Kurt W. Alt
Friedrich W. Rösing
Maria Teschler-Nicola (eds.)

Dental Anthropology

Fundamentals, Limits, and Prospects

SpringerWienNewYork

Priv.-Doz. Dr. Kurt W. Alt
Department of Human Genetics and Anthropology
Freiburg University
Freiburg, Germany

Univ.-Prof. Dr. Friedrich W. Rösing
Department of Human Genetics and Anthropology
Ulm University
Ulm, Germany

Univ.-Doz. Dr. Maria Teschler-Nicola
Department of Anthropology
Museum of Natural History Vienna
Vienna, Austria

Softcover reprint of the hardcover 1st edition 1998

Data conversion: macTypo Werbegraphik, Peter Schmidt, A-1020 Wien

Printed on acid-free and chlorine-free bleached paper

With 224 Figures

ISBN-13:978-3-7091-7498-2 e-ISBN-13:978-3-7091-7496-8
DOI: 10.1007/978-3-7091-7496-8

This book
is dedicated
to our children

Laura Charlotte Alt	** Jan 26, 1967*
Corvin Rösing	** May 28, 1979*
Justin Rösing	** Jul 10, 1981*
Marie Danner	** Sept 10, 1992*
Katharina Teschler	** Oct 11, 1979*
Theresa Teschler	** Nov 4, 1983*

Contents

Foreword

Shelley Saunders

This book offers a welcome diversity of topics covering the broader subjects of teeth and the study of teeth by anthropologists. There is an impressive array of coverage here including the history of anthropological study of the teeth, morphology and structure, pathology and epidemiology, the relationship between nutrition, human behavior and the dentition, age and sex estimation from teeth, and geographic and genetic variation. Most chapter authors have provided thorough reviews of their subjects along with examples of recent analytical work and recommendations for future research. North American researchers should particularly appreciate the access to an extensive European literature cited in the individual chapter bibliographies. Physical anthropologists with even a passing interest in dental research should greet the publication of this book with pleasure since it adds to a growing list of books on how the study of teeth can tell us so much about past human populations.

In addition to the archaeological applications, there is the forensic objective of dental anthropology which the editors refer to in their introduction which is dealt with in this volume. The chapters dealing with methods of sex determination, age estimation of juveniles and age estimation of adults using the teeth are exhaustive and exacting and of critical importance to both "osteoarchaeologists" and forensic anthropologists. Authors Liversidge, Herdeg and Rösing provide very clear guidelines for the use of dental formation standards in juvenile age estimation, recommendations that are so obviously necessary at this time. Authors Rösing and Kvaal point out that histological and chemical techniques of adult age estimation must now be part of standard practice, both for forensic work and archaeological samples. I agree with them. Rapid methods of sample processing and data collection by image analysis are widely available in industry. They should also be more easily available to anthropologists. The chapters by Teschler-Nicola and Prossinger grapple with a variety of mathematical approaches to using tooth size for sex determination, particularly for juvenile skeletons producing some interesting and informative results. Forensic workers can now almost routinely use DNA sex determination for their cases. Those working with archaeological samples are hampered by technological and financial barriers. It is possible to foresee that the next important study on sex determination in juvenile skeletons will include a comparison of documented sex, dental size and DNA determinations. Multiple methods of identification are always better than single methods.

Traditionally, biological anthropology has been concerned with the documentation of human physical variation and its relevance to population relationships and evolutionary trends. These topics are also dealt with in this book. But some of the contributing authors identify the persistence of certain disturbing problems, notably, variability created by inter-observer error and methodological idiosyncracies (see chapters by Schnutenhaus and Rösing and Alt and Vach). I perceive that books like this one can act as catalysts for an increase in major collaborative studies, where a number of dental anthropolgists work together to collect, collate and analyze major bodies of data. Perhaps in this way we can effectively answer some of the sweeping questions about diet, health and behavior throughout human evolutionary history that characterize our discipline.

Foreword

Trinette Constandse-Westermann

It is a great honour and pleasure to be asked to write a foreword to this volume, presenting the state of the art of the physical anthropology of teeth and their surrounding tissues.

When teaching physical anthropology, the dentition is one of the most informative parts of excavated human skeletons, and the understanding to be derived from it relates to a number of different domains. Dentitions are sources of insight into aspects of human phylogeny (evolutionary studies of dental morphology) and anatomy/histology (investigations of gross and microscopic dental structure). They also inform us about individual physical development and, later in life, wear and degeneration (dental development and eruption, dental attrition). Some aspects of human genetics can be elucidated by the study of specific morphological dental traits, whose (partly) hereditary nature can be assessed in the living and thereafter be used in the interpretation of studies of the deceased. Dental pathology comprises a large number of disorders, e.g. caries periapical lesions, ante mortem tooth loss, parodontal disease, calculus and hypoplasia. Finally, aspects of human culture and behaviour can be revealed by the analysis of human dental remains, e.g. the filing of teeth, pipe smoking, dental restoration and occupational hazards. All the foregoing is illustrated in the presented work.

Due to the durability of teeth in their archaeological context, dental analyses also contribute to our insights into processes transpiring at the population level, e.g. demographical studies wherein the dental analysis forms the basis of the age estimations. The occurrence of measurable levels of dental disease in the majority of pre- and/or protohistoric populations lifts most studies of the pathology of teeth, including the tissues around them, above the level of case studies and enables the researcher to perform palaeo-epidemiological analyses. Especially in Section 4 of this work the problems relating to the epidemiology of dental (and other) diseases have been treated.

In addition, the relatively frequent occurrence of teeth in cemetery contexts creates the possibility to go beyond the purely biological interpretation of the data extracted from these human remains. The integrated analysis of dental and other, independently assessed, cultural/archaeological data offers a number of possibilities to gain insight in pre- and/or protohistoric social or socio-economic processes. The study of enamel hypoplasia not only indicates at which specific age each individual was subjected to dietary or pathological stress, but also the archaeological data may make it possible to relate such stress to the age of weaning or to (changes in) dietary patterns. The occurrence of hereditary traits in teeth may be linked to cemetery data, i.e. the placement of the grave within the graveyard, grave structure, grave goods, the treatment of the body of the deceased and its position within the grave. By the integrated analysis of the physical anthropological (dental) and the archeological (cemetery) data sets, possible familial and/or lineage relationships and pattering may be revealed or at least may be ascribed a certain degree of probability. Also the study of social stratification from cemetery or other archeological data may be supported by dental analyses. Observed differences in the level of infectious (dental) disease may be related to differences in dietary patterns and/or occupational stress between hierarchically structured social groups. Especially in relatively simple prehistoric societies, without or with only a minor level of labour specialization, the execution of sex/gender specific tasks may be revealed by the joint analysis of dental, skeletal and archeological data. Social and/or ethnic information my be derived from the analysis of specific intentional treatments of the teeth, e.g. filing.

In all the foregoing fields of research, the data on teeth often surpass other skeletal data in their usefulness. However, studies fully and fruitfully integrating dental and other physical anthropological data on the one hand and cultural/ archeological information on the other, are still scarce in the scientific literature. It is through this sort of cultural and biological processes transpiring in pre- and protohistoric societies will become clearer. This book constitutes a firm platform for the continuing integration of the biological (dental) and the cultural/ archeological research paradigms.

Contributors

Kurt W. Alt, received his dental degree (Dr.med.dent.) from the Freie Universität (West)Berlin, Germany. He completed his studies in anthropology, archaeology and ethnology at the Albert-Ludwigs-University, Freiburg, Germany (Dr. habil. in 1992). Afterwards, he worked as a dental and forensic anthropologist at the Institute of Forensic Medicine, Heinrich-Heine-University, Düsseldorf, Germany. Since 1997, he is asssociate lecturer in historic and biological anthropology at the University of Freiburg. Dr. Alt's research focused on physical, forensic and dental anthropology, forensic odontology, dental evolution, paleopathology, and the history of dentistry, with particular emphasis on macromorphological, micromorphological and metrical variants of teeth and jaws with regard to kinship analysis (determination of genetic relationships in skeletal remains) and to "dental species" (reconstruction of the taxonomic status of hominid fossils and their ecological background; e.g., *Homo erectus* from Dmanisi 1/Georgia; *Australopithecus* (RC 911) and *H. rudolfensis* (UR 501) from Malema and Uraha/Malawi). His numerous publications include *Pulpoalveolar disease: etiology, incidence, and differentiation of periapical lesions* (with Wächter R and Türp JC), Journal of Paleopathology, 1992; *Artificial tooth-neck grooving in living and prehistoric populations* (with Koçkapan C), Homo, 1993; *Prosthetics, periodontal therapy and conservative dentistry in the eighteenth century: archeological findings from Grand Sacconex, Geneva, Switzerland,* Bull Hist Dent, 1994; *Kinship analysis in skeletal remains – concepts, methods, and results* (with Vach W), Forensic Sciece International 1995; *Die Evolution der Zähne. Phylogenie – Ontogenie – Variation* (co-edited with Türp JC), Quintessenz, Berlin, 1997, and *Taxonomische Marginalien zum Dmanisi Unterkiefer aufgrund dentalmorphologischer Vergleichsanalysen* (with Henke W and Rothe H), in: *Beiträge zur Archäozoologie und Prähistorischen Anthropologie,* Kokabi M (Hrsg), 1997. His research has been supported by grants from the Deutsche Forschungsgemeinschaft and the Carlsberg Foundation, Denmark. Dr. Alt is a member of many scientific associations, including the American Association of Physical Anthropologists, the European Anthropological Association, the International Society for Forensic Odontostomatology, the Gesellschaft für Anthropologie (Germany) and the Deutsche Gesellschaft für Zahn-, Mund- und Kieferheilkunde. Address: Department of Human Genetics and Anthropology, Freiburg University, Breisacher Str. 33, D-79106 Freiburg, Germany.

C. Loring Brace received his B.A. (in geology) from Williams College, Williamstown, MA, and his Ph.D. (in Anthropology) from Harvard University in Cambridge, MA. He is Professor of Anthropology in the Department of Anthropology at the University of

Michigan, and he is also Curator of Biological Anthropology at the Museum of Anthropology in the Museum of Natural History at the University of Michigan, Ann Arbor. Dr. Brace's broader involvement has been in studying the course of human evolution from the appearance of the first hominids in the Pliocene over 5 million years ago up through the emergence of "modern" humans including the biological variation in living human populations sometimes subsumed under the term "race". His published documentation of the changes and relationships has concentrated on craniofacial measurements and odontometrics. In addition, he has expanded his interests into the realms of evolutionary theory and the history of science. His publications include *The fate of the "classic" Neanderthals: A consideration of hominid catastrophism,* Current Anthropology, 1964; *Man's Evolution: An Introduction to Physical Anthropology* (with Montagu A), Macmillan, New York, 1965 ([2]1977); *The Stages of Human Evolution,* Prentice-Hall, Englewood Cliffs, NJ, 1967 ([5]1995); *Environment, tooth form and size in the Pleistocene,* Journal of Dental Research, Suppl. No. 5, 1967; *Occlusion to the anthropological eye,* in: The Biology of Occlusal Development, McNamara JA (ed), University of Michigan, Ann Arbor, 1977; *Krapina, "classic" Neanderthals, and the evolution of the European face,* Journal of Human Evolution, 1979; *Australian tooth size clines and the death of a stereotype,* Current Anthropology, 1980; *Oceanic tooth size variation as a reflection of biological and cultural mixing* (with Hinton RJ), Current Anthropology, 1981; *Japanese tooth size past and present* (with Shao X-q, Zhang Z-b), in: The Origins of Modern Humans: A World Survey, Spencer F, Smith F (eds), Liss, New York, 1984; *Gradual change in human tooth size in the late Pleistocene and post-Pleistocene* (with Rosenberg K, Hunt K), Evolution, 1987; *Reflections on the face of Japan: A multivariate craniofacial and odontometric perspective* (with Brace ML, Leonard WR), American Journal of Physical Anthropology, 1989; *What big teeth you had Grandma! Human tooth size, past and present* (with Smith SL and Hunt KD), in: Advances in Dental Anthropology, Kelley MA, Larsen CS (eds), Liss, New York, 1991; *Bio-cultural interaction and the mechanism of mosaic evolution in the emergence of "modern" morphology,* American Anthropologist, 1995; *Trends in the evolution of human tooth size,* in: Aspects of Dental Biology: Paleontology, Anthropology and Evolution, Moggi-Cecchi J (ed), Istituto de Antropologia, Università di Firenze, 1995; *Cro-Magnon and Qafzeh — vive la différence,* Dental Anthropology Newsletter, 1996. Dr. Brace's work has been supported by the National Science Foundation, the LSB Leakey Foundation, the Committee on the Scientific Cooperation with the People's Republic of China of the National Academy of Sciences, and the Museum of Anthropology at the University of Michigan. He is a member of the American Association of Physical Anthropologists, the American Association for the Advancement of Science, the History of Science Society, and he is a member and past-president of the Dental Anthropology Association. Address: University of Michigan, Museum of Anthropology, Ann Arbor MI 48109, USA.

Franz Brandstätter, born 1953, studied mineralogy at the University of Vienna (Dr. phil. 1979). Since 1980 he is with the Department of Mineralogy and Petrography at the Natural History Museum Vienna, Austria. He is specialized in mineral analysis by scanning electron microscopy and electron microprobe techniques. His research interests involve mineralogy and petrology of meteorites. Address: Natural History Museum Vienna, Department of Mineralogy and Petrography, Burgring 7, A-1014 Wien, Austria.

K. H. Peter Caselitz, born 1951 in Bremen. After studying chemistry and physics in Münster he switched to prehistory, anthropology and ethnology in Hamburg (important teachers F. W. Rösing, H. Ziegert and J Jensen). For 5 years he was scientific employee at the University of Hamburg and responsible for editing the journal Hamburger Beiträge zur Archäologie. His Ph.D. dissertation on paleonutrition was published in 1986 in Oxford. He was member of several excavations in Germany, Greece, Egypt, Syria etc. and directed the excavations of some medieval cemeteries in northern Germany as well as for two years a large excavation in the center of Lübeck. His research focuses on the interdisciplinary overlap between archeology and physical anthropology – osteoarcheology, as he prefers to call it – population history of western and northern Germany, analysis of cremated bones and the computer-based analysis of ceramics. At present he teaches computing, he is the expert on osteoarcheology in some federally funded projects, and he is a severe critic of mediocrity in university and science. His bibliography encompasses some eighty titles. PC was an executive editor of the former journal Archaeologica Atlantica. He is a member of the editorial board of Homo and Archivio per l'Antropologia e la Etnologia. Address: Department of Archeology, Hamburg University, Johnsallee 35, D-20148 Hamburg, Germany.

Petra Carli-Thiele, Dr. rer. nat., born 1963 in Kassel, studied zoology and microbiology from 1983 to 1988 at the Georg August University in Göttingen. After completing her Master of Science degree in 1990 at the German Primate Center in Göttingen, she conducted research with the paleopathology team at the Anatomy Center of the University of Göttingen. In 1994 she obtained her Ph.D in human biology at the University of Braunschweig, with additional work in microbiology and physical chemistry. Since 1995 she has participated in a project with the paleopathology team on the etiology and epidemiology of disease in prehistoric and early historic populations, with a focus on deficiency diseases. Publications in paleopathology and prehistoric anthropology. Author of the book *Vestiges of deficiency diseases in Stone Age child skeletons,* Göltze, Göttingen, 1996. Address: Department of Anatomy, Goettingen University, Kreuzbergring 36, D-37075 Goettingen, Germany.

Gisela Grupe (study of Physical Anthropology at the Institute of Anthropology, Göttingen, Ph.D. 1986, Dr.rer.nat.habil. 1990). Since 1991 Professor at the Institute of Anthropology and Human Genetics, Munich, head of the department of Physical Anthropology and Environmental History, and director of the Anthropological Collections in Munich. She is specialized in archeometric investigations of ancient human skeletal remains with the aim of reconstructing ancient environments and human living conditions. Current work focusses on ancient serum proteins, stable carbon and nitrogen isotope ratios in bone collagen and stable strontium isotope rations in bone mineral. Recent publications include: *In vitro decomposition of bone collagen by soil bacteria. Implications for stable isotope analysis in archeometry* (with Balzer A, Gleixner G, Schmidt HL and Turban-Just S), Archeometry 1997; *Molecular preservation and isotopy of mesolithic human finds of the Ofnet cave (Bavaria, Germany)* (with Bocherens H, Mariotti A and Turban-Just, S), Anthropologischer Anzeiger, 1997; *Serum proteins in archeological human bone* (with Turban-Just S), International Journal of Osteoarcheology 1996. Prof. Grupe is member of the Gesellschaft für Anthropologie, the European Anthropological Association, the American Association of Physical Anthropology, the Gesellschaft Deutscher Naturforscher und Ärzte, and the Mediävistenverband. Address: Department of Anthropology, University of Munich, Richard-Wagner Str. 10/I, D-80333 München, Germany.

Winfried Henke (Study of Biology, Anthropology, Geosciences, Philosophy and Pedagogy at the Universities of Kiel and Braunschweig; Dr. rer.nat. at the Christian Albrechts-University of Kiel, 1971; Dr. rer.nat. habil., Johnannes Gutenberg-University Mainz 1990) is currently Professor and Academic Director at the Institute of Anthropology of the Biological Faculty, Mainz University, Germany. He is specialized in paleoanthropology, prehistoric anthropology and the biology and demography of recent populations. His more than 100 scientific publications focus on the evolutionary biology of fossil and subfossil populations, including: *A comparative approach to the relationships of European and non-European Late Pleistocene and Early Holocene Populations,* ERAUL 56, Liège 1992; *Affinities of European Upper Paleolithic Homo sapiens and later Human Evolution,* J Hum Evol 1992; *Paläoanthropologie* (with Rothe H), Springer, Berlin, 1994; *Qualitative and quantitative analysis of the Dmanisi mandible* (with Roth H and Simon C), in: Proccedings of the 10th International Symposium on Dental Morphology, Radlanski R, Renz H (eds), Brünne, Berlin, 1995; *Homo erectus – ein valides Taxon der europäischen Hominiden?,* Bull Soc Suisse Anthrop 1996; *Zahnphylogenese der nicht-menschlichen Primaten,* and *Zahnphylogenese der Hominiden* (with Rothe H) in: *Die Evolution der Zähne,* Alt KW, Türp JC (eds), Quintessenz, Berlin, 1997. Prof. Henke was ERASMUS-docent at the universities of Heraklion (Greece), Complutense/Madrid (Spain), Komotini (Greece), Bordeaux (France) and Firenze (Italy) and has done field work in Iceland and Jordan. He is member of the board of the Gesellschaft für Anthropologie and has been elected as a member of the board responsible for the European Master in Anthropology and Human Biology. Since many years Prof. Henke has been local coordinator of the ERASMUS/SOCRATES networks in Biology and Human Biology. Furthermore he is cooperator and referee of different scientific journals and elected referee for Anthropology in the Deutsche Forschungsgemeinschaft. Address: Department of Anthropology, Mainz University, Saarstr. 21, D-55099 Mainz, Germany.

Berthold Herdeg, born 1965, finished his studies of dentistry in 1991 with the state examination and in 1992 his dental doctor's thesis at the medical school in Ulm, Germany (directed by F.W. Rösing) *Die Zahnentwicklung beim Menschen. Kritische Analyse der bisherigen Zahlen und Zitierreihen.* He is a dental practitioner in Bopfingen. Address: Jenaer Str. 14, D-73479 Ellwangen, Germany.

Horst Kierdorf, born 1956, studied at the University of Cologne, where he received M.Sc. degrees in biology and geography in 1984/85, a doctoral degree (Dr. rer. nat.) in zoology in 1988, and the venia legendi (Habilitation) in biology in 1994. From 1986 to 1990 he undertook research in deer genetics, parasitology and management for the Game Research Station of the federal state of North Rhine-Westphalia in Bonn. Currently Dr. Kierdorf is senior lecturer of zoology at the University of Cologne. His main fields of research are: dental development and dental pathology in mammals, ultrastructure and functional morphology of the dental hard tissues, dental fluorosis and osteofluorosis, wildlife ecotoxicology and genetics, and antler formation in deer. Address: Department of Zoology, Köln University, Weyerstr. 119, D-50923 Köln, Germany.

Uwe Kierdorf, born 1956, studied bio- and geosciences at the University of Cologne, Germany. He received M.Sc. degrees in biology and geography in 1984/85 and a doctoral degree (Dr. rer. nat.) in zoology from the University of Cologne in 1988. From 1985 to 1988 he worked as an ecotoxicologist for the Game Research Station of the federal state of North Rhine-Westphalia in Bonn. From 1988 to 1994 he was lecturer of zoology at the I. Zoological Institute, University of Goettingen, Germany. During 1994/95 he was guest researcher in the Department of Dental Pathology, Operative Dentistry and Endodontics, Royal Dental College, University of Aarhus, Denmark. Currently Dr. Kierdorf works in the Department of Animal Ecology, Institute of General and Systematic Zoology, Justus-Liebig-University of Giessen, Germany. His main research interests are: dental development and dental pathology in mammals, ultrastructure and functional morphology of dental hard tissues, mammalian osteopathology, dental fluorosis and osteofluorosis, wildlife ecotoxicology, and the development, biomineralization and pathology of deer antlers. Address: Department of Zoology, Giessen University, Stephanstr. 24, D-35390 Giessen, Germany.

Michaela Kneissel, born 1966, received her M.Sc. (1990, biology) and her doctoral degree (Dr. rer. nat., 1993, human biology) at the University of Vienna, Austria and is currently working as a specialist for bone histomorphometry in the Bone Metabolism Unit at Novartis Pharma AG, Basel, Switzerland. She has performed studies of cancellous bone structure during growth and aging in historic and prehistoric populations (Department of Anatomy and Developmental Biology, University College London, U.K.; Ludwig Boltzmann Institute for Osteology, Vienna, Austria; Museum of Natural History, Vienna, Austria) and worked on the quantitative characterization of bone structure in the edentulous jaw (with Doz. Dr. C. Ulm, Clinic for Dental Research, University of Vienna, Austria). During her postdoctoral fellowship at the University of Utah, Salt Lake City, U.S.A. she studied changes in the maternal skeleton during pregnancy and lactation. Currently one main focus of her research is the influence of dietary interventions on bone. A second is the characterization of bone structure, mineralization, dynamics, and turnover during and after intermittent administration of parathyroid hormone in animal models. Recent publications in the field of anthropology and dental research include: *Age- and sex-dependent cancellous bone changes in a 4000 BP population.* (Kneissel M, Boyde A, Hahn M, Teschler-Nicola M, Kalchhauser G, Plenk jr H) Bone 15: 539–545, 1994; *Characteristics of the cancellous bone of edentulous mandibles.* (Ulm C, Kneissel M, Hahn M, Matejka M, Watzek G, Donath K) Clin Oral Impl Res Vol 8: 125–130, 1997. *Cancellous bone structure in the growing and aging lumbar spine in an historic Nubian population.* (Kneissel M, Roschger P, Steiner W, Schamall D, Kalchhauser G, Boyde A, Teschler-Nicola M) Calc Tiss Int 61: 95–100, 1997. Address: Bone Metabolism Unit, K-125.10.59, Novartis Pharma AG, CH-4002 Basel, Switzerland.

Thomas Koppe (Study of dentistry at the Ernst Moritz Arndt University Greifswald, 1985 final examination; Dr. med., Ernst Moritz Arndt University Greifswald, 1985; Dr. med. habil., Rostock University, 1991) is currently an Assistant Professor in the Department of Anatomy at the Okayama University Dental School, Japan. Dr. Koppe is member of the cooperative research programme of the Kyoto University, Japan. He is specialized in craniofacial growth and comparative anatomy of the masticatory apparatus. Recently he works on the functional morphology of the paranasal sinuses of higher primates using computed tomography and 3D-image reconstructions. His publications in this research field include: *Investigations on the growth pattern of the maxillary sinus in Japanese human fetuses* (with Yamamoto T, Tanaka O, and Nagai H), Okajimas Folia Anat, 1994; *Growth pattern of the maxillary sinus in orang-utan based on measurements of CT scans* (with Röhrer-Ertl O, Hahn D, Reike R, and Nagai H), Okajimas Folia Anat., 1995; *The pneumatization of the facial skeleton in the Japanese macaque (Macaca fuscata) – a study based on computerized three-dimensional reconstructions* (with Inoue Y, Hiraki Y, and Nagai H), Anthrop. Sci., 1996. Recently Dr. Koppe organized the symposium "Pneumatization of the Vertebrate Skull" in cooperation with Dr. LM Witmer (Ohio, USA), part of the 5th International Congress of Vertebrate Morphology at the University of Bristol, U.K. in July 1997. Dr. Koppe is member of the Anatomische Gesellschaft, Gesellschaft für Anthropologie, the Japanese Association of Anatomists, the Japanese Association of Oral Biology, The Primate Society of Japan, the Anthropological Society of Nippon, the American Association of Physical Anthropology, and the International Society of Vertebrate Morphologists. Address: School of Dentistry, Department of Anatomy, Okayama University, Shikata-cho 2-5-1, J-700 Okayama, Japan.

Kerstin Kreutz, born 1963, studied anthropology and zoology at the Georg August University of Göttingen. 1996 Dr. rer. nat. in anthropology at the Justus Liebig University of Giessen. Title of the dissertation: "Ätiologie und Epidemiologie an den Kinderskeleten der bajuwarischen Population von Straubing". Since 1989 member of the paleopathology team at the Anatomy Center of the Georg August University of Göttingen. Address: Department of Anatomy, Goettingen University, Kreuzbergring 36, D-37075 Goettingen, Germany.

Sigrid I. Kvaal obtained a Bachelor's Degree in Dental Surgery (B.D.S.) at the University of London in 1978. After three years in England, working mainly in oral surgery units, she returned to her homeland, Norway. She then spent one year in north Norway where she was engaged in the Public Health Service, but started private practice when she moved to Oslo. In addition to this, she completed a one-year postgraduate education course in General Biology in 1988, in which year she was appointed a research assistant in the Department of Oral Pathology and Section for Forensic Odontology, Dental Faculty, University of Oslo. In 1995 she obtained the Doctorate of Odontology (dr.odont.) at the same university. Dr. Kvaal's main published research has been in dental age changes with particular reference to the field of forensic odontology and dental anthropology, and she has studied dental conditions in various post Reformation collections in both Norway and Sweden. She has taken part in the identification of a number of single cases and has officially been engaged in indentification work in major disasters including the "Scandinavian Star" ferry disaster (1990) involving 158 deaths and the Russian Tupolev air disaster at Spitsbergen (1996) involving 141 deaths. Dr. Kvaal has also examined the dentist's role in cases of physical child abuse. She is now a clinical instructor at the Dental Faculty in the University of Oslo where she teaches forensic odontology, general and oral pathology, dental radiology and oral surgery. She is a member of the Norwegian Dental Association and other Norwegian, Nordic and international scientific societies. Address: Department of Oral Pathology, Section of Forensic Odontology, Box 1109 Blindern, 0317 Oslo, Norway.

Helen Liversidge grew up and trained as a dental surgeon in South Africa. While working as a general practitioner, she continued her education in London at the Hard Tissue Research Unit, Department of Anatomy and Developmental Biology, University College London with Alan Boyde and Christopher Dean. After ten years in private practice she joined the Pediatric Dentistry Department of the School of Medicine and Dentistry, Queen Mary and Westfield College (formerly London Hospital Medical College) as a lecturer and divides her time between clinical undergraduate and postgraduate teaching as well as ongoing research at the Human Origins Group, Natural History Museum with Thea Molleson and AEW (Loma) Miles. She is a member of the British Dental Association, British Association of Paediatric Dentistry, International Association/British Society for Dental Research, Dental Anthropological Association and the Royal Society of Medicine (Odontology Section). Her contributions to meetings include the American Association of Physical Anthropologists, 9th and 10th International Symposia of Dental Morphology as well as the International Association and British Society of Dental Research. Her publications include quantitative data on the developing human dentition, accuracy of age estimation in early childhood, growth standards and tooth size of deciduous teeth and crown formation times of permanent teeth. Ongoing work includes facial growth as well as comparison of dental maturation between population groups. Address: Department of Paediatric Dentistry, St. Bartholomew's and the Royal London School of Medicine and Dentistry, Turner Street, London, E1 2AD, Great Britain.

Hiroshi Nagai (Ph.D., Kyoto University, 1962) is a Professor of anatomy and head of the Department of Anatomy at the Okayama University Dental School. He specializes in experimental teratology, comparative functional anatomy, and medical history. His publications include: *Effects of transplacental injected alkylating agents upon development of embryos-appearance of intrauterine death and mesodermal malformation* Bull. Tokyo Dental College, 1972; *Disturbance of dentine formation on rats and mice after administration* *of L-azetidine-2-carboxyl acid,* Congenital Anomalies, 1977; *A Study of the history of anatomical copperplates,* Shin-Nippon Publ. Co., 1979; *An ultrastructural study of cytochrome oxidase activity in metaphyseal bone cells,* Acta Histochem. Cytochem., 1991. He is a member of the Japanese Association of Anatomists, the Japanese Association of Oral Biology, the Japanese Society for Bone and Mineral Research, the Japanese Teratology Society, and the Japan Society of Medical History. Address: School of Dentistry, Department of Anatomy, Okayama University, Shikata-cho 2-5-1, J-700 Okayama, Japan.

Aleš Obrez received his dental degree (D.M.D.) from the University of Ljubljana Medical School (Slovenia) in 1981. After two years of postgraduate training in prosthodontics (1985, College of Dentistry), he continued with graduate studies on functional morphology of the stomatognathic systems in the Department of Anatomy and Cell Biology, College of Medicine, The University of Illinois at Chicago (Ph.D., 1992) under the mentorship of Dr. Susan W. Herring. From 1992 to 1994 he served as an Assistant Professor of Dentistry in the Facial Pain Clinic, School of Dentistry, University of Michigan, Ann Arbor. Dr. Obrez is currently an Assistant Professor in the Department of Restorative Dentistry, College of Dentistry, The University of Illinois at Chicago. Dr. Obrez's research interests include functional and evolutionary morphology, physiology, and pathology of the masticatory system. His publications include *Bone growth and periostal migration control masseter muscle orientation* (with Herring SW), Anatomical Record, 1993; *The temporomandibular joint: development, anatomy, physiology, and function,* in: Clark's Clinical Dentistry, Vol. 2, Hardin JF (ed), 1993; *Jaw muscle pain and its effect on gothic arch tracing* (with Stohler CS), Journal of Prosthetic Dentistry, 1996; *The effect of molar teeth on development of power stroke in miniature pig (Sus scrofa),* Acta Anatomica, 1996; *Die odontogenetische und postnatale Entwicklung des menschlichen Kiefergelenks* (with Türp JC, Radlanski RJ), in: Die Evolution der Zähne, Alt KW, Türp JC (eds), Quintessenz, Berlin, 1996. Dr. Obrez is a member of the American Association for Dental Research, the American College of Prosthodontists, and the Association of University TMD and Orofacial Pain Programs. Address: Department of Restorative Dentistry, College of Dentistry, University of Illinois at 801 S. Paulina St., Chicago, IL 60612, USA.

Sandra L. Pichler, studied Prehistoric Archaeology, Physical Anthropology, Geology and English at the Albert-Ludwigs University, Freiburg, Germany, where she obtained her Ph.D. with a thesis on *Paläoökologie des Östlichen Gravettien – Paläoklimatische und kulturökologische Analyse archäologischer Grabungsbefunde aus Mittel- und Osteuropa* (Prof. Dr. W. Schüle) in 1995. She took part in numerous archeological excavations in Germany and abroad, and has extensive experience in the conception and realisation of archaeological and anthropological exhibitions in museums. Dr. Pichler participated in research projects on early hominid subsistence and adaptation, on the reconstruction of biological structures and living conditions in the European paleolithic and mesolithic, and on aDNA analysis. Her research interests are cultural ecology, gender roles and social differentiation in earlier populations, ethology and human exploitation of Quaternary faunas, and adaptive interactions of man and environment. Recent publications include: *Geschlechtsbestimmung anhand der aDNS an einer Bestattung des bandkeramischen Gräberfeldes Ensisheim "les Octrois", Dept. Haut-Rhin* (with Kuntze K) in: Jeunesse C, Campagne de fouille 1996 sur la nécropole rubanée d'Ensisheim "les Octrois" (Haut-Rhin), CNRS, Strasbourg, 1996; *Zähne und Kiefer in der Archäozoologie – Relevanz und Methoden* in: *Die Evolution der Zähne – Phylogenie, Ontogenie, Variation,* Alt KW, Türp JC (eds), Quintessenz, Berlin, 1997; *25,000 year old triple burial from Dolní Věstonice – an ice age family?* (with Alt KW, Vach W, Klima B, Vlček E, and Sedlmeier J), American Journal of Physical Anthropology, 1997. Address: Institute of Prehistory, University of Freiburg, Belfortstr. 22, D-79085 Freiburg, Germany.

Hermann Prossinger (Dr. phil, University of Vienna, 1976) develops mathematical models for anthropological research teams (Natural History Museum, Vienna, and Institute of Human Biology, Vienna University). He has simulated neutron fluxes in nuclear reactor safety models (Argonne.National Laboratory, USA) and physical microclimate models for pest control (FAO, Rome). He has also studied scientific developments in medieval astronomy. As an augmentation to his mathematical modelling research, he is currently reviewing the epistemological implications of precursors of the (positivist) Wiener Kreis for biology and the implications of the visualist approach to morphology in paleobiological studies. Recent publications include: *Suggestions for improving the objectivity of paleodemographic data, exemplified in the analysis of Early Bronze Age cemeteries of the Lower Traisen Valley* (with Teschler-Nicola M) Anthropologie 1992; *Schädelasymmetrien, Denkasymmetrien und Artefakte: zur Genesis der Imagination und der Nachweis ihrer Zeitlichkeit.* In: *Interface II: Weltbilder/Bilderwelten.* Hans Bredow Institut, Hamburg, 1995; *Metrication meets morphology when attempting to reconstruct both permanent from deciduous and maxilla from mandibular tooth dimensions* (with Tescher-Nicola M) In: Proceedings of the 10th International Symposium on Dental Morphology, Berlin, 1995; *Elemente des Übergangs. Bausteine für Kunstwelten im anbrechenden Computerzeitalter* (with Clausberg K), Paragrana – Int Zeitschr Hist Anthrop, 1996; *Die Autorenschaft eines Reliefs am Campanile von Santa Maria del Fiore in Firenze. Untersuchungen zur Krise der Wissenschaftsauffassung am Ausgang des XIII. Jahrhunderts in Italien,* submitted for publication. Dr. Prossinger is a member of the Austrian Physical Society (Österreichische Physikalische Gesellschaft). Address: Natural History Museum Vienna, Department of Anthropology, Burgring 7, A-1014 Wien, Austria.

Ralf J. Radlanski, is Professor and Head of the Department of Oral Structural Biology at the Medical Faculty at Freie Universität Berlin. After education in medicine and dentistry in Göttingen and Minneapolis he was graduate student in the Department of Anatomy at Göttingen University. His doctoral thesis was about functional anatomy of the forelimb of a mongoose species (1985). In the Department of Embryology at Göttingen University he worked about dental embyology and enamel microstructure. He then specialized in Orthodontics at Göttingen University and habilitated (1989) at the Medical Faculty at Göttingen University. His habilitation thesis was a contribution to the development of human deciduous tooth primordia. The findings were computer-aided 3-dimensional reconstructions from serial sections, including the structures surrounding the developing dental primordia. In 1992 he became Professor and Head of the Department of Oral Structural Biology at the Medical Faculty at Freie Universität Berlin. In addition, he works part-time as an orthodontist in a private practice in Berlin. In 1995 he was president of the 10th International Symposium On Dental Morphology in Berlin. Main research interests: presently, he is involved in basic research dealing with the description of human craniofacial development as revealed by computed tomography. From these findings a computer-animated film is under production. Furthermore, computer-aided 3-D-reconstructions from serial sections are in production to cover the 3-dimensional development of the mandible, the maxilla, the temporomandibular joint, and adjacent soft tissues. This collection covers the stages from 18 to 280 mm CRL. Micromorphology of structures of the human stomatognathic system is another field of main interest. The findings dealing with enamel structure are partly presented in the contribution of this volume. At present, the structural development of fetal enamel is under closer investigation. Address: Department of Oral Structural Biology, Medical Faculty at Freie Universität Berlin, Aßmannshauser Str. 4–6, D-14197 Berlin, Germany.

Friedrich Wilhelm Rösing. Born 28 February 1944. After studying human biology, prehistory, sociology and journalism in Mainz, Düsseldorf and Hamburg (important teachers: his mother I. Schwidetzky, R. Knussmann and H. Ziegert), he finished his biology diploma thesis in 1972 in Mainz, his natural science thesis in 1975 in Hamburg and his anthropology habilitation thesis in 1987 in Ulm. Since 1994 he is professor of anthropology at the medical school of Ulm university and responsible for a genetics and forensics ambulance. His scientific fields are taxonomy of populations in Europe and Africa, population history of Egypt and Baden-Württemberg, method development and forensic sciences. Additional teaching fields are human evolution, demography, ethics in science, general biology (game theory, self-regulation, the evolution of sex) and statistics. He directed 26 MD, DD, PhD and master theses, 16 are in progress at the moment. For several years he served in university management positions, the most important were member of the university senate and central university commissioner for continuing education. His present long-term interest focuses on mediaeval population adaptation and above all forensic anthropology, including paternity analysis, identification of skeletons and of living persons and age diagnosis in juvenile offenders. In skeletal identification he is particularly interested in mass graves, particularly when connected to human rights issues. His bibliography encompasses some eighty titles, including five monographs. He is one of the managing editors of the journal Homo and member of the editorial board in some others. Learned society memberships focus on physical anthropology, forensics and architecture history. Address: Institute of Anthropology, Human Genetics, and Clinical Genetics, Department of Human Genetics, Liststr. 3, D-89079 Ulm, Germany.

Jerome C. Rose, born in 1947, received his B.A. in Anthropology from the University of Colorado, Boulder, in 1969 and his Ph.D. in Anthropology (Biological Anthropology) from the University of Massachusetts, Amherst, in 1973. He taught Anthropology at the University of Alabama in Birmingham (1973–1976) before moving to the University of Arkansas, Fayetteville where he is Professor of Anthropology. Prof. Rose's research areas are dental anthropology and bioarcheology. He has published a number of articles and book chapters on microscopic enamel defects and enamel hypoplasia (e.g., *American Journal of Physical Anthropology, Human Biology, Journal of Human Evolution and the Yearbook of Physical Anthropology*). More recently, he is engaged in Cultural Resource Management Bioarcheology where he has conducted a number of skeletal analyses, but specializes in literature syntheses and bioarcheological management overviews for various government agencies. He has participated in the development of bioarcheology overviews covering approximately 55% of the continental U.S.A. This research has been funded by 33 grants and research contracts and has appeared in a number of journal articles, book chapters, and published contract reports (e.g., *International Journal of Osteoarcheology, Journal of Economic History,* and *Plains Anthropologist*). Prof. Rose has conducted bioarcheological excavations in Illinois, Arkansas, Texas, Egypt and Jordan. In addition, he has served as a bioarcheology consultant to the U.S. Government Services Administration and as a forensic anthropologist for several law enforcement agencies. He has held research assignments at Harvard University, Cambridge University, Ohio State University and was visiting Professor of Anthropology at Yarmouk University, Irbid, Jordan. Prof. Rose is a member of the American Association of Physical Anthropologists, Dental Anthropology Association, Paleopathology Association, Sigma Xi, American Center of Oriental Research, and American Research Center in Egypt. Address: Department of Anthropology, Old Main 330, University of Arkansas, Fayetteville, AR 72701, USA.

Sigmar Schnutenhaus, born 19 May 1965. After school, he was admitted to the military academy of the German medical corps. He received his dental doctor's degree (Dr. med. dent.) from the University of Ulm; the thesis was directed by F. W. Rösing. After some clinical positions he is now a major of the medical corps and staff officer in the dental corps of the south-west German military district headquarters. His main clinical field of work is endodontic therapy and periprothodontics. Dr. Schnutenhaus is member in several scientific associations, e.g. Deutsche Gesellschaft für Zahn- Mund- und Kieferheilkunde (DGZMK), Deutsche Gesellschaft für Parodontologie (DGP), German Association for Forensic Odonto-Stomatology and Deutsche Gesellschaft für Wehrmedizin und Wehrpharmazie. Address: Postfach 1135, D-78245 Hilzingen, Germany.

Tyede Helen Schmidt-Schultz, Dr. rer. nat., studied biochemistry, immunology, microbiology, organic chemistry and anthropology. Master of Science degree finished at the beginning of 1989; her Ph. D. thesis, completed April 1992, dealt with regeneration of myelin-producing oligodendrocytes. She worked more than five years in the Max Planck Institute for Experimental Medicine, Göttingen. She is currently working on protein chemistry in the Department of Clinical Biochemistry in the Internal Medicine Center, University of Göttingen. For ten years she has been involved in several research projects in paleopathology. She has investigated archeological skeletal remains from several sites in Eastern Europe, the Near East, North and Middle America. Publications in enzyme chemistry, neurobiology, cell biology, and paleopathology. Address: Department of Clinical Biochemistry, Internal Medicine Center, University of Göttingen, Humboldt-Allee 3, D-37075 Göttingen, Germany.

Michael Schultz, Dr. med., Dr. phil. nat., Dr. med. habil., startet his studies in medicine, anthropology, microbiology, archeology (prehistory and early history), and ethnology at the Johann-Wolfgang-Goethe-University in Frankfurt am Main in 1966. After finishing his diploma thesis in anthropology in 1972 and receiving his Master of Science degree in 1973 he completed state medical examinations in the same year. He was medical assistant at the University hospitals in Frankfurt, and the local hospitals in Northeim und Fürstenzell 1973–74. In 1974 he was approbated as physician. Scientific research and teaching in human anatomy, primatology, and paleopathology at the Centre of Anatomy of the Georg-August-University in Göttingen 1974–1979. Since 1976 he is curator of the Blumenbach Collection. He finished his doctoral thesis in medicine (Dr.med.) in 1977 and his doctoral thesis in natural sciences (Dr.phil.nat.) in 1979. Since 1979 he has been lecturing human gross anatomy, histology, and embryology. Additionally, from 1979–89 he has held a professorship at the Department of Anthropology and Human Genetics for Biologists at the University of Frankfurt teaching physical anatomy, anthropology, primatology, and paleopathology. Since 1980 he has been the director of prosectorship. Since 1985 he is directing the paleopathology research team at the Centre of Anatomy of Göttingen University. Inauguration (Habilitation) and Dr.med.habil. followed in 1988. Additionally, in 1990, he has held a professorship for paleopathology and morphology of mummies at the Department of Egyptology at the University of Heidelberg. Since 1990 he has been director of the School for Medical Assistants at Göttingen University. He has been professor of anatomy since 1993 and since 1995 Academic Director at the Centre of Anatomy of Göttingen University and visiting professor of the University of Santa Cruz at Tenerife. In 1996 he was elected president of the Association of German Anthropologists (Gesellschaft für Anthropologie) for the period 1997–98. In 1997 he was a visiting professor of the University of Mexico at Mexico City. He has carried out research in the USA, Mexico, Union of Soviet Socialist Republics, Italy, Spain, Austria, Switzerland, Greece, Turkey, Iran, Syria, Jordan, Egypt, Tunesia, etc. He has developed special methods and techniques for macroscopic and microscopic research in paleopathology. He has published work on paleopathology, especially on the etiology and epidemiology of diseases in prehistoric and early historic populations in the Old and New Worlds, prehistoric anthropology and paleoanthropology, comparative morphology and functional anatomy of primates, archeology of the North American Southwest, and also recent human anatomy and forensic taphonomy. He is the editor of the book series "Advances in Paleopathology and Osteoarcheology" and member of the editorial board of several national and international scientific journals. He has organized several national and international scientific meetings (e.g., Xth European Meeting of the Paleopathology Association in 1994). Address: Department of Anatomy, Goettingen University, Kreuzbergring 36, D-37075 Goettingen, Germany.

Tore Solheim is professor at the Department of Oral Pathology, Odontologic Faculty, University of Oslo, Norway. After dental education in Oslo in 1966, for two years he worked in the Public Dental Health service in Norway. In 1969 he finished a one year postgraduate education in general biology. In 1971 he started training in forensic odontology, and he in 1974 was attached to the department of Oral Pathology as associated professor with special responsibility for forensic odontology. His main teaching responsibility has been for forensic odontology and general pathology and in addition oral pathology. When the permanent Norwegian Identification Commission was created in 1975, he was appointed a member. Under the Commission he has been responsible for most of the dental identification in single cases and also in a number of major disasters, including the "Alexander L Kielland" oil rig disaster in 1980, where 124 persons died, the "Scandinavian Star" ferry disaster in 1990, where 158 people died, and also the Russian Tupolev disaster at Svalbard 1996, where 141 people died. In 1976 he was acknowledged by the Department of Justice as a forensic medical expert in odontologic questions. In 1986 he was also recongnized by the Department of Health as qualified in Oral Pathology. In 1982 he was on study-leave for forensic odontology in Adelaide, Australia (Dr. Kenneth Brown), in 1983 in Seattle, USA (Dr. Thomas Morton) and in 1995 in Melbourne, Australia (Dr. John Clement). His main research interest has been in identification techniques, especially age estimation in adults. In 1993 he was awarded the degree of dr. odont. for his thesis "Dental age-related regressive changes and a new method for calculating the age of an individual." The new method was based on a systematic study of a number of age related changes in the teeth. Current main research interests are systems of coding dental information and computerization for identification and epidemiology of dental restorations and their discriminatory potential in identification. Address: Department of Oral Pathology, Section of Forensic Odontology, Postboks 1052 Blindern, N-0317 Oslo, Norway.

Thomas F. Strohm was studying to be a dentist at the Free University of (West)Berlin, Germany, and the University of Erlangen-Nürnberg, Germany, where he received his dental degree (Dr. med. dent.). Later on he was employed at the University of Freiburg (Germany) with focal point in periodontology, besides he was a student of archaeology at the Albert-Ludwigs-University of Freiburg, Germany. Now he is established as a dentist. Dr. Strohm is member of the Gesellschaft für Anthropologie, Germany. Address: Urbanstr. 8, D-79104 Freiburg, Germany.

Eugen Strouhal, born 1931, received his MD in 1956, his Dr. degree in archeology and history in 1959 at Charles University in Prague, his PhD in physical anthropology in 1968 at Comenius University in Bratislava and his habilitation in 1992 in Prague. After long-term positions at the Czechoslovak Institute of Egyptology and in the Department of Prehistory and Antiquity of the Náprstek Museum, he became head of the Institute for the History of Medicine at 1st Medical Faculty, Charles University, Prague in 1990. His research concentrates on anthropology, paleopathology and history of medicine, with focus on ancient Egypt and Nubia. Presently, his main interest is in the history and paleopathology of malignant tumours. During 20 Czechoslovakían, American, French and Austrian research expeditions he worked in Egypt. His list of scientific publications contains about 220 entries, including 10 monographs, of which *The Life of the Ancient Egyptians* achieved 16 editions in 8 languages and was elected "Book of the Year" by the Oklahoma University Press in 1992. Currently he teaches medical history, paleopathology and anthropology at the faculties of medicine, sciences and arts of Charles University. He is member of many learned societies and particularly active in the European Anthropological Association. Address: Institute for the History of Medicine. 1st Medical Faculty, Charles Universtiy Prague, Katarinska 32, 12108 Prague 2, Czech Republic.

Wolf-Rüdiger Teegen, born 1961, studied European prehistory, history of medicine, physical anthropology, Near Eastern archaeology, and Italian studies at the Universities of Göttingen and Rome "La Sapienza". 1996 Dr. phil. in European prehistory at the University of Göttingen (*Der Pyrmonter Brunnenfund*). Co-author of *Starigard VI – Die menschlichen Skeletreste* (with Schultz M), Neumünster, in press. Since 1991 member of the paleopathology team at the Anatomy Center of the Georg August University of Göttingen. Address: Department of Anatomy, Göttingen University, Kreuzbergring 36, D-37075 Göttingen, Germany.

Maria Teschler-Nicola (Dr. phil., University of Vienna 1976, habilitation 1993) is currently head of the Department of Anthropology, Natural History Museum Vienna, Austria. Her research fields include paleopathology, skeletal biology and demography of Neolithic and Early Bronze age populations in Austria. She is editor of the volume *Mensch und Umwelt im Neolithikum und der Frühbronzezeit in Mitteleuropa. Ergebnisse interdisziplinärer Zusammenarbeit zwischen Archäologie, Klimatologie, Biologie und Medizin.* Recent publications include *Age- and sex-dependent cancellous bone changes in a 4000 y BP population* (with Kneissel M, Boyde A, Hahn M, Kalchhauser G and Plenk H), Bone 1994; *Suggestions for improving the objectivity of paleodemographic data, as exemplified in the analysis of Early Bronze age cemeteries of the Lower Traisen Valley* (with Prossinger H), Anthropologie 1992; *Stapedial footplate fixation in a 4000-year-old temporal bone from Franzhausen II, Austria* (with Ziemann-Becker B, Pirsig W and Lenders H) J Osteoarcheol 1994; *Priest, hunter, alpine shepherd, or smelter worker?* (with Gössler W, Schlaghaufen C, Irgolie KJ, Wilfing H and Seidler H) In: Der Mann im Eis. Neue Funde und Ergebnisse. Springer, Wien. She is member of the advisory commitee of the German Society of Anthropology and member of Paleopathology Association, the Gesellschaft für Anthropologie, the European Anthropological Association and the American Association of Physical Anthropology. Address: Natural History Museum Vienna, Department of Anthropology, Burgring 7, A-1014 Wien, Austria.

Jens C. Türp received his dental degree (Dr. med. dent.) from the Albert Ludwigs University at Freiburg (Germany) in 1988. He is currently a Visiting Assistant Professor of Dentistry at the Department of Biologic and Materials Sciences, School of Dentistry, University of Michigan, Ann Arbor. Dr. Türp's clinical main research is facial pain. Additional research topics are anatomy and function of the temporomandibular joints, dental pathology, dental anthropology and dental evolution, and history of dentistry. His publications include *Zahnfegen, Zahnpinsel, Zahnputzhölzer: Zur Aktualität traditioneller Formen der Mund- und Zahnhygiene,* Curare, 1990; *Lae kariesvoorkoms in Sub-Sahara-Afrika gesien in die lig van ontoereikende tandheelkundige dienste* (with Carstens IL), Journal of the Dental Association of South Africa, 1991; *Zur histomorphologischen und röntgenologischen Differenzierung von periapikalem Granulom und radikulärer Zyste – mit historischem Exkurs* (with Wächter R, Alt KW), Parodontologie, 1992; Curriculum Prothetik, Vol. I-III (co-authored with Strub JR et al.), Quintessenz, Berlin, 1994; *Designating teeth: The advantages of the FDI's two-digit system* (with Alt KW), Quintessence International, 1995; *Determining condylar and ramus height with the help of an Orthopantomogram – a valid method?* (with Alt KW et al.), Journal of Oral Rehabilitation, 1997; Die Evolution der Zähne (co-edited with Alt KW), Quintessenz, Berlin, 1996; *Richard Owen and the Comparative Anatomy of Teeth* (with Brace CL, Alt KW), Bulletin of the History of Dentistry, 1997; *Morphology of the temporomandibular joint of primates – with special consideration of Pongidae and Hominidae* (with Alt KW, Picq PG), in: The paranasal sinuses of higher primates. Development, function, and evolution, Koppe T, Nagai H (eds), University Press, Tokyo, 1997. Dr. Türp is a member of several professional organizations, among them the American Association of Physical Anthropologists, the Dental Anthropology Association, and the International Association of Dental Research. Address: Department of Biologic and Materials Sciences, School of Dentistry, University of Michigan, Facial Pain Clinic, Ann Arbor, MI 48109–1078, USA.

Peter S. Ungar, born 1963, received his B.A. in Anthropology from the State University of New at Binghamton in 1985 and his Ph.D. in Anthropological Sciences from the State University of New York at Stony Brook in 1992. He was a Postdoctoral Fellow in the Department of Cell Biology and Anatomy at the Johns Hopkins University School of Medicine between 1992 and 1993, and was a Research Associate in the Department of Biological Anthropology and Anatomy at the Duke University Medical Center between 1993 and 1995. He is now an Assistant Professor in the Department of Anthropology at the University of Arkansas in Fayetteville. Dr. Ungar's main research interests have focused on the reconstruction of diet and tooth use in early hominids and other fossil primates through the comparative study of the relationships between teeth and feeding behavior in living primates in the wild. He has participated in paleontological field work projects in the United States, Hungary and Jordan, and primatological field work projects in Venezuela, Costa Rica, and Indonesia. He has associated the diets and ingestive behaviors of numerous living higher primates to aspects of dental morphology and patterns of microscopic tooth wear. Dr. Ungar's research also involves the development of computer-assisted methods of analysing tooth morphology and microscopic wear, and he has studied tooth shape and wear in australopithecines from South Africa, and Miocene apes from Europe and Africa. This work is published in book chapters and journals including the *American Journal of Physical Anthropology, American Journal of Primatology, Anthropological Közlemények, Cota Zero, Folia Primatologica, Journal of Human Evolution, Proceedings of the National Academy of Sciences (USA),* and *Scanning.* Dr. Ungar's research has been funded by the US National Science Foundation, the Boise Fund, the LSB Leakey Foundation, the Andrew Mellon Foundation, Sigma Xi, the Arkansas Space Grant Consortium, and the University of Arkansas. Dr. Ungar is a member of the American Association of Physical Anthropologists, the Dental Anthropology Association and Sigma Xi. Address: Department of Anthropology, Old Main 330, University of Arkansas, Fayetteville, AR 72701, USA.

Werner Vach, received his PhD from the Department of Statistics, University of Dortmund, Germany. He was working at the Center for Data Analysis and Model Building and the Institute of Medical Biometry and Medical Informatics at the University of Freiburg and is currently professor at the Odense University, Denmark (Department of Statistics and Demography). Dr. Vach has written several papers on the application of statistical methods in medicine, biology, anthropology and archaeology with special emphasis on incomplete data problems. His contributions (all together with Alt KW) relevant to anthropology include *Zur statistischen Analyse der horizontalstratigraphischen Verteilung eines odontologischen (epigenetischen) Merkmals,* Fundber Baden-Württemberg, 1990; *The reconstruction of genetic kinship in prehistoric burial complexes – problems and statistics,* in: Classification, Data Analysis, and Knowledge Organization, edited by Bock HH, Ihm P, Springer 1991; *Non-spatial analysis of genetic kinship in skeletal remains,* in: Analysing and Modeling Data and Knowledge, edited by Schader M, Springer, 1992; *Detection of kinship structures in prehistoric burial sites,* in: Computing the Past – Computer Applications and Quantitative Methods in Archaelogy, edited by Andresen J, Madsen T, Scollar I, Aarhus University Press, 1993; *Detection of blocks in a binary matrix – a Bayesian approach,* in: From Data to Knowledge: Theoretical and Practical Aspects of Classification, Data Analysis and Knowldege Organization, edited by Gaul W, Pfeiffer D,

Springer, 1996. Address: Department of Statistics and Demography, Odense University, Hestehaven 201, DK-5220 Odense 50, Denmark.

Rüdiger Wächter, M.D., D.D.S., born 1958 in Herrenberg, Germany. Studies: medicine and dentistry in Heidelberg, Freiburg and Tübingen. Visiting clinical courses in Austria (Linz), Scotland (Edinburgh), Australia (Sydney). Internship: in Stuttgart and Tübingen: Department of Oral and Maxillofacial Surgery (Prof. Dr. Dr. N. Schwenzer). Post graduate education: Scientific assistant in the Pathological Department of the University of Erlangen (Nürnberg) (Prof Dr. V. Becker): *Histomorphometric study on the lipomatosis of the pancreatic gland.* Since 1989 scientific assistant at the Department of Oral and Maxillofacial Surgery of University of Freiburg (Prof. Dr. W. Schilli). He is currently established as dental surgeon. Member in several oral and maxillofacial associations. Speaker in many national and international congresses and meetings. Topics of main research: Paleopathological and histomorphological research. Morphology and differentiation of periapical cystic bone lesions. Osseointegration of dental implants under the circumstances of irradiation (experimental and clinical studies). Dental implants in vascularized and non-vascularized bone grafts. Alloplastic mandibular reconstruction in tumor surgery and after complex mandibular fracture. Mandibular reconstruction using microvascular bone grafts; value for dental implantation. Characterisation of new titanium surfaces (experimental and clinical studies). Evaluation of 3D-CI-reconstructions in patients suffering from panfacial fractures. Ultrasonographic study of the stomatognatic system. His numerous publications include *Pulpoalveolar disease: Etiology, incidence and differentiation of periapical lesions* (with Alt, K.W. and Türp, J.C.), J. Paleopathol 1992; *Bilaterale Unterkieferfraktur an einem Schädel aus dem 17. Jahrhundert – Möglichkeiten und Grenzen der paläopathologischen Diagnostik* (with Alt, K.W. and S. Ullrich-Bochsler), Quintessenz 1993. Address: Heinrichstr. 35, D-36037 Fulda, Germany.

Christoph Willms (Dr. phil., University of Münster/Westfalia, 1978) is involved in interdisciplinary research concerning archeological metallurgy; this research is supported by grants from the Volkswagen-Stiftung, Hannover. He has been on the staff of the University of Frankfurt/Main, where he investigated ecological aspects of human impact on the disappearance of flora and fauna, primarily during the Neolithic. Subsequently, he was involved in archeological research for the Archäologischer Dienst in Bern, Switzerland. A further Aspect of his research involves studies in human nutrition during the Neolithic. Recent publications include *Zwei Fundplätze der Michelsberger Kultur im westlichen Münsterland, gleichzeitig ein Beitrag zum neolithischen Silexhandel in Mitteleuropa,* 1986; *Der Hausesel nördlich der Alpen,* 1990; *Mensch und Natur in der Jungsteinzeit. Über Eingriffe des Menschen in seine Umwelt in der Zeit von 5600–2400 v.Chr.,* 1991; *Getreide im europäischen Frühneolithikum,* 1991. Address: Institute of Prehistoric Archeology, University of Münster, Domplatz 20–22, D-48143 Münster, Germany.

1 Dental Anthropology – An Introduction

Kurt W. Alt, Friedrich W. Rösing, and Maria Teschler-Nicola

Dental Anthropology provides an excellent view into biological, ecological and cultural aspects which help to detect and understand individuality, human behavior, living conditions, and environments. Teeth are used to separate fossil hominids, demonstrate trends in hominid dentition, reflect individual and group patterns of demography, biological relationships in the context of affinity and kinship, aspects of diet and cultural adaptation, and supply information on dental health, art, cult, and custom in fossil and archeological series. In forensic odontology and anthropology, they permit the identification of unknown bodies in the context of mass disasters, and the evaluation of bitemarks in corpse or objects.

Teeth and jawbones are used to address questions in numerous disciplines including paleoanthropology, paleontology, prehistoric anthropology, archeology, dentistry, comparative anatomy, genetics, embryology, and forensic medicine. Which are the main advantages of dental remains to make them an object of study in so many disciplines? Jaws and teeth are more durable compared to skeletal remains (less post-mortem decomposition, best represented part of skeleton, record of fossil species, past and recent population), they possess a high degree of morphological individuality representing personal, familial, and population characteristics, and they can be directly observed and evaluated in both living and past populations. Furthermore, because of their high heritability they are useful in assessing evolutionary and population origins, developments and dynamics, they reflect dietary and cultural behavior and environmental effects. And finally, the non-genetic characteristics of teeth such as wear and disease make them well suited for research of dietary adaptations, regional variation in disease manifestations, epidemiological status and others. Dental anthropology makes ample use of this research potential. This discipline has a unique holistic view of teeth, striving to place them in every possible context.

One objective of Dental Anthropology is the reconstruction of the phylogenetics of humans and primates. Our understanding of primate evolution is ultimately based on patterns of phyletic relationships and morphological change in the fossil record. In this field, teeth are a prime source of information ("key structures") to reconstruct the form and life history of early hominids, to understanding biological adaptation and patterns in human evolutionary ecology. If, as is frequently the case, only teeth have survived, taxa are defined as *dental species*. New analytical developments and conceptual advances – especially in dental anthropology – have produced an enormous number of new answers to the main questions of paleoanthropology, i.e. the relationships of the hominoids and the split of the hominid lineage from those of other primates, the morphological changes in the hominid phylogeny, the ecological niches of the fossil hominids, and finally the interrelationships of the various fossil species. In this research sector, comparative anatomical, macroscopic, microscopic, phylogenetic, biochemical, molecular, and ecological methods and results dominate.

Another objective of Dental Anthropology is the biological reconstruction of early populations (prehistoric anthropology), using the ontogenetic and populational variability of teeth. In this field, teeth are decisive for understanding biological developments and dynamics as well as cultural and economic processes. The teeth of our ancestors are useful for the reconstruction of life history by demographic parameters (estimation of age and sex), morphological (anatomical) variants, individual features, nutritional patterns, origin and population history, identification of familial relationships (for reconstructing social structures in past populations), accidental and intentional cultural behaviour (artificial dental modifications), and dental diseases. Apart from classical methods in this field, many innovative techniques such as extraction of ancient DNA (aDNA), trace element and stable isotope analyses are used in this context. Three applications of aDNA analysis are of interest: access to genetic information at the individual, at the infrapopulation, and at the interpopulation level. Trace element and stable isotope analysis is helpful in the detection of subsistence strategies, endogamy versus exogamy, migration, social differentiation, ontogenetic trends, toxic accumulation of elements such as Pb or As, and paleopathological features.

Finally there is the forensic objective of Dental Anthropology. In forensic medicine teeth play an essential role in the personal identification of unknown bodies, of victims of crimes and natural or civil mass disasters, and in cases of individuals in mass graves, victims of armed conflict or of political terror. This objective, however, rather forms a discipline of its own, forensic odontology. Apart from routine analyses, dental findings are used in investigations concerning estimation of least number of individuals, population or ancestry of the individuals, reconstruction of nutritional status and health history, occupational markers or features caused by habitual activities, trauma, or other lifetime events. In case of isolated parts of bodies or skeletons, death by fire etc., oral findings are often the only evidence for the identity of the victim. In addition to the number and distribution of teeth, restorations, dentures, congenital anomalies and other dental characteristics may aid the identification. In forensic medicine, teeth, like (DNA) fingerprints, are individual, but as they resist the ravages of time far better than other parts of the body, they represent an unsurpassed record of the individual.

This spectrum of research objectives shows on the one hand that dental anthropology is mainly rooted in the framework of biology, and on the other that Dental Anthropology forms a strong bridge to paleontology, dentistry, genetics, ontogenetics, and to the humanities. Generally the results of dental anthropology may be incorporated into the body of knowledge of more than one science.

These considerations lead to the concept of this volume, which was initially formed some years ago. Starting with aspects of *Teeth in History*, the book continues with chapters on *Dental Morphology, Structure, and Evolution*; *Dental Pathology and Epidemiology*; on teeth in the context of *Nutrition and Human Behaviour*; *Age and Sex Estimation* by dental parameters; and *Geographical and Familial Tooth Variation*. The 33 authors of the 26 contributions are specialists in many fields: dental anthropology, paleoanthropology, prehistoric anthropology, molecular anthropology, paleopathology, forensic odontology, forensic anthropology, archeology, anatomy, embryology, history of medicine, dentistry, and statistics.

Consequently most aspects of the book are new, particularly the syntheses. This is now a volume which provides an introduction to the field as well as a reference both for specialists and students in anthropology, paleontology, ecology, dentistry, and the cultural sciences. The basic literature and experience is of a broad international origin and not limited to English language sources only. Moreover, the book can also contribute to further research on selected topics of the field. Numerous illustrations and tables help to clarify the statements given in the text.

The production of this book has been generously supported by: Bundeszahnärztekammer Köln (Präsident: Dr. F. J. Willmes), Gesellschaft von Freunden und Förderern der Heinrich-Heine-Universität Düsseldorf e.V. (Präsident: R. Schwarz-Schütte), Naturhistorisches Museum Wien (Generaldirektor: Prof. Dr. B. Lötsch), and Universitätsgesellschaft Ulm (Vorsitzender: E. Ludwig). The English language revision was performed by Dr. Valery Lee, Ann Arbor, Michigan, USA and Heidi Asmus, Ulm, Germany. Finally we thank the Springer-Verlag Wien, New York, particularly R. Petri-Wieder, for the efficient, flexible, and highly motivated cooperation.

Kurt W. Alt, Freiburg, Germany March 1998
Friedrich W. Rösing, Ulm, Germany
Maria Teschler-Nicola, Vienna, Austria

2 Teeth in History

2.1 Johann Wolfgang von Goethe, Weimar, and Dental Anthropology

Kurt W. Alt

Introduction

The coincidence of several facts gave rise to this short report, which we are placing at the beginning of this volume. *Weimar* is the town where the symposium was held which finally gave rise to this volume. At the same time, Weimar is also the place where the German poet and naturalist *Johann Wolfgang von Goethe* (1749–1832), one of the town's greatest sons, lived and worked for most of his life (Fig. 1). His studies in the field of skeletal morphology touch on the main theme of our 1993 symposium: *Dental Anthropology*.

Fig. 1. Johann Wolfgang v. Goethe; copperplate by Heinrich Lips. Goethe-Nationalmuseum. Stiftung Weimarer Klassik. Photo: S. Geske.

Goethe won fame not only by his literary accomplishments, but also by his studies in the natural sciences, where he made important discoveries. In biology and dentistry his name is mentioned primarily in two contexts (see Worm 1922). In 1784 *Goethe* detected and described the premaxillary bone in man (Goethe 1820), and he was the first to report a case of ectopic eruption of a tooth into the nasal cavity in 1797 (Goethe 1978).

Autobiographical aspects relating to the discovery of the intermaxillary bone are encoded in *Goethes* dramatic poem "Faust", in which Dr. Faust is confronted with the figure of Mephistopheles. According to Hellmich and Hellmich, there are certain indications "that the person of *J. H. Merck* can be recognized in this figure, and that *Goethe's* discovery of the human intermaxillary bone had its part in creating the figure of Mephistopheles as we know it today" (1982, 553).

In *Goethe's* life and works, health, treatment, and healing are ever-recurring themes (Nager 1992; Nechwatal 1992). Especially the dental problems he was frequently afflicted with are often mentioned in his correspondence (Goethe 1786). Yet to begin with, we will take a closer look at some of his scientific accomplishments.

Premaxilla (os intermaxillare, os incisivum)

Even though the human os intermaxillare, resp. a suture between the os intermaxillare and the os maxillare had already been described in Antiquity (*Galen* 129–199), the existence of such a bone was denied by leading anatomists of the 16th to 18th centuries like *Andreas Vesalius* (1514–1564), *Pieter Camper* (1722–1789) and *Samuel Thomas Sömmering* (1755–1830) as well as by the "father of anthropology", *Johann Friedrich Blumenbach* (1752–1840). The intermaxillary bone was therefore believed to represent a distinguishing feature between humans and apes until the end of the 18th century (Camper 1778; see also Trefz 1989).

The human maxilla is a bone separated from its mate by the intermaxillary suture and is derived from at least two components, the maxilla proper, forming in the region of the canine to the molar teeth, and the premaxilla (an autonomous bone), arising in the incisal region. In man, both premaxilla and os maxillare resp. the sutures between the os intermaxillare and the surrounding bones merge at an early ontogenetic stage whereas they partially persist in primates (Vogel 1965), "so that, in the adult, the facial aspect at least appears both macroscopically and microscopically as a continuous bony mass. An occasional adult skull and all skulls at birth show an incisive suture between the premaxilla and maxillary components, particularly in the palatine process" (Chase 1942, 1991).

In most animals other than man, these two bones remain more or less distinctly separated by sutures throughout life. It bears the incisors in most vertebrates, and is also present in animals lacking the incisory teeth (cervides). The reduction in the incisory region is often accompanied by a reduction of the os intermaxillare (e.g. in microchiroptera, xenarthra). Yet this process is not inevitable as the presence of the os intermaxillare in the tubulidentata and artiodactyla demonstrates, which also lack incisors (Starck 1979).

Goethe had been conducting studies on the collection of the anatomist *Loder* in Jena from 1781 till 1782. In the course of these anatomical studies, which aroused his interest in osteology and comparative investigations of animals and humans, he discovered the disputed intermaxillary bone in 1784 (Fig. 2). *Goethe* was highly pleased with his discovery as he was attempting to demonstrate the presence of all mammalian bones in man. In a letter dated March 27th of the same year he wrote to *J. G. Herder*:

> "I have found – neither gold nor silver but still something which gives me unspeakable joy – the human os intermaxillare! I compared human and animal skulls with Loder, caught the scent, and look, there it is. [...] because it is like the keystone of man, it is not missing, it is actually there." (Jena 1784)

Goethe's report on the os intermaxillare dating to the year of the discovery bears the title:

> "Both man and animals are to be ascribed an intermediate bone of the upper jaw." (Jena 1786)

It was initially available to a limited number of people as a manuscript until it was published (although without illustrations) in his second booklet on the natural sciences, "Zur Morphologie", in 1820.

When *Goethe* discovered the os intermaxillare he was not aware of the fact that the French anatomist *Vicq d'Azyr* had already reported on it in the year 1780 (Ashley-Montagu 1935). *Goethe's* merit in the debate on the intermaxillary bone lies in the application of methodical criteria and their application in comparative

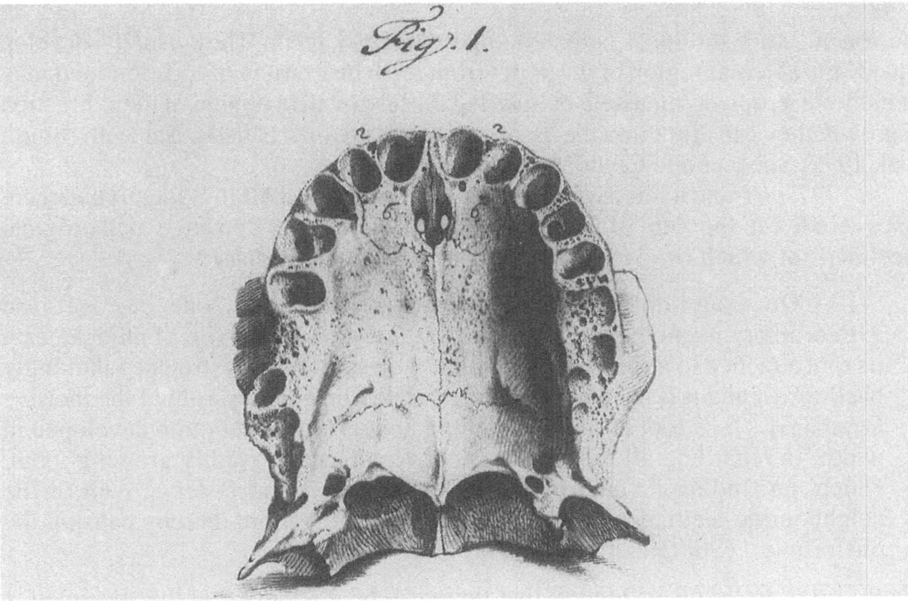

Fig. 2. Human os intermaxillare, palatal view; ink drawing by J. C. W. Waitz 1784. GSA Stiftung Weimarer Klassik, Weimar. Photo: S. Geske.

biological studies (Franz 1933; Vogel 1965). In this context he wrote about the incomprehensible rejection of the finding by contemporary anatomists:

> "Here now we had the strange case, that one meant to attribute the difference between apes and man to the fact that the first were ascripted an intermaxillary bone, whereas the second were not; however, as the aforementioned bone is most noteworthy because it provides the setting for the upper incisor teeth, it was all the more incomprehensible that humans should have incisors but lack the bone wherein they are set." (Jena 1819)

In allusion to *Camper* the lack of recognition of his findings by the emminent authorities in the field again prompted him to the following lines in 1819:

> "It certainly shows a particular lack of worldly wisdom, a kind of youthful self-confidence, when a mere amateur apprentice dares to contradict the masters of the guild, and, even more foolishly, tries to convince them of this argument." (Jena 1819)

Goethe's statements concerning the os intermaxillare touch on central, recurring questions in morphology studies: first, research in homologies is nearly exclusively constituted by comparative anatomy, second, form is determined by function (cf. Vogel 1965). His inferences are based on well-defined methodical concepts, especially on evaluation criteria called for in anatomical studies. That was his actual credit in the discovery of the os intermaxillare.

Nasal Tooth

Mesiodentes are the most common supernumerary teeth. They usually develop outside the alveolar region of the upper front teeth in a variety of locations and may be multiple, erupted, impacted, or inverted. In case of an inversion of their direction of growth they can erupt into the nasal cavity and are then called nasal teeth (Smith et al. 1979; Spencer and Couldery 1985).

In 1797, en route to his Swiss journey *Goethe* stopped off in Stuttgart and there discovered, on the 6th of September with *Rapp*, the "strange osteological specimen" of which he gives a detailed description in his diary:

> " [...] On examining the nasal cavities on the skull, one discovers an extraordinary phenomenon: there is a tooth just under the rim of the eye, with its root attached to a small, round, wrinkly mass of bone. It stretches slantingly backwards, and it is as if it pierced the palate immediately behind the incisive foramen. [...]; its root is simple and long and its crown not quite developed in width. In regarding all this it appears to be a healthy, readily growing tooth, which, on finding its path barred by the irregular and faster growth of the neighbouring teeth, developed in a backwards direction, thereby causing the misfortune." (Stuttgart 1797)

Goethe's description is so precise that there can be no doubt that the specimen is a mesiodens, more precisely a nasal tooth. He himself believed that it was "the missing molar" which had become retained (Goethe 1797, 144). The honour of the first description of the finding thus befits *Goethe* in spite of its misinterpretation.

The incidence of nasal teeth is very small (0.01%; see Wood and Mackenzie 1987), yet there are some observations in prehistoric populations (Alt 1990) and in anthropoid primates (Schwartz 1984). The etiology of mesiodentes is unclear. Whereas in former times phylogenetic theories were discussed concerning the pathogenesis, it is today believed that they are caused by remnants of the dental laminae or accessory laminae. Complication such as obstruction, headache, nasal discomfort, epistaxis, purulent rhinorrhea, and rhinitis caseosa are prevented by the surgical removal of intranasal teeth (Wood and Mackenzie 1987).

Johann Wolfgang von Goethe: Homo patiens

Like other mortals, *Goethe* was not spared from disease (Nager 1992) and, as we know from many letters to his contemporaries, for all his life his dental problems were especially troublesome to him (Nechwatal 1992). In these correspondences he often complains about "swollen cheek", "toothache" and "gumboils" which often confined him to bed (Dietz 1931). On the 7th of April 1786 he writes to Duke *Carl August* from Weimar:

> "I am most unhappy that I am unable to take up your invitation, but must remain at home instead. A small swelling of the tooth which caused me so much trouble last year in Neustadt, and which I have been trying for a week to suppress, has developed now in to a gumboil, feels so tight and throbs so much that I feel I may be overcome at any minute." (Weimar 1786)

And on the following day:

> "I am most sorry to spoil your outing and impede your plans. I fear that my trouble has developed as I predicted. My cheek is swollen and I am forced to pack my face about with herbal pouches." (Weimar 1786)

Goethe's case history presents a succession of dental problems, which in retrospect comprised purulent, febrile dental infections, gingivitis, dentitio difficilis and abscesses (Nechwatal 1992). Other than the "opening of gumboils", treatment in *Goethe's* time consisted of oral rinses, herbal packs and in the end the inevitable extraction of affected teeth.

Yet all of *Goethe's* health problems did not prevent him to gain immortality in the literary world. Neither were they a hindrance in *Goethe's* pursuit of the natural sciences as various theoretical and comparative studies impressively demonstrate (Bräuning-Octavio 1982; Krätz 1995).

Acknowledgements

I wish to extend thanks to Dr. S. Pichler, Prehistoric Institute, Albert Ludwigs University, Freiburg and A. F. Bulmer BA, English Seminar, Heinrich Heine University, Düsseldorf for help in the translation, and to Dr. F. Bartelt, University Library, Heinrich Heine Universität,

Düsseldorf for her invaluable assistance in unearthing works of Goethe. The author is especially grateful to the Herzogin Anna Amalia Bibliothek, Stiftung Weimarer Klassik, for the generous consent to the reproduction of the figures.

References

Alt KW (1988) Mesiodens in der Nasenhöhle bei einer frühmittelalterlichen Bestattung. Quintessenz 39: 1075–1081.

Alt KW (1990) Nasal teeth: Report of a historic case. Int J Anthrop 5: 245–249.

Ashley-Montagu MF (1935) The premaxilla in the primates. Quart Rev Biol 10: 32–59; 181–208.

Bräuning-Octavio H (1982) Goethes naturwissenschaftliche Schriften und die Freiheit von Forschung und Lehre. Jahrbuch des Freien Deutschen Hochstifts. Niemeyer, Tübingen.

Camper P (1778) Kurze Nachricht von der Anatomie verschiedener Orang-Utangs. In: Kleinere Schriften Bd.1, Leipzig.

Chase SW (1942) The early development of the human premaxilla. J Am Dent Assoc 29: 1991–2001.

Dietz MA (1931) Goethes Zahnleiden und Zahnärzte. Med Diss, Würzburg.

Franz V (1933) Goethes Zwischenkieferpublikation nach Anlaß, Inhalt und Wirkung. Mit Ausblicken auf Goethes Morphologie überhaupt. Ergebn Anat Entwicklungsgesch 30: 469–543.

Goethe JW von (1784) Goethes Briefe, 6. Band: 1. Juli 1782 – 31. Dezember 1784 (Weimarer Ausgabe). Böhlau, Weimar 1890.

Goethe JW von (1786) Goethes Werke, Band 13: Naturwissenschaftliche Schriften 1. Textkritisch durchgesehen und kommentiert von D. Kuhn und R. Wankmüller (Hamburger Ausgabe). Beck, München 1981.

Goethe JW von (1797) Reise in die Schweiz 1797. Gedenkausgabe der Werke, Briefe und Gespräche, Bd 12. Herausgegeben von E. Beutler. Artemis, Zürich 1949.

Goethe JW von (1819) Morphologische Hefte. Erster Band, Zweites Heft. Die Schriften zur Naturwissenschaft. Abtl. 1, Band 9. Bearbeitet von D. Kuhn. Böhlau, Weimar 1954.

Hellmich S, Hellmich J (1982) Der Zwischenkieferknochen und Goethes Mephistopheles. Laryng Rhinol Otol 61: 552–556.

Krätz O (1992) Goethe und die Naturwissenschaften. Callwey, München.

Nager F (1989) Goethe – homo patiens. Swiss Dent 10: 25–32.

Nager F (1992) Der heilkundige Dichter. Goethe und die Medizin. Artemis, Zürich.

Nechwatal N (1992) Zahnweh – Die Zahnheilkunde in der Dichtung. Wittal, Wiesbaden.

Reichenbach E (1968) Goethe und die Stomatologie. Dtsch Stomat 11: 556–568

Schwartz JH (1984) Supernumerary teeth in anthropoid primates and models of tooth development. Arch Oral Biol 29: 833–842.

Smith RA, Gordon NC, De Luchi SF (1979) Intranasal teeth. Oral Surg Med Pathol 47: 120–122.

Spencer MG, Couldery AD (1985) Nasal tooth. J Laryngol Otol 99: 1147–1150.

Starck D (1979) Vergleichende Anatomie der Wirbeltiere, Bd. 2. Springer, Berlin Heidelberg New York

Trefz H (1989) Goethe et l'os intermaxillaire. Actualités Odonto-Stomatol 167: 407–415.

Vicq d'Azyr F (1780) Observationes anatomiques sur trois singes [...]. Hist Acad Roy Sci Paris, p. 478–493.

Vogel C (1965) Varianten des Zwischenkiefers und ihre Abhängigkeit von der Gebißdifferenzierung bei den Primaten einschließlich der Hominiden. Fortschr Kieferorthop 26: 357–376.

Wood GD, Mackenzie I (1987) A dento-nasal deformity. Oral Surg Med Pathol 63: 656–657.
Worm W (1922) Stomatologisches bei Goethe. Berlinische Verlagsanstalt, Berlin.

2.2 The History of Dental Anthropology

Kurt W. Alt, C. Loring Brace, and Jens C. Türp

A tooth! A tooth!
My kingdom for a tooth!

Thomas Huxley (1858)

Introduction

Among the diverse tasks of the human dentition, the uptake and crushing of food is doubtlessly the most characteristic. The English Victorian naturalist Richard Owen (1804–1892) underscored the importance of this function in the introduction to his classic treatise *Odontography* (1840–1845): "Teeth are firm substances attached to the parities of the beginning of the alimentary canal, adapted for seizing, lacerating, dividing and triturating the food, and are the chief agents in the mechanical part of digestive function". This function, however, did not emerge until a late phase of evolution. Before that time, primitive, tooth-like structures or elements such as gill traps had facilitated food uptake (Gutmann 1997). Other utilizations of the teeth are secondary, for instance the use of teeth as weapon or tool, as a structure for the characterization of age and sex, or as ornamental objects (Kanner and Remy 1924; Alt et al. 1990; Kelley and Larsen 1991; Alt and Pichler, this volume). Additionally, among humans teeth have great significance for vocal articulation (Schumacher et al. 1990) and for the esthetics of the appearance (Alt 1994b).

Teeth and jaws usually abound among paleontological and archeological finds because of their resistance to postmortal influences. As a consequence, many phylogenetic concepts are based solely on the interpretation of tooth forms (Maier 1978). Since after completion of amelogenesis tooth crowns do not undergo further changes (except pathological [e.g., caries], and age-dependent processes [e.g., attrition]), teeth play an important role for comparative anatomical investigations and for the reconstruction of phylogenetic mechanisms in the evolution of mammals: "Tooth form varies with taxonomy and phylogeny and so can be used to reconstruct evolutionary patterns" (Foley and Cruwys 1986, 1). Teeth have therefore become "index fossils" in paleontology, paleozoology, and paleo-

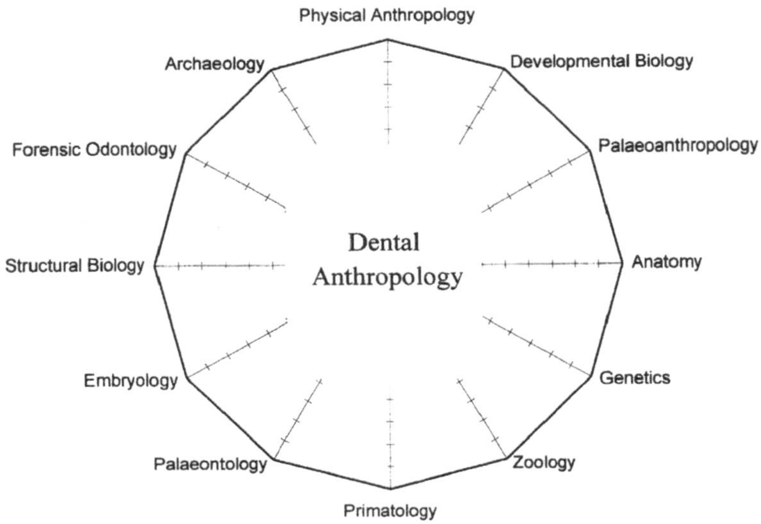

Fig. 1. Spectrum of scientific disciplines related to Dental Anthropology.

anthropology (Hillson 1986; Thenius 1989; Henke and Rothe 1994; Alt and Türp 1997a). In addition, teeth provide valuable information about environmental parameters and diet as well as answers to biostratigraphic questions (Kay 1978; Maier 1984).

Teeth and dentition are today the focus of specific investigations in anthropology, archeology, zoology, and other fields of the natural sciences and medicine (Fig. 1). In recent decades, dental anthropology has had a great impact on phylogenetic research in paleoanthropology and has contributed to advances in related disciplines such as primatology, osteology, and population biology. Yet dental anthropology is not only useful for the exploration of the past, but it also influences clinical basic research. For example, the recognition of evolutionary trends, such as the size reduction of teeth and jaws, has important implications for clinical dentistry.

Odontology, the precursor of dental anthropology, was the classical scientific discipline dealing with fundamental questions about the development and structure of teeth (Peyer 1968; Würtz 1985; Alt and Türp 1997a). During the 20th century, the interdisciplinary significance of odontology increased as innovative methods were introduced. This process was closely linked to concepts and developments in other scientific disciplines, such as comparative anatomy, zoology, paleontology, embryology, and physiology, and it was influenced by rapid advances in general and population genetics (cf. Vogel and Motulsky 1996). At the same time, methods derived from biomechanics, biochemistry, and statistics, as well as technical procedures, such as the identification of microscopical structures, provided valuable impulses for basic research in odontology.

In the 1960s, odontology became incorporated in the rapidly developing discipline of dental anthropology. With few exceptions (e.g., in "forensic odontology"), the term "odontology" has been completely replaced by "dental anthropology" (Brothwell 1963; Scott and Turner 1988). Today, dental anthropology is an important subdiscipline of physical anthropology. It interdisciplinarily combines all those fields outside clinical dentistry which are concerned with odontological questions. The historical development of dental anthropology, however, has only recently become a subject of interest (Foley and Cruwys 1986; Scott and Turner 1988; Dahlberg 1991; Alt 1997 b, c).

Origin and Development of Basic Research in Odontology

Researchers have been engaged in the study of dentitions since antiquity. The *Corpus* of Hippocrates (460–375 B. C.) is among the oldest written records on teeth. Medically related observations prevail in antique literature. Statements about the anatomy of teeth and jaws, however, are also present, and are in part based on comparative investigations. One of the oldest sources to include comparative anatomical reflections about teeth is *De generatione animalium* by Aristotle (384–322 B. C.). Considered the founder of the biological sciences, Aristotle compiled the knowledge of his time, relying on anatomical knowledge derived from animal dissections as well as on his own observations. The encyclopedic work *Natura historia* by Plinius (23–79 A. D.) also includes anatomical comments about teeth. The most important description of antique anatomy and medicine, however, is considered to be Galen's (129–189 A. D.) *De anatomicis administrationibus*, as he was frequently cited as a major source of anatomical knowledge during the following centuries.

The scientific and medical knowledge of classical antiquity was later also compiled by Islamic authors, such as Rhazes (865–923) and Avicenna (980–1037). As a result of these compilations, classical ideas were kept alive until the end of the Middle Ages, and so fundamentally shaped the development of science and medicine in Europe (Hoffmann-Axthelm 1985).

After the era of natural philosophy, an orientation toward scientific ideas was observable in the 18th and early 19th centuries which also favored purposeful research on the origin, development, morphology, and structure of teeth. However, for the time being research was limited to basic medical disciplines, such as anatomy. The establishment of independent disciplines, such as dentistry, zoology, comparative anatomy, embryology, and paleontology, and the recognition of the scientific importance of teeth formed the basis for comparative anatomical investigations on odontologic and phylogenetic questions, which were carried through on a broad basis during this period (Cole 1949).

During the 19th century, important contributions were made to basic dental science by anatomists, histologists, and dentists, such as C. H. T. Schreger (1768–1833), A. A. Retzius (1796–1860), R. Owen (1804–1892), J. Tomes (1815–1895), and O. Hertwig (1849–1922). Research was focused mainly on the histology of structural elements of teeth (e.g., Hunter Schreger striae of enamel prisms, Retzius lines, Tomes fibers; Schreger 1800; Retzius 1837; Schwann 1839; Tomes 1859; Kölliker 1863; Waldeyer 1864; Hertwig 1874). These studies also initiated investigations in comparative anatomy and zoology.

Systematic investigations on dental morphology began in the first half of the 19th century. Under the supervision of J. E. Purkinje, Raschkow (1835) and Fränkel

(1835) presented theses about the development and structure of mammalian teeth in which they revised traditional concepts on amelogenesis and contributed much to the theory of the development and structure of dental enamel (Würtz 1985). A few years later, Owen published the first comprehensive monograph on the comparative dental anatomy of living and fossil animals (Fig. 2). His *Odontography* (1840–45) remained the basis for all comparative investigations for a long time and

ODONTOGRAPHY;

OR, A

TREATISE

ON THE

COMPARATIVE ANATOMY OF THE TEETH;

THEIR PHYSIOLOGICAL RELATIONS, MODE OF DEVELOPMENT,

AND

MICROSCOPIC STRUCTURE,

IN THE

VERTEBRATE ANIMALS.

BY RICHARD OWEN, F.R.S.

CORRESPONDENT OF THE ROYAL ACADEMY OF SCIENCES OF PARIS, BERLIN, &c.&c, HUNTERIAN PROFESSOR TO THE ROYAL COLLEGE OF SURGEONS, LONDON.

VOLUME I.

TEXT.

LONDON:
HIPPOLYTE BALLIERE, PUBLISHER,
FOREIGN BOOKSELLER TO THE ROYAL COLLEGE OF SURGEONS.
219, REGENT STREET.
Paris: J. B. BALLIÈRE LIBRAIRE DE L'ACADÉMIE DE MÉDECINE.
LEIPZIG: T. O. WEIGEL
1840-1845

Fig. 2. First monograph on comparative anatomy of teeth.

stimulated various investigations in the areas of odontogeny and dental histology (Türp et al. 1997).

The reciprocal stimulation of the different disciplines had a positive effect on early research on the phylogeny of teeth. In France, odontology remained under the influence of G. Cuvier (1769–1832), the founder of paleontology. French odontology was therefore predominantely oriented towards comparative anatomy and systematics (Cuvier 1835–1846). A similar situation existed in North America, where comparative anatomical studies were primarily limited to fossils (cf. Wortman 1886; for comprehensive descriptions see Zittel 1899 and Schindewolf 1948). Yet as a result of these developments, odontology had gained a certain independence from clinical dentistry by the end of the 19th century.

Theories and Concepts on the Phylogeny of Teeth

In the 19th century it was realized that teeth can serve as a source of information for a multitude of ontogenetic and phylogenetic questions. This marked the beginning of investigations on the phylogenetic development of mammalian teeth. Studies pertaining to this subject were carried out in different scientific disciplines and were based both on fossil material and on embryologic findings. The first theories on the phylogenetic development of teeth and the morphogenesis of tooth forms were developed in paleontology around the turn of the century. Important pioneers were, among others, E. D. Cope (1840–1897), H. F. Osborn (1857–1935), and W. K. Gregory (1876–1970). Odontology, the science of the comparative morphology of teeth, is rooted in the observation made by Cope (1874, 1883) that the tritubercular type was ancestral to many if not all of the higher types of molar teeth. Cope's hypothesis was further advanced by Osborn (1888, 1907) and Gregory (1916, 1922), and it was considered "one of the most important generalizations ever made in mammalian comparative anatomy" (Osborn 1907, 3). Under the terms "differentiation theory" or "tritubercular theory" it was the basis of all theories to follow.

Extensive, well-stratified fossil finds form the material basis of the differentiation theory, which claims a uniform origin for all mammalian teeth. Due to their functionally and morphologically differentiated forms, and very different evolutionary tendencies concerning their shape, size, and structure, the various types of molars became the taxonomically most important dental category (Thenius 1989). The haplodont reptile tooth with its sharp, cone-shaped tip and its single root is considered the basic tooth form from which molars evolved. According to this theory, differentiation and further accentuation of two marginal projections are the basis for the development of a triconodont tooth with separated roots among early mammals (*Triconodonta*) in the Upper Triassic. Through changes in the position of the cusps, the triconodont tooth is believed to have evolved to a tritubercular tooth whose cusps form a triangle (see Simpson 1936; Starck 1982; Thenius 1989). From today's point of view, the tribosphenic (trituberculo-sectorical) tooth, which is perfectly suited for the crushing of food by pressure and friction of antagonist teeth, is the tooth type which all specialized molar forms can

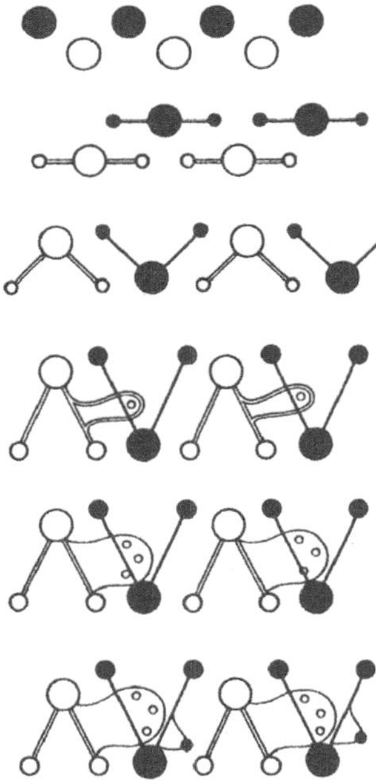

Fig. 3. Relations between maxillary (black) and mandibular (white) molars in the course of mammalian phylogenetic development; 1: haplodont reptiles – staggered positioning of teeth; 2: triconodont tooth-form – newly occuring cusps (small dots) improve shearing effects; 3: tritubercular molars – development of occlusal surfaces, efficiency low because of staggered positioning; 4 & 5: development of talonid (loop) and talonid cusps (dots inside loop) make possible efficient chewing; 6: later hypocone develops on maxillary molars (from Schulze 1987).

be derived from (Starck 1982). From the tribosphenic tooth evolved the most important basic patterns of tooth crowns as well as their combinations.

Osborn (1892) employed the original cusp terminology derived from the tritubercular tooth as a basis for a nomenclature of maxillary and mandibular molar cusps which is still used today (Fig. 3). As a denotation for the triconodont tooth form, Osborn (1895) chose terms that had been used for the description of cusp tips of the protodont tooth type: protoconus (maxilla) and protoconid (mandible) for the main cusps, and paraconus and metaconus (maxilla) as well as paraconid and metaconid (mandible) for the two marginal projections. Among other morphologically important structures, the fissure pattern of the mandibular molars known as the Dryopithecus pattern (Gregory 1916) has attained considerable significance in dental anthropology. On lower molars, the Y-like fissure pattern characteristically divides three outer cusps from two inner cusps. According to Remane (1960), the Dryopithecus pattern is an important distinguishing feature between Pongidae and Cercopithecidae. It is considered to be the original crown pattern from which hominid molars can be derived (Butler 1978; Sakai 1982).

Objections against Cope's and Osborn's differentiation theory came above all from embryologists. Advocates of the concrescence theory, such as W. Kükenthal

(1861–1922) and C. Röse (1864–1947), believed that today's molars and premolars evolved by fusion of several simple conoform teeth. Embryological studies had indicated that tooth anlagen of the same dentition in tandem arrangement as well as anlagen of different dentitions lying side by side are able to fuse. According to this theory, due to lateral fusion of the originally multiple tooth rows of lower vertebrates, only two dentitions are left among mammals (Kükenthal 1892; Röse 1892; Schwalbe 1894). The odontogenetically founded dimer (two element) theory of anatomist L. Bolk (1866–1930) represents a combination of the differentiation concrescence theories (Bolk 1913, 1914). In Bolk's view, two protodont tooth germs fuse to form a tooth with three cuspids. Two three-cuspid teeth fuse forming a tooth with six cusps. Bolk (1914) called the phylogenetically older part of the dimer tooth protomer (buccal), and the younger part deuteromer (lingual).

Resistance against the differentiation theory grew, now from paleontologists, giving rise to the premolar analogy theories of Wortman (1902) and Gidley (1906). These postulate that premolars are simple, underdeveloped molars (Maier 1978). The premolar analogy theory attributes the leading role in genesis to the simple incisor, degrading the protoconus from the first main cusp to a secondary development (Remane 1960). Though there was a high degree of consensus that with the advancing development of the numerous and originally homogeneous teeth their number was reduced, no agreement about the kind of adaptation patterns regarding tooth morphology could be reached. Researchers such as Adloff (1916) believed that tooth forms evolved in consequence of function. In contrast, champions of the descendence theory (Aichel 1917) denied any influence of function, but assumed selection to be the determining factor. Aichel was convinced that teeth influenced the selection of food, and not *vice versa*.

After the emergence of the classic theories on odontogenesis, investigations on diverse topics of dental anatomy and phylogeny continued, and knowledge and understanding of phylogeny, classification, and systematics improved (Simpson 1936; Bourdelle et al. 1937; Weidenreich 1937; Butler 1941; Remane 1960). Yet the descriptive and casuistic approach to the study of fossils laid down at the outset was critisized already in 1923 by Eidmann. He pointed out gaps that remained in the understanding of developmental processes in animal as well as human dentitions. Eidmann's objection that, because of these shortcomings, phylogenetic theories were based on unfirm grounds, has not lost its validity, as illustrated by current discussions about the *Ardipithecus ramidus* finds from Aramis, Ethiopia (White et al. 1994, 1995).

Research in Dentistry and Its Influence on Dental Anthropology

The establishment of dentistry as a scientific discipline originated in the 18th century mainly under the influence of scientists from England, Germany and France (Hoffmann-Axthelm 1985). In the beginning, major contributions were made by researchers from medical disciplines such as anatomy and histology. A major role in the emergence of scientific dentistry is adjudged to J. Hunter (1728–1793), whose internationally reknowned *Natural History of the Human Teeth* (1771) is considered to be the most important contemporary publication

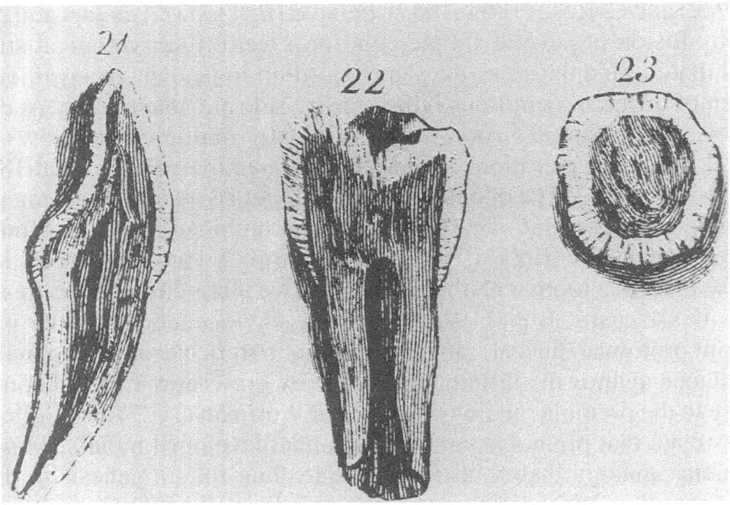

a

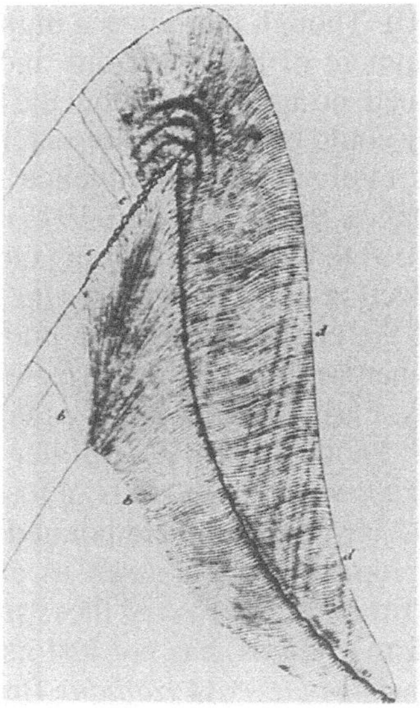

b

Fig. 4a, b. Historic specimens with (a) Hunter-Schreger-bands and (b) Retzius-lines (after Hoffmann-Axthelm 1985).

on this subject (King 1994). New insights into the development and structure of teeth (cf. Retzius 1837; Hertwig 1874) as well as studies in embryology (Baume 1882) and dental anatomy (Mühlreiter 1870) represent important advances in the field of dentistry and iniated more detailled studies in dental anthropology (Fig. 4 a, b).

Around the turn of the century one can observe a cumulation of publications on the development of mammalian dentitions on the part of dentistry, as well as on other selected topics, such as micro- and macromorphology of tooth structures and odontogenesis (v. Ebner 1890/91; Schlosser 1890/91; Walkhoff 1894; Leche 1895, 1902; Dependorf 1898; Adloff 1908, 1916; Ahrens 1913; De Jonge 1917, 1926; Eidmann 1923). All of these works gave impetus to the dental sciences and considerably affected the orientation of future research. The construction of the scanning electron microscope by Ruska in 1927 made possible new investigations on dental ultrastructure (Shobusawa 1952; Helmcke 1953; Boyde 1964; Miles 1967). These studies laid the foundations for the current success of that field of research (Rensberger 1978; Boyde and Martin 1984; v. Koenigswald et al. 1993; Pfretzschner 1993).

After the fundamentals of population genetics had been developed in the 1930's, genetic problems (e.g. family and twin studies) were also gaining importance in dentistry (Korkhaus 1930; Keeler 1935; Ritter 1937). This trend is undiminished through the present (Steward and Prescott 1976; Schulze 1987; Townsend et al. 1986, 1990, 1994; Richards et al. 1990). The strict heredity of many dental traits is the main reason for their importance in population studies. The variation in the phenotype is reflected by a multitude of individual and population characteristics which are successfully used for infra- and interpopulation studies (Turner 1986; Scott 1991; Alt 1997 a, b).

The bulk of publications in dentistry concerns dental anatomy (cf. Türp and Alt, this volume). Comprehensive studies on dental anomalies began with the investigations of Colyer and Sprawson (1938) and Euler (1939), whose works initiated further research (Brabant et al. 1958; Brabant 1967; Pindborg 1970). These investigations improved knowledge on odontogenesis as well as on ontogenetic development by elucidating etiologic and pathogenetic questions. Contemporary studies on dental anomalies have lately been published by Schulze (1987), and Schroeder (1991). In contrast, until recently little interest has been devoted to anatomical studies of the tooth roots (Visser 1948; Kovacs 1964). Additional fields of interest in dentistry that influence dental anthropology are related to research on growth and development (Kraus and Jordan 1965; Dahlberg and Graber 1977; Osborn and Ten Cate 1983; Radlanski 1993; Osborn 1995), craniofacial development (Brown and Molnar 1992), and on macroscopic and microscopic anatomy (Boyde 1964; Miles 1967; Taylor 1978, Carlsen 1987).

"The overlap between dentistry and anthropology in relation to dental research has long been recognized". This observation made by Cruwys and Foley (1986) in the foreword to their monograph *Teeth and Anthropology* underscores the close and long-lasting scientific connections between clinical research and dental anthropology. As an example, Cruwys and Foley (1986, foreword), mention population-specific investigations on the variability of dental traits suitable for "defining treatments or understanding the aetiology of dental disease or abnormalities". The common interests of both fields are related to specific questions concerning structural and functional characteristics of teeth, e.g., the biomechanics of tooth wear (Teaford and Walker 1984; Teaford and Runestad 1992), as well as pathological and epidemiological studies (on temporomandibular joints, parodontopathies, caries, dental genetics etc.). Both disciplines cooperate to validate findings on a broad basis, and to promote the transfer of knowledge through scientific exchange.

Scope of Dental Anthropology

Dental anthropology has evolved as a subdiscipline of physical anthropology. Today, it embraces all disciplines engaged in the study of teeth outside clinical dentistry (cf. Alt 1997c). According to Scott and Turner (1988), the term "Dental Anthropology" was first used around the turn of the century. The beginnings of the independent development of dental anthropology are marked by papers which introduced the idea of the "population" as the standard reference in morphological and pathological studies (Campbell 1925; Krogman 1927; Hellman 1928).

A. Hrdlička (1869–1943) was doubtlessly one of the most influential promotors of dental anthropology in the first half of the century (Hrdlička 1911, 1920, 1921). Much earlier, numerous publications on dentition and teeth appeared mainly in the French speaking countries of Europe from the 1870's onward (e.g. Lambert 1877; Broca 1879; Topinard 1892; Regnault 1894). These papers on racial and ethnic differences may have paved the way for the growing interest in population biology observable in Europe, North America, Japan, and other regions in the 20th century. Medical historians and ethnographers also began to work on the masticatory apparatus before the era of dental anthropology. The links thus established between anthropology and both the cultural sciences and medicine continued into the 20th century (Thompson 1903; Schröder 1906; Guerini 1909; Kanner 1928).

Fig. 5. Albert A. Dahlberg, the "nestor" of Dental Anthropology (from Hylander and Mayhall 1996).

In Europe, physical anthropology was an established discipline by the turn of the century. By translating systematized results into theories, odontology contributed considerably to the rapid development of certain branches of anthropology (e.g. paleoanthropology). After the importance of morphology for understanding biological processes in populations had been recognised and their genetic basis established, there was a rapid increase of investigations on dental morphology on the population level from the 1950s onward. The works of A. A. Dahlberg (1908–1993; Fig. 5), who is often referred to as the "nestor" of dental anthopology (Alexandersen 1994), reflect the beginning of this era of anthropological research (Dahlberg 1945, 1951).

The rapid advance of modern dental anthropology, the aims and methods of which were described for the first time by Klatsky and Fisher (1953), was initiated by researchers from dentistry more than by anthropologists (Pedersen 1949; Lasker 1950; Riethe 1955; Moorrees 1957; Brabant and Twiesselmann 1964; Brabant 1965, 1967). Distinct changes occured when the introduction of palaeodemography led to a fundamental re-orientation of approaches in anthropology. The "biologic reconstruction" of populations now becomes "a major task" in prehistoric anthropology, and morphology serves to "infer biological processes on a population level" (Acsádi and Nemeskéri 1957; Schwidetzky 1988).

International symposia on dental anthropology have been held since the 1960s, which, because of their interdisciplinary character, had a great impact both within and beyond the discipline. The starting point was the London symposium on *The Scope of Physical Anthropology and Human Population Biology and Their Place in Academic Studies* organized by the Ciba Foundation. Contributions concerning dental anthropology were published in a volume titled *Dental Anthropology* (Brothwell 1963), which is today considered "the" classic in dental anthropological literature. Already in 1958, the *Society for the Study of Human Biology* had been founded in the British Museum, in which dental anthopology

Fig. 6. Participants of the First International Symposium on Tooth Morphology at Fredensborg, Denmark, September 27–30, 1965. At the discussant's table are F. R. Parrington, United Kingdom; G. H. R. von Koenigswald, Netherlands; A. A. Dahlberg, United States, and P. O. Pedersen, Denmark (photo from Journal of Dental Research 46: 775 in 1967).

was also appropriately represented. The international establishment of dental anthropology in the scientific world was achieved at the latest in 1965. In that year, the first international symposium on *Dental Morphology* initiated by A. A. Dahlberg, P. O. Pedersen, V. Alexandersen, and P. M. Butler took place in Fredensborg, Denmark (Dahlberg 1967; Alexandersen 1994; Fig. 6).

Although initially dental anthropology was more or less limited to hominid questions, it rapidly incorporated all scientific fields interested in dental research. Considering the growing trend towards specialization and fission within scientific disciplines, the integrating feature of dental anthropology becomes the more noteworthy. It may be due to the fact that up to that point there had been no scientific platform for groups doing research in odontology. Against this background, the success of the first symposium on *Dental Morphology* (Dahlberg 1967), a new subject of central importance in dental anthropology up to the present, becomes all the more understandable.

This symposium provided the impetus for establishing the growth of several directions of dental research within the field of anthropology as a whole. Among these were studies dealing with the evolution of the basic dimensions of tooth crown size in prehistoric and recent human representatives, and a consideration of dental metrics in living human populations (Brace 1979, 1995; Brace et al. 1987, 1991). In parallel fashion, systematic assessments of dental morphology (Turner 1990, 1992; Turner et al. 1991) and patterns of human tooth wear (Molnar 1972; Smith 1984) were extended as a consequence of that stimulus. Even further, the human pattern of tooth crown and root formation and dental eruption has been investigated from the perspective of mammalian development in general and comparative primate development in particular (Smith 1989, 1991 a, b, 1994 a, b; Smith et al. 1994).

Within dental anthropology, classic odontology has found its scientific forum in the international congresses which take place periodically. So far, nine subsequent symposia on dental morphology – London 1968 (Dahlberg 1971), Brussels 1971, Cambridge 1974 (Butler and Joysey 1978), Turku 1979 (Kurtén 1982), Reykjavik 1983, Paris 1986 (Russel et al. 1988), Jerusalem 1989 (Smith and Tchernov 1992), Florence 1992 (Moggi-Cecchi 1995), and Berlin 1995 (Radlanski and Renz 1995) – have attracted growing numbers of scientists. Main topics at these meetings were ontogenetic and phylogenetic questions on morphology, embryology, and genetics. By its broad, multidisciplinary incorporation of different disciplines, dental anthropology has been receiving an ever increasing attention.

A further constitutional step was the foundation of the *Dental Anthropological Association* (DAA) in Albuquerque, New Mexico, with the aim to create a general forum for co-operation and communication of all scientists working in the field (Işcan 1989). On the foundation of the DAA in 1986, three major goals were set down in the association's by-laws: 1) promoting the exchange of educational, scientific, and scholarly knowledge in the field of dental anthropology, 2) stimulating interest in dental anthropology, and 3) publishing in the *Dental Anthropology Newsletter* (now: Dental Anthropology). Now, after ten years have elapsed, Philip L. Walker, current president of the DAA, reflected that these goals have been achieved fully (Walker 1996). Today, *Dental Anthropology* still remains

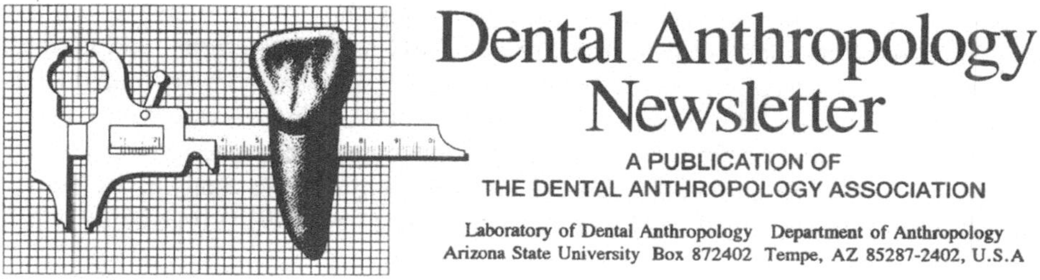

Fig. 7. Logo and head of the DAA *Dental Anthropology Newsletter*.

the only periodical which focusses purely on dental anthropolgy, and it has evolved into a valuable source of information and communication (Fig. 7). In 1996, the DAA, a subsidiary of the American Association of Physical Anthropologists (AAPA), had 324 members, two thirds of them from the United States (Bailey 1996).

Current Situation and Perspectives

The many facets and multidisciplinary relevance of dental anthroplogy are reflected by a diverse body of literature, which has steadily grown during the past thirty years. Research activities in the field of dental anthropology continue to be of great importance, as teeth "represent a permanent record of ontogeny and evolution [...]. They offer information about phyletic relationships, evolutionary history, and adaption" (Larsen and Kelley 1991, 1). Nevertheless, the objectives of dental anthropology are subject to constant change due to new tasks and research trends. The developments in paleontology, zoology, and paleoanthropology are representative of such changes. In the 1970s, the focus of research in this discipline shifted to concepts and hypotheses on evolutionary morphology with regard to dynamic components (temporomandibular joints, chewing mechanisms, diet). This re-orientation came about as researchers realized that for decades morphological studies had been mostly typological and had disregarded the complex relations of dental form and function. This had resulted in many inconsistent findings and concepts, and was increasingly criticised. Today's concepts are mainly based on the theory of the tribosphenic tooth (Simpson 1936), and are entirely embedded in constructional and functional morphological considerations (Crompton and Hiiemae 1970; Kay 1973; Rensberger 1973; Gantt 1977; Janis 1979; Maier 1980; Daegling et al. 1992; Spencer and Demes 1993; Hylander and Johnson 1994).

In the meantime, dental anthropology has produced a number of important hypotheses and well-founded theories on the phylogenetic development of mammals, especially of hominids. The study of occlusal relations and dental attrition has provided new insights into masticatory processes with regard to evolutionary morphology and ecology. In combination with the investigation of metric parameters as well as data on enamel thickness and attritional phenomena, these findings led to a new understanding of evolutionary processes. An example of this are the evolutionary changes in mammal and hominid molars, which are now interpreted as adaptive processes, respectively as constructional adaptations to dietary changes (Wolpoff 1971; Covert and Kay 1981; Maier 1984; Teaford and Walker 1984; Janis 1988; Teaford and Runestad 1992; Turner and Wood 1993; Spears and Crompton 1995). These developments owe much to new methods and approaches made possible by technical progress, e.g., in the analysis of dental ultrastructures by improved SEM techniques (Pfretzschner 1992; von Koenigswald et al. 1993).

After the publication of several standard works on odontology in the 1960s and 1970s (Keil 1966; Peyer 1968; Miles 1972; Poole 1976), only brief summaries appeared (Berkovitz et al. 1980; Schumacher et al. 1990). With the notable exeption of Hillson's monograph on *Teeth* (1986), the latest developments in odontological research could only be gathered from published proceedings of the "Dental Morphology" symposia. This lack of an extensive, up-to-date overview has been attempted to be remedied by a lately published volume on *The Evolution of Teeth* (in German; Alt and Türp 1997a), in which a number of established scientists from various different disciplines, such as anatomy, zoology, anthropology, and archeology, have compiled a comprehensive survey of the field.

Questions on taxonomy, ontogeny, and phylogeny have always played an essential role in dental anthropological research. Authors like Frisch (1965), Wolpoff (1971), Miles (1972), Swindler (1976), Lavelle et al. (1977) and Frayer (1978) reported on the significance of teeth in primate evolution and hominisation. As far as dentitions are concerned, classic approaches (odontogeny, phylogeny) still dominate in disciplines such as paleoanthropology, paleontology, zoology, genetics, embryology, and dentistry. Comprehensive and systematic overviews on teeth and dentitions can be found in basic disciplines such as zoology (Thenius 1989), paleoanthropology (Henke and Rothe 1994), and others (Osborn 1973; Walker 1981; Aiello and Dean 1990; Mayhall 1992).

The symposia on dental morphology represent an adequate forum for interdisciplinary communication and discussion in the field (cf. Smith and Tchernov 1992; Moggi-Cecchi 1995; Radlanski and Renz 1995). For other fields of dental anthropology (e.g., prehistoric anthropology, paleoanthropology) such periodic international meetings are still lacking. The annual meetings of the DAA, which have taken place since 1986, are no equivalent for multi-national events at changing locations. Yet, such organizational difficulties have not negatively affected these branches of dental anthropology, which is witnessed by the quality and number of monographs (Hillson 1986; Kieser 1990; Lukacs 1993), masters or doctoral theses (e.g., Haeussler 1985; Hollander 1992; Irish 1993; Rieger 1993; Schnutenhaus 1993), and the host of publications in anthropological, archeological, biological, anatomical, dental, and paleontological journals (Scott and Turner 1988; Scott 1991; Alt and Koçkapan 1993; Smith 1994a; Smith and Zilberman 1994; Van Reenen and Reid 1995), as well as in the number of congress volumes (Reddy 1985; Isçan 1989; Kelley and Larsen 1991).

As "specific morphological traits occur with greater frequency in some population than in others" (Foley and Cruwys 1986, 13), studies on the variability of morphognostic and metric characteristics of teeth are invaluable for infra- and interpopulation studies. In the past thirty years, dental characteristics helped to unravel processes of ethnohistoric differentiation and population dynamics (Lukacs 1983; Turner 1986; Scott 1991; Haeussler and Turner 1992; Scott and Turner 1997), and innovative procedures in odontologic kinship analysis on skeletal remains opened new perspectives for sociocultural interpretations of earlier populations (Alt and Vach 1991, 1995; Vach and Alt 1993; Alt 1997a). In these contexts, knowledge on dental genetics is of vital importance (Witkop 1962; Goose 1971; Stewart and Prescott 1976; Garn 1977; Townsend and Brown 1978 a, b; Potter et al. 1981; Schultze 1987; Nichol 1989; Richards et al. 1990; Townsend et al. 1986, 1990; Dempsey et al. 1995). Further investigations focussed on the biochemical status of teeth, on methodological problems of age and sex determination in skeletons, on normal and pathological tooth wear, on periodontal disease and other pathological findings (see Hillson 1996). The state of the art in dental anthropology has been sketched several times during the last few years (Foley and Cruwys 1986; Dahlberg 1991; Larsen and Kelley 1991; Scott 1991; Lukacs 1993; Alt 1997c; Alt and Türp 1997b).

Cultural aspects have always been of interest to dental anthropologists as well. Since the turn of the century, archaeological specimens (Milner and Larsen 1991; Scott 1991; Alt and Pichler, this volume), ethnographic findings (Heymer 1986), and objects of medico-historical interest (Whittaker and Hargreaves 1991; Alt and Koçkapan 1993; Alt 1994a) have time and again been the subject of research in dental anthropology. These investigations focus on the use of teeth as tools, as body decoration, medical instruments, amulets, relics, as drugs and remedies, prophylactic agents, trophies, souvenirs, and as objects of value (Kanner and Remy 1924; Verger-Pratoucy 1970; Alt et al. 1990).

The present survey of the *History of Dental Anthropology* is intended to illustrate the many roots and the broad scientific basis of modern dental anthropology. In the course of the last two decades, dental anthropology provided new, well-founded explanations and arguments for better insights into the phylogenetic development of teeth. It also produced a wealth of methodological innovations which are now indispensable in the reconstruction of the living conditions and subsistence of earlier populations. In a surprisingly large number of disciplines, teeth have become key objects for the investigation of numerous and diverse questions, and dental anthropology developed into a *sine qua non* for the understanding of evolutionary processes.

Acknowledgements

We would like to thank the publishing companies John Wiley and Sons, Inc., New York, and Quintessenz GmbH, Berlin, for their consent to reproduce photographs from their academic publishing program. We also owe thanks to V. Alexandersen, Copenhagen, Denmark, W. Hylander, Durham, U.S.A., J. Mayhall, Toronto, Canada, W. Hoffmann-Axthelm, Perleberg, Germany, and Ch. Schulze, Berlin, Germany, for their cooperation.

References

Acsádi G, Nemeskéri J (1957) Paläodemographische Probleme am Beispiel des frühmittelalterlichen Gräberfeldes von Halimba-Cseres. Homo 8: 133–148.

Adloff P (1908) Das Gebiß des Menschen und der Anthropomorphen. Vergleichend anatomische Untersuchungen. Springer, Berlin Heidelberg New York
Adloff P (1916) Die Entwicklung des Zahnsystems der Säugetiere und des Menschen. Eine Kritik der Dimertheorie von Bolk. Meusser, Berlin.
Ahrens H (1913) Die Entwicklung der menschlichen Zähne. Bergmann, Wiesbaden.
Aichel O (1917) Die Beurteilung der rezenten und prähistorischen Menschen nach der Zahnform. Z Morph Anthrop 20: 457–550.
Aiello L, Dean C (1990) An Introduction to Human Evolutionary Anatomy. Academic Press, London.
Alexandersen V (1994) Al Dahlberg: Thoughts from Denmark. Dent Anthrop Newsletter 8 (2): 9.
Alt KW (1994a) Prothetics, periodontal therapy and conservative dentistry in the eighteenth century. Bull Hist Dent 42: 67–70.
Alt KW (1994b) Die historische Entwicklung der zahnärztlichen Prothetik. In: Strub JR, Türp JC, Witkowski SWJ, Hürzeler MB, Kern M (Hrsg) Curriculum Prothetik, Bd. 1. Quintessenz, Berlin, p. 37–61.
Alt KW (1997a) Odontologische Verwandtschaftsanalyse. Individuelle Charakteristika der Zähne in ihrer Bedeutung für Anthropologie, Archäologie und Rechtsmedizin. Fischer, Stuttgart.
Alt KW (1997b) Kategorien und Konzepte dentalanthropologischer Studien. In: Alt KW, Türp JC (eds) Die Evolution der Zähne. Phylogenie – Ontogenie – Variation. Quintessenz, Berlin, p. 679–687.
Alt KW (1997c) Die historische Entwicklung der Odontologie und Dental-Anthropologie. In: Alt KW, Türp JC (eds) Die Evolution der Zähne. Phylogenie – Ontogenie – Variation. Quintessenz, Berlin, p. 47–76.
Alt KW, Koçkapan C (1993) Artificial tooth-neck grooving in living and prehistoric population. Homo 44: 5–29.
Alt KW, Parsche F, Pahl MW, Ziegelmayer G (1990) Gebißdeformation als Körperschmuck – Verbreitung, Motive und Hintergründe. Zahnärztl Mitt 80: 2448–2456.
Alt KW, Türp JC (eds) (1997a) Die Evolution der Zähne. Phylogenie – Ontogenie – Variation. Quintessenz, Berlin.
Alt KW, Türp JC (1997b) Vergleichende Odontologie und Dental-Anthropologie. Bedeutung – Positionen – Strategien. In: Alt KW, Türp JC (eds) Die Evolution der Zähne. Phylogenie – Ontogenie – Variation. Quintessenz, Berlin, p. 31–46.
Alt KW, Vach W (1991) The reconstruction of genetic kinship in prehistoric burial complexes – problems and statistics. In: Bock HH, Ihm P (eds) Classification, Data Analysis, and Knowledge Organisation. Springer, Berlin Heidelberg New York, p. 299–310.
Alt KW, Vach W (1995) Odontologic kinship analysis in skeletal remains – Concepts, methods, and results. Forensic Sci Int 74: 99–113.
Bailey SE (1996) Minutes from the 1996 Annual Meeting of the Dental Anthropology Association (DAA), Raleigh-Durham, April 11, 1996. Report from the Secretary-Tresurer. Dent Anthrop Newsl 10 (3), 20–21.
Baume R (1882) Odontologische Forschungen I. Versuch einer Entwicklungsgeschichte des Gebisses. Felix, Leipzig.
Berkovitz BKB, Holland GR, Moxham BJ (1978) A Colour Atlas and Textbook of Oral Anatomy. Wolfe, London.
Berkovitz BKB, Holland GR, Moxham BJ (1992) Color Atlas and Text of Oral Anatomy, Histology, and Embryology. Mosby, St. Louis.
Bolk L (1913) Die Ontogenie der Primatenzähne. Fischer, Jena.
Bolk L (1914) Die Morphogenie der Primatenzähne. Fischer, Jena.
Bourdelle E, Bennejeant C, Wicart A (1937) Anatomie et physiologie bucco-dentaires. Anatomie Dentaire Comparee. Baillière, Paris.

Boyde A (1964) The Structure and Development of Mammalian Enamel. Phil Thesis, London.
Boyde A, Martin L (1984) The microstructure of primate dental enamel. In: Chivers DJ, Wood BA, Bilsborough A (eds) Food Aquisition and Processing in Primates. Plenum Press, New York, p. 341–368.
Brabant H (1965) Observations sur l' évolution de la denture temporaire humaine en Europe occidentale. Bull Group Int Rech Sc Stomat 8: 235–302.
Brabant H (1967) Comparison on the characteristics and anomalies of the deciduous and the permanent dentition. J Dent Res 46: Suppl 5, 897–902.
Brabant H, Klees L, Werelds RJ (1958) Anomalies, Mutilations et Tumeurs des Dents Humaines. Prélat, Paris.
Brabant H, Twiesselmann F (1964) Observations sur l'évolution de la denture permanente humaine en Europe occidentale. Bull Group Int Rech Sc Stomat 7: 11–84.
Brace CL (1979) Krapina, "classic" Neanderthals, and the evolution of the European face. J Hum Evol 8: 527–550.
Brace CL (1995) Trends in the evolution of human tooth size. In: Moggi-Cecchi J (ed) Aspects of Dental Biology: Palaeontology, Anthropology and Evolution. Istituto di Anthropologia, Florence, p. 437–446.
Brace CL, Rosenberg KR, Hunt KD (1987) Gradual change in human tooth size in the late Pleistocene and post-Pleistocene. Evolution 41: 705–720.
Brace CL, Smith SL, Hunt KD (1991) What big teeth you had Grandma! Human tooth size, past and present. In: Kelley MA, Larsen CS (eds) Advances in Dental Anthropology. Wiley-Liss, New York, p. 33–57.
Broca P (1879) Instructions relatives á l'étude anthropologique du système dentaire. Bull Soc Anthrop Paris 2: 128–163.
Brothwell DR (ed) (1963) Dental Anthropology. Pergamon Press, Oxford.
Brothwell DR (ed) (1990) Dental Anthropology. Macmillan, New York.
Brown T, Molnar S (eds) (1992) Craniofacial Variation in Pacific Populations. Univ Press, Adelaide.
Butler PM (1941) A theory of the evolution of mammalian molar teeth. Am J Sci 239: 421–450.
Butler PM (1978) Molar cusp nomenclature and homology. In: Butler PM, Joysey KA (eds) Development, Function and Evolution of Teeth. Academic Press, London.
Butler PM, Joysey KA (eds) (1978) Development, Function, and Evolution of Teeth. Academic Press, London.
Campbell TD (1925) Dentition and Palate of the Australian Aborigines. Hassell Press, Adelaide.
Carlsen O (1987) Dental Morphology. Munksgaard, Copenhagen.
Cole S (1949) A History of Comparative Anatomy. From Aristotle to the Eighteenth Century. Macmillan, London.
Colyer JF, Sprawson E (1938) Dental Surgery and Pathology. Longmans-Green, London.
Cope ED (1874) On the homologies and origin of the types of molar teeth of mammalia Educabilia. J Acad Nat Sci Philadelphia NS 8, Pt I: 71–89.
Cope ED (1883) Note on the trituberculate type of superior molar and the origin of the quadrituberculate. Am Naturalist 17: 407–408.
Covert HH, Kay RF (1981) Dental microwear and diet: Implications for determining the feeding behaviours of extinct primates, with a comment on the dietary pattern of Sivapithecus. Am J Phys Anthrop 55: 331–336.
Crompton AW, Hiemae K (1970) Molar occlusion and mandibular movements during occlusion in the American oppossum, *Didelphis marsupialis* L. Zool J Linn Soc 49: 21–47.
Cruwys E, Foley R (eds) (1986) Teeth and Anthropology. BAR Int Ser 291, London.

Cuvier G (1835–1846) Leçons d'Anatomie Comparée de Georges Cuvier. Crochard, Paris.
Daegling DJ, Ravosa MJ, Johnson KR, Hylander WL (1992) Influence of teeth, alveoli, and periodontal ligaments on torsional rigidity in human mandibles. Am J Phys Anthrop 89: 59–72.
Dahlberg AA (1945) The changing dentition of man. Am J Phys Anthrop 8: 15–25.
Dahlberg AA (1951) The dentition of the American Indian. In: Laughlin WS (ed) The Physical Anthropology of the American Indian. Viking Fund, New York, p. 138–176.
Dahlberg AA (ed) (1967) Proceedings on the International Symposium on Dental Morphology. J Dent Res 46: 769–992.
Dahlberg AA (ed) (1971) Dental Morphology and Evolution. Univ Press, Chicago.
Dahlberg AA (1991) Historical perspective of dental anthropology. In: Kelley MA, Larsen CS (eds) Advances in Dental Anthropology. Wiley-Liss, New York.
Dahlberg AA, Graber TM (eds) (1977) Orofacial Growth and Development. Hague, Mouton.
Dempsey PJ, Townsend GC, Martin NG, Neale MC (1995) Genetic covariance structure of incisor crown size in twins. J Dent Res 74: 1389–1398.
Dependorf T (1898) Zur Entwicklungsgeschichte des Zahnsystems der Marsupialia. Denkschr Med Naturwiss Ges Jena, Bd. 6, Jena.
Ebner V von (1890/91) Histologie der Zähne mit Einschluß der Histogenese. In: Scheff's Handbuch für Zahnheilkunde. Bd. 1. Hölder, Wien, p. 209–262.
Eidmann H (1923) Die Entwicklungsgeschichte der Zähne des Menschen. Meusser, Berlin.
Euler H (1939) Die Anomalien, Fehlbildungen und Verstümmelungen der menschlichen Zähne. Lehmanns, München.
Foley R, Cruwys E (1986) Dental anthropology: Problems and perspectives. In: Cruwys E, Foley R (eds) Teeth and Anthropology. BAR Int Ser 291, London, p. 1–20.
Fränkel M (1835) De penetori dentium humaniorum structure observationes (Beobachtungen über die innere Struktur der Zähne). Nat Diss, Preßburg.
Frayer DW (1978) The Evolution of the Dentition in Upper Paleolithic and Mesolithic Europe. Lawrence, Kansas.
Friant M (1933) Contribution à l'étude de la differenciation des dents jugale chez les mammifères. Publ Mus Natn Hist Nat Paris 1: 1–132.
Frisch JE (1965) Trends in the Evolution of the Hominoid Dentition. Bibliotheca Primatologica, Fasc 3. Karger, Basel.
Gantt DG (1977) Enamel of Primate Teeth: Its Thickness and Structure with Reference to Functional and Phyletic Implications. PhD Thesis, St Louis.
Garn SM (1977) Genetics of dental development. In: McNamara JA (ed) The Biology of Occlusal Development. Center for Human Growth and Development, Ann Arbor, p. 13–28.
Gidley J (1906) Evidence bearing on tooth-cusp development. Proc Wash Acad Sci 8: 91–110.
Giebel CG (1855) Odontographie. Vergleichende Darstellung des Zahnsystems der lebenden und fossilen Wirbeltiere. Abel, Leipzig.
Goose DH (1971) The inheritance of tooth size in British families. In: Dahlberg AA (ed) Dental Morphology and Evolution. Univ Press, Chicago, p. 144–149.
Gregory WK (1916) Studies in the evolution of primates. Bull Am Mus Nat Hist 35: 239–355.
Gregory WK (1922) The Origin and Evolution of the Human Dentition. Williams and Wilkins, Baltimore.
Guerini V (1909) History of Dentistry. Lea and Febinger, Philadelphia.
Gutmann W (1997) Chordaten-Konstruktionen und ihre Evolution. Von der Kiemenreuse zum zahnbesetzten Kieferapparat. In: Alt KW, Türp JC (eds) Die Evolution der Zähne. Phylogenie – Ontogenie – Variation. Quintessenz, Berlin, p. 363–390.

Haeussler A (1985) Dental Morphology of New World, Eastern Siberia, and Soviet Central Asia Population. MA Thesis, Tempe AR.
Haeussler A, Turner CG (1992) The dentition of Central Asia and the quest for New World ancestors. In: Lukacs JR (ed) Culture, Ecology, and Dental Anthropology. J Hum Evol Special Issue 2: 273–297.
Hellman M (1928) Racial characters in human dentitions. Proc Am Philos Soc 67: 157–174.
Helmcke HJ (1953) Atlas des menschlichen Zahnes im elektronenmikroskopischen Bild. Transmare-Photo, Berlin.
Henke W, Rothe H (1994) Paläanthropologie. Springer, Berlin Heidelberg New York.
Hertwig O (1874) Über Bau und Entwicklung der Placoidschuppen und der Zähne der *Selachier*. Z Naturwiss (Jena) 8: 331–404.
Heymer A (1986) Öko-physiologische Zahnabrasion bei den Bayaka-Pygmäen in Zentralafrika. Curare 9: 51–58.
Hillson S (1986) Teeth. Cambridge Univ Press, Cambridge.
Hillson S (1996) Dental Anthropology. Cambridge Univ Press, Cambridge.
Hoffmann-Axthelm W (1985) The History of Dentistry. Quintessence, Chicago.
Hollander GA (1992) Dental anthropology: A view into biological and cultural factors. PhD Thesis Fullerton/CA.
Hrdlička A (1911) Human dentition and teeth from evolutionary and racial standpoint. Dominion Dent J 23: 403–421.
Hrdlička A (1920) Shovel-shaped teeth. Am J Phys Anthrop 3: 429–465
Hrdlička A (1921) Further studies of tooth morphology. Am J Phys Anthrop 4: 141–176.
Hunter J (1771) The Natural History of Teeth. London.
Huxley TH (1853) On the development of the teeth and on the nature and importance of Nasmyths "persistent capsule". Quart J Micr Sci 1: 149–164.
Hylander Wl, Johnson KR (1994) Jaw muscle function and wishboning of the mandible during mastication in macaques and baboons. Am J Phys Anthrop 94: 523–547.
Hylander WL, Mayhall JT (1996) In memoriam Albert A. Dahlberg (1908–1993). Am J Phys Anthrop 99: 627–633.
Irish JD (1993) Biological Affinities of Late Pleistocene Through Modern African Aboriginal Population. The Dental Evidence. PhD Diss, Tempe AR.
Isçan MY (1989) The emergence of Dental Anthropology. Am J Phys Anthrop 78: 1.
Isçan MY (ed) (1989) Current directions in Dental Anthropology Symposium. Am J Phys Anthrop. 78: 1–122.
Janis CM (1979) Aspects of the Evolution of Herbivory in Ungulate Mammals. PhD Thesis, Harvard Univ.
Janis CM (1988) An estimation of tooth volume and hypsodonty indices in ungulate mammals and the correlation of these factors with dietary preference. In: Russell DE, Santoro JP, Sigogneua-Russell D (eds) Teeth revisited. Mem Mus Natn Hist Nat C 53, Paris.
de Jonge T (1917) Die Kronenstruktur der unteren Prämolaren und Molaren. Ein Beitrag zur Morphologie des menschlichen Gebisses. Med Diss, Zürich.
de Jonge TE (1926) Die Dimerie der Frontzähne. Z Ges Anat Z Anat Entwickl Gesch 78: 277–307.
Kanner L (1928) Folklore of the Teeth. MacMillan, New York.
Kanner L, Remy CE (1924) Die Verwendung der Zähne außerhalb des Mundes bei den verschiedenen Völkern. Dtsch Mschr Zahnheilk 42: 573–578.
Kay RF (1973) Mastication, Molar Tooth Structure and Diet in Primates. Phil Diss, Yale Univ.
Kay RF (1978) Molar structure and diet in extant Cercopithecidae. In: Butler PM, Joysey KA (eds) Development, Function, and Evolution of Teeth. Academic Press, London, p. 309–339.

Keeler CE (1935) Heredity in dentistry. Dent Cosmos 77: 1147–1163.
Keil A (1966) Grundzüge der Odontologie. Borntraeger, Berlin.
Kelley MA, Larsen CS (eds) (1991) Advances in Dental Anthropology. Wiley-Liss, New York.
Kieser JA (1990) Human Adult Odontometrics. Univ Press, Cambridge.
King R (1994) John Hunter and the natural history of the human teeth: dentistry, digestion, and the living principle. J Hist Med Allied Sci 49: 504–520.
Klatsky M, Fisher RL (1953) The Human Masticatory Apparatus: An Introduction to Dental Anthropology. Dental Items of Interest Publ, Brooklyn.
Koenigswald W von, Martin T, Pfretzschner HU (1993) Phylogenetic interpretation of enamal structures in mammal teeth: possibilities and problems. In: Szalay S, Novacek MJ, McKenna MC (eds) Mammal Phylogeny. Springer, Berlin Heidelberg New York, p. 303–314.
Kölliker A von (1863) Die Entwicklung der Zahnsäckchen der Wiederkäuer. Z Wiss Zool 12: 455–460.
Korkhaus G (1930) Die Vererbung der Kronenform und -größe menschlicher Zähne. Z Anat 91: 594–617
Kovacs I (1964) Contribution à l'étude des rapports entre le dévelopment et la morphologie des racines des dents humaines. Bull Group Int Rech Sc Stomat 7: 85–120.
Kraus BS, Jordan RE (1965) The Human Dentition Before Birth. Lea and Febiger, Philadelphia.
Krogman WM (1927) Anthropological aspects of the human teeth and dentition. J Dent Res 7: 1–108.
Kükenthal W (1892) Über den Ursprung und die Entwicklung der Säugetierzähne. Jenaische Z Naturwiss 26: 469–489.
Kurtén B (1982) Teeth: Form, Function, and Evolution. Columbia Univ Press, New York.
Lambert E (1877) Morphologie du système dentaire des races humaines. Bull Acad Roy Sci Belg 43: 559–603.
Larsen CS, Kelley MA (1991) Introduction. In: Kelley MA, Larsen CS (eds) Advances in Dental Anthropology. Wiley-Liss, New York, p. 1–5.
Lasker GW (1950) Genetic analysis of racial traits of the teeth. Cold Spring Harbor Symp Qunat Biol 15: 191–203.
Lavelle CLB, Shellis RP, Poole DFG (1977) Evolutionary Changes to the Primate Skull and Dentition. Thomas, Springfield.
Leche W (1895) Zur Entwicklungsgeschichte des Zahnsystems der Säugetiere, Teil I, Ontogenie. Bibliotheca Zoologica, Heft 17.
Leche W (1902) Zur Entwicklungsgeschichte des Zahnsystems der Säugetiere, Teil II, Phylogenie. Bibliotheca Zoologica, Heft 37.
Lukacs JR (1983) Human dental remains from early Neolithic levels at Mehrgarh, Baluchistan. Curr Anthrop 24: 390–392.
Lukacs JR (1993) Culture, Ecology and Dental Anthropology. Kamla-Raj, Delhi.
Maier W (1978) Zur Evolution des Säugetiergebisses – typologische und konstruktionsmorphologische Erklärungen. Natur und Museum 108: 288–300.
Maier W (1980) Konstruktionsmorphologische Untersuchungen am Gebiß der rezenten Prosimiae (Primates). Abh Senckenb Naturf Ges 538: 1–158.
Maier W (1984) Tooth morphology and dietary specialization. In: Chivers DJ, Wood BA, Bilsborough A (eds) Food Aquisition and Processing in Primates. Plenum Press, New York, p. 303–330.
Mayhall JT (1992) Techniques for the study of dental morphology. In: Saunders SR, Katzenberg MA (eds) Skeletal Biology: Research Methods. Wiley-Liss, New York, p. 59–78.

Miles AEW (1972) Teeth and Their Origins. Oxford Biology Readers. Nr. 21. Oxford Univ Press, London.
Miles AEW (ed) (1967) Structural and Chemical Organisation of Teeth, Vol I. Academic Press, London.
Milner GR, Larsen CS (1991) Teeth as artifacts of human behavior: intentional mutilation and accidental modification. In: Kelley MA, Larsen CS (eds) Advances in Dental Anthropology. Wiley-Liss, New York, p. 357–378.
Moggi-Checchi JM (ed) (1995) Aspects of Dental Biology: Palaeontology, Anthropology, and Evolution. Pontecorboli, Florenz.
Molnar S (1972) Tooth wear and culture: A survey of tooth functions among some prehistoric populations. Current Anthrop 13: 511–526.
Molnar S, Richards L, McKee J, Molnar I (1989) Tooth wear in Australian aboriginal populations from the Murray Valley. Am J Phys Anthrop 79: 185–196.
Moorrees CFA (1957) The Aleut Dentition. Harvard Univ Press, Cambridge MA.
Mühlreiter E (1870) Anatomie des menschlichen Gebisses. Felix, Leipzig.
Nichol CR (1989) Complex segregation analysis of dental morphological variants. Am J Phys Anthrop 78: 37–59.
Osborn HF (1888) The evolution of mammalian molars to and from the tritubercular type. Amer Nat 21: 1067.
Osborn HF (1892) The history and homologies of the human molar cusps. Anat Anz 7: 740–747.
Osborn HF (1895) The history of the cusps of the human molar teeth. Int Dent J 7: 55–65.
Osborn HF (1907) Evolution of mammalian molar teeth to and from the triangular type. Macmillan, New York.
Osborn JW (1973) Variations in structure and development of enamel. Oral Sci Rev 3: 3–83.
Osborn JW (1995) A mechanical model stimulates the folding of the internal dental epithelium during cusp formation. In: Radlanksi RJ, Renz H (eds) Proc 10th Int Symp Dental Morphology. Brünne, Berlin, p. 22–26.
Osborn JW, Ten Cate AR (1983) Advanced Dental Histology. Wright, Bristol.
Owen R (1840–45) Odontography; Or, a Treatise on the Comparative Anatomy of the Teeth; Their Physiological Relations, Mode of Development, and Microscopic Structure, in the Vertebrate Animals. Bailliere, London.
Pedersen PO (1949) The East Greenland Eskimo Dentition: Numerical variations and anatomy. Medd Grønl 142: 1–256.
Peyer B (1968) Comparative Odontology. Univ Press, Chicago.
Pfretzschner HU (1992) Enamel microstructure and hypsodonty in large mammals. In: Smith P, Tchernov E (eds) Structure, Function and Evolution. Freund, London, p. 147–162.
Pfretzschner HU (1993) Enamel microstructure in the phylogeny of the Equidae. J Vertebrate Paleont 13: 342–349.
Pindborg JJ (1970) Pathology of the Dental Hard Tissue. Musgaard, Kopenhagen.
Poole DFG (1976) Evolution of mastication. In: Anderson DJ, Matthews B (eds) Mastication. Wright, Bristol.
Potter RHY, Corruccini RS, Green LJ (1981) Variance of occlusion traits in twins. J Craniofac Genet 1: 217–227.
Radlanski RJ (1993) Contributions to the Development of Human Deciduous Tooth Primordia. Quintessence, Chicago.
Radlanski RJ, Renz H (eds) (1995) Proceedings of the 10th International Symposium on Dental Morphology. Brünne, Berlin.
Raschkow I (1835) Meletemata circa mammalium dentium evolutionem (Untersuchungen über die Entwicklung der Säugetierzähne). Med Diss, Preßburg.

Reddy VR (ed) (1985) Dental Anthropology: Application and Methods. Inter-India Publ., New Delhi.
Regnault F (1894) Variations dans la forme des dents suivant les races humaines. Bull Soc Anthrop Paris 4: Ser 5, 14–18.
Remane A (1960) Zähne und Gebiss. In: Hofer H, Schultz AH, Starck D (Hrsg) Primatologia. Bd. III/2, Karger, Basel, p. 637–846.
Rensberger JM (1973) An occlusion model for mastication and dental wear in herbivorous mammals. J Paleont 47: 515–528.
Rensberger JM (1978) Scanning electron microscopy of wear and occlusal events in some small Herbivores. In: Butler PM, Joysey KA (eds) Development, Function, and Evolution of Teeth. Academic Press, London, p. 415–438.
Retzius A (1837) Bemerkungen über den inneren Bau der Zähne mit besonderer Berücksichtigung auf den im Zahnknochen vorkommenden Röhrenbau. Arch Anat Phys, Berlin, Anat Abt 486–566.
Richards LC, Townsend GC, Brown T, Burgess VB (1990) Dental arch morphology in south Australian twins. Arch Oral Biol 35: 983–989.
Rieger S (1993) Untersuchungen zur Geschlechtsbestimmung der Individuen einer merowingerzeitlichen Skelettserie aus Baden-Württemberg anhand der Zahnmaße. Med Diss, Freiburg.
Riethe P (1955) Zur Phylogenie des Primatengebisses. Verlag Kunst und Wissenschaft, Baden-Baden.
Ritter R (1937) Über die Frage der Vererbung von Anomalien der Kiefer und Zähne. Meusser, Leipzig.
Röse C (1892) Über die Entstehung und Formabänderungen der menschlichen Molaren. Anat Anz 7: 392–421.
Russell DE, Santoro JP, Sigogneua-Russell D (1988) Teeth revisited. Mem Mus Natn Hist Nat C 53, Paris.
Sakai T (1982) Morphogenesis of molar cusps and tubercles in certain primates. In: Kurtén B (ed) Teeth: Form, Function, and Evolution. Columbia Univ Press, New York, p. 307–322.
Schindewolf OH (1948) Wesen und Geschichte der Paläontologie. Springer, Berlin Heidelberg New York.
Schlosser M (1890/91) Die Differenzierung des Säugetiergebisses. Biol Centralbl 10: 238–252, 264–277.
Schnutenhaus S (1993) Weltverteilung der Zahngrößen. Überprüfung der Eignung von Zahnmaßen für eine Taxonomie des Menschen. Med Diss, Ulm.
Schreger BG (1800) Beitrag zur Geschichte der Zähne. Beitr Zergliederungskunst 1: 1–7.
Schröder H (1906) Die künstliche Deformation des Gebisses. Abel, Greifswald.
Schroeder HE (1991) Pathobiologie oraler Strukturen. Karger, Basel.
Schulze C (1987) Anomalien und Mißbildungen der menschlichen Zähne. Quintessenz, Berlin.
Schumacher GH, Schmidt H, Börnig H, Richter W (1990) Anatomie und Biochemie der Zähne. Fischer, Stuttgart.
Schwalbe G (1894) Über Theorien der Dentition. Anat Anz Jena, Erg. Heft 9: 5–45.
Schwann T (1839) Mikroskopische Untersuchungen über die Übereinstimmung in der Struktur und dem Wachsthum der Thiere und Pflanzen. Sander, Berlin.
Schwidetzky I (1988) Geschichte der Anthropologie. In: Knussmann R (Hrsg) Anthropologie, Bd I, 1: Wesen und Methoden der Anthropologie. Fischer, Stuttgart, p. 47–127.
Scott GR (1991) Dental anthropology. In: Dulbecco R (ed) Encyclopedia of Human Biology, Vol 2, Academic Press, San Diego, p. 789–804.
Scott GR, Turner CG (1988) Dental anthropology. Ann Rev Anthropol 17: 99–126.

Scott GR, Turner CG (1997) The Anthropology of Modern Human Teeth. Dental Morphology and Its Variation in Recent Human Populations. Cambridge Univ Press, Cambridge.
Shobusowa M (1952) Vergleichende Untersuchungen über die Form der Schmelzprismen der Säugetiere. Okajimas Folia Anat Jap 24: 371–392.
Simpson GG (1936) Studies of the earliest mammalian dentitions. Dent Cosmos 78: 791–800.
Smith BH (1984) Patterns of molar wear in hunter-gatherers and agriculturalists. Am J Phys Anthrop 63: 39–56.
Smith BH (1989) Growth and development and its significance for early hominid behavior. Ossa 14: 63–96.
Smith BH (1991a) Dental development and the evolution of life history in the hominidae. Am J Phys Anthrop 86: 157–174.
Smith BH (1991b) Standards of human tooth formation and dental age assessment. In: Kelley MA, Larsen CS (eds) Advances in Dental Anthropology. Wiley-Liss, New York, p. 143–168.
Smith BH (1994a) Patterns of dental development in *Homo*, *Australopithecus*, *Pan*, and *Gorilla*. Am J Phys Anthrop 94: 307–325.
Smith BH (1994b) Sequence of emergence of the permanent teeth in Macaca, Pan, Homo, and Australopithecus: Its evolutionary significance. Amer J Hum Biol 6: 61–76.
Smith BH, Crummett TL, Brandt K (1994) Ages of eruption of primate teeth: A compendium for aging individuals and comparing life histories. Yearbook of Phys Anthrop 37: 177–231.
Smith P and Zilberman U (1994) Thin enamel and other tooth components in Neanderthals and other hominids. Am J Phys Anthrop 95: 85–87.
Smith P, Tchernov E (1992) Structure, Function, and Evolution of Teeth. Freund, London.
Spears J, Crompton R (1995) Mechanics of food reduction in hominoid molars. In: Radlanksi RJ, Renz H (eds) Proc 10th Int Symp Dental Morphology. Brünne, Berlin, p. 387–390.
Spencer MA, Demes B (1993) Biomechanical analysis of masticatory system configuration in Neandertals and Inuits. Am J Phys Anthrop 91: 1–20.
Starck D (1982) Vergleichende Anatomie der Wirbeltiere auf evolutionsbiologischer Grundlage, Bd 3. Springer, Berlin Heidelberg New York.
Stewart RE, Prescott GH (1976) Oral Facial Genetics. Mosby, St. Louis.
Swindler DR (1976) Dentition of Living Primates. Academic Press, London.
Taylor RMS (1978) Variation in Morphology of Teeth. Thomas, Springfield.
Teaford MF, Runestad JA (1992) Dental microwear and diet in Venezuelan primates. Am J Phys Anthrop 88: 347–364.
Teaford MF, Walker A (1984) Qualitative differences in dental microwear between primate species with different diet and a comment on the presumed diet of Sivapithecus. Am J Phys Anthrop 64: 191–200.
Thenius E (1989) Zähne und Gebiß der Säugetiere. Hb Zoologie Bd 8, Mammalia. De Gruyter, Berlin.
Thompson A (1903) Ethnographic Odontology: The Inca Peruvians. [s. i. ;s. l.].
Tomes C (1872) On the nature of the cuticula dentis (Nasmyth's membrane). Quart J Micr Sci 12: 321–326.
Tomes J (1887) A System of Dental Surgery. Lindsay and Blakiston, Philadelphia.
Topinard P (1892) De l'évolution des molaires et prèmolaires chez les primates. Anthropologie Paris 3: 641–710.
Townsend GC, Brown T (1978 a) Inheritance of tooth size in Australian aboriginals. Am J Phys Anthrop 48: 305–314.
Townsend GC, Brown T (1978 b) Heritabilities of permanent tooth size. Am J Phys Anthrop 49: 497–502.

Townsend GC, Yamada H, Smith P (1990) Expression of the entoconulid (sixth cusp) on mandibular molar teeth of an Australian aboriginal population. Am J Phys Anthrop 82: 267–274.
Townsend GC, Brown T, Richards LC, Rogers JR, Pinkerton SK, Travan GR, Burgess VB (1986) Metric analyses of the teeth and faces of South Australian twins. Acta Genet Med Gemellol 35: 179–192.
Townsend GC, Dempsey P, Brown T, Kaidonis J, Richards L (1994) Teeth, genes and the environment. Perspect Hum Biol 4: 35–46.
Turner A, Wood B (1993) Comparative palaeontological context for the evolution of the early hominid masticatory system. J Hum Evol 24: 301–308.
Turner CG (1986) Dental evidence for the peopling of the Americas. Nat Geo Soc Res Rep 19: 573–596.
Turner CG (1990) Major features of Sundadonty and Sinodonty, including suggestions about East Asian microevolution, population history, and late Pleistocene relationships with Australian aboriginals. Am J Phys Anthrop 82: 295–317.
Turner CG (1992) Microevolution of East Asian and European populations: A dental perspective. In: Akazawa T, Aoki K, Kimura T (eds) The Evolution and Dispersal of Modern Humans in Asia. Hokusen-Sha, Tokyo, p. 415–438.
Turner CG, Nichol CR, Scott GR (1991) Scoring procedures for key morphological traits for the permanent dentition: The Arizona State University dental anthropology system. In: Kelley MA, Larsen CS (eds) Advances in Dental Anthropology. Wiley-Liss, New York, p. 13–31.
Türp JC, Brace CL, Alt KW (1997) Richard Owen and the comparative anatomy of teeth. J Hist Dent 45: 11–16.
Vach W, Alt KW (1993) Detection of kinship structures in prehistoric burial sites based on odontological traits. In: Andresen J, Madsen T, Scollar I (eds) Computer Applications and Quantitative Methods in Archaeology. Univ Press, Aarhus, p. 287–292.
Van Reenen JF, Reid C (1995) The configuration of maxillary molars. In: Radlanksi RJ, Renz H (eds) Proc 10th Int Symp Dental Morphology. Brünne, Berlin, p. 167–171.
Verger-Pratoucy JC (1970) Recherches sur les mutilations maxillo-dentaires préhistoriques. Bull Group Int Rech Sc Stomat 13: 133–310.
Visser JB (1948) Beitrag zur Kenntnis der menschlichen Zahnwurzelformen. Med Diss, Zürich.
Vogel F, Motulsky AG (1996) Human Genetics: Problems and Approaches. Springer, Berlin Heidelberg New York.
Waldeyer W (1864) Untersuchungen über die Entwicklung der Zähne. Königsberger Med Jb, 1 Abt 4: 236–299.
Walker A (1981) Dietary hypotheses and human evolution. Philos Trans Roy Soc London B 292: 57–64.
Walker PL (1996) Presidential address. Dent Anthrop Newsl 10 (3): 1.
Walkhoff O (1894) Mikrophotographischer Atlas der normalen Histologie menschlicher Zähne. Risel, Hagen.
Weidenreich F (1937) The dentition of Sinanthropus Pekinensis: A comparative odontography of the hominids. Paleont Sinica, NS D 1: 1–132.
White TD, Suwa G, Asfaw B (1994) *Australopithecus ramidus*, a new species of early hominid from Aramis, Ethiopia. Nature 371: 306–312.
White TD, Suwa G, Asfaw B (1994) Corrigendum. Nature 375: 88.
Whittaker DK, Hargreaves AS (1991) Dental restorations and artificial teeth in a Georgian population. Br Dent J 171: 371–376.
Witkop CJ (ed) (1962) Genetics and Dental Health. McGraw-Hill, New York.
Wolpoff MH (1971) Metric Trends in Hominid Dental Evolution. Studies in Anthropology 2. Case Western Reserve University Studies Anthrop, Cleveland.

Wortman JL (1886) The comparative anatomy of the teeth of vertebrates.
Wortman JL (1902) Studies of Eocene Mammalia in the marsh collection, Peabody Museum. Am J Sci 13: 39–46.
Würtz G (1985) Zahnentwicklung: Kenntnisstand im Ablauf der Geschichte. Ferber, Gießen.
Zittel KA von (1899) Geschichte der Geologie und Paläontologie. Oldenbourg, München.

2.3 Roll Call: Thirty-Two White Horses on a Red Field. The Advantages of the FDI Two-Digit System of Designating Teeth

Kurt W. Alt and Jens C. Türp

Thirty white horses on a red hill,
First they champ,
Then they stamp,
Then they stand still.

JRR Tolkien, The Hobbit (1937)

Introduction

Every scientific field evolves its own terminology to accomodate (more or less) its particular interests and requirements (Swales 1990). In order to facilitate a dialog within and among disciplines, it is necessary to use an accepted nomenclature as well as standardized abbreviations and signs. Nonetheless, an "officially" implemented nomenclature is not always recognized by all members of the scientific community. This leads to the unsatisfying situation wherein several different terminologies are in parallel use, thus often causing misunderstandings. In anatomy, for example, a standard Latin nomenclature was compiled for the first time in 1895 (Baseler Anatomische Nomenklatur [Basle Nomina Anatomica, B.N.A.]). In an attempt to improve upon the B.N.A., the Anatomical Society of Great Britain and Ireland presented a Birmingham Revision in 1933 (in Latin and English), which had been widely used in the English-speaking world, while the German Anatomical Society (Deutsche Anatomische Gesellschaft) adopted its own revision in 1935 (in Latin) which became known as the Jena Nomina Anatomica (J.N.A.) (Kopsch 1941).

Twenty years later, in 1955, a new nomenclature was introduced (in Latin), the Nomina Anatomica Parisiensia (N.A.P.) (First Edition), which was "a rather conservative revision of the B.N.A." (Woerdeman 1957, 6). The latest (sixth) edition of the Nomina Anatomica was published in 1989, together with the third editions of Nomina Histologica and Nomina Embryologica. However, inconsistencies, errors and omissions, as well as minor or major changes in nomenclature between different editions of the P.N.A., continue to produce "confusion, rage and – in medical students who have to 'get the names right' for their examinations – something very like despair", as pointed out by Staubesand and Steel (1988, 265). Furthermore, in contrast to the German anatomical literature, which in most cases employs Latin expressions, this is almost never the case in English language journals; instead, preference is given to English translations of the "official" anatomical terms. This, and the fact that each medical specialty has its "own" anatomical terminology, shows clearly that internationally the Nomina Anatomica are far from being accepted in daily use.

Tooth numbering systems, which are important for scientific fields dealing with human (or animal) teeth, share some of the problems mentioned above. While, unlike the anatomical nomenclature, dental connotation systems in use today are free of errors per se (unless made by the individuals who are using them), there is no recommended tooth designating method that has gained full international acceptance. This is also true for the two-digit system proposed by the Fédération Dentaire Internationale (FDI) (1971), despite the fact that both the FDI and the World Health Organization (WHO) (1977, 1987) have recommended this method and that it has been recognized and used by international agencies like the International Organization for Standardization (ISO 3950) (1984, 1995) and Interpol (1989).

Furthermore, disciplines outside dentistry, such as forensic odontology, palaeoanthropology, palaeontology or zoology, often choose systems different from any of those used in dental science. This heterogeneity among different disciplines and countries can be the cause of misinterpretations. The aim of this chapter is to encourage the use of the FDI two-digit system in all fields that are involved with human teeth in some form.

The History of Tooth Designating Systems

For a long time, different methods of designating and encoding teeth have been in use (Tab. 1). The oldest known system was proposed by Adolph Zsigmondy (Vienna, Austria) in 1861 (Symbolic System). He suggested numbering the teeth of the permanent dentition consecutively, starting with *1* for the central incisors, and ending with *8* for the third molars. In order to determine if the tooth in question is located in the upper right, upper left, lower left or lower right quadrant of the mouth, a grid symbol (┘, └, ┌, ┐) is placed around the number. Therefore, the system has sometimes also been called the Set-Square-System (FDI 1983). For numbering deciduous teeth, the Roman numerals *I* to *V* are used. However, instead of Roman numerals, several variations for recording deciduous teeth have been proposed, such as the upper case characters *A* through *E* (Lyons 1947); the lowercase letters *a* through *e*; the letter *D* or *d* (for "deciduous") placed before or after the number of the tooth; or the letter *m* (for "milk teeth") after the tooth number (Frykholm and Lysell 1962).

The same method of recording teeth had also been described by Corydon Palmer (Warren, Ohio) at the tenth annual meeting of the American Dental Association in 1870. Obviously not aware of Zsigmondy's earlier publications,

Palmer claimed the authorship of this tooth numbering system (1891). Therefore, in English-speaking countries it is generally known as "Palmer's Notation" (Paquette 1960; Frykholm and Lysell 1962).

Table 1. Examples of tooth designations under different systems

Tooth	Tooth numbering system				
	Zsigmondy/Palmer	Mühlreiter	Haderup	Universal	FDI
Permanent maxillary left lateral incisor	$\lfloor 2$	I^2	+2	10	22
Permanent mandibular right second molar	$7\rceil$	$_2M$	7–	31	47
Primary maxillary left canine	$\lfloor$III or $\lfloor$c	dC^1 or c^1	+03 or +III	H or 48	63

Eduard Mühlreiter (Salzburg, Austria) (1870) combined the upper case letters *I*, *C*, *P*, and *M* (abbreviations for permanent incisors, canines, premolars, and molars, respectively) with numerals (*1* for canines; *1* or *2* for incisors and premolars; *1*, *2*, or *3* for molars) whereby the position of the letter relative to the numeral indicates if the tooth is maxillary or mandibular, and left or right. Deciduous teeth are indicated by placing a *d* immediately before the upper case letter. An alternative method of designating the primary teeth was introduced by L. Bolk (Amsterdam, Netherlands) who suggested using the lower case letters *i*, *c*, *p*, and *m*.

Victor Haderup (Copenhagen, Denmark) (1887) proposed a system which omitted angular symbols placed around numerals and introduced "+"-signs (for indicating maxillary teeth) or "–"-signs (for mandibular teeth). Left and right sides are indicated by placing the sign before (left side) or after (right side) the numeral. Decidous teeth were originally indicated by adding an *l* immediately before the numeral; after a few years, however, the *l* was replaced by a *0*. Variations in the notation of primary teeth are the use of Roman numerals. Haderup's system has been very popular in Scandinavia (Frykholm and Lysell 1962).

The Universal System (American System, National System) proposed by Julius Parreidt (Leipzig, Germany) in 1882, has been widely used throughout the U.S.A. (American Dental Association 1967, 1968). It assigns the numerals *1* to *32* consecutively to singular permanent teeth in a clockwise sequence. Numbering starts at the maxillary right third molar (*1*), follows the dental arch until the maxillary left third molar (*16*), continues at the mandibular left third molar (*17*), and ends at the mandibular right third molar (*32*). The deciduous teeth are designated in a similar matter, but with the letters *A* through *T*. In place of letters, Goodman (1967) proposed the use of the numerals *41* through *60* for the twenty primary teeth.

Language-adapted methods have been also used. They are combinations of characters (for the designation of quadrants) and digits or characters (for the identification of teeth). Thus, in the English-speaking world the upper right, upper left, lower left, and lower right quadrants have been identified by *UR*, *UL*, *LL*, and

LR respectively. The French system differentiates between the right and the left sides of the tooth arches. An uppercase *D* (for "droite" = "right") means (in the permanent as well as the deciduous dentition) "upper right", a lowercase *d* "lower right"; an uppercase *G* (for "gauche" = "left") means "upper left", a lowercase *g* "lower left". The numbers *1* to *8* are allotted to the teeth in the permanent dentition, the Roman numerals *I* to *V* to those of the primary dentition. For additional tooth numering systems, see Schwartz and Stege (1977) and Goaz (1981).

Unfortunately, none of these methods complies with the five basic requirements set by the FDI (1971). Following these standards, a tooth-designating system should be simple to understand and to teach, easy to pronounce in conversation and dictation, readily communicable in print and by wire, easily adaptable to typewriter or data-processing keyboards, and easily adaptable to standard charts used in general practice.

If these requirements are considered, many traditional systems have disadvantages. For example, the use of grid signs, like in Zsigmondy's method, is a major obstacle in fast communication and data processing. While arabic numerals (*0* to *9*) are used even in countries which do not have the Roman alphabet, this is neither the case for alphabetical characters (*A*, *B*, *C*, etc.) nor for Roman numerals (*I*, *II*, *III*, etc.). Language-dependent methods, on the other hand, are limited to certain countries. The disadvantages of the Universal System have been summarized by Hrabowsky and Sim: "Its major drawback is the necessity for memorizing 32 digits and 20 characters and associating these 52 unrelated symbols with individual teeth. Not only can this prove to be a source of confusion, but it also precludes instant recognition of particular teeth and quadrants" (1971, 197). Interestingly enough, at an international dental meeting in Paris in 1890, a commission appointed to consider different notation methods came to the conclusion "that any system of numbering [the permanent teeth, eds.] by thirty-two was inconvenient, confusing, and difficult to memorize. It was therefore unanimously rejected" (Palmer 1891, 198).

The FDI Two-Digit System for Designating Teeth

Because of the shortcomings of the existing tooth-numbering systems, the General Assembly of the FDI, at its 58th annual session which took place in Bucharest (Romania) from September 26 to October 1, 1970, accepted unanimously a resolution proposing that the two-digit system of designating teeth be adopted worldwide (thirty-eight out of fifty-six representatives voted in favour and eleven against the resolution, with seven abstaining [FDI 1971]). The FDI believed that only this system seemed to comply with the requirements mentioned earlier (Keiser-Nielsen 1971). Since it was subsequently accepted by the International Organization for Standardization (ISO), it is sometimes also referred to as the "ISO/FDI Two Digit System" (ADA 1995).

In the two-digit system, which was originally described by Jochen Viohl (Berlin, Germany) in 1966 – and in slightly different versions by Pirquet in 1924 and by Denton in 1963 – each tooth is identified by a unique two-digit combination. The first digit specifies one of the four quadrants of the mouth, starting with the upper right segment and proceeding in a clockwise sequence. Because the permanent teeth are the main concern within dentistry, the corresponding quadrants are allotted the digits 1 through 4, while for the deciduous teeth the quadrants 5 through 8 have been chosen:

Maxillary right quadrant, permanent dentition:	1
Maxillary left quadrant, permanent dentition:	2
Mandibular left quadrant, permanent dentition:	3
Mandibular right quadrant, permanent dentition:	4
Maxillary right quadrant, deciduous dentition:	5
Maxillary left quadrant, deciduous dentition:	6
Mandibular left quadrant, deciduous dentition:	7
Mandibular right quadrant, deciduous dentition:	8

The second digit indicates the actual tooth within the quadrant. In every quadrant, the (permanent) teeth are numbered mesial to distal, from 1 to 8, beginning with the middle incisor and ending with the third molar:

Permanent central incisor:	1
Permanent lateral incisor:	2
Permanent canine:	3
Permanent first premolar:	4
Permanent second premolar:	5
Permanent first molar:	6
Permanent second molar:	7
Permanent third molar:	8

The division of quadrants and descriptions of single permanent teeth is summarized in Fig. 1.

Upper jaw right	8	7	6	5	4	3	2	1	1	2	3	4	5	6	7	8	Upper jaw left
Lower jaw right	8	7	6	5	4	3	2	1	1	2	3	4	5	6	7	8	Lower jaw left

Fig. 1. Division of the quadrants and descriptions of single permanent teeth.

To describe a single tooth, two digits are used – the first for the quadrant, the second for the particular tooth (Fig. 2).

18	17	16	15	14	13	12	11	21	22	23	24	25	26	27	28
48	47	46	45	44	43	42	41	31	32	33	34	35	36	37	38

Fig. 2. Notation for permanent teeth by the FDI system (the grid is included for clarification).

The two-digit system described by Clemens Pirquet (1924) differed in two points from Viohl's FDI version:

- ♦ The quadrants 1 through 4 denoted the deciduous teeth while the quadrants 5 through 8 indicated the permanent teeth.
- ♦ The third and seventh quadrants corresponded to the mandibular right quadrant, and the fourth and eighth quadrants to the mandibular left quadrant.

In the system proposed by George B. Denton (1963), the right quadrants in the permanent dentition were allotted the first digits 2 and 3, while the left quadrants were indicated by the numbers 1 and 4. With regard to the deciduous teeth, in contrast, the first "digit" was an upper case *b* or *c* (for the upper and lower right quadrant), and *a* and *d* (for the upper and lower left quadrant respectively).

The two digits of the FDI System should be pronounced separately, e.g., "one-eight" (18) [not: "eighteen"], "one-seven" (17), "four-one" (41), etc. This makes translation into other languages easy because it requires only the mastery of the digits from *one* to *eight* in the required language (Hrabowsky and Sim 1971). Furthermore, by pronouncing the digits independently, confusion with the Universal System can be prevented.

55	54	53	52	51	61	62	63	64	65
85	84	83	82	81	71	72	73	74	75

Fig. 3. Notation for the deciduous teeth by the FDI system (the grid is included for clarification).

The same principle of notation applies for the deciduous teeth (Fig. 3). Teeth within the same quadrant are allotted the digits 1 (first incisor) through 5 (second molar):

Deciduous central incisor:	1
Deciduous lateral incisor:	2
Deciduous canine:	3
Deciduous first molar:	4
Deciduous second molar:	5

The permanent mandibular right first molar is therefore described as 46 and the deciduous mandibular right first molar as 84.

The FDI System has two features in common with the notation systems described by Zsigmondy-Palmer and Haderup:

- ♦ Teeth which are anatomically similar in dentition are characterized by the same number (e. g., lateral incisors: 2, canines: 3, first premolars: 4);
- ♦ Counting starts with the central incisor (1) and ends with the third molar (8).

In place of an angular symbol or a plus or minus sign, however, the quadrant is indicated by a digit that precedes the one designating the specific tooth. By determining a clockwise sequence of the four quadrants, the FDI showed consideration for the Universal System. In doing so, the FDI's binomial system represents a compromise between the principles of notation shared by the Zsigmondy-Palmer and Haderup systems, and those of the Universal System.

Other suggestions which were debated – and rejected – during the 1970 FDI session were to give all left, or all right side quadrants odd digits (1, 3, 5, 7, 9, etc.), or to allot the deciduous teeth the quadrant digits 1 to 4, and the permanent teeth the digits 5 to 8 (FDI 1971).

Possible Modifications of the Two-Digit System and Its Use in Anthropology

Because the Universal System uses in part the same written numbers as the two-digit system (although they should at least be pronounced in a different manner), this similarity can be a source of confusion in those countries where the Universal System is used. Tooth *11* ("one-one"), for example, indicates in the two-digit system the permanent maxillary right central incisor, whereas tooth *11* ("eleven") in the Universal System refers to the permanent maxillary left canine. To avoid possible misinterpretations, e.g., of patient records or accident victims, Sharma and Wadhwa (1977) proposed a slight modification of the FDI two-digit system by placing a hyphen between the two digits in written communications (*1-1*). However, this suggestion, which could facilitate a transition from the Universal System to the FDI two-digit system, was not adopted.

Fortunately, in anthropology and paleopathology the Universal System is not used. On the other hand, neither did the two-digit system have great importance until now. In comparison with other notation methods, the FDI System has many outstanding advantages. For the documentation and evaluation of (pre-) historic skeletons, it is useful not only because every tooth is clearly and logically described but also because it is possible to feed the data into a computer without difficulty. In addition, it should be considered that anthropologists and dentists are professionally involved with one another through common problems (Alt 1989). For this reason, the FDI nomenclature should be favored as it is understood by members of both professions. A great advantage of the FDI System is that it holds the possibility of being extended into a three-digit system, as suggested by Kryszinski (1986) and Villa Vigil (1989). This is extremely helpful in order to designate supernumerary teeth, which have been fairly common among certain ethnic groups and subpopulations (Ducka-Karska 1983; Hurlen and Humerfelt 1984; O'Dowling 1989), and are thus of special importance in anthropology and epidemiological dental surveys.

The three-digit system as proposed by Kryszinski (1986) differentiates three possibilities as to how teeth in excess of the normal development can be designated:

1 "If the supernumerary tooth resembles a normal one, it is represented, in addition to the first digit [which indicates the quadrant, *authors*], by the following two digits: the second digit, which indicates the normal tooth that the extra tooth resembles in anatomical form; and the third digit, *0*.

2 If the accessory tooth does not have the normal morphology, it is represented, in addition to the first digit, by a second digit, which indicates the preceding [i. e., mesial, *authors*] normal tooth if the accessory tooth is arranged in the dental arch. If the extra tooth is placed outside the dental arch, the second digit indicates the normal tooth at which is projected the crest of the curvature on the mesial crown

surface of the extra tooth when viewing the occlusal (on radiographs, buccal) aspect. Third digits, *1* to *3*, are also used. These indicate, in clockwise succession, the actual accessory tooth or teeth.

3 All the supernumerary elements in or near the midline which lack the normal shape, are included under the classification *Mesiodens*. The notation is expressed *101* (*102*, *103*) or *301* (*302*, *303*), the first digit indicating the maxillary or mandibular tooth, respectively. The third digit indicates, in a clockwise sequence, the extra teeth (up to three) located between the central incisors. The primary mesiodens designation is *501* for the maxillary tooth and *701* for the mandibular tooth." (p. 127–128).

Examples:

- A supplementary permanent mandibular left fourth molar whose morphology resembles the one of the existing third molar, is expressed *380* (Fig. 4).
- A right supplementary deciduous maxillary molariform tooth would be designated as *540* if it resembles the deciduous first molar, and *550* if it resembles the second molar (Fig. 5).

Fig. 4. Supplementary permanent mandibular left fourth molar (*380*).

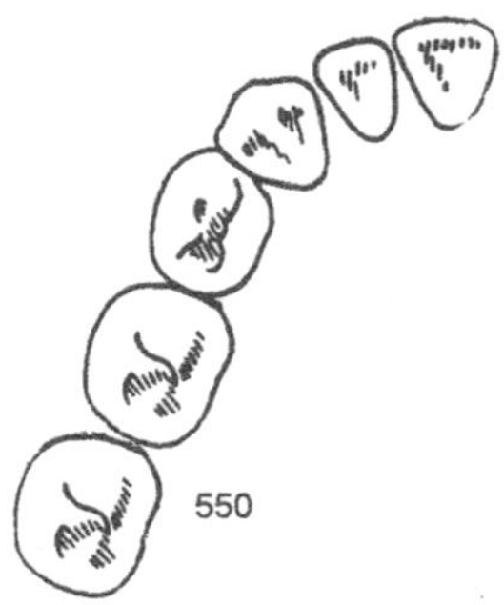

Fig. 5. Supplementary deciduous maxillary right molariform tooth that resembles the second molar (*550*).

- A supplementary permanent maxillary posterior tooth with an abnormal morphology whose preceeding (normal) tooth is *15* or, in the case that it is not arranged in the dental arch, whose crest of curvature on the mesial surface projects to the tooth *15*, is indicated *151* (Fig. 6 a).
- If there is another supplementary tooth with abnormal morphology located further distally in this quadrant, e.g., a supernumerary tooth preceeded by tooth *18*, the tooth distal to *18* would be called *181*, and the extra tooth preceeded by tooth *15* would be indicated *152* (because in this case it is the second supernumerary tooth with an abnormal form in the tooth arch when counted in a clockwise sequence starting at the last molar of the patient's maxillary right side) (Fig. 6 b).
- A supplementary permanent cone-shaped tooth (mesiodens) in or near the midline of the upper jaw (between the central incisors) becomes *101*, a second one *102* (Fig. 7).

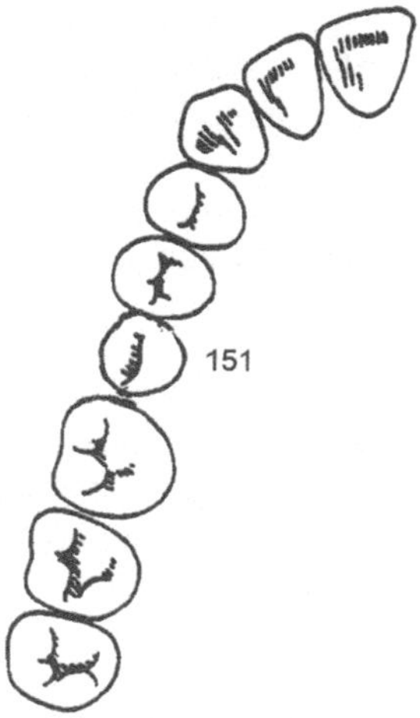

Fig. 6 a: Supplementary permanent maxillary posterior tooth with an abnormal morphology (*151*).

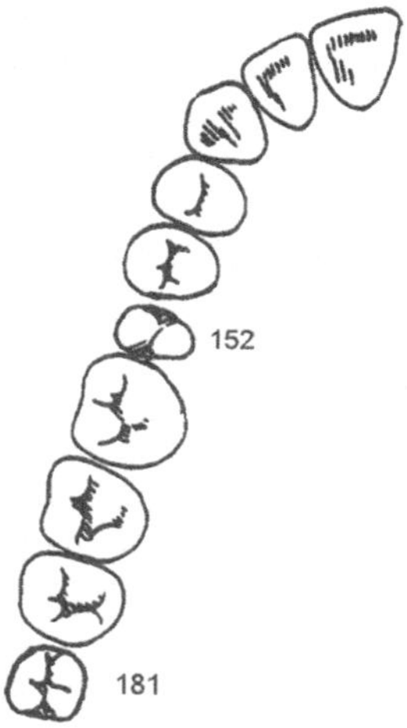

Fig. 6 b: Supplementary permanent maxillary posterior teeth with abnormal morphology (*181* and *152*).

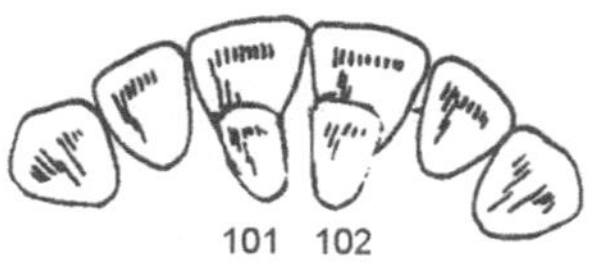

Fig. 7. Supplementary permanent cone-shaped maxillary teeth (*101* and *102*).

Villa Vigil et al. (1989), also choosing a three-digit system, proposed a different method to designate supernumerary teeth, and to give information about the form, topography, and possible anatomic abnormalities of a tooth.

- Supernumerary teeth are indicated by a letter of the alphabet which can be uppercase (i.e., the extra tooth resembles a normal tooth) or lowercase (i.e., the extra tooth has an abnormal shape or size).
- In the case of extra teeth placed within the dental arch – proximal (beneath) to a normal tooth or a tooth of reference – for mnemotechnical reasons the letters *P*, *R*, *O*, and *X* (derived from the Latin word proximalis) are used. Depending on the position of the tooth with regard to the tooth of reference, the letter can appear as a third or a first "digit": "The letter is placed as a third 'digit', when the supernumerary tooth is distal to the tooth of reference. The letters *P*, *R*, *O*, and *X* are used respectively to designate successive teeth in the mesiodistal direction. The appropriate letter is placed as a first 'digit', when the supernumerary tooth is mesial to the tooth of reference (e.g., in the case of a mesiodens). In this case, the notations used are *P*, *R*, *O*, and *X*, respectively, in the distomesial direction" (p. 300).

In contrast, for extra teeth which are placed outside the dental arch, Villa Vigil et al. (1989, 300) recommended the following notation rules:

- "A *V* (or a *v* in the case of an abnormal tooth) is used if the supernumerary tooth is positioned buccally (vestibular); if there is more than one, the notations are, successively, from mesial to distal sides: *V*, *E*, *S*, and *T* (the first four letters of the Latin term vestibularis).
- An *L* (or an *l*) is used if the extra tooth is positioned lingually; if there are several, the letters *L*, *I*, *N*, and *G* (from the term lingualis) will be used successively in the mesiodistal direction".

Examples:

- A supplementary permanent mandibular left fourth molar whose morphology resembles the one of the existing third molar, is expressed *38P* when it is placed within the dental arch (Fig. 8 a), and *3V8* or *3L8* when it is located bucally or lingually of the arch respectively (Fig. 8 b).

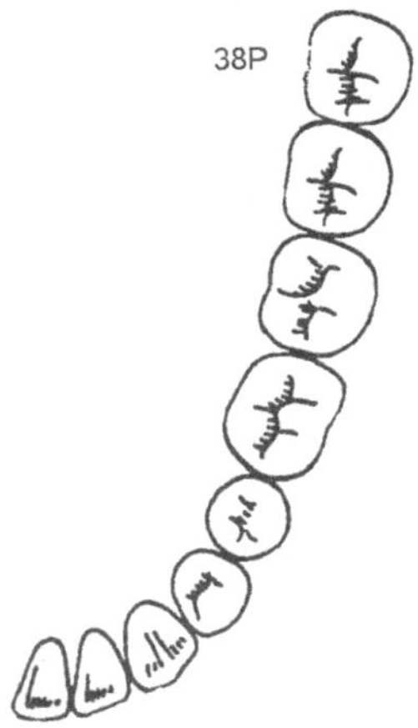

Fig. 8 a: Supplementary permanent mandibular left fourth molar within the dental arch (*38P*).

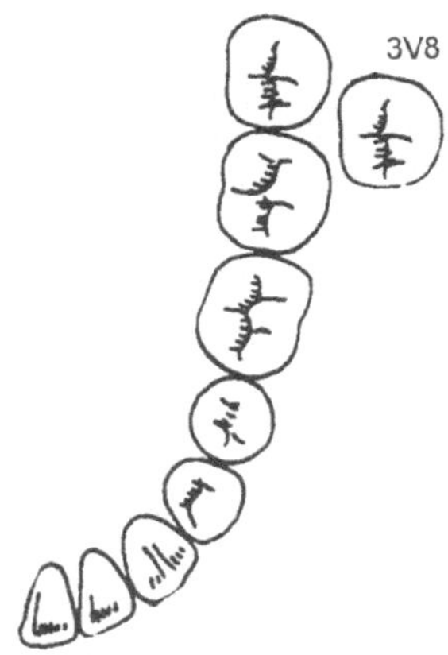

Fig. 8 b: Supplementary permanent mandibular left fourth molar outside the dental arch (*3V8*).

- A right supplementary deciduous maxillary molariform tooth which is placed bucally of the dental arch is indicated as *5v4* if it resembles the deciduous first molar, and *5v5* if it resembles the second molar (Fig. 9).

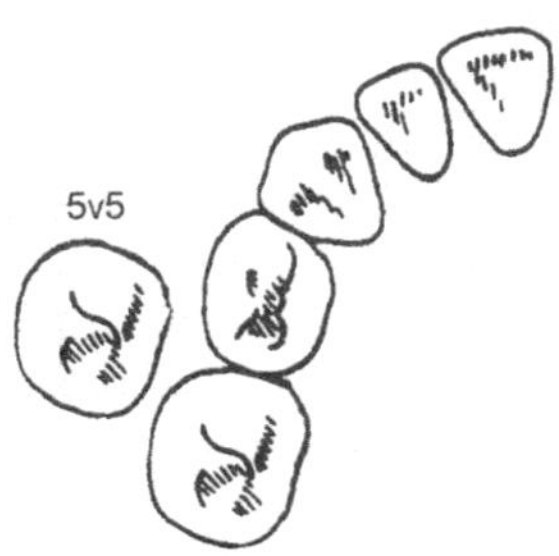

Fig. 9. Supplementary deciduous maxillary molariform tooth outside the dental arch (*5v5*).

- A supplementary permanent maxillary posterior tooth with an abnormal morphology whose preceeding (normal) tooth is tooth *15*, is indicated *15p* (Fig. 10 a). If there is another supplementary tooth with abnormal morphology located further distally in this quadrant, e.g. a supernumerary tooth preceded by tooth 18, the tooth distal to *18* would be called *18r* (Fig. 10 b).

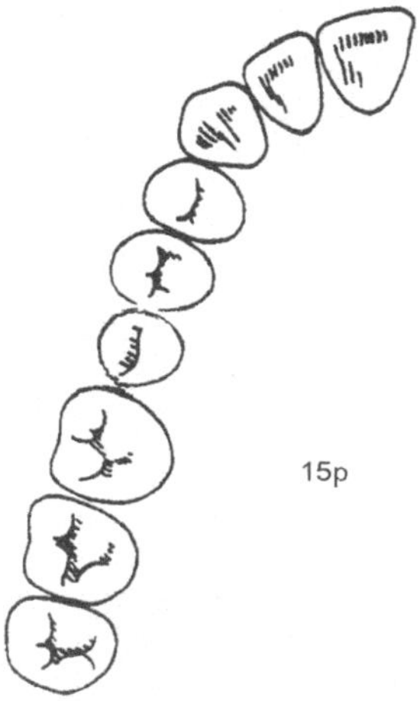

Fig. 10 a: Supplementary permanent maxillary posterior tooth with an abnormal morphology within the dental arch (*15p*).

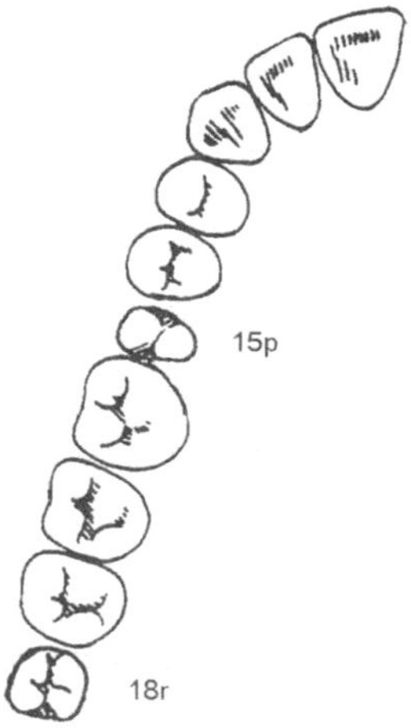

Fig. 10 b: Supplementary permanent maxillary posterior teeth with an abnormal morphology within the dental arch (*15p* and *18r*).

- A supplementary permanent cone-shaped tooth in or near the midline of the upper jaw (between the central incisors), but within the dental arch becomes *p11* or *p21* (as it is proximal to tooth *11* as well as to tooth *21*) (Fig. 11 a). In the case of a second cone-shaped tooth next to the first one, the one mesial to tooth *11* is allotted *p11*, and the one mesially to tooth *21*, *p21*. If these teeth are placed lingually to instead of within the dental arch, the notations are *1l1* and *2l1*, respectively (Fig. 11 b).

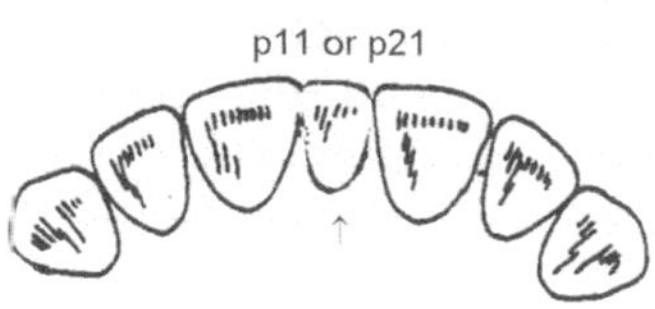

Fig. 11 a: Supplementary permanent cone-shaped tooth between the central incisors within the dental arch (*p11* or *p21*).

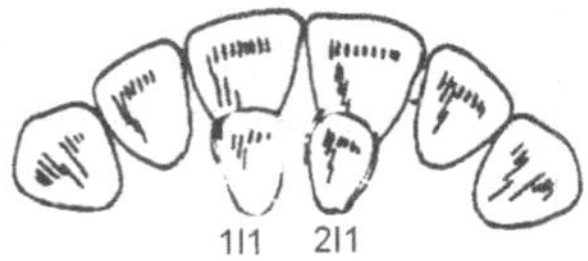

Fig. 11 b: Supplementary cone-shaped teeth outside dental arch (*1l1* and *2l1*).

Villa Vigil et al. also proposed notations for fused and geminated teeth. In the case of fused normal teeth, "the second digit indicates the mesial tooth that participates in the fusion, and the third digit indicates the distal tooth of the fusion" (1989, 299).

Geminations – the result from the fusion of a normal tooth and a supernumerary one – are indicated in such a way that "the second and the third digits, respectively, are the number of the normal tooth participating in the gemination and *9*, which represents the fused supernumerary" (1989, 300).

Examples:

- A fusion of a permanent mandibular left lateral incisor (*32*) with its adjacent canine (*33*) would be designated *323* (Fig. 12 a).
- A gemination between a permanent mandibular left lateral incisor (*32*) and a supernumerary incisor would be indicated as *329* (Fig. 12 b).

Fig. 12 a: Fusion of a permanent mandibular left lateral incisor with its adjacent canine (*323*).

Fig. 12 b: Gemination between a permanent left lateral incisor and a supernumerary incisor (*329*).

Obstacles to the World-Wide Adoption of the FDI Two-Digit System

When proposing the new binominal system in 1970, the FDI representatives realized that it would take a number of years until full recognition could be achieved – despite the fact that the "new system seemed so simple to learn that, once officially introduced, difficulties were unlikely to persist beyond a relatively short transition period" (FDI 1971). The FDI Director's appeal in 1988 that more countries adopt the two-digit system (N.N. 1988) shows that obstacles with regard to world-wide acceptance remain. In 1988, there were still about "40 different systems in use due to uncoordinated development" (N.N. 1988) (p 49). Frykholm and Lysell (1962) gave an explanation for the apparent obstacles associated with the introduction of a new tooth-numbering method, such as the FDI system, when they described the most common of the many tooth designating systems at the time: "Due to a certain conservatism, there is always a certain tendency for the method that one uses oneself to become the 'only possible method' [...]" (p 204).

Particularly in the United States, where the Universal System is still ubiquitous, the dental community has been reluctant to adopt the FDI tooth-designating method. The concerns of many American dentists are reflected in a remark made by Schwartz and Stege: "In almost every country in the world, whole numbers are expressed and written in a customary manner: ten (10); eleven (11); twelve (12); ... twenty eight (28), etc. Usurping this worldwide standard for the written twenty eight (28) as two-eight ... is very confusing. It is difficult to believe that anything but professional chaos will result" (1977, 105). Errors, however, are more likely to evolve from the Universal System, which has been correctly described as "a language without logic" (Peck and Peck 1993, 645). Parreidt himself was frank enough to admit that after using his 1-to-32 method for two months in the clinical courses at the University of Leipzig he had to abandon it because the students made too many mistakes (Parreidt 1882).

In order to foster the international application of the FDI two-digit system, many editorials and articles in international dental journals have encouraged dentists for its adoption (Sharma and Wadhwa 1977; O'Connor 1983; Sandham 1983; Thurow 1986; Peck and Peck 1993, 1996; Simonsen 1995; Türp and Alt 1995). Finally, in 1994, on their 135th Annual Session, the American Dental Association (ADA) recognized the usefulness of the FDI two-digit system. The ADA adopted resolutions in which they supported the integration of the FDI method of designating teeth into clinical computer systems and encouraged all dental schools in the United States to teach students this method along with the Universal System (American Dental Association 1995).

Conclusion

An internationally accepted nomenclature which is widely used in odontology and related fields enhances the understanding among all disciplines dealing with teeth and, at the same time, reduces the risk of misinterpretation. Because of its simplicity, accuracy, safety, applicability in modern technology, and expandibility, the FDI two-digit system is most useful (Türp and Alt 1995). In our view, it seems to be the best method for designating teeth and for ensuring that all who have an interest in designating the thirty-two teeth have a common language.

References

Alt KW (1989) Empfehlung zur einheitlichen Kennzeichnung der Zähne bei der 2-1-2-3 Zahnformel. Anthropologie (Brno) 27: 5–12.

American Dental Association (1995) Joint report of the council on dental benefit programs, the council on dental materials, instruments and equipment and the council on dental practice. Electronic data interchange and designating systems for teeth. Supplement to Annual Reports and Resolutions. ADA 135th Annual Session. New Orleans, Louisiana, October 22–26, 1994, p. 288–291.

Denton GB (1963) [News and notes.] Dent Abstr 8: 66.

Ducka-Karska K (1983) Certaines anomalies dentaires dans la population noire de Lubumbashi (Zaïre). Odontostomatol Trop 6: 127–130.

Fédération Dentaire Internationale (FDI) (1971) Two-digit system of designating teeth. Int Dent J 21: 104–106.

Frykholm KO, Lysell L (1962) Different systems for the recording of teeth and teeth surfaces. Int Dent J 12: 194–207.

Goaz PW (1981) Dental Charting Systems. American Dental Association.

Goodman P (1967) A universal system for identifying permanent and primary teeth. J Dent Child 34: 312–315.

Haderup V (1887) Vorschlag zu einer internationalen Bezeichnung der Zähne. Korrespbl Zahnärzte 16: 314–315.

Hrabowsky I, Sim J (1971) FDI approved system for encoding the human dentition. J Can Dent Assoc 37: 196–198.
Hurlen B, Humerfelt D (1984) Hyperodontia in the 14th – 18th century Norwegian populations: A radiographc study on skulls. Dentomaxillofac Radiol 13: 135–139.
Interpol (1989) Interpol-Disaster-Victim-Identification. Lyon.
International Organization for Standardization (1984) ISO 3950. Dentistry – Designation System for Teeth and Areas of the Oral Cavity. 2nd ed.
International Organization for Standardization (1995) ISO 3950. Dentistry – Designation System for Teeth and Areas of the Oral Cavity. Corrected and reprinted 2nd ed.
Keiser-Nielsen S (1971) Two-digit-system of designating teeth. Dent Pract Dent Rec 21: 347.
Kopsch F (1941) Die Nomina anatomica des Jahres 1895 (BNA) nach der Buchstabenreihe geordnet und gegenübergestellt den Nomina anatomica des Jahres 1935 (JNA). Thieme, Leipzig.
Krysinski Z (1986) The three-digit system of designating supernumerary teeth. Quintessence Int 17: 127–128.
Lyons H (Committee on Nomenclature, American Dental Association) (1947) Committee adopts official method for the symbolic designation of teeth. J Am Dent Assoc 34: 647.
Mühlreiter E (1870) Anatomie des menschlichen Gebisses. Felix, Leipzig.
NN (1988) FDI Director calls on more countries to adopt the FDI two-digit tooth-numbering system. J Ir Dent Assoc 34: 49.
Nomina Anatomica – Sixth Edition (1989). Livingstone, Edinburgh.
O'Connor JT (1983) Let's really standardize our tooth numbering system. Operative Dent 8: 73–74.
O'Dowling IB (1989) Hypo-hyperdontia in an Irish population. J Ir Dent Assoc 35: 114–117.
Palmer C (1870) Dent Cosmos 12: 523.
Palmer C (1891) Palmer's dental notation. Dent Cosmos 33: 194–198.
Paquette NJ (1960) Quo vadis? Can dentistry agree on a uniform tooth designation? Dent Surv 36: 37–41.
Parreidt J (1882) Zahnärztliche Mittheilungen aus der chirurgischen Universitätspoliklinik zu Leipzig. Felix, Leipzig, p. 13.
Peck S, Peck L (1993) A time for change of tooth numbering systems. J Dent Educ 57: 643–647.
Peck S, Peck L (1996) Tooth numbering progress. Angle Orthod 66: 83–84.
Pirquet C (1924) Nummerierung der Zähne. Wien Klin Wochenschr 37: 566.
Sandham JA (1983) The FDI two-digit system of designating teeth. Int Dent J 33: 390–392.
Schwartz S, Stege D (1977) Tooth numbering systems: A final choice. Ann Dent 36: 99–106.
Sharma PS, Wadhwa P (1977) Evaluation of the FDI two-digit system of designating teeth. Quintessence Int 8: 99–101.
Simonsen RJ (1995) It just makes sense. Quintessence Int 26: 441.
Staubesand J, Steel F (1988) A note on degenerative changes in anatomical terminology. Acta Anat 133: 265–268.
Swales JM (1990) Genre Analysis: English in Academic and Research Settings. Univ Press, Cambridge.
Türp JC, Alt KW (1995) Designating teeth: The advantages of the FDI's two-digit system. Quintessence Int 26: 501–504.
Thurow RC (1986) Babble or Babel? [editorial] Angle Orthod 56: 266–267.
Villa Vigil MA, Alvarez Arenal A, Rodríguez González MA (1989) Notation of numerical abnormalities by an addition to the FDI system. Quintessence Int 20: 299–302.
Viohl J (1966) Dokumentation mit Maschinenlochkarten in der konservierenden Zahnheilkunde an Universitätskliniken. Dtsch Zahn Mund Kieferheilk 46: 345–364.

WHO (1987) Oral Health Surveys. Basic Methods. World Health Organization, Geneva.
Woerdeman MW (1957) Nomina Anatomica Parisiensia (1955) et B. N. A. (1895). Oosthoek, Utrecht.
Zsigmondy A (1861) Grundzüge einer praktischen Methode zur raschen und genauen Vormerkung der zahnärztlichen Beobachtungen und Operationen. Dtsch Vierteljahrsschr Zahnheilk 1: 209–211.

2.4 A Recently Discovered Etruscan Dental Bridgework

Maria Teschler-Nicola, Michaela Kneissel, Franz Brandstätter, and Hermann Prossinger

Introduction

Human skeletal remains from the distant past are not only biological source material – being particularly relevant in the absence of historic documents – but also documents of cultural and medical importance. Therapeutic attempts, such as the care of bone injuries or the restauration of teeth or jaws, can be documented even for epochs in which written documents do not exist. Early authentic archeological finds of tooth prostheses are extraordinarily rare.

Except for two tooth replacements and/or fastening mechanisms from ancient Egypt (Junker 1929, Harris and Iskander 1975, Hoffmann-Axthelm 1985, Alt 1994, Puech 1995) – the interpretation of their primary implementation remaining controversial –, the only genuine dental work that can be documented stems from historic antiquity (Asbell 1948, Jackson 1988, Walter 1989). Among such works, those of the Etruscans must be considered exceptionally good. Nearly twenty finds of tooth replacements and examples of parodontal bridgework (now kept in various museums; see Waarsenburg (1991) and Becker (1994)) certify the high technical expertise of their makers. They show various technical styles, the most common being the use of a gold band in order to hold loose teeth or – in the case of dental loss – a tooth replacement; other, more sophisticated ones were made from small separate gold loops.

The exceptional rarity of such early dental work is the reason why we here present an isolated dental bridgework found a few years ago in an Etruscan tomb near Lake Bracciano, north of Rome, and now stored in the Department of Anthropology, Natural History Museum, Vienna. Waarsenburg (1991) bemoans the problems resulting from the "non-accounted and usually fictive dating of dental specimens" presented in the literature. In the case of this "Bracciano bridge" we have incomplete information as to the arrangement of the grave goods in the tomb. However, some accompanying archeological artefacts have been preserved (now in the Department of Prehistory, Natural History Museum, Vienna), which enable us to date this bridgework between 700–600 BC. These same artefacts also indicate that the wearer was female, probably from a higher social class.

In order to elucidate the *in situ* circumstances, the origin of the tooth replacement, the functionality of the implementation, the metal used, as well as the construction technique employed by the ancient goldsmith, we used a variety of methods; we investigated the artefact macroscopically, radiologically and with both light and scanning electron microscopes.

Tooth Features

This Etruscan dental bridgework consists of three frontal teeth of the upper jaw (right lateral, right central and left central incisor) and an approximately 4 mm broad band of gold, which is riveted to the right medial incisor and surrounds the adjacent teeth (Figs. 1–4).

Fig. 1. Bracciano bridge (Etruscan; NHMW Inv. Nr. 24.286): Labial view.

The tooth crowns of the outer teeth (teeth 12 and 21) are completely preserved, including the neck regions, while the major parts of the roots have deteriorated. The horizontally and transversally orientated root-fractures must be recent fracture lines, as there are no smoothing effects due to influences incurred during extended post-mortem deposition. These fractures could be a consequence of the postmortal removal of the dental device, which possibly occurred during the excavation by an amateur. On both teeth, a moderate shoveling is evident. In the case of the left central incisor this feature is accompanied by the presence of a lingual tuberosity.

By contrast, the right central incisor shows a different degree of preservation: This tooth consists of the crown and a small root remnant. Based on macroscopic, light and scanning electron microscope investigations, all parts seem to have been

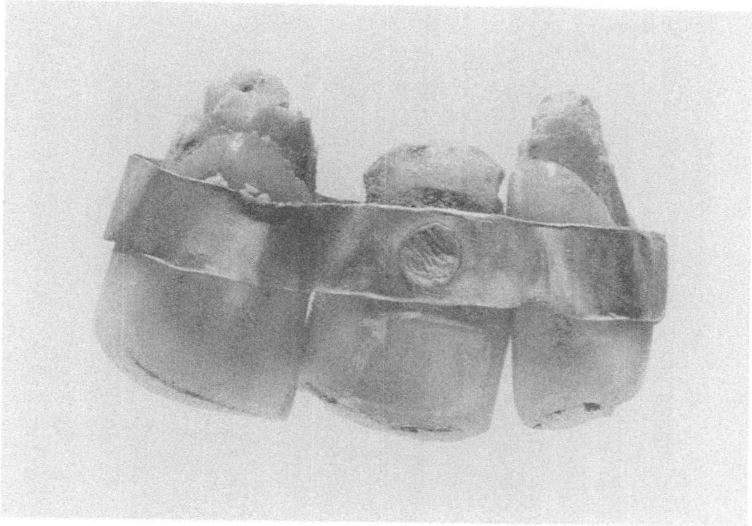

Fig. 2. Bracciano bridge (Etruscan; NHMW Inv. Nr. 24.286): Lingual view.

artificially altered. The lingual side of the crown has been partly ground down (Fig. 2) in order to achieve a flat surface. For the fixation of the gold band, a careful preemptive drilling of the tooth was necessary. (Otherwise, either driving a pin through the tooth would have destroyed it or attempting to use the gold rivet for the fixation of the replaced tooth would have been impossible.) The metric dimensions, which are summarized in Tab. 1 document this manipulation. The morphological features of the remnant proximal part of the root indicate a special

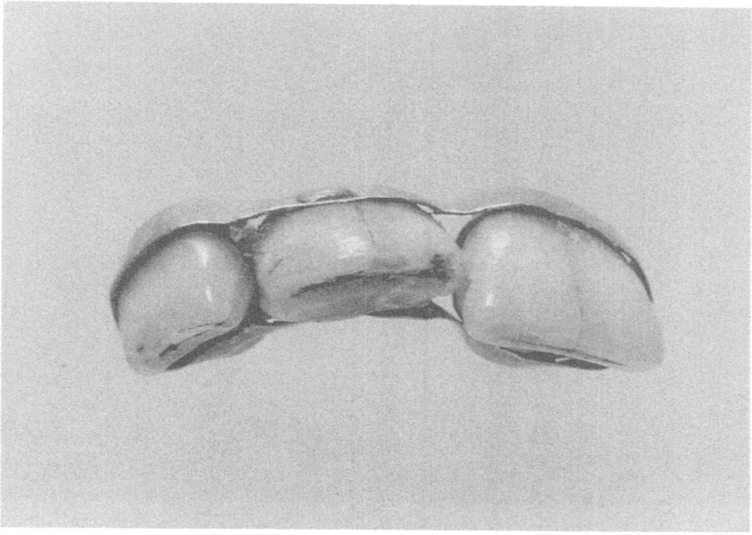

Fig. 3. Bracciano bridge (Etruscan; NHMW Inv. Nr. 24.286): Incisal view.

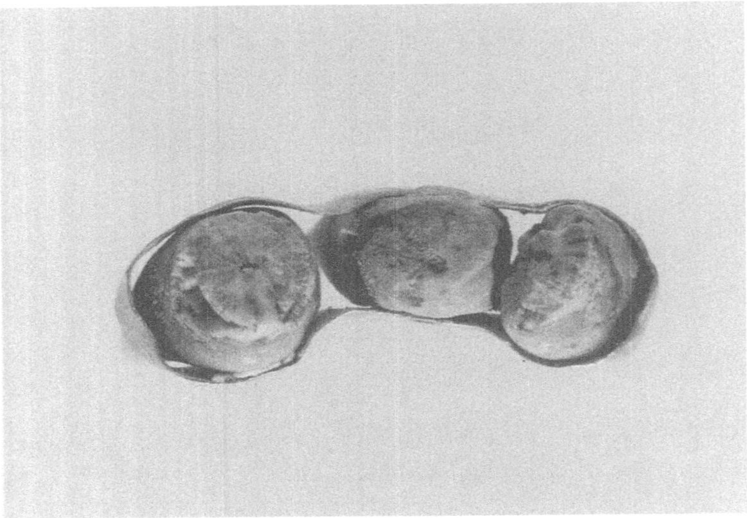

Fig. 4. Bracciano bridge (Etruscan; NHMW Inv. Nr. 24.286): Alveolar view.

treatment too, resulting in a rounded, smooth surface (Figs. 4, 5). All these features, and the riveting technique used for the fixation of the gold band indicate that the right central incisor must be the replaced one. The form, size and color of any tooth is so unique that we can conclude that the artificially anchored tooth was most certainly the wearer´s own. This conclusion is also supported by the similar extent

Fig. 5. Bracciano bridge (Etruscan; NHMW Inv. Nr. 24.286): Superior (alveolar) view of the right central incisor: structure of the remnant root (SEM).

Table 1. Dental measurements (MD = mesiodistal diameter, BL = buccolingual diameter, CH = height of tooth crown, H = height of tooth crown and root fragment; values in parentheses are actual measurements, not corrected for losses, etc)

	MD	BL	CH	H
I1 L	8.5	6.5	(10.4)	(10.7)
I1 R	(8.0)	(4.6)	10.9	11.9
I2 R	5.9	6.2	10.2	(13.3)

of dental abrasion; even though the anchored incisor – due to its heavier use – resulted in more uncovering of dentine. However, such a differentiation could have originally been caused by a malposition of this incisor in the jaw, so that it was subject to higher strain.

Since other parts of the skeleton, especially the jaws, are missing, it is – unfortunately – impossible to determine the cause for the loosening of the central incisor. It could have been extracted because of possible periodontal inflamation or loosening by a blow, although tooth or jaw fractures – especially in women – are rare in prehistoric populations, as evidenced in the extensive Bronze Age populations from Lower Austria (Teschler-Nicola 1994).

From the degree of wear, which we ought to investigate very carefully (there are no diagenetic changes visible), we conclude that the person appears to have died between 25 and 40 years of age. A comparison of the metric dimensions with those of Bronze Age populations (Teschler-Nicola 1992) suggests that this device was probably worn by a woman. These metric dimensions, along with other known examples of such pontics (Becker 1994) corroborate the archeological evidence.

The Bridge Construction

The gold bridge consists of a single long band with a rivet (Fig. 6). The band, overlapping by about 8 mm between the right medial and lateral incisor (Figs. 1, 7), has been forged to a thickness of 0.24–0.30 mm (up to 0.39 mm in the overlapping region). The metric dimensions of this "Bracciano bridge" are shown in Figs. 8 a–8 c. To obtain further clues about the technology used by the ancient goldsmith (the high manufacturing quality supports the view that goldsmiths – not physicians – are the makers), chemical analyses were performed using a scanning electron microscope (JEOL–6400) equipped with a KEVEX energy dispersive system (EDS) were performed (Hartmann 1970, 1982, Echt and Thiele 1987, Brandstätter 1988/1989). The results of analysis from ten different regions of the gold band, including the rivet, show the use of a relatively inhomogeneous material (Tab. 2, Fig. 9). The proportion of silver varies between 0.7 and 7.0% by weight and copper, which is present only in a few regions, varies between < 0.2 and 0.8% by weight. According to Hartmann (1982) gold with a copper content in concentrations less than 1% represents naturally occuring gold, gained from secondary deposits, such as panned gold. The dental bridge was certainly produced by a – widespread – cold forging technique. The fabrication marks, caused by the forging tools, can be

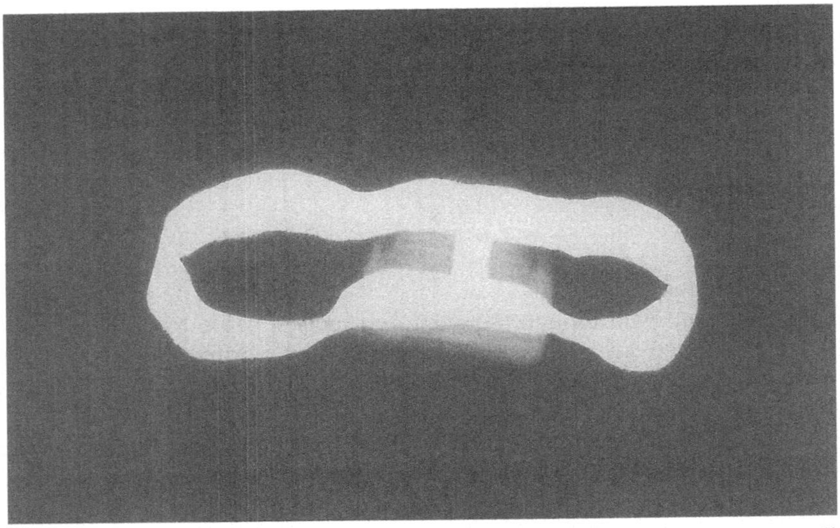

Fig. 6. Bracciano bridge (Etruscan; NHMW Inv. Nr. 24.286): X-ray image of the appliance.

observed over the whole band (Figs. 7, 10). One of the widest scratches shows signs of repair, obviously done by the goldsmith himself. Fig. 10 shows this part of the gold band with the broad scratch covered by a small amount of material added later.

The functionality of this early dental bridgework is indicated by tiny chips, situated on the occlusal margin along the gold band, and very small (0.5 mm) scratches in varying directions (Figs. 2, 11). These scratches are quite different from those which were produced by the goldsmith's tools.

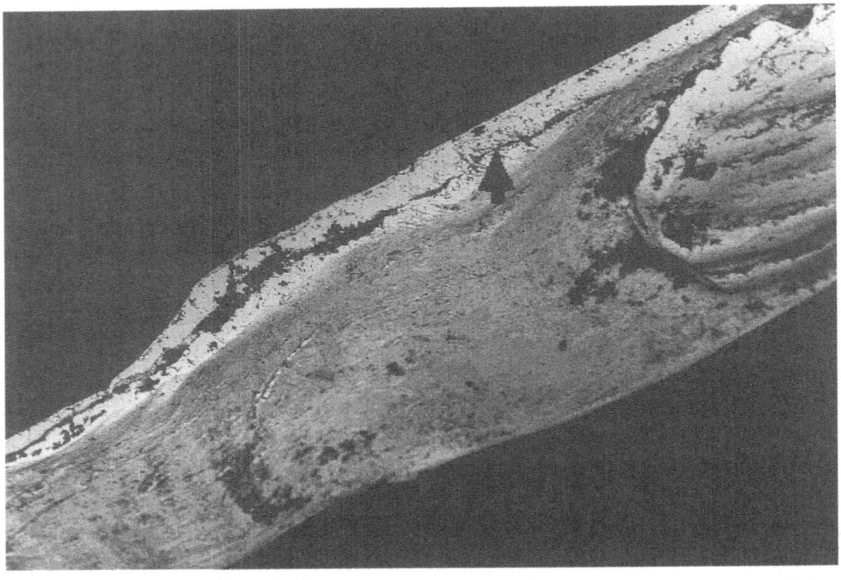

Fig. 7. Bracciano bridge (Etruscan; NHMW Inv. Nr. 24.286): Superior margin of the gold band in the region of overlap (SAM).

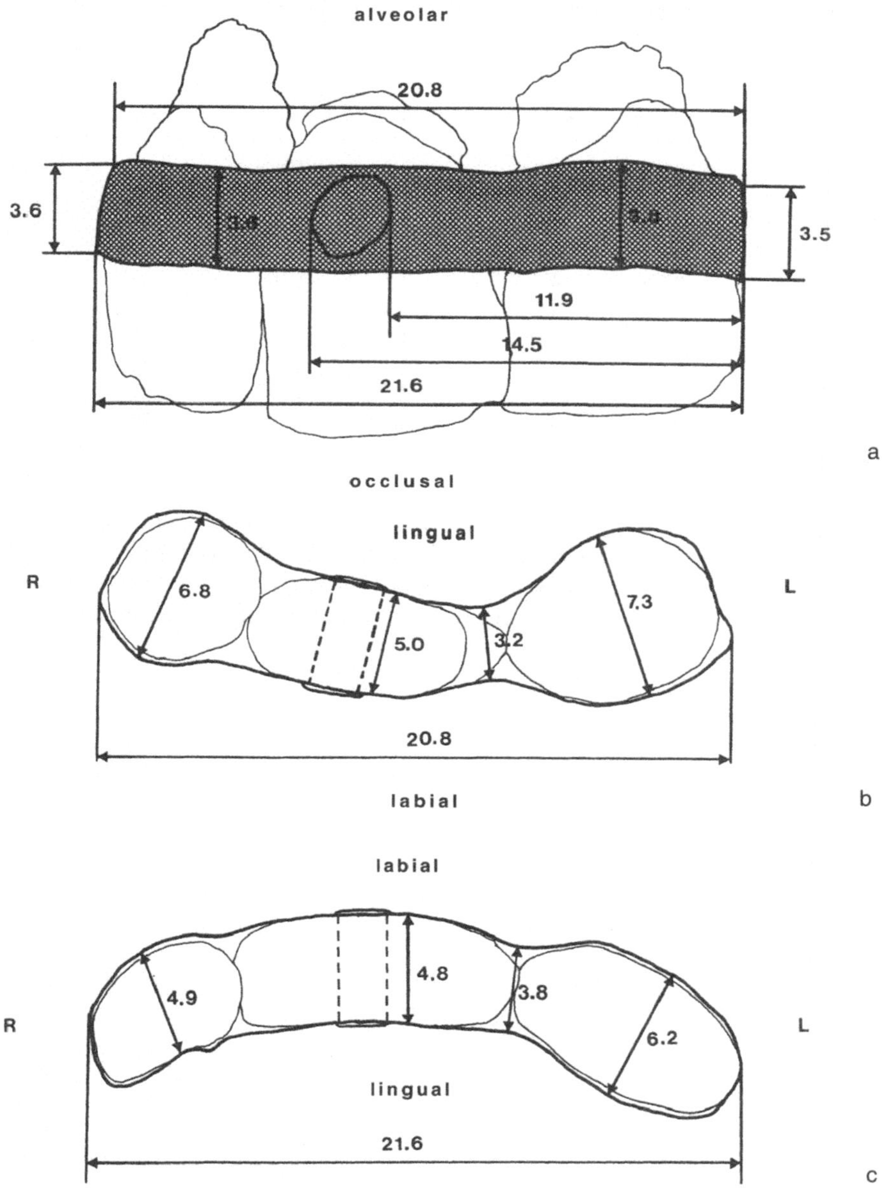

Fig. 8. Bracciano bridge (Etruscan; NHMW Inv. Nr. 24.286): Line drawing, showing the dimensions in mm; **a:** labial view, **b:** superior view, **c:** incisal view.

Discussion

The "Bracciano bridgework" is further evidence for the high technical standard of dental prostheses, as has been known from other specimens found in Etruscan

Table 2. EDS-analyses on ten different locations (in wt.-%, S. D. = standard deviation)

Location	Au		Ag		Cu	
	wt.-%	S. D.	wt.-%	S. D.	wt.-%	S. D.
A1	97.2	3.3	2.0	0.1	0.8	0.1
A2	97.8	2.9	1.5	0.1	0.7	0.1
A3	98.2	3.2	1.3	0.1	0.5	0.1
A4	92.2	3.0	7.0	0.1	0.8	0.1
A5	97.2	3.2	2.2	0.1	0.7	0.1
A6	93.8	3.0	5.4	0.1	0.8	0.1
A7	95.3	3.0	4.1	0.1	0.6	0.1
A8	97.6	3.2	2.1	0.1	0.3	0.1
A9	97.6	3.0	2.4	0.1	<0.2	—
A10	98.6	3.1	0.7	0.1	0.6	0.1

tombs (Waarsenburg 1991, Becker 1994, Capasso and Di Totta 1993, Johnstone 1932a, 1932b, Pot 1985, Corrucini and Pacciani 1989, Terzioglu and Uzel 1987). In order to produce such appliances, special tools were needed (and used), which could not be recognised as such in ancient grave ensembles of medical tools (Como

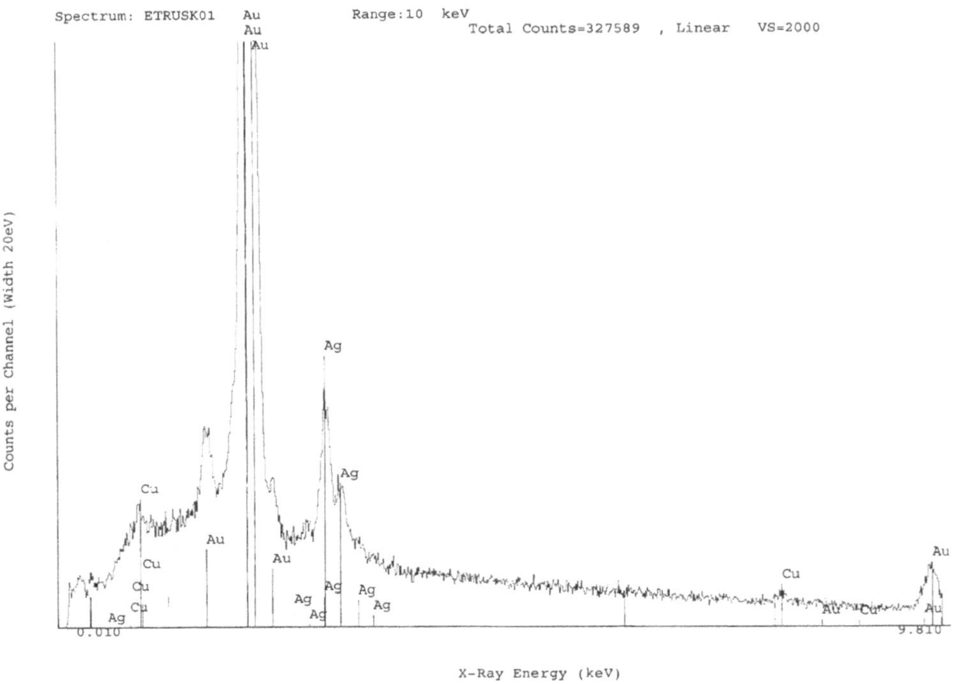

Fig. 9. Bracciano bridge (Etruscan; NHMW Inv. Nr. 24.286): Energy dispersive x-ray spectrum of area 1 (A1) of the gold appliance.

Fig. 10. Bracciano bridge (Etruscan; NHMW Inv. Nr. 24.286): Parts of the dental appliance showing fabrication marks; the band with scratches and signs of repair (SEM image).

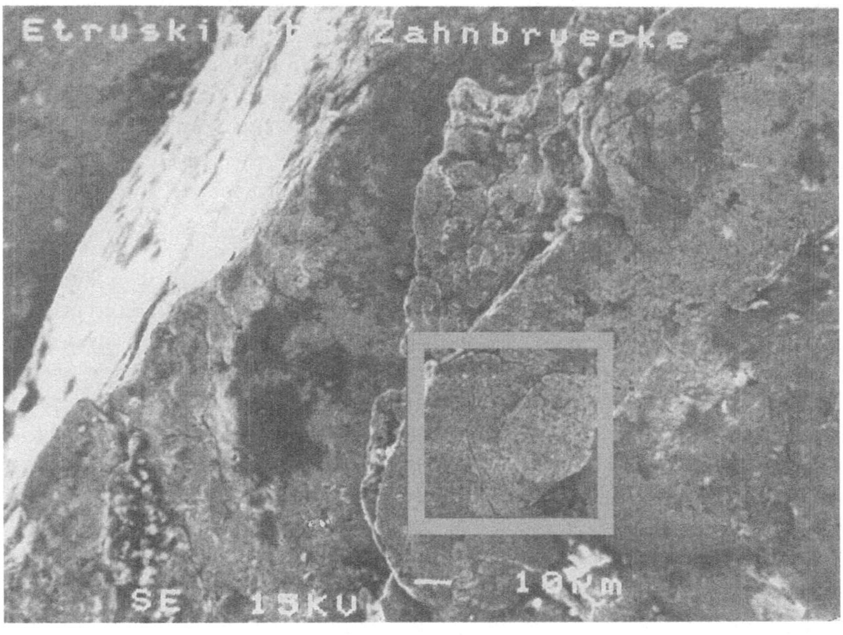

Fig. 11. Bracciano bridge (Etruscan; NHMW Inv. Nr. 24.286): Details of the appliance; very small scratches (SEM image).

1925, Künzl 1983). Because of the specific characteristics of the material used for the manufacture of dental bridgework, these gold appliances were most probably produced by ancient goldsmiths. Their professionalism is further documented by their knowledge of different techniques for the fixation of granulation on jewelry work (Echt and Thiele 1987).

The method of tooth fixation with gold was so common that the Roman "Twelve Table Law", which intended to constrain luxurious burial rites, was obliged to tolerate the use of dental gold. The edict specifically refers to dental gold when it forbids the inclusion of gold artefacts in graves, yet allows the gold fastened to teeth to remain during burial or cremation (Krug 1984).

A review of the published appliances shows that women were their primary users (Becker 1995). Nonetheless, we cannot ascertain the reason for their application with confidence. Capasso and Di Totta (1993) present the results of paleopathological examinations of an Etruscan sample, consisting of 119 individuals, which could perhaps supply an explanation for the need of dental bridgeworks. They documented a rather high frequency of alveolar bone infection or inflammatory lesions (apical granulomas, ca. 28%) in the molar region of the jaw. They could furthermore demonstrate that "about 5% of all central incisors were lost ante-mortem." Because of such circumstances – that central incisors were rarely affected by caries and seldom by severe attrition – "trauma may have been a contributing factor to this loss." Thus, it would be highly relevant to shed light on the degree of periodontal alterations in this population – especially alveolar bone reduction – as they could be a possible cause for ante-mortem tooth loosening.

Becker (1995), who has extensively studied Etruscan and other dental appliances, suggests another – possibly cultural – phenomenon. Since maxillary central incisors, as opposed to side teeth, are rarely lost, he formulates the difficult-to-substantiate hypothesis "that tooth evulsion was practiced in Etruria and that the gold pontics were used as replacements and as ornaments." The bridgework presented here can – at best, if at all – only be considered marginal support for this view.

Conclusion

An isolated dental bridgework in the Museum of Natural History in Vienna (Inv. Nr. 24.286) represents a further example of an early Etruscan gold band prosthesis, presumably worn by a female, a conclusion corroborated by the archeological artefacts and the tooth dimensions. It consists of three frontal teeth, a gold band and a rivet, fixed to the right central incisor. As the analyses of macro- and microscopic features show, this must have been the wearer´s own reused tooth – perhaps having previously been loosened by periodontal disease or by a sharp blow – and replaced by a talented goldsmith who employed techniques of a high standard in order to stabilize it.

Since the dental bridge is isolated and no other parts of the skeleton have been preserved, there is no possibility to examine the status of the alveolar margin and to find evidence for the cause of loosening.

Although considerations of decorative and cosmetic aspects as well as physiognomic appearances cannot be ignored, this construction seems to be primarily functional. Mechanical clues, such as small scratches and grooves on the gold band document the use of this prosthesis over a time span of least several years.

References

Asbell MB (1948) Specimens of the dental art in ancient Phoenicia (5th–4th Century B.C.). Bulletin of the History of Medicine 22 (6), 812–821.

Alt KW (1994) Die historische Entwicklung der zahnärztlichen Prothetik. In: Strub JR, Türp J Ch, Witkowski S, Hürzeler MB, Kern M (eds) Curriculum Prothetikum. Quintessenz, Berlin, p. 37–61.

Becker MJ (1994) Etruscan gold dental appliances: Origins and functions as indicated by a example from Orvieto, Italy, in the Danish National Museum. Dental Anthropology Newsletter 8 (3), 2–8.

Becker MJ (1995) Tooth evulsion among the ancient Etruscans: Recycling in Antiquity. Dental Anthropology Newsletter 9 (3), 8–9.

Brandstätter F (1988/1989) Ein Beitrag zur Kenntnis der chemischen Zusammensetzung frühbronzezeitlicher Goldfunde: Elektronenstrahl-Mikrosondenanalysen von Funden aus dem Gräberfeld von Hainburg-Teichtal, VB Bruck an der Leitha, Niederösterreich. Mitt. Anthrop. Ges. Wien 118/119, 151–153.

Capasso L, Di Totta G (1993) Etruscan teeth and odontology. Dental Anthropology Newsletter 8 (1), 4–7.

Como J (1925) Das Grab eines römischen Arztes in Bingen. Germania 9, 152–162.

Corrucini RS, Pacciani E (1989) "Orthodontistry" and dental occlusion in Etruscans. Angle Orthodontists 59 (1), 61–64.

Echt R, Thiele WR (1987) Etruskischer Goldschmuck mit gelöteter und gesinterter Granulation. Arch. Korrespondenzblatt 17, 213–222.

Hartmann A (1970) Prähistorische Goldfunde aus Europa. In: Bittel K, Hartmann A, Otto H, Sangmeister E, Schickler H, Schöder M (eds) Studien zu den Anfängen der Metallurgie, Bd. 3, Gebrüder Mann Verlag, Berlin.

Hartmann A (1982) Prähistorische Goldfunde aus Europa II. In: Bittel K, Hartmann A, Otto H, Sangmeister E, Schickler H, Schröder M (eds) Studien zu den Anfängen der Metallurgie, Bd. 5. Gebrüder Mann Verlag, Berlin.

Harris JE, Iskander Z, Farid S (1975) Restorative dentistry in ancient Egypt: An archaeological fact. J Michigan Dent Ass 57, 401–404.

Hoffmann-Axthelm W (1985) Die Geschichte der Zahnheilkunde (2nd ed.). Quintessenz, Berlin.

Jackson R (1988) Doctors and Disease in the Roman Empire. British Museum Publications, London.

Johnstone MA (1932a) The Etruscan collection in the free public museums of Liverpool. Annals of Archaeology and Anthropology 19, 1221–137.

Johnstone MA (1932b) The Etruscan Collection in the Public Museum of Liverpool. Studi Etruschi 6, 443–452.

Junker H (1929) Giza I. Die Mastabas der IV. Dynastie auf dem Westfriedhof. Denkschriften der Akademie der Wissenschaften, Phil.-hist. Klasse 69, 1. Abhandlung. Wien, Leipzig.

Krug A (1984) Heilkunst und Heilkult: Medizin der Antike. Beck, Nördlingen.

Künzl E (1983) Medizinische Instrumente aus Sepulkralfunden der römischen Kaiserzeit. Führer des Rheinischen Landesmuseums Bonn Nr. 115. Habelt, Bonn.

Pot T (1985) Two Etruscan gold dental appliances, found in 19th century excavations at Satricum and Praeneste. Mededelingen van het Nederlands Instituut te Rome 47 (N.S. 11), 35–39.

Puech PFF (1995) Dentistry in ancient Egypt: Junkers' teeth. Dental Anthropology Newsletter 10 (1), 5–7.

Terzioglu A, Uzel I (1987) Die Goldbandprothese in etruskischer Technik. Ein neuer Fund aus Westanatolien. Phillip Journal für restaurative Zahnmedizin 4, 109–112.

Teschler-Nicola M (1992) Sexualdimorphismus der Zahnkronendurchmesser. Ein Beitrag zur Geschlechtsdiagnose subadulter Individuen anhand des frühbronzezeitlichen Gräberfeldes von Franzhausen I, Niederösterreich. Anthrop. Anz. 50, 51–65.

Teschler-Nicola M (1994) Bevölkerungsbiologische Aspekte der frühen und mittleren Bronzezeit. In: Neugebauer JW (ed.) Bronzezeit in Ostösterreich. Verlag Niederösterreichisches Pressehaus, St. Pölten, Wien, p. 167–183.

Waarsenburg DJ (1991) Auro dentes iuncti. An Inquiry into the Study of the Etruscan Dental Prosthesis. Stips votiva, Festschrift für C. M. Stibbe. Amsterdam, p. 241–247.

Walter C (1989) Die Geschichte der zahnärztlichen Prothetik. Swiss Dent 10 (7–8), 15–31.

3 Dental Morphology, Structure, and Evolution

3.1 Anatomy and Morphology of Human Teeth

Jens C. Türp and Kurt W. Alt

Introduction

The macromorphology of human teeth is of great importance to the scientific disciplines of anatomy, dentistry, physical anthropology, and forensic medicine. Morphological studies concerning the interrelation between the form and the function of teeth are essential to the understanding of ontogenetic and phylogenetic processes. The task of odontography is to allow the precise identification of each tooth, whether it be deciduous (primary) or permanent (secondary), maxillary (upper jaw) or mandibular (lower jaw), left or right, and to give its exact position within the dentition (e.g. permanent mandibular left second molar).

A knowledge of the variation within the human dentition is indispensible for population studies and forensics. Although, at a superficial level, the teeth of different individuals may appear similar on closer examination they exhibit great variation in both form and size. Descriptions of normal tooth form are the product of subjective observation. All forms within a certain tolerance can be described as variants of a norm (the variation is normal) (Alt 1997). But findings which go beyond biological variation are abnormalities. Extreme cases are regarded as malformations (Schulze 1987).

Topography and Nomenclature

The international nomenclature for dental anatomy and for the description of a tooth's position within the dental arch is used inconsistently and is sometimes at variance with anatomic nomenclature. The terminology used for the surfaces of the crowns of teeth generally follows the *Nomina Anatomica* (1989) whereas the cusps and roots are named according to their position in the dental arch.

The part of the tooth covered by enamel is the crown, that covered by cementum is the root. The transition from crown to root takes place at the cervix or neck of the tooth in a sinuous outline, and is called the cementoenamel junction

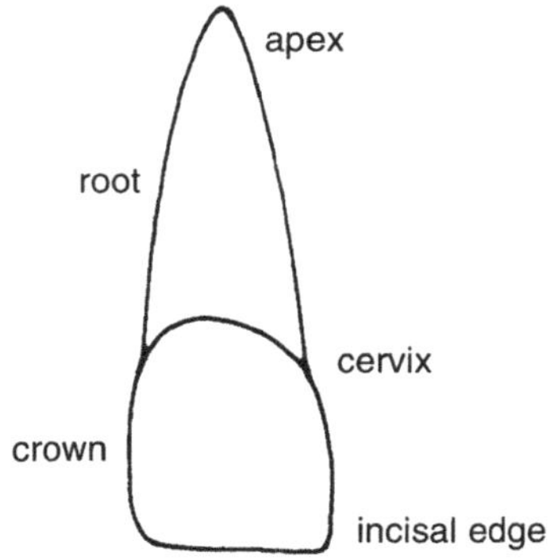

Fig. 1. Topographical regions of an incisor.

or the cervical line. The different regions of the tooth are described with terms derived from Latin (Fig. 1).

The dental arches divide the mouth into two areas, that inside the dental arch (the oral cavity proper), and that outside. The oral cavity can be further divided lingual in the mandible, palatal in the maxilla, but is often generally described as lingual. The facial area is often subdivided into a labial and a buccal part (from the Latin words *labium:* lip, and *bucca:* cheek).

The crown of each tooth has five surfaces. In addition to the facial and lingual surfaces every tooth has either an incisal edge or ridge (incisors and canines) or an

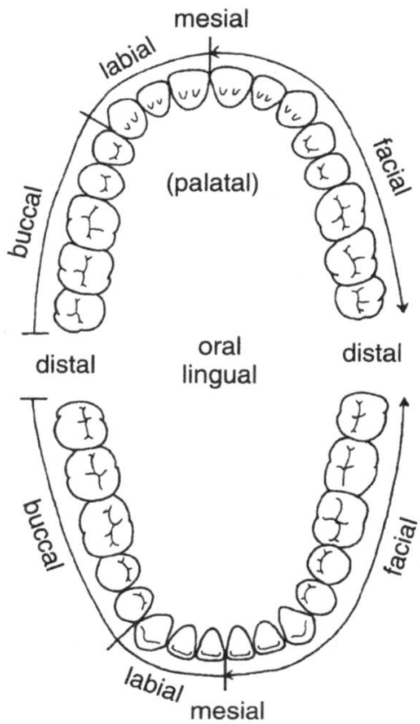

Fig. 2. Upper and lower jaw: Directions and tooth surfaces.

occlusal surface (premolars and molars) as well as two proximal surfaces (mesial and distal), i.e. contact areas between adjacent teeth. Mesial refers to the surface which faces toward the center line of the dental arch and distal refers to the surface which faces away from the center line of the dental arch (Fig. 2). This nomenclature remains even if teeth are rotated, tilted, or displaced. Notations used in anthropology, paleontology, biology, and dentistry aim to represent clearly in abbreviations, teeth and types of teeth using formulae and notation systems (see Chapter 2.3, this volume, also Türp and Alt 1995).

Tooth Structure

Human teeth are composed of four tissues, the soft tissue of the pulp and the three calcified tissues dentin, enamel and cementum. The cementum, also called root cementum, is simultaneously a part of the periodontium, the supporting structure of the teeth (Fig. 3).

The dental pulp occupies the pulp cavity of the tooth. The pulp can be further divided into coronal pulp, which occupies the pulp chamber, and radicular pulp, which fills the pulp or root canals. The pulp contains cells (odontoblasts, fibroblasts, undifferentiated mesenchymal cells), nerve fibers, and blood and lymph vessels. The pulp cavity is lined by a layer of cells called odontoblasts.

Dentinal tubules give dentin its characteristic structure. Chemically, dentin is composed of, by weight, 70% inorganic (mainly calcium and phosphate in hydroxyapatite crystals), 20% organic matrix (mainly collagen), and 10% water (Schroeder 1991). The density and diameter of the tubules decreases the further they are from the pulp cavity. Typical values for the permanent teeth of a young adult for areas close to the pulp cavity (0.1–0.5 mm from the pulp-dentin-border) and areas farther away (3.1–3.5 mm from the pulp-dentin-border) are represented in Table 1.

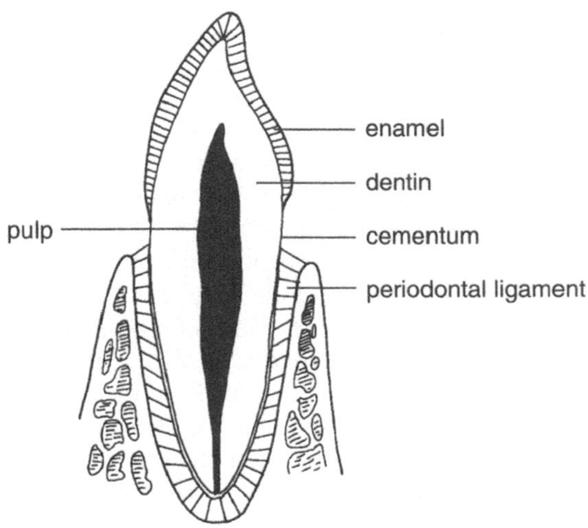

Fig. 3. Buccolingual section of a lower anterior tooth.

Table 1. Average density and average diameter of dentin tubules (Garberoglio and Brännström 1976)

	0.1 – 0.5 mm from the pulp	3.1 – 3.5 mm from the pulp
Mean density (number/mm²)	43,000	19,000
Tubule diameter (μm)	1.9	0.8

Dentinal tubules contain the odontoblast processes (Tomes' processes). The periodontoblastic space around the processes is filled with a liquid tissue (dentin liquor). The dentinal tubules are surrounded by peritubular dentin which is characterized by a high degree of mineralization. The majority of the dentin is formed by intertubular dentin.

Dental enamel consists of 95% inorganic substance by weight (organic matrix: 1%; water: 4%) (Schroeder 1991). Because enamel contains neither cells nor cell processes some authors regard it not as a hard tissue but as a crystalline structure (Schroeder 1983). Nevertheless it is formed by cells, the ameloblasts. Human tooth enamel consists of enamel prisms (rods) (density: 20,000–30,000/mm^2; average diameter: 5 μm) which are constructed of hydroxyapatite crystals.

The superficial layer of enamel is aprismatic in all deciduous teeth and in 70% of permanent teeth. The formation of enamel prisms and their variation in primates can be used to decide questions of taxonomy and phylogeny (Gantt 1983).

Structure of the Periodontium

The periodontium consists of those tissues which anchor and support the tooth in the socket (alveolus) in the alveolar process of the jaw. They comprise of the cementum, the supra alveolar gingival fibre bundles, the periodontal ligament fibres, the junctional epithelium, and the alveolar bone (Fig. 3).

The root cementum, which anatomically belongs to the tooth and functionally to the periodontium, surrounds the root dentin in a thin layer. The cementum layer is thinnest at the cervix (50–150 μm) and increases towards the apex (200–600 μm) (Rateitschak et al. 1989). Chemically, cementum consists, by weight, of about 61% inorganic matrix, 27% organic matrix, and 12% water (Schroeder 1991). Depending on whether the cementum is cellular and the presence of intrinsic or extrinsic fibers of the desmodont (periodontal ligament), four kinds of cementum can be distinguished (see Schroeder 1993). The main components of the periodontal ligament are collagen fibers which are interwoven in a three-dimensional array (Sharpey fibers). The tooth is elastically suspended in the alveolus by these fibers which extend from alveolar bone to the cementum in the periodontal crevice (periodontal space). Depending on the course and topography of the bundles, the fibers are distinguished as crestal, horizontal, diagonal (main part) and apical, as well as interradicular in the case of multirooted teeth (Fig. 3) (Rateitschak et al. 1989). In addition to Sharpey fibers, the periodontal crevice contains dense networks of blood and lymph vessel, cells (fibroblasts, osteoblasts, osteoclasts, cementoblasts, osteoprogenitor cells and cementoprogenitor cells, blood cells and epithelial cells), neural networks, nociceptors (pain) and mechanoreceptors (pressure), as well as tissue fluid.

The collagen fiber bundles of the gingiva are inserted coronally to the periodontal crevice. They attach the gingiva to the tooth surface (the so-called supraalveolar fiber apparatus) (Fig. 4). The alveolar processes are dependent upon the existence of teeth; they are largely resorbed in edentulous persons. The alveolar processes consist of three parts: that nearest the tooth, the alveolar bone, is part of the periodontium, the outermost part is made up of compact bone. In between is trabecular bone filled with fatty marrow (Figs. 3 and 4). The alveolar bone is approximately 0.1 to 0.4 mm thick and has perforations (Volkmann canals) through which blood and lymph vessels as well as nerves enter the periodontal crevice.

Morphology and Variation of Tooth Form and Size in Permanent Teeth

Human teeth show great variability with regard to form and size. Whereas in anthropology and forensic medicine anatomic peculiarities such as the form

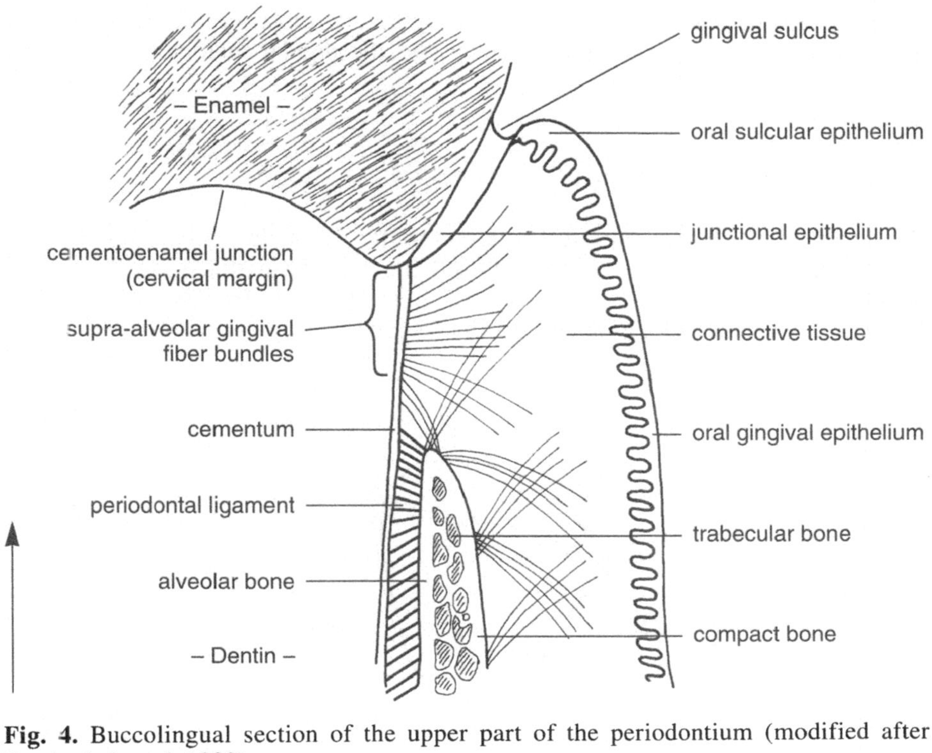

Fig. 4. Buccolingual section of the upper part of the periodontium (modified after Rateitschak et al. 1989).

and prominence of marginal crests, cusps and roots are important, from the dentist's viewpoint characteristics such as crown contour and curvature, the functional details which play a role when teeth are rehabilitated have greater relevance.

The range of variation of human teeth was first depicted by Mühlreiter (1870). Around the turn of the century, famous scientists were dealing with the anatomy and morphology of human teeth and comparative anatomy (Zuckerkandl 1891; Black 1897; Tomes 1898; de Terra 1905; Adloff 1908; Bolk 1913, 1914; de Jonge-Cohen 1917; Hrdlička 1921; Sicher and Tandler 1928). From around the 1950's on, studies centered on macromorphology, which is still important today (Visser 1948; Häupl 1949; de Jonge 1958; Remane 1960; Sicher 1960; Marseillier 1965; Keil 1966; Kraus et al. 1969; Wheeler 1969; Fujita 1973; Tallec 1975; Scott and Simons 1977; Berkovitz et al. 1978; Taylor 1978). Standard works have since been published (DuBrul 1980; Schumacher 1980, 1984; Kurtén 1982; Ash 1984; Carlsen 1987; Williams et al. 1989; Matsumura 1995).

Tooth-Distinguishing Characteristics

For paleodontological and forensic purposes, it can be very important to classify single teeth very precisely. As a rule, every tooth shows certain characteristics that allow its precise identification as left or right. The most well-known of these general features are the curvature characteristic, the angular characteristic, and the root characteristic (Mühlreiter 1870).

Curvature Characteristic (Arch Characteristic)

The evaluation of the curvature characteristic is made while viewing the tooth in question from the incisal or occlusal direction. From this point of view the height of contour of the facial plane of the root is located mesially, which means that the tooth seems to be bulkier mesially (Fig. 5).

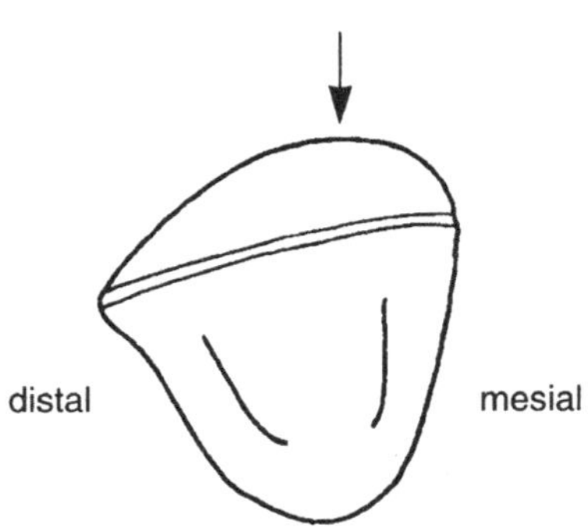

Fig. 5. Tooth determining characteristics: permanent maxillary right central incisor: curvature characteristic.

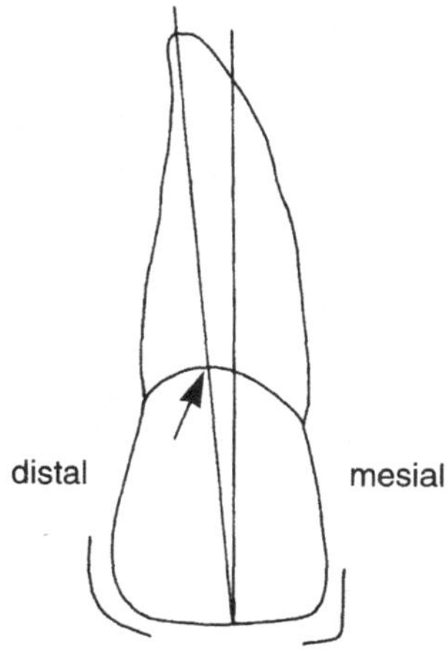

Fig. 6. Tooth distinguishing characteristics: permanent maxillary right central incisor: angular characteristic, root feature, and cervical characteristic.

Angular Characteristic

The angular characteristic is assessed from the facial direction. In the evaluation of the angles made up by the incisal edge and both the mesial and distal side planes of the tooth crown, one can recognize that these are more accute toward the mesial and more rounded toward the distal (Fig. 6). Angularity is especially well-developed in maxillary incisors. As a result of the angular characteristic, the area which is surrounded by the external outline of the tooth and straight lines set on the incisal edge and the respective approximal surface of the tooth shows a smaller volume mesial than distal.

Root Characteristic

The root characteristic is assessed by viewing the tooth from its facial aspect. One can recognize that the root is bent towards the distal (Fig. 6). This feature is found in all teeth except on the central mandibular incisors.

Further Distinguishing Features

A further distinguishing characteristic is the cervical characteristic. It is assessed from the facial direction. In this case, the most apical point is found distally (Fig. 6). This characteristic is found on all incisors and is especially developed on the central maxillary incisors.

Mandibular incisors often show a concavity on the distal surface of the root. By using this characteristic, the very symmetrically built mandibular incisors can usually be assigned to their respective sides.

Morphology and Variation of the Permanent Tooth Crown

As a result of their cutting function, incisors are either spatula- or chisel-shaped. The canines, on the other hand, are designed to hold food and to tear pieces off. This is why they are pointed. Morphologically, canines represent the transition from incisor to premolar (Berkovitz et al. 1980). Tables 4 and 5 show the variation concerning the number of cusps (and roots) of permanent teeth.

Crown Morphology of Permanent Anterior Teeth

Maxillary Central Incisor

The maxillary central incisors are the largest human incisors (Fig. 7a). They possess curvature, angular, root, and cervical characteristics. The labial and lingual surfaces may be either smooth or furrowed. Variations of the lingual surface include a

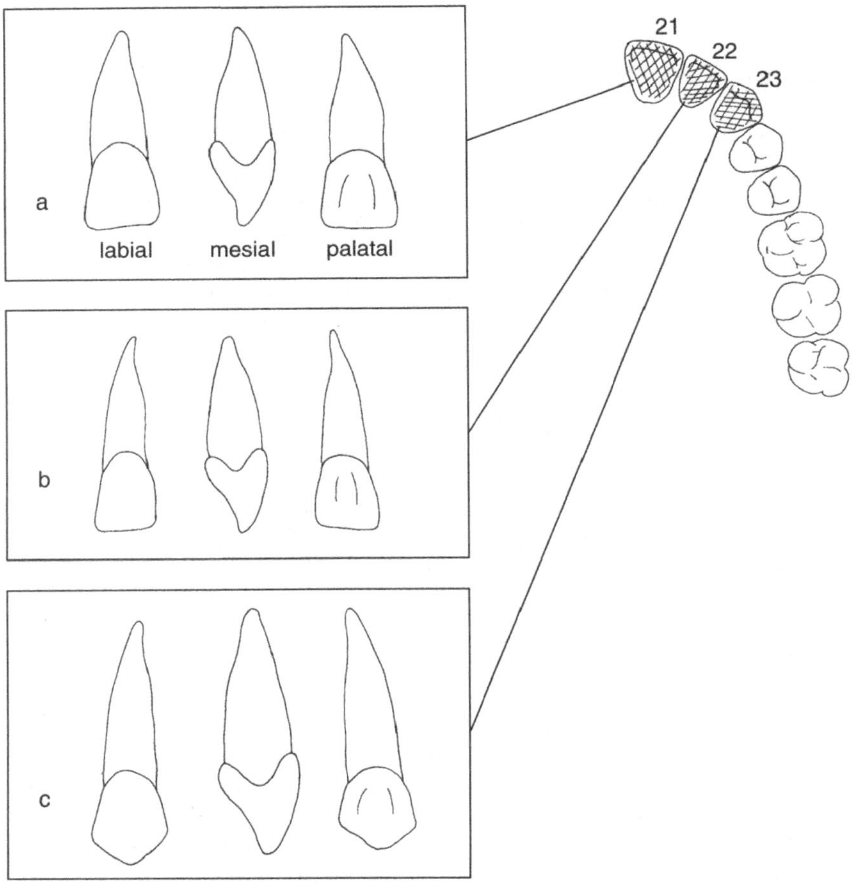

Fig. 7 a–c. Permanent maxillary left front teeth: **a:** central incisor; **b:** lateral incisor; **c:** canine.

lingual fossa, a cingulum (lingual lobe), and marginal and accessory ridges (narrow linear elevations).

Maxillary Lateral Incisor

The maxillary lateral incisors are smaller and slimmer than their mesial neighbours (Fig. 7b). Angular characteristic, root feature and cervical characteristic are typical, whereas the curvature characteristic is less obvious. On the palatal surface, one often finds a foramen caecum, i.e. a blind-ending impression. Variations in form are frequently found in this tooth. The main variations are mostly dilacerations, a caniniform appearance, and a reduction in size (atrophy). The latter feature can be regarded as a micromanifestation of aplasia.

Maxillary Canine

The maxillary canines are the longest human teeth (Fig. 7 c). The outline of the labial surface of the tooth is strongly marked by the form of the mesial and distal edges. The tooth has a form close to that of two isosceles triangles. The mesial edge which descends from the point angle is shorter and slightly flatter than the distal edge. At the transition from the distal edge to the distal approximal surface the tooth widens a little. Marginal ridges are usually found on the lingual surface. If a central ridge occurs, two shallow grooves (sulci) can be found mesially and distally (lingual fossae). The tooth has curvature, angular, root, and cervical characteristics. Variations from the norm most frequently concern the form of the lingual tuberculum (cingulum), the marginal ridges, and the incisal tip.

Mandibular Central Incisor

The mandibular central incisors are the smallest incisors (Fig. 8 a). The teeth are built very symmetrically and as seen from the facial side show a rectangular to triangular contour. The incisal edge is sometimes divided into three equally large cusps by nicks. Marginal ridges, grooves and the tuberculum dentis are often poorly developed. The teeth usually have neither angular, root, nor cervical characteristics, but laterality may be decided upon by the distal concavity in the root.

Mandibular Lateral Incisor

The mandibular lateral incisors, like the centrals, are less variable than their antagonists in the maxilla (Fig. 8 b). The teeth have a poorly developed angular characteristic and a slight root characteristic. Variations of the incisal edge can give the tooth an appearance similar to that of a canine. The incisal edge can be divided into cusps by nicks like the mandibular central incisor. Features such as ridges, grooves etc. seldom occur.

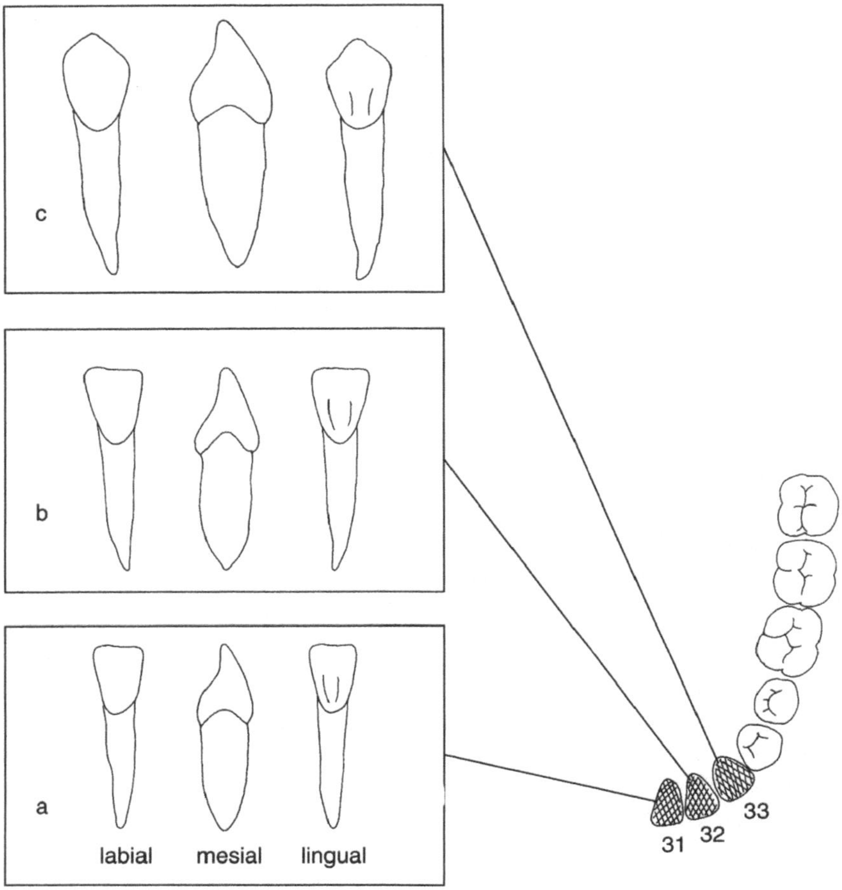

Fig. 8 a–c. Permanent mandibular left front teeth: **a:** central incisor; **b:** lateral incisor; **c:** canine.

Mandibular Canine

The mandibular canines are slimmer than the maxillary canines (Fig. 8 c). From the facial aspect, the tooth shows almost parallel ridges and a rectangular form with a triangle set on top to form a point. Like the mandibular premolars and molars, the crowns of the mandibular canine show a slight lingual inclination. Curvature and root characteristics are obvious. Lingual structures are less developed than in the maxillary canine; the marginal ridges are the only exception.

Crown Morphology of Permanent Posterior Teeth

The main function of the permanent posterior teeth is to crush food. Typically they consist of two cusps (bicuspid). The maxillary premolars are much more simple than their antagonists in the mandible. In many primates, the form of the premolar can resemble that of a canine (caninization), while the first premolar may resemble a molar (molarization). For this reason the premolars can be seen morphologically as a transition from the canine to the molar.

Maxillary First Premolar

The maxillary first premolar usually shows a reverse curvature characteristic, i.e. as seen from the occlusal aspect the height of contour has moved toward the distal (Fig. 9 a). While the tooth has a trapezoid form, if the tooth has a mesial concavity, it has a more kidney-like appearance. The buccal cusp is slightly higher and larger than the palatal one, and the buccal cusp tip is set distally with respect to the palatal one. The fissure pattern of the tooth is very variable. The basic shape of the longitudinal fissure is usually straight or "S"-shaped and ends in a groove in front of the marginal ridge. The main occlusal fissure runs from mesiopalatal to distopalatal and typically shows a drop toward the distal as is typical for molars. Unlike in the mandible, the crowns of the maxillary premolars are clearly defined mesially and distally by marginal ridges. This definition is accentuated by variable diagonal fissures which separate the cusps toward the marginal ridges.

Maxillary Second Premolar

This tooth is somewhat smaller than the maxillary first premolar (Fig. 9 b). Its crown

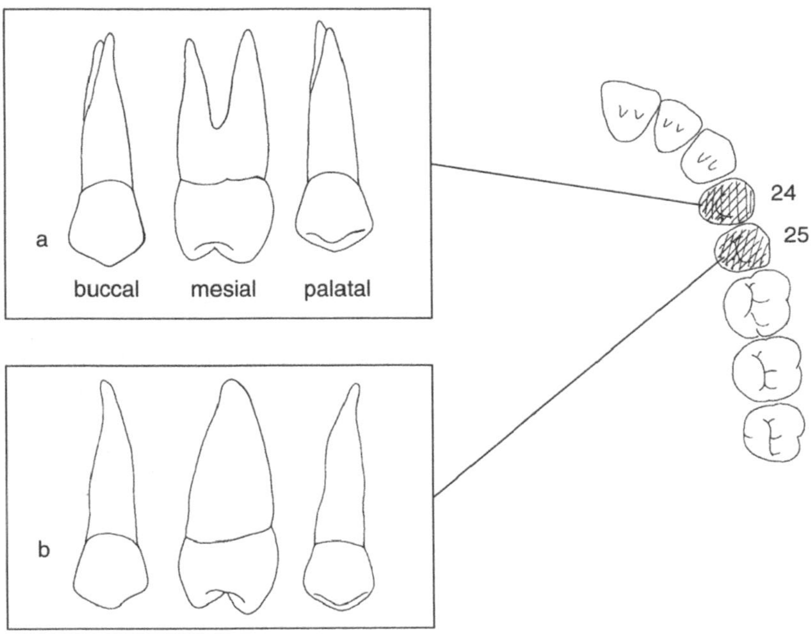

Fig. 9 a, b. Permanent maxillary left premolars: **a:** first premola; **b:** second premolar.

does not differ significantly in form from that of the first premolar. The curvature characteristic is only weakly developed. Both cusps are almost equal in height and size. Unlike the first premolar, the kidney-formed concavity of the mesial marginal ridge is usually missing. The form and size of the second premolar can vary.

Mandibular First Premolar

The mandibular first premolar is the smallest of all premolars (Fig. 10 a). It shows the greatest variation concerning the number and form of its cusps and the course of its fissures. The crown of the tooth is set lingually relative to the root, with the buccal cusp tip in line with the root axis. There is an obvious height difference between the buccal and the lingual cusp. The lingual cusp(s) do not reach the occlusal plane. The tooth has a curvature and a root characteristic.

Mandibular Second Premolar

Like the maxillary first premolar, the crown of the mandibular second premolar is lingually inclined, and shows angle and curvature characteristics (Fig. 10 b). This tooth is also very variable regarding its crown morphology. Usually it possesses three cusps (buccal, mesiolingual, and distolingual). The fissure course is closely

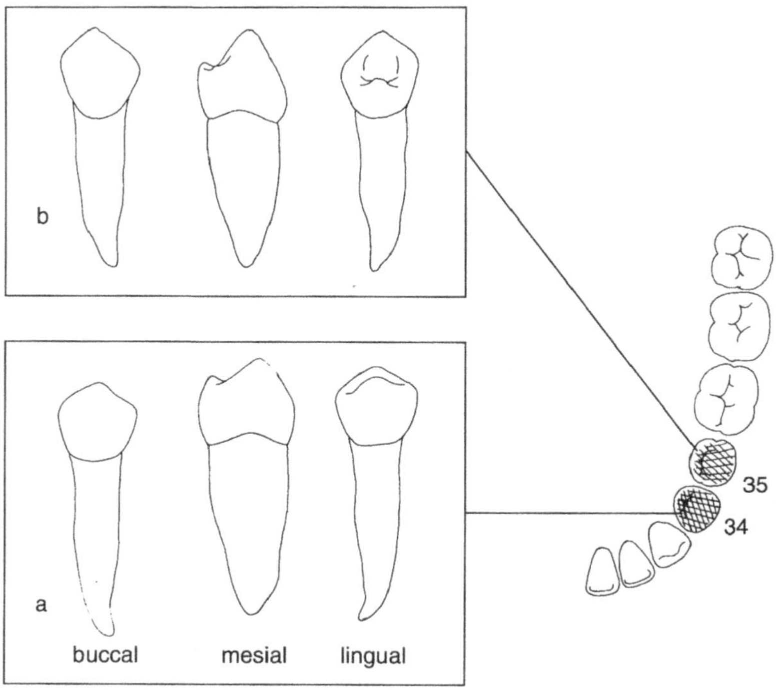

Fig. 10 a, b. Permanent mandibular left premolars: **a:** first premolar; **b:** second premolar.

related to the number and definition of the cusps. If the cusps are strongly developed the fissure course is usually Y-shaped, if they are less developed the course is half round, and if the tooth has two cusps it is straight. The root is triangular to round in cross-section.

Maxillary First Molar

The maxillary first molar is basically rhomboid in form (Fig. 11 a). The two mesial cusps are larger and higher than the distal ones: The mesiopalatal cusp (protoconus) is the largest, followed by the mesiobuccal (paraconus), the distobuccal (metaconus) and the distopalatal (hypoconus; for comparative terms for cusps used in paleontology and those customary in dentistry see Table 2).

The occlusal surface of the first molar possesses two main fissures, one mesial and one distal. The mesial fissure originates between the two buccal cusps, runs diagonally, then turns mesially and runs longitudinally. The distal fissure has its origin between the palatal cusps. It travels distobuccally and ends in front of the marginal ridge. The transverse ridge which is very variable in its pronounciation, runs from the tip of the mesiopalatal cusp to the distobuccal cusp. It separates the two main fissures. If the distopalatal cusp is missing, the transverse ridge takes on the character of a marginal ridge. A great number of secondary fissures may be present as branches or ramifications of the main fissures. The maxillary first molar shows angle, curvature and root characteristics.

Maxillary Second Molar

The basic form of the maxillary second molar is comparable to that of the first (Fig. 11 b). Nevertheless, the second molar is slightly smaller and shows greater

Table 2. Paleontological and odontological nomenclature of molar cusps

Jaw	Paleontological term	Odontological term
Maxilla	Paraconus	Mesiobuccal cusp
	Metaconus	Distobuccal cusp
	Protoconus	Mesiopalatal cusp
	Hypoconus	Distopalatal cusp
	(Protoconulus)	(Mesial intermediate cusp)
	(Metaconulus)	(Distal intermediate cusp)
Mandible	Protoconid	Mesiobuccal cusp
	Hypoconid	Mediobuccal cusp
	Metaconid	Mesiolingual cusp
	Entoconid	Distolingual cusp
	Hypoconulid	Distobuccal cusp
	(Paraconid)	(Mesial cusp)

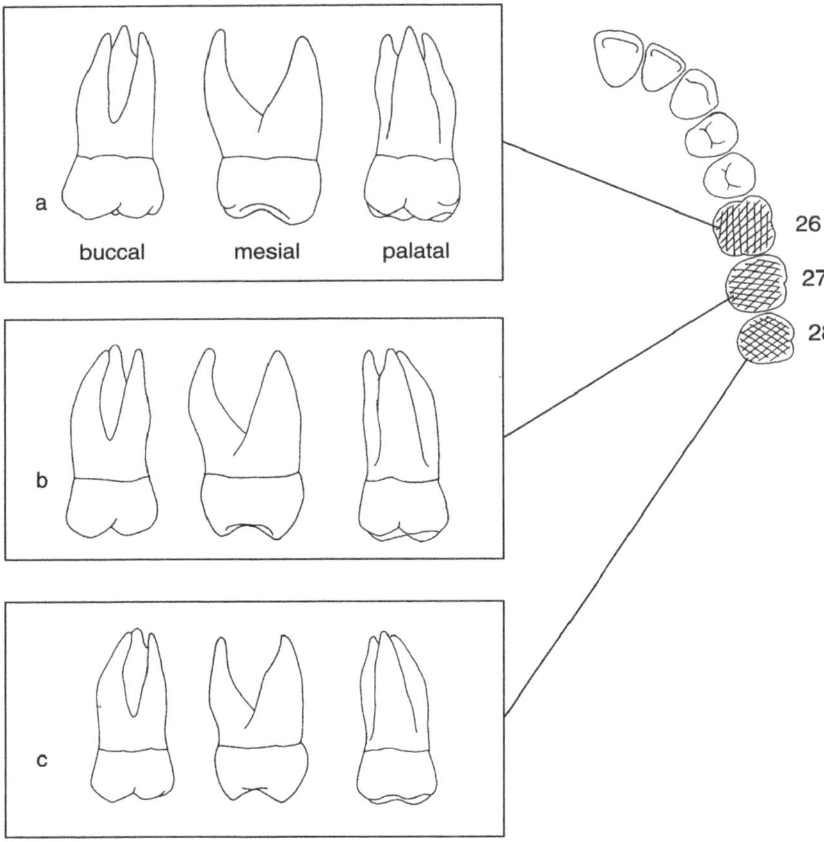

Fig. 11 a–c. Permanent maxillary left molars: **a:** first molar; **b:** second molar; **c:** third molar.

variation. The rhomboid shape, which is typical for all maxillary molars, is most apparent in the second maxillary molar. In almost 70% of cases the tooth has four cusps; in these cases the distopalatal cusp (hypoconus) is missing. 20% of the teeth have a reduced hypoconus which suggests a tendency for the cusp to disappear.

Maxillary Third Molar

Wisdom teeth are for the most part the smallest molars (Fig. 11 c). They generally have an irregular form. This also applies to the roots which are often fused.

Mandibular First Molar

The basic form of this tooth is a rectangle (Fig. 12 a). Curvature and angle

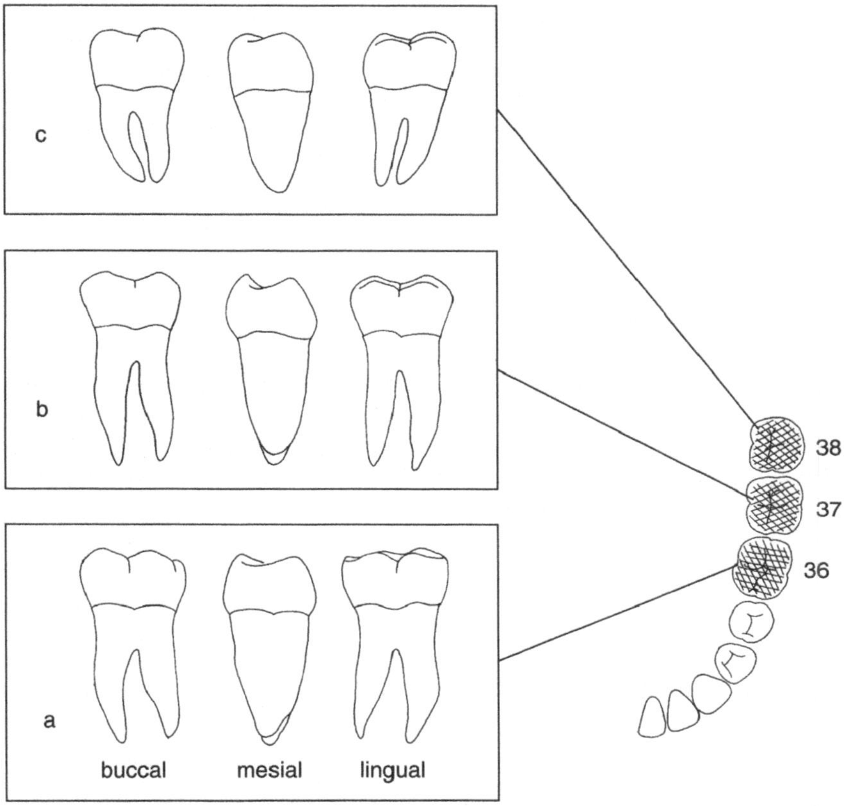

Fig. 12 a–c. Permanent mandibular left molars: **a:** first molar; **b:** second molar; **c:** third molar.

characteristics are present as well as a lingual inclination of the crown. The tooth typically has five cusps: three buccal (mesiobuccal, mediobuccal, distobuccal) and two lingual (mesiolingual, distolingual). Whereas the buccal cusps show a decrease in height from mesial to distal, the lingual ones maintain theirs. The occurrence of four cusps always coincides with the loss of the hypoconulids. In addition to the longitudinal fissure (main fissure), which runs from mesial to distal and usually ends in the region of the marginal ridge, three diagonal fissures branch off from the center. In the first molar, one diagonal fissure runs toward the lingual and two run toward the buccal side. Depending on the form and number of cusps – as with the other mandibular molars – a Y-shaped fissure pattern, or an X-pattern exists (Tab. 3).

The "Y"-pattern derives from the original *Dryopithecus*-pattern while the other two variations are adaptations. In Middle European people the Y-5-pattern, followed by the "+"-5-pattern is the most common in the first mandibular molar. Internationally, the distinction of three fissure characteristics ("Y", "+" or "X") has not generally been maintained. The "X"-variation and the cross-shaped variation are seen as one by many examiners.

Table 3. Fissure patterns of mandibular molars

Y-pattern	Cusp contact between metaconid and hypoconid; protoconid and entoconid are separated by a fissure and do not come into contact
+-pattern	Point contact of metaconid, protoconid, entoconid, and hypoconid in the fovea centralis
X-pattern	Cusp contacts between protoconid and entoconid; metaconid and hypoconid are separated by a fissure

Mandibular Second Molar

The crown of the mandibular second molar is lingually inclined (Fig. 12 b). The tooth also has a curvature and a root characteristics. Its basic form is similar to that of the first molar although it is smaller. The two buccal and two lingual cusps are equal in size and shape. The second mandibular molar has four cusps in 90% of all cases (Twiesselmann and Brabant 1967), the "+"- variation predominates. In rare cases it has only three cusps.

Mandibular Third Molar

The mandibular wisdom tooth like the maxillary one is very irregularly formed (Fig. 12 c). Its roots are often fused.

The Roots of Permanent Teeth and Their Variability

Usually only molars and maxillary first premolars have more than one root. The mandibular molars usually possess two roots, and the maxillary molars three roots (cf. Tables 4 and 5).

In contrast to the crowns, roots have so far only been examined in large-scale surveys. Since the knowledge gained from radiography is limited, information gained from skulls and from extracted teeth is of far greater value (Visser 1948). As with crowns, in the root area normal variations exist, especially in the form and number of roots.

Where there are more roots than normal, the extra root is formed, either by splitting the main root or by formation of another independent root. The latter are accessory roots and are discussed with other rare root variations (for example: taurodontism, pyramidalism) in Chapter 3.2. In principle, the only roots which are regarded as multiple roots are partially – at the apex of the root – or completely separated.

Certain teeth are affected more often by multirootedness than others. The incisors and canines of permanent teeth seldom have extra roots through differentiation; instead, a tendency toward separation, i.e. division (furrow formation), is more likely to be found (Keil 1943; Sykaras 1972). The mandibular canines are an exception (Alexandersen 1963; Bruszt 1963; Le Huche 1954;

Table 4. Number of roots and cusps of permanent maxillary human teeth (Schumacher 1991)

Teeth	Roots	Cusps
11+21	1	–
12+22	1	–
13+23	1	–
14+24	2 (>60%) or 1	2
15+25	1 (>85%) or 2	2
16+26	3	4
17+27	3	4

Table 5. Number of roots and cusps of permanent mandibular human teeth (Schumacher 1991)

Teeth	Roots	Cusps
41+31	1	–
42+32	1	–
43+33	1	–
44+34	1 (74%) or 2 (26%)	2 (75%) or 3 (25%)
45+35	1 (85%) or 2 (15%)	3
46+36	3	5
47+37	3	4

Sassenberg 1939) because they show a second root with a frequency of up to 8% (Visser 1948).

Although the existence of an appendiciformed root is rarely expressed in percentages, it is still found as a surplus root developed through differentiation (Schulze 1987). During differentiation of the root, splitting follows certain regularities: the maxillary incisors split in a mesio-distal direction, the mandibular incisors – primarily lateral incisors – split in a facial-lingual direction. The permanent canines always differentiate in a facial-lingual direction.

In premolars, singlerootedness is generally the norm, although the first maxillary premolars often possess two roots. In contrast to molars, premolars are characterized by a far greater constancy of root formation. Multirootedness arises primarily from root differentiation. Of all teeth, the molars are most often effected by extra roots (and are most often the site of anomalies). In maxillary molars, fusion of the normally three divergent roots is often observed. Root divergence is greatest in the maxillary first molar.

The Root Size of Permanent Teeth and Their Variability

Form and size, the most frequently noted variations, are determined by separate hereditary factors. Complimentary tooth form and size are important for a correct occlusion (Eismann 1980). The deciding factor is not the size of the tooth itself, but its proportion within the dentition (Graf und Koch 1988; Klein 1982). Discrepancies in form and size lead to malocclusion, which is intensified by a discrepancy between tooth and jaw size (Gegenfurtner 1982 a, b; Harzer and Müller 1986; Schulze 1987). For anthropologists, anatomists, and biologists the main themes when studying the metric characteristics of the teeth are: evolutionary biological events (e.g., questions of evolutionary trends in reduction of tooth size [Brace et al. 1987; Lavelle 1971, 1972 a, b; Frayer 1977; Matsumura 1995; Smith 1976, 1982]), comparative population studies and gender detemination (e.g. the diagnoses of the sex of children by means of the discriminating-analytic separating function [Henke 1990; Langenscheidt 1985; Paul 1990]).

Usually the determination of tooth size is limited to two measurements, namely the mesio-distal and the buccal-lingual crown diameters (Herren 1974, Doris et al. 1981; Matsumura 1995).

A large degree of agreement has been achieved at present concerning existing sex dimorphism and the size of teeth (cf. Kieser 1990). The meaning for anthropology was pointed out by Henke (1990) and Paul (1990) among others. Studies from Miethke (1972) as well as Heckerodt und Köhnen (1983) have shown that the teeth of a male individual tend to be broader than those of a female. Within the groups of teeth, differences are especially marked in canines and mandibular

Table 6. Variations in entire length of permanent maxillary human teeth (in mm)

Teeth	Entire length (min–max)	Entire length (min–max)
11+21	18.0–32.0	17.2–27.8
12+22	17.5–28.0	19.8–26.4
13+33	19.0–37.0	19.9–40.7
14+24	16.2–28.2	18.3–26.2
15+25	17.5–27.0	16.3–28.4
16+26	17.5–29.0	17.9–28.9
Investigator	Mühlreiter (1891)	Alt (1997)

Table 7. Variations in entire length of permanent mandibular human teeth (in mm)

Teeth	Entire length (min–max)	Entire length (min–max)
41+31	18.0–27.0	19.2–21.4
42+32	19.0–29.0	19.0–26.3
43+33	20.0–34.0	18.1–28.7
44+34	18.5–27.0	18.8–23.4
45+35	19.0–27.5	17.7–23.8
46+36	18.3–26.0	17.2–24.4
Investigator	Mühlreiter (1891)	Alt (1997)

Table 8. Variations in crown length of permanent maxillary human teeth (in mm)

Teeth	Crown length (min–max)	Crown length (min–max)
11+21	8.5–14.5	10.2–13.3
12+22	7.8–12.0	8.8–11.7
13+33	7.5–13.0	9.2–13.0
14+24	7.0–10.8	7.4–10.2
15+25	6.2–10.2	6.9– 9.7
16+26	6.8– 9.0	6.2– 8.3
Investigator	Mühlreiter (1891)	Alt (1997)

Table 9. Variations in crown length of permanent mandibular human teeth (in mm)

Teeth	Crown length (min–max)	Crown length (min–max)
41+31	7.0–11.5	7.8– 9.4
42+32	8.2–11.8	7.7–10.3
43+33	8.5–14.5	9.8–12.2
44+34	7.5–11.0	7.3–10.2
45+35	6.9–10.0	7.0– 9.7
46+36	7.0– 9.0	5.4– 7.8
Investigator	Mühlreiter (1891)	Alt (1997)

molars (Rieger 1993). Normally, the differences in size are 0.2 to 0.3 mm. With canines, differences are between 0.4 and 0.6 mm and with mandibular molars 0.4 mm or more.

In Tables 6 and 7, variations of the total lengths of the maxillary and mandibular teeth are collected. The chart shows a comparison of results which Mühlreiter (1870) obtained from teeth of volunteers from the Salzburg area (Austria) and results which Alt (1997) gathered from teeth of the Schwarz tooth collection (School of Dentistry, University of Basel/Switzerland).

Results for crown lengths alone are shown in Tables 8 and 9.

Occlusal Relationships

The term occlusion has two meanings. The first is the act or process of closure or of being closed or shut. The second is the static relationship between the incising or masticating surfaces of the maxillary or mandibular teeth (or tooth analogs). The interdigitation of cusps of opposing teeth is described with the term intercuspation. The maximum interdigitation of opposing teeth (independent of the position of the mandibular condyles) is called maximum intercuspation. Articulation refers to the relationship between the occlusal surfaces of the teeth in contact during function.

In the incisor area, during maximum intercuspation a distinction can be made between vertical and horizontal overlap. Vertical overlap (overbite) is the distance the incisal edges of the maxillary incisors extend below those of the mandibular incisors when the teeth are in maximum intercuspation (Fig. 13: VO). Horizontal overlap (overjet) is the distance between the labial surface of the maxillary incisors and the labial surface of the mandibular incisors (Fig. 13: HO).

As a rule, the mandibular arch is smaller than the maxillary arch and mandibular incisors are slimmer than those of the maxilla. Therefore, with maximum intercuspation, and provided that the row of teeth is not discontinuous, posterior teeth have contact with not one but with two teeth of the opposite jaw (exception: the most distally placed lower molar has only one antagonist). Accordingly, a main antagonist can be distinguished from a secondary antagonist (example: the mandibular first molar has the maxillary first molar as a main antagonist and the maxillary second premolar as a supporting antagonist).

In the case of regular intercuspation of the posterior teeth, the maxillary canine occludes between the mandibular canine and the mandibular first premolar, while the maxillary first molar with its mesiobuccal cusp contacts the main buccal fissure

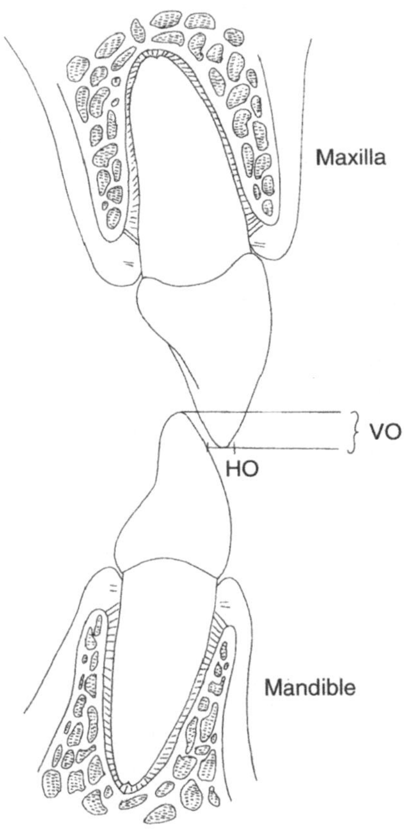

Fig. 13. Normal occlusal relationship of the anterior teeth showing two types of overlap: VO: vertical overlap (overbite), HO: horizontal overlap (overjet).

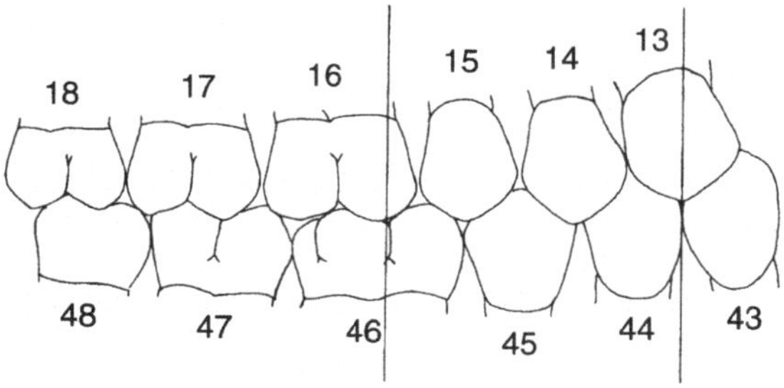

Fig. 14. Normal occlusal relationship of the first molars and the canines: The mesiobuccal cusp of the maxillary first molar is situated over the groove between the mesiobuccal and mediobuccal cusp of the mandibular first molar. The maxillary canine occludes between the mandibular canine and the mandibular first premolar.

of the first mandibular molar, which lies between the mesio- and mediobuccal cusps (Fig. 14).

As a rule, maxillary posterior teeth are more buccally placed than mandibular ones (normal occlusion). As a result, the palatal cusps of maxillary posterior teeth and the buccal cusps of mandibular teeth are centric (or supporting) cusps. The buccal cusps of the upper jaw and the lingual cusps of the lower jaw, on the other hand, are described as noncentric (or guiding) cusps. For further definitions concerning occlusal relations refer to The Glossary Prosthodontic Terms (1994).

References

Adloff P (1908) Das Gebiß des Menschen und der Anthropomorphen. Vergleichend-anatomische Untersuchungen. Springer, Berlin Heidelberg New York.

Alexandersen V (1963) Double-rooted human lower canine teeth. In: Brothwell DR (ed): Dental Anthropology. Pergamon Press, Oxford, p. 235–244.

Alt KW (1989) Empfehlung zur einheitlichen Kennzeichnung der Zähne bei der 2-1-2-3 Zahnformel. Anthropologie 27: 5–12.

Alt KW (1997) Odontologische Verwandtschaftsanalyse. Fischer, Stuttgart.

Berkovitz BKB, Holland GR, Moxham BJ (1978) A Colour Atlas and Textbook of Oral Anatomy. Wolfe Medical Publications, London, p. 24, 26, 185, 199, 217.

Berkovitz BKB, Moxham B (1981) Tooth morphology. In: Osborn JW (ed): Dental Anatomy and Embryology, Vol. 1, Book 2: A Companion to Dental Studies. Blackwell, Oxford, p. 118–154.

Black GV (1897) Descriptive Anatomy of the Human Teeth. White Dental Manufac., Philadelphia.

Bolk L (1912) Odontologische Studien I. Fischer, Jena.

Bolk L (1913) Die Ontogenie der Primatenzähne. Versuch einer Lösung der Gebißproblematik. Fischer, Jena.

Bolk L (1914) Die Morphogenie der Primatenzähne. Eine weitere Begründung und Ausarbeitung der Dimertheorie. Fischer, Jena.

Brace CL, Rosenberg KR, Hunt KD (1987) Gradual change in human tooth size in the late Pleistocene and post-Pleistocene. Evolution 41: 705–720.

Bruszt P (1963) Über seltene Form-, Zahlen- und Lageanomalien der Eckzähne (I, II). Stoma 16: 172–180, 208–213.
Carlsen O (1987) Dental Morphology. Munksgaard, Copenhagen.
de Jonge-Cohen TE (1917) Die Kronenstruktur der unteren Prämolaren und Molaren. Ein Beitrag zur Morphologie des menschlichen Gebisses. Med Diss, Zürich.
de Jonge T (1958) Anatomie der Zähne. In: Meyer W: Die Zahn-, Mund- und Kieferheilkunde. Vol. I. Urban & Schwarzenberg, München, p. 169–232.
de Terra M (1905) Beiträge zu einer Odontographie der Menschenrassen. Berliner Verlagsanstalt, Berlin.
Doris JM, Bernhard BM, Kuftinec MM, Stom D (1981) A biometric study of tooth size and dental crowding. Am J Orthodont 79: 326–336.
DuBrul EL (1980) Sicher's Oral Anatomy. Mosby, St. Louis, p. 242–272.
Eismann D (1961) Untersuchungen zur Variation der Kronendiameter. In: Forschungs- und Entwicklungsbericht 05263007 der Medizinischen Akademie Erfurt, Erfurt.
FDI (Fédération Dentaire Internationale) (1971) Two-digit system of designating teeth. Int Dent J 21: 104–106.
Frayer, DW (1977) Metric dental change in the European upper paleolithic and mesolithic. Am J phys Anthrop 46: 109–120.
Fujita T (1973) Dental Anatomy [in Japanese]. Kinbara Shuppan, Tokyo.
Gantt DG (1983) The enamel of neogene hominoids. Structural and phyletic implications. In: Ciochon RL, Corruccini RS (eds) New Interpretations of Ape and Human Ancestry. Plenum Press, New York, p. 249–298.
Garberoglio R, Brännström M (1976) Scanning electron microscopic investigation of human dentinal tubules. Arch Oral Biol 21: 355–362.
Gegenfurtner H (1982a) Zahngrößen und deren Beziehungen zum Zahnengstand. Med Diss, München.
Gegenfurtner R (1982b) Kiefergrößen und deren Beziehungen zum Zahnengstand. Med Diss, München.
Graf H, Koch EM (1988) Gebißparameter und Korrelationen. Ergebnisse, Überlegungen, Schlußfolgerungen. Zahn Mund Kieferheilkd 76: 456–458.
Harzer W, Müller S (1986) Zur Variation ausgewählter Gebißparameter in Zwillingsfamilien. Nova Acta Leopoldina NF 58 262: 521–524.
Häupl K (1949) Lehrbuch der Zahnheilkunde, Vol I. Urban & Schwarzenberg, Wien, p. 48–66.
Heckerodt M, Köhnen EJ (1983) Mesiodistale Distanzen von Zähnen und Zahnbogensegmenten bei Geschwistern. Med Diss, Hannover.
Henke W (1990) "De discrimine sexuum". Möglichkeiten und Grenzen der Geschlechtsbestimmung menschlichen Skelettmaterials. In: Referat Gemeinschaftssymposium der AG Historische Anthropologie der Sektion Anthropologie der Biologischen Gesellschaft der DDR und des Landesmuseums für Vorgeschichte Halle. Anthropologisches Institut, Mainz.
Henke W, Rothe H (1994) Paläoanthropologie. Springer, Berlin Heidelberg New York.
Herren, P (1974) Die summierten mesio-distalen Kronenbreiten des anterioren Zahnbogens und ihre Differenz zwischen Ober- und Unterkiefer. Schweiz Monatsschr Zahnheilk 84: 853–872.
Hillson S (1986) Teeth. Cambridge Univ Press.
Hrdlička A (1921) Further studies of tooth morphology. Am J Phys Anthrop 4: 141–176.
Keil A (1943) Über einen zweiwurzeligen mittleren oberen Schneidezahn im bleibenden Gebiß. Dtsch Zahnärztl Wochenschr 46: 24–25.
Keil A (1966) Grundzüge der Odontologie. Borntraeger, Berlin.
Kieser JA (1990) Human Adult Odontometrics. The Study of Variation in Adult Tooth Size. Cambridge Studies in Biological Anthropology. Cambridge University Press, Cambridge.

Klein G (1982) Zur Vererbung von Schneidezahngrößen und Schneidezahnformen. Med Diss, Erfurt.
Kraus BS, Jordan RE, Abrams L (1969) Dental Anatomy and Occlusion. Williams and Wilkins, Baltimore.
Krysinski Z: The three-digit system of designating supernumerary teeth. Quintessence Int 17: 127–128.
Kurtén B (ed) (1982) Teeth: Form, Function, and Evolution. Columbia Univ Press, New York.
Langenscheidt F (1983) Diskriminanzanalytische Geschlechtsbestimmung an Hand von Zahnmaßen – unter Verwendung von Verfahren zur angenähert unverzerrten Schätzung der Trennstärke. Homo 34: 22–27.
Langenscheidt F (1985) Methodenkritische Untersuchungen zur Paläodemographie am Beispiel zweier fränkischer Gräberfelder. In: Materialien zur Bevölkerungswissenschaft. Sonderheft 2. Bundesinstitut fiir Bevölkerungsforschung, Wiesbaden.
Lavelle CL (1971) Mandibular tooth configuration in different racial groups. J Dent Res 50: 1353–1358.
Lavelle CL (1972a) The incisors of men and apes. Bull Group Int Rech Sci Stomatol 15: 285–301.
Lavelle CL (1972b) Maxillary and mandibular tooth size in different racial groups and in different occlusal categories. Am J Orthodont 61: 29–37.
Le Huche R (1954) Quelle est la fréquence de la bifidit, de la racine de la canine inférieure? Rev Stomat 55: 656–663.
Marseillier E (1965) Les Dents Humaines – Morphologie. Gauthier-Villars, Paris.
Matsumura H (1995) A microevolutinal history of the Japanese people as viewed from dental morphology. Nat Sci Mus Monogr No 9, Tokyo.
Miethke RR (1972) Zahnbreiten und Zahnbreitenkorrelationen. Berlin.
Mühlreiter E (1870) Anatomie des menschlichen Gebisses. Felix, Leipzig.
Nomina Anatomica (1989) Chuchill Livingstone, Edinburgh.
Paul G (1990) Geschlechtsbestimmung von Skeletten mit Diskriminanzfunktionen fiir bleibende Zähne. Med Diss, Ulm.
Rateitschak KH, Rateitschak EM, Wolf HF (1989) Periodontology. Color Atlas of Dental Medicine 1. Thieme, Stuttgart, p. 4–9.
Remane A (1960) Zähne und Gebiß. In: Hofer H, Schultz AH, Starck D (ed) Primatologia. Handbuch der Primatenkunde III/2. Karger, Basel, p. 637–846.
Rieger S (1993) Untersuchungen zur Geschlechtsbestimmung merowingerzeitlicher Individuen aus Baden-Württemberg anhand der Zahnmaße. Med Diss, Freiburg.
Sassenberg F (1939) Über einen Fall von beiderseitig zweiwurzeligen, unteren Eckzähnen. Dtsch Zahnärztl Wochenschr 42: 163.
Schroeder HE (1983) Pathobiologie oraler Strukturen. Karger, Basel.
Schroeder HE (1991) Orale Strukturbiologie. Thieme, Stuttgart.
Schroeder HE (1993) Human cellular mixed stratified cementum: A tissue with alternating layers of acellular extrinsic- and cellular intrinsic fiber cementum. Schweiz Monatsschr Zahnmed 103: 550–560.
Schulze C (1987) Anomalien und Mißbildungen der menschlichen Zähne. Quintessenz, Berlin.
Schumacher GH (1980) Odontographie. Barth, Leipzig.
Schumacher GH (1984) Anatomie für Stomatologen. Lehrbuch und Atlas, Bd. 1. Barth, Leipzig.
Schumacher GH (1991) Anatomie. Lehrbuch und Atlas, Bd. 1. Barth, Leipzig.
Scott JH, Symons NBB (1977) Introduction to Dental Anatomy. Churchill Livingstone, Edinburgh.
Sicher H, Tandler J (1928) Anatomie für Zahnärzte. Springer, Wien.

Smith P (1973) Variation in dental traits within population. In: Dahlberg AA, Graber TM (eds): Orofacial Growth and Development (World Anthropology). Mouton, The Hague, p. 171–181.

Smith P (1982) Dental reduction. Selection or drift? In: Kurtén B (ed) Teeth, Form, Function, and Evolution. Columbia University Press, New York, p. 366–379.

Sykaras SN (1972) A two-rooted maxillary lateral incisor. Oral Surg Med Pathol 34: 349.

Tallec P (1975) Morphologie Dentaire. Paris.

Taylor RMS (1978) Variation in Morphology of Teeth. Thomas, Springfield.

The Academy of Prosthodontics (1994) The glossary of prosthodontic terms. J Prosthet Dent 71: 41–116.

Tomes CS (1898) A Manual of Dental Anatomy, Human and Comparative. Blackinston's Sons, Philadelphia.

Türp JC, Alt KW (1995) Designation teeth: The advantages of the FDI's two-digit system. Quintessence Int 26: 501–504.

Twiesselmann F, Brabant H (1967) Nouvelles observations sur les dents et les maxillaires d'une population ancienne d'âge Franc de Coxyde (Belgique). Bull Group Int Rech Sc Stomat 10: 5–180.

Visser JB (1948) Beitrag zur Kenntnis der menschlichen Zahnwurzelformen. Med Diss, Zürich.

Wheeler RC (1969) An Atlas of Tooth Form. Saunders, Philadelphia.

Wheeler RC (1974) A Textbook of Dental Anatomy, Physiology, and Occlusion. Saunders, Philadelphia.

Williams PL, Warwick R, Dyson M, Bannister LH (1989) Gray's Anatomy. Churchill Livingstone, Edinburgh.

Zuckerkandl E (1891) Makroskopische Anatomie. In: Scheff J (ed) Handbuch der Zahnheilkunde, I. Band. Hölder, Wien, p. 1–208.

3.2 Hereditary Dental Anomalies

Kurt W. Alt and Jens C. Türp

Introduction

Studies of dental morphology are quite frequent due to their importance for comparative anatomy, phylogeny, and dentistry (Taylor 1978; Hillson 1986; Carlsen 1987; Schumacher et al 1990; Matsumura 1995). Dental anomalies, by contrast, are seen as curiosities, and have been investigated only rarely (Brabant et al. 1958; Thoma and Goldman 1960; Pindborg 1970; Ash 1986; Schulze 1987; Schroeder 1991; Regezi and Sciubba 1993). The study of the anomalies of human teeth increases our knowledge of odontogeny and aids our understanding of ontogenetic processes by answering questions of etiology and pathogeny (Smith and Tchernov 1992). Furthermore, the knowledge obtained enables us to evaluate important aspects of evolutionary processes more precisely (Frisch 1965; Hillson 1986; Henke and Rothe 1994). Comparative studies of fossil and recent animal teeth, especially of higher primates, are therefore indispensable (Colyer 1936; Remane 1960; Alt and Türp 1997).

Morphological variants and anomalies of teeth are highly significant: for the study of human evolution, the skeletal biology of ancient populations (especially for inter- and intragroup investigations), and in paleopathology and forensic odontology (Butler and Joysey 1978; Kurtén 1982; Hillson 1986, 1996; Scott and Turner 1988; Kieser 1990; Kelley and Larsen 1991; Alt 1997 a, b). Normal dental variation in humans has been discussed previously (see Türp and Alt, this volume). This present contribution gives a short systematic review of dental anomalies, paying special attention to those of permanent teeth. While hereditary anomalies also occur in deciduous teeth, there are considerable differences in the frequency of their occurrence. Discussions of pathogeny, etiology, trait frequencies etc. are largely omitted (for this topic, see Pindborg 1970; Stewart and Prescott 1976; Schulze 1987). An international classification of diseases of the digestive system has been made available by Med-Index Publications (1993), and clinical dental terminology has been outlined by Zwemer (1993).

An investigation of anomalies requires that the limits of normal variation are well established. Simply describing anomalies as deviations from the norm is insufficient. Schellong and Pfeiffer (1963) define anomalies to be morphological and/or structural changes of all or part of an organ during individual development, and which lie outside normal variability but do not impede normal function. As such, dental anomalies are abnormalities in which a tooth or several teeth have deviated from normal form, function, or position (Zwemer 1993). This can occur as part of a more generalized disorder (syndrome) or it can exist independently. The transition from normal variation to anomaly is always fuzzy, and affects diagnostic evalution. Population-specific differences in the frequencies of variants result in differing classifications of traits, as in the case of shovel-shaped incisors, which are considered to be a normal variation in some groups, but in others are classified as anomalies.

This chapter focuses on the following ontogenetic disturbances: tooth shape (form), concrescence/gemination/twinning/fusion, size and number, tooth position and malocclusion, tooth structure, disturbances of tooth eruption, and congenital malformations and disturbances of bone formation and growth of the skeleton affecting jaws and teeth. The description is generally presented with respect to tooth groups, since in the majority of cases single disturbances occur on all teeth of a given tooth group. Like most morphological traits, dental anomalies show continuous variability, which results in a gradual distinction and classification of their expression. Anomalies of both tooth shape and tooth number can be differentiated as plus and minus variants. Plus variants may include supplemental cusps and accessory roots, whereas minus variants may differ in shape, size and number of cusps and roots.

Abnormalities of the Tooth Shape (Form)

Tooth Crowns

Anomalies of the shape of the upper incisors include shovel-shape, double (lingual and buccal) shovelling, and barrel and peg shape in the case of the maxillary second incisors ("peg lateral"). Further variants of the incisors or canines include accessory cusps, tubercles, ridges, as well as supplemental grooves (only rarely on the buccal sides), talon cusps and dilated composite odontomas (Hillson 1986; Alt 1997 a) (Fig. 1).

Morphologically, shovel-shaped incisors are a characteristic feature in mongoloid populations, in which they occur nearly universally (Carbonell 1963). However, their frequency among Caucasians is so low that they are classified as an anomaly. The occurrence of marginal ridges without concavity of the palatal surface should not be included in the shovel-shaped type. Ethnic differences in shovelling indicate that this trait is controlled by genetic factors. This hypothesis has been corroborated by twin and family studies (Blanco and Chakraborty 1976).

The most frequent anomalies of the shape of incisors among Caucasian populations are to be found on the dental tubercle. These anomalies appear as singly or multiply grooved cingula or cusplets. Occasionally, the dental tubercle is enlarged into an independent cusp (de Jonge-Cohen 1935). Increased frequencies of multiply-grooved cingula in families indicate a genetic origin for this trait. The talon cusp is a dental anomaly which is manifested as an accessory cusp-like

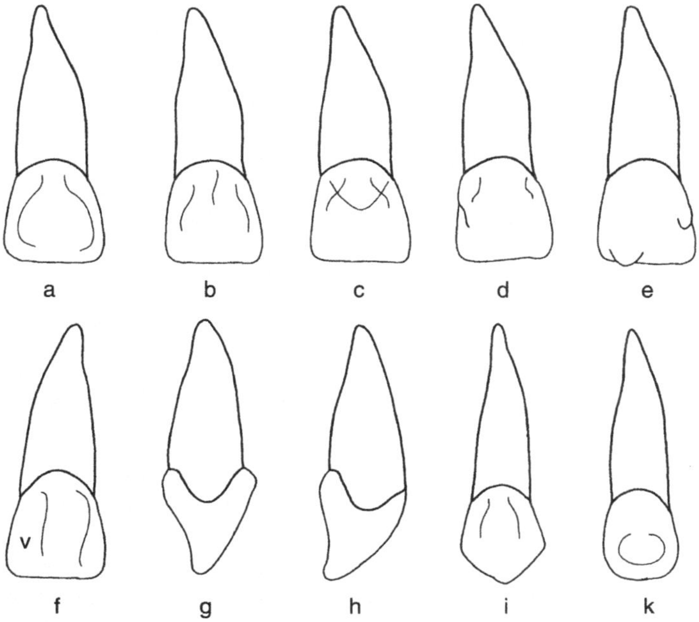

Fig. 1. Anomalies of the shape of the upper incisors; palatal view (**a–e**); buccal view (**f**); mesial view (**g–h**); **a:** shovel-shaped incisor; **b:** grooved cingulum; **c:** accessory tubercle; **d:** grooved lingual marginal ridges; **e:** accessory cusp on incisal or marginal ridge; **f:** buccal ridge, fissure or cusp; **g:** buccal cingulum; **h:** extreme curvature of buccal surface; **i:** caniniform incisor; **k:** barrel-shaped incisor.

structure on the tooth crown (Fig. 2). As far as the incisal edge of the incisors is concerned, the crown contour may assume a Y- or T-shape (Kimball et al. 1970; Hattab et al. 1995).

The dilated composite odontome *(dens invaginatus coronalis/radicularis)* is the result of a folding process of the enamel organ during ontogenesis, with coronal and radicular variants (Hallet 1953) (Fig. 3). The frequency of dilated composite

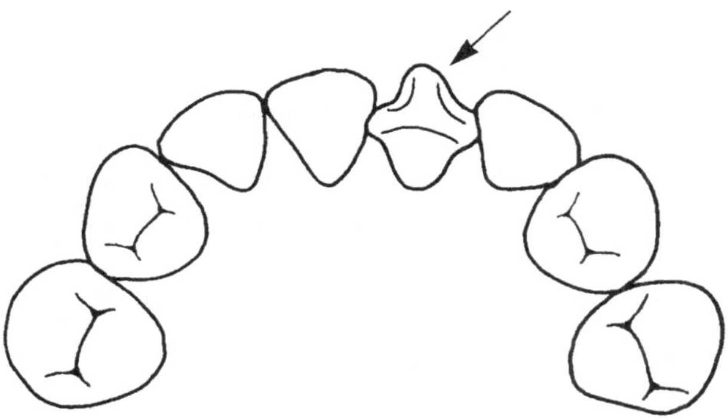

Fig. 2. Talon cusp (mandibular incisor).

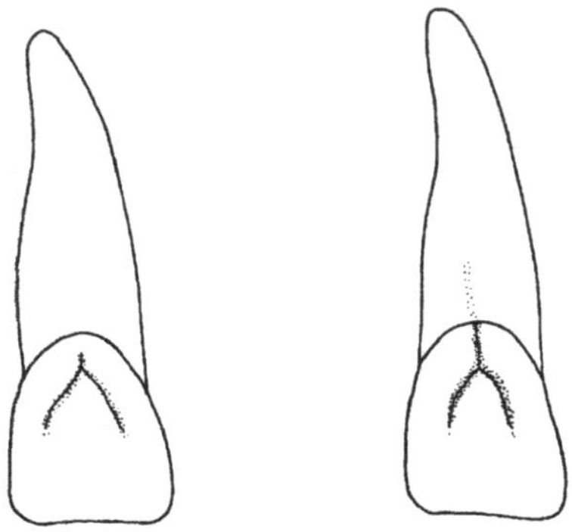

Fig. 3. Invagination (*dens in dente*); **a:** coronal invagination; **b:** radicular invagination.

odontomes shows high interpopulation variety. Family studies carried out by Grahnén et al. (1958) clearly indicate the participation of hereditary factors in the occurrence of this trait.

Premolars show fewer anomalies than do incisors or molars (Kutscha 1985; Burnett 1996). Rare anomalies of the lower premolars include accessory ridges (Keil 1966), prominent marginal ridges, as well as bucco-lingually compressed crowns (sectorial form) (Hitchin and Ferguson 1958). In both the mandible and the maxilla there may be additional cusps (molarisation) (Lunt 1976) or accessory cusps, such as the paramolar tubercle on the buccal surface (primarily upper premolars) (Fujita et al. 1963) and on the occlusal surface (tuberculated premolars; syn. *dens evaginatus*) (Fig. 4). The latter is more frequent in lower than in upper premolars, and its most marked development can be found in mongoloid populations (Reichart et al. 1982). In upper premolars, the palatal cusp may be reduced or absent. There are clear indications of a hereditary basis for most involved traits (see Alt 1997a).

Typical manifestations of molar anomalies are commonly found on all molars, but the frequencies of the different traits vary from tooth to tooth. Certain cusps of both upper and lower molars show either reduction or extension, and a number of additional cusps or cusplets can occur. Extension occurs most frequently in the hypocone and protocone cusps of upper molars (Fig. 5), and in the protoconid and metaconid cusps of lower molars (Fig. 6) (Schulze 1987). More common is the underdevelopment of the molar crown, although most authors consider both hypocone and metacone reduction in upper molars and metaconid or entoconid reduction in lower molars to be within the limits of normal variation (Hofmann 1985).

Of greater importance for population studies are additional cusps or cusplets (Scott 1980; Hofmann 1985; Hillson 1986; Kanazawa et al. 1990; Turner et al. 1991; Matsumura 1995) (Fig. 7). In maxillary teeth they can be present as:

- the Carabelli tubercle (CAT), on the palatal base of the protocone;
- the paramolar tubercle (parastyl), on the buccal base of the paracone (PAT);
- the mesial paracone tubercle (MPT);
- the mesial accessory tubercle (MAT);
- the protoconule (PL; c-6), at the mesial marginal ridge;

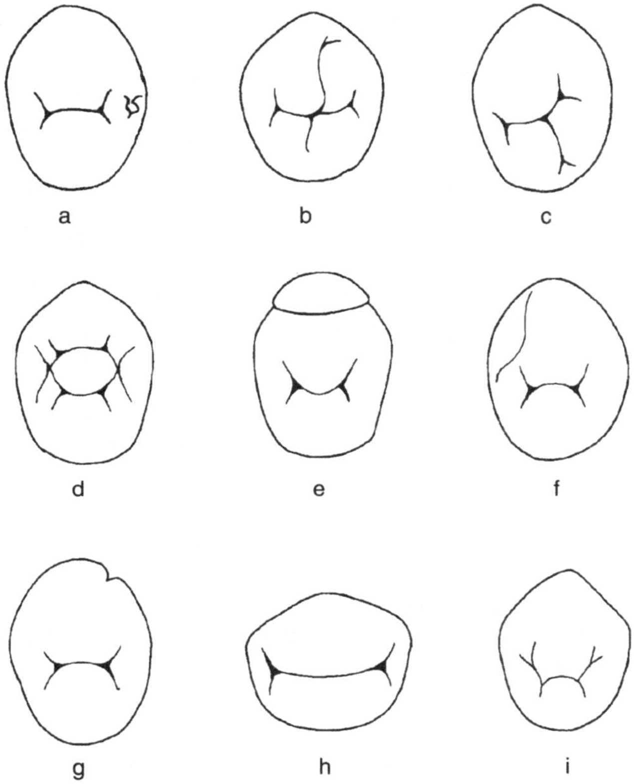

Fig. 4. Anomalies of premolar crowns; **a:** accessory cusp of marginal ridge; **b:** molariform mandibular second premolar (4 cusps); **c:** molariform maxillary first premolar (3 cusps); **d:** tuberculated premolar; **e:** paramolar tubercle; **f:** accessory occlusal ridge (from buccal cusp tip to distal marginal ridge); **g:** "Uto-Aztecan" premolar (fossa in distobuccal part of paracone); **h:** sectorial premolar (buccolingually compressed crown); **i:** caniniform premolar.

- the lingual paracone tubercle (LPT), toward the center;
- the distal accessory tubercle (DAT), at the distal marginal ridge;
- the metaconule (ML; c-5), between hypocone and metacone.

Kanazawa et al. (1990, 177) state that "the terminology of the anomalous cuspules on the human upper molar is confusing" which complicates comparative studies.

Additional cusps in mandibular teeth (see Mizoguchi 1977; Schulze 1987) (Fig. 8) are:

- the tuberculum intermedium (c-7; metaconulid [MC]), between metaconid and entoconid;
- the paramolar tubercle (PAT), at the base of the protoconid (protostylid);
- the Citroën tubercle (CIT), at the base of the metaconid (metastylid);
- the entoconulid (c-6; tub. sextum [EC]), between entoconid and hypoconulid;
- the paracone tubercle (PCT), at the mesial marginal ridge.

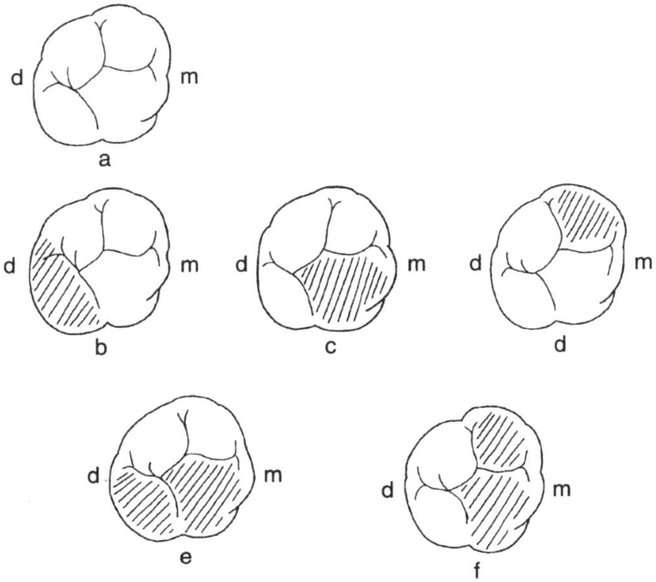

Fig. 5. Overdeveloped maxillary cusps; m = mesial; d = distal; **a:** regular molar crown; **b:** overdeveloped hypocone; **c:** overdeveloped protocone; **d:** overdeveloped paracone; **e:** combination of **b** and **c**; **f:** combination of **c** and **d**.

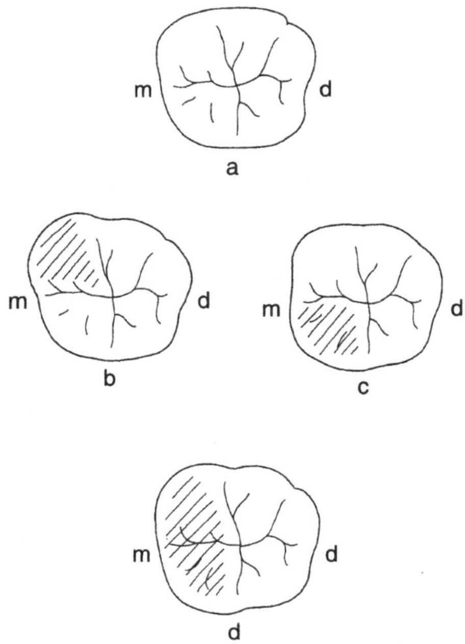

Fig. 6. Overdeveloped mandibular cusps; m = mesial; d = distal; **a:** regular molar crown; **b:** overdeveloped protoconid; **c:** over-developed metaconid; **d:** combination of **b** and **c**.

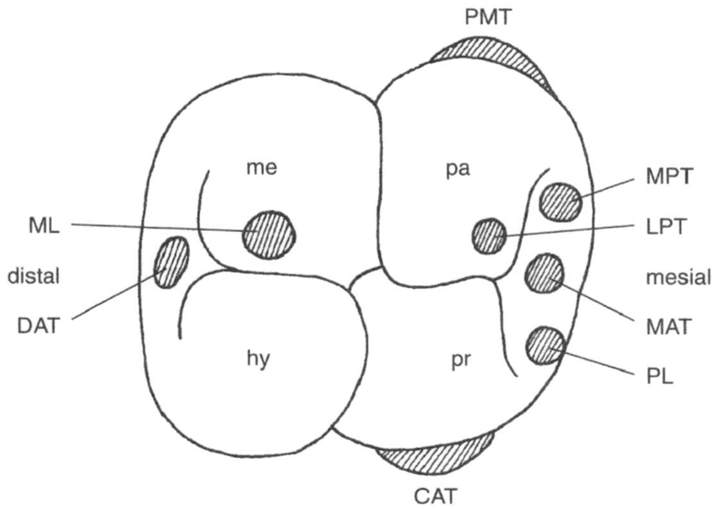

Fig. 7. Accessory maxillary cusplets (schematic representation); me = metacone; pa = paracone; pr = protocone; hy = hypocone; CAT: Carabelli cusp; PAT: paramolar tubercle; DAT: distal accessory tubercle; ML: metaconule; MPT: mesial paracone tubercle; LPT: lingual paracone tubercle; MAT: mesial accessory tubercle; PL: protoconule.

A special formation of the median ridge of the metaconid is known as a "deflecting wrinkle" (Hanihara et al. 1964) (Fig. 8). Twin studies showed that this trait is to a high degree genetically determined (Mizoguchi 1977).

All cusps vary continuously in shape and size. Furthermore, additional cusps may be present on the palatal or buccal surfaces, while microforms like grooves, ridges, or merely a depression may replace an actual cusp (Hillson 1986). All of these features seem to belong to a single variant, now called the Carabelli or

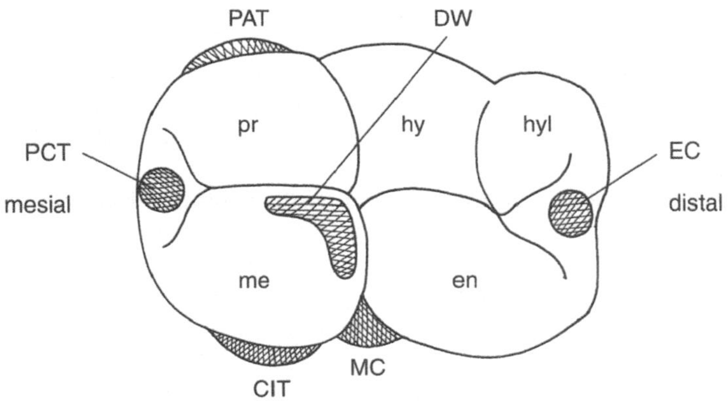

Fig. 8. Accessory mandibular cusplets (schematic representation); pr = protoconid; hy = hypoconid; hyl = hypoconulid; me = metaconid; en = entoconid; PCT: mesial paraconid tubercle; PAT: paramolar tubercle; DW: deflecting wrinkle; EC: entoconulid; CIT: Citroën tubercle; MC: intermedium tubercle.

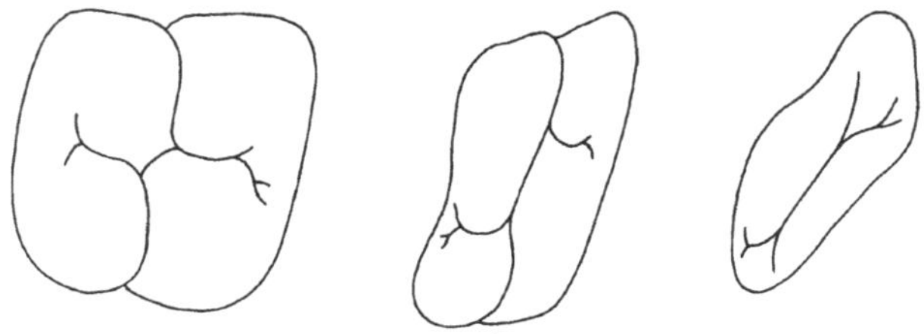

Fig. 9. Compressed crown of maxillary second and third molars; **a:** regular form; **b:** semi-compressed crown; **c:** compressed crown.

paramolar structure. In population studies, the scoring of all features is especially important, as the microforms are far more frequent than the cusps themselves (in the case of the Carabelli tubercle in a 3:1 ratio). Differences in the frequency of their occurrence suggest a genetic basis of the additional cusps (cusplets). Twin studies with a 100% concordance in identical twins and approximately a 50% concordance in dizygotic twins confirm a marked hereditary trend for traits like the Carabelli tubercle (Scheffler 1976).

A distinctive feature of the second and third molars in the maxilla is the mesiodistal compression of the crown (compressed crown) (Fig. 9). Bilateral occurrence and an increased familial frequency are strong indications for the participation of genetic factors (Schulze 1987).

Tooth Roots

Findings on tooth roots are generally less conclusive than on tooth crowns due to the longer and more intense influence of exogenous factors during root growth. As they are not as easily accessible, anomalies of tooth roots have not been studied very intensively (Visser 1948; Kovacs 1967). Twin studies conducted for an estimation of the heredity of root traits (Korkhaus 1930; Townsend et al. 1992) showed a surprisingly high degree of agreement in the shape and size of tooth roots. In monozygotic twins, characteristic agreements were found even in details, whereas these were largely absent in dizygotic probands. This strongly supports the genetic determination of these traits.

Apart from supernumerary roots caused by the splitting of a main root (most frequent in lower canines), there also exist genuine accessory roots (Fig. 10). *Radiculae appendiciformes* are small accessory roots which may occur buccally or lingually on all teeth (Alt 1990) (Fig. 10). They develop from folding processes or by excessive growth of the dental enamel. In the literature, the two manifestations are often not differentiated.

If accessory cusps are present, accessory roots are often found, and the roots can usually be easily identified. The *radix paramolaris*, for example, is bound to the presence of the paramolar tubercle. This tubercle may occur on all teeth, whereas the accessory root appears only next to the regular mesiobuccal root of the corresponding upper and lower molars. The horse-tail shaped *radix praemolarica*

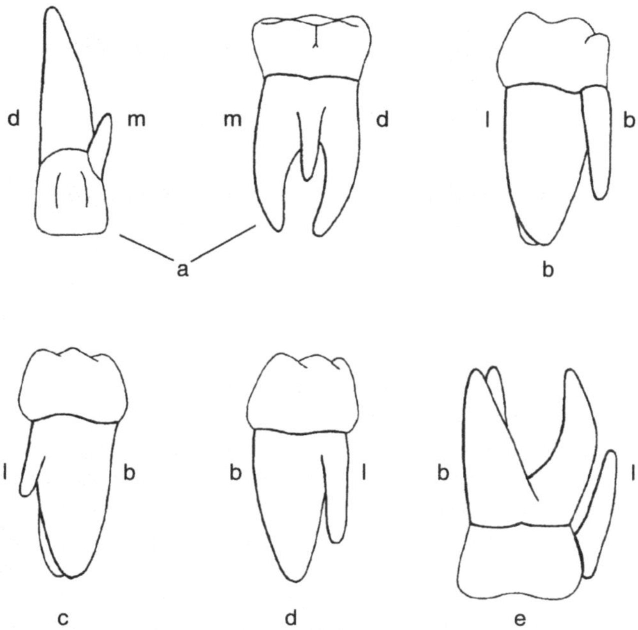

Fig. 10. Supernumerary roots; m = mesial; d = distal; b = buccal; l = lingual; **a:** examples of *radices appendiciformes;* **b:** *Radix paramolaris;* **c:** *Radix entomolaris* (syn. *R. praemolaris*); **d:** *Radix Citroën;* **e:** *Radix Carabelli.*

(entomolaris) is located lingually on lower molars. Other cusps with accessory roots are the Carabelli tubercle (*radix Carabelli*) and the Citroën tubercle (*radix Citroën*) (for details see Schulze 1987) (Fig. 10).

Pyramidalism is a more or less complete fusion of all roots in permanent molars (Fig. 11). Hoffmann (1985) states that second molars are primarily affected. Root fusion caused by an excess deposition of cementum between roots of the same tooth

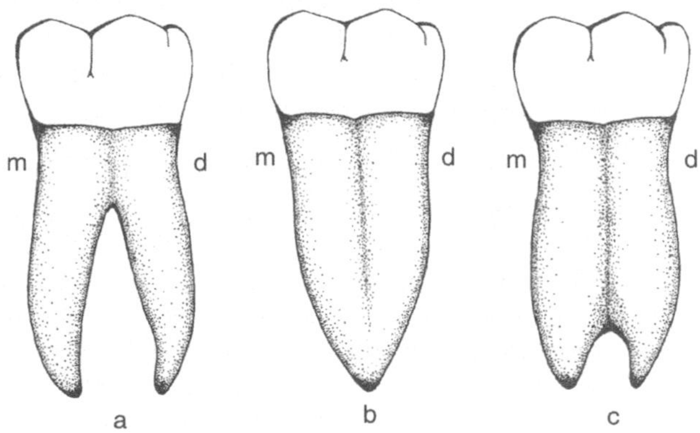

Fig. 11. Types of root fusion; m = mesial; d = distal; **a:** regular form; **b:** pyramidalism; **c:** taurodontism.

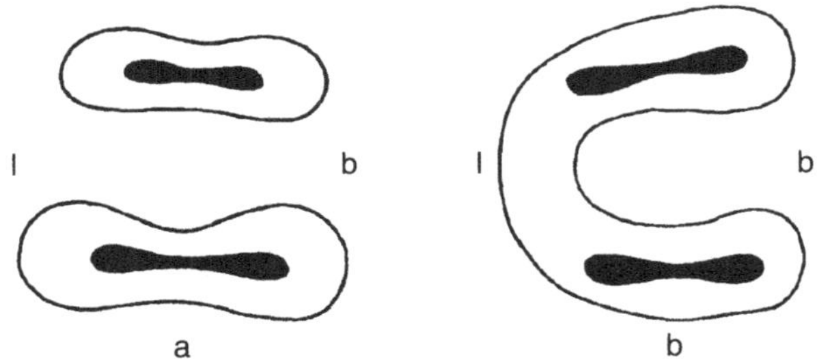

Fig. 12. Special form of apex formation; b = buccal; l = lingual; **a:** regular formation; **b:** horseshoe reduction form.

(cemental bridging), or a partial joining of tooth roots is not scored as pyramidalism. Studies by Kovacs (1967) support the assumption that the trait is primarily genetically determined. Pyramidal roots of molars which show a longitudinal folding of the whole root can have a shape (in the lower third of the root) which is sometimes referred to as "horseshoe reduction form" (Fig. 12).

Taurodontism is characterized by an abnormally large pulp chamber in molars in association with an apical displacement of the bifurcation or trifurcation of affected teeth (Hofmann 1985) (Fig. 11). Anthropologically, the diagnosis should be limited to two distinct variants, mesotaurodontism and hypertaurodontism. The increased occurrence of taurodontism in specific populations (e.g., the Krapina neanderthals) indicates the important role of genetic factors in the occurrence of this dental anomaly (Fischer 1961).

In the case of abnormal root deviations, a distiction between genetic and exogenous disturbances is easily achieved in most cases. Marked root curvatures caused by exogenous factors are characteristically located at the apical third of the root, whereas genetically caused deviations affect the whole length. A special form of root deviation, termed dilaceration, is characterized by a severe crescent-shaped angular curvature. The curvature starts at the tooth neck, but does not affect the crown (Fig. 13).

"Cemental bridging" stems from the fusion of the roots of multirooted teeth (Visser 1948). The bridges are most frequently located between the palatal and distobuccal roots of upper molars (Fig. 14). The most extreme expression is a complete fusion of all components into one pyramid-shaped root, i.e. there has been no differentiation of the epithelial root sheath (Schulze 1987).

Concrescence, Gemination, Twinning and Fusion

The union of two (or more) teeth after eruption or after completion of root formation caused by fusion of their cementum surfaces is called acquired concrescence (Pindborg 1970; Zwerner 1993) (Fig. 15). If such a union occurs during tooth development, the condition is termed a true concrescence, and is most often seen between maxillary second and third molars (Pindborg 1970).

Gemination is the attempt of a tooth bud or follicle to divide, resulting in the formation of double crowns on a single tooth root (Pindborg 1970; Zwerner 1993)

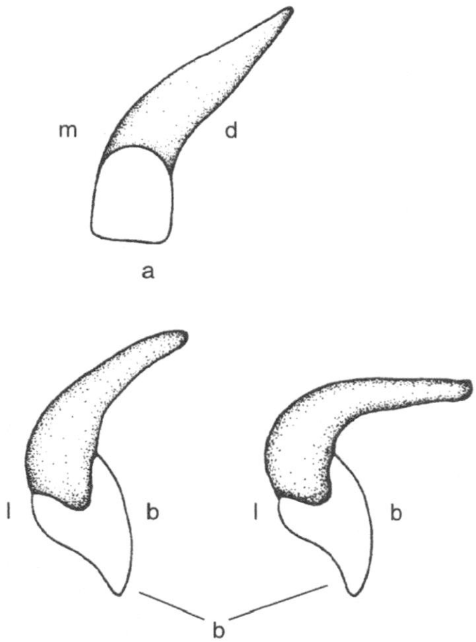

Fig. 13. Abnormal root deviations; m = mesial; d = distal; b = buccal; l = lingual; **a:** abnormal root curvature; **b:** examples of root dilacerations.

(Fig. 15). Gemination mostly appears among incisors and canines; however, premolars can also be affected (Pindborg 1970). Clinically, an incisal groove or depression can be distinguished. The total number of teeth is not affected in the case of gemination (Tannenbaum and Alling 1963).

Twinning (schizodontia) is caused by a complete cleavage of the tooth bud "resulting in the formation of an extra tooth in the dental arch which is usually a mirror image of its adjacent partner" (Tannenbaum and Alling 1963, 886) (Fig. 15).

Fusion (synodontia [de Jonge 1955], or false gemination [Ennis 1949]) is "a union between the dentin and/or enamel of two or more separate developing teeth" (Pindborg 1970, 51) (Fig. 15). It can affect the coronal part of the tooth, the radicular part, or both. In most cases, it is found in primary teeth (incisors and canines) (Pindborg 1970). It has been reported that fused teeth ("double tooth") have been

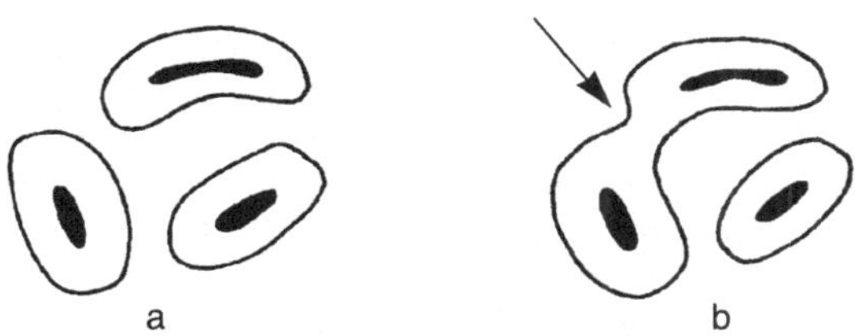

Fig. 14. Special fusion of molar roots; **a:** regular formation; **b:** cemental bridging.

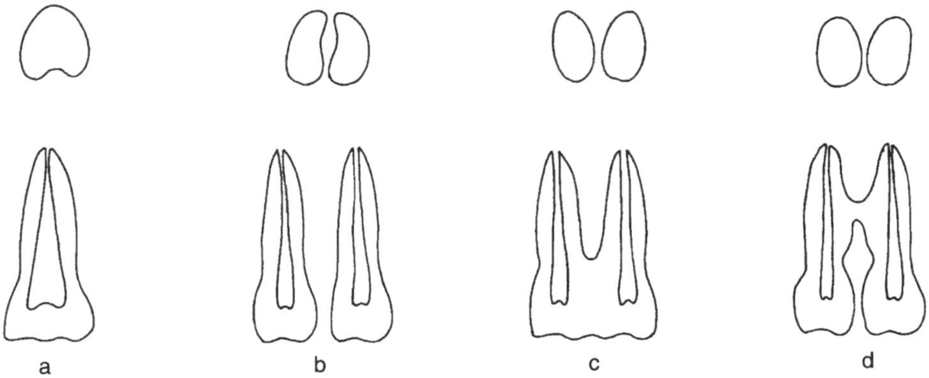

Fig. 15. Schematic representation of tooth union; **a:** gemination; **b:** twinning; **c:** fusion; **d:** concrescence.

found with a higher frequency in certain families (Moody and Montgomery 1934). Fusion in the primary dentition is frequently accompanied by fusion or agenesis of the smaller of the two affected teeth in the secondary dentition (Gysel 1965). Fusion can be differentiated from gemination by the fact that in the case of fusion the adjacent tooth is congenitally absent (Tannenbaum and Alling 1963).

Teeth involved in twinnig or fusion may either be entirely separate, or may have only one pulp chamber and a depression running down the center to mark the division (Schulze 1987). The mandibular incisors are affected most frequently.

Abnormalities of Tooth Size

Tooth size, i.e. the diameters of a tooth in its mesio-distal, bucco-lingual, and vertical directions, varies world-wide among populations and between the sexes (sexual dimorphism) (Kieser 1990; Schnutenhaus and Rösing, this volume). Sexual dimorphism particularly affects the permanent maxillary central incisors as well as the maxillary and mandibular canines and first molars. Therefore, these teeth are useful in determining the sex of an individual with the help of discriminant analysis (e.g. Alt et al. 1995).

Deviations from anormal length are manifest in either a lengthening or a shortening of dental roots (Fig. 16). Here, one has to distinguish between a general occurrence of this anomaly and its development in groups of teeth only. In the upper jaw, abnormally long roots can be observed in all teeth except the central incisors. However, the upper canines are affected most frequently. As for mandibular teeth, long roots are most often observed in canines and premolars. Abnormally short roots in maxillary teeth occur most frequently in the first incisors, while in the mandible in premolars. Abnormally short roots of central maxillary incisors are probably due to hereditary causes. Lind (1972) discovered that in cases of affected individuals, siblings and – in most cases – also one parent had short dental roots.

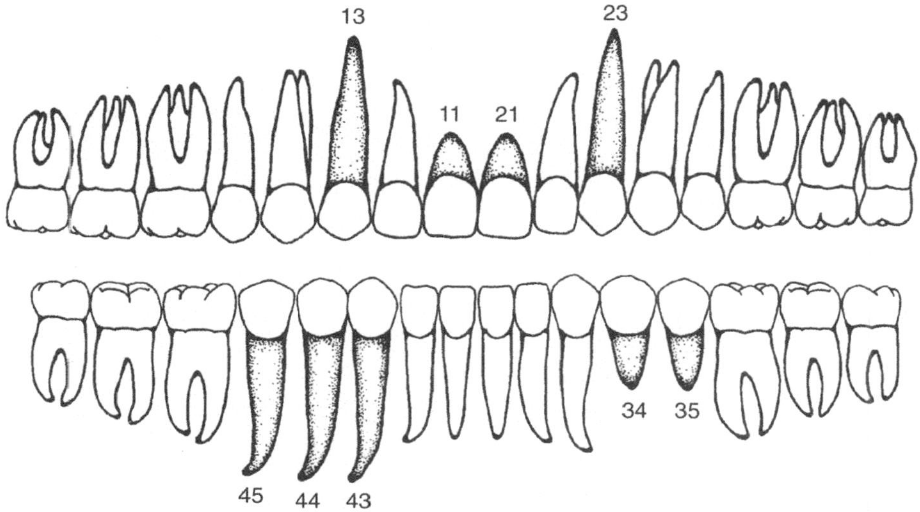

Fig. 16. Abnormalities of tooth size; abnormally long roots: teeth 13, 23, 45, 44 and 43; abnormally short roots: teeth 11, 21, 34 and 35. (FDI tooth numbering system).

Extreme variations in the size of teeth can be regarded as abnormalities (Schulze 1987). Abnormally large tooth size is called macrodontia (megadontism), whereas remarkably small size is referred to as microdontia (Schroeder 1991). In both cases, one, several, or all teeth of an individual can be affected. Microdontia is more common than macrodontia. Microdontia occurs most often for single teeth. After the wisdom teeth, the maxillary second incisors are most frequently affected and often show a peg-shaped form ("peg lateral") (Ash 1986). Microdontia is a characteristic sign present in the case of hypodontia (partial anodontia) and that can affect all remaining teeth.

Disorders of Tooth Number

Anomalies of tooth number can be arranged in three groups: supernumerary teeth (excluding twinning, fusion), special supplementary dental formations, and missing teeth.

Supernumerary Teeth

Supernumerary teeth (hyperodontia, polydontia, polygenesis) occur in excess of the normal number of 20 deciduous or 32 permanent teeth. The extra teeth may be eumorph (supplementary tooth) or heteromorph; they may represent single formations, or may occur in the form of twinning (schizodontia) (Pindborg 1970; Schulze 1987).

Supernumerary teeth can generally occur anywhere in the maxilla and mandible (Fig. 17). In the permanent dentition there are three preferred regions: the anterior part of the maxilla, the molar region in the maxilla, and the premolar region in the mandible. Based on topography, they occur with decreasing frequency

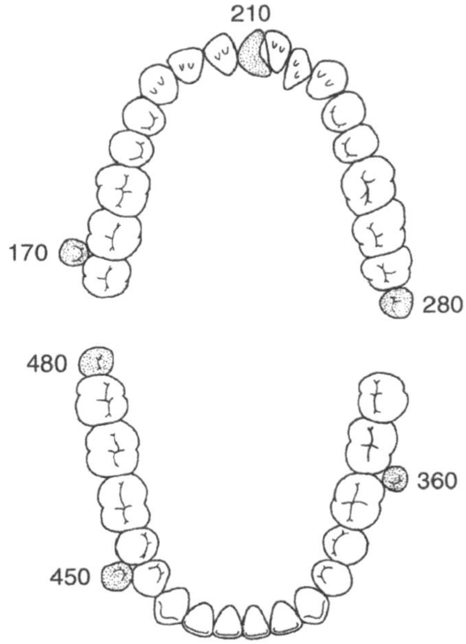

Fig. 17. Supernumerary teeth; 210: mesiodens; 360 and 170: paramolars; 480 and 280: distomolars; 450: supernumerary regularly formed premolar (three-digit system of designating supernumerary teeth after Krysinski 1987, Quintessenz Int 17: 127–128).

in the following order: mesiodentes, paramolars and distomolars (all irregulary shaped), and, in the region of premolars in the mandible, supplemental teeth (Sumiya 1959; Lavelle and Moore 1973).

The frequencies of supernumerary teeth range from 0.2% to 1.0% in the deciduous dentition and from 0.5% to 3.5% in the permanent dentition. On average, 2% of all permanent teeth are supernumerary. The upper and lower jaws are affected in an 8:1 ratio. This is primarily due to the mesiodentes, which account for about 50% of all supernumerary teeth. The next most frequently encountered supernumerary teeth are paramolars and distomolars in the maxilla, followed by extra premolars in the mandible. Twin and family studies substantiate the hereditary genesis of supernumerary teeth (Gysel 1971).

Multiple supernumerary teeth have up to now only been observed in the permanent dentition. Large numbers of supernumerary teeth do not occur frequently. Optionally, impaction may be present. Multiple supernumerary teeth may occur within the context of certain syndromes (for example dysostosis cleidocranialis). A hereditary cause for this anomaly is suggested by its increased frequency within families (Mercuri and O'Neill 1980).

Supplementary Dental Formations

Special supplementary formations include accessory teeth of abnormally small size (microteeth), interradicular teeth (mostly fused to the roots of

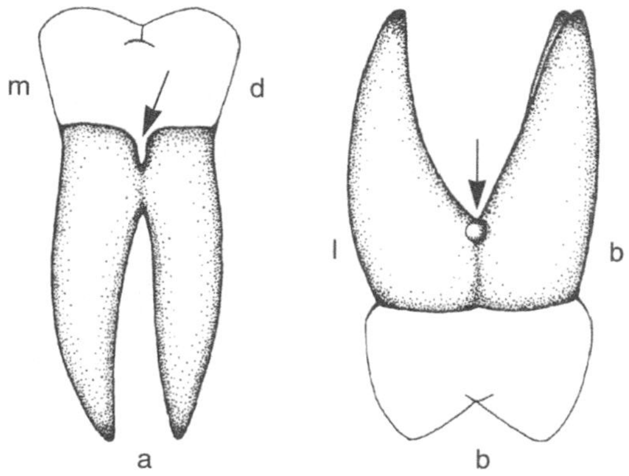

Fig. 18. Supplementary dental formations; m = mesial; d = distal; b = buccal; l = lingual; **a:** enamel pearl; **b:** cervical enamel projection.

relevant teeth), enamel pearls (enamelomas) and cervical enamel projections (Fig. 18).

Microteeth may occur anywhere in the jaws, while interradicular teeth are located between the roots of the molars (Schulze 1987). Enamel pearls are spherical surplus productions of the dental lamina, and are attached to the surface of roots, most frequently at the furcation of molars (Risnes 1974a). The upper and lower jaws are affected in an 8:1 ratio, and the frequency on certain molars varies considerably. In skeletal material of known demographic composition an increased frequency of enamel pearls has been observed within families (Kaiser 1994). As in the case of enamel pearls, cervical enamel projections have been diagnosed primarily on molars, mostly on the buccal sides (Risnes 1974b). In extreme variants, enamel projections can extend all the way to the furcation. Both enamel pearls and cervical enamel projections show considerable population-specific differences in frequency.

Missing Teeth

Anodontia (syn. agenesia, aplasia of teeth) describes any congenital absence of teeth normally present in either the permanent or deciduous dentition (Schulze 1987). Whereas these terms are used to indicate the general fact that teeth are missing, the terms hypodontia, partial anodontia (oligodontia) and total anodontia denote characteristic expressions and degrees of aplasia. Total anodontia occurs very rarely, while partial anodontia (usually the absence of eleven teeth) is observed more frequently (see Henkel 1963).

Because of its frequency and hereditary genesis, hypodontia plays a more important role than the above mentioned anomalies affecting tooth number. Hypodontia describes the absence of several characteristic teeth: the third molars (wisdom teeth) and second premolars in maxilla and mandible, the second incisors in the maxilla and the first incisors in the mandible (Fig. 19). Absence of these teeth accounts for more than 95% of all aplasias (Alt 1989).

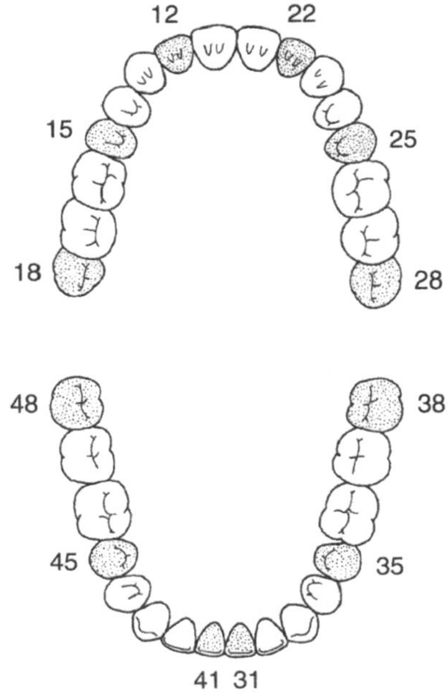

Fig. 19. Missing teeth (hypodontia): maxillary and mandibular third molars (18, 28, 38, 48); second premolars (15, 25, 35, 45); second maxillary incisors (12, 22); first mandibular incisors (41, 31) (FDI tooth numbering system).

All of the characteristic hypodont teeth may exhibit a number of microsymptoms, which are regarded to be micromanifestations of aplasia (Sollich 1974). In addition to reduction of form and size, such as the peg-shape of the wisdom teeth and the second incisors in the maxilla, premolars frequently exhibit symptoms indicating disturbances of normal development (Fig. 20). These include retarded mineralization, retarded or abnormal sequence of eruption, false direction of eruption, and impactions (Fig. 21). All of these symptoms have recently been addressed as microsymptoms because they all are a reaction of the tooth germ to

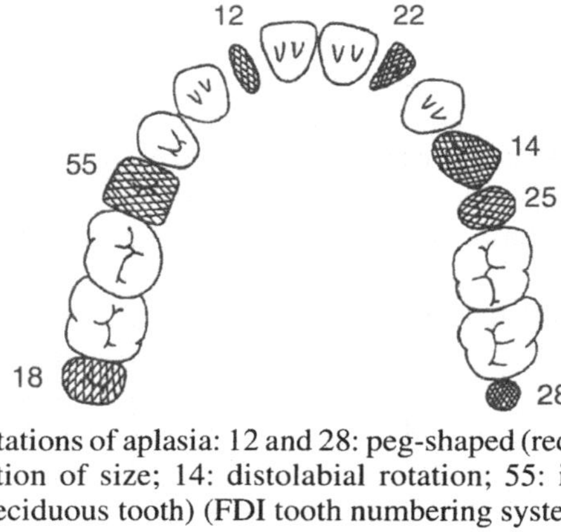

Fig. 20. Micromanifestations of aplasia: 12 and 28: peg-shaped (reduction of form and size); 22, 25 and 18: reduction of size; 14: distolabial rotation; 55: impaction of permanent premolar (persistent deciduous tooth) (FDI tooth numbering system).

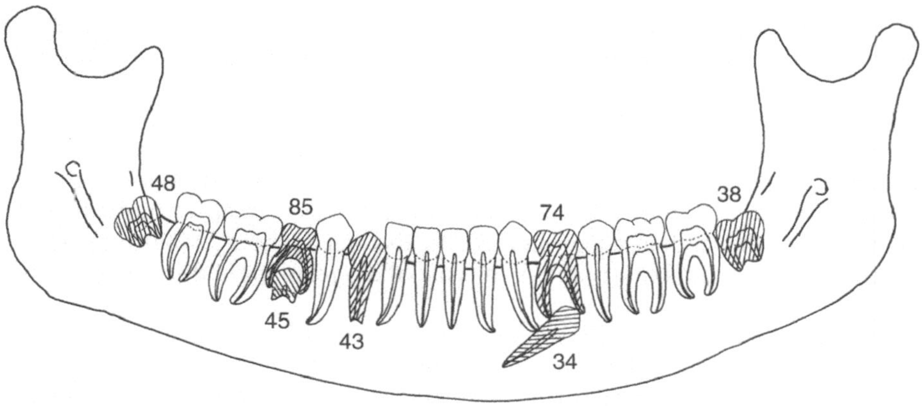

Fig. 21. Micromanifestations of aplasia (radiological findings): 48 and 38: several forms of impaction; 45: retarded mineralization; 43: abnormal sequence of dental eruption; 34: impaction as a result of false direction of eruption; 85 and 74: persistence of the deciduous teeth (FDI tooth numbering system).

an etiologic factor which only in its fullest expression causes aplasia (Schulze 1987, 346).

A characteristic trait of microsymptoms is their co-occurrence with aplasia or the simultaneous occurrence of several microsymptoms in one individual. The most common modifications concern the shape and size of affected teeth. Schulze (1987) interprets such reductions as a first step towards aplasia. Berry and Berry (1967) reported similar phenomena in strains of mice. Grüneberg (1951) regards aplasia to be the ultimate variant of the continuously varying expression of tooth size.

According to all reports, there is a high frequency of hypodontia in the families of affected individuals. Grahnén (1956), in a study of a group of individuals with hypodontia of the wisdom teeth, showed that among brothers and sisters of affected individuals, agenesis of the third molar was twice as high as in a control group without agenesis. A comparison of the subjects with and without hypodontia and their families revealed that the families of unaffected individuals showed a 6% rate of aplasia, which corresponds to the average population frequency, whereas parents or siblings of persons with aplasia showed a frequency of 44%. The heredity of hypodontia is well-established (Alt 1989). Vastardis et al. (1996) suggest that the normal, highly conserved Arg 31 residue of the homeodomain is necessary for MSX1 functions that are required for second premolar and third molar formation (found in all affected family members). Hypodontia is the most frequent hereditary dental anomaly and therefore of special importance in intrapopulation studies.

Anomalies of Tooth Position (Malalignments) and Dental Arch Relationship (Malocclusion)

Anomalies of tooth position are difficult to diagnose in skeletal material (Alt 1997a). The main reason is that good conservation of both jaws is an essential prerequisite and that the recognition and diagnosis of such anomalies require much experience. Above all, the question arises as to how an anomaly of tooth position is defined.

Complementary to the term "eugnathia" which describes the normal and harmonious relationship of the jaws (including the teeth) to one another, the term "dysgnathia" or "dysgnathic anomaly" describes the abnormalities of one or both jaws (including the teeth). A eugnathic anomaly, in contrast, is limited to abnormalities of the teeth and their immediate alveolar supports. Malocclusion is a deviation in the intramaxillary and/or intermaxillary relations of the teeth (Zwemer 1993).

The diagnosis and classification of dental anomalies requires criteria which establish the appearance of the eugnathic dentition. The position of the teeth during intercuspidation is the basis of all investigations of eugnathic or dysgnathic anomalies. In addition to abnormal positions of singular teeth, such as tipping, displacement, rotation, crowding, spacing (abnormal), and transposition, abnormalities concerning the relationships of both jaws in the sagittal, horizontal or vertical plane can appear. Typically, in the case of specific dysgnathic anomalies, complex malpositions of teeth and tooth groups can be found.

In orthodontics, treatment planning requires a comprehensive diagnostic procedure based on radiographic and occlusal evaluations of the patient. The occlusal evaluation is composed of functional (evaluation of the patient) and static parts (analysis of dental models). The latter consists of an intra-arch analysis, followed by an analysis of the inter-arch relations and an assessment of tooth size. Still, the standard values and indices developed and applied in orthodontics are not entirely suitable for investigations of skull material since they do not take population or evolution specific needs into consideration.

Anthropological investigations should be limited to abnormalities of the relationship of the upper and lower jaws and to the identification of localized eugnathic anomalies. Examples are the assessment of the intercuspidation in the sagittal plane, of the vertical relationship of the front teeth to each other, of the evaluation of dysgnathic anomalies in the horizontal plane, of the relationship between the size of the teeth and the jaws, and of impactions and transpositions of teeth.

The amount of tooth divergence from a eugnathic situation is most commonly expressed in terms of premolar widths. Regarding the intercuspidation in the sagittal plane two forms of malocclusion, namely a disto-occlusion and a mesio-occlusion, can be distinguished (Fig. 22). This description is based on the assumption that the lower molars are responsible for the discrepancies between the jaws. Following this assumption, a disto-occlusion is identified when the lower teeth occlude distally to their normal relationship to the upper teeth, whereas in the case of a mesio-occlusion the lower teeth are positioned mesially (Zwemer 1993). Of course, both forms of malocclusion also affect the relations of the anterior teeth (Fig. 23).

The two most important forms of malocclusion in the anterior region of the dental arches in the vertical plane are an open bite and a deep bite (Fig. 24). While the normal range of vertical overlap of the upper front teeth over their lower antagonists is between 2 and 4 mm, in an anterior open bite the anterior teeth do

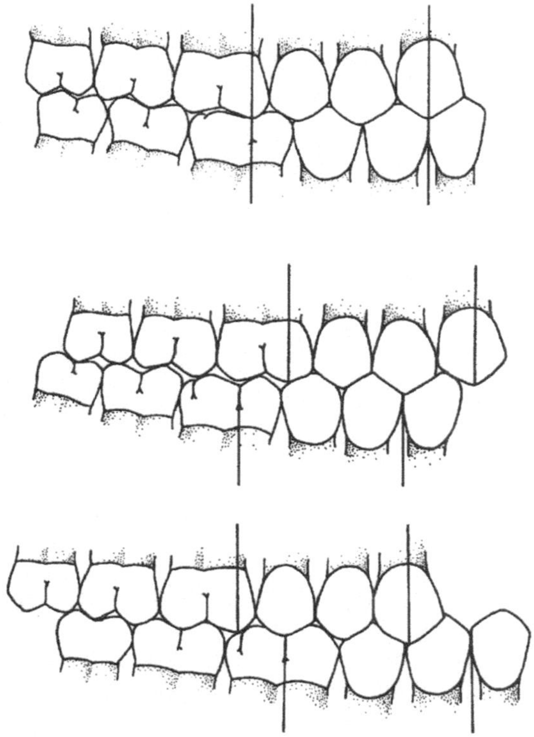

Fig. 22. Relationship of the dental arches (malocclusion); **a:** Angle class I (neutrocclusion, normal anterposterior relationship); **b:** Angle class II (distocclusion; [division 1: with labioversion of the maxillary incisors; division 2: with linguoversion of the maxillary incisors]); **c:** Angle class III (mesiocclusion).

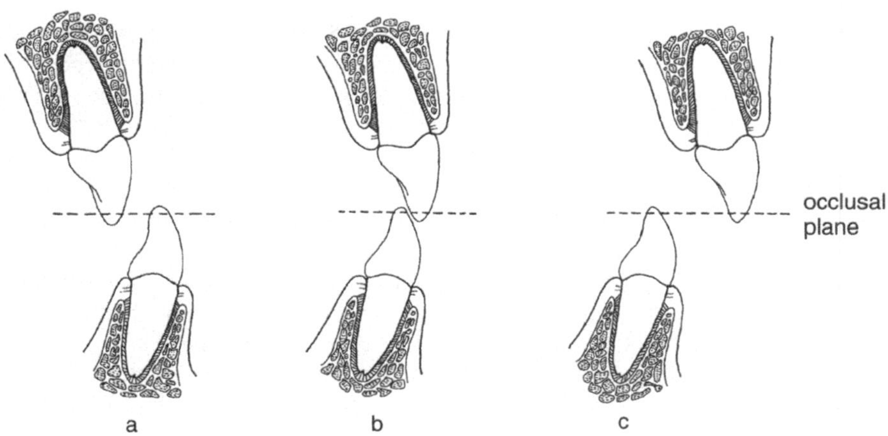

Fig. 23. Malocclusions in the anterior region of the dental arches in the occlusal plane (positive, negative); **a:** anterior crossbite; **b:** regular sagittal relationship; **c:** strong horizontal overlap (overjet).

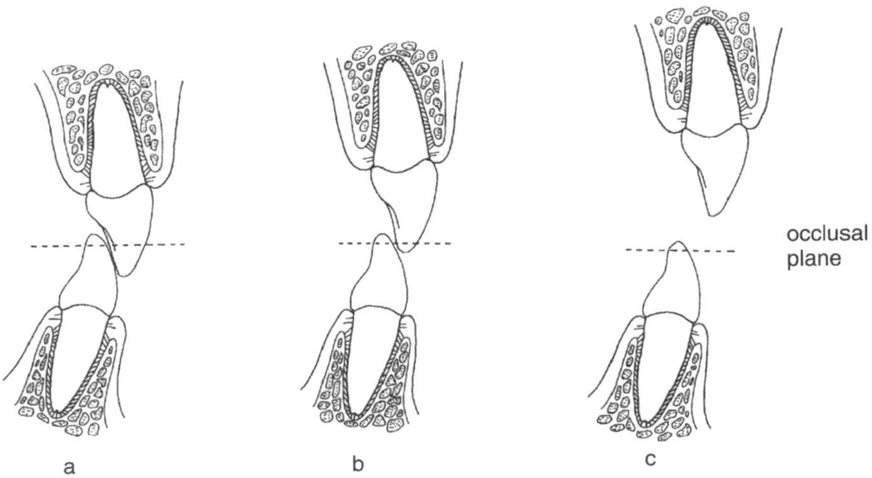

Fig. 24. Malocclusions in the anterior region of the dental arches in the vertical plane; **a:** deep bite (overbite); **b:** regular vertical relationship; **c:** anterior openbite.

not overlap or contact when the posterior teeth are brought into occlusion. This means that an edge-to-edge bite, in which the incisal edges of the maxillary and of the mandibular incisors meet, is not an open bite. In the case of a deep bite (deep overbite), in contrast, the vertical overlapping exceeds 4 mm. As a rule, a deep bite is accompanied by malpositions in the sagittal and horizontal planes. Also seen in cases of deep bite is a steep position of the dental axis and/or a great overjet.

The term "crossbite" is a collective term for malocclusions in the anterior or posterior region of the dental arches in which the normal labio-lingual or bucco-lingual relationship between the teeth of the upper and lower dental arch is reversed, i. e., the lower teeth overlap the upper ones (Fig. 25). In the case of an anterior crossbite, the maxillary incisors are positioned lingually in relation to the mandibular incisors (see Fig. 23). In a posterior crossbite, the maxillary posterior teeth are in a lingual position in relation to their maxillary antagonists. Lateral crossbite is a symptom of many dysgnathias; in most cases, singular teeth are affected.

Characteristic of specific dysgnathias is their syndrome-like character. Certain irregularities appear regularly. As a rule, there is a typical combination of different malpositions of teeth and tooth groups. The causes of this phenomenon are genetic and functional (Harzer 1988). Specific dysgnathias are characterized by the position of the antagonist teeth in the sagittal plane (normal occlusion, distocclusion, mesiocclusion) and the position of the incisors (extrusion, intrusion).

In the case of distocclusion accompanied by a large overjet (horizontal overlap), the missing anterior occlusal contact leads to a deep bite (functional adaptation) and to an anterior maxillary compression (narrowing of the anterior part of the palate). Based on the results of studies carried out on twins and families there is no doubt about the hereditary character of this dysgnathia (Tammoscheit 1971).

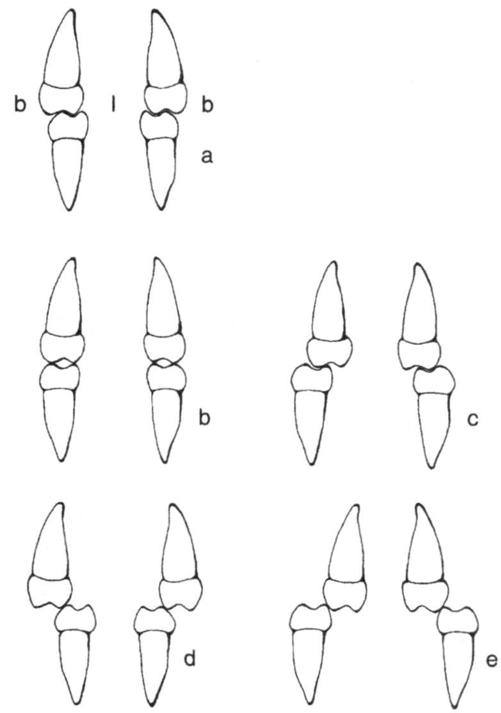

Fig. 25. Transverse occlusal deviations of the posterior teeth; b = buccal; l = lingual; **a:** regular transverse occlusion; **b:** edge-to-edge bite (posterior crossbite tendency); **c:** posterior crossbite; **d:** buccal non-occlusion; **e:** lingual non-occlusion.

Typical signs of an Angle class II, division 2 malocclusion ("Deckbiss") are inverted first incisors in the upper jaw (linguoversion) and a crowding of the mandibular anterior teeth. The latter is accompanied by an abridgement of the lower dental arch. As a consequence of the inversion, a deep bite covering the lower teeth is most likely to develop. In contrast to the flat rise of the palatal curve in the case of a distocclusion, the rise here is very steep. The form of alterations in the anterior region of the mandibular front teeth (crowding, retrusion) depends on the bite. More than 90% of all individuals with an Angle Class II, Division 2 malocclusion have at the same time a deep bite, whereas an open bite never occurs. Korkhaus, on a basis of twin studies, in 1928 affirmed the heredity nature of a Class II, Division 2 malocclusion. The probability of a manifestation is very high (93 %), indicating that exogenous factors play only a very small part (see also Christiansen-Koch 1981).

The third specific dysgnathia is mandibular prognathism. In this situation, the mandible is in a forward relationship relative to the maxilla resulting in an inverted overbite between the front teeth of both jaws (anterior crossbite). As mentioned before, the difference of the tooth positions compared to the eugnathic situation is given in premolar widths. The anomaly can be accompanied by a sagittal discrepancy; however, this is not mandatory.

Whereas in earlier times mandibular prognathism was considered to be caused predominantly by exogenous factors (with only some recognized exceptions like the kinship of the Austrian Habsburg dynastry), twin studies prove that hereditary influences are the determinative factors. In an investigation carried out by Schulze (1979) the degree of concordance was 90% for identical twins, and only 12.2% for dizygotic twins. The ratio of concordance to discordance of about 7 : 1 can be explained only by polygenic factors, and not by dominant hereditary patterns. There is strong evidence that the discussed autosomal hereditary pattern of the anomaly of the Habsburg dynastry (Sander 1990), which has been transmitted from generation to generation (penetrance > 90%), is caused by marriages between relatives (homocygotia). Werner (1979) showed that when relatives are considered, the number of the affected individuals increases by a factor of 2.

Open bite, excessive overbite (e.g, deep bite) and crossbite are considered to belong to the group of specific dysgnathic anomalies. Although they are mainly additional symptoms of typical dysgnathia, they can also be caused by exogenous factors. With the exception of deep bite, they can additionally be a genetically caused anomaly with an independent character (compare Schulze 1987; Alt 1997a). Frequencies occurring in families show that hereditary factors are part of the etiology of these anomalies (Hausser 1961).

Regarding malpositions of single teeth, tipping, displacement, rotation, crowding, diastema, spacing, transposition, and paraxial dystopia (parallel displacement of a tooth with correct position of its length axis) can be distinguished (Fig. 26). Transposition is a change of position between two teeth in a dental arch, i.e. the positions of the first premolar and canine can be reversed. It seems that dental malpositions are mostly caused by dysgnathia and exogenous factors. However, several observations give reason to believe that in part hereditary factors also play a role (Iizuka 1976). Incisors are much more affected than premolars and molars.

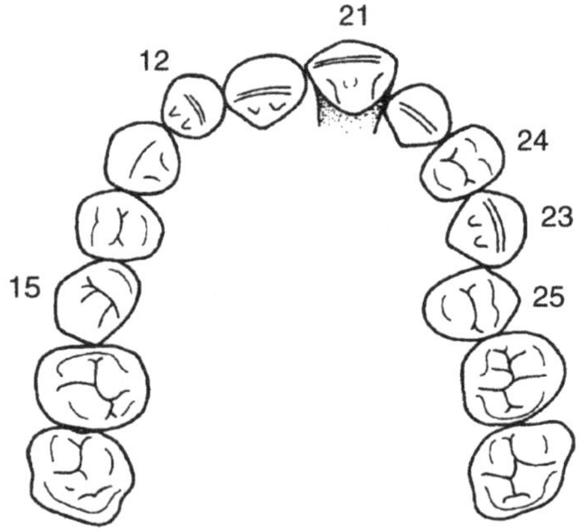

Fig. 26. Several forms of anterior and posterior malposition of single teeth (maxilla); 12: mesiolabial rotation; 15: distolabial rotation; 21: inclination; 24 and 23: transposition (interchange in position of two adjacent teeth); 25: linguoversion (FDI tooth numbering system).

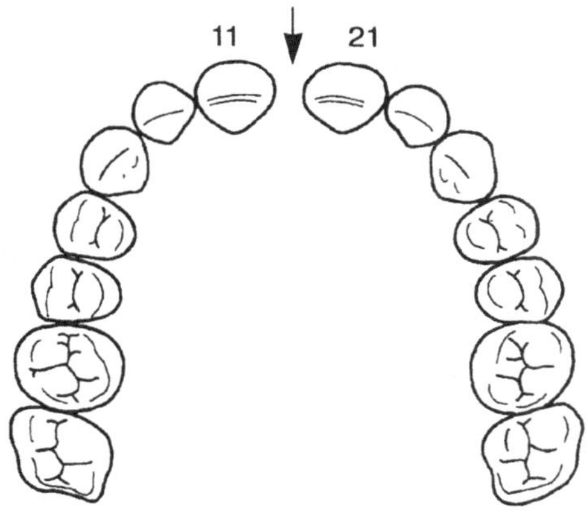

Fig. 27. Diastema (upper midline diastema [trema]).

A medial diastema is a space which typically can be found between the maxillary central incisors of the permanent dentition (Fig. 27). The width of the space varies from individual to individual. In contrast to the physiologic diastemata that appear within the pongides ("primate space", trema), a medial diastema is considered to be an anomaly, provided that it remains for life after the eruption of the canines and does not result from other clinical causes. Terwee pointed out the hereditary character of this eugnathic anomaly in 1922.

Crowding and abnormal spacing of permanent teeth are considered to be indicators for a disproportion between the size of the jaw and the teeth. In the case of spacings, of course, possible extractions of teeth or hypodontia must be excluded. The typical location of crowding or spacing is the frontal region of the dental arch (Fig. 28). A disproportion between the size of the jaw and the teeth can be recognized in the form of paraxial malpositions, and, in the case of macrodontia or micrognathia, of rotations and/or tippings of the anterior teeth or of spaces between the anterior teeth. For diagnostic purposes one can speak of a disproportion between the size of the jaw and the teeth because both signs vary quantitatively and are genetically very independant (Schulze 1982).

An impaction (syn. ankylosed tooth) is a situation where an unerupted tooth is wedged against another tooth or teeth or is otherwise located in a way that it cannot erupt normally (Zwemer 1993) (see Fig. 21). A delayed eruption of the teeth (dentitio tarda) must be excluded, and root formation must be finished before a diagnosis is made. Retention means that the tooth has erupted only partially but is not able to erupt fully because of an obstacle (mostly another tooth or teeth). In living individuals, the affected tooth can either be fully covered by oral mucosa or it can be partially erupted (Pindborg 1970). Impaction can be accompanied by an ectopia. Ectopia means that a tooth is located in a place remote from the normal site.

According to Kwon "the impacted tooth is one where eruption has been interfered with by other teeth, overlying bone or fibrous tissue" (1987, 130). Impaction is seen most frequently among wisdom teeth, followed by canines, maxillary central incisors and premolars. Impacted (retented) or embedded teeth

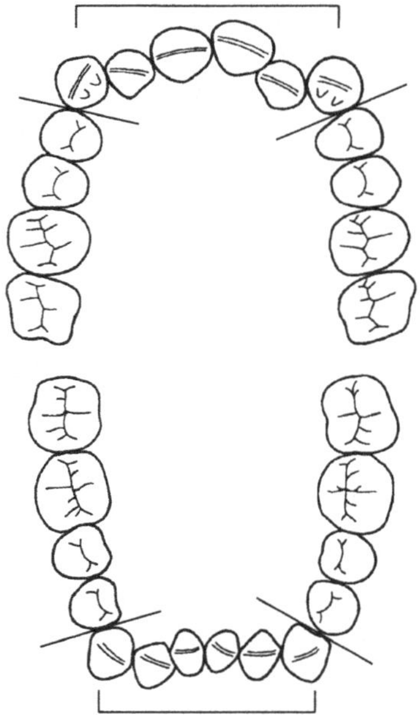

Fig. 28. Anterior crowding in mandible and maxilla.

may be discussed considering three separate theories: orthodontic (exogenous) theory (e.g. infection, trauma, malocclusion), phylogenic theory (civilization effect), and Mendelian theory (heredity). The most common etiologic factors in impaction are heredity and local/mechanical disturbance. Family studies show that hereditary factors play an important role (Kurol and Bjerklin 1982). The prevalence of impaction in living populations is about 2% (Baumer 1985; Alt 1991).

Hereditary Disturbances in Tooth Structure

Hereditary anomalies of tooth structure concern faults in the structure/construction and/or the mineralization of enamel and dentine which occurred during the development of the tooth. Aside from genetically determined disturbances of the hard tissues, there are also a number of acquired, i.e. non-hereditary causes for such defects, but these will not be discribed. The anthropological literature contains only a few reports of hereditary anomalies of tooth structure (see e.g. Berger 1992).

Hereditary anomalies of tooth structure are divided into anomalies of enamel and anomalies of dentine (Witkop 1989). In the presence of anomalies of structure, additional anomalies such as aplasia, impaction, open bite or taurodontism can frequently be observed. All of the hereditary structural anomalies have in common that their expression varies not only within families and between the sexes, but also in the deciduous and permanent dentitions. This makes their diagnosis difficult. The two most common structural anomalies of teeth are amelogenesis imperfecta and dentinogenesis imperfecta.

Amelogenesis Imperfecta

Amelogenesis imperfecta is a hereditary defect in enamel formation which affects all teeth in both dentitions. Clinically, four groups can be distinguished (Witkop 1989): hypoplasia (type I), hypomaturation (type II), hypocalcification (type III), and partial hypomaturation and hypocalcification combined with taurodontism (type IV). Groups I-III contain several subclassifications. Hypocalcification and hypomaturation represent a disturbance of the secondary function of ameloblasts, the mineralization and maturation of the enamel, leaving matrix formation unaffected. In contrast, enamel hypoplasia represents a disturbance of matrix formation, which is later affected by the disturbance in mineralization.

The terms "hypoplasia" and "dysplasia" denote two different defects in enamel formation. Hypoplasia indicates both a qualitative and quantitative deficiency in enamel formation, whereas dysplasia results in only a qualitative deficiency due to hypocalcification. The slight differences between aplasia and hypoplasia are made obvious in clinical and roentgenological findings. Aplasia denotes the absence of dental enamel (which is so thin it can only be distinguished microscopically). In the case of hypoplasia, the enamel cover is visible, but greatly reduced in thickness, and the affected tooth surface is generally very irregular (Fig. 29).

Hereditary disturbances in enamel formation (amelogenesis) have been the object of numerous family studies, which form the basis for their differentiation (for reviews see Witkop and Sauk 1976, Schulze 1987). It has been possible to

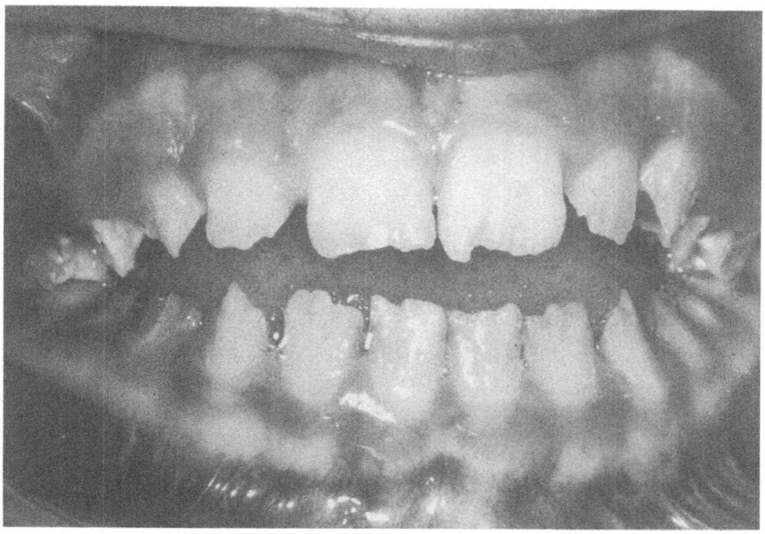

Fig. 29. Amelogenesis imperfecta – Hypomineralized type. This 12-year-old girl has yellowish-brown, very soft enamel. The surfaces of the teeth are dull and covered with ridges and grooves. Noticeable amounts of enamel have fractured away. The enamel dysplasia is combined with an open bite. (Photo: I. Jonas; from Rakosi T, Jonas I, Graber TM (1993) Orthodontic Diagnosis. Thieme, Stuttgart, Fig. 163).

diagnose a special form of enamel hypoplasia affecting only the lower third of the dental crowns in 5 probable family members of 13 individuals from a Late Roman skeletal series (Alt et al. 1992; Berger 1992). Further anomalies like taurodontism and disturbances in tooth number were also observable in that case.

Dentinogenesis Imperfecta

Anomalies of the structure of the dentin are classified into two main groups (Jorgenson 1989): isolated disturbances, and anomalies associated with other diseases. Dentinogenesis imperfecta is an isolated disturbance of the dentin and is of genetic origin. It is characterized by a defective (early) calcification of the dentin leading to fast attrition of the hard tissues. Four variants may be distinguished by different hereditary mechanisms deduced from family studies (Mars et al. 1976; Jorgenson 1989). An older classification of dentin dysplasia had been proposed by Shields et al (1973).

The most frequent type of dentinogenesis imperfecta (type I) results in a "mushroom-shaped" crown and is characterized by an opalescent appearance of the teeth and obliteration of the pulp chamber (Fig. 30). The main finding of type II are rootless teeth (types after Jorgenson 1989, see also Schulze 1987).

Dentinal dysplasia may be associated with an array of syndromes, e.g. pseudohypoparathyroidism (Albright's syndrom), Ehlers-Danlos syndrome, and many others (Jorgenson 1989; Schroeder 1991).

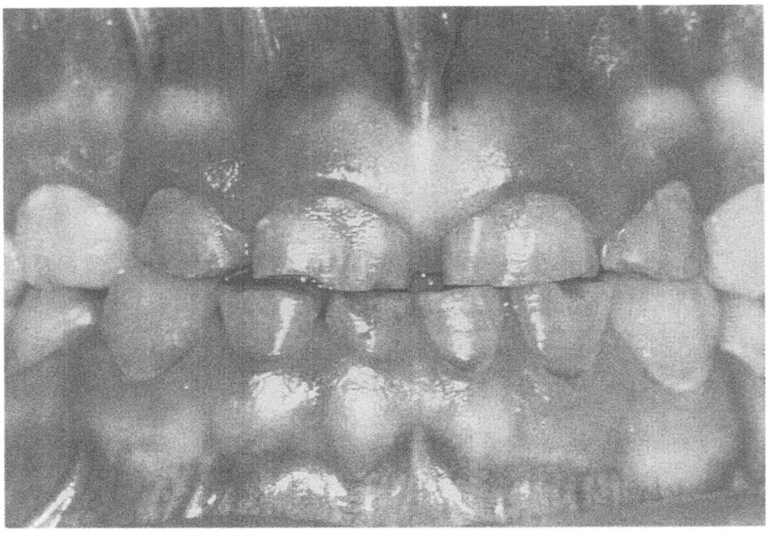

Fig. 30. Dentinogenesis imperfecta. Hereditary dysplasia of the dentin in a 19-year-old patient. Short, amber-colored upper and lower anterior teeth, with splintered enamel and marked attrition of the incisal edges are typical clinical findings in this disturbance in tooth formation. (Photo: I. Jonas; from Rakosi T, Jonas I, Graber TM (1993) Orthodontic Diagnosis. Thieme, Stuttgart, Fig. 160).

Criteria for the Diagnosis of Hereditary Disturbances in Tooth Structure

In anthropological investigations, a differentiation of dental hard tissues requires criteria different from those used in clinical studies. The investigator must be able to recognize the presence of macroscopically visible structural disturbance. A differentiation of hereditary and non-hereditary disturbances is of secondary importance, as this can usually be achieved by answering a simple question: Does the defect occur on isolated teeth or within tooth groups, or is it generalized?

If isolated teeth are affected, one can commonly assume that a local effect, such as an inflammation or trauma, is the cause. A systematic occurrence on teeth or tooth groups (more rarely on all teeth) suggests the presence of endogenous disturbances, which took effect as the affected teeth developed (see Schultz et al., this volume). In this case the duration of the disturbance determines the extent of the anomaly. In contrast to such localized disturbances, hereditary structural anomalies are always generalized and can be found in both the deciduous and permanent dentitions. This rather simple mode of distinguishing structural anomalies is the first and basic step in their differential diagnosis.

Disturbances in Tooth Eruption

Disturbances of tooth eruption can be found in both the deciduous and permanent dentitions. Retarded eruption *(dentitio tarda)* may be diagnosed if the eruption of the teeth occurs beyond the normal variation (van der Linden and Duterloo 1976). Delayed eruption need not be of pathological nature (obstructed eruption), but can represent local or general deficiencies of the organism.

Delayed eruption is attributed to a hereditary delay in the anlage of certain teeth. In premature newborns, *dentitio tarda* is due to the discrepancy between postnatal age and degree of maturation. Transpositions of tooth germs, anomalies of tooth form, obstacles of tooth eruption and acute inflammatory processes are symptoms of local disturbances. *Dentitio tarda* is almost always observed when cleidocranial dysostosis is diagnosed. Retarded eruption can also represent generalized disturbances of ossification. The most common of these are rickets as well as deficiencies of the thyroid gland or the hypophysis (Hausser 1960; Dausch-Neumann and Bierich 1989).

Dentitio praecox denotes the premature eruption of teeth. It occurs only rarely in the deciduous dentition. In the permanent dentition it is usually found as a result of premature loss of deciduous teeth (Pindborg 1970).

Congenital Malformations and Disturbances of Bone Formation and Growth of the Skeleton Affecting Jaws and Teeth

Congenital malformations of the skull and disturbances of bone formation and growth occasionally involve jaws and teeth (David et al. 1982; Smith 1988). The non-uniform use of the nomenclature of craniofacial disorders may lead to problems of communication (Maroteaux 1986; Jorgenson 1989). Congenital malformations of the skeleton are dysontogenetically caused changes of normal structure which develop in the embryonic period and are most commonly found in the region of the skull (Prokop 1974; Gorlin et al. 1976). In contrast, disturbances of bone formation and growth belong to the group of systemic diseases, which represent general disturbances of the differentiation and growth of the bones (Burkhardt 1970).

According to their localization, craniofacial dysplasia are differentiated into dysplasia of the cerebral part of the skull (*cranium cerebrale*) and dysplasia of the facial part of the skull (*cranium faciale*). In this context, only those types of craniofacial dysplasia involving teeth and jaws are considered (Tab. 1). These include hypoplasia, hyperplasia and cases of cleft palate. The specialist literature contains numerous specific descriptions and systematic reviews. We are therefore giving only a short description of the findings on the jaws and skeleton in the following charts (see also Alt 1997a).

Table 1. Congenital malformations of the skeleton affecting jaws and teeth

Malformation	Skeletal findings	Findings on jaws and teeth
Faciomandibular (mandibulofacial) dysostosis	development disturbance of the cranial bones; hypoplasia of the zygomatic bone	hypoplasia of the mandibular body; hyperplasia of the ramus; tooth malalignments
Median cleft of the mandible	(–)	cleft of mandible
Maxillofacial dysostosis	hypoplasia of the zygomatic bone; decrease in size of the frontal cranial base	hypoplasia of the upper jaw; open bite; mandibular protrusion
Pierre Robin's syndrome	(–)	hypoplasia of the lower jaw (retrognatism); cleft mandible; isolated cleft palate (inconstantly)
Cleft of lip, jaw, and palate	(–)	cleft of upper jaw and palate; anomalies of tooth number; tooth malalignments
Craniofacial dysostosis (Crouzon)	acro(turri)cephaly; hypertelorism; wide root of the nose	underdevelopment of the maxilla; underdeveloped palate; tooth malalignments
Cleidocranial dysostosis (Sainton's disease)	agenesis/hypoplasia of the clavicles; supernumerary bone formation on the skull (brachycephaly); "Impressio et cyphosis basilaris"; delayed closure of the sutures and fontanels; hyposomia	hypoplasia of the upper and lower jaw; persistent primary teeth; multiple impaction/retention; formation of supernumerary teeth; delayed eruption of teeth

The more important congenital disturbances of bone formation and growth which include malformations of jaws and teeth are shown in Table 2:

Table 2. Congenital disturbances of bone formation and growth

Disturbance	Skeletal findings	Findings of jaws and teeth
Osteogenesis imperfecta (Lobstein's diseases)	failure of ossification: fragility and deformation of bones, peculiar structure	dentinogenesis imperfecta; translucent teeth; hypomineralized enamel
Fibrous (fibro-osseous) dysplasia (Morbus Jaffé-Lichtenstein)	systemic enlargement and deformation of the bone; mostly affected cranial base, frontal bone, occipital bone	enlarged alveolar processes without involvement of the chin and the condylar processes; radiopacity of the maxillary sinus
Cherubism (familial fibrous dysplasia of the jaws; familial intraosseous swelling)	frontal dysplacement of the frontal wall of the maxillary sinus; raised eyes	symmetric polycystic expansion; enlargement of upper and lower jaw (swelling); deformations and displacements of the teeth; retarded eruption; osteomas and osteofibromas

Congenital malformations of the skeleton and disturbances of bone formation and growth affecting jaws and teeth are also evident in earlier humans (Brothwell and Powers 1968; Turkel 1989).

Conclusion

More than any other organ, teeth offer great potential for variation. In addition, "given their nature and function, teeth are used to address several kinds of questions" (Scott 1991, 789). Their extreme degree of hardness and their better ability to resist against decomposition, compared to other body tissues, make them a research object for a variety of scientific disciplines (anthropology, paleontology, biology, dentistry, anatomy, forensic sciences). Studies investigate problems concerning evolutionary and growth factors, phyletic relationships and development, adaptation, comparative anatomy, structural analyses, forensic odontology, pathology, genetics and others (see also Kelley and Larsen 1991; Lukacs 1992; Hillson 1996; Alt and Türp 1997; Scott and Turner 1997).

Teeth are relatively easy to observe in living as well as past populations. Most dental variables have a strong genetic component. Variation of normal tooth morphology, especially dental anomalies of morphology, size, number, structure, and position of teeth, are of great interest for research in inter- and intrapopulation studies because they provide information about the genetic background of a population (Scott 1991). Many dental anomalies described in the present chapter, such as variants of crown morphology and missing teeth (see Scott 1973; Lukacs 1987; Turner 1987; Scott and Turner 1997), have a long tradition in interpopulation studies, and recently in infrapopulation research (kinship analysis) and in forensic anthropology/ odontology as well (Alt 1995, 1997b; Alt and Vach 1995).

The aim of this chapter was to give an overview of hereditary dental anomalies which are important for population studies and epidemiological investigations (palaeopathology) in dental anthropology. Pathological findings which are caused by exogenous factors (e.g., environmental defects; see Hillson 1986) were deliberately excluded. Findings of this type, which can be observed quite frequently in archaeological samples, are summarized in other chapters of this volume (e.g. caries, attrition and others). We hope that this review will help to enlarge the number of dental traits found and recorded in population investigations. Because of the usually poor preservation of teeth found in anthropological material, findings are often limited. Therefore, the number of possible traits to be investigated in past populations should be as large as possible. Most of the traits described in this chapter have proved to be very helpful in giving detailed information (see Alt and Vach, this volume).

Acknowledgements

The authors would like to thank Prof. Dr. I. Jonas, Dental Clinic, University of Freiburg, for photographs, G. Thieme Verlag, Stuttgart and New York for the possibility to reproduce the Figure 29 and 30, and Prof. Dr. U. Wolf, Institute of Human Genetics and Anthropology, University of Freiburg, for financial support. Original drawings were designed by Dr. E. Faulstich, Freiburg.

References

Alt KW (1989) Zur Problematik odontologischer Verwandtschaftsanalysen in der prähistorischen Anthropologie am Beispiel der Aplasie/Hypodontie. Z Morph Anthrop 78: 43–71.

Alt KW (1990) Radicula appendiciformis an einem lateralen Oberkieferschneidezahn – Fallbericht über eine seltene Anomalie. Quintessenz 41: 105–110.

Alt KW (1991) Zur Definition, Häufigkeit und Ätiologie retinierter Zähne. Anthrop Anz 49: 261–272.

Alt KW (1995) Dental anthropology in forensic odontology and archaeology. A review. In: Jacob B, Bonte W, Alt KW, Pieper P (eds) Advances in Forensic Sciences, Vol. 7. Köster, Berlin, p. 255–262.

Alt KW (1997a) Odontologische Verwandtschaftsanalyse. Fischer, Stuttgart.

Alt KW (1997b) Kategorien und Konzepte dentalanthropologischer Studien. In: Alt KW, Türp JC (Hrsg) Die Evolution der Zähne. Phylogenie – Ontogenie – Variation. Quintessenz, Berlin, p. 679–687.

Alt KW, Vach W (1995) Odontologic kinship analyses in skeletal remains – concepts, methods, and results. Forensic Sci Int 74: 99–113.

Alt KW, Türp JC (Hrsg) (1997) Die Evolution der Zähne. Phylogenie – Ontogenie – Variation. Quintessenz, Berlin.

Alt KW, Rieger S, Krekeler G, Vach W (1995) Odontometrische Geschlechtsbestimmung. Evaluierung frühmittelalterlicher Bestattungen. Z Rechtsmed 5: 82–87.
Ash MM (1986) Kerr and Ash's Oral Pathology. Lea & Febiger, Philadelphia.
Baumer K (1985) Die Retention bleibender Zähne. Med Diss, Erlangen.
Berger T (1992) Amelogenesis imperfecta an kaiserzeitlichen Skelettresten aus Regensburg-Harting – eine differentialdiagnostische Studie. Med Diss, Tübingen.
Berry AC, Berry JR (1967) Epigenetic variation in the human cranium. J Anat 101: 361–379.
Blanco R, Chakraborty R (1976) The genetics of shovel shape in maxilla central incisivi in man. Am J Phys Anthrop 44: 233–236.
Brabant H, Klees L, Werelds RJ (1958) Anomalies, mutilations et tumeurs des dents humaines. Prélat, Paris.
Brothwell DR, Powers R (1968) Congenital malformations of the skeletons in earlier man. In: Brothwell DR (ed) The Skeletal Biology of Earlier Human Populations. Pergamon Press, Oxford.
Burkhardt L (1970) Pathologische Anatomie des Schädels. In: Uehlinger E (Hrsg) Handbuch der speziellen pathologischen Anatomie und Histologie IX/7. Springer, Berlin, p. 1–352.
Burnett SE (1996) A new look at premolar trait variation: maxillary premolar accessory ridges (MxPAR). Dent Anthrop Newsl 10 (3): 17–18.
Butler PM, Joysey KA (eds) (1978) Development, Function and Evolution of Teeth. Academic Press, London.
Carbonell VM (1963) Variations in the frequency of shovel-shaped incisors in different populations. In: Brothwell DR (ed) Dental Anthropology. Pergamon Press, Oxford. p. 21–234.
Carlsen O (1987) Dental morphology. Munksgaard, Kopenhagen.
Christiansen-Koch S (1981) Familien- und Sippenuntersuchungen zur Frage der Vererbung des Deckbisses. Med Diss, Berlin.
Colyer F (1936) Variations and Diseases of the Teeth of Animals. London.
David DJ, Poswillo D, Simpson D (1982) The Craniosynostoses. Causes, Natural History, and Management. Springer, Berlin Heidelberg New York.
Dausch-Neumann D, Bierich J (1989) Wirken sich wachstumsfördernde Hormone bei Minderwuchs auf die Gebißentwicklung aus? Fortschr Kieferorthop 50: 54–64.
Ennis LM (1949) Dental Roentgenology. Lea & Febiger, Philadelphia.
Fischer H (1961) Die prismatischen Molaren von Krapina/Kroatien im Lichte rezenter Funde. Dtsch Zahnärztl Z 16: 8–15.
Frisch JE (1965) Trends in the Evolution of the Hominoid Dentition. Karger, Basel.
Fujita T, Sumiya Y, Sakamoto K (1963) Paramolar tubercle in the premolar region. Acta Anat Nippon 38: 109–119.
Gorlin RJ, Pindborg JJ, Cohen MM (1976) Syndromes in the Head and Neck. McGraw-Hill, New York.
Grahnén H (1956) Hypodontia in the permanent dentition. Odont Revy 7: Suppl 3.
Grahnén H, Lindahl B, Omnell KA (1958) Palatal invaginations (dens in dente) of the second maxillary permanent incisors. Odont Revy 9: 163–166.
Grüneberg H (1951) The genetics of a tooth defect in the mouse. Proc Roy Soc Biol 139: 437–451.
Gysel C (1965) La fusion et le gémination des dents temporaires au regard des definitives. Rev Belge Méd Dent 20: 35–86.
Gysel C (1971) Gemination, mesiodens et genetique. Rev Belge Med Dent 26: 145–150.
Hallet GEM (1953) The incidence, nature, and clinical significance of palatal invaginations in the maxillary incisor teeth. Proc Roy Soc Med 46: 491–499.
Hanihara K, Kuwashima T, Sakao N (1964) The deflecting wrinkle on the lower molars in recent man. J Anthrop Soc Nippon 72: 740–743.

Harzer W (1988) Zur Bedeutung und Wertigkeit von genetischen und Umweltfaktoren für die Ätiologie und Pathogenese von Dysgnathien. Zahn- Mund Kieferheilkd 76: 439–443.
Hattab FN, Yassin OM, Al-Nimri KS 81995) Talon cusp – clinical significance and management: Case reports. Quintessence Int 26: 115–120.
Hausser E (1960) Schilddrüse, Nebenschilddrüse, Hypophyse und Kauorgan. Fortschr Kieferorthop 21: 176–182.
Hausser E (1961) Erbeinfluß und Gebissanomalien. Fortschr Kieferorthop 22: 72–82.
Henke W, Rothe H (1994) Paläoanthropologie. Springer, Berlin Heidelberg New York.
Henkel G (1963) Die Vererbung der hochgradigen Zahnunterzahl (Oligodontie und Anodontie) beim Menschen. Med Diss, Göttingen.
Hillson S (1986) Teeth. Cambridge Univ Press, Cambridge.
Hillson S (1996) Dental Anthropology. Cambridge Univ Press, Cambridge.
Hitchin AD, Ferguson HW (1958) A compression odontome and its possible relationship to aberrant crown forms. Dent Pract Dent Rec 8: 309–311.
Hofmann KH (1985) Morphologische Variationen und Anomalien im Bereich der ersten und zweiten Molaren bei 1.400 kieferorthopädischen Patienten unter besonderer Berücksichtigung des Taurodontismus und Pyramidalismus. Med Diss, Berlin.
Iizuka A (1976) The bilateral mesiopalatal rotation of the upper central incisors. J Anthrop Soc Nippon 84: 31–47.
Jonge TE de (1935) Die Dimerie der Frontzähne, ein weiterer Beitrag zur Morphologie des menschlichen Gebisses. Schweiz Mschr Zahnheilk 45: 95–122.
Jonge TE de (1955) Beschouwingen over synodontie. Tijdschr Tandheelkd 62: 828–834.
Jorgenson RJ (1989) Problems in nomenclature of craniofacial disorders. J Craniofac Genet Dev Biol 9: 7–20.
Kaiser KG (1994) Schmelzperlen und Schmelzsporne und ihre paläodontologische und parodontologische Bedeutung. Med Diss, Freiburg.
Kanazawa E, Sekikawa M, Ozaki T (1990) A quantitative investigation of irregular cuspules in human maxillary permanent molars. Am J Phys Anthrop 83: 173–180.
Keil A (1966) Grundzüge der Odontologie. Bornträger, Berlin.
Kelley MA , Larsen CS (eds) (1991) Advances in Dental Anthropology. Wiley, New York.
Kieser JA (1990) Human Adult Odontometrics. The Study of Variation in Adult Tooth Size. Cambridge Univ Press, Cambridge.
Kimball M, Ripa LW, Rochester NY (1970) Talon cusp: a clinically significant anomaly. Oral Surg 29: 225–228.
Korkhaus G (1930) Die Vererbung der Kronenform und -größe menschlicher Zähne. Z Anat 91: 594–617.
Kovacs I (1967) Contribution to the ontogenetic morphology of roots in the human teeth. J Dent Res 46: 865–847.
Kurol J, Bjerklin K (1982) Ectopic eruption of maxillary first permanent molars: Familial Tendencies. J Dent Child 49: 35–38.
Kurtén B (ed) (1982) Teeth: Form, Function, and Evolution. Columbia Univ Press, New York.
Kutscha B (1985) Morphologische Variationen und Anomalien im Prämolarenbereich bei 1.400 kieferorthopädischen Patienten. Med Diss, Berlin.
Kwon JH (1987) Impacted teeth. In: Waite DE (ed) Textbook of Practical Oral and Maxillofacial Surgery. Lea & Febiger, Philadelphia, p. 130–148.
Lavelle CLB, Moore WJ (1973) The incidence of agenesis and polygenesis in the primate dentition. Am J Phys Anthrop 38: 671–680.
Lind V (1972) Short root anomaly. Scand J Dent Res 80: 85–93.
Linden FPGM van der, Duterloo HS (1976) Development of the Human Dentition. An Atlas. Harper & Row, New York.

Lukacs JR (1987) Biological relationships derived from morphology of permanent teeth: recent evidence from prehistoric India. Anthrop Anz 45: 97–116.
Lukacs JR (1992) Culture, ecology, and dental anthropology. J Hum Evol: Spec Issue No 2. Kamla-Rai, Delhi.
Lunt DA (1976) Molarization on of the mandibular second premolar. J Dent 4: 83–86.
Maroteaux P (1986) International nomenclature of constitutional diseases of bones. Birth Defects 22 (4): 1–54.
Mars M, Farrant S, Roberts GJ (1976) Dentinogenesis imperfecta. Br Dent J 140: 206–209.
Matsumura H (1995) A microevolutional history of the Japanese people as viewed from dental morphology. Nat Sci Mus Monogr No 9, Tokyo.
Med-Index Publications (1993) ICD 9 CM. International Classifcation of Diseases. Wiley, Salt Lake City.
Mercuri LG, O'Neill R (1980) Multiple impacted and supernumerary teeth in sisters. Oral Surg 50: 293.
Mizoguchi Y (1977) Genetic variability of permanent tooth crowns as ascertained from twin data. J Anthrop Soc Nippon 85: 301–309.
Moody E, Montgomery LB (1934) Hereditary tendencies in tooth formation. Am Dent Assoc 21: 1774–1776.
Pindborg JJ (1970) Pathology of the Dental Hard Tissue. Munksgaard, Kopenhagen.
Prokop O (1974) Genetik erblicher Syndrome und Mißbildungen. Springer, Berlin Heidelberg New York.
Rakosi T, Jonas I, Graber TM (1993) Orthodontic Diagnosis. Thieme, Stuttgart.
Regezi J, Sciubba J (1993) Oral Pathology. Clinical-Pathologic Correlations, 2nd ed. Saunders, Philadelphia.
Reichart PA, Metah D, Sukasem M (1982) Morphological findings in dens evaginatus. Int J Oral Surg 11: 59–63.
Remane A (1960) Zähne und Gebiss. In: Hofer H, Schultz AH, Starck D (Hrsg) Primatologia. Hb der Primatenkd III/2. Karger, Basel, p. 637–846.
Risnes S (1974a) The prevalence and distribution of cervical enamel projections reaching into the bifurcation on human molars. Scand J Dent Res 82: 413–419.
Risnes S (1974b) The prevalence, location, and size of enamel pearls on human molars. Scand J Dent Res 82: 403–412.
Sander H (1990) Die Vererbung der Progenie im Hochadel Europas. Med Diss, Freiburg.
Scheffler B (1976) Häufigkeit und Formvariabilität des Tuberculum Carabelli. Med Diss, Berlin.
Schellong G, Pfeiffer RA (1963) Mißbildungen im Kopfbereich aus der Sicht des Kinderarztes. Dtsch Zahnärztebl 17: 231–244.
Schroeder HE (1991) Oral Structural Biology. Thieme, Stuttgart.
Schulze C (1979) Zur Ätiologie der Progenie. Fortschr Kieferorthop 40: 87–104.
Schulze C (1982) Lehrbuch der Kieferorthopädie, Bd. 3. Quintessenz, Berlin.
Schulze C (1987) Anomalien und Mißbildungen der menschlichen Zähne. Quintessenz, Berlin.
Schumacher GH, Schmidt H, Börnig H, Richter W (1990) Anatomie und Biochemie der Zähne. Fischer, Stuttgart.
Scott GR (1973) Dental Morphology: A Genetic Study of American White Families and Variation in Living Southwest Indians. Ph D Diss, Arizona State University, Tempe.
Scott GR (1980) Population variation of Carabelli's trait. Hum Biol: 52: 63–78.
Scott GR (1991) Dental anthropology. In: Dulbecco R (ed) Encyclopedia of Human Biology. Academic Press, San Diego.
Scott GR, Turner CG (1988) Dental anthropology. Ann Rev Anthropol 17: 99–126.
Scott GR, Turner CG (1997) The Anthropology of Modern Human Teeth. Cambridge Univ Press, Cambridge.

Shields ED, Bixler D, El-Kafrawy AM (1973) A proposed classification for heritable human dentine defects with a description of a new entitiy. Arch Oral Biol 18: 543–553.
Smith DW (1988) Recognizable Patterns of Human Malformations. Saunders, Philadelphia
Smith P, Tchernov E (1992) Structure, Function and Evolution of Teeth. Freund, London.
Sollich A (1974) Zur Aplasie bleibender Zähne unter Berücksichtigung ihrer Mikrosymptome. Med Diss, Berlin.
Stewart RE, Prescott GH 81976) Oral Facial Genetics. Mosby, Saint Louis.
Sumiya Y (1959) Statistic study on dental anomalies in the Japanese. J Anthrop Soc Nippon 67: 215–233.
Tammoscheit UG (1971) Zur Ätiologie und Pathogenese der sogenannten Angle-Klasse II, 1. Habil-Schrift, Berlin.
Tannenbaum KA, Alling EE (1963) Anomalous tooth development. Case reports of gemination and twinning. Oral Surg 16: 883–887.
Taylor RMS (1978) Variation in Morphology of Teeth. Thomas, Springfield.
Terwee PJ (1922) Die Vererbung des Diastemas. Med Diss, Bonn.
Thoma KH, Goldman HM (1960) Oral Pathology. Mosby, St. Louis.
Townsend GC, Richards LC, Brown T, Burgess VB, Travan GR, Rogers JR (1992) Genetic studies of dental morphology in south Australian twins. In: Smith P, Tchernov E (eds) Structure, Function, and Evolution of Teeth. Freund, London, p. 501–518.
Turkel SJ (1989) Congenital abnormalities in skeletal populations. In: Işcan MY, Kennedy KAR (eds) Reconstruction of Life From the Skeleton. Liss, New York, p. 109–127.
Turner CG (1987) Late pleistocene and holocene population history of east Asia based on dental variation. Am J Phys Anthrop 73: 305–321.
Turner CG, Nichol CR, Scott GR (1991) Scoring procedures for key morphological traits of the permanent dentition: The Arizona State University dental anthropological system. In: Kelley MA, Larsen CS (eds) Advances in Dental Anthropology. Wiley-Liss, New York, p. 13–31.
Vastardis H, Karimbux N, Guthua SW, Seidman JG, Seidman CE (1996) A human *MSX1* homeodomain missense mutation causes selective tooth agenesis. Nature Genetics 15: 417–421.
Visser JB (1948) Beitrag zur Kenntnis der menschlichen Zahnwurzelformen. Med Diss, Zürich.
Werner C (1979) Sippenuntersuchungen zur Frage der Vererbung der Progenie. Med Diss, Berlin.
Witkop CJ, Sauk JJ (1976) Heritable defects of enamel. In: Stewart RE, Prescott GH (eds) Oral Facial Genetics. Mosby, St. Louis, p. 151–226.
Witkop CJ (1989) Amelogenesis imperfecta, dentinogenesis imperfecta, and dentin dysplasia revisited: Problems in classification. J Oral Pathol 17: 547–553.
Zwemer TJ (1993) Boucher's Clinical Dental Terminology. Mosby, St. Louis.

3.3 Micromorphological Features of Human Dental Enamel

Ralf J. Radlanski

Introduction

Dental enamel is the hardest tissue produced by the human body. For this reason teeth are usually well preserved in fossil and prehistoric material. Enamel consists of 95% inorganic substance (calcium and phosphate in the form of hydroxyapatite), only 1% organic substance and ca. 4% water, and is the product of secretory cells (ameloblasts). Its structure can only be understood when the developmental processes active during tooth development are described.

Embryological Basis of Enamel Structure

Dental development begins with a thickening and local invagination of the oral epithelium in an embryo of about 6 weeks of age. According to the number of teeth to be developed there arise discrete swellings which change their form until a bell stage is reached (Figs. 1, 2). This epithelial tooth bell embraces the underlying mesenchyme, the dental papilla, in which the outer cells differentiate to form dentine and the underlying dental pulp. The thin, cuffed margin of the tooth bell is called *Hertwig*'s epithelial sheath. Here a rapid increase in cell number is realized by frequent mitosis. Thus the primordium grows and, by downgrowth of *Hertwig*'s sheath, the roots are formed. After initial dentin formation, the cells of the inner vault of the bell-stage primordium differentiate into secretory ameloblasts. At their secretory end the ameloblasts bear a pyramidal protrusion, the process of *Tomes*. They produce a matrix rich in glycoproteins, from which water and organic substances are re-resorbed by the ameloblasts and calcium and phosphate are incorporated to form hydroxyapatite crystals. Fig. 4 shows impressions the processes leave at the fetal enamel surface, and Fig. 3 depicts a simplified, schematic view of the orientation of the crystals

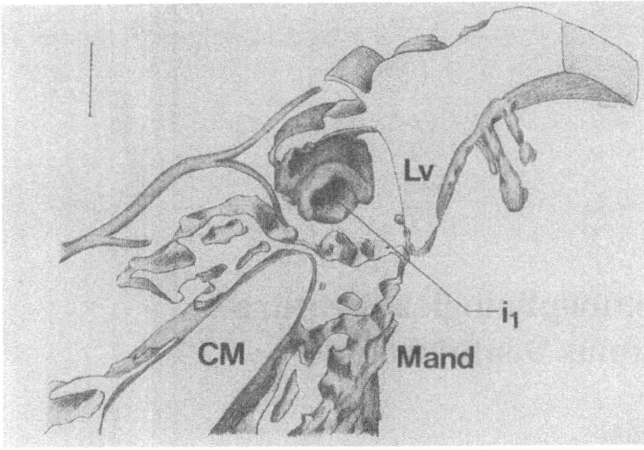

Fig. 1. Epithelial primordium of a lower right central deciduous tooth of a human fetus of 64 mm CRL (Crown-rump-length). Graphical reconstruction from a histological serial section to show the spatial arrangement of the primordium (i_1) with selected surrounding structures such as vestibular lamina (Lv), Meckel's cartilage (CM) and some vessels. View from the right aspect, scale bar: 250 μm.

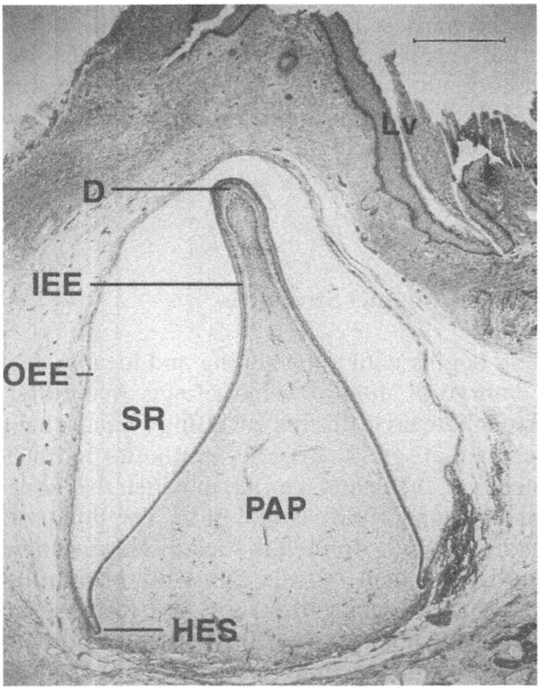

Fig. 2. Sagittal section through the incisor primordium in a late bell stage of a human fetus of 250 mm CRL. The formation of dentine and enamel has begun. IEE: inner enamel epithelium, OEE: outer enamel epithelium, SR: reticular stratum, PAP: mesenchymal papilla, **d:** dentine with a thin layer of enamel, HES: Hertwig's epithelial sheath, Lv: Vestibular lamina. Scale bar: 500 μm.

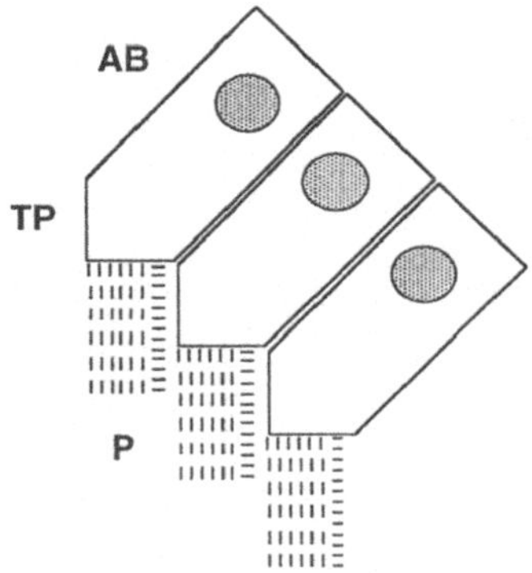

Fig. 3. Schematic illustration of ameloblasts (AB) and the orientation of the hydroxylapatite crystallites. Ameloblasts are characterized by a Tomes process (TP) that bears a secretory and a sliding surface which influence the orientation of crystallites. The prism structure of human enamel is the result of the pattern of crystallite orientation. P: Prism.

in relation to the form of the secretory end of the ameloblasts. More detailed knowledge concerning this topic can be found in Warshawsky et al. (1981), Suga (1983), Boyde (1989), and Radlanski (1993).

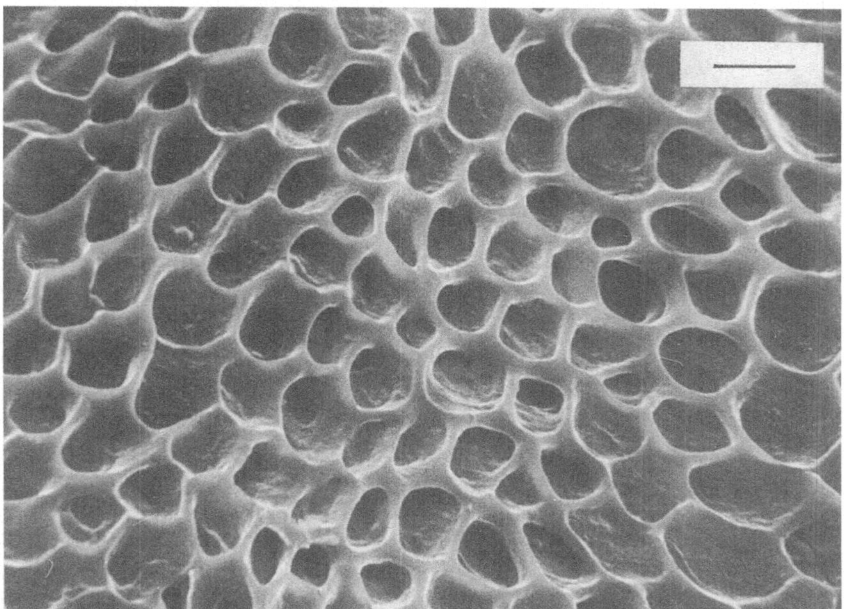

Fig. 4. Scanning electron microscopical reproduction of the fetal enamel matrix surface, and after the ameloblast epithelium has been removed, the impressions of the Tomes processes with their variable forms are revealed. Scale bar: 4 μm.

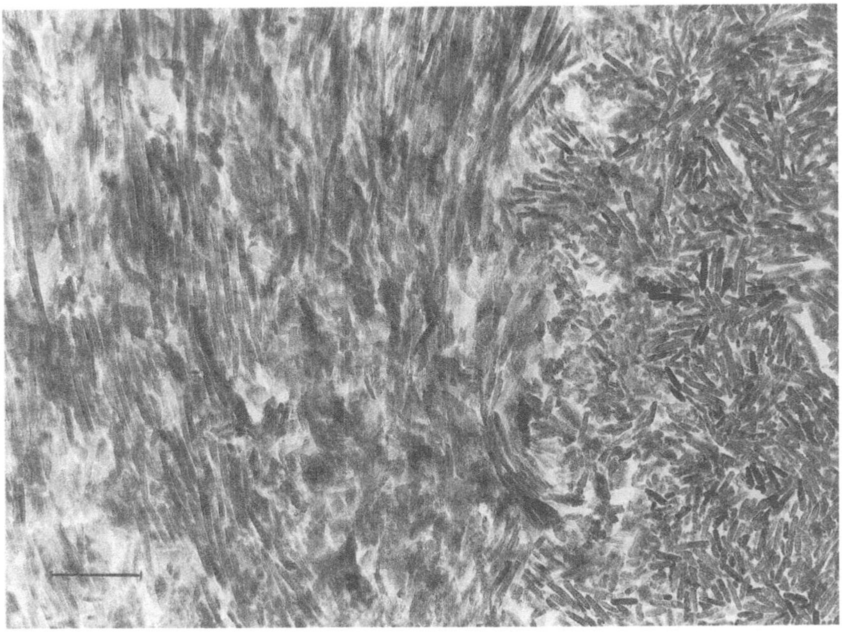

Fig. 5. Transmission electron microscopical representation of an ultrathin section to show the hydroxylapatite crystallites in human enamel. The different crystallite orientation pattern is responsible for the metastructure made out of prisms. Scale bar: 0.1 μm.

Prisms as a Metastructure of Human Enamel

Due to the alternating orientation of the hydroxyapatite crystals (Figs. 5, 6 a) a „metastructure" of enamel can be found, which is usually described as enamel prisms (Schumacher et al. 1990; Schroeder 1992). A clear interdependence exists between the enamel prisms and the ameloblasts from which they were formed. A quarter of the surfaces of 4 ameloblasts are required to form one complete prism (Boyde 1989; Suga 1983; Wakita and Kobayashi 1983).

Outlines of the impressions of the secretory processes are very variable, as are (therefore) outlines of the prisms in ground enamel (Holtgrave and Hennes 1995) and at the fetal enamel surface (Quast et al. 1995; Radlanski 1997). However, prisms of horseshoe and keyhole outlines are predominantly described (Schroeder 1992).

It can be roughly stated that prisms start at the dentino-enamel junction and run centrifugally towards the enamel mantle surface without interruption. Their path is straight in some regions, while in other regions they show a wavy, groupwise interwoven, undulated pattern where they deviate back and forth (Fig. 9 a). In ground sections, prisms are cut either longitudinally or transversally, leading to different optical phenomena (Fig. 9 b). Prisms cut transversally appear darker and are called diazonia, and regions in which prisms are cut longitudinally are called parazonia (Preiswerk 1895). These alternating dark and light regions (Fig. 8 a) are

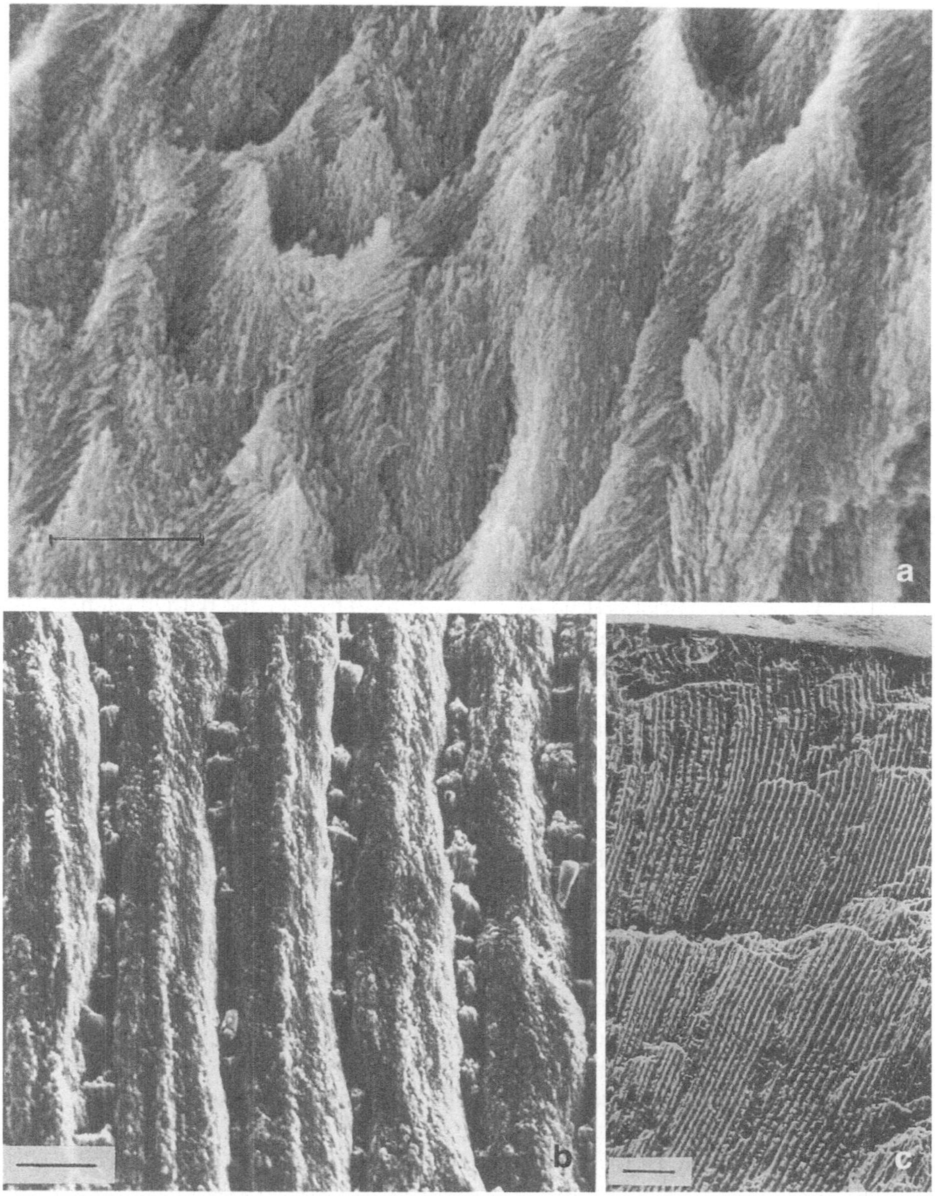

Fig. 6 a–c. a: Scanning electron microscopical reproduction of a ground and acid-etched surface within the enamel mantle. This specimen shows the composition of prisms made out of crystallite needles. Scale bar: 5 µm; **b:** Scanning electron microscopical reproduction of a fractured specimen of human enamel to show the enamel prisms. At certain distances the prisms reveal nodular swellings and constrictions, although their general diameter remains constant while passing from the dentino-enamel junction to the enamel surface. Between the prisms there (can be seen) remnants of fractured adjacent prisms. Scale bar: 4 µm; **c:** Scanning electron microscopical reproduction of a larger area of fractured human enamel. At the upper margin the enamel surface is reached. Scale bar: 40 µm.

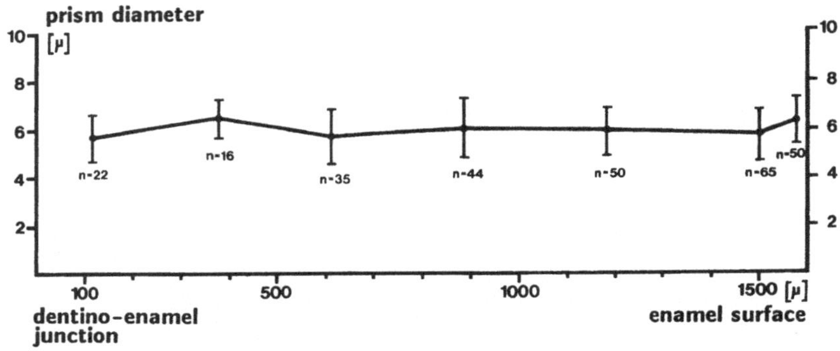

Fig. 7. The diameter of the prisms remains constant from the dentino-enamel junction towards the enamel surface.

usually called *Hunter-Schreger*-Bands (Kawai 1953; Beyer 1983; Schumacher et al. 1990).

Due to the fact that prisms are the product of living secretory cells, any alterations of their metabolic changes lead to nodular swellings and constrictions of the thickness of the prisms (Fig. 6 b). It is difficult to trace the path of each single prism, because the prisms are relatively small subunits of the enamel mantle. With the human enamel mantle being up to 4 mm thick while prism diameters range from 5–7 μm, a prism can be 660 times longer than it is thick.

There have been attempts to gain insight into prism orientation within the enamel mantle by reconstruction of their paths obtained from serial ground sections (Süss 1939; Helmcke 1967). For technical reasons, however, only a limited number of sections could be followed. Osborn (1967) managed to follow the path of prisms for at least a distance of 100 μm with his elegant focus-through technique.

Applying scanning electron microscopy to fractured and ground preparations of enamel (Fig. 6 c) it was possible to follow prisms in larger regions and for a greater distance (Radlanski et al. 1995). The longest path of a single prism that we were able to follow was 500 μm. Measurements of the diameter of prisms (Fig. 7) in different enamel regions from the dentino-enamel junction to the enamel surface were found to be constant (Radlanski et al. 1986). The question of how the increasing volume of the enamel mantle is created with prisms characterized by a constant diameter will be described after the following paragraph.

Incremental Lines of the Enamel Mantle and Enamel Surface

In addition to other structural features polished ground sections reveal incremental lines of enamel (Fig. 8 a). Incremental lines are the result of the growth pattern of the enamel organ in its bell stage, which is gradual rather than continuous. The growth rate varies between 20 and 28 μm per day (Deutsch 1982; Risnes 1986) in deciduous teeth. Phases of more accentuated rest lead to structural changes in the structure of the prisms, visible as lines of Retzius in ground sections (Fig. 8 a). When enamel is fractured, incremental layers can be made visible as terrace-like structures

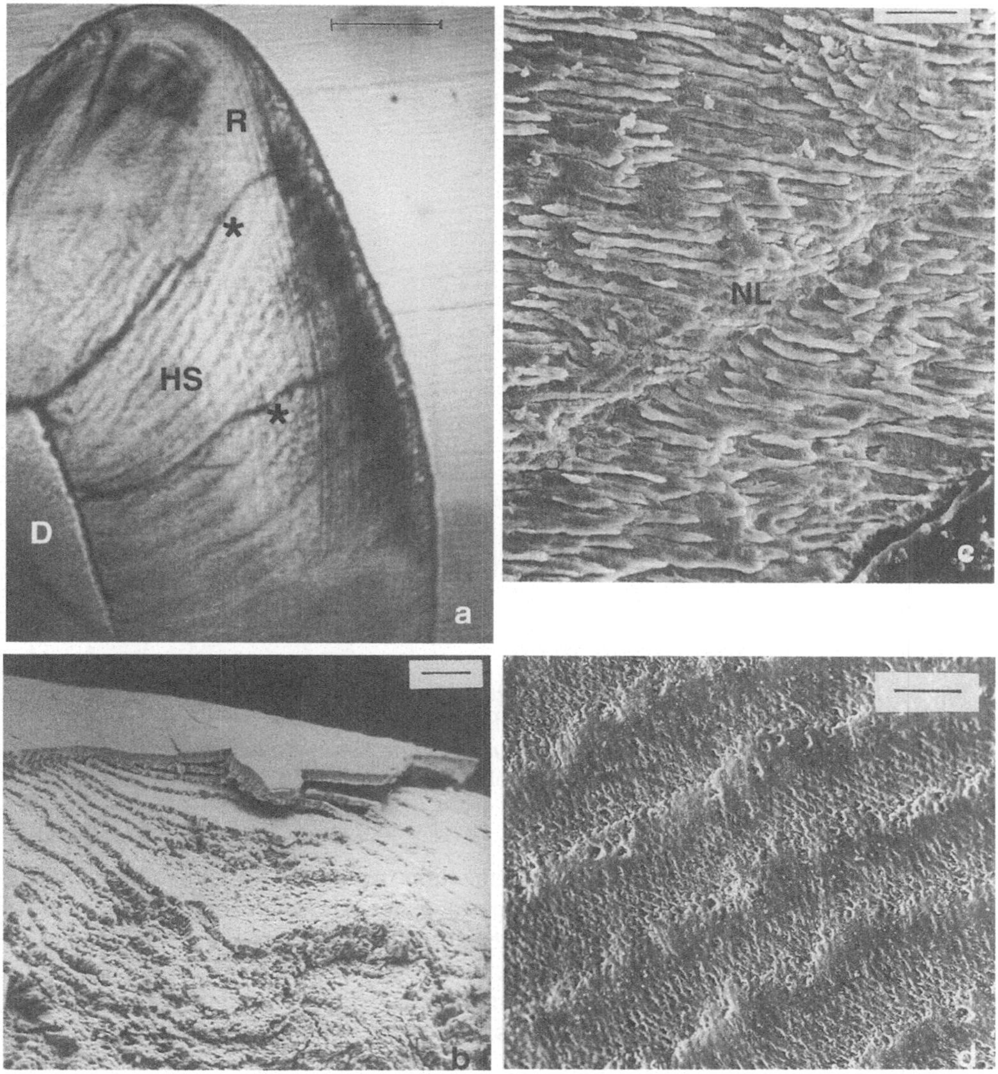

Fig. 8 a–d. a: Ground section of a human molar cusp to visualize the incremental lines (Lines of *Retzius*, R) and the *Hunter-Schreger*-Bands (HS). D: Dentine. The extent of the enamel cracks (*) is artificial. Scale bar: 1 mm; **b:** Scanning electron microscopical reproduction of a fractured specimen of human enamel to show the incremental layers in increasing depths of the enamel mantle. Top: natural enamel surface. Scale bar: 200 µm; **c:** Scanning electron microscopical reproduction of a ground and acid-etched plane below the enamel surface. At NL there is shown a special incremental line, the neonatal line, where the prism's paths deviate due to the nutritional changes after birth. Scale bar: 20 µm; **d:** Scanning electron microscopical reproduction of the enamel surface showing perikymates, where the incremental layers reach the surface. Scale bar: 40 µm.

(Fig. 8 b). A more marked arrested increment line is the neonatal line (Whittaker and Richards 1978), which results from metabolic changes in connection with

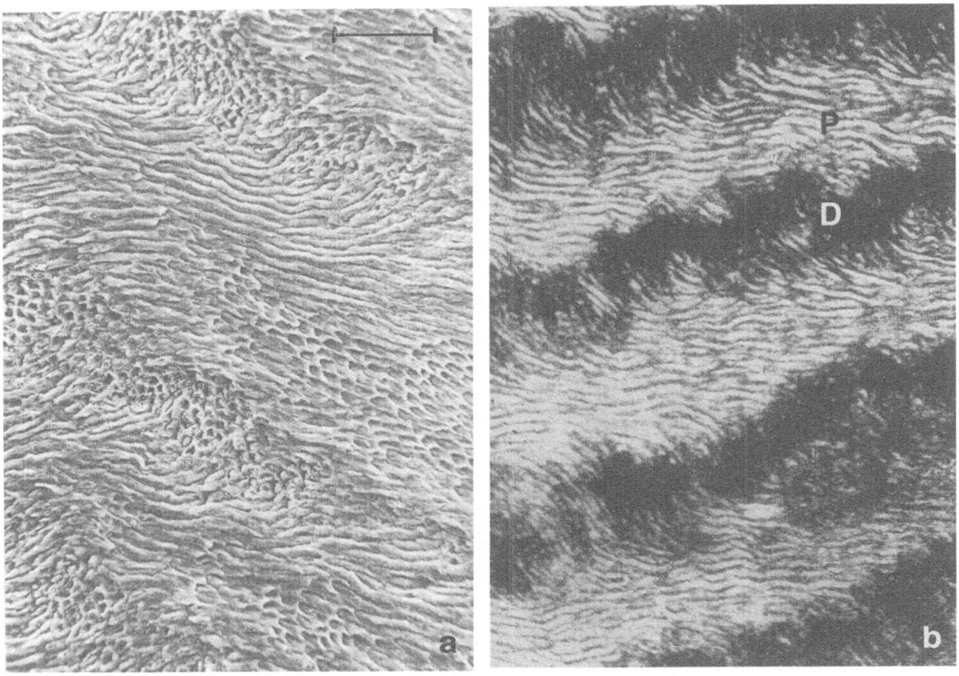

Fig. 9 a–b. a: Scanning electron microscopical reproduction of a ground and acid-etched plane showing the alternating and interwoven path of enamel prisms. In ground sections (Fig. 8 a, 9 b) this leads to the optical phenomenon of *Hunter-Schreger*-bands and Para- and Diazonia. Scale bar: 50 µm; **b:** Ground section of enamel with Parazonia (P) and Diazonia (D). From Meyer (1932).

perinatal nutrition (Fig. 8 c) when mineralization may cease for some days. This line can be found in primary teeth and in permanent teeth which are mineralized at birth (first molars and incisors). When the incremental layers reach the enamel surface they form the perikymates at the circumference of the tooth (Fig. 8 d).

Although most prisms reach the enamel surface, there are areas of prismless enamel, where crystallite orientation does not lead to a prismatic enamel structure (Fig. 10). This, however, is confined only to the outer 10µm of the enamel (Schroeder 1992).

Prism Orientation in Enamel

Prism Diameter

Most descriptions of human dental enamel rely on an increase of prism diameter to explain the increased volume of the enamel mantle (see Fig. 11). Other descriptions proposed artifacts of the 2-dimensional histology of ground sections (Fig. 12), assuming ramified Mummery (1916, 1924) or additional prisms (Dewey 1914; Lewis and Stöhr 1914; Andrews 1919; Broomell and Fischelis 1922;

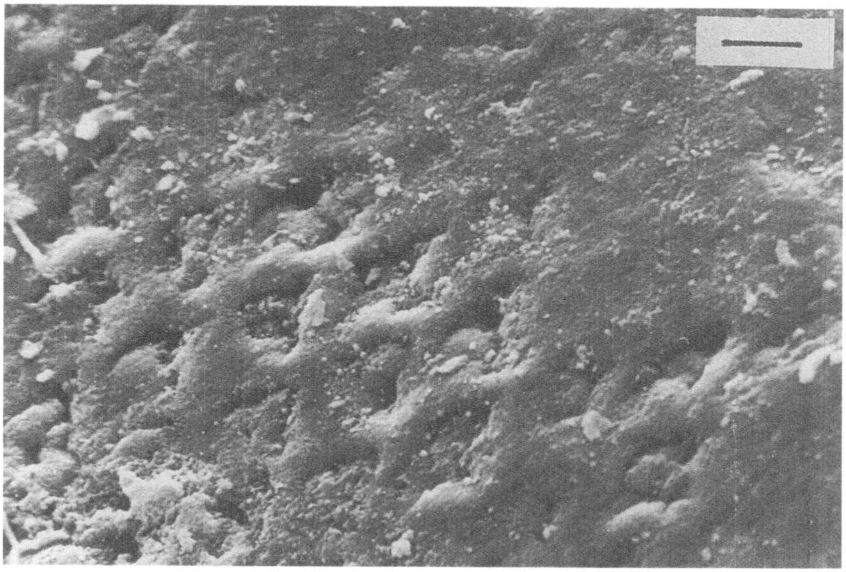

Fig 10. Scanning electron microscopical reproduction of the enamel surface. In the left and lower half the prismatic structure reaches up to the surface where impressions of the *Tomes*' processes can be seen, while in the upper and right half the enamel surface remains prismless.

Hopewell-Smith 1926). Only a few authors believed in the constancy of prism diameter (Noyes and Thomas 1921; Wolf 1942; Quigley 1959; Schumacher et al. 1990). These authors were well aware that a specific arrangement of the prisms must be responsible for creating the form of the enamel mantle. An example is Wolf's *Umreihungstheorie* (1942).

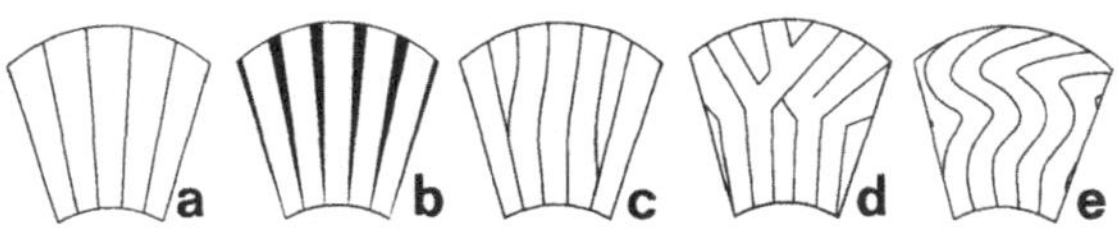

Fig. 11 a–e. Assumptions about the prism arrangement within the enamel in schematic representations: **a:** The prism diameter increases on its way to the enamel surface and thus creates the volume of the enamel mantle (Pickerill 1913; Williams 1923; Chase 1924; Meyer 1932; Yosida 1938; Blechschmidt 1942; Heuser 1956; Fosse 1968; Swancar et al. 1970; Skobe and Stern 1980); **b:** In addition to the increase in prism diameter a so-called „interprismatic substance" increases towards the periphery (Meyer 1932; Shobusawa 1952). **c:** Towards the enamel surface there arise new, additional prisms in order to fill the volume of the enamel mantle (Dewey 1914; Lewis and Stöhr 1914; Andrews 1919; Broomell and Fischelis 1922; Hopewell-Smith 1926); **d:** Ramified prisms are described by Mummery (1916, 1924); **e:** The diameter of the prisms is constant, the surface increment of the enamel mantle is created by the angulation of the prisms, towards the surface, and in addition, the volume of the enamel mantle is created by the meandering path of the prisms within the enamel mantle (Radlanski et al. 1986; Radlanski et al. 1989; 1990).

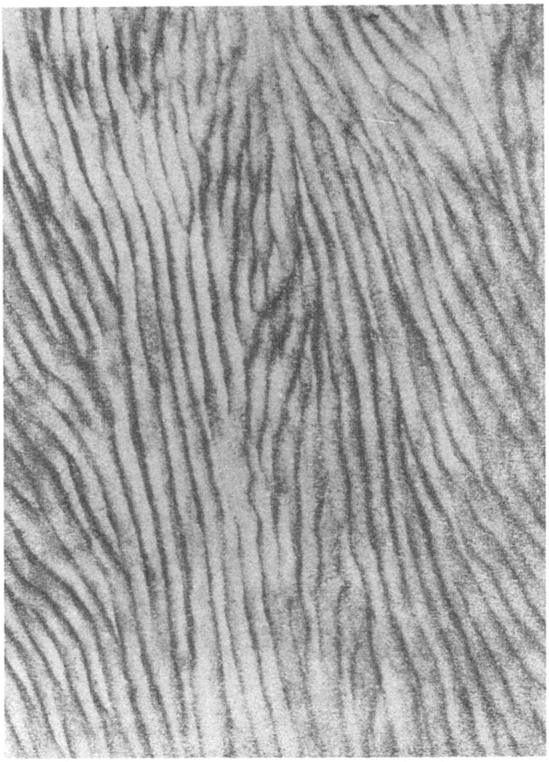

Fig. 12. A ground section of human enamel shows artifacs like „ramified“ and „additional“ prisms. From Meyer (1932).

Angulation of the Prisms at the Enamel Surface

It is well known that prisms only rarely reach the enamel surface at a perpendicular angle (Noyes and Thomas 1921; Wolf 1942; Osborn 1968; Skobe and Stern 1980). Black (1924) had this in mind when describing the design of cavity preparation for

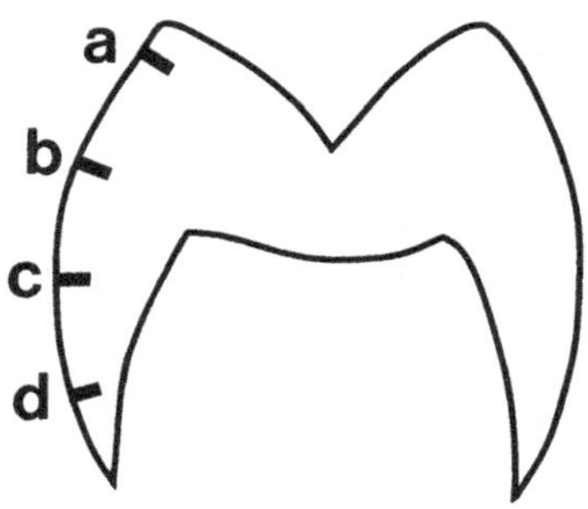

Fig. 13. The outline of the outer enamel surface is longer than that of the inner surface at the dentino-enamel junction. In addition, the localisation of the sections of Figs. 14 a–d is given.

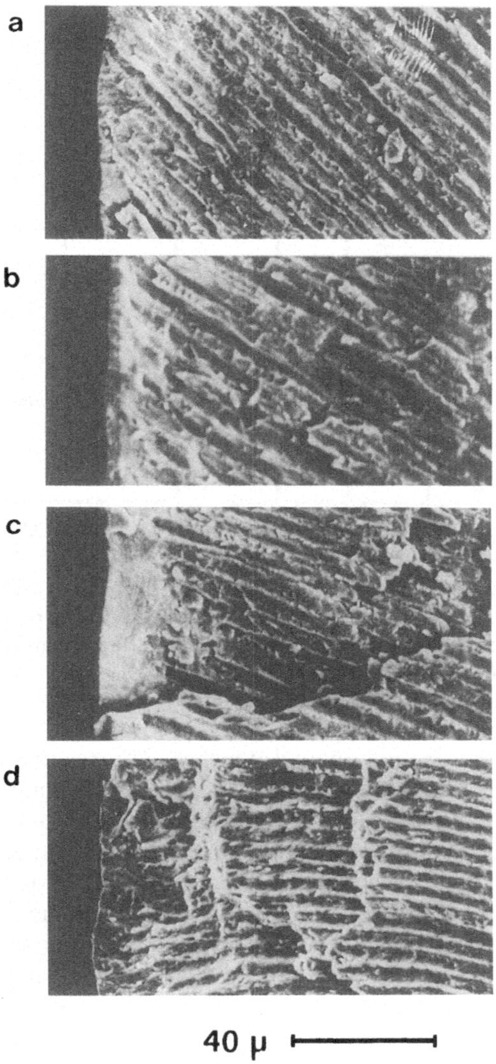

Fig. 14 a–d. Scanning electron microscopical reproductions of fracture preparations show that the angulation that prisms obtain towards the surface is variable in different regions (location of sections see Fig. 13).

dental fillings. In cervical regions prisms tend to reach the surface mostly perpendicularly, while their angulation increases towards the occlusal region (Figs. 13, 14). The result is that the thicker the enamel, the more oblique the prisms which reach the surface (Fig. 15).

The angulation that a prism obtains with the enamel surface leads to an increase in its *effective* diameter. We found factors between 1 (perpendicular ending) and 2.92 (angulation at 70°), the mean factor being 1.3 (Radlanski et al. 1989). This would lead to a mean increase in the contour-line length at the enamel surface of

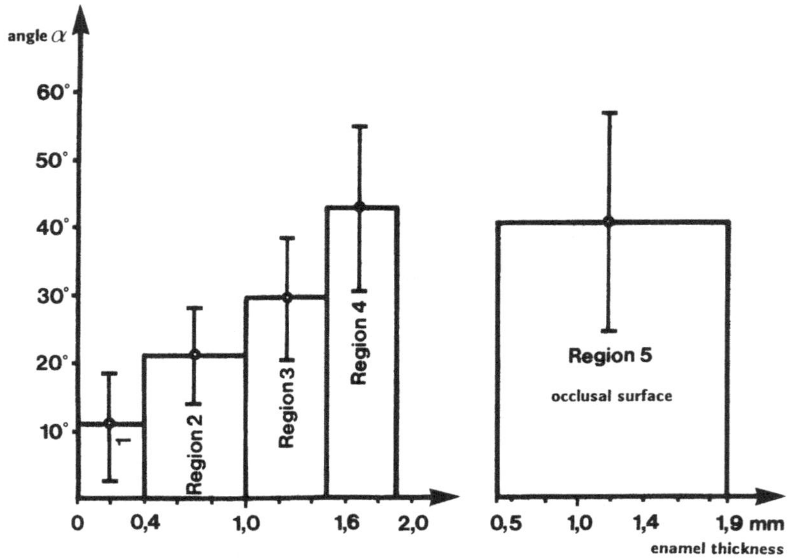

Fig. 15. The interdependence between the thickness of the enamel and the corresponding angulations of the prisms, represented by the angle a. As the occlusal region is cut by several fissures, the angle-values show a wider distribution. Therefore, the mean value for a in the occlusal region 5 is smaller than that of region 4.

30% as compared to the dentino-enamel junction. This result comes close to the geometrical conditions found in premolars (Hugel 1970; Schaaf 1971).

It is interesting to re-evaluate the measurements that e.g. Chase (1924) obtained from ground sections. He published values of prism diameters at the dentino-enamel junction and at the enamel surface, but neglected the angulation of prisms at the enamel surface. These light-microscopic findings led him to the conclusion that the prism diameter increased. We, however, assume a constant prism diameter, and if we use his increment factors to calculate the angulations that prisms must have obtained, we reach values ranging between 18° and 54°, which sufficiently match our findings (Radlanski et al. 1989). Chase (1924) interpreted his findings as real thickenings of the prisms. Although he mentions a sometimes oblique path of the prisms, Chase did not believe that this would be responsible for surface enlargement. On the contrary, our findings show that the prisms contribute to the enlargement of the enamel surface almost threefold, and as a mean value, depending on the region, at least one and one-half times the actual diameter. Therefore the angulation of prisms at the surface contributes an important, if not the decisive contribution to enlargement of the surface of the enamel mantle.

Volume Increase

The answer to the question of whether prisms with constant diameters can completely fill the enamel mantle requires a description of the specific arrangement

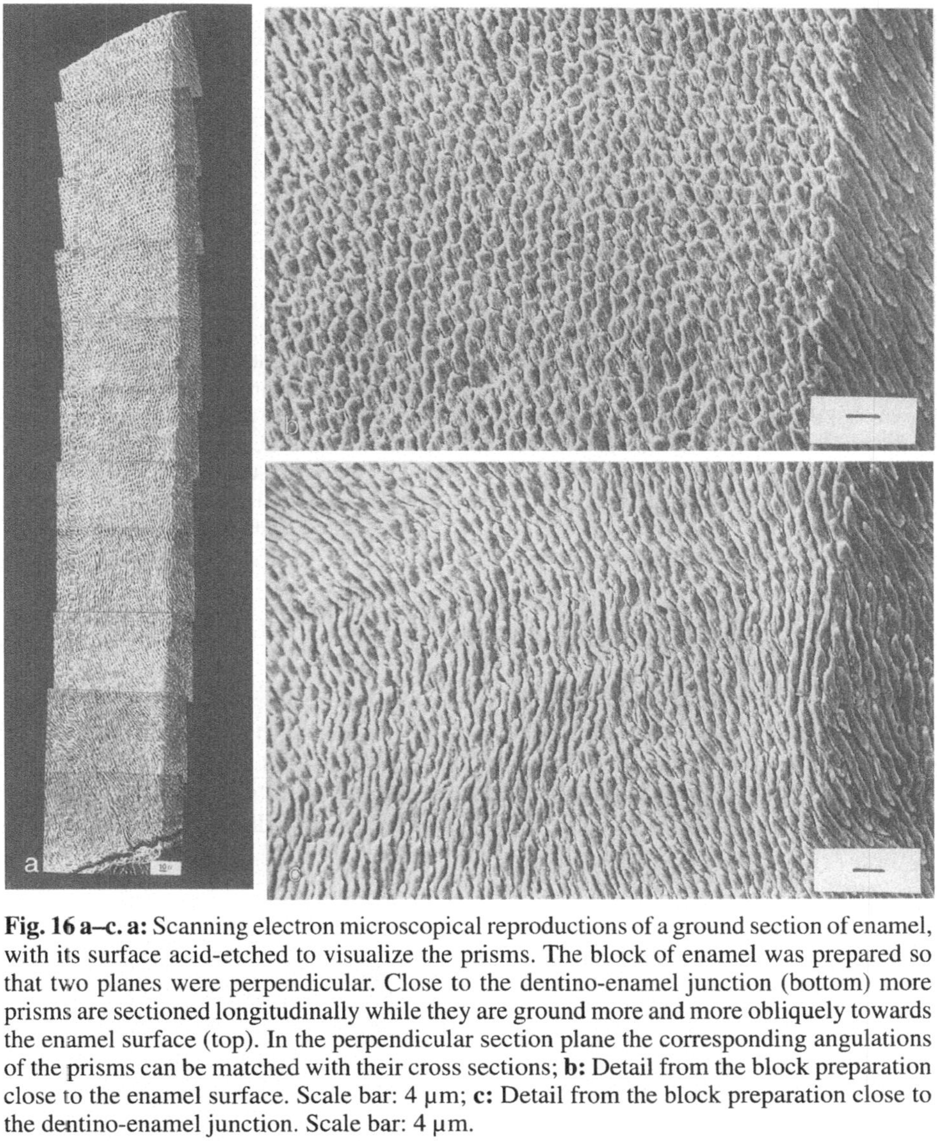

Fig. 16 a–c. a: Scanning electron microscopical reproductions of a ground section of enamel, with its surface acid-etched to visualize the prisms. The block of enamel was prepared so that two planes were perpendicular. Close to the dentino-enamel junction (bottom) more prisms are sectioned longitudinally while they are ground more and more obliquely towards the enamel surface (top). In the perpendicular section plane the corresponding angulations of the prisms can be matched with their cross sections; **b:** Detail from the block preparation close to the enamel surface. Scale bar: 4 μm; **c:** Detail from the block preparation close to the dentino-enamel junction. Scale bar: 4 μm.

of the prisms characterized by an increasing deviation from the perpendicular path (Figs. 16 a–c). From these findings a schematic, pattern-like diagram can be developed (Fig. 17). The interdependence between enamel thickness and prism orientation found at the enamel surface is transferred into the inner bulk of enamel. Then the angulation of a prism can be seen as a function of its distance from the dentino-enamel junction (Radlanski et al. 1990, 1995). Although with this model the contradiction between constant prism diameter and volume increase of the enamel mantle can be explained, it must be stated that regional compensation of different prism orientation is possible and meets the micromorphological findings.

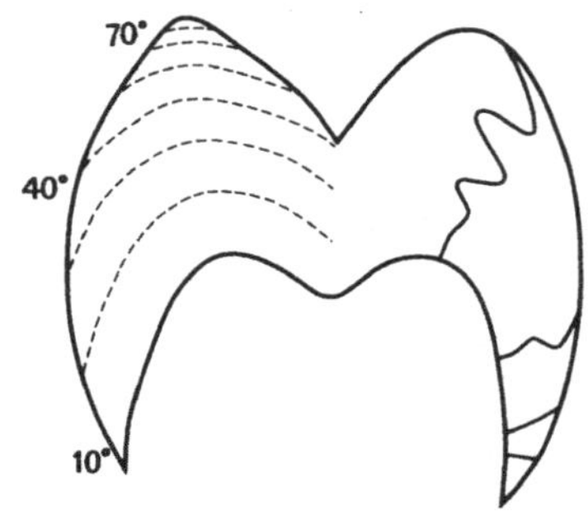

Fig. 17. Schematic, pattern-like diagram of the path of the prisms in the right half of the figure: In the cervial region the prisms run mostly straight and reach the enamel surface in a perpendicular orientation. In the cuspal region the prisms obtain a more wavy path and they reach the surface at a more acute angle. In the left half of the diagram the angulations of the prisms, as they are found at the surface, are continued into the inner part of the enamel mantle. According to this model an almost constant angulation of the prisms at the enamel surface as well as in the inside of thicker enamel should be expected to be found in a constant distance from the dentino-enamel junction along the dotted lines (isogonial lines). The increment of the angulation of the prisms is not linear.

Artifacts on Ground Sections and Surfaces

Whenever a plane is sectioned through enamel, one is never sure of the angle under which the prisms are cut. This is due to the impossibility of predicting their orientation at a particular spot. Any oblique section distorts the appearance of the outline of the prisms (Figs. 18 a, b). Therefore measurements of prism diameter in ground sections can lead to false results (Fosse 1968). In addition, it must be stated

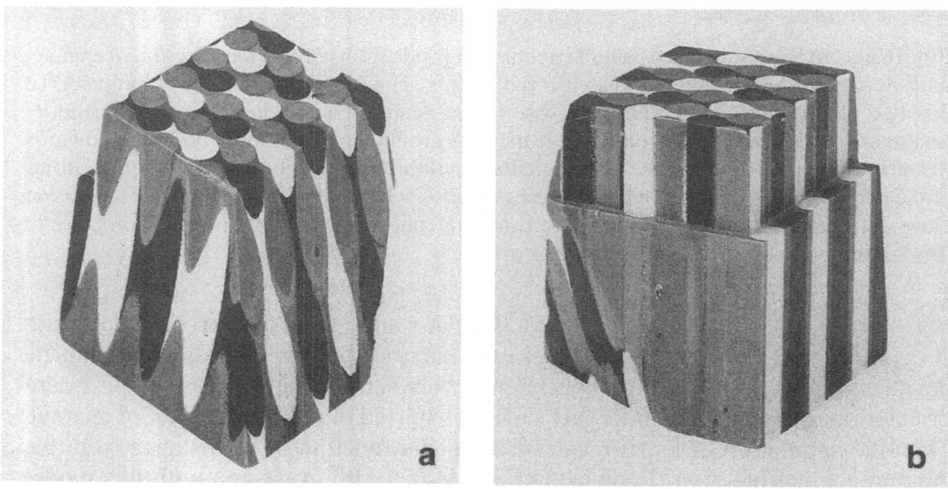

Fig. 18 a–b. Photographic reproduction of a plastic model of schematic enamel prisms. Sections are carried out at different planes with different orientation and location. The interdependence between the appearance of the prism outline and the section plane is obvious.

that evaluations of prism outline types are prone to be erroneous when ground sections are used. Preparations of enamel blocks with two perpendicular planes help to reduce the problem of misinterpretation.

Conclusions

Human enamel is a tool to understand the developmental mechanisms of tooth formation, because in mature enamel the prisms represent the unchanged pathways of the migration of the ameloblasts. In addition, enamel may serve for comparative anatomical research, focused on crown morphology as well as on its prism structure. As far as prism arrangement is concerned, however, the risk of artifacts in ground sections has to be kept in mind.

References

Andrews RR (1919) The development of the teeth and some of the contested points in regard to their development and structure. J Dental Res 1: 353–385.

Black GV (1924) Operative Dentistry. London, Kensington.

Blechschmidt E (1942) Die Schmelzarchitektur der menschlichen Zähne. Z Anat Entw-Gesch Berlin 112: 141–184.

Beier K (1983) Hunter-Schreger-Bänder im Zahnschmelz von Känguruhs (Macropodinae, Marsupialia). Zool Anz Jena 210: 315–332.

Boyde A (1989) Enamel. In: Berkovitz BKB, Boyde A, Frank RM, Höhling HJ, Moxham BJ, Nalbadian J, Tonge CH (eds) Teeth. Springer, Berlin Heidelberg New York, p. 309–473.

Broomell JN, Fischelis P (1922) Anatomy and Histology of the Mouth and Teeth. Blakiston, Philadelphia.

Chase SW (1924) The absence of supplementary prisms in human enamel. Anat Rec 28: 79–89.

Deutsch D (1982) Development of enamal in human fetal teeth. J Dent Res 61: 1543.

Dewey M (1914) The enamel and its relation to cavity preparation. Western Dental J Kansas City 28: 4–9.

Fosse G (1968) A quantitative analysis of the numerical density and distributional pattern of prisms and ameloblasts in dental enamel and tooth germs, V. The number of prisms per unit area in the outer surface of human permanent canines. Acta Odont Scand 26: 501–543.

Helmcke JG (1967) Ultrastructure of enamel. In: Miles AEW (ed) Structural and chemical organisation of teeth, Vol. II. Acad Press, New York, p. 135–144.

Heuser H (1956) Oberflächenhistologische Untersuchungen über die Größe der Schmelzprismen in den einzelnen Schmelzschichten am menschlichen Zahn. Dtsch Zahnärztl Z 11: 705–711.

Holtgrave EA, Hennes T (1995) A morphological investigation in the enamel structure of different deciduous teeth. In: Radlanski RJ, Renz H (eds) Proceedings of the 10th International Symposium on Dental Morphology. Brünne, Berlin, p. 216–221.

Hopewell-Smith A (1926) Concerning human enamel: Facts, explanations and applications. Dent Cosmos 68: 639–667.

Hugel R (1970) Untersuchungen über die Wandstärken von Schmelz und Dentin an oberen Frontzähnen und Prämolaren. Med Diss, Würzburg.

Kawai N (1953) Comparative anatomy of the bands of Schreger. Okajimas Folia Anat 27: 115–131.
Lewis FT, Stöhr D (1914) A Textbook of Histology Arranged upon an Embryological Basis. Blakiston, Philadelphia.
Meckel AH, Griebstein WJ, Neal RJ (1965) Structure of mature human dental enamel as observed by electron microscopy. Arch Oral Biol 10: 775–783.
Meyer W (1932) Normale Histologie und Entwicklungsgeschichte der Zähne des Menschen. Lehmanns, München.
Mummery JH (1916) On the structure and arrangement of the prisms, especially as shown in enamel of the elephant. Proc Roy Soc Med 9: 121–138.
Mummery JH (1924) The Microscopic and General Anatomy of the Teeth. Oxford Univ Press, London.
Noyes B, Thomas NG (1921) A Textbook of Dental Histology and Embryology, Including Laboratory Directions. Lea & Febiger, Philadelphia.
Osborn JB (1967) Three-dimensional reconstructions of enamel prisms. J Dent Res 46: 1412–1419.
Osborn JB (1968) Directions and interrelationships of prisms in cuspal and cervical enamal of human teeth. J Dent Res 47: 395–402.
Pickerill HP (1913) The structure of enamel. Dent Cosmos 55: 969–988.
Preiswerk G (1895) Beiträge zur Kenntnis der Schmelzarchitektur bei Säugern. Nat Diss, Basel.
Quast K, Renz H, Radlanski RJ (1995) Outline and distribution of prism forms at the fetal enamel surface. In: Radlanski RJ, Renz H (eds) Proceedings of the 10th International Symposium on Dental Morphology. Brünne, Berlin, p. 95–99.
Quigley MB (1959) Electron microscopy of developing enamel matrix in the Syrian hamster. J Dent Res 38: 180–187.
Radlanski RJ, Seidl W, Steding G (1986) Über das Dimensionsverhalten der Schmelzprismen in verschiedenen Tiefen und Regionen des Schmelzes menschlicher permanenter Zähne. Anat Anz Jena 162: 317–324.
Radlanski RJ, Seidl W, Steding G, Jäger A (1989) Über die Richtung der Schmelzprismen an der Schmelzoberfläche. Anat Anz Jena 168: 405–412.
Radlanski RJ, Seidl W, Steding G, Jäger A (1990) Über die Ausrichtung der Prismen im Zahnschmelz menschlicher permanenter Zähne. Anat Anz Jena 170: 329–337.
Radlanski RJ (1993) Contributions to the Development of Human Deciduous Tooth Primordia. Quintessence International, Chicago.
Radlanski RJ, Seidl W, Steding G (1995) Prism arrangement in human dental enamel. In: Moggi-Cecchi J (ed) Aspects of Dental Biology, Palaentology, Anthropology and Evolution. International Institute for the Study of Man, Florence, p. 33–49.
Radlanski RJ (1997) Morphogenetic and morphological aspects of prism structure in human dental enamel. Acta Med Dent Helv 2: 86–95.
Risnes S (1986) Enamel apposition rate and the prism periodicity in human teeth. Scand J Dent Res 94: 394–404.
Schaaf R (1971) Untersuchungen über die Ausmaße von Schmelz und Dentin an unteren Frontzähnen und Praemolaren. Med Diss, Würzburg.
Schroeder HE (1992) Orale Strukturbiologie. Entwicklungsgeschichte, Struktur und Funktion normaler Hart- und Weichgewebe der Mundhöhle und des Kiefergelenks. Thieme, Stuttgart.
Schumacher GH, Schmidt H, Börning H, Richter W (1990) Anatomie und Biochemie der Zähne. Volk und Gesundheit, Berlin.
Shobusawa M (1952) Vergleichende Untersuchungen über die Form der Schmelzprismen der Säugetiere. Okajimas Folia Anat 24: 371–392.
Skobe Z, Stern S (1980) The pathway of enamel rods at the base of cusps of human teeth. J Dent Res 59: 1026–1032.

Süss W (1939) Über die Architektur des Schmelzes. Zeitschr Zellforsch Mikr Anat Abt A Berlin 30: 171–193.
Suga S (ed) (1983) Mechanisms of Tooth Enamel Formation. Quintessence International, Tokyo.
Swancar JR, Scott DB, Njemirowskij Z (1970) Studies on the structure of human enamel by the replica method. J Dent Res 49: 1025–1033.
Wakita M, Kobayashi S (1983) The three dimensional structure of Tomes' processes and the development of the microstructural organization of tooth enamel. In: Suga S (ed) Mechanisms of Tooth Enamel Formation. Quintessence International, Tokyo.
Warshawsky K, Josephsen, K, Thylstrup A, Fejerskov O (1981). The development of enamel structure in rat incisors as compared to the teeth of monkey and man. Anat Rec 200: 371–378.
Williams JL (1923) Disputed points and unsolved problems in the normal and pathological histology of enamel. J Dent Res 5: 27–115.
Whittaker DK, Richards D (1978) Scanning electron microscopy of the neonatal line in human enamel. Arch Oral Biol 23: 45–50.
Wolf J (1942) Der Einfluß der Ameloblastenverschiebungen auf die Gestalt und den Verlauf der Schmelzprismen. Dtsch Zahn-Mund-Kieferheilkd Leipzig 9: 488–522.
Yosida K (1938) Vergleichende Untersuchungen über den Zahnschmelz. Kokub Gakkai Zass Tokyo 12: 1–17.

3.4 The Temporomandibular Joint

Aleš Obrez and Jens C. Türp

Introduction

The two temporomandibular joints (TMJs) form the bilateral articulation of the mandible with the cranium. Together with the neuromuscular system, the anatomy of the TMJ contributes to specific mandibular functions. Studying the TMJ exclusively on dry skulls may lead to erroneous conclusions regarding both mandibular function and joint biomechanics. It is for this reason that in addition to the bony components of the TMJ, the cartilage and soft tissues demand close attention.

Morphology and function of the temporomandibular joint are strongly related throughout development. Understanding the relationship between form and function may therefore help in reconstructing mandibular function, especially in cases where evidence is limited to skeletal remains. The purpose of this chapter is to introduce the reader to the anatomy of the temporomandibular joint, its growth and remodelling, and its function.

Anatomy

The temporomandibular joint (TMJ) is formed by the squamous temporal, tympanic, and mandibular bones (Fig. 1). It is the only diarthrodial joint in the craniofacial region (Fick 1904). The articular surfaces of the joint are separated from each other by an articular space, which is divided into upper and lower compartments by an articular disk, and from the surrounding tissues by an articular capsule. The joint cavities thus formed contain synovial fluid, making the TMJ a synovial joint.

The articular surfaces of the majority of mammalian synovial joints are covered by a hyaline cartilage. The exceptions are the TMJ and sterno-clavicular joints. The differences in histology between the TMJ and more typical joints has been attributed to the unique characteristics of the TMJ's phylogeny and ontogeny.

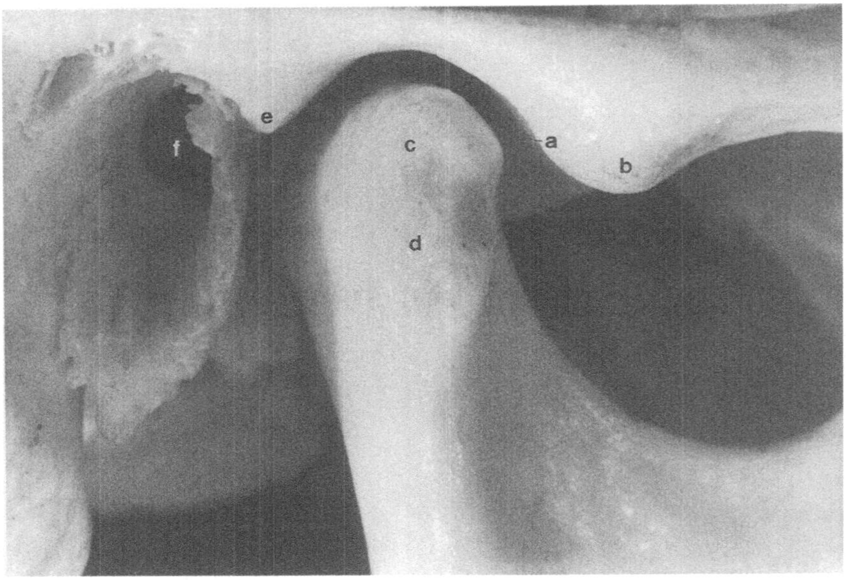

Fig. 1. Lateral view of the right side of a human skull with the components of the TMJ. **a:** Articular eminence; **b:** Articular tubercle; **c:** Condylar process; **d:** Condylar neck; **e:** Postglenoid process; **f:** External auditory meatus.

Mandibular Condyle

The condylar process of the mandible *(Processus condylaris mandibulae)* forms the mandibular part of the TMJ. The process is a postero-superior extension of the mandibular ramus. It consists of a head *(Caput mandibulae)* and a neck *(Collum mandibulae)* (Fig. 1). The position of the condyle relative to the temporal components of the TMJ is in part influenced by the relationships of the teeth.

The articulating areas of the condyle are the superior, anterior and medial surfaces of the condylar head. A significant intra- and interindividual variation in form and size of the condylar head and in the outline of the articular fossa and the articular eminence exists (Meyer 1865; Öberg et al. 1971; Costa 1986; Wendler et al. 1989; Koppe et al. 1993; Türp et al. 1995). The shape of the condylar head varies from rounded to flat at its superior surface (Yale et al. 1966; Lubsen et al. 1985; Dibbets and van der Weele 1991). Younger individuals have more rounded condylar heads than adults. It is important to acknowledge that the actual form of the condylar head depends on the thickness of the connective tissue layers covering it (Pullinger et al. 1993a, b).

The subchondral bone of the condylar process and the articular fossa is covered by three tissue layers (Fig. 2): (1) a superficial avascular fibrous connective tissue layer (modified bilaminar periosteum), which contains fibroblasts and collagen fibers; (2) a proliferative layer of undifferentiated mesenchymal cells and fibroblasts (zone of maturation); and (3) the hypertrophic layer of the cartilage

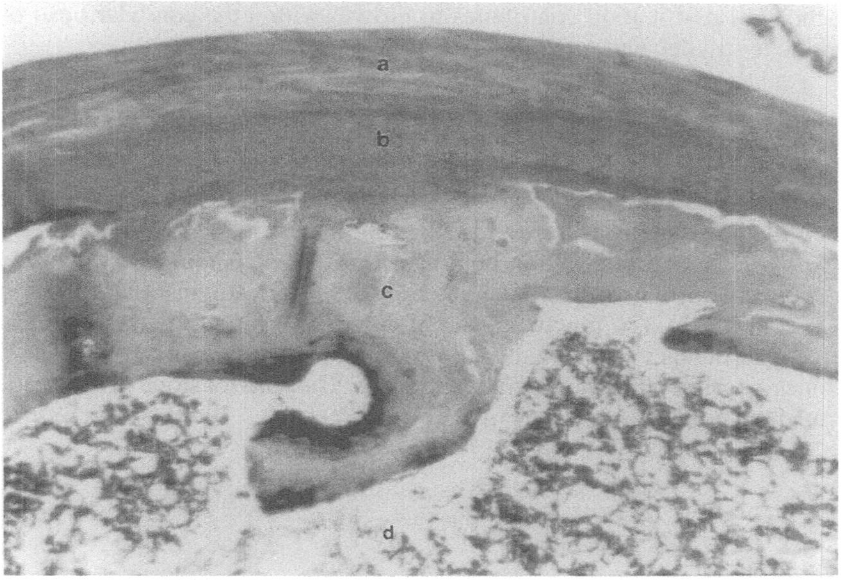

Fig. 2. The tissue layers of the condyle and the temporal components of the TMJ; **a:** Fibrous connective tissue layer; **b:** Proliferative layer; **c:** Hypertrophic layer; **d:** Subchondral bone (modified from Türp et al. (1997). Reprinted by permission of Quintessenz, Berlin).

containing chondrocytes and collagen fibers (Blackwood 1966; Carlsson and Öberg 1974; Wright and Moffett 1974; Thilander et al. 1976; Carlson et al. 1978; Luder et al. 1988). The periosteum covering the condylar neck divides into superficial and deep layers as it approaches the articular surface of the condylar head. While the superficial layer contributes to the articular capsule, articular disk and the periosteum of the temporal bone, the deeper layer forms the fibrous articular tissue of the condylar head. The latter is characterized by an abundance of collagen fibers (type I) oriented in the antero-posterior direction and parallel to each other (Luder et al. 1988). In addition, the articular tissue contains fibrocytes and elastic fibers. The proportion of individual components in this layer changes during growth as the number of cellular and vascular components reduces significantly (Thilander et al. 1976).

Underneath the fibrous layer covering the condylar head is a layer of undifferentiated mesenchymal cells (zone of maturation). These cells provide a source of cells for the underlying bone and/or secondary cartilage, evidently depending on the loads applied to the surface of the condylar head. Normal TMJ loading during growth in younger individuals results in chondrogenesis, while a reduction in loading predominant in older individuals induces osteogenesis (Carlson et al. 1978; Stutzman and Petrovic 1979, 1982; Silberman et al. 1987; Copray et al. 1989). As the chondrocytes eventually die (within the zone of endochondral ossification) the osteoblasts arising from the medullary part of the condylar head fill the newly formed empty spaces. The extent to which endochondral ossification contributes to growth within the condylar head of the mandible is still debated (see below) (Enlow 1994).

Glenoid Fossa and Articular Eminence

The glenoid fossa, or mandibular fossa *(Fossa mandibularis),* is part of the inferior surface of the squamosal portion of the temporal bone *(Pars squamosa ossis temporalis).* Though concave in both directions, this anatomical structure is of greater dimension in its width (mediolaterally) than in its length (anteroposteriorly) (Öberg et al. 1971). The functional part of the glenoid fossa, i.e., the region that is covered by articular tissue, is called the articular fossa (Hylander 1992). The fossa continues anteriorly into the articular eminence. The anterior slope of the eminence is termed the preglenoid plane (Fig. 3).

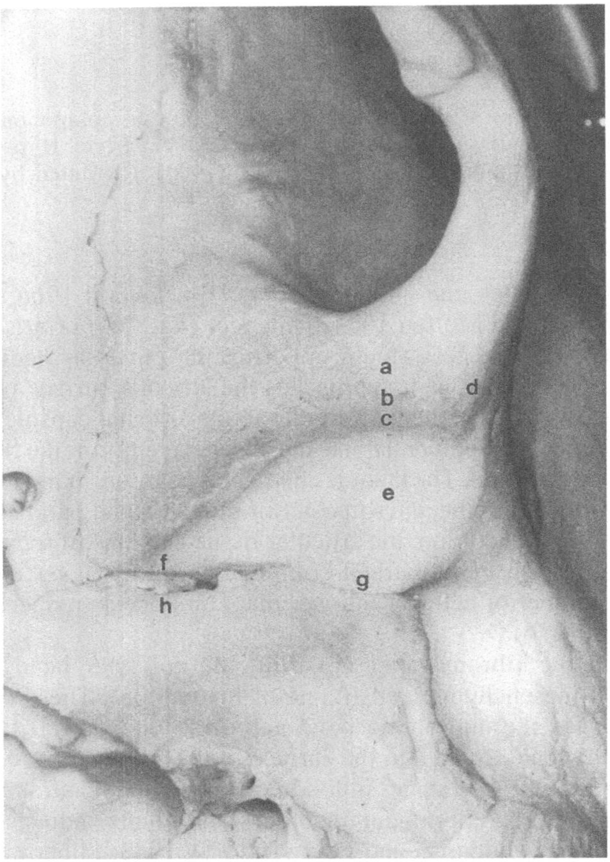

Fig. 3. Left side of the cranial base of a human skull; **a:** Preglenoid plane; **b:** Crest of the articular eminence; **c:** Articular eminence; **d:** Articular tubercle; **e:** Articular fossa; **f:** Petrosquamous fissure; **g:** Tympanosquamous fissure; H: Petrotympanic fissure.

As with the condylar head, there is significant variation in the form of the glenoid fossa and articular eminence in both males and females (Koppe 1993). This variation in form seems to be independent of the shape of the condylar head (Solberg et al. 1985). As in the case of the condylar head, the functional form of the temporal joint surfaces depends not only on the contour of the bony foundation but also on the thickness of the connective tissue layers covering it (Pullinger et al. 1993a, b). For example, the layers are thickest on the lateral and posterior parts of the articular eminence (Hansson et al. 1977).

The bony roof of the glenoid fossa is very thin. As it continues posteriorly, the glenoid fossa is traversed by the tympanosquamosal fissure *(Fissura tympanosquamosa)* (does not always exist as a separate entity) which separates medially into the petrosquamosal fissure *(Fissura petrosquamosa)* (medio-ventrally), and the petrotympanic fissure *(Fissura petrotympanica)* (medio-dorsally) (Fig. 3) (Dauber 1987). These fissures divide the glenoid fossa into a larger anterior and a smaller posterior part. The former continues anteriorly without a sharp border into the articular eminence, the preglenoid plane and the infratemporal fossa. Both the glenoid fossa and the eminence contribute to the articulating surface *(Facies articularis)* of the TMJ. When viewed in a parasagittal plane they form an S-shaped profile. The steepness of the articular eminence develops as mandibular function increases during human development. At birth, the eminence is flat (Wright and Moffett 1974). However, as development progresses it becomes more prominent. Similarly, with loss of the teeth the eminence flattens (Granados 1979).

The small tubercle lateral to the articular eminence, not part of the articular surface, is called the articular tubercle (DuBrul 1988; Hylander 1992; Obrez 1993; van Rensburg 1995) (Figs. 1 and 3). The official anatomic nomenclature (Nomina Anatomica 1989) does not make a distinction between the two and calls the entire complex the articular tubercle *(Tuberculum articulare)*. The articular tubercle is an attachment site of the lateral ligament of the TMJ (DuBrul 1988).

The posterior part of the articular fossa supports loose retroarticular tissue composed of connective tissue, fat, veins, and nerves (Zenker 1956; Dauber 1987). Between the mandibular fossa and the tympanic plate of the external auditory meatus *(Meatus acusticus externus)* is a bony ridge (posterior articular lip of the postglenoid process) that extends caudally (Fig. 4). This is the site of attachment of the articular capsule. In many individuals this ridge is wider laterally. In that case the bony element is called a postglenoid tubercle or postglenoid (retroarticular) process (Sicher 1949; DuBrul 1988; Hylander 1992). Compared to other primate TMJs, the human postglenoid process is small (Carlson et al. 1980; Hinton 1981, 1983; Henke and Rothe 1994).

Associated Connective Tissues

Articular Disk

In healthy TMJs the articular disk *(Discus articularis)* is interposed between the condylar head and the articular eminence. The disk divides the joint cavity into

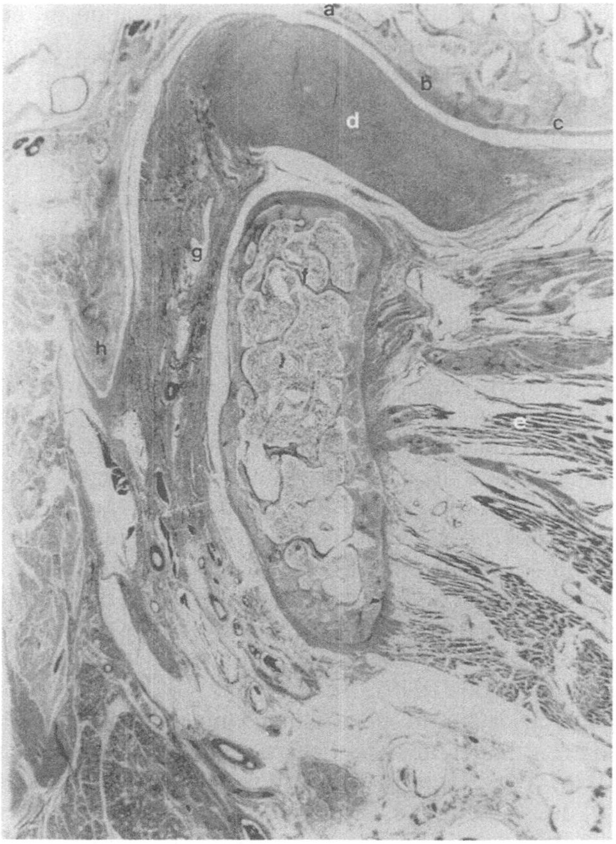

Fig. 4. Parasagittal histological section through a right human temporomandibular joint. (Goldner, thickness of cut: 12 mm.); **a:** Articular fossa; **b:** Articular eminence; **c:** Preglenoid plane; **d:** Articular disk; **e:** Lateral pterygoid muscle; **f:** Condylar process; **g:** Retroarticular tissue; **h:** Postglenoid process (modified from Türp et al. 1997. Reprinted by permission of Quintessenz, Berlin).

two completely separated compartments, the upper disko-temporal and the lower disko-mandibular space (Fick 1904; Tandler 1919) (Fig. 4). With the exception of the TMJ, articular disks are only found in the sternoclavicular joint and, incompletely, in the proximal wrist (Tillmann and Töndury 1987). The articular disk is the eutherian mammal characteristic, and is also found in some marsupials, but is absent in monotremes (Sprinz 1965; Goose and Appleton 1982). The TMJ disk is assumed to compensate for the incongruities between the temporal and mandibular components of the joint during functional movements of the mandible. The articular disk has sometimes been erroneously called a meniscus (Meyer 1865; Macalister 1954; Rees 1954). The difference is that a meniscus only partially divides an articular joint space into two joint compartments, allowing the joint spaces to communicate with each other through an aperture. An example of a joint containing a meniscus is the knee joint.

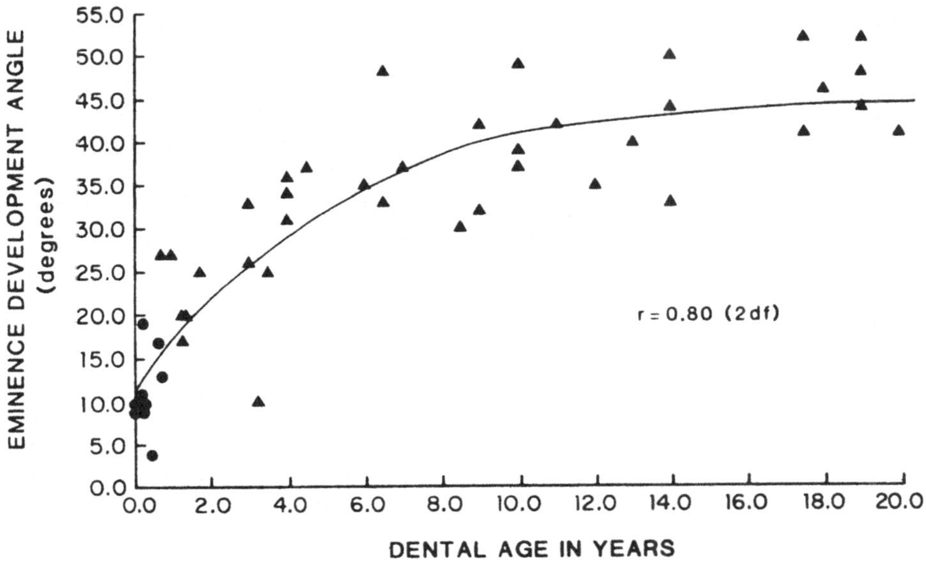

Fig. 5. Development of the articular eminence in dependence of dental age. Triangles and circles refer to skull samples originatings from two different collections (from: Nickel et al. 1988. Reprinted by permission of The American Association for Dental Research, Washington D.C.).

The articular disk can be divided morphologically into three parts: an anterior band, a thinner intermediate zone and a (broader) posterior band (Rees 1954) (Fig. 5). The posterior band of the articular disk continues posteriorly into the richly innervated and vascularized retrodiskal tissue, containing the bilaminar zone. The superior lamina *(Lamina superior, Stratum superior)* contains elastic fibers and inserts onto the postglenoid process posteriorly and into the petrosquamosal and tympanosquamosal fissures medially, unless the latter is missing (Dauber 1987). The inferior lamina *(Lamina inferior, Stratum inferior)* is attached to the posterior region of the condylar neck. Between the two laminae there is richly innervated and vascularized retroarticular connective tissue (Zenker 1956). The blood vessels from this area supply the majority of the articular disk except for the avascular intermediate zone (Wright and Moffet 1974). The articular disk is firmly attached to the medial and lateral sides (poles) of the condylar head via collateral ligaments. Laterally the disk only approximates the articular capsule, while anteriorly the disk continues into the articular capsule, and in many cases through the anterior extension (prediskal lamina) into the fibers from the superior belly of the lateral pterygoid muscle (Meyenberg et al. 1986; Heylings et al. 1995).

During early human postnatal growth the disk is vascularized abundantly (Moffett 1957; Thilander et. al 1976). However, during growth the central part of the articular disk becomes avascular and lacking in innervation (Boyer et al. 1964; Keith 1982). Histologically the articular disk is a dense fibrous connective tissue containing type I collagen and glycosaminoglycans comprised of chondroitin sulfate, hyaluronic acid and keratan sulfate (Axelsson et al. 1992; Kobayashi

1992). The collagen fibers within the intermediate zone of the articular disk are directed both antero-posteriorly and medio-laterally (Scapino 1983; Mills et al. 1988, 1994). This specific arrangement of the fibers becomes pronounced as the temporomandibular joint matures (Goose and Appleton 1982). Healthy articular disks in individuals under 30 years of age contain only a small amount of cartilage cells, predominantly around the periphery. Whereas the presence of cartilage cells in the central part of the articular disk seems to be related to the dental condition and function of the TMJ, this is not true for the diskal periphery (Carlsson et al. 1973).

Articular Capsule

The TMJ is surrounded by the articular capsule *(Capsula articularis)* containing blood vessels and nerve fibers. The articular capsule is a modified fibrous (external) layer of the bilaminar periosteum that continues from the condylar neck. The capsule is attached to the squamous part of the temporal bone, the postglenoid process, the tympanosquamosal fissure and to the area between the condylar head and neck. The height of the capsular attachment onto the condylar neck may depend on age (eg., more superior in children, Öberg et al. 1971) and mobility of the TMJ (eg., more inferior in hypermobile TMJs, Solberg et al. 1985). Anteriorly and posteriorly, the boundaries of the articular capsule are not well defined (Dauber 1987; Wendler et al. 1989; Luder 1991). The connective tissues covering the preglenoid plane and the fibrous layer of the periosteum of the condylar neck approach each other and the connective tissue from the superior head of the lateral pterygoid muscle to become posteriorly continuous with the articular disk (Meyenberg et al. 1986; Hylander 1992).

The inner surface of the articular capsule with its recesses is lined by a synovial membrane *(Membrana synovialis, Stratum synoviale)*. The membrane is continuous with the osteogenic (internal) layer of the bilaminar periosteum covering the condylar neck. The membrane secretes synovial fluid that serves as lubrication for the articular cartilage. Though the specific composition of the synovial fluid varies, it is generally considered to be a plasma-like fluid containing hyaluronate. Whether the synovial fluid is also a source of nutrition is controversial (Ten Cate 1994).

The articular capsule is reinforced by the lateral and medial temporomandibular ligaments. The former *(Ligamentum laterale)* originates from the lateral part of the articular tubercle and inserts onto the lateral aspect of the condylar neck (DuBrul 1988; Hylander 1992). However, this ligament does not always exist as a separate entity (Savelle 1988; Hylander 1992). The delicate medial ligament *(Ligamentum mediale)* is located in the medial part of the capsule. Both ligaments are assumed to provide limitations to condylar movements. Two other craniomandibular ligaments are the stylomandibular *(Ligamentum stylomandibulare)* and the sphenomandibular ligament *(Ligamentum sphenomandibulare)*, both extending from the mandible to the base of the cranium. The sphenomandibular ligament is the remnant of the posterior part of Meckel's cartilage. It is questionable if the latter two ligaments are of any important functional significance (Ten Cate 1994).

Growth and Remodelling

The TMJ is the unique characteristic of a mammal. In reptiles, the joint between the lower and upper jaws is formed between the articular bone (the posteriorly located ossified part of Meckel's cartilage) and the quadrate bone (the posteriorly located ossified part of the palatoquadrate cartilage). During evolution from mammal-like reptiles to mammals the two bones forming the reptilian jaw joint assumed a new function as part of the sound transmission complex in the middle ear (malleus, incus) (Barghusen and Hopson 1970; Crompton and Parker 1978). The articular disk developed later than the bony joint components, namely during the evolution of some marsupials and the vast majority of eutherian mammals (Sprinz 1965; Moffet 1966). The evolutionary transition from mammal-like reptiles to mammals is also paralleled during mammalian ontogeny.

The human true (secondary) TMJ starts to develop during the 7th prenatal week (Richany et al. 1956; Furstman 1963; Keith 1982) independently from other parts of the mandible (ramus, corpus). At this stage of ontogeny, the primary craniomandibular joint consists of the anlagen of the incus and malleus, analogous structures to those found in reptiles (Hertwig 1888). The secondary TMJ develops later, lateral to the primitive primary joint, as one of the last joints in the body (Moffett 1966; O'Rahilly and Gardner 1978; Keith 1982). For a short time, until the 18th week, both joints function simultaneously (Avery 1988). Subsequently, malleus and incus are integrated into the hearing complex of the middle ear. All of the essential components of the secondary TMJ are present and functional during the 12th week (Moffett 1957). The articular fossa and eminence are still flat. The histomorphology of the subcortical condylar and temporal areas reveals all the features that are indicative of bony growth: a thick proliferative zone, an increased mitotic activity of undifferentiated mesenchymal cells, and local vascularity (Carlson et al. 1978; Hinton and Carlson 1983).

The mandible and the squamous part of the temporal bone initially ossify directly from mesenchymal tissue without cartilaginous templates, unlike the base of the cranium and the long bones. The mandibular ossification continues lateral to Meckel's cartilage, the primitive cartilagenous lower jaw. As ossification progresses, cartilageneous islands appear within mesenchymal tissue in the areas of the future condylar, coronoid and the angular processes of the mandible. The appearance of these secondary cartilages in association with ossifying mesenchymal tissue is not limited to the developing mandible but is also observed in the midpalatal suture, the pterygoid plate, and the sagittal suture of the cranium (Hall and Hanken 1985). Secondary cartilage differs significantly from the primary (hyaline) cartilage found in other bones that ossifies endochondrally (Stutzman and Petrovic 1982). The difference is most pronounced in the spatial arrangement of the cartilage and the biochemical composition of its extracellular matrix (Bollen et al. 1989, 1990; Milam et al. 1991). In the area of the condylar process, secondary cartilage remains covered by the periosteum that is continuous with periosteum covering the rest of the ossifying mandible.

The secondary cartilage of the TMJ contributes to the underlying subchondral bone. However, the presence of secondary cartilage is not essential for growth of the condylar process. Instead, the growth of the secondary cartilage is regarded as being adaptive to various influences around the TMJ and not intrinsic to the cartilage itself, as for example is the case with the epiphyseal growth plate of long bones (interposed between the epiphysis and the diaphysis). In long bones, primary cartilage is regarded as the primary growth center. The major components of newly forming long bones, for example the epiphyseal center of ossification and growth plate, are thus missing in the developing mandible and remain so throughout growth and adult life. As function of the TMJ develops, periostea that cover the condylar head and the articular portion of the temporal bone become modified into fibrous articular layers, contributing to cartilage formation that subsequently undergoes endochondral ossification.

The most active growth sites of the mandible are the posterior border of the mandibular ramus, the condylar neck and the condylar head. As a result of the growth process, the mandible is displaced downward and forward. Articulation at the TMJ is maintained by condylar growth in the supero-posterior and lateral directions, and bone resorption in its anterior and medial parts (Enlow and Harris 1964). As TMJ loading increases, the superficial articular layer undergoes change such that its cells come to resemble both chondrocytes and fibroblasts. This layer is often termed *fibrocartilage* irrespective of its developmental stage, that is, whether cartilage cells are present in the layer (later in life) or not (early in life; see above). As the thickness of the articular layer increases during growth, the thickness of the subarticular cartilage layer decreases (Wright and Moffet 1974). The decrease of the latter is primarily attributed to the progressive increase of mineralization observed in the deeper layers of the hypertrophic zone. Similarly, the zone of maturation continues to be present until the late teens, when it is replaced by mineralized cartilage (Thilander et al. 1976). These layers eventually surround the newly forming bony cap of the condylar head. Even when the latter is completely formed, condylar growth continues until midlife, but it slows markedly (Carlsson and Öberg 1974).

Similar growth changes are observed in the temporal component of the TMJ. Initially flat, the articular eminence becomes increasingly steeper. Its development is fastest within the first three years of age. After eruption of the primary teeth, more than fifty percent of its mature size and morphology has been reached. Around the age of five, the rate of increase of articular eminence steepness is progressively reduced until reaching zero during the middle or late teens (Fig. 5) (Nickel et al. 1988). Changes in its form are primarily attributed to bone deposition on its anterior and posterior sides and bone resorption of the articular fossa (Wright and Moffet 1974). The thickness of the individual zones changes similarly to the condylar head. During this same period the postglenoid process elongates and contributes to the S-shaped profile of the temporal component of the adult TMJ when viewed in a parasagittal plane.

The components of the temporomandibular joint, as in any other joint, are continuously responding adaptively to their environment (Johnson 1964; Moffett et al. 1964; Blackwood 1966; Öberg et al. 1971; Mongini 1972, 1977; Carlsson and Öberg 1974; Hinton 1981). The capacity to adapt includes the period after the TMJ reaches its adult form (around the age of 20) (Schroeder 1992). The mechanism most often associated with biological adaptation is remodelling. Remodelling is a process where the structure and form of tissue change in response to external stimuli, most often biomechanical stress (increased mechanical loading) (Meikle 1992). For example, there is evidence that changes in form of the articular eminence are related to loss of the posterior teeth with resulting loss of occlusal support (Moffett et al. 1964; Furstman 1965; Blackwood 1966; Öberg et al. 1971; Mongini 1972, 1977; Bergman and Hansson 1979; Granados 1979; Whittaker et al. 1985; Whittaker 1989; Sheridan et al. 1991) (exceptions: Ericson and Lunberg 1968; Toller 1973). Similarly, associations are made between findings of occlusal attrition (enamel loss caused by mastication movements, i.e. by tooth contacts; Schroeder 1991) and abrasion (enamel loss caused by foreign bodies, such as food ingredients, toothpaste, toothbrushes, or by ritual manipulations; Schroeder 1991) and changes in form of the condylar head (Blackwood 1966; Mongini 1975; Seward 1976; Wedel et al. 1978; Bergman and Hansson 1979; Hinton 1981; Richards and Brown 1981) (exceptions: Whittaker et al. 1985, 1990; Sheridan et al. 1991). The sites where remodelling takes place most frequently are the posterior and lateral aspects of the articular eminence and the anterior-superior and lateral parts of the mandibular condyle (Öberg et al. 1971; Carlsson and Öberg 1974; Hylander 1979, 1981, 1992). These tissue alterations result in an increase in thickness of the corresponding articular layers (Hansson et al. 1977; Carlson et al. 1978; Baldioceda et al. 1990a, b) (Fig. 5).

Remodelling can be regarded as a way to maintain the equilibrium between form and function (Hansson 1992). It leads to a change in cellular composition of the fibrous articular layer. The appearance of cartilage cells within the articular layer, which initially contained only fibrocyte-like cells, is regarded as an adaptation to increased biomechanical stress. In addition, the proliferative layer covering the surface of the condylar head and the articular fossa has the ability to produce cells that are capable of either chondrogenesis or osteogenesis. The cartilage layer thus produced is more compliant in the distribution of compressive forces than bone during increased loading. This form of adaptation of the fibrous articular tissue is not unique to the temporomandibular joint, but can be found anywhere where membranous bone is under intermittent loads. If these areas are devoid of pressure for longer periods, bone formation is initiated as shown by experiments wherein the joints are immobilized. The degree of vascularization also influences the differentiation of the mesenchymal cells, produced by the proliferative zone (Hall 1970). An increase in vascularization leads to osteogenesis, while its decrease results in chondrogenesis. As the applied loads on the temporomandibular joint increase, the thickness of the articular tissue changes accordingly. Thus, the histology of the articular layer of these sites supports the theory of an association between thickness and the loading regime (Moffet 1966; Carlson et al. 1978). One proposed mechanism responsible for this tissue response is an increase in the water content within the superficial layer of cells, similar to the mechanism known for the increase in thickness of the hyaline cartilage under load (Radin et al. 1972; Carlson et al. 1980). As the cartilage layer increases in its thickness, so does the rate of its mineralization and resorption within its deeper layers with subsequent replacement by underlying bone (endochondral ossification) (Meikle 1992). This process is associated with a complementary process whereby the subchondral bone is resorbed. Both processes occur simultaneously within the same joint, but at different sites (Schroeder 1987). The overall form of the condylar head and/or the articular fossa thus changes. The remodeled joint morphology ensures continuous function in spite of changes in the amount and location of biomechanical stress.

If the biomechanical stress exceeds the capability of the TMJ to adapt, a breakdown of the articular tissues results (Moffett et al. 1964; Radin 1972). However, the differences between adaptive remodelling and definite pathological changes (osteoarthrosis) are not always clearcut, and both processes may take place at the same time.

Functional Aspects of the Temporomandibular Joint

Since the right and left mandibular condyles are connected through the mandibular body and its rami, every mandibular movement involves both TMJs. Depending on the extent of the movement itself and on whether or not food is present between the teeth, three general types of mandibular movements can be distinguished: border movements, empty movements, and masticatory movements. Border movements of the mandible are extreme movements, and the corresponding positions of the mandible are border positions. Both the border movements and border positions are restrained by anatomical structures (TMJ, masticatory muscles, ligaments, and teeth). Empty (free) movements occur within the boundaries of border movements without the presence of food between the teeth, as for example during speaking. Masticatory movements take place during biting and chewing.

The mandibular condyle is able to carry out two types of movements, namely rotation and translation. Rotation and translation very rarely occur completely independently of each other, but are generally combined (Hylander 1992). Mandibular movements are subject to wide interindividual variability (Merlini and Palla 1988). The following part of this chapter describes condylar movements during the most common empty and border movements (opening and closing, protrusion, retrusion, and lateral shifts) and during typical masticatory movements of the mandible.

Mandibular Opening and Closing

Mandibular opening is initiated primarily by a rotation of the mandibular condyle against the articular disk about a transverse axis. The axis of initial rotation differs among and within subjects. The center of rotation can be located within the outline of the condyle or its immediate vicinity, or even considerably far outside the skull. Even during minimal mandibular opening, the position of the transverse axis of mandibular rotation changes continuously relative to the base of the cranium (Gibbs and Lundeen 1982). As mandibular opening continues, translation of the disk-condyle complex, with simultaneous ongoing condylar rotation, occurs along the posterior slope of the articular eminence in a downward and anterior direction. The average distance of the condylar translation during maximal opening is 13 to 15 mm, while the disk moves anteriorly 5 to 9 mm (Rees 1954; Finlay 1965).

Mandibular opening is primarily accomplished by the combined activity of the anterior belly of the digastric, the mylohyoid and geniohyoid muscles (rotation) – the so-called mandibular depressors – and the inferior head of the lateral pterygoid (translation).

The closing movement is similar to that of opening, but in the opposite direction. However, the paths described by the condylar centers are not identical in both movements (Gibbs and Lundeen 1982). Muscles responsible for mandibular elevation are the masseter, the medial pterygoid, the anterior part of the temporalis and the upper head of the lateral pterygoid.

Mandibular Protrusion and Retrusion

Mandibular protrusion occurs as the mandibular teeth pass their maxillary counterparts in the anterior direction, while maintaining minimal separation. The movement of the condyle-disk complex is primarily translatory. Whereas the inferior head of the lateral pterygoid is responsible mainly for the forward and downward movement of the condyles, the masseter and medial pterygoid muscles as well as the depressors stabilize the mandible in the proper vertical relation to the maxilla.

The posterior part of the temporalis muscle has been acknowledged to be the most important muscle producing a backward movement of the mandible (retrusion) (Hylander 1992). In many subjects, the centric relation of the mandible is posterior to the position of the mandible with the teeth in maximum intercuspation. Only in a relatively small percentage of subjects do these condylar positions coincide. Centric relation is not considered to be a functional position, but rather a border position of the mandible (Fig. 6). It is in part for this reason that this mandibular position is used in dentistry as a reference mandibular position in patients with complete loss of teeth.

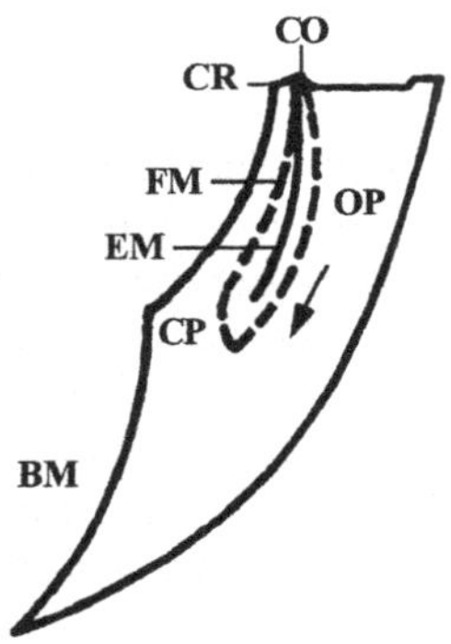

Fig. 6. Paths of the lower central incisors traced in the sagittal plane during functional movements (FM; dashed line), border movements (BM; solid line), and empty movements (EM) of the mandible (after Posselt 1968). CO: centric occlusion; CR: centric relation; OP: opening phase; CP: closing phase; The arrow indicates the direction of the movement.

Lateral Mandibular Movements

During a lateral movement, the mandible exhibits rotation about a vertical axis and bodily translation toward the working side. More specifically, the condyle-disk complex of the contralateral (balancing; non-working) side translates anteriorly, inferiorly and medially. Simultaneously, the ipsilateral (resting; working) condyle rotates about a vertical axis which is positioned immediately posterior to it, accompanied by a small lateral shift (Bennett movement) (Fig. 7).

Major muscles involved in lateral movements of the mandible are the contralateral lateral pterygoid (balancing side) and the ipsilateral middle and posterior part of the temporalis (working side).

Masticatory Movements

Masticatory movements occur within the range of the mandibular movements described above. The basic unit of masticatory action is a chewing cycle. A chewing cycle can typically be divided into four consecutive phases (Hiiemae 1978). After completion of the first two phases – slow mouth opening, followed by fast opening (both phases are also referrred to as the opening stroke) – the lower incisors, which serve here as a reference point for the description of the mandibular movements, move laterally and anteriorly toward the chewing side. In the following phase (fast closing, also called the closing or fast stroke) the mandible is still in an anterior and lateral position and does not shift back toward the midline until the last, or slow closing phase ("power stroke"). Tooth contact is not a requirement before the following chewing cycle is initiated. The shape of the chewing cycle, traced and viewed in the frontal plane, changes significantly during the chewing process as the food bolus is softened and prepared to be swallowed.

Concluding Remarks

The purpose of this chapter was to introduce the reader to the characteristics of the human TMJ anatomy, growth, development, and function. In spite of its uniqueness, it should not be forgotten that the TMJ shares common features with other joints of the human body. As such, it does not react differently to biomechanical stress than do, for example, the knee or the hip joints. In addition, like most other joints, the TMJ morphology is characterized by a considerable degree of intra- and interindividual variation. In an attempt to explain this variability, various studies have postulated a direct association between TMJ morphology and dental parameters. However, factors unrelated to the dental apparatus have to be taken into consideration as well. As Richards (1990, 383) proposed, due to the complex relationships among structures of the craniofacial system, "many questions about the determinants of joint morphology are as yet unanswered". Therefore, the investigation of this complex subject remains a challenging and important task in anthropology and dentistry.

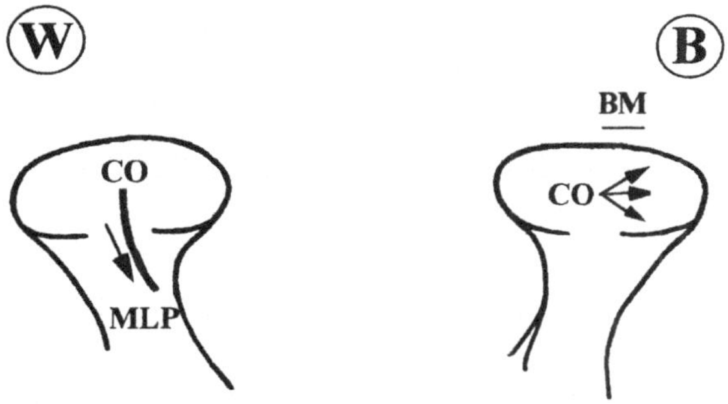

Fig. 7. Tracing of the condylar path during a lateral shift of the mandible to the left side. W: Working side; B: Balancing side; CO: centric occlusion; MLP: Maximum lateral position BM: Bennett movement.

Acknowledgements

The authors are grateful to Susan W. Herring, PhD, University of Washington, Seattle, and Robert P. Scapino, DDS, PhD, University of Illinois, Chicago, who have provided valuable suggestions in earlier drafts of the manuscript.

References

Avery JA (1988) Development of cartilages and bones of the facial skeleton. In: Avery JA (ed) Oral Development and Histology. Williams & Wilkins, Baltimore, p. 42–52.

Axelsson S, Holmund A, Hjerpe A (1992) Glycosaminoglycans in normal and osteoarthrotic human temporomandibular joint disks. Acta Odont Scand 50: 113–119.

Baldioceda F, Bibb CA, Pullinger AG (1990a) Distribution and histological charater of osseous concavities in mandibular condyles of young adults. J Craniomand Disord Facial Oral Pain 4: 147–153.

Baldioceda F, Pullinger AG, Bibb CA (1990b) Relationship of condylar bone profiles and dental factors to articular soft tissue thickness. J Craniomand Disord Facial Oral Pain 4: 71–79.

Barghusen HR, Hopson JA (1970) Dental squamosal joint and the origin of mammals. Science 168: 573.

Bergman LE, Hansson T (1979) Hard tissue changes of the temporomandibular joint in an archaeo-osteological material from the 11th century. Swed Dent J 3: 149–155.

Blackwood HJJ (1966) Adaptive changes in the mandibular joints with function. Dent Clin N Amer 10: 559–566.

Bollen AM, Makinen KK, Makinen PL, Carlson DS (1989) Collagenolytic and phosphate activity in the rat mandible after functional protrusion. Arch Oral Biol 34: 267–273.

Bollen AM, Carlson DS, Dziewiatkowski DD (1990) Aggregation of proteoglycans in cephalic, epyphyseal, and articular cartilages in growing sheep. Conn Tiss Res 24: 121–127.

Boyer CC, Williams TW, Stevens FH (1964) Blood supply of the temporomandibular joint. J Dent Res 43: 224–228.

Carlson DS, McNamara JA Jr., Jaul DH (1978) Histological analysis of the growth of the mandibular condyle in the rhesus monkey (Macaca mulata). Am J Anat 151: 103–117.

Carlson DS, McNamara JA Jr., Graber LW, Hoffman DL (1980) Experimental studies of growth and adaptation of TMJ. In: Irby WB (ed) Current Advances in Oral Surgery, Vol. III. Mosby, St. Louis, p. 28–39.

Carlsson GE, Hassler O, Öberg T (1973) Microradiographic study of human temporomandibular disks obtained at autopsy. J Oral Pathol 2: 265–271.
Carlsson GE, Öberg T (1974) Remodelling of the temporomandibular joints. Oral Sci Rev 6: 53–86.
Copray JCVM, Johnson PM, Decker JD, Hall SH (1989) Presence of osteonectin/SPARC in mandibular condylar cartilage of the rat. J Anat 162: 43–51.
Costa RL Jr (1986) Asymmetry of the mandibular condyle in Haida Indians. Am J Phys Anthrop 70: 119–123.
Crompton AW, Parker P (1978) Evolution of the mammalian masticatory apparatus. Am Zool 66: 192–201.
Dauber W (1987) Die Nachbarschaftsbeziehungen des Discus articularis des Kiefergelenks und ihre funktionelle Deutung. Schweiz Monatsschr Zahnmed 97: 427–437.
Dibbets JMH, van der Weele LT (1991) Flattened condylar projection in children: Reflection of seasonal growth? Eur J Orthod 13: 161–165.
Dibbets JMH, van der Weele LT (1992) Prevalence of structural bony change in the mandibular condyle. J Craniomandib Disord Facial Oral Pain 6: 254–259.
DuBrul EL (1988) Sicher and DuBrul's Oral Anatomy. Ishiyaku EuroAmerica, St. Louis.
Enlow DH, Harris DB (1964) A study of the postnatal growth of the human mandible. Am J Orthod 50: 25–50.
Enlow DH (1994) The condyle and facial growth. In: Sarnat BG, Laskin DM (eds): The Temporomandibular Joint: A Biological Basis for Clinical Practice. Saunders, Philadelphia, p. 48–59.
Ericson S, Lundberg M (1968) Structural changes in the finger, wrist and temporomandibular joints. Acta Odontol Scand 26: 111–126.
Fick R (1904) Handbuch der Anatomie und Mechanik der Gelenke. Erster Teil: Anatomie der Gelenke. Fischer, Jena.
Finlay IA (1965) The movements of the mandibular inter-articular disc. Proc Roy Soc Med 58: 671–675.
Furstman L (1963) The early development of the human temporomandibular joint. Am J Orthod 49: 672–682.
Gibbs CH, Lundeen HC (1982) Jaw movements and forces during chewing and swallowing and their clinical significance. In: Gibbs CH, Lundeen HC (eds) Advances in Occlusion. Wright, Boston, p. 2–32.
Goose DH, Appleton J (1982) Human Dentofacial Growth. Pergamon Press, Oxford.
Granados JI (1979) The influence of the loss of teeth and attrition on the articular eminence. J Prosthet Dent 42: 78–85.
Hall BK (1970) Cellular differentiation in skeletal tissue. Biol Rev 45: 455–484.
Hall BK, Hanken J (1985) Foreword. In: De Beer G (ed) The Development of the Vertebrate Skull. Univ Press, Chicago.
Hansson T, Öberg T (1977) Arthrosis and deviation in form in the temporomandibular joint. A macroscopic study on human autopsy material. Acta Odont Scand 35: 167–174.
Hansson T, Öberg T, Carlsson GE, Kopp S (1977) Thickness of the soft tissue layers and the articular disk in the temporomandibular joint. Acta Odont Scand 35: 77–83.
Hansson TL (1992) Pathological aspects of arthritides and derangements. In: Sarnat BG, Laskin DM (eds): The Temporomandibular Joint: A Biological Basis for Clinical Practice. Saunders, Philadelphia, p. 165–182.
Henke W, Rothe H (1994) Paläoanthropologie. Springer, Berlin Heidelberg New York.
Hertwig O (1888) Lehrbuch der Entwicklungsgeschichte des Menschen und der Wirbelthiere. 2. Aufl. Fischer, Jena.
Heylings DJA, Nielsen IL, McNeill C (1995) Lateral pterygoid muscle and the temporomandibular disc. J Orofac Pain 9: 9–16.

Hiiemae KM (1978) Mammalian mastication: A review of the activity of the jaw muscles and the movement they produce in chewing. In: Joysey KA, Butler PM (eds): Development, Function and Evolution of Teeth. Academic Press, London, p. 1359–1398.
Hinton RJ (1981) Changes in articular eminence morphology with dental function. Am J Phys Anthrop 54: 439–455.
Hinton RJ, Carlson DS (1983) Histological changes in the articular eminence and mandibular fossa during growth of the rhesus monkey (Macaca mulata). Am J Anat 166: 99–116.
Hinton RJ (1985) Adaptive response of the articular eminence and mandibular fossa to altered function of the lower jaw: An overview. In: Carlson DS, McNamara JA Jr., Ribbens KA (eds): Developmental Aspects of Temporomandibular Joint Disorders, Craniofacial Growth Series, No. 16. Univ of Michigan, Ann Arbor, p. 207–234.
Hylander WL (1979) An experimental analysis of temporomandibular joint reaction force in macaques. Am J Phys Anthrop 51: 443–456.
Hylander WL (1981) Patterns of stress and strain in the macaque mandible. In: Carlson DS (ed) Craniofacial Biology. Center for Human Growth and Development, Univ of Michigan, Ann Arbor, p. 1–36.
Hylander WL (1992) Functional Anatomy. In: Sarnat BG, Laskin DM (eds) The Temporomandibular Joint: A Biological Basis for Clinical Practice. Saunders, Philadelphia, p. 60–92.
Johnson LC (1964) Morphologic analysis in pathology: The kinetics of disease and general biology of bone. In: Frost HM (ed) Bone Biodynamics. Little, Brown and Company, Boston, p. 543–654.
Keith DA (1982) Development of human temporomandibular joint. Brit J Oral Surg 20: 217–224.
Kobayashi J (1992) Studies on matrix components relevant to structure and function of the temporomandibular joint. Kokubyo Gakkai Zasshi 59: 105–123 (in Japanese).
Koppe T, Autrum B, Krczal P (1993) Physiologische Formvarianten am Kiefergelenk des Menschen und der Pongiden. In: Benner K, Fanghänel J, Kowalewski R, Kubein-Meesenburg D, Randzio J (eds) Morphologie, Funktion und Klinik des Kiefergelenks. Quintessenz, Berlin, p. 101–109.
Lubsen CC, Hansson TL, Nordstrom BB, Solberg WK (1985) Histomorphometric analysis of cartilage and subchondral bone in mandibular condyles of young human adults at autopsy. Arch Oral Biol 30: 129–136.
Luder HU, Leblond CP, von der Mark K (1988) Cellular stages in cartilage formation as revealed by morphometry radiography and type III collagen immunostaining of the mandibular condyle from weanling rats. Am J Anat 182: 197–214.
Luder HU, Bobst P (1991) Wall architecture and disc attachment of the human temporomandibular joint. Schweiz Monatsschr Zahnmed 101: 557–570.
Luder HU, Bobst P, Schroeder HE (1993) Histometric study of synovial cavity dimensions of human temporomandibular joints with normal and anterior disc position. J Orofac Pain 7: 263–274.
Macalister AD (1954) A microscopic survey of the human temporomandibular joint. N Z Dent J 50: 161–172.
Merlini L, Palla S (1988) The relationship between condylar rotation and anterior translation in healthy and clicking temporomandibular joints. Schweiz Monatsschr Zahnmed 98: 1191–1199.
Meikle M (1992) Remodeling. In: Sarnat BG, Laskin DM (eds) The Temporomandibular Joint: A Biological Basis for Clinical Practice. Saunders, Philadelphia, p. 93–107.
Meyenberg K, Kubik S, Palla S (1986) Relationships of the muscles of mastication to the articular disc of the temporomandibular joint. Helv Odont Acta 30: 1–20. In: Schweiz Monatsschr Zahnmed 96: 815–834.
Meyer H (1865) Das Kiefergelenk. Arch Anat Physiol Wiss Med: 719–731.

Milam SB, Klebe RJ, Triplett RG, Herbert D (1991) Characterization of the extracellular matrix of the primate temporomandibular joint. J. Oral Maxillofac Surg 49: 381–391.
Mills DK, Daniel JC, Scapino R (1988) Histological features and in-vitro proteoglycan synthesis in the rabbit craniomandibular joint disk. Arch Oral Biol 33: 195–202.
Mills DK, Fiandaca DJ, Scapino RP (1994) Morphologic, microscopic, and immunohistochemical investigations into the function of the primate TMJ disc. J Orofac Pain 8: 136–154.
Moffett BC (1957) The prenatal development of the human temporomandibular joint. Carnegie Contr Embryol 36: 19–28.
Moffett BC, Johnson LC, McCabe JB, Askew HC (1964) Articular remodeling in the adult human temporomandibular joint. Am J Anat 115: 119–142.
Moffett BC (1966) The morphogenesis of the temporomandibular joint. Am J Orthod 52: 401–415.
Mongini F (1972) Remodelling of the mandibular condyle in the adult and its relationship to the condition of the dental arches. Acta Anat 82: 437–453.
Mongini F (1975) Dental abrasion as a factor in remodelling of the mandibular condyle. Acta Anat 92: 292–300.
Mongini F (1977) Anatomic and clinical evaluation of the relationship between the temporomandibular joint and occlusion. J Prosthet Dent 38: 539–551.
Nickel JC, McLachlan KR, Smith DM (1988) Eminence development of the postnatal human temporomandibular joint. J Dent Res 67: 896-902.
Nomina Anatomica (1989) Churchill Livingstone, Edinburgh, p. A18.
Öberg T, Carlsson GE, Fajers CM (1971) The temporomandibular joint. A morphologic study on human autopsy material. Acta Odont Scand 29: 349–384.
Obrez A (1993) The Temporomandibular Joint: Development, Anatomy, Physiology, and Function. Clark's Clinical Dentistry, Vol 2, Chap 34.
O'Rahilly R, Gardner E (1978) The embryology of movable joints. In: Sokoloff L (ed) The Joints and Synovial Fluid, Vol I. Academic Press, New York, p. 49–103.
Pullinger AG, Bibb CA, Ding X, Baldioceda F (1993a) Contour mapping of the TMJ temporal component and the relationship to articular soft tissue thickness and disk displacement. Oral Surg Oral Med Oral Path 76: 636–646.
Pullinger AG, Bibb CA, Ding X, Baldioceda F (1993b) Relationship of articular soft tissue contour and shape to the underlying eminence and slope profile in young adult temporomandibular joints. Oral Surg Oral Med Oral Path 76: 647–654.
Radin EL, Paul IL, Rose RM (1972) Role of mechanical factors in pathogenesis of primary osteoarthritis. Lancet 4: 519–521.
Rees LA (1954) The structure and function of the mandibular joint. Br Dent J 96: 125–133.
Rensburg BGJ van (1995) Oral Biology. Quintessence, Chicago.
Richany SF, Bast TH, Anson BJ (1956) The development of the first branchial arch in man and the fate of Meckel's cartilage. Q Bull Northwestern U Med School 30: 331–355.
Richards LC, Brown T (1981) Dental attrition and degenerative arthritis of the temporomandibular joint. J Oral Rehabil 8: 293–307.
Richards LC (1990) Tooth wear and temporomandibular joint change in Australian Aboriginal populations. Am J Phys Anthrop 82: 377–384.
Savelle WPM (1988) Some aspects of the morphology of the human temporomandibular joint capsule. Acta Anat 131: 292–296.
Scapino RP (1983) Histopathology associated with malposition of the human temporomandibular joint disk. Oral Surg 55: 382–397.
Schroeder HE (1987) Orale Strukturbiologie. Thieme, Stuttgart, p. 336–348.
Schroeder HE (1991) Pathobiologie oraler Strukturen. Karger, Basel, p. 60–61.

Seward FS (1976) Tooth attrition and the temporomandibular joint. Angle Orthodontist 46: 162–170.
Sheridan SG, Mittler DM, van Gerven DP, Covert HH (1991) Biomechanical association of dental and temporomandibular pathology in a Medieval Nubian Population. J Phys Anthrop 85: 201–205.
Sicher H (1949) Oral Anatomy. Mosby, St. Louis.
Silbermann M, Reddi AH, Hand AR, Leapman RD, von der Mark K, Franzen A (1987) Further characterization of the extracellular matrix in the mandibular condyle in neonatal mice. J Anat 151: 169–188.
Solberg WK, Hansson TL, Nordström B (1985) The temporomandibular joint in young adults at autopsy: a morphologic classification and evaluation. J Oral Rehabil 12: 303–321.
Sprinz R (1965) A note on the mandibular intra-articular disc in the joints of Marsupialia and Monotremata. Proc Zool Soc (London) 144: 327–338.
Stegenga B (1991) Temporomandibular Joint Osteoarthrosis and Internal Derangement. Van Denderen, Groningen.
Stutzman JJ, Petrovic AG (1979) Intrinsic regulation of the condylar cartilage growth rate. Eur J Orthod 1: 41–54.
Stutzman JJ, Petrovic AG (1982) Bone cell histogenesis: The skeletoblast as a stem-cell for preosteoblasts and for secondary-type prechondroblasts. In: Dixon AD (ed) Factors and Mechanisms Influencing Bone Growth. Liss, New York, p. 29–44.
Tandler J (1919) Lehrbuch der systematischen Anatomie, 1. Band. Vogel, Leipzig.
Ten Cate AR (1994) Temporomandibular joint. In: Ten Cate AR (ed) Oral Histology. Development, Structure, and Function. Mosby, St. Louis, p. 432–455.
Tillmann B, Töndury G (eds) (1987) Rauber/Kopsch – Anatomie des Menschen. Bewegungsapparat. Thieme, Stuttgart.
Thilander B, Carlsson GE, Ingervall B (1976) Postnatal development of the human temporomandibular joint, I. A histological study. Acta Odont Scand 34: 117–126.
Toller PA (1973) Osteoarthrosis of the mandibular condyle. Br Dent J 134: 223–230.
Türp JC, Radlanski RJ, Obrez A (1997) Ontogenetische und postnatale Entwicklung des menschlichen Kiefergelenks. In: Alt KW, Türp JC (Hrsg.) Die Evolution der Zähne. Quintessenz, Berlin, p. 647–675.
Türp JC, Alt KW, Vach W, Harbich K (1995) On asymmetries of mandibular condyles and rami. In: Radlanksi R, Renz H (eds) Proceedings of the 10th International Symposium on Dental Morphology, Berlin, Sept 6–10, 1995. Brünne, Berlin, p. 444–448.
Wedel A, Carlsson GE, Sagne S (1978) Temporomandibular joint morphology in a medieval skull material. Swed Dent J 2: 177–187.
Wendler D, Bergmann M, Schumacher GH, Kunz G (1989) Das Kiefergelenk – Struktur und Funktion im Entwicklungsgang. Anat Anz 169: 1–5.
Whittaker DK, Davies G, Brown M (1985) Tooth loss, attrition and temporomandibular joint changes in a Romano-British population. J Oral Rehabil 12: 407–419.
Whittaker DK (1989) Surface and form changes in the temporomandibular joints of 18th century Londoners. J Dent Res 68 (Spec Iss): 1018 (Abstr 1211).
Whittaker DK, Jones JW, Edwards PW, Molleson T (1990) Studies on the temporomandibular joints of an eighteenth-century London population (Spitalfields). J Oral Rehabil 17: 89–97.
Wright DM, Moffett B (1974) The postnatal development of the human temporomandibular joint. Am J Anat 141: 235–249.
Yale SH, Allison BD, Hauptfuehrer JD (1966) An epidemiological assessment of mandibular condyle morphology. Oral Surg Oral Med Oral Pathol 21: 169–177.
Zenker W (1956) Der retroartikuläre plastische Polster des Kiefergelenkes und seine mechanische Bedeutung. Z Anatomie Entwicklungsgeschichte 119: 375–388.

3.5 The Maxillary Sinus of Extant Catarrhine Primates

Thomas Koppe and Hiroshi Nagai

Introduction

Compared with other structures of the skull such as the teeth, the temporomandibular joint or the cranial base, the paranasal sinuses have received much less attention among comparative anatomists and anthropologists. Although it is assumed that an enlarged maxillary sinus is a characteristic feature of the Hominoidea (Andrews and Martin 1987), the role of the paranasal sinuses in evolution is still uncertain. This situation is somewhat surprising in light of the absence of any accepted theory to explain the functions or even the great variability of the pneumatic cavities. As pointed out by Blanton and Biggs (1969) and as recently discussed by Blaney (1990), many of the functions ascribed to the sinuses are based merely on speculation.

Since the still widely recognized comparative anatomical studies of the paranasal sinuses by Seydel (1891) and Paulli (1900), numerous researchers have contributed to our understanding about the development, growth and variability especially of the human paranasal sinuses (Schaeffer 1920; Leicher 1928; Szilvássy et al. 1987). However, little knowledge is available about the primate paranasal sinuses, and most authors have dealt only with the pneumatic cavities of the great apes (Wegner 1936; Cave and Haines 1940; Blaney 1986; Lund 1988; Koppe et al. 1995).

This obvious lack of information about the primate paranasal sinuses is partially because studies of the internal structures of the skull involve destruction of the specimens and therefore are unsuitable with regard to precious museum collections (Cave 1973). On the other hand, plain radiographic examinations allow the complicated shape of these three-dimensional (3D) objects to be evaluated only partially. CT investigations of internal structures of the skull as well as 3D reconstructions based on CT scans conducted since the 1980s are likely to overcome the above mentioned problems. Although these methods have been increasingly applied by both anthropologists and primatologists (for review see Zonneveld et al. 1989) the primate pneumatic cavities have only been dealt with in a few studies (Ward and Pilbeam 1983; Conroy and Vannier 1987; Koppe et al. 1996).

The largest among the primate pneumatic cavities, the maxillary sinus is in more or less close proximity to the roots of the maxillary posterior teeth. Comparing the maxillary sinus topography in living great apes and humans with a number of skeletal remains of Miocene hominoids and Pliocene hominids, Ward and Pilbeam (1983) hold that the pneumatization of the postcanine alveolar recess as well as the size of the maxillary sinus are a function of body size. However, the pneumatization of the maxillary alveolar recess varies considerably and is only partially understood for humans and the great apes. Thus, this chapter tries to fill this gap by studying the pneumatization of the maxillary sinus not only in Hominoidea but also in different species of Old World monkeys. Furthermore, the present study is designed to discuss the hypothesis that the proportional enlargement of the maxillary sinus during primate evolution is mainly related to increasing body size.

Pneumatization of the Facial Skeleton in Vertebrates

Paranasal sinuses are not limited to mammals, and are seen in other vertebrates as well. Recently, Witmer (1995) showed evidence that the antorbital cavity of birds and the crocodilian caviconchal sinus are homologous structures. Although the old hypothesis that the maxillary sinus of placental mammals is homologous with the paranasal air sinus of Archosaurs (e.g., Bertau 1935) gains support because of certain similarities regarding their development and topography, this hypothesis clearly fails the congruence test of homology (Witmer 1995).

While any kind of pneumatization is lacking in montremes (Paulli 1900), pneumatic cavities in marsupials are rare and until now have been reported only for *Phascolarctus cinerus* (Paulli 1900) and *Thylacinus cynocephalus* (Weinert 1927). In contrast, pneumatization is a distinct characteristic of placental mammals and the maxillary sinus is generally regarded as a primitive eutherian feature (Moore 1981).

The degree of pneumatization varies considerably among the placental mammals. Pneumatic cavities are absent in aquatic mammals and in small bats but are extremely enlarged in size, number and complexity among large ungulates and the great apes (Seydel 1891; Paulli 1900; Wegner 1936; Cave and Haines 1940; Negus 1958; Moore 1981).

The Primate Paranasal Sinuses

Because the morphology of the nasal cavity of most quadrupeds differs drastically from that of primates, which is especially evident regarding the interorbital region as well as the morphology of the turbinals, Cave (1973) holds that the primate paranasal sinuses should be regarded separately.

Novacek (1993) claims that lower primates show a very poor development of any opening besides the maxillary sinus, and only African great apes and humans possess additional sinuses. However, next to the maxillary sinus, a pneumatization of the frontal and sphenoid bones is already present in prosimians such as lemurs and galagos (Seydel 1891; Weinert 1927; Wegner 1955). Similar to quadrupeds, the pneumatic cavities are sometimes partially or almost completely occupied by parts of the turbinals. Thus, the homology of these cavities with those of anthropoids is questionable (Moore 1981).

Tarsius already shows remarkable changes of the nasal cavity which includes the development of an interorbital septum as well as a simplification of the turbinals (Cave 1968). The only pneumatic cavity of *Tarsius* is the maxillary sinus.

The paranasal sinuses of New World monkeys are poorly understood and this is reflected in the lack of agreement regarding their etiology and function. The maxillary sinus is usually well developed and may extend into neighboring bones. Although a pneumatization may also occur within the frontal and sphenoid bones of *Cebus*, *Alouatta* and *Lagothrix* (Seydel 1891; Weinert 1927; Wegner 1955; Cave 1968), their homology with those cavities of the Hominoidea is questionable (Cave 1968).

The only true pneumatic cavity of extant Cercopithecidae is the maxillary sinus (Seydel 1891; Cave 1968; Moore 1981; Lund 1988; Koppe et al. 1996). Certain Old World monkeys lack any pneumatic cavities (Table 1). The virtual absence of a maxillary sinus in *Victoriapithecus*, a middle Miocene cercopithecoid, is probably a derived cercopithecoid feature (Benefit and McCrossin 1993).

Within the catarrhine primates, only the Hominoidea develop pneumatic cavities in addition to a maxillary sinus. The gibbon and the orang-utan already show a sphenoidal sinus. Although ethmoidal air cells have been reported for the orang-utan (Weinert 1927; Wegner 1955), it is still generally accepted that ethmoidal air cells and frontal sinuses are restricted to African great apes and to humans (Cave and Haines 1940; Moore 1981; Blaney 1986). Furthermore, the presence of a frontal sinus in *Proconsul*, *Afropithecus* and *Dryopithecus* suggests that this pneumatic cavity is an ancestral hominoid characteristic (Andrews 1992).

Maxillary Sinus Pneumatization of Extant Catarrhini

We studied the maxillary sinus of macerated skulls of humans, the great apes, as well as of different species of Cercopithecoidea (Tab. 1), trying to analyze always an equal number of male and female skulls. For all skulls, coronal CT scans were done with a Siemens *Somatom* DR CT or a General Electric HiSpeed Advantage RP CT. The distances between the CT scans ranged from 0.5 to 4 mm depending on the size of the skulls. The volume of the maxillary sinus was then calculated from the CT scans. In addition, 3D images were reconstructed for selected CT scans using the *Allegro* medical imaging workstation (ISG Technologies Inc., Toronto).

Humans

The maxillary sinus develops as the first of the paranasal sinuses in the fetal period after the 2nd month of gestation. Not bigger than 7–10 mm in length, 4 mm in height, and 3 mm in width in newborns, the maxillary sinus increases at 2 mm per year vertically and 3 mm horizontally (e.g. Schaeffer 1920; Anderhuber et al. 1992). Studying Indian children, Tanaka (1983) reports that the capacity of the maxillary sinus enlarged from 4.4 cm^3 after the completion of the primary dentition to 8.2 cm^3 in children after the eruption of the first premolar. Although not finished in growth, the maxillary sinus of children at the age of 12 years resembles the shape of the adult sinus (Tillier 1977) and has already enlarged to the level of the nasal floor (McGowan et al. 1993).

The adult human maxillary sinus appears as a three- or four-sided pyramid with the base towards the nasal cavity (Fig. 1a; McGowan et al. 1993). However, as pointed out by Anagnostopoulou et al. (1991), the pyramidal shape of the maxillary sinus is an idealization. They suggest a classification of maxillary sinus shapes into 4 categories: semi-ellipsoid (15%); paraboloid (30%); hyperboloid (47%) and cone (8%). There is a certain amount of sexual dimorphism in the sinus size, with a bigger sinus in males than in females. The size and shape of the maxillary sinus vary among living human populations (Leicher 1928; Tillier 1977).

On coronal CT scans, the sinus reaches its biggest spatial extension at the orbital entrance. Although the sinus is mainly restricted to the boundaries of the maxillary bone, the sinus may also extend into the palate and zygomatic bone. The arrangement of the human maxillary sinus floor shows a big variability (detailed descriptions by Keith 1902; Runge 1928; Uemura 1974; Taylor 1980; Eberhardt et al. 1992). Although the mesiobuccal root of the maxillary second molar usually shows the closest relation to the sinus floor, there are also skulls with a sinus floor clearly seen above the level of the maxillary posterior teeth.

Gibbon and Great Apes

In the gibbon, the sinus is a relatively long and narrow chamber which reaches anteriorly almost to the canine root. Root apices of the second premolar and of all maxillary molars are usually exposed into the sinus floor and make its shape rather irregular. In contrast to humans, the maxillary sinus of the great apes is enormous in size (Fig. 1c). Not only does it occupy almost the whole maxillary bone, but it also usually pneumatizes neighboring bones (for details see Wegner 1936; Cave and Haines 1940). In *Gorilla* and *Pongo*, this cavity posteriorly reaches the anterior wall of the sphenoidal sinus.

Typical of *Pongo*'s maxillary sinus is a capacious frontal recess, which pneumatizes partially the interorbital septum (Koppe et al. 1995). In the gorilla, the anterior part of the maxillary sinus is partially displaced by an enormously dilated nasolacrimal canal (Fig.1c). A characteristic feature of the chimpanzee is the extension of the maxillary sinus into the hard palate. Thus, in the chimpanzee both maxillary sinuses are sometimes only separated by a thin bony septum, which contains also the incisive canal (Fig. 1b).

Typical for the great apes is a multisepted maxillary sinus floor. Furthermore, the sinus floor of the great apes is usually advanced to the root apices of both

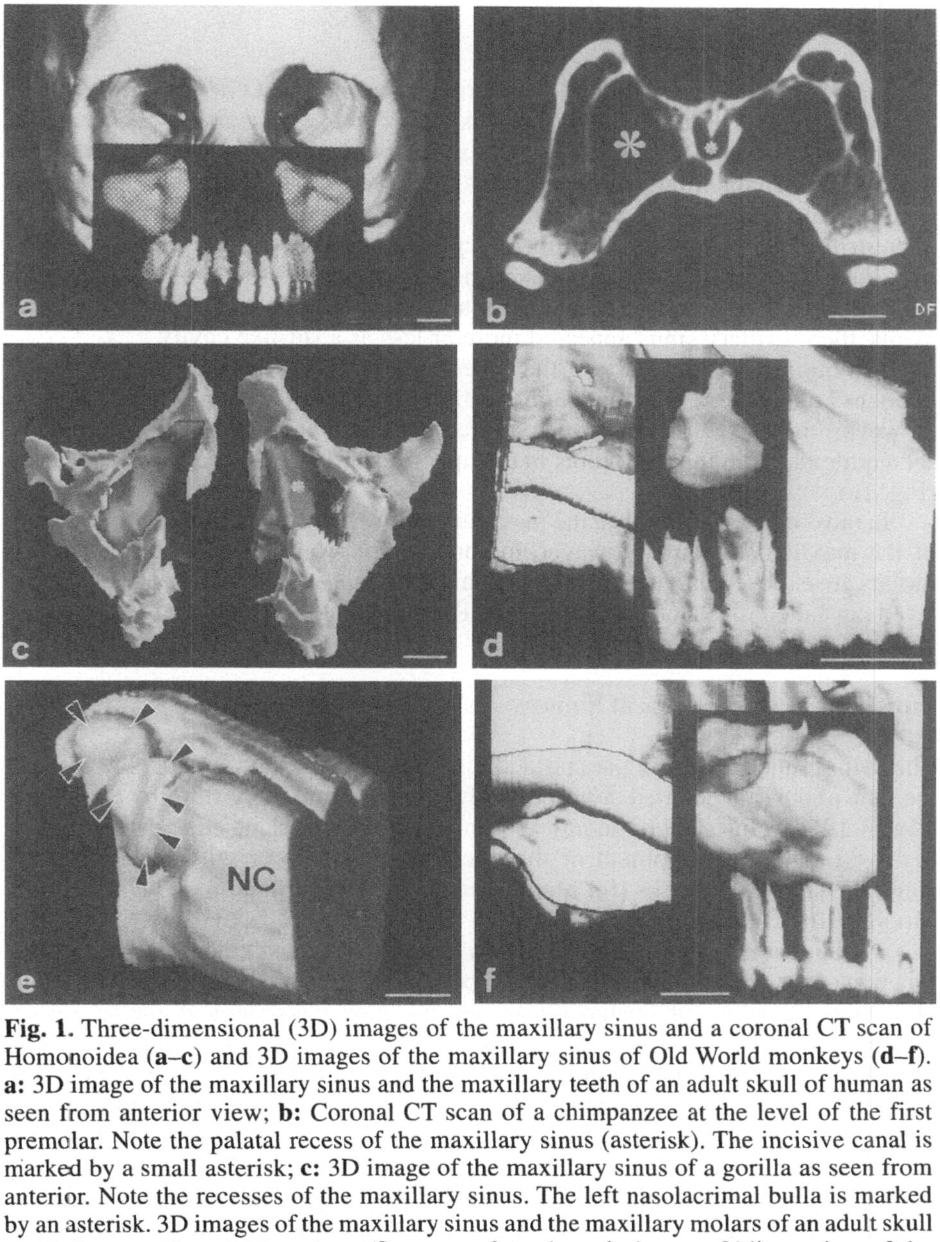

Fig. 1. Three-dimensional (3D) images of the maxillary sinus and a coronal CT scan of Homonoidea (**a–c**) and 3D images of the maxillary sinus of Old World monkeys (**d–f**). **a:** 3D image of the maxillary sinus and the maxillary teeth of an adult skull of human as seen from anterior view; **b:** Coronal CT scan of a chimpanzee at the level of the first premolar. Note the palatal recess of the maxillary sinus (asterisk). The incisive canal is marked by a small asterisk; **c:** 3D image of the maxillary sinus of a gorilla as seen from anterior. Note the recesses of the maxillary sinus. The left nasolacrimal bulla is marked by an asterisk. 3D images of the maxillary sinus and the maxillary molars of an adult skull of *M. fuscata* (**d**) and *M. mulatta* (**f**) as seen from lateral view; **e:** Oblique view of the reconstruction of a part of the nasal cavity of *P. hamadrys*. The lateral recess (arrowheads) is seen in the upper posterior region of the nasal cavity (NC). Bars: 1 cm.

molars and premolars. However, similar to humans, there are ape skulls with a sinus floor clearly seen above the root apices. In *Pongo* the maxillary sinus only occasionally reaches the first premolar (Koppe et al. 1995).

Old World Monkeys

Compared with the Hominoidea, the maxillary sinus of the Old World monkeys is relatively small. While no maxillary sinus was found in *Procolobus badius* and in *Cercopithecus aethiops*, a maxillary sinus is observed in all macaques. In those monkeys with a maxillary sinus, the sinus is restricted to the maxillary bone in the molar region (Fig. 1 d, f). The maxillary sinus of some larger skulls of *Macaca nemestrina* may pneumatize the maxillary bone anteriorly to the first molar. The maxillary sinus of the macaque species is a hemispherical or kidney-like cavity and sometimes also shows a small frontal recess (Koppe et al. 1996). In some skulls, the maxillary sinus appeared more or less as a slit-like cavity.

Although the region above the alveolar process of the very long snouted baboons is concave, a kind of pneumatization of the maxillary bone occurs as well. However, these pneumatic cavities, more properly named lateral recesses, were seen quite far from the teeth roots in the upper posterior region of the nasal cavity (Fig. 1e).

In most macaque samples, the maxillary sinus floor has no relation to the roots of the maxillary molars. However, in some skulls, root apices of the maxillary molars are exposed into the sinus floor. This is not only seen in the relatively large-sized pig-tailed macaque, but also in the Rhesus monkey (Fig. 1f).

Functions of the Paranasal Sinuses

The actual function of the paranasal sinuses has remained enigmatic since 1651, the time of the first detailed description of the human maxillary sinus by Highmore (Gysel 1967). Since then, countless publications have dealt either with factors involved in the development of the pneumatic cavities or with the functional significance of the sinuses (for review see Negus 1958; Blanton and Biggs 1969; Moore 1981; Blaney 1986).

Although it is impossible to discuss or even to mention all the proposals for the function of the pneumatic cavities, two aspects can be distinguished. The first group of theories claims that the paranasal sinuses may have physiological functions such as conditioning of the inspired air. In this context, it is noteworthy that the complex maxilloturbinals of certain mammals are important to reduce the respiratory water loss, which is considered important in terms of endothermy (Hillenius 1992). Assuming that Hillenius's approach is correct, the question arises as to how the primate nasal cavity, which is characterized by a radical simplification of the turbinals, is involved in this process. Answering this question could be of interest considering the functions of the primate paranasal sinuses because the morphological changes of the primate nasal cavity are associated with obvious alterations of the pneumatic cavities. Shea (1977), who studied the maxillary sinus volume in different Eskimo populations, suggested that the decrease of the sinus size in colder areas may be due to structural peculiarities of the nasal cavity, especially of the inferior nasal concha (maxilloturbinal) and the inferior nasal meatus.

According to a second group of authors, the paranasal sinuses are basically functionless and only present to replace the unnecessary bone between the bony

Table 1. Maxillary sinus volume of Catarrhini analyzed by different methods*

Taxon (N)	Computed tomography Mean (cm^3)	S.D.	Plain radiography[a]	Visual inspection
Homo sapiens (10)	12.51	5.27	(–)	(–)
Australopithecus africanus (1)[b]	3.5	(–)	(–)	(–)
Pan troglodytes (10)	18.98	4.09	(–)	(–)
Gorilla gorilla (10)	39.84	15.40	(–)	(–)
Pongo pygmaeus (11)	24.73	11.05	(–)	(–)
Hylobates lar (4)	2.45	0.79	(–)	(–)
Hylobates syndactylus (1)	(–)	(–)	length – 3.0 cm	(–)
Papio hamadrys (4)	lateral recess	(–)	(–)	(–)
Papio anubis (1)	lateral recess	(–)	(–)	(–)
Papio cynocephalus (1)	(–)	(–)	lateral recess	(–)
Mandrillus sphinx (1)	(–)	(–)	lateral recess	(–)
Theropithecus gelada (1)	(–)	(–)	lateral recess	(–)
Macaca nemestrina (6)	5.65	4.01	(–)	(–)
Macaca mulatta (8)	1.15	0.24	(–)	(–)
Macaca fuscata (10)	0.88	0.46	(–)	(–)
Macaca fascicularis (8)	1.51	0.77	(–)	(–)
Macaca assamensis (4)	1.34	0.53	(–)	(–)
Macaca radiata (1)[c]	(–)	(–)	(–)	present
Cercopithecus aethiops (4)	not present	(–)	(–)	(–)
Cercocebus torquatus (1)	(–)	(–)	lateral recess	(–)
Precolobus badius (1)	not present	(–)	(–)	(–)
Nasalis larvatus (1)	(–)	(–)	lateral recess	(–)
Victoriapithecus (5)[d]	(–)	(–)	(–)	not present

* Because measurements are not available, the maxillary sinus pneumatization of extinct hominoids is not included, except for the Taung skull. N = number of skulls; a = taken from Lund (1988); b = Conroy and Vannier (1987); c = Seydel (1891); d = Benefit & McCrossin (1993)

pillars of the facial skeleton. Thus, the presence of the paranasal sinuses is seen in close relation to the growth of the skull and the strains applied to the bone.

Even though growth and development of the paranasal sinuses are closely linked to both growth of the skull and dentition, it has been demonstrated that the pneumatic cavities possess a developmental potential of their own (e.g., Libersa et al. 1981). This assumption has gained support from recent clinical studies which have demonstrated that the effect of gross malformations such as cleft palate (e.g., Robinson et al. 1982) on the pneumatization process is obviously very small. Furthermore, according to Farkas et al. (1966) the relations between the size of the maxillary sinus and the external dimensions of the maxillary bone in humans are weak.

Evolution of the Maxillary Sinus Size

We performed a simple regression analysis using the length of the skull (prosthion – opisthocranion) as the independent variable and the volume of the maxillary sinus

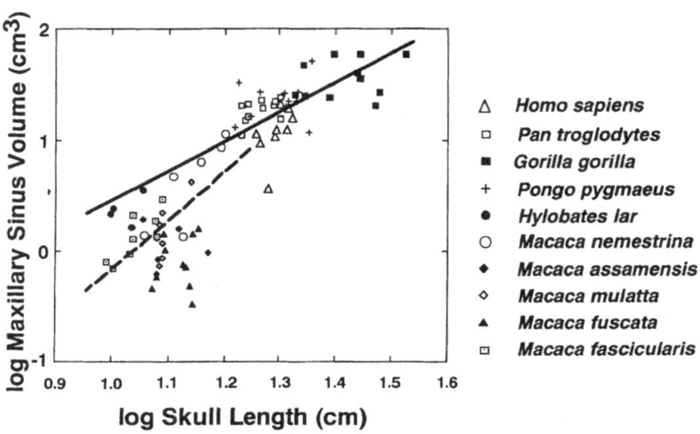

Fig. 2. Bivariate plot of the maxillary sinus volume against the skull length for catarrhine primates. The regression lines (least-squares linear regression) are included for Hominoidea (solid line): $r = 0.816$, $y = -2.141 + 2.615x$; and for Cercopithecoidea (dashed line) $r = 0.679$, $y = -4.586 + 4.427x$. The difference in slope of the regression lines was significant at $p < 0.05$. Because of the evaluation of the residuals, the values of the *M. fuscata* were treated as outlier. Thus, *M. fuscata* is not included in the regression analysis.

as a dependent variable for both Hominoidea and Cercopithecoidea. The differences in the slope of the regression lines were tested by employing the t-test ($p < 0.05$). Whereas the size of the maxillary sinus increased with increasing skull size, the statistical analysis revealed significant differences in the slopes of the regression lines of Hominoidea and Cercopithecoidea (Fig. 2).

Considering that the skull is a very complex structure which consists of numerous more or less functionally independent components, it is impossible to explain the variability of the pneumatic cavities by a single factor such as body size. Apart from the multitude of interactions between the different components of the skull during growth (for review see Herring 1993), the radical changes of craniofacial morphology during primate evolution, such as flexion of the cranial base, shortening of the jaws and reduction of teeth size, certainly affected the pneumatization process and the function of the pneumatic cavities as well (Takahashi 1984). Regarding the diversity of the paranasal sinuses between the different orders of mammals, Moore (1981) claims that it is by no means certain that the sinuses serve the same functions in each of these orders.

Conclusion

The primate maxillary sinus is probably the only sinus which is homologous with their non-primate counterpart. Thus, the maxillary sinus can be considered as a primitive eutherian feature. Whereas the maxillary sinus is restricted in Old World monkeys mainly to the maxillary bone in the molar region, the sinus increases in size within the Hominoidea, especially in the great apes. Furthermore, the relations between the maxillary sinus floor and the maxillary posterior teeth seem to be closer in hominoid species. However, the present review suggests that these differences between Old World monkeys and Hominoidea cannot be explained by only a single factor such as body size. Although this does not say much about the actual function of the maxillary sinus, we suppose that the pneumatic cavities are functional and they are far from being simple spaces between the mechanically essential bony pillars of the facial skeleton. However, to resolve the controversy regarding whether the role of the paranasal sinuses is "structural" or "functional" (or both) more research is needed in a wide range of extant and extinct primates. For this purpose, CT is a very valuable nondestructive method to visualize the engimatic pneumatic cavities.

Acknowledgements

This study was supported by the Cooperation Research Program of the Primate Research Institute of Kyoto University, Japan. We are grateful to Prof. T. Kimura (Tokyo), Dr. Y. Hamada, Dr. T. Mouri and Dr. Y. Kunimatsu (Inuyama) for valuable advice and support. Prof. W. Schmid (Greifswald) and Prof. Y. Hiraki (Okayama) kindly provided examination time on the CT scanners. Furthermore, the authors are especially grateful to Prof. Y. Hiraki for allowing access to all computer facilities in his department. Prof. J. Fanghänel (Greifswald) and Prof. T. Sugimoto (Okayama) kindly allowed to examine skulls of humans and great apes of their departments. We thank Mr. Y. Ohkawa (Okayama) for his enthusiastic help during the whole study and Mr. P. Randolph for kind editorial assistance.

References

Anagnostopoulou S, Venieratos D, Spyropoulos N (1991) Classification of human maxillary sinuses according to their geometric features. Anat Anz 173: 121–130.

Anderhuber W, Weiglein A, Wolff G (1992) Cavitas nasi und Sinus paranasales im Neugeborenen- und Kindesalter. Acta Anat 144: 120–126.

Andrews P (1992) Evolution and environment in the hominoidea. Nature 360: 641–646.

Andrews P, Martin L (1987) Cladistic relationship of extant and fossil hominoids. J Hum Evol 16: 101–118.

Benefit BR, McCrossin M (1993) Facial anatomy of *Victoriapithecus* and its relevance to the ancestral cranial morphology of Old World monkeys and apes. Am J Phys Anthrop 92: 329–370.

Bertau M (1935) Zur Entwicklungsgeschichte des Geruchsorgans der Krokodile. Z Anat Entwicklungsgesch 104: 168–202.

Blaney SPA (1986) An allometric study of the frontal sinus in *gorilla, pan* and *pongo*. Folia Primatol 47: 81–96.

Blaney SPA (1990) Why paranasal sinuses. J Laryngol Otol 104: 690–693.

Blanton PL, Biggs NL (1969) Eighteen hundred years of controversy: The paranasal sinuses. Am J Anat 124: 135–148.

Cave AJE (1968) Observations on the Platyrrhine nasal fossa. Am J Phys Anthrop 26: 277–288.

Cave AJE (1973) The primate nasal fossa. Proc Biol J Linn Soc London 5: 377–387.

Cave AJE, Haines RW (1940) The paranasal sinuses of the anthropoid apes. J Anat 72: 493–523.

Conroy GC, Vannier MW (1987) Dental development of the Taung skull from computerized tomography. Nature 329: 625–627.

Eberhardt JA, Torabinejad M, Christiansen EL (1992) A computed tomographic study of the distances between the maxillary sinus floor and apices of the maxillary posterior teeth. Oral Surg Oral Med Oral Pathol 73: 345–346.
Farkas LG, Feiglova B, Hajnis Y, Kustra T (1966) Contribution to the anatomy of the sinus maxillaris in Central-European skulls. Arch Jpn Chir 35: 1054–1065.
Gysel C (1967) Les descriptions du sinus maxillaire au seizième siècle. Janus 54: 262–269.
Herring SW (1993) Epigenetic and functional influences on skull growth. In: Hanken J, Hall BK (eds) The Skull, Vol. 1. Chicago Univ Press, Chicago, p. 153–206.
Hillenius WJ (1992) The evolution of nasal turbinates and mammalian endothermy. Paleobiology 18: 17–29.
Keith A (1902) The relationship of the eruption of the permanent molar teeth to the expansion of the maxillary sinus (Antrum of Highmore). Br J Dent Sci 16: 530–544.
Koppe T, Röhrer-Ertl O, Hahn D, Reike R, Nagai H (1995) Growth pattern of the maxillary sinus in orang-utan based on measurements of CT scans. Okajimas Folia Anat 72: 37–43.
Koppe T, Inoue Y, Hiraki Y, Nagai H (1996) The pneumatization of the facial skeleton in the Japanese macaque *(Macaca fuscata)* – a study based on computerized three-dimensional reconstructions. Anthropol Sci 104: 31–41.
Leicher H (1928) Vererbung anatomischer Variationen der Nase, ihrer Nebenhöhlen und des Gehörgangs. Bergmann, München.
Libersa C, Laude M, Libersa JC (1981) The pneumatization of the accessory cavities of the nasal fossae during growth. Anat Clinica 2: 265–273.
Lund VJ (1988) The maxillary sinus of higher primates. Acta Otolaryngol 105: 163–171.
McGowan DA, Baxter PB, James J (1993) The Maxillary Sinus and Its Dental Implications. Wright, Oxford.
Moore WJ (1981) The Mammalian Skull. Cambridge Univ Press, Cambridge.
Negus V (1958) The Comparative Anatomy and Physiology of the Nose and the Paranasal Sinuses. Livingstone, Edinburgh.
Novacek MJ (1993) Patterns of diversity in the mammalian skull. In: Hanken J, Hall BK (eds) The Skull, Vol. 2. Chicago Univ Press, Chicago, p. 438–545.
Paulli S (1900) Über die Pneumaticität des Schädels bei den Säugetieren. Eine morphologische Studie, I, II, III. Gegenbaurs Morphol Jahrb 28: 147–178, 179–251, 483–564.
Robinson HE, Zerlin GK, Passy V (1982) Maxillary sinus development in patients with cleft palates as compared to those with normal palates. Laryngoscope 92: 183–187.
Runge H (1928) Die Beziehungen der Zahnkeime und Zahnwurzeln zur Oberkieferhöhle während des Kindesalters. Z Anat Entwicklungsgesch 85: 734–761.
Schaeffer JP (1920) The Nose, Paranasal Sinuses, Nasolacrymal Passageways, and Olfactory Organ in Man. Blakiston, Philadelphia.
Seydel O (1891) Über die Nasenhöhle der höheren Säugethiere und des Menschen. Gegenbaurs Morphol Jahrb 17: 44–99.
Shea BT (1977) Eskimo craniofacial morphology, cold stress and the maxillary sinus. Am J Phys Anthrop 47: 289–300.
Szilvássy J, Kritscher H, Vlcek E (1987) Die Bedeutung röntgenologischer Methoden für die anthropologische Untersuchung ur- und frühgeschichtlicher Gräberfelder. Ann Naturhist Mus Wien A89: 313–352.
Takahashi R (1984) The formation of the human paranasal sinuses. Acta Otolagryngol Suppl 408: 1–28.
Tanaka M (1983) Study on the maxillary sinus in human cranium in deciduous and mixed dentition periods. Shikwa Gakuho 83: 803–839.
Taylor RMS (1980) The maxillary sinus and molar teeth. Aspects of varying anatomical relations. J Dent Ass South Africa 35: 51–56.

Tillier AM (1977) La pneumatisation du massif cranio-facial chez les hommes actuels et fossiles. Bull Mém Soc Anthrop Paris XIII, 4: 177–189.
Uemura J (1974) Morphological studies on the maxilla of the edentulous skulls and the skulls with teeth, 1. On the sinus of the maxilla. Shikwa Gakuho 74: 1860–1889.
Ward SC, Pilbeam DR (1982) Maxillofacial morphology of miocene hominoids from Africa and Indo-pakistan. In: Cichion RL, Corruccini RS (eds) New Interpretations of Ape and Human Ancestory. Plenum Press, New York, p. 211–238.
Wegner RN (1936) Sonderbildungen der Kieferhöhle bei Anthropoiden. Anat Anz 83: 161–193.
Wegner RN (1955) Studien über Nebenhöhlen des Schädels. Wiss Z Ernst-Moritz-Arndt-Univ Greifswald 5: 1–111.
Weinert H (1927) Die Ausbildung der Stirnhöhlen als stammesgeschichtliches Merkmal. Z Morphol Anthropol 25: 243–357.
Witmer LH (1995) Homology of facial structures in extant archosaurs (birds and crocodilians), with special reference to paranasal pneumaticity and nasal conchae. J Morphol 225: 269–327.
Zonneveld FW, Spoor CF, Wind J (1989) The use of CT in the study of the internal morphology of hominid fossils. Medicamundi 34: 117–128.

3.6 Current Aspects of Dental Research in Paleoanthropology

Winfried Henke

Introduction

From the beginning of paleoanthropological research until the mid-nineteenth century, dental evidence was dominant in discussions on primate evolution in general (Butler 1963, 1986; Chivers et al. 1984; Fleagle 1988; Gregory 1920; Gregory and Hellman 1926; Osborn 1907; Owen 1859; RD Martin 1990; Remane 1921, 1960), and the reconstruction of human evolution in particular. This is understandable since teeth constitute the most enduring component of the body (overview in Brothwell 1963; Butler and Joysey 1978; Szalay and Delson 1979; Cruwys and Foley 1986; Fleagle 1988; Grine 1988; Aiello and Dean 1990; P Smith and Tchernov 1992; Henke and Rothe 1994, 1997 a, b; Alt and Türp 1997). Teeth are often the only preserved biological substratum or trace of a vertebrate organism. Useful not only for taphonomic purposes, teeth are an extremely valuable source of palaeontological information. Since teeth develop early in life embedded in the jaws and are protected from external environmental influences, they provide much information about the genetic constitution and development of an individual. On the other hand worn teeth and their special macro- and microstructures disclose much about food acquisition and processing.

This survey aims to review the current dental studies in paleoanthropology, focusing on diachronic changes in perspectives and describing innovative fields of recent and future dental research in paleoanthropology. Still, the scope of this review is limited, and this paper is far from comprehensive. Rather, its main intent is to demonstrate how dental anthropology and paleoanthropology interlink methodologically, bringing together comparative morphology of living apes and humans with that of early fossil hominoids for a better understanding of our evolutionary relationships and the patterns in human evolutionary ecology.

The Scope of Paleoanthropology

Knowledge of our evolutionary past has undergone breathtaking advances during the last decades. This progress cannot be explained merely by the enormous increase of fossil evidence, which of course caused radical revisions of earlier interpretations. Far more important for the present dynamic, expansionary phase of paleoanthropology was a fundamental methodological change in its evaluation. While the traditional approach has for the most part been a detailed, purely morphological description and consideration of the phylogenetic change of morphological form, current studies are multidisciplinary projects from various fields of biological and medical sciences as well as biochemistry, physics, earth sciences and archeology. One decisive aim of this integrated approach is the explanation of evolutionary adaptation. As morphology alone does not provide a sufficient foundation on which to analyse the phenomenon of biological adaptation, in the 1960s v. Bertalanffy (1960), Bock and v. Wahlert (1965) and other biologists proposed the study of two dimensions of phenotypic features – i.e. form and function. A current understanding of morphology regards form and function as "two inseparable components of biological features and must always be considered together" (see Bock and v. Wahlert 1965, 271). A simplified scheme illustrating the hierarchy and relationships between the components of the organism and the environment which are pertinent to the understanding of biological adaptation is shown in Fig. 1 (additional comments in Henke and Rothe 1994).

The indicated theoretical approach demonstrates a new understanding of paleoanthropology insofar as fossil evidence from dental and skeletal remains is a prime source of information from which to reconstruct the form and lifestyle of early hominids, i.e. "to define the problems that early hominids faced, and to relate the problems to the evolution of hominid adaptive strategies" (Foley 1987, xxi).

In addition to these aspects of human evolutionary ecology, current paleoanthropology is characterized by an intensive search for reliable principles and methods of classification and phylogenetic reconstruction (e.g. Hennig 1950; Remane 1952; Mayr 1969; Ax 1984; Willmann 1985; RD Martin 1990; Henke and Rothe 1994). While classification infers the allocation of species to groups and the construction of a nomenclature – a discipline termed taxonomy by

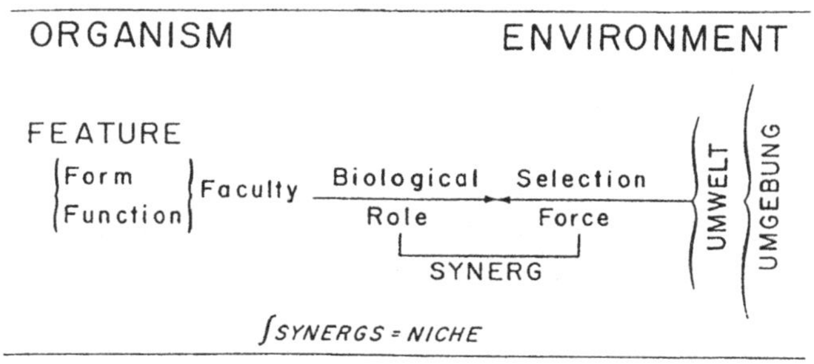

Fig. 1. Scheme of hierarchy and relationships between the components of the organism and of the environment, illustrating adaptation as a compromise between the demands of all the synergetical relationships in which it occurs (from Bock and v. Wahlert 1965, Fig. 1).

Simpson (1961) – phylogenetic reconstruction concerns the pattern of relationships within and among fossil and extant members of those groups. It is essential to recognize that classification and phylogenetic reconstruction have distinct objectives although they are interdependent. Even the recent literature displays the confusion concerning the underlying theoretical principles. There are three competing concepts:

- First, the cladistic school (alternatively labeled phylogenetic systematics; see Eldredge and Cracraft 1980);
- second, evolutionary systematics, which is founded on the more traditional thoughts of Simpson (1931, 1961, 1963) and Mayr (1942, 1969) on a grade-based classification, and
- third, the school of numerical taxonomy (Sneath and Sokal 1973), which does not recognize the need for a classification to reflect evolutionary relationships among the taxa included.

These different methods of phylogenetic reconstruction are – obviously unavoidably – based on the assessment of probabilities and do not depend on certainties. *We are just modelling*! The presently adequate comprehensive approach to the reconstruction of the phylogenetic history of a group such as the primates is shown in a flow chart (Fig. 2; see RD Martin 1990).

Human Evolution: Evidence From Comparative Dental Morphology

Our understanding of primate evolution is ultimately based on patterns of phyletic relationships and morphological change in the fossil record, but in addition to this source of information, there are of course biochemical and molecular data that provide new information about the interrelationships of living primate species and the probable times of divergence of their lineages from a common ancestor (Goodman 1982; Cronin 1983; RD Martin 1990). Though these new innovative disciplines, which have been named paleogenetics, have gained increasing importance over the last decades, they have not rendered the traditional disciplines obsolete. New analytical developments and conceptual advances – especially in dental anthropology – have produced an enormous number of new answers to the main questions of paleoanthropology, i.e. the relationships of the hominoids and the split of the hominid lineage from those of other primates, the morphological changes in the hominid phylogeny and the ecological niches of the fossil hominids, and finally the interrelationships of fossil hominids (e.g. Remane 1960; Dahlberg 1971; Butler and Joysey 1978; Kurtén 1982; Ciochon and Corruccini 1983; Foley and Cruwys 1986; Scott and Turner 1988; RD Martin 1990; Kelley and Larsen 1991; Smith and Tchernov 1992; Henke and Rothe 1994).

All primates have teeth in both the maxilla and the mandible, and within the upper and lower jaws primate teeth are bilaterally symmetrical. Their permanent dentitions are made up of four types of teeth, in mesial to distal order: incisors (I), canines (C), premolars (P) and molars (M). While the hypothetical ancestral primate had three I, one C, four P and three M (dental formula 3.1.4.3), there is a

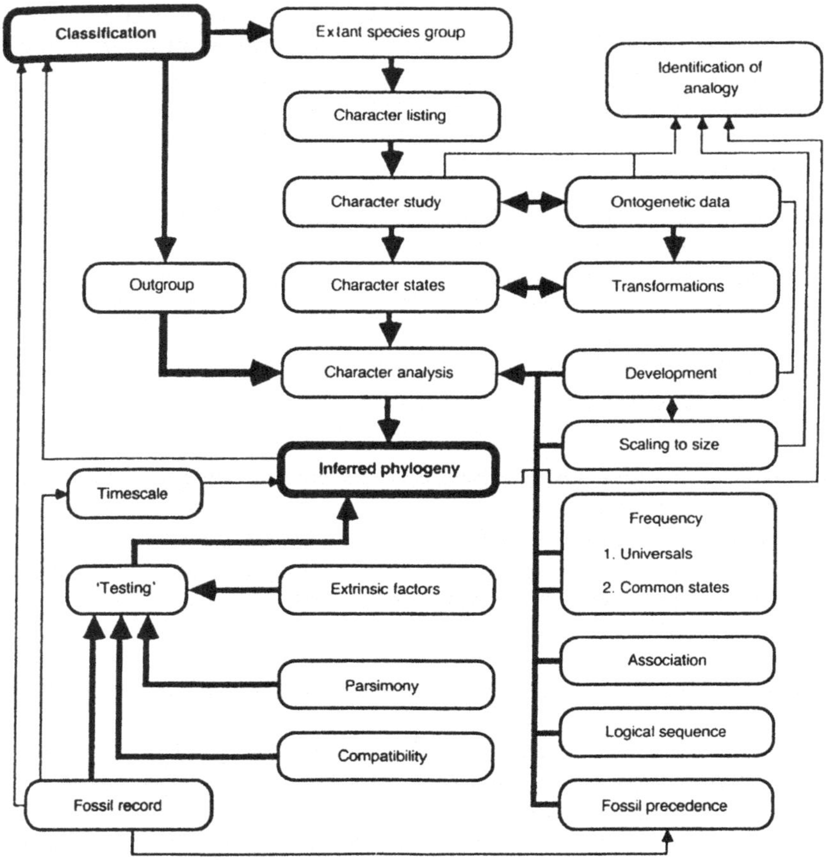

Fig. 2. Basic flow diagram indicating the main procedures advocated for phylogenetic reconstuction using traditional biological evidence (after RD Martin 1990, Fig. 3.12).

progressive reduction in the number of teeth (see Fig. 3) and simultaneously a trend towards a more complex structure in the remaining teeth. Old world monkeys (Catarrhini), apes (pongids and hylobatids) and hominids all have the dental formula 2.1.2.3, hence a reduction by one incisor (probably the lateral) and two mesial premolars (P1 and P2). The premolars in the Catarrhini are usually called P3 and P4 (see Remane 1960). As all other mammals primates are diphyodont, i.e. there is an initially deciduous dentition which is replaced by the successional or permanent dentition. Hominoids have two deciduous incisors, one deciduous canine and two deciduous premolars (called milk molars). Though there is a large and continually expanding body of information on various aspects of the deciduous dentition of primates (e.g. Remane 1960; Grine 1985; Aiello and Dean 1990; RD Martin 1990), this article focuses primarily on paleoanthropological conclusions from the permanent dentition.

Incisors of prosimians have evolved from very simple single rooted conical teeth. They are very variable in lower primates; e.g. *Daubentonia madagascariensis,*

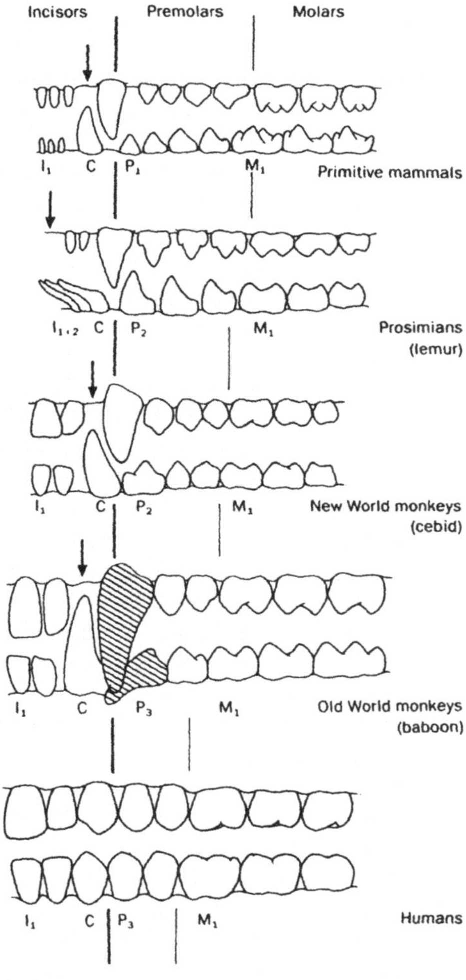

Fig. 3. The teeth of a primitive mammal and the special features of the main primate groups (after Dean 1992, 57). The vertical lines are oriented through the centre of the upper canine and the contact point between the last upper premolar and the first upper molar and show the different proportions of each part of the dentition; arrows indicate a gap or diastema. In the prosimians (lemur) the lower canine is incorporated into the comb (stippled). Lemurs and lorises have a median diastema. The molar series reduces in size from mesial to distal in all groups except Old World monkeys and apes (represented by baboons). The honing complex formed by the upper canine and lower sectorial premolar is hatched in the baboon.

commonly called Aye-Aye, possesses lifelong growing incisors like rodents, but in simians incisors are blade-like teeth for cutting and shearing food. Odontometrical studies demonstrate a continuous reduction in relative size through Old World monkeys, apes and humans (Remane 1960).

The canines, which are single-rooted like the incisors, are large teeth that can pierce food. Aside from their masticatory function they are important in social life, especially in monkeys and apes, where they are highly sexually dimorphic. There are obvious differences in the crown size and shape between pongids and hominids indicating a functional change in the chewing apparatus (Hiiemae and Kay 1973). While the upper canines tend to become more and more similar to the premolars (premolarization), the lower canines show an "incisification". Diminuative and morphologically altered canines have long been recognized as one of many autapomorphic human characteristics (Huxley 1863; Darwin 1871), but there is still a broad debate on this issue (RD Martin 1990).

The premolars are commonly bicuspid teeth with cutting and grinding functions. In forms with large canines – e.g. apes and monkeys – the first lower premolar is sectorial, i.e. one cusp is very large. When the jaws are closed, the convex-curved anterior face shears against the back of the upper canine. While *Australopithecus afarensis* has lower premolars that closely resemble the sectorial pattern of the pongids, all other hominids have very similar P_3 and P_4 which are non-sectorial. Some of them show a tendency toward additional cusps called molarization (Remane 1960; Swindler 1976; Aiello and Dean 1990; Henke and Rothe 1994). The differences between pongids and hominids are gradual, since the traits are intraspecifically highly variable, and there is interspecific overlapping.

Molars show the most complicated morphology of all tooth types. They have more cusps than premolars and enlarged occlusal surfaces for crushing and grinding food. These multicusped teeth have a complex topography. The cusps of the upper molars correspond to hollows in the lower molars, and vice-versa, functioning like a mortar and pestle. The basic design of mammalian molars is a triangle, a tooth pattern that has been described as tribosphenic. The upper molars are a simple triangle (or trigon) with only three main cusps. The apices are oriented toward the palate. The lingual/palatinal cusp in the apex is the protocone, buccomesial lies the metacone, and buccodistal the paracone. Small cusps adjacent and lingual to these major cusps are the paraconule and the metaconule (see Fig. 4). Furthermore, there are accessory folds of enamel on the buccal surface of the tooth (styles) and a collar of enamel around the base of the tooth which has been described as a cingulum.

The lower molars form the same basic triangle but with the apex pointing buccally. To distinguish the cusps of the upper and lower molars, the suffix *-id* is, added to their names i.e. protoconid, paraconid, and metaconid. This mesially-positioned basic triangle (trigonid) is fused at its distal end to a talonid which is formed by two or three cusps (hypoconid [buccal], entoconid [lingual]), and a small, most distal cusp between them, the hypoconulid (for further discussions on Osborn's (1888, 1907) system of nomenclature see Hershkovitz 1971; Butler 1978; Starck 1979; RD Martin 1990).

This basic pattern in molars is recognizable in recent tarsiers, while in other extant primates there is commonly an additional fourth cusp buccolingually (inside of the rear) of the upper molars as well as a trend to lose the leading cusp on the main triangle of the lower molar. This trend is illustrated in Fig. 4. Due to the fact, that the patterns of cercopithecoids and hominoids developed in different ways, these tooth characteristics are excellent for a differential diagnosis of both taxa. Old World monkeys have developed a cusp pattern known as bilophodonty (four cusps linked in pairs by transverse ridges); hominoids show a pattern which has been called the Y-5- or *Dryopithecus* pattern (Gregory 1916, 1921; Gregory and Hellman 1926, 1927). The evolutionary significance of the "*Dryopithecus* pattern"

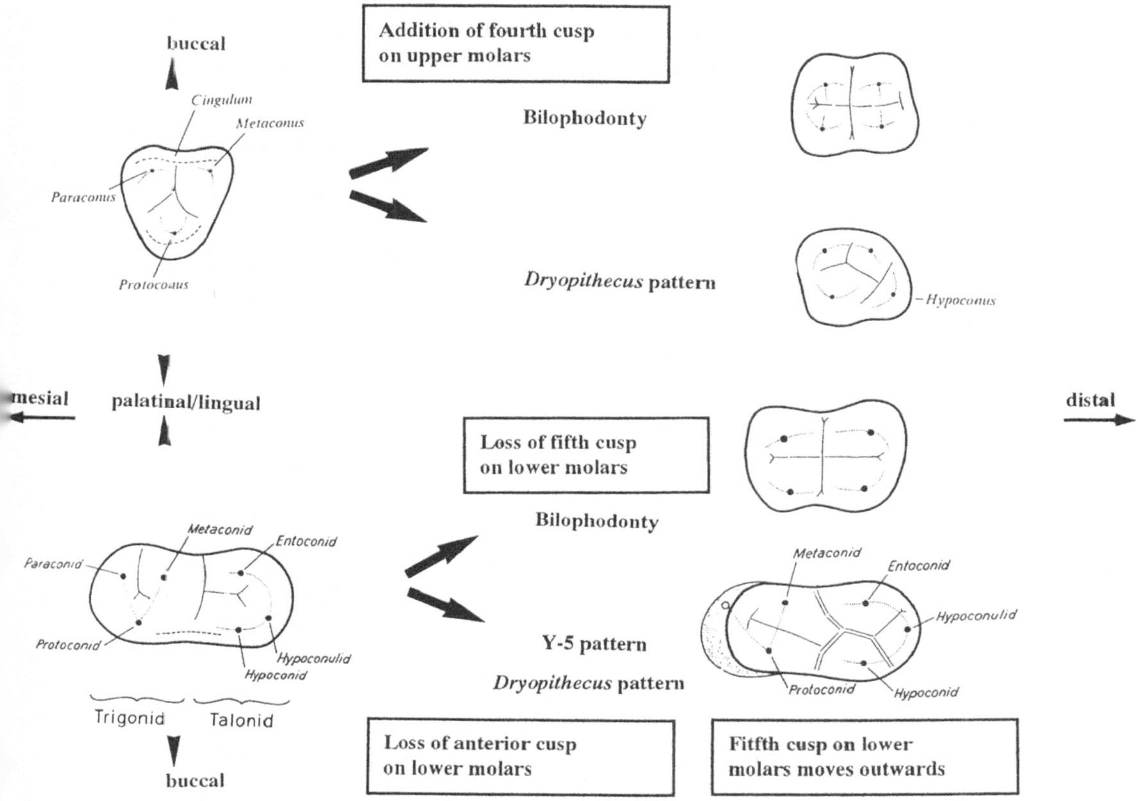

Fig. 4. Evolution of bilophodonty and *Dryopithecus*-pattern (combined based on Starck 1979; Dean 1992; Henke and Rothe 1994).

is, that it is characteristic of all Hominoidea and therefore a useful diagnostic feature for establishing phylogenetic affinities. Maier and Schneck (1981, 127) emphasize that "the '*Dryopithecus* molar-type' is by no means 'archaic', but represents a derived structural complex" (Fig. 5). Since the "*Dryopithecus* pattern" has undergone considerable alteration in its configuration, a number of sequences of change have been proposed (e.g. Remane 1960; Erdbrink 1965; Johanson 1974). Based on studies of human molars, a sequence from a Y-5 pattern to +-5, to +-4, and to the most advanced stage Y-4 is regarded as quite plausible (Johanson 1979).

Based on the morphognostic features of the simian dentition and their diagnostic value described above, it is, for example, possible to rule out all cercopithecoid monkeys from the line of ascent leading to the anthropoid apes and man due to their bilophodonty (see Gregory 1920). Furthermore, the differential structures of the canine-premolar-complex in pongids and hominids enable us to separate both taxa, although in this case the transition is rather obscure. Dart (1925) used these dental features [of the deciduous dentition] to claim the hominid status of the Taung child, and Johanson et al. (1978) recognized *Australopithecus afarensis* as a new species by careful evaluation of the dentition (in addition to cranial and postcranial features).

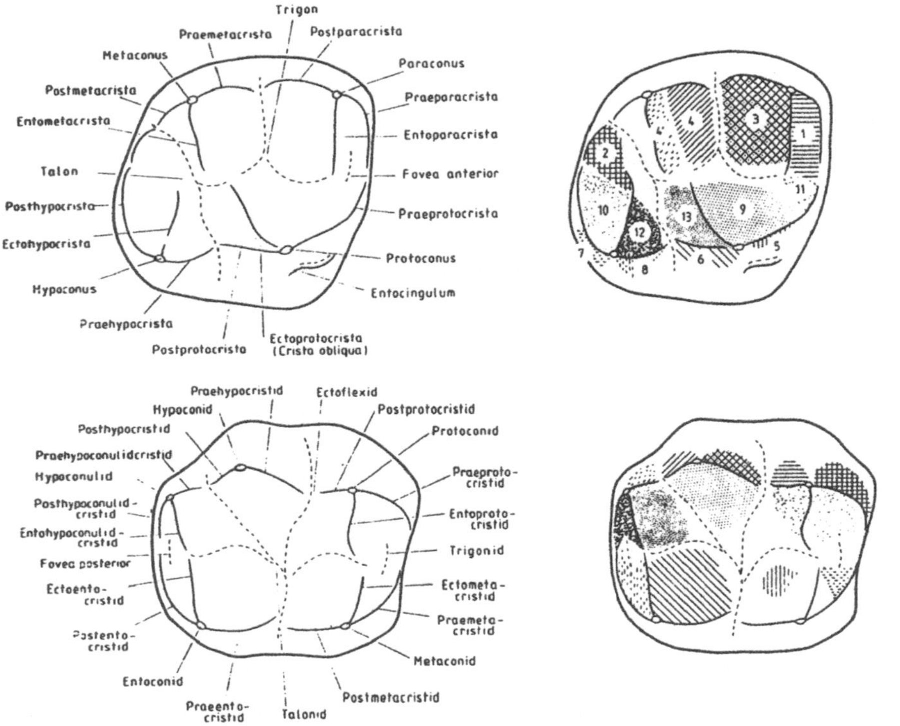

Fig. 5. Descriptive (left) and functional morphological (right) nomenclature (after Maier and Schneck 1981). The descriptive terms correspond to the conventional odontological-paleontological terminology. The functional morphology is demonstrated by the complementary pairs of facets of the upper and lower antagonists; the different elements of construction are numbered after Crompton (1971), Kay (1977), and Maier (1977). The equivalent pairs of facets are characterized by the same hatches. The relevant shearing edges and guiding tracks are shown; the valleys are dotted; the apices of the cusps are marked by small ovals.

Johanson and White (1979) mention, for example, that the large, asymmetric, pointed, lower canines project only slightly above the tooth row as well as beyond the upper ones. When worn, they often bear an exposed strip of dentine at the distal occlusal edge, and apical wear is often present as well. Together with the mandibular C/P3 complex, which is not functionally analogous to the pongid condition, and several other traits and states of the teeth and dentition (see Tobias 1991; Skelton and McHenry 1992), these constituted good arguments to regard *A. afarensis* as the oldest hominid species discovered until September 1994. At this time there was collected an adult, 4.1 million years old mandible at Kanapoi (Kenya). The strait parallel tooth rows and the receding symphysis beneath the front teeth distinguish this mandible from other early hominid species, while the configuration of the teeth are very similar to formerly described Australopithecines (Coffing et al. 1994; Leakey 1995). This is not the case concerning the *Species nova Ardipithecus ramidus* from Aramis (Ethiopia) which has been described firstly as *Australopithecus ramidus* by White et al. (1994, 1995). The fossils are still older than *A. afarensis* and *A. ramidus* but obviously do not belong to the Australopithecines. Its phylogenetic status has been recently discussed by Rothe et al. (1997). The discovery of a further *Species nova* from Koro Toro (Chad) by Brunet et al. (1995, 1996; see also Alt et al. 1996), which has been named *Australopithecus bahrelghazali* demonstrates that the early hominids occupied even habitats in the northern zones of Africa.

When considering relationships within the hominid lineage itself, we must explain dramatic adaptive changes in the configuration of the skull and the teeth of plio-pleistocene hominids. Though there have never been doubts, that the main task of the dentition of various primate taxa consists of the mechanical treatment of ingested foodstuff, a refined morpho-functional analysis of molar crown patterns was started for the first time in the 1970s by Kay (1973, 1977) and Maier (1977, 1978 a, b; Maier and Schneck 1981). Ethological field studies on food acquisition in primates (e.g. Clutton-Brock 1977; Chivers et al. 1984) and detailed functional studies on food processing (e.g. Hiiemae and Kay 1973; Hylander 1979) increased the availability of comparative data on the diets and tooth morphology of living species (Chivers et al. 1984). This essential knowledge allowed an integrated morphological approach to prove reliable functional hypotheses on skull configuration, dentition and crown pattern of fossil primates, and applies especially to current biomechanical models of the ecological niche of the australopithecines (Rak 1983; Demes and Creel 1988; Preuschoft 1989; Henke and Rothe 1994). Maier and Schneck (1981,131) stated in regard to the principles of functional tooth morphology: "Erst ein hinreichendes Verständnis biotechnischer, ethologischer und ökologischer Zusammenhänge wird die Formulierung befriedigender Evolutionsmodelle ermöglichen", and Preuschoft (1989, 421) postulated: "What we need is an explaining theory [...], which helps us to understand all evolutionary changes from ape-like forms through the earliest hominids to modern man."

Features of the australopithecine face, especially of the "robust" species, have been interpreted by DuBrul (1977), Grine (1981, 1988), Rak (1983), Demes and Creel (1988), and Preuschoft et al. (1989) to be adaptations for masticating a mechanically resistent diet that demands the generation of powerful bite forces. This interpretation is indicated by the large postcanine teeth (molarization of the premolars, enlargement of the occlusal surfaces of the molars), thick enamel as well as expanded and anteriorly-positioned chewing muscle insertions, higher mid-facial skeletons and extremely massive jaws (Preuschoft 1989; Henke and Rothe 1994). Demes and Creel (1988) derived estimates of bite forces from cranial measurements on extant hominoid species and hominid fossils and used these data to test the hypotheses which have been advanced about bite force, occlusal pressure and diet in fossil hominids. As can be seen from Fig. 6 there exists a roughly isometric relationship between bite force and molar crown area. Because neither the gracile nor the robust australopithecines deviate substantially from the regression line based on living hominoid species, the authors suggest: "..that the high bite force of the robust australopithecines was a necessary consequence of the enlargement of the molar crown surface" (Demes and Creel 1988). The authors concluded that Walker's interpretation, that the australopithecine chewing apparatus was adapted to low-energy food, that had to be processed in great quantities (Walker 1981), is in agreement with their results. Former "dietary hypotheses that postulate the processing of hard objects can be maintained only under the assumption that the

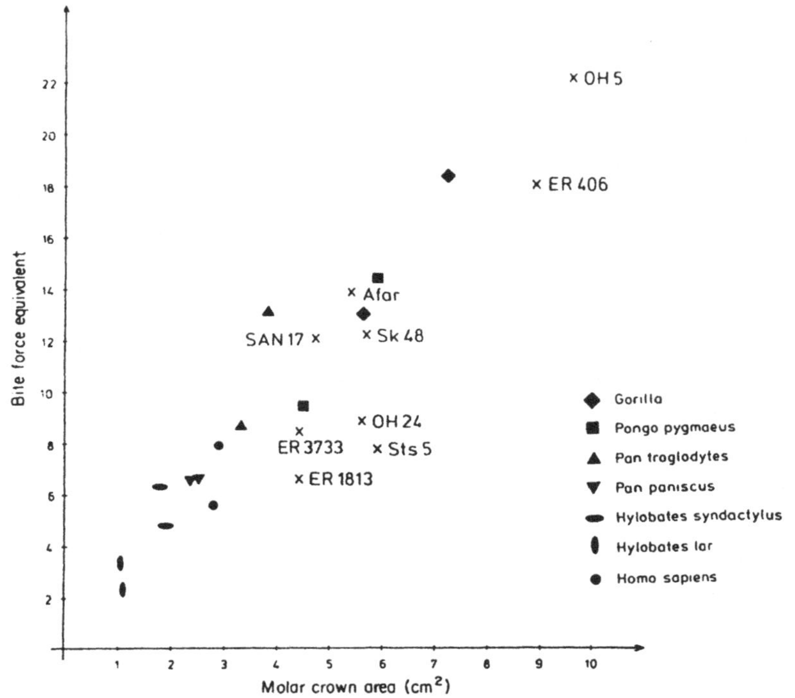

Fig. 6. Bite force estimates versus molar crown area (after Demes and Creel 1988).

area of contact between teeth and food was small" (Demes and Creel 1988, 667). Small or hard and round-shaped food objects like seeds or nuts fit their model, if only a small mouthful of food was triturated at a time. Scanning electron microscope analysis of occlusal events, i.e. dental microwear, by Grine (1981) suggests that the "robust" australopithecines from South Africa employed more crushing and puncture crushing activity than the "gracile" form. The "robust" hominids habitually, or at least seasonally, triturated harder, more resistant and perhaps smaller food objects than were masticated by the "gracile" australopithecines.

Representative SE micrographs of the occlusal surfaces of teeth from a cheetah (carnivore), an orang-utan (frugivore), and a spotted hyena (scavenger, bone eater) in comparison to a "robust" *Australopithecus* show that the fossil hominid is very dissimilar to the carnivores *s.l.* and resembles the herbivore form (Fig. 7). Walker (1981) claimed that *A. boisei* seems to have been a fruit-eater of fairly hard seasonal fruit, while *H. erectus*' microwear patterns are indicative of a more omnivorous and eclectic diet.

Experimental approaches have been carried out by analysing microwear of laboratory animals (*Didelphis marsupialis*, oppossum) fed different diets (insectivorous and herbivorous). Covert and Kay (1981, 331) came to the conclusion that "diets of extinct forms cannot always be deduced by the analysis of microwear", but noted that microwear analysis can provide information on whether the animal's food included components of exogenous grit or silicate-

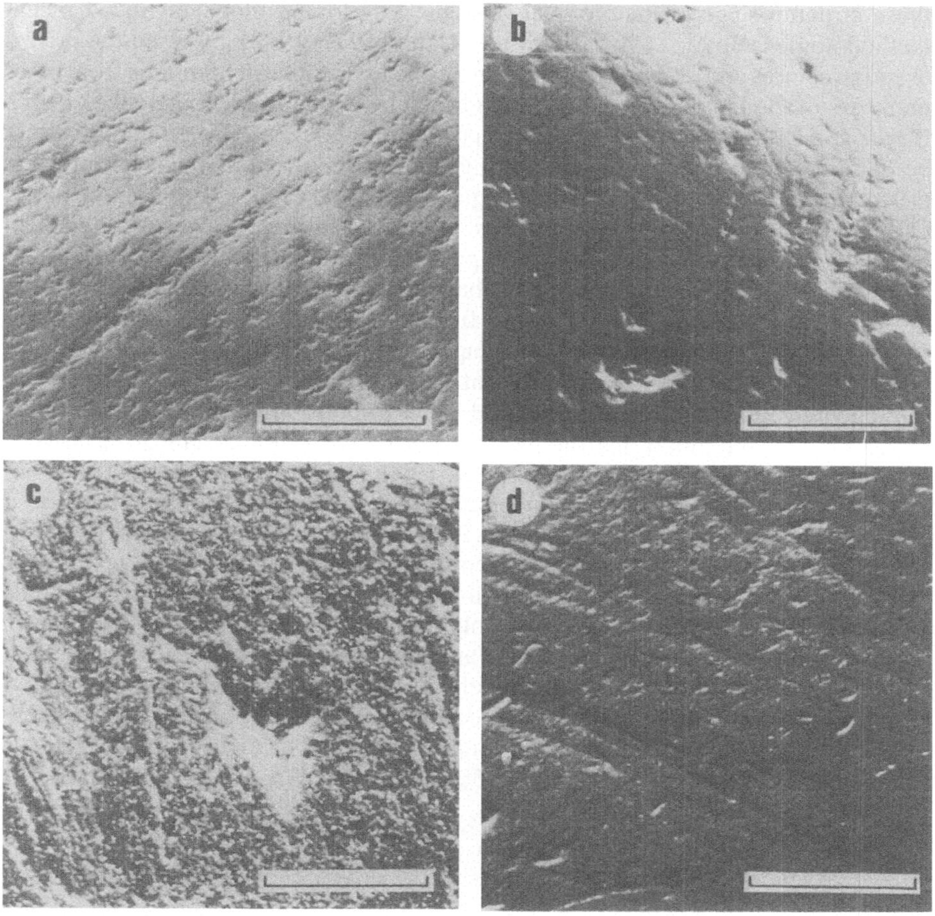

Fig. 7. Micrographs (SEM) of the occlusal surfaces of (**a**) cheetah, (**b**) orang-utan, (**c**) spotted hyena (**d**) "robust" *Australopithecus*; scale bars: 200 μm (after Walker 1981, Plate 2c–f).

containing plants. Further studies on dental microwear of chimpanzees by Gordon (1982) suggested that microwear is not only caused by diet, but also by masticatory movement, position of the tooth or the facet in the mouth and even the individual age of the specimen examined. As of now the question has not been resolved of whether the scars, pits and scratches on teeth are reliable indicators of the normal diet of an individual or whether they indicate only the food eaten shortly before death. These results show that microwear studies are extremely difficult and require further improvement to attain reliable results.

Apart from microwear, other scanning and optical miscroscopic aspects of the teeth yielded a wealth of new information. Enamel, the best preserved and most highly mineralized portion of the tooth, was expected to yield much information about human evolution. Though the high variability of the enamel structure has been mentioned by Tomes (1848). Carter (1922) was probably the first to study these structures for primatological reasons. Employing more advanced recent methodologies, Boyde (1964, 1975) and Gantt (1977, 1980) analysed the prisms or rod structures and claimed that it was possible to distinguish hominids from other members of the Hominoidea. Vrba and Grine (1978), Xirotiris and Henke (1981), Boyde and LB Martin (1984), Radlanski (this volume) pointed out that one should be extremely cautious when interpreting the results taxonomically, because the prism pattern is highly variable in a single species and even within a single tooth. Only the higher taxa among the primates (e.g. superfamilies) may be distinguished by relative frequencies of different rod patterns and by the extent of decussation. The pitch or angle of Hunter-Schreger bands in early *Homo* does not differ from those of modern humans, but does differ from those of *A. boisei* (Beynon and Wood 1986). Strong decussation may be an adaptation to resist enamel fracture and may have evolved independently in different taxonomic lineages. The objections to phylogenetic information from enamel ultrastructure are supported by results indicating the likelihood that the prism pattern affects biomechanical properties (Foley and Cruwys 1986; Aiello and Dean 1990; Henke and Rothe 1994).

Tooth enamel thickness and structure analyses of hominoid fossils have been carried out by JT Robinson (1956), Jolly (1970), Simons (1972, 1976), Gantt (1977), Kay (1981, 1985), LB Martin (1985), P Smith and Zilberman (1994). It was shown that the teeth of *Gorilla* and *Pan* have a thinner layer than do those of fossil hominids and fossil apes. The enamel in *Pongo* is only slightly thicker than that in African apes, but does not wear away as rapidly. There are good reasons to believe that tough food might favour the evolution of thick enamel, but allometry and the ontogenetic period of functioning over a lifetime may also affect the thickness of the layer. The postcanine teeth of the "robust" australopithecines have [corrected to tooth size] the thickest enamel of any primate (LB Martin 1985; Beynon and Wood 1986; Grine and LB

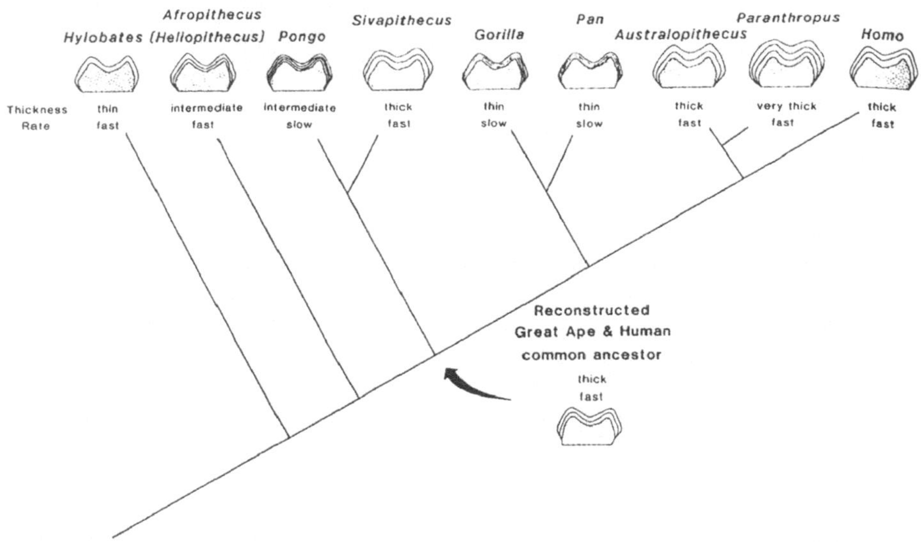

Fig. 8. Cladogram of molar enamel thickness and rates of enamel deposition in hominoids (after LB Martin 1985, from Fleagle 1988, Fig. 13.21)

Martin 1988). Even modern humans have thicker enamel than apes, obviously an indicator of a tough diet. Taking thickness rates for enamel deposition in fossil and extant apes into account, LD Martin (1985) suggested that thick enamel is best interpreted as the ancestral condition for great apes and humans, with chimpanzees and gorillas showing a secondary reduction in thickness (Beynon and Wood 1986; Fleagle 1988; Grine and Martin 1988). Fig. 8 demonstrates that tooth enamel thickness on its own does not give reliable phylogenetic information, but may be a useful indicator for the reconstruction of the diet niche. P Smith and Zilberman (1994) claim that thin enamel appears to characterize all Neanderthals. They guess that the thin enamel of Neanderthal teeth may be due to slow rates of enamel deposition or early cessation of ameloblast activity. This pattern separates the Neanderthals from *Homo sapiens sapiens* and shows them to be a highly specialized group.

The prisms of enamel can be described as tightly packed bundles of apatite crystals that transverse the tooth from the surface to the junction with the dentine. Along the length of the rods are incremental markings known as the brown striae of Retzius. Their outcroppings form a pattern of ridges and grooves, the perikymata. The striae of Retzius seem to occur in a circaseptan rhythmic manner. It is possible to count these markings and use them as chronological indicators of development. Bromage and Dean (1985) and Dean (1988) analysed these structures in pongids and hominids. The first molar of the famous Taung child, assigned to *A. africanus*, seems to have erupted at about 3 to 4 years, a pongid-like rate of tooth formation which implies – if the calibration is correct – a pongid-like pattern of development, a further step to the "dehominization" of the australopithecines (see Henke and Rothe 1994).

Only sparse results exist concerning the microscopic structure of dentine and cementum in fossil hominids (Aiello and Dean 1990). Morris (1978) analysed the reparative dentine which prevents the exposition of the pulp in cases of extreme attrition. His approach of using the banding of secondary dentine as an indicator of ontogenetic development was not successful because the underlying causes of this process were not yet understood. We are faced with similar problems in regard to cementum deposition. Foley (1986) and Foley and Cruwys (1986) pointed out the necessity of further research in this field "so that cementum banding can be interpreted, as it should be, in terms of an individual's life history, rather than general external conditions" (Foley and Cruwys 1986, 7).

Dental calculus occurs in relative abundance archaeologically but is of even greater interest in terms of the reconstruction of past human diet and oral ecology. The identification of microscopic food inclusions (e.g. phytoliths) is a highly problematic field of research involving botany, zoology and archaeology and is now only in its earliest stages (Rovner 1971; Newman 1980; Dobney and Brothwell 1986; Fox and Pérez-Pérez 1994).

In additon to the above-mentioned morphological and histological results which contributed essentially to a new understanding of the phylogeny and ecology of the early hominids, we must further mention the large amount of metrical dental studies on the plio-pleistocene and holocene hominids. Brace (1962), Wolpoff (1971, 1978), Frayer (1978), Brace et al. (1991) and others have demonstrated the relevant trends in tooth size and sexual dimorphism and the influence of culture on human dentition from plio-pleistocene to recent times based on crown diameters, indices and areas of occlusal surfaces (see Fig. 9). Others like Wood and Abbott (1983), Wood et al.

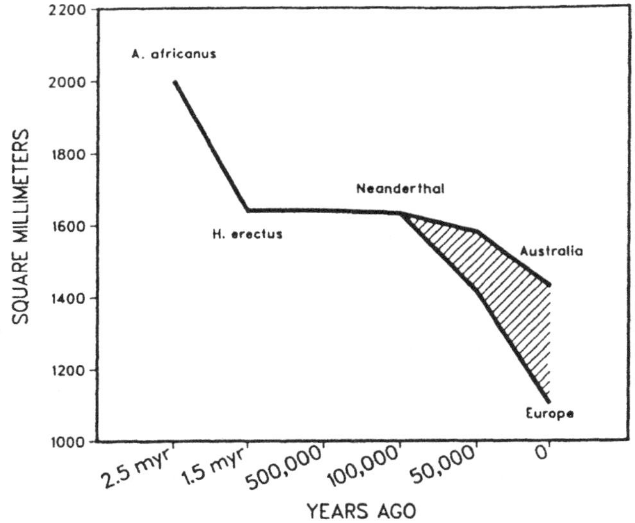

Fig. 9. Hominid tooth size from the australopithecines to the spectrum visible in living populations of *Homo sapiens* (after Brace et al. 1991, 49).

(1983), Wood and Uytterschaut (1987), Wood and Engleman (1988), Grine (1988), Grine and Martin (1988), and Suwa (1988) developed continually more sophisticated techniques. They established and documented detailed and precise morphometrical criteria to distinguish early hominid taxa by dental characteristics

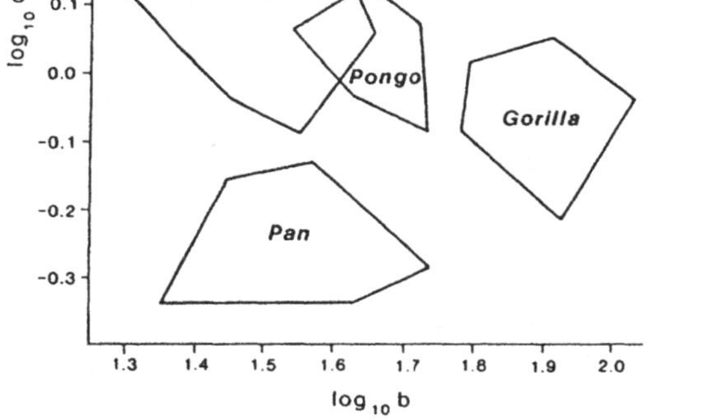

Fig. 10. Plot of average enamel thickness (Y axis) versus the dentine component of tooth size (X axis) The polygons encompass all specimens compromising the extant hominoid sample (after Suwa 1988, Fig. 1.19).

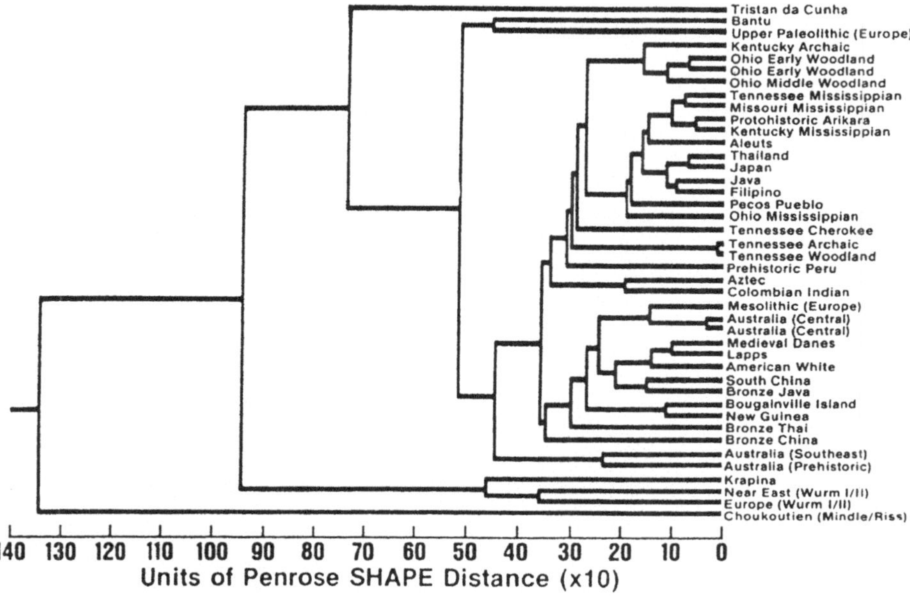

Fig. 11. Dendrogram based on Penrose shape coefficients computed from dental crown measurements of fossil and extant populations (Perzigian 1984, Fig. III).

and used the characteristics of these reference populations as a guide for assessing the affinities of specimens whose taxonomic designation is controversial (Fig. 10; see Wood et al. 1983; Suwa 1988). On the one hand there is a trend to include more informative measurements of the crown, the root or internal structures (dentine and enamel layers, pulpa measures, wear angles, etc.) of the teeth. On the other hand there is a shift from uni- and bivariate statistical methods to multivariate approaches (e.g. Penrose distances [Fig. 11], MMD, DFA, FA, PCA).

Within the *Homo* lineage there is evidence suggesting that bite force reduction paralleled by a reduction of the dental occlusal area is a late trend in hominid evolution. Furthermore, a steady decrease in sexual dimorphism has been described. It is necessary to mention that odontometrical studies give rise to many methodological problems, for example the exact measurement of the variables, the standardization to overall body size of the organism and the complications caused by abrasion and attrition.

Macrowear, especially, may give indications about age, diet, patterns of mastication and paramasticatory habits of an individual. An exemplary study by BH Smith (1984) on diachronic Nubian samples, dating from 5000 to 2000 B.P., showed more steeply angled wear of the molar crown through time, suggesting a trend towards an increased reliance on agricultural products (Fig. 12). Further biochemical analysis may be important for proving a possible change in food habits of human populations and the reconstruction of an individual's life history. The promising results from chemical analysis on bones (e.g. Grupe and Herrmann 1988; Price 1989) demonstrate that biochemical research of teeth can provide

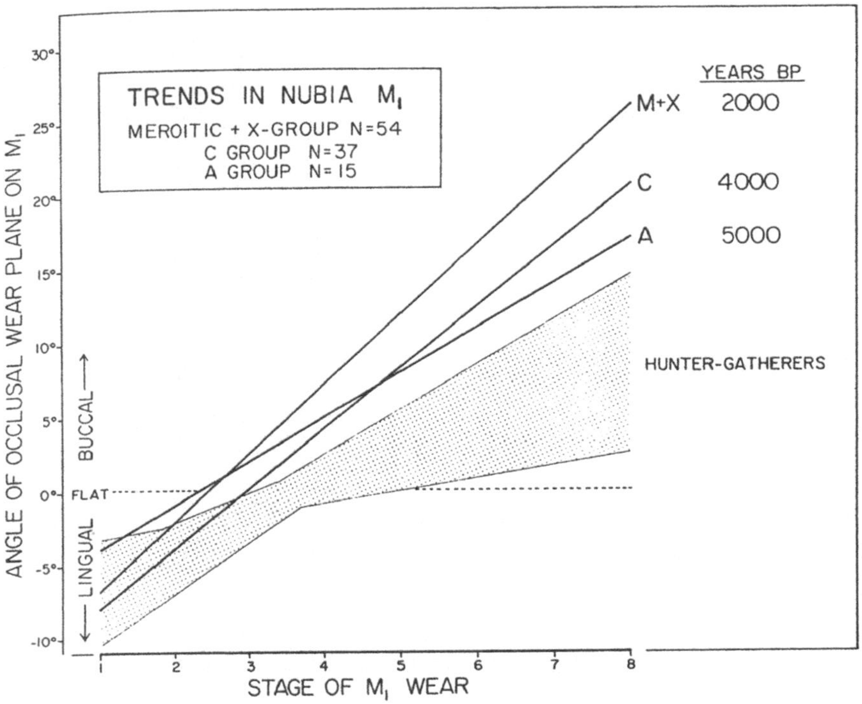

Fig. 12. Least squares lines from regression of M_1 wear plane angle on stage of wear for three time periods in Nubia. Range of least squares lines of hunter-gatherer is shaded (BH Smith 1984, Fig. 7).

valuable information to paleoanthropology in the near future (see C Robinson et al. 1986; Molleson 1988; Lubell et al. 1994; Grupe this volume).

Finally, non-metrical traits, as for example winging, shoveling, and double-shoveling (I^1), interruption grooves (I^2), tuberculum dentale (I^2), mesial ridge (C^-), distal accessory ridge (C^-), hypocone (M^2), root number of the premolars and molars, Carabelli trait, various extra cusplets, ridges and dental tubercles, and innumerable further traits have been studied in fossil, prehistorical and recent populations. Though the variation in human dentition is immense and the functional significance of many traits obscure, reasonable hypotheses of modern human origin have been established. Turner (1983, 1986, 1987) developed the Sundadont dental hypothesis for anatomically modern man's origin. He described two different dental types in East Asian populations, Sundadonty and Sinodonty. The Sundadonty, which is not found in Africa, Europe, or Australo-Melanesia and is characteristic of populations peripheral to East Asia (Southeast Asia, offshore achipelagos, Ainus, Pacific Island) is in all probability older than Sinodonty, a complex of dental traits characteristic of mainland East Asia and American Indians. Comparison of the dental morphology of circumpacific populations suggests that neither Europids [Caucasians] nor Mongolids participated in the first settlement of Northeast Asia and, after that, of the Americas. Sinodonty has obviously preserved pre-mongolid archeomorphic features (Turner 1986).

Conclusions

This short review shows that teeth are an essential source of information in paleoanthropology. Recent methodologies have employed a wealth of new techniques, e.g. sophisticated odontometry, comparative morphology and histology at transspecific, intraspecific, and individual levels, optical and scanning microscopy and – in its infancy – biochemical analyses. Since then a common agreement has evolved that much basic research has yet to be done. There is no doubt that current dental paleoanthropology – which focuses on a multi- and interdisciplinary functional approach to explain the selective advantage and survival value of particular features on the one hand, and on a better understanding of ecological structures and phylogenetic patterns of our forerunners on the other hand – is an encouraging and innovative discipline.

Acknowledgements

I am grateful to Dr. Maria Teschler-Nicola for her invitation to what proved to be an excellent and constructive workshop. I also thank the editors for publishing my contribution, and Dr. Ariane Kemkes-Grottenthaler and Heidi Asmus for improving my English translation.

References

Aiello L, Dean C (1990) An Introduction to Human Evolutionary Anatomy. Academic Press, San Diego.

Alt KW, Henke W, Türp JC (1996) Zähne und Kiefer – Schlüsselstrukturen zum Verständnis der Evolution. Quintessenz 47: 1711–1724.

Alt KW, Türp JC (1997) Die Evolution der Zähne. Phylogenie – Ontogenie – Variation. Quintessenz, Berlin.

Ax P (1984) Das phylogenetische System. Systematisierung der lebenden Natur aufgrund ihrer Phylogenese. Fischer, Stuttgart.

Bertalanffy L von (1960) Problems of Life. Harper and Brothers, Reprint, New York.

Beynon AD, Wood BA (1986) Variations in enamel thickness and structure in East African Hominids. Am J Phys Anthrop 70: 177–195.

Bock WJ, Wahlert G von (1965) Adaptation and the form-function complex. Evolution 19: 217–227.

Boyde A (1964) The Structure and Development of Mammalian Enamel. PhD Thesis, London.

Boyde A (1975) A method for the preparation of cell surfaces hidden within bulk tissue for examination in the SEM. In: Johari O (ed) Scanning Electron Microscopy. Illinois Inst Tech Res, Chicago, p. 295–303.

Boyde A, Martin LB (1984) The micro-structure of primate dental enamel. In: Chivers DJ, Wood BA, Bilsborough A (eds) Food Acquisition and Processing in Primates. Plenum Press, New York, p. 341–367.

Brace CL (1962) Cultural factors in the evolution of the human dentition. In: Montagu MFA (ed) Culture and the Evolution of Man. Oxford Univ Press, New York, p. 343–354.

Brace CL, Smith SL, Hunt KD (1991) What Big Teeth You Had, Grandma! Human Tooth Size, Past and Present. In: Kelly MA, Larsen CS (eds) Advances in Dental Anthropology. Wiley-Liss, New York, p. 33–57.

Bromage TG, Dean MC (1985) Re-evaluation of the age at death of immature fossil hominids. Nature 317: 525–527.

Brothwell DR (1963) (ed) Dental Anthropology. Symposia of the Society for the Study of Human Biology, Vol V. Pergamon Press, Oxford.
Brunet M, Beauvilain, Coppens Y, Heintz E, Moutage AHE, Pilbeam D (1995) The first australopithecine 2.500 kilometres west of the Rift Valley (Chad). Nature 378: 273–275.
Brunet M, Beauvilain, Coppens Y, Heintz E, Moutage AHE, Pilbeam D (1996) *Australopithecus bahrelghazali,* une nouvelle espèce d'hominidé ancien de la région de Koro Toro (Tchad). Compt Rds làcad Sci, Ser IIa, 322: 907–913.
Butler PM (1963) Tooth Morphology and Primate Evolution. In: Brothwell DR (ed) Dental Anthropology,Vol V. Pergamon Press, Oxford, p. 1–13.
Butler PM (1978) Molar cusp nomenclature and homology. In: Butler PM, Joysey KA (eds) Development, Function, and Evolution of Teeth. Academic Press, London, p. 439–454.
Butler PM (1986) Problems of dental evolution in the higher primates. In: Wood BA, Martin LB, Andres PA (eds) Major Topics in Primate and Human Evolution. Cambridge Univ Press, Cambridge, p. 89–106.
Butler PM, Joysey KA (eds) (1978) Development, Function and Evolution of Teeth. Academic Press, London.
Carter JT (1922) The structure of the enamel in the primates and some other mammals. Proc R Zool Soc Lond 1922: 599–608.
Chivers DJ, Wood BA, Bilsborough A (eds) (1984) Food Acquisition and Processing in Primates. Plenum Press, New York.
Ciochon RL, Corruccini RS (eds) (1983) New Interpretations of Ape and Human Ancestry. Plenum Press, New York.
Clutton-Brock TH (ed) (1977) Primate Ecology. Academic Press, London.
Coiffing K, Feibel C, Leakey M, Walker A (1994) Four-million-year-old hominids from east Lake Turka. Am J Phys Anthrool 93: 55–65.
Covert HH, Kay RF (1981) Dental microwear and diet: Implications for determining the feeding behaviours of extinct primates, with a comment on the dietary pattern of *Sivapithecus*. Am J Phys Anthrop 55: 331–336.
Crompton AW (1971) The origin of the tribosphenic molar. In: Kermack DM, Kermack A (eds) Early Mammals. Zool J Linn Soc London 50 (Suppl), London, p. 65–87.
Cronin JE (1983) Apes, humans and molecular clocks. A reappraisal. In: Ciochon RL, Corruccini RS (eds) New Interpretations of Ape and Human Ancestry. Plenum Press, New York, p. 115–149.
Cruwys E, Foley RA (1986) Teeth and Anthropology. BAR Intern. Series 291, Oxford.
Dahlberg AA (1971) (ed) Dental Morphology and Evolution. Univ Press, Chicago, p. 81–94.
Dart RA (1925) *Australopithecus africanus*: The ape-man of South Africa. Nature 115: 195–197.
Darwin C (1871) The Descent of Man and Selection in Relation to Sex. Murray and Sons, London.
Dean MC (1988) Growth of teeth and development of the dentition in *Paranthropus*. In: Grine FE (ed) Evolutionary History of the "Robust" Australopithecines. De Gruyter, New York, p. 43–54.
Dean C (1992) Jaws and teeth. In: Jones S, Martin R, Pilbeam D, Bunney S (eds) The Cambridge Encyclopedia of Human Evolution. Cambridge Univ Press, Cambridge, p. 56–59.
Demes B, Creel N (1988) Bite force, diet and cranial morphology of fossil hominids. J Hum Evol 17: 657–670.
Dobney K, Brothwell D (1986) Dental calculus: Its relevance to ancient and oral ecology. In: Cruwys E, Foley RA (eds) Teeth and Anthropology. BAR International Series 291, p. 55–82.
DuBrul LE (1977) Early hominid feeding mechanisms. Am J Phys Anthrop 47: 305–320.
Eldredge N, Cracraft J (1980) Phylogenetic Patterns and the Evolutionary Process: Method and Theory in Comparative Biology. Columbia Univ Press, New York.

Erdbrink DP (1965) A quantification of the *Dryopithecus* and the lower molar patterns in man and in some apes. Z Morph Anthrop 57: 70–108.
Fleagle JG (1988) Primate Adaptation and Evolution. Academic Press, San Diego.
Foley RA (1986) Cementum deposition among tropical African ungulates: implications for paleoecological studies. In: Cruwys E, Foley RA (eds) Teeth and Anthropology. BAR Int Ser 291, Oxford, p. 83–100.
Foley RA (1987) Another Unique Species. Patterns in Human Evolutionary Ecology. Longman, Harlow.
Foley RA, Cruwys E (1986) Dental anthropology: Problems and perspectives. In: Cruwys E, Foley RA (eds) Teeth and Anthropology. BAR Int Ser 291, Oxford, p. 1–20.
Fox CL, Pérez-Pérez A (1994) Dietary information through the examination of plant phytoliths of the enamel surface of human dentition. J Archeol Science 21: 29–34.
Frayer D (1978) Evolution of the Dentition in Upper Palaeolithic and Mesolithic Europe. Univ Kansas Publ Anthrop, Lawrence.
Gantt DG (1977) Enamel of Primate Teeth: Its Thickness and Structure with Reference to Functional and Phyletic Implications. PhD Diss, St. Louis.
Gantt DG (1980) The enamel of neogene hominoids. Structural and phyletic implications. In: Ciochon RL, Corruccini RS (eds) New Interpretations of Ape and Human Ancestry. Plenum Press, New York, p. 249–298.
Goodman M (1982) Biomolecular evidence on human origins from the standpoint of Darwinian theory. Hum Biol 54: 247–264.
Gordan KD (1982) A study of microwear on chimpanzee molars: Implications for dental microwear analysis. Amer J Phys Anthrop 59: 195–215.
Gregory WK (1916) Studies on the evolution of Primates, I. The Cope-Osborn "theory of trituberculy" and the ancestral molar pattern of the Primates. Bull Am Mus Nat Hist 35: 239–257.
Gregory WK (1920) The origin and evolution of the human dentition: A palaeontological review, Part IV. The dentition of the higher primates and their relationships with man. J Dent Res 2: 607–717.
Gregory WK (1921) The origin and evolution of the human dentition: A palaeontological review, Part V. Later stages in the evolution of the human dentition; with a final summary and a bibliography. J Dent Res 3: 87–228.
Gregory WK, Hellman M (1926) The crown pattern of fossil and recent human molar teeth and their meaning. Nat Hist NY 26: 300–309.
Gregory WK, Hellman M (1927) The dentition of *Dryopithecus* and the origin of man. Anthrop Pap Am Mus Nat Hist 28: 1–123.
Grine FE (1981) Trophic differences between "gracile" and "robust" australopithecines: A scanning electron miscroscope analysis of occlusal events. South Afr J Science 77: 203–230.
Grine FE (1985) Australopithecine evolution: The deciduous dental evidence. In: Delson E (ed) Ancestors: The Hard Evidence. Liss, New York, p. 153–167.
Grine FE (ed) (1988) Evolutionary History of the "Robust" Australopithecines. De Gruyter, New York.
Grine FE Martin LB (1988) Enamel Thickness and Development in *Australopithecus* and *Paranthropus*. In: Grine FE (ed) Evolutionary History of the "Robust" Australopithecines. De Gruyter, New York, p. 3–42.
Grupe G, Herrmann B (1988) (eds) Trace Elements in Environmental History. Springer, Berlin.
Henke W, Rothe H (1994) Paläoanthropologie. Springer, Berlin.
Henke W, Rothe H (1997a) Zahnphylogenese der nichtmenschlichen Primaten. In: Alt KW, Türp JC (eds) Die Evolution der Zähne – Phylogenie, Ontogenie, Variation. Quintessenz, Berlin, p. 223–272.

Henke W, Rothe H (1997b) Zahnphylogenese der Hominiden. In: Alt KW, Türp JC (eds) Die Evolution der Zähne – Phylogenie, Ontogenie, Variation. Quintessenz, Berlin, p. 273–353.
Hennig W (1950) Grundzüge einer Theorie der Phylogenetischen Systematik. Deutscher Zentralverlag, Berlin.
Hershkovitz P (1971) Basic crown patterns and cusp homologies of mammalian teeth. In: Dahlberg AA (ed) Morphology and Evolution. Chicago Univ Press, Chicago, p. 95–150.
Hiiemae KM, Kay RF (1973) Evolutionary trends in the dynamics of primate mastication. Symp 4th Int Congr Primatol 3: 28–64.
Huxley TH (1863) Evidences as to Man's Place in Nature. Williams and Norgate, London.
Hylander WL (1979) The functional significance of primate mandibular form. J Morphol 160: 223–239.
Johanson DC (1974) Morphological and Metrical Variability in the Chimpanzee Molar Dentition. PhD Diss, Chicago.
Johanson DC (1979) A consideration of the "*Dryopithecus* pattern". Ossa 6: 125–135.
Johanson DC, White TD (1979) A systematic assessment of early African hominids. Science 203: 321–330.
Johanson DC, White TD, Coppens Y (1978) A new species of the genus *Australopithecus* (Primates, Hominidae) from the Pliocene of eastern Africa. Kirtlandia 28: 1–14.
Jolly CJ (1970) The seedeaters: A new model of hominid differentiation based on a baboon analogy. Man NS 5: 5–26.
Kay RF (1973) Mastication, Molar Tooth Structure, and Diet in Primates. PhD Thesis, Ann Arbor.
Kay RF (1977) Diet from early Miocene African hominoids. Nature 268: 628–630.
Kay RF (1981) The nut-crackers: A new theory of the adaptation of the Ramapithecinae. Am J Phys Anthrop 455: 141–152.
Kay RF (1985) Dental evidence for the diet of *Australopithecus*. Ann Rev Anthrop 14: 315–343.
Kelley MA, Larsen CS (eds) (1991) Advances in Dental Anthropology. Wiley-Liss, New York.
Kurtén B (ed) (1982) Teeth: Form, Function, and Evolution. Columbia Univ Press, New York.
Leakey M (1995) The farthest horizon. National Geographic Magazine 188: 38–51.
Lubell D, Jackes M, Schwarcz H, Knyf M, Meiklejohn C (1994) The mesolithic-neolithic transition in Portugal: Isotopic and dental evidence of diet. J Archeol Science 21: 201–216.
Maier W (1977) Die Evolution der bilophodonten Molaren der Cercopithecoidea. Eine funktionsmorphologische Untersuchung. Z Morph Anthrop 68: 26–58.
Maier W (1978a) Zur Evolution des Säugetiergebisses – typologische und konstruktionsmorphologische Erklärungen. Natur u. Museum 108: 288–300.
Maier W (1978b) Die Evolution des tribosphenischen Säugetiermolaren. Sonderb Naturwiss Verh Hamburg 3: 41–60.
Maier W, Schneck G (1981) Konstruktionsmorphologische Untersuchungen am Gebiß der hominoiden Primaten. Z Morph Anthrop 72: 127–169.
Martin LB (1985) Significance of enamel thickness in hominoid evolution. Nature 314: 260–263.
Martin RD (1990) Primate Origins and Evolution. A Phylogenetic Reconstruction. Chapman and Hall, London.
Mayr E (1942) Systematics and the Origin of Species. Columbia Univ Press, New York.
Mayr E (1969) Principles in Systematic Zoology. McGraw-Hill, New York.
Molleson T (1988) Trace elements in human teeth. In: Grupe G, Herrmann B (eds) Trace Elements in Environmental History. Springer, Berlin Heidelberg New York, p. 67–82.

Morris P (1978) The use of teeth for estimating the age of wild animals. In: Butler PM, Joysey KA (eds) Development, Function, and Evolution of Teeth. Academic Press, London, p. 483–494.
Newman HN (1980) Dental Plaque: The Ecology of the Flora on Human Teeth. Springfield, Illinois.
Osborn HF (1888) The nomenclature of the mammalian dental cusps. Am Nat 22: 926–928.
Osborn HF (1907) Evolution of the Mammalian Molar Teeth. Macmillan, New York.
Owen R (1859) On the orang, chimpanzee and gorilla. In: N.N. (ed) On the Classification and Geographical Distribution of the Mammalia, Appendix B. Parker, London, p. 64–103.
Perzigian AJ (1984) Human odontometric variation: An evolutionary and taxonomic assessment. Anthropologie (Brno) 22: 193–198.
Preuschoft H (1989) Biomechanical approach to the evolution of the facial skeleton of hominoid primates. In: Splechtna H, Hilgers H (eds) Trends in Vertebrate Morphology. Progress in Zoology, Vol 35. Fischer, Stuttgart, p. 421–431.
Price TD (1989) The Chemistry of Prehistoric Human Bone. Cambridge Univ Press, Cambridge.
Rak Y (1983) The Australopithecine Face. Academic Press, New York.
Remane A (1921) Beiträge zur Morphologie des Anthropoidengebisses. Wiegmann Arch Naturgesch Abt 1 87: 1–179.
Remane A (1952) Die Grundlagen des natürlichen Systems, der vergleichenden Anatomie und der Phylogenetik. Akad Verlagsges Geest & Portig, Leipzig.
Remane A (1960) Zähne und Gebiß. In: Hofer H, Schultz AH, Starck D (eds) Primatologia. Handbuch der Primatenkunde, Vol III, Teil 2. Karger, Basel, p. 637–846.
Robinson JT (1956) The dentition of the Australopithecinae. Transvaal Mus Mem 9: 1–79.
Robinson C, Kirkham J, Weatherell JA, Strong M (1986) Dental enamel – a living fossil. In: Cruwys E, Foley RA (eds) Teeth and Anthropology. BAR Int Ser 291, Oxford, p. 31–54.
Rothe H, Wiesemüller B, Henke W (1997) Phylogenetischer Status des fossilen Neulings *Ardipithecus ramidus:* eine kritische Evaluation gegenwärtiger Konzepte. In: König V, Hohmann H (eds) Bausteine der Evolution. Edition Archaea, Gelsenkirchen, Schwelm, p. 159–168.
Rovner I (1971) Potential of opal phytoliths for use in palaeoecological reconstruction. Quaternary Research 1: 343–359.
Scott GR, Turner CG (1988) Dental anthropology. Ann Rev Anthropol 17: 99–126.
Simons EL (1972) Primate Evolution: An Introduction to Man's Place in Nature. Macmillan, New York.
Simons EL (1976) The nature of the transition in the dental mechanism from pongids to hominids. J Hum Evol 5: 511–528.
Simpson GG (1931) A new classification of the mammals. Bull Am Mus Nat Hist 59: 259–293.
Simpson GG (1961) Principles of Animal Taxonomy. Columbia University Press, New York.
Simpson GG (1963) The meaning of taxonomic statements. In: Washburn SL (ed) Classification and Human Evolution. Aldine, Chicago, p. 1–31.
Skelton RR, McHenry (1992) Evolutionary relationships among early hominids. J Hum Evol 23: 309–349.
Smith BH (1984) Dental development in *Australopithecus* and early *Homo*. Nature 323: 327–330.
Smith P, Tchernov E (1992) Structure, Function, and Evolution of Teeth. Freund, London.
Smith P, Zilberman U (1994) Thin enamel and other tooth components in Neanderthals and other hominids. Am J Phys Anthrop 95: 85–87.

Sneath PAH, Sokal RR (1973) Numerical Taxonomy. The Principles and Practice of Numerical Classification. Freeman, San Francisco.

Starck D (1979) Vergleichende Anatomie der Wirbeltiere auf evolutionsbiologischer Grundlage, Vol 3. Springer, Berlin.

Suwa G (1988) Evolution of the "Robust" Australopithecines in the Omo succession: Evidence from mandibular premolar morphology. In: Grine FE (ed) Evolutionary History of the "Robust" Australopithecines. De Gruyter, New York, p. 199–222.

Swindler DR (1976) Dentition of Living Primates. Academic Press, London.

Szalay FS, Delson E (1979) Evolutionary History of the Primates. Academic Press, New York.

Tobias PV (1991) Olduvai Gorge, Vol 4, parts V-IX. The Skulls, Endocasts, and Teeth of *Homo habilis*. Cambridge Univ Press, Cambridge.

Tomes J (1848) Lectures on Dental Physiology and Surgery. Parker and Son, London.

Turner CG II (1983) Sinodonty and Sundadonty: A dental anthropological view of Mongoloid microevolution, origin, and dispersal into the Pacific basin, Sibiria, and the Americas. In: Vasilievsky RS (ed.) Late Pleistocene and Early Holocene Cultural Connections of Asia and America. USSR Acad Sci, Siber Branch, Novosibirsk, p. 72–76.

Turner CG II (1986) The first Americans: The dental evidence. Natl Geogr Res 2: 37–46.

Turner CG II (1987) Late Pleistocene and early Holocene population history of East Asia based on dental variation. Am J Phys Anthrop 73: 305–321.

Vrba ES, Grine FE (1978) Australopithecine enamel prism patterns. Science 202: 890–892.

Walker A (1981) Diet and teeth. Dietary hypotheses and human evolution. Phil Trans R Soc Lond B 292: 57–64.

White TD, Suwa G, Asfaw B (1994) *Australopithecus ramidus,* a new species of early hominid from Aramis, Ethiopia. Nature 371: 306–312.

White TD, Suwa G, Asfaw B (1995) *Australopithecus ramidus,* a new species of early hominid from Aramis, Ethiopia. Corrigendum. Nature 375: 88.

Willmann R (1985) Die Art in Raum und Zeit. Das Artkonzept in Biologie und Paläontologie. Parey, Hamburg.

Wolpoff MH (1971) Metric Trends in Hominid Dental Evolution. Case Western Reserve Univ Stud Anthrop, Vol 2.

Wolpoff MH (1978) Some aspects of canine size in the australopithecines. J Hum Evol 7: 115–126.

Wood BA, Abbott SA (1983) Analysis of the dental morphology of Plio-Pleistocene hominids, I. Mandibular molars: Crown area measurements and morphological traits. J Anat 136: 197–219.

Wood BA, Engleman CA (1988) Analysis of the dental morphology of Plio-Pleistocene hominids, V. Maxillary postcanine tooth morphology. J Anat 161: 1–35.

Wood BA, Uytterschaut H (1987) Analysis of the dental morphology of Plio-Pleistocene hominids, III. Mandibular premolar crowns. J Anat 154: 121–156.

Wood BA, Abbot SA, Graham SH (1983) Analysis of the dental morphology of Plio-Pleistocene hominids, II. Mandibular molars – study of cusp areas, fissure pattern, and cross sectional shape of the crown. J Anat 137: 287–314.

Xirotiris NI, Henke W (1981) Enamel prism patterns of European hominoids – and their phylogenetical aspects. In: Chiarelli B, Corruccini RS (eds) Primate Evolutionary Biology. Selected Pap (Part A), VIIIth Congr Int Primatol Soc, Florence, 1980. Springer, Berlin, p. 109–116.

4 Dental Pathology and Epidemiology

4.1 Caries – Ancient Plague of Humankind

Peter Caselitz

Introduction

Humans are all too often haunted by their decayed teeth. Caries is not the largest plague of humankind, but certainly a daily nuisance. The aim of this chapter is to review the current theories of etiology, methods of scientific analysis, and particularly the evolution of caries in a historical dimension. Caries, or *caries dentium,* is the common name for tooth decay. It is a local disease characterized by an irreversible and progressive destruction of the hard dental tissue. Caries starts at the enamel surface of the tooth or – in case of recessed gingiva – at exposed parts of the neck. It works its way progessively through the dentine into the pulp cavity. The crown can be totally destroyed and other periodontal difficulties such as abscess may follow (for pulpoalveolar and periodontal diseases see Strohm and Alt, this volume). Finally – in more favourable cases – the affected tooth will fall out and the alveolus will be closed naturally. Alternatively, inflammation will spread into the surrounding bone or – specifically in the case of the upper teeth – the maxillary sinus will suppurate, leading to further, sometimes even deadly complications (for periapical diseases see Alt et al., this volume).

Etiology

Different theories of the etiology of caries have been discussed (for a historical review: Greve 1949; Cootjans 1955; Nikiforuk 1985). According to the most commonly accepted theory, the destruction of the hard dental tissue results from acids produced by micro-organisms of the oral flora, especially in an adherent gelatinous film on teeth and gingiva, called plaque. Plaque consists of food debris, salivary proteins, micro-organisms and polysaccharides of bacterial origin built up on the teeth in the absence of efficient oral hygiene (Moore and Corbett 1983, 140). Plaque is found mostly in fissures, small pits, and on the sides of the crown, and especially in the small area where the gingiva and tooth are in contact. It sometimes forms a supragingival calculus called dental tartar which is visible in archeological finds as a band of hard material surrounding the crown like a ribbon. The occlusal surfaces are normally not affected because of the abrasive effect of rough diet components.

The conditions for the development of caries depend upon the chemical balance between acid and alkaline produced by plaque bacteria's metabolism of different food components left over in the mouth as micro-debris. The metabolization of protein and carbohydrates produces alkaline waste products and lactic acid. The chemical balance switches from acid to alkaline several times a day. Particularly high acidity occurs when the diet contains large quantities of sugar. Sugars are metabolized much more rapidly than are other carbohydrates and as a consequence more lactic acid is produced more quickly (Hillson 1979).

Lactic acid carries out the initial attack on a tooth, resulting in demineralization of the enamel of the crown or of the cementum of an exposed part of the neck. The effect can be promoted by bacteria, such as *Streptococcus mutans*, *Streptococcus sanguis*, or *Staphylococcus albus* (Clarke 1924; Menzel 1925; Clement 1961; Shklair 1973; Hite 1975; review in Keene 1981). In case of caries in the neck area of a tooth, *Actinomyces viscosus* and *Actinomyces naeslundi* seem to be involved in the process. However, the acid theory has never reached total acceptance. Caries may also depend on genetic factors (Stewart and Spence 1976 versus Grundgeir 1987) as well as on socio-cultural influences. Even environmental factors may have an influence as well as may pathological conditions of saliva formation or nutritional disturbances during tooth formation manifestated as enamel hypoplasias. It can be summarized that "caries is a complex multifactorial disease caused by the interaction between a susceptible tooth, the presence of certain cariogenic micro-organisms, and a suitable oral environment" (Patterson 1984, 62). Yet caries seems to depend largely on the sort and manner of eaten food (Horster 1939; Somogyi 1964; Bibby 1966; Hartles and Leach 1975). Another theory claims a possible transmission of caries-promoting bacteria from parents to children when they are fed with a licked spoon or baby's bottle (Köhler and Bratthall 1978; Berkowitz and Jones 1985).

Caries in Past Populations

Caries is a very ancient and most widely spread disease. It already existed in the Australopithecines of South Africa (Robinson 1952; Clement 1956) as well as in the *Homo erectus* species from Java/Indonesia (Brodrick 1948). In the Neanderthals of Mount Carmel/Palestine caries is as common as in French upper paleolithic humans (e.g. in Aurignacian and Solutrean; see Praeger 1925; Vallois 1936; Krogman 1938; Brothwell 1963). Unfortunately, with the exception of two Neanderthal series the number of observations is too small for a statistical analysis. Caries is not limited to humans. Modern apes are affected (Colyer 1936; Schultz 1935;1967) as well as Pliocene animals and Pleistocene mammals (Clement 1958). Caries is also common in domesticated animals, especially when infected with *Streptococcus mutans* (Berger 1937; Zuhrt 1967; Baker and Brothwell 1980 versus Shklair 1981).

According to current scientific knowledge, the rate of caries increased notably with the shift from hunters and gatherers to farmers in Neolithic times when food became rich in carbohydrate. The fibre content of the diet has decreased continuously until the present time. There are two main factors for the extremely high caries increase in postmedieval history: the consumption of sugar and potatoes. Sugar is the simplest example of carbohydrate. Sugar was first gained from sugar cane and imported from America to Europe after 1550 AD, where beet-sugar has been produced since 1753 AD. These new sugars promoted caries much better than did honey, which had been used for sweetening in prehistoric times. Another promotor of the caries rate appears to be the shift from bread – with a proportionally higher degree of fibre and protein – to a potato-based diet during the 17th/18th century AD in Europe. More proteins and less carbohydrate, as paradoxically consumed in times of famine, are more favorable for the health of teeth especially when a high rate of rough and abrasive components of the diet help to scrub plaque off.

A common opinion in the literature is that in "primitive" tribes caries occurs less often than in "acculturated" societies. The rise of culture seems to bring an increase in tooth decay. Transferred into a historical dimension this would indicate a low caries rate in the hunters and gatherers of the Palaeolithic and Mesolithic periods and higher caries rates from Neolithic times onwards, when food became more refined and sophisticated, up to the still high caries frequency in present day urban populations. This opinion is supported by several surveys on caries in relation to the process of acculturation (e.g. Australian Aborigines: Cran 1959; Inuits: Waugh 1937; Pedersen 1939; Baaregaard 1949; Russell et al. 1961; Mayhall 1970; natives of the Alps: Seiler 1931). Some authors even believe in a caries immunity in primitive societies (see Clement 1961).

Caries in Research

Different parameters can be taken into consideration in an analysis of carious lesions in populations but unfortunately most of them are not comparable with one another. Widely differing opinions in literature (see Caselitz 1986, Fig. 14) are the result of this incomparability and their only common conclusion is the increase in caries from prehistoric to present times. The easiest approach in caries research is to either count the individuals with and without carious lesions or to count the carious and the healthy teeth. The DMF-Index (*Decayed, Missing, Filled*) is often used in the analysis of living populations, but DMF-data are not comparable with the data from research on (pre-)historic populations, since the DMF method includes medically treated, filled teeth which were extremely uncommon in ancient times. So, DMF based data give a too optimistic view of the real dental conditions. Without medical treatment, a small carious lesion will normally grow until it destroys the crown or finally causes the loss of the tooth. It is therefore more convenient to count missing teeth that were lost during lifetime as well as – in archeological material – in postmortal times, which allows the reconstruction of the number of teeth still in use during lifetime.

The Most Reliable and Tested Index

In order to compare populations it is advisable to devise a numerical term, which only needs one figure to describe the caries stress in a given quantity of individuals. The easiest procedure is to count the number of teeth still in function, those lost during lifetime or postmortally, as well as those affected with caries. These parameters form an index of the frequency of caries (abbreviated I-CE: index of caries et extractio). For this index the number of carious teeth is used, including the number of alveolae closed during lifetime which are considered to indicate loss of a seriously decayed tooth. Next, the sum of observed and postmortally lost teeth gives the number of teeth which were still in use during the lifetime of the individuals. The number of unhealthy teeth is divided by the number of observed tooth positions. All of the needed parameters are macroscopically easy to observe. Only the third molar can turn out to be a problem due to its irregular occurrence. This procedure allows a comparison of most of the analyses of archaeological series.

Considering only the I-CE, an increase of caries is evident over time (Fig. 1). The analysis was originally based on 603 series mostly cited in Caselitz (1986, Tab. 16) and completed here. The present sample consists of 518 series with at least 851,656 tooth positions, 1,311,637 teeth, 118,386 teeth lost during lifetime, 136,941 teeth lost after death, 80,030 analysed individuals with 28,302 caries-infected individuals. Unlike in our former study (Caselitz 1986), the arithmetic average of the dating span of a series is not used to obtain the time position. Instead,

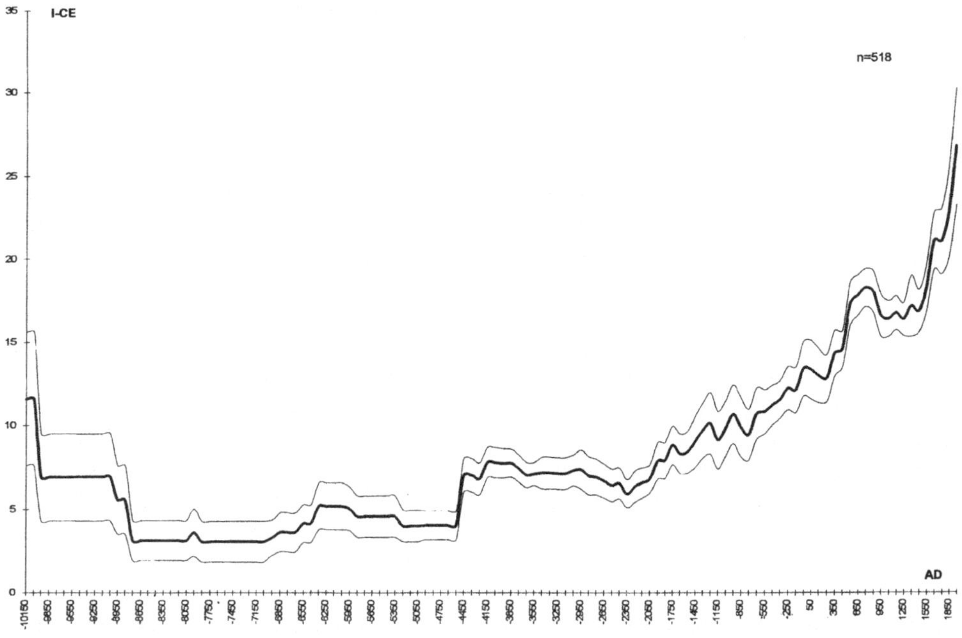

Fig. 1. Evolution of caries demonstrated by I-CE in a worldwide sample of 518 series. The small lines show the range of one standard deviation around the I-CE.

now the I-CE of a series is included in the summarizing count when its dating span falls into the concerned century. Historical reality can be closely traced in this way because dating spans of an archeological series are often quite large and there is no way of obtainig an exact time position for each individual. The formerly used mean value of time simplifies this circumstance a bit too much.

Only published findings were taken into account for a first view on evolution of caries in historical dimension. A series is usable when a minimum of 100 tooth positions (healthy teeth plus intravitally and postmortally lost teeth) or a minimum of 10 analysed persons is given and there were no selections for age or sex. If possible, only teeth of the permanent dentition are considered. Before taking a view on the evolution of caries in history there remains a statistical problem to be resolved. A series – especially an archeological one – is sometimes accidentally biased and the reality for dental health may be distorted and not correctly seized and interpreted. Therefore, following Grimm and Oehmisch (1956) the confidence of the relative rate of caries was calculated for each series. Series with greater than a value of 5.0 were excluded. The remaining sample is large enough to neglect the interobserver reliability mentioned by Rudney et al. (1983). However, the quality of the series in our sample is quite different. Occasionally one or more parameters are not listed. These omissions particularly concern the number of teeth lost during lifetime or after death and the mean age of the individuals of a series. In some cases the dating span is upgraded from actual knowledge.

Worldwide Evolution of I-CE

Five hundred eighteen series are used for this worldwide review of caries evolution. The oldest series are the Palaeolithic groups of Skhul/Mount Carmel (Israel) and a European Neanderthal collection. Their average I-CE is 5.3. The relationships appear to deteriorate in epipaleolithic times, but the only two series of Taforalt and the Algerian Iberomaurusian are used. Unfortunately the author omitted listing the teeth lost postmortally. So this I-CE rate seems too high and most probably obscures historical reality. In addition, not every tooth lost during lifetime is a result of caries, so I-CE rates have a slight tendency to be a bit too high. This may be a diachronical problem in the study of caries evolution.

Taking an overall look at the evolution of caries from 10,000 BC to the present time in century steps (Fig. 1), low caries rates are noted in Mesolithic and early Neolithic times. The rate of caries demonstrated here by I-CE remains relatively constant between the middle of the ninth to the middle of the fifth millenium BC. The I-CE then increases dramatically in the span of only one century, rising around 4.500 BC from 4.0 to 7.0 points. This phenomenon may be due to neolithisation. The sample in this century consists of 15 series. Two of them are hunter and gatherer groups, while the others are agricultural societies. The majority of the series come from Europe, two are from northern Africa and one from China. Hypothetically, a diet rich in meat and poor in cereals before the middle of the fifth century may be assumed. Caries-promoting conditions could be linked to the spreading use of grains. The caries rate then remains nearly stable up to the Middle Bronze Age and increases continuously from the middle of the second millenium to late Roman times or the middle of the first millenium AD. A small peak is found around 750 AD followed by a phase of relative steadiness in the Middle Ages. A second dramatic increase in the I-CE rate begins in the 16th century. The increase may be a result of the introduction of new foodstuffs into Europe after the discovery of the Americas. Present caries conditions are a record.

Continental View on I-CE

This overview of caries evolution brings up many questions: does the described evolution represent a worldwide phenomenon, does it depend on statistical insufficiency, or is it caused by an altogether different phenomenon, perhaps a biological as well as a sociocultural one? Splitting up the worldwide sample into regions yields only three: Europe, the Mediterranean including the Arabian peninsula, and the Americas (without Inuit populations; see Fig. 2). The European region is the biggest subsample (n = 358). Here, the evolution of the I-CE differs only slightly from the worldwide sample. Low caries rates can be noticed in very early times. A dramatic increase is still present around 4500 BC. A more continuous increase starts in Europa around 1200 BC, which is about one millenium later than in the worldwide sample. In the Mediterranean and Arabia region (n = 71) the prehistoric increase of caries rates starts as early as in the 7th millenium and shows a dramatic effect around 6000 BC. Up to Roman times the I-CE is generally higher than in Central Europe. The oldest populations in the Americas (n = 46) analysed with regard to caries phenomena dated from 7000 BC onwards. In these populations the I-CE starts at an already high level and decreases to a good, healthy level around

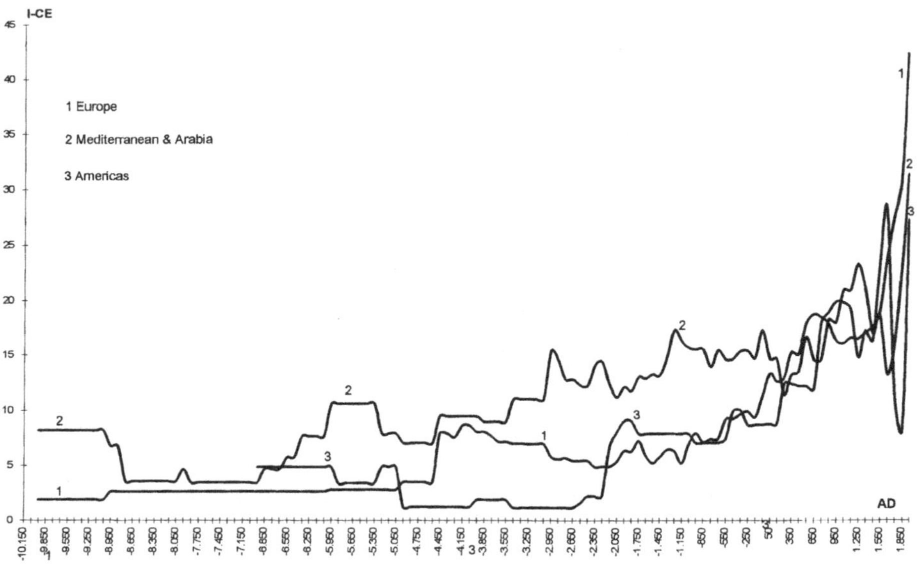

Fig. 2. Evolution of caries demonstrated by I-CE in continental subsamples.

5000 BC. This low caries rate span ends around 2300 BC, and a dramatic increase is noticed in the next centuries. The evolution of the I-CE in the times following is then similar to that seen in Europe. The low number of observations prevents further continental or regional analyses of caries (sample sizes: Africa: 6; Central Asia with India: 8; East Asia: 14; Australia: 11).

The dramatic increases in the I-CE may be due to neolithisation. This process occurs in the three regions at different times. The earliest horticulturalists are found in the Near East and the Mediterranean area. The shift from hunters and gatherers to agriculturalists in Central Europe appears to follow after 1500 years. The American pattern is different. Even so hunters and gatherers lived here side by side with horticulturalists up to subrecent times. The increase of the I-CE in the Americas in the last third of the third millenium corresponds well with archaeological contexts. However, since the number of observations for the American and Mediterranean samples per century is quite low, these interpretations can only be approximate.

A strong correlation (r_{xy} = + 0.809) exists between time and I-CE evolution in the worldwide material as well as in the subsamples (Europe: r_{xy} = + 0.781; Mediterranean and Arabia: r_{xy} = + 0.857; Americas: r_{xy} = + 0.743). It may be summarized that caries shows an undeniable tendency to increase with the standard of living. However, there may be a statistical problem with regard to the most ancient times. While the number of observations per century increase with dating time (worldwide r_{xy} = + 0.800; Europe r_{xy} = + 0.723; Mediterranean and Arabia r_{xy} = + 0.547; Americas r_{xy} = + 0.721), there is a lack of observations for the Mesolithic and early Neolithic periods. Not until after the sixth millenium BC do the number of observations reach a statistically useful level.

I-CE and Life Span

A common opinion is that human life span increases throughout history. Teeth are therefore in use longer and have an increased chance of being affected by caries. Is the mentioned increase of caries a consequence of this sociobiological phenomenon? To answer this question we correlate the average life span of the individuals of each dental analysis to its I-CE value. Astonishingly, the relationship between these two parameters is very weak (r_{xy} = + 0.146; n = 260), perhaps a result of the differences in the quality of data. As mentioned above, the I-CE is formed by the relative number of carious teeth plus intravitally lost teeth. Analysing each parameter separately also produces a weak correlation (carious teeth: r_{xy} = – 0.060; intravital loss: r_{xy} = + 0.230). The dispersion will be wide in graphical presentations (Fig. 3). Taking into account the I-CE only of series in which both parameters exist, the correlation is still weak (r_{xy} = + 0.162; see Fig. 3c).

Do life spans of the samples really increase with time? The average life spans per century are calculated by the same method as are the I-CE. The correlation between life spans and dating centuries are strong for the worldwide material (r_{xy} = + 0.725) as well as for subsamples of the Old World (Europe [n = 190] r_{xy} = + 0.685; Mediterranean and Arabia [n = 29] r_{xy} = + 0.631). Surprisingly, there is a

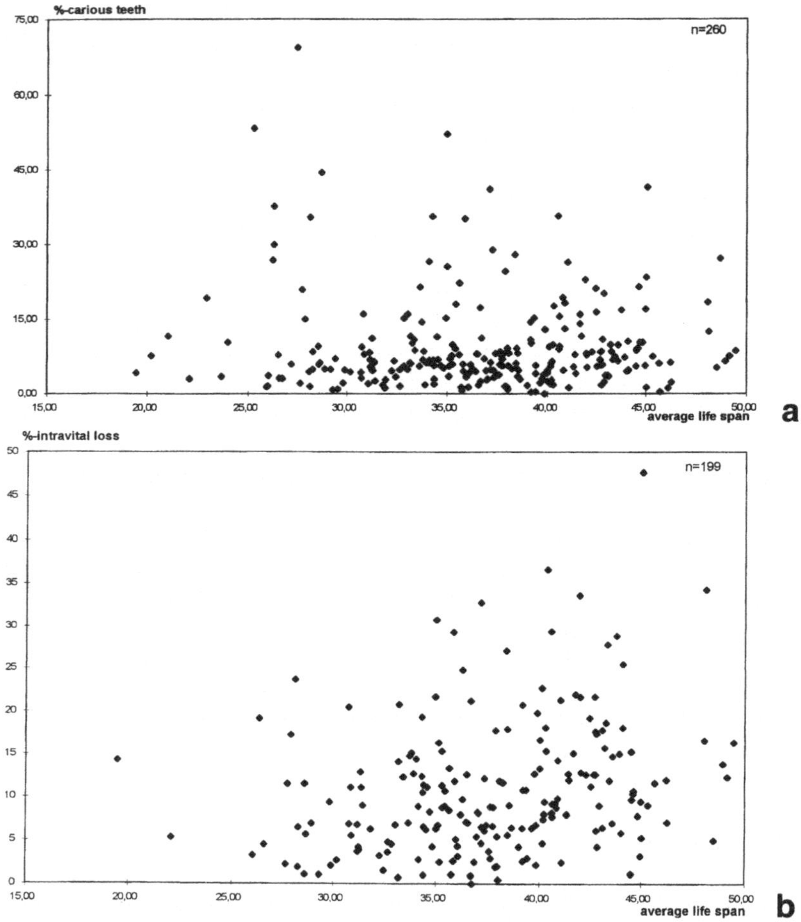

Fig. 3 a, b. Scattergram of average life span in relation to percentage of carious teeth (**a**) and percentage of intravital loss (**b**).

strong negative correlation in the American subsample ([n = 20] $r_{xy} = -0.879$). Here, life spans seem to decrease with historical time. Even if life span and I-CE value increase – at least in the Old World – throughout time, why do I-CE and life span not correlate more closely? The only answer is a wide variation in both parameters. Populations with low as well as high I-CE values and short average life spans are to be found, as well as are populations with the same I-CE but long life spans.

I-CE and Its Sociological Aspects

Maybe the increase in caries – as shown by the I-CE – is not a biological but a sociocultural phenomenon. This assumption seems to be confirmed by the sociological context of each series (see Tab. 1). The lowest caries rate – and even

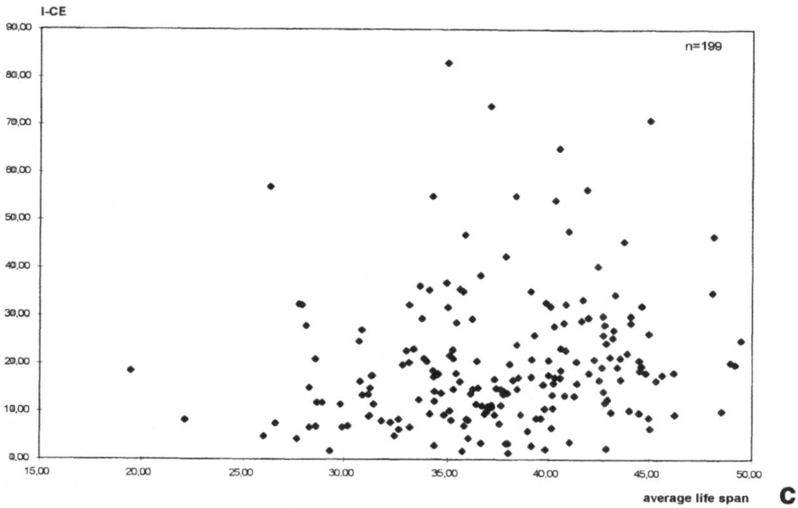

Fig. 3 c. Scattergram of average life span in relation to percentage of I-CE.

the lowest average life span – is noted in hunters and gatherers, whereas agriculturalists, although their average life span differs by only two years, show an average I-CE rate that is more than twice as high. A comparably medium I-CE rate is noted in the group of monks and nuns. Their lifespan is similar to that of urbanites and aristocrats, which show the highest I-CE rate. It may be summarized that monks live as long as urbanites but their teeth appear to be in better condition. This may be the result of a less opulent diet and perhaps also of more careful dental hygiene.The effect of culturisation is visible in a small group of acculturated Inuits (n = 228). Including their dental findings, the I-CE rate of hunters and gatherers jumps to 7.89 and the average life span decreases by one year. However, the four analysed groups cannot be followed through the complete span between Paleolithic and recent times. Urbanites and monks are a relatively young phenomenon in the history of humankind.

The I-CE rate increases with dating time in agriculturalists (r_{xy} = + 0.925) as well as in urbanites (r_{xy} = + 0.590). The number of observed series increases as well (r_{xy} = + 0.789 resp. r_{xy} = + 0.760). In hunters and gatherers the I-CE seems to increase by only a small amount (r_{xy} = + 0.468). Looking at the total dating span of monks there appears a steady rate of I-CE (r_{xy} = – 0.148). Differences show up when the monks are split into older (AD 775 – 1599) and younger groups (≥ AD 1600). In the older group a strong negative correlation exists between time and I-CE and the number of observed persons increases (r_{xy} = + 0.830). In the younger group the rate of I-CE decrease is less sharp (r_{xy} = – 0.449), but the number of analysed sets of teeth decreases notably (r_{xy} = – 0.727). This phenomenon is easily accounted for by a historical process. The time of prosperity of monasteries was before AD 1600. The rules of daily life faded away in the younger group as the monasteries changed more and more into hospitals and old people's homes.

Caries Rate in Individuals

Another parameter in the analysis of caries is the percentage of affected individuals. Unfortunately, some of the series lack relevant data. The following discussion is based on a reduced sample size of 340 series (268 European, 39 Mediterranean and Arabian, and only 13 American). As far as the worldwide evolution of the percentage of affected individuals is concerned (Fig. 4), a statistical problem is caused by the lack of a sufficient number of analysed series before BC 4500. The observations before this date are only suggestive for a trend as noted by the area of standard deviation. Statistical relevance ($n > 5$ series per century) begins in the middle of the fifth millennium BC, at which point 29 percent of adult individuals were affected by dental caries, a rate nearly constant for a millennium. The rate rises above this in the last half of the fourth millenium followed by a phase of 28 percent. A slight intermediate minimum is found around 2250 BC. In the middle and late Bronze Age the rate of affected individuals increases continuously to three intermediate maxima of 36 percent. From the middle of the last half of the second millenium to 300 BC the rate decreases. From this date onward the percentage of affected individuals increases dramatically to 56 percent in the 7th century AD. Conditions seem to improve during the Middle Ages and deteriorate again after AD 1300. In the recent series of our sample, more than two thirds of individuals are affected by caries. In actual clinical observations, this value should be over 95 percent.

An amazingly low rate of 0.38 percent is found in a subrecent Inuit series from Greenland followed by more realistic, but also low rates (< 10%), in two Viking age series (Trelleborg and Haithabu) and in a Goth population of Roman Age. A series from Nebira/Papua New Guinea as well as a population of St. Catherines Island complete this group. Surprisingly, there is only one hunter and gatherer series in this sample of little-affected populations. At the negative end of the variation

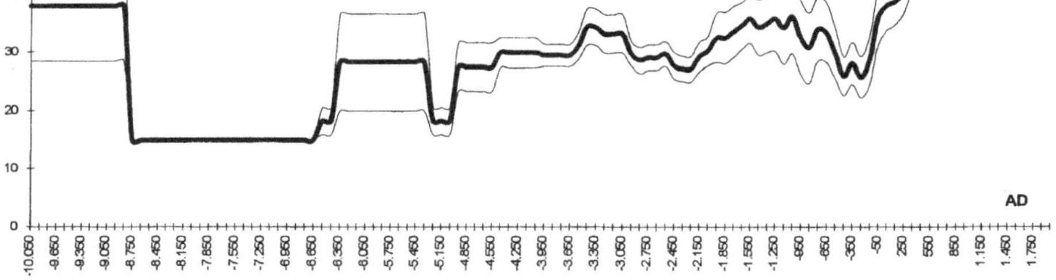

Fig. 4. Evolution of the percentage of individuals affected by caries in a worldwide sample of 340 series. The small lines give the range of one standard deviation around the parameter.

(> 95%) many 20th century series from Europe are to be found, including two Merowingian populations (Eppstein-Frankenthal and Griesheim, Germany). Considering the overall timespan, nearly every second human was affected by caries ($\bar{x}$ = 48.3 percent). The percentage of affected individuals correlates with the I-CE (r_{xy} = + 0.637). It correlates well also with the dating time of the series (r_{xy} [10,000 BC–AD 2,000] = + 0.639 resp. r_{xy} [4,500 BC–AD 2,000] = + 0.788).

Splitting the worldwide sample into regions the evolution of the percentage of affected individuals shows a rather confusing picture (Fig. 5). American values vary extremely as a result of the low number of observations, and are neglected in this discussion. In Europe the evolution begins with a phase of ups and downs around 27 percent but approximately 200 BC figures increase dramatically, reaching 57 percent in the seventh century AD. Medieval times bring a slight improvement, but from 1300 AD on the percentage of affected individuals increases again to recent very high caries rates. In Mediterrean and Arabian areas the rate appears higher than in Europe throughout up to 2000 BC. At this point, the number of observations in this region reaches statistical relevance. The general decreasing tendency is similar to that in Europe. After levels of a relatively low percentage of affected individuals in the third century BC as well as in the third and fourth centuries AD an increase to 62 percent in the 13th century AD is noted. The improvement in subrecent times may be influenced again by low numbers of observations.

The sociological aspects of the percentage of caries-affected individuals are similar to those discussed regarding the index of decay and extraction mentioned above. A low average is found in hunters and gatherers (Tab. 1). Nearly every second individual in agriculturalists and monks was affected with caries. The average of this rate in urbanites and aristocrats is 62 percent, but is low in early medieval times and rises strongly with the dating position of the urban series (r_{xy} = + 0.934). An increase with time is observed in agriculturalists (r_{xy} = + 0.757) as well as in monks and nuns (r_{xy} = + 0.750) but not in hunters and gatherers

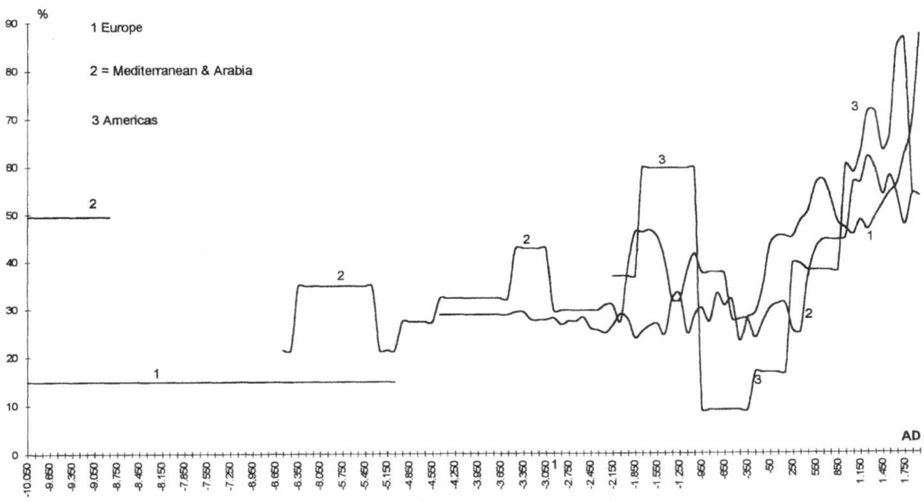

Fig. 5. Evolution of percentage of caries-affected individuals in continental subsamples.

Table 1. Caries parameters in relation to sociocultural context

Minimum number of	Hunters and gatherers*	Horti- and agriculturalists	Urbanites and aristocrats	Monks and nuns
series	46	404	43	18
positions of teeth	38697	570741	185953	22894
teeth	62427	865895	300036	49067
intravital loss	2385	76676	30763	2966
postmortal loss	8685	105711	15894	1099
carious teeth	2417	92955	72071	3237
individuals	3063	46102	25473	3004
individuals with caries	361	19113	6380	1375
average of life span	34.40	36.62	38.76	38.71
average of I-CE	6.92	15.67	26.32	14.98
average percentage of carious teeth	4.18	8.16	15.86	7.65
average percentage of intravital loss	6.29	10.91	13.79	13.60
average percentage of carious infected individuals	31.43	47.81	61.84	48.86

* Without the acculturated Inuit series (see Caselitz 1986, 164)

(r_{xy} = + 0.143). The main difference in the evolution exists between I-CE and the percentage of affected individuals among urbanites. The increase appears to be lower in the I-CE in this social group whereas the spreading rate of caries in individuals rises sharply. In conclusion, there is a notable increase in modern times in the percentage of affected individuals as well as in the I-CE (for a detailed discussion of further parameters like number of carious teeth, intravital losses etc. see Caselitz 1986, 166).

Localization of Caries in the Teeth

A carious lesion is not the simple entity as it appears. For the individual it is quite important where the lesion is located and how far it has developed. A common opinion is that the first molar is generally the most affected tooth. Since it is the first tooth of the permanent dentition to erupt, it must withstand large chewing pressures, and it has the greatest surface area of all teeth. These reasons seem to predestine the first molar for a carious attack. This common opinion, however, is without any extensive support of relevant data. Its validity can be determined by examining the 518 series once again. Only 220 series have useful data which explicitly mention the localization of caries. Some analyses neglect the data for intravitally lost teeth, so our sample is reduced to 146 series worldwide.

The number of carious teeth per mandible, maxilla, and both jaws together were converted to percentages. The number of teeth affected with caries were summed for the left and right sides together. The result was transformed to a relative value with respect to tooth type. Low averages are noted in anterior teeth (incisors and

Table 2. Average of the relative occurrence of carious lesions in the different types of teeth based on maximally 146 worldwide series. (1st = central incisor, 2nd = lateral incisor ... 8th = third molar).

Type of tooth	1st	2nd	3rd	4th	5th	6th	7th	8th
maxilla (n = 141 series)	5.39	5.09	5.71	10.58	12.80	25.37	21.64	12.90
mandible (n = 141 series)	3.95	3.14	2.84	5.26	11.40	31.88	25.95	15.58
maxilla and mandible	4.79	4.27	4.19	7.76	12.37	28.47	23.84	14.31

canine), whereas high rates were found for first and second molars (Tab. 2). More than half of the carious lesions were found in posterior teeth. The common opinion seems to be corroborated. However, the rate of caries differs in the maxilla and the mandible. Lower molars are more affected than are their upper opponents, while the anterior teeth – especially from second incisor to first premolar – are less affected.

Are these relationships constant in historical dimensions? Taking an overall look at the evolution of the occurrence of caries in upper and lower teeth from 10.000 BC to the present time in steps of 250 years, some changes are noted (Fig. 6). The description starts at 4500 BC due to the lack of older series in the sample. At least two phenomena are conspicuous. The first and second molars have the largest number of carious lesions over all time. The percentage of caries in anterior teeth increases from AD 1000 onward. Anterior teeth especially are considered to be nearly caries-free. Two additional dates in

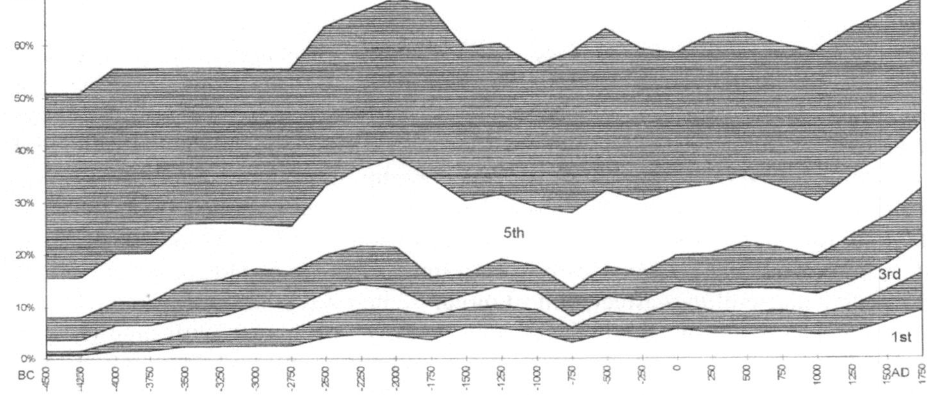

Fig. 6. Frequency distribution of the relative occurrence of caries in the different types of tooth summarizing mandible and maxilla as well as left and right side from first incisor (below) to third molar (at top).

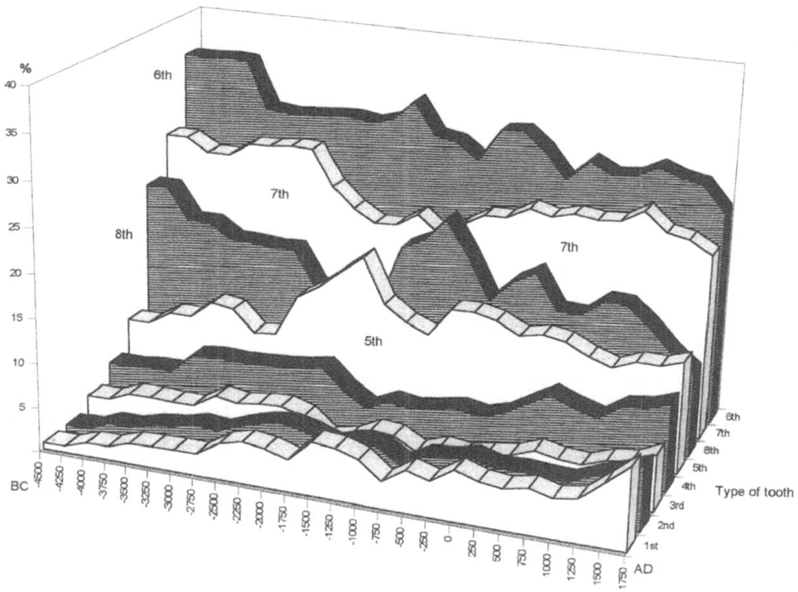

Fig. 7. Three-dimensional frequency polygon of the relative occurrence of caries in the different type of tooth. Note the switched order in molars.

evolution are remarkable: 2750 BC and 1750/1500 BC. Around 2750 BC the percentage of carious second premolars increased dramatically, whereas the caries rate in the second molars decreased at the same time. These changes may be a result of alterations in diet caused by a shift from an economy based on crop cultivation to one dominated by herding. This shift may have been caused by climatic changes.

The interdependent changes in the occurrence of caries are more easily discernible when a different, but methodologically slightly less correct kind of graphical presentation is chosen (Fig. 7). A generally significant tendency to increase is observed in the incisors and is strongly correlated with the historical time ($r_{xy[1st]}$ = + 0.830; $r_{xy[2nd]}$ = + 0.818). The increase in the canine and the first premolar is weakly correlated with time ($r_{xy[3rd]}$ = + 0.470; $r_{xy[4th]}$ = + 0.607), and may be caused by the intermediate phase of high rates between middle Neolithic and early Bronze Age (c. 3500 – 1750 BC). The evolution of the percentage of caries in the second premolar is very inconsistent (r_{xy}[5th] = + 0.386). The phase of maximal rate starts around 2500 BC and ends about 750 BC. Around 1750/1500 BC nearly one-fifth of carious lesions are found in this tooth. This phase correlates well with a phase of less affected third molars. The first molar is continuously the most affected tooth, but there is a strong negative correlation with time ($r_{xy[6th]}$ = – 0.813). Whereas the second molar appears to be more highly affected by caries from 1250 BC onward, seen over the total time span, there is a weak tendency to decrease ($r_{xy[7th]}$ = – 0.497). Two phases of decrease are noted for the third molar ($r_{xy[8th]}$ = – 0.502). Starting at 4500 BC a high level of occurrence of carious lesions decreases to a minimum around 2000/1750 BC followed by a strong increase to a maximum at 750 BC. Again a phase of decrease is noted, eventually reaching the relatively low rate in recent times. Extension of the total evolution predicts further decrease in all three molars and an increase in second premolars and especially in the first incisors.

The relationships differ in the lower and upper dentitions as expected from the separate data given in Tab. 2. These differences are easily demonstrated by subtracting the average data of the maxilla from those of the mandible for each time interval separately (Fig. 8). The relationships in anterior teeth – up to the first premolar – appear to be more or less constant. The lower second premolar is more affected by caries than is its upper opponent until 1750 BC. The relationships then reserve to a higher degree of affection in the upper second premolar with a maximum around 500 BC. There follow relatively constant relationships up to the present time. Remarkable are the differences in the first molar. Between 4500 and 3500 BC this tooth is extremely more affected with caries in the maxilla. From 3250 BC onward caries in first molars occurs consistently more often in the lower teeth. The phenomenon of a greater caries rate in lower molar teeth is also observable in the second molar and – except for the time span from 3250 to 1250 BC – in the third molar. Generally, the lower anterior teeth (1st to 4th) are less affected with caries than are their upper opponents, whereas caries occurs more frequently in the molars of the maxilla. The second premolars show an intermediate relationship. From 1500/1250 BC onward they are better preserved against caries in the mandible. These results of paleostomatology may produce further consequences for clinical dentistry concerning dental prostheses.

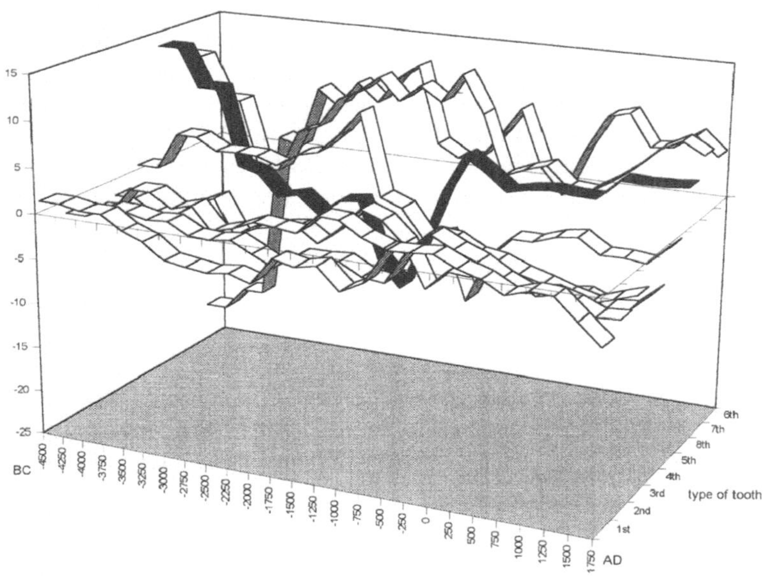

Fig. 8. Three-dimensional surface histogram of differences in the relative occurrence of caries between maxilla and mandible. Negative data stand for higher affection in the upper teeth. The relationships of the third molar are emphasized. Note again the switched order in molars.

Approach to Intensity

When studying a series it is advisable to number the carious and intravitally lost teeth and to compare both to healthy teeth. For epidemiological considerations it is also of interest to take the intensity of carious lesions into account. We prefer a system of five gradations to classify the pathological condition of each tooth (see Tab. 3). The grades are added up for each tooth type and finally divided by the number of observations. In the end there is just one numerical term to describe the intensity of caries for each tooth type (for an example see below Tab. 5).

Only a few series, mostly coming from Northern Germany, are suited for an analysis of the intensity of carious lesions. Only the teeth of the permanent dentition are taken into consideration and differentiated by the eight types of teeth (first incisor to third molar). The data for left and right sides are summarized. Up to now it has been convenient to add the data from the mandible and the maxilla to avoid statistical confusion caused by the low number of observations. Data of teeth from the deciduous dentition were neglected. This procedure is also usable on children's teeth and for a detailed analysis of age or sex groups. It is advisable to differentiate between teeth of the first and second dentitions and to avoid too low numbers of observations per tooth type.

A totally healthy tooth is given by a degree of 1.0, but is very rarely observed (see Fig. 9). This situation is generally found in the anterior teeth such as in the friarly dominated series from Bad Iburg, Schleswig and the medieval subseries of Stade. The worst conditions prevail in the first incisors of the Palmyrenian population, but this is due to a very low number of preserved anterior teeth in those skeletons. Very poor dental conditions are found in the samples from Lüchow-Dannenberg and the Frenswegen cemetery. Both historical populations originated from countrysides. First molars of the Frenswegen group have the highest rate ever found ($\bar{x} = 3.22$ points), followed by second molars of the series ($\bar{x} = 3.19$ points) and the molars of the Lüchow-Dannenberg sample ($\bar{x} = 3.14$ resp. 3.16 points). Both series are very similar in their total dental conditions ($\bar{x} = 2.62$ points).

Relatively good dental health is attributed to the Dominican friars from Schleswig ($\bar{x} = 1.20$ points). Remarkable are their very healthy molars, comparable

Table 3. Gradations of carious lesions

Grade	Description
1	Healthy teeth. No caries found.
2	Small pit or small fissure caries. Not to be confused with normal anatomical pits on the enamel surface (especially on first molars like buccal pits or foramen caecum hypoplasia).
3	Large carious lesion, often bigger than a pinhead – but with less than two thirds of the tooth crown destroyed.
4	Nearly complete destruction of the tooth crown, with only the roots remaining.
5	Tooth lost, alveola closed or still closing. No remains of the root.

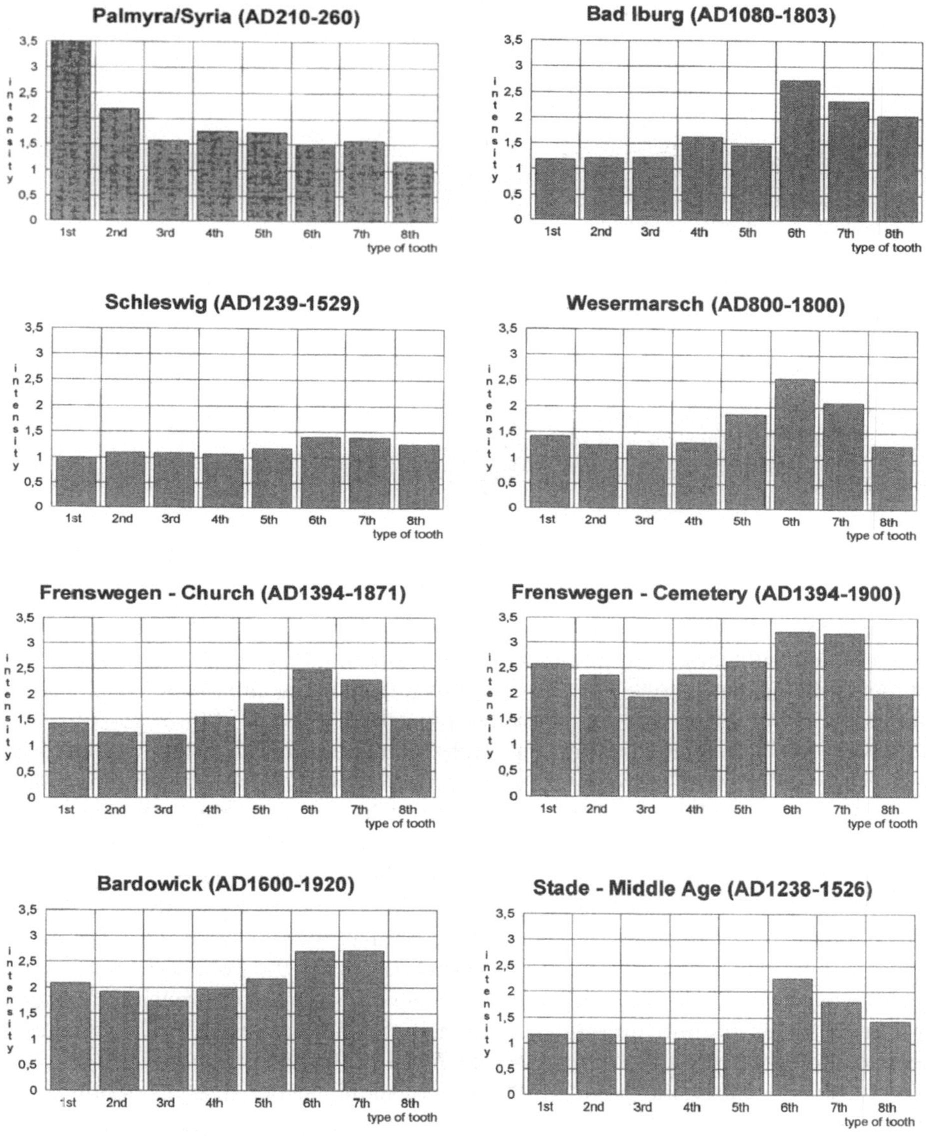

Fig. 9. Intensity of carious lesions per type of tooth in historical populations of Northern Germany and in a Roman Age series of Palmyra/Syria (based on Caselitz 1986, Fig. 19).

only with the good molars of the Palmyrenians (totally = 1.72 points) and of the Stade – Middle Ages series ($\bar{x}$ = 1.41 points). Generally poor dental conditions prevail in the samples from Lüchow-Dannenberg and Frenswegen cemetery (for both: = 2.62 points) and perhaps also in Bardowick series ($\bar{x}$ = 2.16 points).

The two series from Stade as well as both groups from Frenswegen allow a sideview on the intensity of caries under different social conditions. Aristocrats and

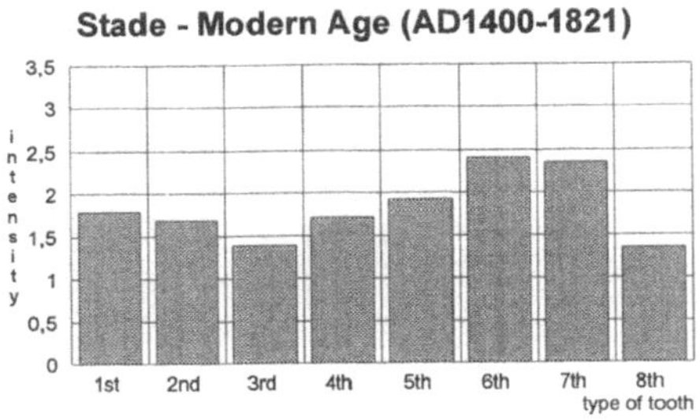

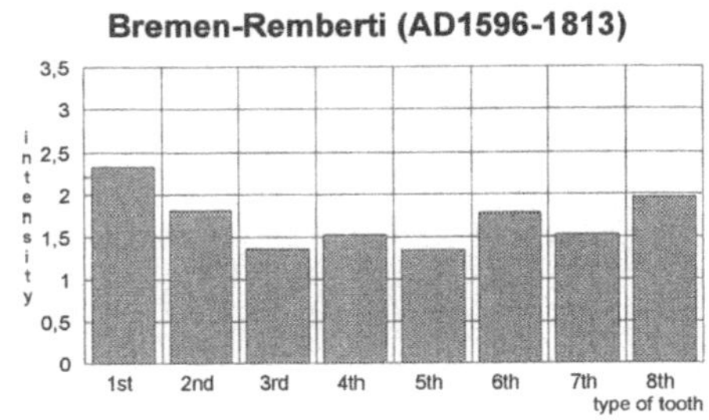

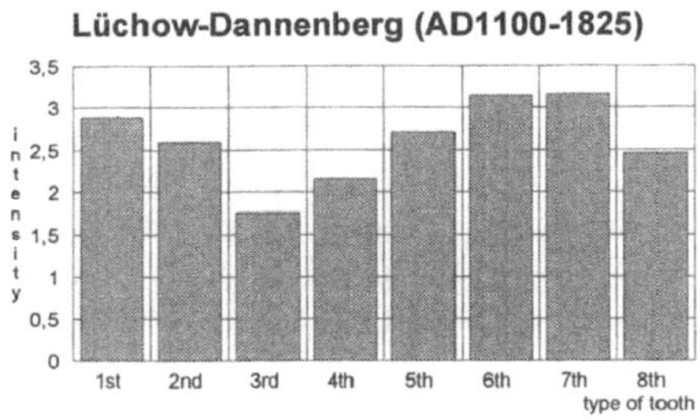

Fig. 9 (continued)

perhaps some friars were buried inside the Frenswegen church, whereas local farmers and villagers were inhumated in the surrounding outside cemetery. The intensity of carious lesions in the two groups differs in the anterior teeth as well as generally. Two series originated from Stade town. The Middle Age subseries consisted of friars whereas poorhouse inhabitants are found in the Modern Age subseries. Differences in the intensity of caries exist in the second molars and in anterior teeth. Parallel to the historical trend of increasing caries with time there appears a tendency of increasing caries intensity. The conditions for the development of caries seem to depend on the dietary habits, but it is astonishing that the food of the Frenswegian farmers should be more likely to promote caries. The diet of aristocrats and friars – in context with a probably better oral hygiene – appears to hinder caries. A comparable social parallel to the Frenswegen villagers is given by the Bardowick and Lüchow-Dannenberg series. The intensity of carious lesions is lower, but the amplitude of intensity is similar – especially in first and second molars as well as the number of highly affected incisors. This may be typical for a countryside population. The constantly better dental conditions in friars must be considered in the context of the more favourable ingredients of daily food. Perhaps this phenomenon is a result of the better economic conditions of monasteries as well as of fasting periods.

Discussion

Caries is a very prevalent disease, widely spread over time and continents. It appears to be easily analysed only by the naked eye and a dental diagnostic probe. However, not every pit on the enamel is caused by caries. Some defects may depend on genetics (e.g. the buccal pits on first molars), whereas others – especially in an archeologically derived sample – may originate from normal decomposition from the influences of soil. Scientists must carefully distinguish "true" from "false" caries to exclude pseudopathological phenomena from the data base (Werner 1937). Some methods of analysis adapted from the clinical literature are less reliable in skeletal populations (Rudney et al. 1983).

The majority of enamel defects are still due to caries. The problems in paleostomatology begin with data processing and the variety of methods used to record lesions. "Some authors estimated the number of carious teeth as a percentage of the total number of teeth available or as a percentage of each tooth type [further indices e.g. in Bornstein 1952]. Others have sought to add correction factors and take account of ante-mortem or post-mortem loss" (Kerr et al. 1988, 143). Third molars were taken into account or neglected. Some authors mix data from the first and second dentitions. Others give only a percentage of caries-affected individuals. This great confusion reduces the number of suitable paleostomatological analyses, just as the archeological material itself limits the available data. It is often fragmentary, incomplete and the teeth broken postmortally. Experience shows that a greater number of diagnosed tooth positions ($n \geq 300$) seems to minimize this statistical irregularity.

When analysing a new series, a lesion should be taken into account only if it is an obvious defect in the integrity of the enamel or root surface. Doubtful phenomena should be neglected. If more than one carious defect is found in one tooth, they should count as one and the more serious lesion accounted for the statistics of intensity. Data from the first and second dentitions are considered separately. The numbers of affected as well as of healthy teeth are listed seperately for each tooth type in the maxilla and mandible. Findings of postmortally lost teeth are counted separately in order to reconstruct the total number of teeth present during lifetime. Intravitally lost teeth are a great problem in paleostomatology. Most were destroyed by caries. We prefer to consider intravitally lost teeth even though their number could be a little high. Another problem arises with regard to third molar. This tooth is often unerupted and its pathological relationship to the other teeth could be distorted. There is no evidence of divergence in enamel composition of the third molars in relation to other types of teeth. Therefore, pathological findings on third molars are to be taken into account. The estimation of the percentage of caries in the group of postmortally lost teeth should be neglected because it lacks methodological relevance. In the presentation of results the absolute number of findings is preferable. The relative number of lesions is – in case of need – easy to calculate. Finally, experience has shown that it is wise to crosscheck the data because there are many obvious miscalculations in the literature.

A model of data account and graphical presentation is shown (see Tab. 4 and Fig. 10); to complete the presented upgrade of the Bad Iburg series it should be considered that 72 percent of the 39 analysed individuals are affected with caries (incl. intravitaly lost teeth). Our exemplary schedule for the calculation of intensity is also shown (Tab. 5). These models can be adapted for any differential analysis of teeth from the left and right sides as well as – with the same calibration (e.g. labelling of the types of tooth) – for the first dentition. In addition, this method can be stressed for more detailed analyses with regard to localization of caries in the

Table 4. Example of data presentation of a paleostomatological analysis: the Bad Iburg series (AD 1080–1803) (enlarging Caselitz 1981)

				maxilla					
tooth present	40	38	45	42	45	32	35	15	292
postmortally lost	10	8	6	5	2	4	6	13	54
intravitally lost	2	2	2	6	4	16	10	9	51
total number of tooth positions	52	48	53	53	51	52	51	37	397
carious teeth	0	2	2	4	2	7	5	2	24
percentage of carious teeth	0.00	4.35	3.92	8.51	4.26	19.44	12.20	7.14	6.94
percentage of intravital loss	19.23	16.67	11.32	9.43	3.92	7.69	11.76	35.14	13.60
I-CE	19.23	21.01	15.24	17.94	8.18	27.14	23.96	42.28	20.54
type of tooth	**1**	**2**	**3**	**4**	**5**	**6**	**7**	**8**	**sum**
tooth present	43	50	56	53	53	36	32	30	353
postmortally lost	17	11	6	5	2	0	3	0	44
intravitally lost	2	2	2	6	7	22	19	8	68
total number of tooth positions	62	63	64	64	62	58	54	38	465
carious teeth	0	0	1	4	1	8	0	4	18
percentage of carious teeth	0.00	0.00	1.61	6.90	1.82	22.22	0.00	13.33	4.53
percentage of intravital loss	27.42	17.46	9.38	7.81	3.23	0.00	5.56	0.00	9.46
I-CE	27.42	17.46	10.99	14.71	5.04	22.22	5.56	13.33	14.00
				mandible					

tooth (e.g. occlusal, labial, lingual, mesial, and distal; for the nomenclature see e.g. White 1991, Fig. 5.2) as well as for differentiation between crown or root-manifested caries.

Conclusions

Carious teeth have been haunting humankind worldwide and for all time. Different theories of the etiology are under discussion. In all probability, caries is the result of the interdependence of oral chemical balance due to bacterial infection (especially with sorts of streptococcus), dietary habits, and oral hygiene as well as a certain mild influence of genetic predisposition. Caries is well related to social conditions and it appears possible to obtain background information

Table 5. Example of presentation of the intensity of carious lesions: the Bad Iburg series(AD 1080–1803) (enlarging Caselitz 1981). Caries degrees are described in Tab. 3

				maxilla					
degree 1	2	2	2	6	4	16	10	9	51
degree 2	52	–	–	–	–	–	–	–	–
degree 3	–	–	–	–	–	–	–	–	–
degree 4	–	–	–	3	1	2	1	1	8
degree 5	2	2	2	6	4	16	10	9	51
intensity	1,19	1,28	1,26	1,73	1,43	2,58	2,11	2,71	1,74
type of tooth	**1**	**2**	**3**	**4**	**5**	**6**	**7**	**8**	**$\bar{x}$**
degree 1	43	–	55	49	52	28	32	26	335
degree 2	–	–	–	1	1	1	–	1	4
degree 3	–	–	–	1	–	1	–	2	4
degree 4	–	–	1	2	–	6	–	1	10
degree 5	2	2	2	6	7	22	19	8	68
intensity	1,18	1,15	1,19	1,56	1,48	2,88	2,49	2,05	1,75
				mandible					

concerning culture from the study of teeth. The present study was based on stomatological data of more than 600 published series mostly derived from archeological excavations. An index of carious and intravitally lost teeth in relation to the number of healthy teeth (I-CE) is presented, which is used to look at the evolution of caries in a historical dimension. The results present two unsuspected developments in the caries rate in historical time. First, a dramatic increase in I-CE about 4500 BC appears to depend on the phenomenon of neolithisation, as pointed out on continental subsamples. Second, an increase in the caries rate from about AD 1500 to recent still very prevalent conditions may be caused – especially in Europe – by the introduction of new foodstuffs originating in the Americas. The social context of the series correlates well with the caries rate, whereas astonishingly the average of age at death of analysed individuals is only weakly correlated to the I-CE. The lowest caries rate is observed in hunters and gatherers, whereas very poor conditions existed in urbanites and aristocrats. The frequency of caries-affected individuals reaches a formerly unknown intermediate maximum in the 7th century AD and increases dramatically after AD 1300.

The discussion of the localization of carious lesions in the types of teeth confirms their prevalence in molars, but the differentiation between the upper and lower dentitions brings – in addition to some significance in historical tendency – some relevance to clinical dentistry. Notes for further research on caries as well as an example of data presentation – enlarging a study on a medieval series from Northwestern Germany – close this study.

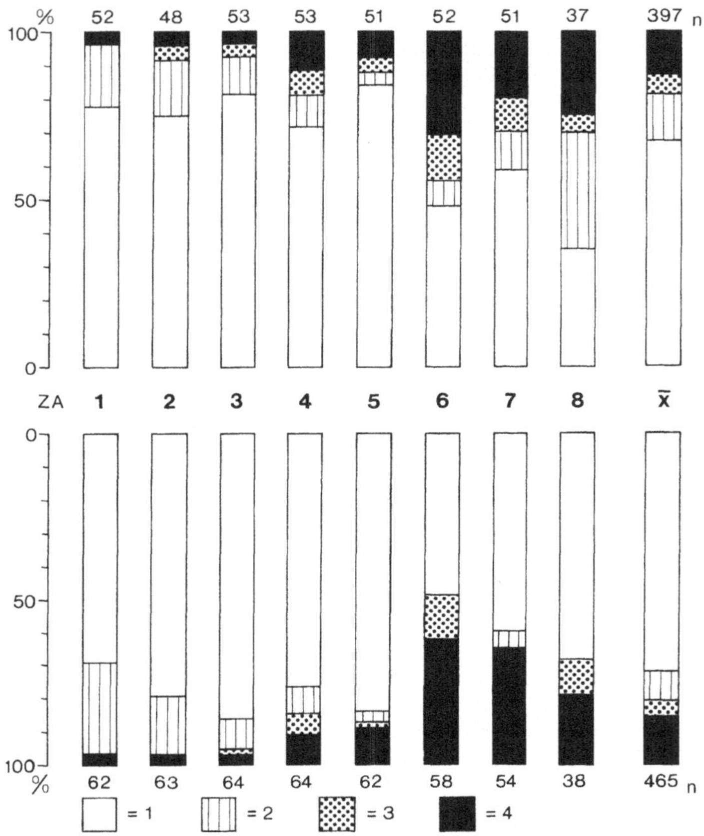

Fig. 10. Example of graphical presentation of intensity of caries: the Bad Iburg series (AD 1080–1803) (adapted from Caselitz 1981, Fig. 1). Signatures: 1 = healthy teeth; 2 = postmortally lost; 3 = carious teeth; 4 = intravitally lost. Above = maxillary teeth; below = mandibular teeth. Note the position of postmortally lost teeth, which clearly indicate the "twilight zone" of this phenomenon.

Acknowledgements

We are grateful to PD Dr. Kurt W. Alt, Düsseldorf, and Prof. Dr. Friedrich W. Rösing, Ulm, for their advice on literature as well as to Rainer B. Michl M.A., Hamburg, for his alltime willingness to discuss hypotheses and statistical problems as well as especially to Rona v. Reinersdorff M.A., Hamburg, for help with the translation.

References

Prenote: the bibliographical evidence for the used series is mostly given in Caselitz 1986. A completion as well as a necessary complete listing of used series would unfortunately exceed the limits extent of this publication.

Baaregaard A (1949) Dental conditions and nutrition among natives in Greenland. Oral Surg Oral Med Oral Pathology 2: 995–1007.

Berger F (1937) Die Entwicklung der Zahn- und Kieferkrankheiten, insbesondere der Karies, vom Tertiär bis zur Bronzezeit. Med Diss, Bonn.

Berkowitz RJ, Jones P (1985) Mouth-to-mouth transmisson of the bacterium *Streptococcus mutans* between mother and child. Arch Oral Biol 30: 377–379.

Bibby BG (1966) The cariogenicity of different foodstuffs. In: Nizel AE (ed) The Science of Nutrition and Its Application in Clinical Dentistry, 2nd Ed. Saunders, Philadelphia, p. 318–330.

Bornstein EJ (1952) Die Kariesintensität verschiedener Völker und Volksgruppen. Med Diss, München.

Brodrick AH (1948) Early Man; a Survey of Human Origins. Hutchinson, London.

Brothwell DR (1963) The macroscopic dental pathology of some earlier human populations. In: Brothwell DR (ed) Dental Anthropology. Macmillan, New York, p. 271–288.

Caselitz P (1981) Untersuchungen zur Zahnkaries und deren Folgen am Beispiel der in der Klosterkirche zu Bad Iburg Bestatteten. Osnabrücker Mitteilungen 87: 9–18.

Caselitz P (1986) Ernährungsmöglichkeiten und Ernährungsgewohnheiten prähistorischer Bevölkerungen. BAR Int Ser 314, Oxford.

Clarke JK (1924) On the bacterial factor in the aetiology of dental caries. Brit J Exp Pathol 5: 141–147.

Clement AJ (1956) Caries in the South African ape-man: Some examples of undoubted pathological authenticity believed to be 800,000 years old. Brit Dent J 102: 4–7.

Clement AJ (1958) The antiquity of caries. Brit Dent J 104: 115–123.

Clement AJ (1961) Caries Immunity in Primitive Peoples. MA Thesis, Johannesburg.

Colyer F (1936) Variations and Diseases of the Teeth of Animals. Bale, Sons and Danielsson, London.

Cootjans G (1955) Le problème de la carie dans l'antiquité. Rev Belge Stomatol 52: 677–688.

Cran JA (1959) The relation of diet to dental caries. Austr Dent J 4: 182–190.

Greve HC (1949) Die Zahnkaries in geschichtlicher, anatomisch-biologischer, anthropologisch-soziologischer, allgemein-pathologischer und hygienischer Hinsicht. Schattauer, Stuttgart.

Grimm H, Oehmisch E (1956) Statistische Bemerkungen zur Karieshäufigkeit in vorgeschichtlicher und geschichtlicher Zeit. Dtsch Zahn- Mund- Kieferheilk 23: 283–291.

Grundgeir WM (1987) Soziale und geographische Differenzierungsmuster von Ernährung und Lebensweise in prähistorischen Zeiten, untersucht am Beispiel der extrem hohen Kariesfrequenz der Alemannen von Kirchheim/Teck. Zulassungsarbeit zur wissenschaftlichen Prüfung für das Lehramt an Gymnasien, Ulm.

Hartles RL, Leach SA(1975) The effect of diet on dental caries. Brit Med Bull 31: 137–141.

Hillson SW (1979) Diet and dental disease. World Archaeol 11: 147–162.

Horster H (1939) Gebißverfall und Ernährungsweise in Mainfränkischen Notstandsgebieten. Verhandl Dtsch Ges Innere Med 51. Kongress, Wiesbaden 1939: 652–657. München.

Keene HJ (1981) History of dental caries in human populations: The first million years. In: Tanzer JM (ed) Animal Models in Cariology. Information Retrieval, Washington, p. 23–40.

Kerr NW, Bruce MF, Cross JF (1988) Caries experience in the permanent dentition of the late Medieval Scots (1300–1600 A.D.). Arch Oral Biol 33: 143–148.

Köhler B, Bratthall B (1978) Intrafamilial levels of *Streptococcus mutans* and some aspects of the bacterial transmisson. Scand J Dent Res 86: 35–42.

Krogman WM (1938) The role of urbanization in the dentitions of various population groups. Z Rassenkde 7: 41–72.

Mayhall JT (1970) The effect of culture change upon the Eskimo dentition. Arctic Anthrop 7/1: 117–121.

Menzel G (1925) Beiträge zur Kariesforschung. Künstliche Karies des Dentiums. Vjschr Zahnheilkd 41: 72–75.
Moore WJ, Corbett E (1983) Dental and alveolar infection. In: Hart GD (ed) Disease in Ancient Man. Clark, Toronto, p. 139–155.
Nikiforuk G (1985): Understanding Dental Caries, Vol. 1: Etiology and Mechanisms. Basic and Clinical Aspects. Karger, Basel.
Patterson DK (1984) A Diachronic Study of Dental Palaeopathology and Attritional Status of Prehistoric Ontario Pre-Iroquois and Iroquois Populations. Archaeol Survey of Canada 122.
Pedersen PO (1939) Ernährung und Zahncaries primitiver und urbanisierter Grönländer. Verhandl Dtsch Ges Innere Med, 51. Kongr., Wiesbaden 1939. Urban & Schwarzenberg, München, p. 661–668.
Praeger W (1925) Das Gebiß des Menschen in der Altsteinzeit und die Anfänge der Zahnkaries. Dtsch Zahnärztl Wschr 28: 88–99, 112–122.
Robinson JT (1952) Some hominid features of the ape-man dentition. J Dent Assoc South Afr 7: 102–113.
Rudney JD, Katz RV, Brand JW (1983) Interobserver reliability of methods for paleopathological diagnosis of dental caries. Am J Phys Anthrop 62: 243–248.
Russell AL, Consolazio CF, White CL (1961) Dental caries and nutrition in Eskimo scouts of the Alaska National Guard. J Dent Res 40: 594–603.
Schultz AH (1935) Eruption and decay of permanent teeth in primates. Am J Phys Anthrop 19: 489–581.
Schultz AH (1967) Diseases and healed fractures of wild apes. In: Brothwell DR, Sandison AT (eds) Dieseases in Antiquity. A Survey of the Diseases, Injuries and Surgery of Early Populations. Thomas, Springfield, p. 47–55.
Seiler J (1931) Frühere Ernährungsweise und die Zähne älterer Taleinwohner im Bezirk Goms. Schweizer Monatsschr Zahnheilk 41: 489–524.
Shklair IL (1973) *Streptococcus mutans* and the epidemiology of dental caries. *Streptococcus Mutans* and Dental Caries: 7–13. National Institute of Dental Research, DHEW Publication 74–286.
Shklair IL (1981) Natural occurence of caries in animals. Animals as vector and reservoirs of cariogenic flors. In: Tanzer JM (ed) Animal Models in Cariology. Information Retrieval, Washington, p. 41–48.
Somogyi JC (ed) (1964) Zusammenhang zwischen Ernährung und Zahnkaries. Karger, Basel.
Stewart RE, Spence MA (1976) The genetics of common dental diseases. In: Stewart RE, Prescott GH (eds) Oral Facial Genetics. Mosby, St. Louis, p. 81–104.
Vallois H (1936) La carie dentaire à la chronologie des hommes pré-historiques. L'Anthropologie 46: 201–212.
Waugh LM (1937) The influence of diet on the jaws and face of American Eskimo. J Am Dent Assoc 24: 1640–1647.
Werner H (1937) Scheinbare und wirkliche Karies an prähistorischen Zähnen. Z Rassenkde 5: 70–79.
White GE (1975) Dental Caries. Thomas, Springfield.
White TD (1991) Human Osteology. Academic Press, San Diego.
Zuhrt R (1967) Die Entwicklung der Zahnkaries von der Vorzeit bis heute. Proc 7th Int Congr Nutrition, Hamburg 1966, Vol 1. Braunschweig, p. 217–224.

4.2 Periodontal Disease – Etiology, Classification and Diagnosis

Thomas F. Strohm and Kurt W. Alt

Introduction

Periodontal disease reaches far back into the history of humankind, and can be verified by periodontal lesions in skeletal material of the earliest humans (Loe et al. 1992). Hillson states that diseases are a "part of the ecology of an individual. It represents the impact of the environment and part of the body's reaction to it [...]. Disease of the teeth reflect much of what is in the diet. Teeth are in direct contact with all of the diet, but most of the dental disease are related to interactions between diet and microorganisms that live in the mouth" (1986, 283). Although anthropological study has been devoted to periodontal disease for a long time, opinions diverge concerning the progress of periodontopathy through the millenia as well as regarding the diagnosis and findings of periodontal disease (see Lavigne and Molto 1995).

Some anthropologists are of the opinion that periodontal disease has increased since the palaeolithic period, with relative frequency in the Neolithic period and further increases up to the present time (Schranz 1962; Moore and Corbett 1983; Clarke et al. 1986). In the present day, the prevalence of the disease ranges from 75% and 96% (Curilovic et al. 1972; Mutschelknauss 1977). Other anthropologists state that periodontal disease cannot be regarded as an illness produced by civilization, but was already existent in former times (Newman 1974; Bach et al. 1975; Brothwell 1981; Costa 1982; Lange 1983; Hildebolt and Molnar 1991; Loe et al. 1992). Rathbun (1984) reports that the constant part of "periodontoclasia" is about 50–60% (see also Costa 1982). Recently Periera et al. (1994) described the oral condition of three Yanomama Indian tribes of South America. The frequency of periodontal disease (after Ramfjord's 1967 disease index) is of a usual level in the Yanomama groups (frequencies range between 77 and 94%).

The aim of this paper is to present an overview of the manifestations of marginal periodontal disease from the anthropological point of view. For an understanding and diagnosis of historical findings a specific knowledge of present day clinical results is imperative. Therefore a description of the anatomic basis as well as etiological questions and clinical results are an essential part of this paper. Only on the basis of modern scientific results can dental/anthropological examinations of periodontal disease make sense and be successful.

Gingival, or beginning periodontal disease, can be easily recognized in the clinical patient, but cannot be verified in osteological material. At first glance this implies that the frequency of periodontopathy was lower in former times (Brothwell 1981; Moore and Corbett 1983), but is additionally the reason why living and historical examinations can only be compared in a limited way.

Many difficulties exist regarding the results of periodontal disease in osteological material: first, postmortal damage caused by ground storage may be found in sensitive alveolar bone (manifestation of decomposition); second, it is difficult to draw conclusions relating bone and soft tissue; third, insufficient findings exist concerning the effect of abrasion on the periodontium; and fourth, standardization regarding the recording of results is lacking (Moore and Corbett 1983; Hildebolt and Molnar 1991). However, "alveolar recession has been used by archaeologists as an indication of the severity of periodontal disease found in populations of former times" (Watson 1986, 123).

Examination of skeletal material offers many advantages: Because of the lack of soft tissue, bone surfaces and the alveolar crest can be interpreted in a precise way; fenestrations, dehiscences, bifurcation and trifurcation involvement and the loss of bone are easier to diagnose and the corresponding measurements are easier to make (Hildebolt and Molnar 1991).

Anatomy of Healthy Periodontium

When examining skeletal material, the clinical connection should never be neglected. Knowledge of the normal biological structure and morphology of periodontal tissue is a precondition for the understanding and recognition of pathological changes.

Under the term "periodontium" four hard and soft tissues are contained: the cementum of the root surface, the alveolar bone, the periodontal ligament and the gingiva. These four elements of the periodontium anchor the tooth in its osseous socket, the alveolus. The periodontium can adapt to functional and topographical changes and forms the contact area with the oral cavity (Schroeder 1992). Unfortunately, in historical material only two of the four elements of the periodontium can be analysed, the alveolar crest and the root cementum.

The cortical plates of the lamina externa of the maxilla and the mandible enclose cancellous bone which is rich in marrow. The alveolar crest, which encloses the alveoli, is a part of the maxilla. The corresponding structure in the mandible is called the pars alveolaris. Tooth sockets (alveoli dentales) are embedded in trabecular cancellous bone and separated by septa (septa interalveolaria). Within an alveolus further subdivisions for multirooted teeth are made by septa interradicularia. At the exterior surface the roots result in prominences called juga alveolaria. The lamina dura (lamina cribriformis) lines the inner surface of the alveolus and consists of cancellous bone perforated like a sieve. On radiographs the lamina dura appears as a line not transparent to X-rays (Benninghoff and Goerttler 1975; Rateitschak et al. 1978; Strahan and Waite 1978; Ramfjord and Ash 1979; Spranger 1980; Hildebolt and Molnar 1991; Schroeder 1992; Flemmig 1993).

The alveolar bone as a functional pillar is a tooth-dependent structure which develops with the formation and eruption of the teeth and regresses after their loss, until the structure is no longer visible. In contrast, the alveolar crest remains partly preserved for a long time (Rateitschak et al. 1978; Ramfjord and Ash 1979; Lindhe 1983; Schroeder 1992). In toothless parts of the jaw the annual loss of bone is about 0.2 mm (Tallgreen 1972).

Etiology of Periodontal Disease

Over the course of life the periodontium is subjected to various changes through a number of causes (Cripps 1984). The principal cause of periodontal disease are bacterial tooth deposits (dental plaque) which accumulate as a result of inadequate oral hygiene and evoke an inflammation of the periodontium. The mineralization of this dental plaque leads to the formation of dental calculus or subgingival calculus. Changes in the periodontium also arise from physiological or functional causes such as malposition of teeth, loss of supporting area and traumatic occlusal forces. However, infectious disease, iatrogenous causes (habits) or endogenous factors (for example hormonal disturbances) as well as inadequate nutrition (hypo- or avitaminosis) can produce or contribute to periodontal disease.

Rathbun (1984) claims that diet and environmental factors can be the source for periodontal disease, although Held (1989) denies that nutrition has any causal effect. According to Kennedy (1984) the special nutrition of hunters and gatherers in contrast to the rural population had no detectable effect on atrophy of the alveolar crest.

Oral Hygiene

For many centuries toothpicks of twigs, stalks or roots from certain types of trees or bushes were used by different people to care for the teeth (Türp 1989). The use of instruments for tooth care goes considerably further back into the past, as evidenced by cavities which were probably caused by toothpicks (Alt and Koçkapan 1993). Held (1989) evaluated the finding of toothpicks as an indirect proof for the atrophy of the alveolar crest.

Malocclusion

Malocclusions accompanied by traumatic occusal forces can lead to tooth migration or looseness of teeth. In the periodontium they result in a widening of the periodontal ligament space and a cribration of the lamina interna. Taylor (1963) traced the findings of tilted molars with denudation of the complete root to malocclusions. Experiments proved that occlusal forces cannot cause periodontitis, but the continuation of an existing periodontitis can be accelerated (Lindhe and Ericsson 1982; Hanamura et al. 1987; Renggli 1990; Merte 1992).

Clinical Appearance and Classification of Periodontal Disease

The terms "periodontal disease" or "periodontopathy" serve as collective terms for all disease of the periodontium, except for apical periodontitis resulting from pulp disease. Periodontal disease appears as an inflammatory or non-inflammatory, recessive or degenerative disease of the periodontium. These processes lead to varying degrees of destruction of the periodontium and finally to the loss of teeth (Cripps 1984). Sex or ethnic grouping does not have effects on the disease of the periodontium (Ramfjord and Ash 1979; Held 1989; Merte 1992), though with age the frequency and degree of periodontal disease accelerates. In youth the main reason for extraction is caries, but after approximately 30 years of age the loss of teeth caused by periodontal disease increases considerably. After about age 35 or 40 most tooth loss is the result of periodontal disease (Allen 1944; Krogh 1958; Mutschelknauss 1977; Power 1985). Data concerning historical populations are contradictory: on the basis of the low rate of caries, Brothwell (1981) mentions periodontopathy as a main factor for tooth loss, in contrast to Lange's opinion (1983) that caries is the main source for the loss of teeth.

Regarding osteological material, periodontopathy can be categorised as follows:

- ♦ Inflammatory forms
  - ♦ Gingivitis
  - ♦ Periodontitis
- ♦ Involutive forms
- ♦ Manifestations of systemic disease

Inflammatory Forms

Gingivitis

In gingivitis an inflammation is found of the gingiva without recognizable bone destruction. This phenomenon generally cannot be diagnosed in dry material.

Periodontitis Marginalis

All forms of periodontitis are caused by an anaerobic mixed infection which leads to inflammation, gingival or intraosseous pockets and abscesses (Rateitschak et al. 1989) and forces the organism to destroy parts of itself in order to fight the infection. The range and therefore the effect of this destruction to the alveolar crest caused by plaque is approximately 1.5 – 2.5 mm (Tal 1984). Periodontitis is chronic and differs in speed of progression.

Further forms of periodontitis are:

- ♦ prepubertal periodontitis (PP)
- ♦ localized juvenile periodontitis (LJP)
- ♦ adult periodontitis (AP)
- ♦ rapidly progressive periodontitis (RPP)

Special forms of periodontitis manifest in the prevalence of particular teeth: for example localized juvenile periodontitis (LJP), which in puberty attacks the upper anterior teeth and the first molars with heavy vertical bone destruction (Rateitschak et al. 1989).

Subdivision of Periodontitis Marginalis into Horizontal and Vertical Bone Destruction

Two different forms of periodontitis marginalis can be described according to the characteristics: horizontal bone destruction can be described as a pure atrophy of the absolute height of the alveolar septum. Vertical bone destruction can be described as lateral bone destruction with formation of angular intraosseous pockets (Mittermayer 1984; Hildebolt and Molnar 1991; White and Folkens 1991). Horizontal bone resorption with a symmetrical loss of bone represents the most widespread form. Vertical bone resorption is rarer and most often isolated. It correlates with serious forms of periodontitis. According to Alt (1987), horizontal loss of bone was also the dominant form in the past. Vertical bone resorption was predominantly found with isolated teeth. In these cases it was mostly circular. Henkel (1961) reports that until the age of 30, horizontal bone loss predominates and that after age 40, vertical bone loss increases.

Vertical loss of bone can show different forms (Goldman and Cohen 1975):

- ♦ the infraosseous pocket with three osseous walls and one dental wall. Karn et al. (1984) call this arrangement a "crater";
- ♦ the infraosseous pocket, limited by two osseous and two dental walls (interdental crater). The combination of two osseous walls, one dental wall and soft tissue an the fourth side is also possible;
- ♦ the infraosseous pocket with one osseous wall, limited by two dental walls, one osseous wall and soft tissue (Karn et al. 1984 named this a "trench");
- ♦ the combined infraosseous pocket, the basin-like defect: this defect surrounds the whole tooth (Karn et al. 1984 labelled this defect a "moat").

In addition, Karn et al. (1984) distinguish between "ramp" (bone loss bevelled at the outer side) and "plane" (bone loss with even levels at the edges).

Subdivision of Periodontitis Marginalis According to the Degree of Bone Destruction

With periodontitis marginalis superficialis a relatively uniform minor loss of bone of less than a third of the length of the root is detectable. This loss of bone is a pure loss of height from alveolar bone. Concerning periodontitis marginalis profunda an uneven loss of bone of more than a third of the length of the root with or without bifurcation and trifurcation involvement takes place. This loss of bone is also a loss of height of the alveolar bone, perhaps in combination with lateral bone resorption (Spranger 1980; Lindhe 1983; Flemmig 1993).

Involutive Forms

Periodontal Recession

The manifestations of periodontal recession are predominantly dehiscence or fenestration. A systematic description of these forms will be given later (p. 235).

Atrophy of the Alveolar Crest

Inflammatory (periodontitis) and non-inflammatory changes (periodontal atrophy, periodontosis) can be differentiated in osteological material. However, in palaeopathology the terms "periodontitis/periodontal atrophy, periodontosis" do not describe a clinical phenotype, but the morphological condition of the pure dry bone. Periodontal atrophy is more probably caused physiologically as a matter of age than as a matter of pathology, since all aggressive signs characteristic of periodontitis are lacking. Bone destruction may be advanced as well, but the surface of the bone is smooth or only slightly wavy (Mühlemann et al. 1975; Schultz 1988) (see Fig. 1 and Tab. 1).

Manifestations of Systemic Disease

Manifestations of systemic disease are rarely found and are difficult to diagnose, therefore in the anthropological context they are not of importance. Three groups can be differentiated:

- metabolic disease (for example: diabetes, nutritional disturbance, avitaminosis). A vitamin C deficiency for example, can cause flat, porous osseous "exostoses" along the limbus alveolaris; these "exostoses" are calcified haematomas (Schultz 1988)
- haematological disease (for example leukaemia)
- parts of genetically-conditioned syndromes (for example Albright syndrome, Down syndrome, Rathbun syndrome)

Supragingival and Subgingival Calculus

Dental calculus "is a dead material, but is usually covered by a layer of active plaque" (Hillson 1986, 301) and occurs not only in humans, but also in various animals. From old Israel, Egypt, Greece, India and China come reports about the

Table 1. Distinguishing features between periodontitis and atrophy in osteological material

	Periodontitis	*Atrophy*
loss of attachment	+	+
horizontal bone loss	+	+
vertical bone loss possible	+	–
uneven bone loss possible	+	–
fast temporal progression possible	+	–
agressive progression possible	+	–
special forms possible	+	–
probable physiological incident	–	+
probable pathological incident	+	–
signs of inflammation at the bone	+	–
roughened surface of the bone	+	–
boardlike border of limbus alveolaris	+	–
secondary renewal of bone possible	+	–

harm of calculus and instructions with regard to oral hygiene (Held 1989). It is striking that enormous functional demands and abrasions undoubtedly connected with very extensive cleaning have been shown to correlate with substantial calculus deposition (Henkel 1961; Alexandersen 1967; Bach et al. 1975; Newman 1980; Brothwell 1981; Dobney and Brothwell 1986).

With increasing age calculus deposition becomes heavier and documents an obvious extensive lack of oral hygiene (Alexandersen 1967; Alt and Pichler 1991). Other possible causes for heavy calculus deposits – especially where deposits are uneven – could be heavy pain, loss of teeth or muscular disorder (Alexandersen 1967).

"Based on its position on the tooth surface there are two recognised categories differentiated only by position, colour and degree of calcification:

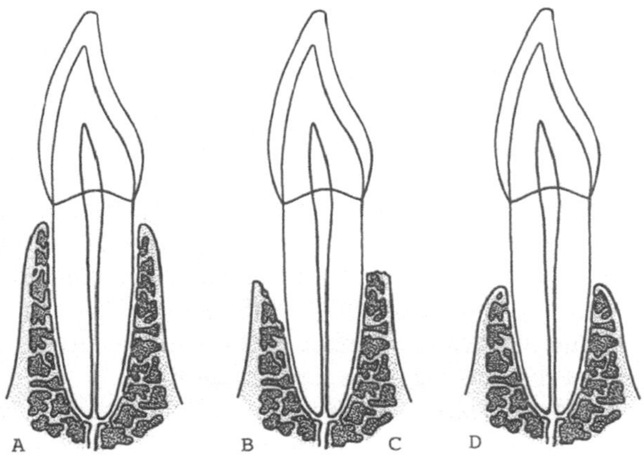

Fig. 1. Classification of periodontal disease. **A:** healthy periodontium; **B:** periodontitis marginalis with vertical bone destruction – infraosseous pockets; **C:** periodontitis marginalis with horizontal bone destruction; **D:** atrophy.

supragingival calculus situated above the gingival margin and, more common, and subgingival calculus usually found isolated in gingival pockets" (Dobney and Brothwell 1986, 55). Supragingivally-located calculus is caused by precipitation of mineral elements from saliva and mineralization of dental plaque, chiefly in the area of the salivary gland openings (Ramfjord and Ash 1979; Lindhe 1983; Hillson 1986; Herrmann et al. 1990). As a consequence, periodontal pocket calculus, located subgingivally, can be produced. Subgingival calculus is a strongly adherent calcified deposit, coloured by side-products of blood from the periodontal pocket (Ramfjord and Ash 1979; Herrmann et al. 1990; Schroeder 1992). In contrast to subgingival calculus, supragingival calculus does not have a pathological source.

When recovering and reconstructing the jaws and the teeth, care is required in order to save the calcified layer of the teeth (Alexandersen 1967; Brothwell 1981; Dobney and Brothwell 1986; Schultz 1988; Brasili Gualandi 1992). Most often the calcified deposits only lightly coat the surface of the teeth, and can be easily split off. Schultz (1988) therefore recommends fixing the supra- and subgingival calculus during excavation with an aqueous solution of glue.

Loss of Attachment

Attachment describes the functional condition, or the quality of the periodontium. The loss of attachment indicates the degree of decomposition taking place in the periodontium which results in exposure of tooth roots through loss of the limbus alveolaris. The distance CEJ-AC (cementoenamel junction-alveolar crest) for a periodontally healthy person is 1–2 mm (Alexandersen 1967; Ramfjord and Ash 1979; Lange 1981; Watson 1986; Schroeder 1992; Flemmig 1993), but, with protruded teeth, could be higher. An increase in this value to more than 2 mm is an indication of periodontal disease (Clarke and Hirsch 1991), though the measurement increases physiologically with age (Hildebolt and Molnar 1991).

The reason for the loss of attachment is an inflammatory process caused by the bacterial stimulus of plaque and the demands on space which are connected with this process. The inflammatory process leads first to gingivitis and later to loss of bone and to periodontitis. Therefore the loss of attachment is the main symptom and the most striking event concerning periodontitis (Rateitschak et al. 1989).

Frequently the loss of attachment runs parallel with abrasion; the relationship between the length of the clinical crown and the length of the remaining resilient root remains the same. In this case one could speak of an elongation of the teeth affected by abrasion (Alexandersen 1967). Philippas (1952) mentions a distance of 7.5 mm between the masticatory surface and the limbus alveolaris of the first mandibular molar that remains the same throughout life.

Clarke et al. (1986) and Clarke and Hirsch (1991) do not consider periodontitis alone to be responsible for the gradual increase of the exposed root of the teeth, but in addition two physiological and two pathological processes operate. From a physiological point of view the tooth erupts continually and cranial structures grow continually until old age; from a pathological point of view gingivitis progresses to periodontitis and abscesses are caused by the inflamed pulpa – here, the connections between pulpa and periodontium, located everywhere in the dentin, are the cause.

Symptomatology of Bone Caused by Periodontal Disease

Boardlike Border of the Limbus Alveolaris

According to Zuhrt (1956) the boardlike border of the limbus alveolaris, which can often be discovered in the molar region (Schranz 1962), is regarded to be a protective reaction of the periodontium when overtaxed by masticatory use. This border, however, distinctive and built like a board, can probably be regarded as a typical characteristic of periodontitis (Lange 1983; Watson 1986; Hedemann 1988).

Roughening of the Bone Surface

Periodontitis is characterized by bone loss caused by inflammation. The bone surface is no longer smooth, but torn and shows groovelike or irregular porous structures. Fine secondary bone plugs at the peripheral zone of the diseased alveolus or porous osseous deposits caused by inflammatory reactions of the periosteum are frequently observable (Zuhrt 1956; Alexandersen 1967; Schultz 1988; Lukacs 1989; Herrmann et al. 1990).

Supporting structures in the bone become enlarged by increased metabolism and can be easily observed. In the case of continuing resorption of the cortex, the underlying trabecular structure of the spongious bone will be exposed. Only if either a loss of cortical surface at the limbus alveolaris or a boardlike border in place of the normally sharp border of the limbus alveolaris is observed, do Clarke et al. (1986) consider periodontal disease to be present.

In serious forms of periodontitis, scurvy should be considered (Herrmann et al. 1990; Hildebolt and Molnar 1991). In the case of a pure bone atrophy, clearly visible changes resulting from inflammation are lacking. After the loss or extraction of a tooth, a regular, smooth, rounded healing of bone does not always occur. In some cases, alveoli remain open for a long time or will form irregularly as the bone reorganizes (Czarnetzki et al. 1982).

Fenestration

In dystopical location of single teeth, or in regions with prominent roots, the remaining bone structure can be very thin. In many cases there is no spongious structure between the lamina externa of the alveolar bone and the lamina interna of the concerned periodontium. In this region an atypical resorption with resulting fenestration of bone can appear, caused by reorganization of bone (Larato 1970; Carranza 1979; Spranger 1980; Grzimek 1986, Merte 1992; Schroeder 1992).

Fenestration describes little windowlike defects in the region of the root, located at the biggest curvature of the juga alveolaria, although still located an osseous bridge incisally. Only when this bridge is lacking one can speak of dehiscence (Lange 1981; Tal 1983; Rateitschak et al. 1989). Elliott and Bowers (1963) regard fenestration to be only a transitional stage on the way to dehiscence. In teeth with fenestrated bone facets, the osseous bridge can completely disappear through continual intermittent forces (Spranger 1980). Hildebolt and Molnar (1991) regard fenestrations and dehiscences to be normal variations of osseous architecture and not to be described as periodontal defects; but this opinion contradicts the clinical phenomenon.

To speak of dehiscence or fenestration at all, the remaining alveolar crest at the transition of the bone to the root must be very thin and should show a flowing transition (Davies et al. 1974; Tal 1983; Grzimek 1986). In most cases damage can be detected if the border of bone appears thick and suggests an artefact; in this case a stereoscopic microscope or at least a magnifying glass should be used.

Causes of bone changes are prominent roots with a pronounced convex shape, pressure of lips or cheeks, disturbance of functions, anomaly of positions and trauma from occlusion or other mechanical reason – in each case in combination with a thin bone plate (Elliott and Bowers 1963; Stahl et al. 1963; Larato 1970; Carranza 1979; Ramfjord and Ash 1979; Tal 1983; Grzimek 1986; Rateitschak et al. 1989; Schroeder 1992; Flemmig 1993). As Volchansky and Vieira (1981) state the proximity to the processus zygomaticus maxillae also plays an important role when the first upper molar is concerned.

The frequency of fenestrations is an estimated 4% (Larato 1970; Tal 1983), 5 to 16% (Volchansky and Vieira 1981; Grzimek 1986; Schroeder 1992) or even 20% (Carranza 1979; Ramfjord and Ash 1979). Fenestrations are observed most frequently at incisors, canines and premolars and rarely at molars (Elliott and Bowers 1963; Davies et al. 1974; Strahan and Waite 1978; Rateitschak et al. 1989). The correlation between the higher incidence of fenestrations and increasing age is partly confirmed (Stahl et al. 1963; Grzimek 1986) and partly refuted (Larato 1970; Davies et al. 1974; Tal 1983).

Dehiscence

Dehiscences describe the mostly vestibularly located atrophic process of the limbus alveolaris. In this case the vestibular CEJ-AC-distance is considerably larger than approximal. Dehiscences are particularly found at anomalies of position at prominent roots and in areas with especially thin vestibular bone (Spranger 1980; Tal 1983; Rateitschak et al. 1989). In patients 3D surface imaging allows a first qualitative orientation, but no quantification of dehiscences. At present, CT cannot be used with adequate safety to identify a thin alveolar lamella (0.1–0.2 mm) and differentiate it from dehiscences (Fuhrmann et al. 1993).

The increased appearance of dehiscences in the mandible and of fenestrations in the maxilla are described by Elliott and Bowers (1963) and Grzimek (1986). The explanation is a matter of anatomy: in the upper jaw, the radius of the osseous base is smaller than the radius of the tooth row, leading to fenestrations; in the mandible the situation is reversed. Consequently a dehiscence is more likely.

Dehiscence is most frequently found vestibularly at the incisors, canines and premolars, but rarely at the upper molars – and then usually palatally (Davies et al. 1974; Mühlemann et al. 1975; Strahan and Waite 1978; Lange 1981; Rateitschak et al. 1989; Hildebolt and Molnar 1991).

Volchansky and Vieira (1981) estimated the presence of dehiscences to be 2% of teeth, Larato (1970) estimated 3%, Tal (1983) 7% and Grzimek (1986) 13%. The finding that dehiscences decrease with increasing age (Davies 1974; Grzimek 1986), is contested by Larato (1970).

Table 2 describes distinguishing features between fenestration and dehiscence at osteological material (Fig. 2).

Table 2. Distinguishing features between fenestration and dehiscence

	Fenestration	*Dehiscence*
incisal bone bridge	+	–
gingival recession	–	+
possibility of an artefact	+	+
location rather in the mandibula	–	+
location rather in the maxilla	+	–
rather in coronal third of root	–	+
rather in middle third of root	+	–

Confluencing Lesions Between Endodontium and Periodontium

The endodontium and the periodontium are connected through the foramen apicale and frequently through lateral canals. A pulp disease can have consequences for

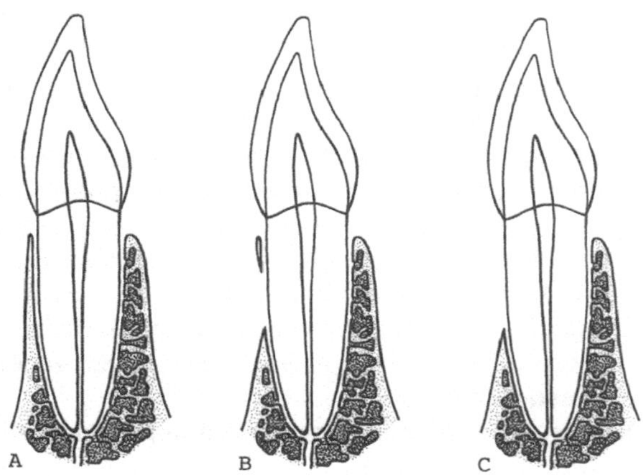

Fig. 2. Distinguishing features between fenestration and dehiscence. **A:** osseous structure without spongious portion; **B:** fenestration; **C:** dehiscence.

the periodontium, just as a disease of the periodontium, in the form of advanced periodontitis, can lead to an inflammation of the pulp (Guldener 1982; Lindhe 1983; Clarke et al. 1986; Rateitschak et al. 1989). Furthermore, independent endodontic and periodontal problems can exist at the same time.

Among diseases of the periodontium that are caused by a pathological condition of the pulp, Lindhe (1983) distinguishes between lateral lesions (by lateral canals), acute periapical lesions (by apex) and combined endodontic-marginal lesions.

Concepts and Procedures in the Description and Evaluation of Periodontal Disease

To describe periodontal processes, indices are used (Clarke et al. 1986; Lavigne and Molto 1995). Various authors have proposed a number of different criteria and examination methods. Requirements for indices are (Hazen 1974):

- simple application
- clear and understandable criteria for examination, in order to support exactness and consistency
- accurate description of the clinical events
- support for statistical evaluation.

Loss of Attachment

To measure the loss of attachment up to six measurements can be taken: mesiobuccal, mesiolingual, distobuccal and distolingual as well as oral and vestibular (the two last measurements always in the middle of the tooth (Hildebolt and Molnar 1991). Concerning the buccal or lingual measurement care must be taken because dehiscences could already exist.

The manifestation of decomposition of the periodontium as well as excavation artefacts can easily be confused with periodontal disease. In this way pathological processes can be falsely identified or even ignored (Schranz 1962; Alt 1987; Hildebolt and Molnar 1991). Because the vestibular bone wall is endangered, an evaluation of interdental height of the septum alone is proposed, in order to avoid pseudopathological interpretations (Valentin 1990).

If the loss of attachment is expressed only in absolute numbers, it is possible that the real condition is not accurately represented. For example, shorter roots are more endangered than longer roots. On the other hand, a percentage measurement could be misleading, since in tooth with long roots, a considerable remainder of attachment could exist. Furthermore a cylindrical root has more remaining surface than does a conical tapering root; a multirooted tooth similarly has – in addition to the more robust support of two or three roots – a larger remainder of attachment than does a single-rooted tooth.

For the measurement of teeth and bones, a dental periodontal probe which is equipped with appropriate calibration is especially suitable: for example the CPITN probe, Williams probe, CP 12 probe or the GC-American probe (Rateitschak et al. 1989).

Examples for the Percentage Measurement of Attachment Loss

- slight degree: initial loss of attachment up to a maximum of a third of the root length;
- moderately severe degree: loss of attachment of approximately half the root length. In addition to horizontal bone loss vertical bone loss can be found;
- severe degree: striking loss of attachment of more than half the root length; it often appears as vertical bone loss (Brabant 1960; Hillson 1979; Schultz 1988; Lukacs 1989; Rateitschak et al. 1989).

Examples for Absolute Measurement of Loss of Attachment

Alexandersen (1967) classifies bone loss in relation to the tooth which is most affected, O'Leary (1967) analyzes the most severely attacked tooth of every sextant of the jaws, Hildebolt and Molnar (1991) assess each tooth and Ramfjord (1967) only the following teeth: 16, 21, 24, 36, 41, 44. These teeth are taken as representative of the whole set. The measurements are given in millimetres, although classifications are often made.

Radiographic Measurement of Attachment Loss

Hildebolt and Molnar (1991) as well as Hildebolt et al. (1992) and Lavelle and Moore (1969) propose to measure and document the loss of attachment by radiographical means with standard X-rays in the mandibular masticatory center. The loss is measured in relation to the length of the root. Digital radiography, involving digital recording and evaluation of X-rays, is more modern (Hildebolt and Molnar 1991).

Bifurcation and Trifurcation Involvement

Bifurcation and trifurcation involvement is a special form of attachment loss in which bone loss exposes the furcation of the molars. Different degrees of horizontal furcation attack can be distinguished. The most important result in this connection is the possibility of complete penetration (Goldman and Cohen 1975; Hamp et al. 1975; Lindhe 1983; Rateitschak et al. 1989). Vertical bifurcation and trifurcation involvement can also be classified. Measurements are given in millimetres from the roof of furcation (Tarnow and Fletcher 1984).

Fenestration

Stahl et al. (1963) make qualitative distinctions in the size of the fenestration, ranging from just visible up to the maximum possible extension; Grzimek (1986) records the size of defects in millimetres.

Dehiscence

The measurement of a dehiscence can be carried out in the following way: a mean value for the general level of bone loss in the approximal area is calculated. Subtraction of this mean value from the measurement of the dehiscence results in the effective loss of substance due to dehiscence itself (Grzimek 1986). Davies et al. (1974) define the dehiscence as a defect, at which the cervical bone lies at least 4 mm apical of the point of the interproximal bone.

Attack by Calculus

Remainders of calculus can often be found on the surface of teeth. The following classification is possible:

- the first visible, spotlike deposit, covering less than one third of the surface;
- deposits surrounding the neck of a tooth in a bandlike way, covering up to two thirds of the surface;
- solid deposits reaching the masticatory surface (Brabant 1960; Greene and Vermillion 1964; O'Leary 1967; Ramfjord 1967; Hillson 1979; Brothwell 1981).

Every abnormal, possibly local formation of calculus should be noted separately, e.g. particularly large quantities of calculus on the masticatory surface or the incisal margin. Suomi et al. (1969) recommend the separate registration of supragingivally located calculus and subgingivally located calculus.

Attack of Subgingival Calculus

Subgingival calculus is situated further apically than is supragingival calculus. The following classification can be made:

- punctual subgingival calculus, covering not more than one third of the cervical tooth in the horizontal direction;
- punctual or marked subgingival calculus, covering more than one third of the cervical tooth in the horizontal direction;
- continuous bandlike subgingival calculus, covering more than two-thirds of the cervical tooth in the horizontal direction (Suomi et al. 1969).

Surface of the Bone After Loss of Teeth

Hillson (1979) and Lukacs (1989) evaluate the bone surface after loss of teeth as follows:

- healed areas are visible as osteogenic healing after loss of teeth, but areas of destruction remain clearly recognizable
- the areas of bone destruction are covered with new bone
- remodelling is complete – the bone surface is solid and smooth.

Criteria for the Registration of Periodontal Disease on Skeletons

Loss of Attachment

Absolute measurements in metric units are to be recommended; conversion into percentage sizes, which is a source of error as well as additional work, can then be dropped. Variations in tooth length are only a few millimetres, and therefore the absolute distance CEJ-AC is sufficient. It should be noted that a distance of up to 2 mm CEJ-AC is healthy – vestibularly the measurements can be a little greater. A distance greater than 2 mm is considered to be a loss of attachment.

To allow later comparision, two measurements should be taken for each tooth: mesial and distal. The information should be in millimetres, not in arbitrary classifications which can be mentioned in the analysis or discussion.

The classification scheme could be applied:

- degree 0 = CEJ-AC between 0 and 2 mm
- degree 1 = CEJ-AC between 2 and 4 mm
- degree 2 = CEJ-AC between 4 and 6 mm
- degree 3 = CEJ-AC more than 6 mm

Ramfjord's (1967) method, which restricts measurements to six fixed teeth, offers an interesting technique for examining extensive material.

Radiographic Measurement of Loss of Attachment

This method is not suited for general use in dental anthropology, because it is too expensive and time-consuming. It is easier and more accurate to examine the osteological materials directly. Only in the case of comparative investigations with living material one can speak of an objective method, especially in combination with digital evaluation. The evaluation of X-rays with computer technology might offer new perspectives.

Fenestration

It is most useful to record the size of defects in millimetres, both in horizontal and vertical directions. By multiplying these measurements one arrives at an area. For the sake of simplicity the defect is treated as if it were rectangular, in contrast to its rather elliptical original form. By indicating the distance between the upper margin of the fenestration and the limbus alveolaris, further observations concerning the progression of fenestrations to dehiscences or concerning specific family or population groupings can be made. In this case preservation condition plays a very important role.

Dehiscence

It is useful to record results according to the method of Grzimek (1986). The approximal result is included given in the measurement CEJ-AC (loss of attachment), so only the measurement of the oral or vestibular defect is necessary. The condition of preservation plays a very important role.

Evaluation

In anthropological investigations, the following measurements should be registered:

- distance CEJ-AC mesially and distally in millimetres;
- attack of furcation horizontally in degrees, vertically in millimetres;
- fenestration in square millimetres;
- vestibular or oral dehiscence, distance CEJ-AC in millimetres;
- calculus or subgingival calculus in degrees;

and supplementary information such as:

- ascertainable traumatic occlusal forces;
- confluencing lesions between endodontium and periodontium.

An exact description of the working method is important to facilitate later comparative investigations. In presenting and evaluating data, measurements should be as precise as possible and preferably in millimetres, although standardized, widely-used classifications may be used. The evaluation of measurements may distinguish between "incidence" (frequency over a period of time) and "prevalence" (frequency at a particular point in time) (Ramfjord and Ash 1979).

Conclusion

According to epidemiological investigations in both extinct and contemporary populations, periodontal disease has been a major cause for loss of teeth in early and late adulthood. Periodontal disease, which is characterised by vertical and horizontal loss of alveolar bone, is multifactorial in origin. Genetic, environmental, diet, and dental hygiene factors, along with influences by culture and ecology, play important roles in its etiology and frequency. As far as ancient populations are concerned, however, these factors are difficult to reconstruct. Unfortunately, no standardized system is available for the measurement of periodontal disease in past populations, which increases the difficulty of gaining precise knowledge regarding its frequency. It would, therefore, be of great advantage if paleopathologists would finally establish and agree upon an internationally accepted method that allows the quantification of alveolar bone loss.

References

Alexandersen V (1967) The pathology of the jaws and the temporomandibular joint. In: Brothwell D, Sandison AT (eds) Diseases in Antiquity. Thomas, Springfield, p. 551–595.

Allen EF (1944) Statistical study of the primary causes of extractions. J Dent Res 23: 453–458.

Alt KW (1987) Zahnerkrankungen sind schon Tausende von Jahren alt. Ein Beitrag zur Paläodontologie. Zahnärztl Mitt 77: 2274–2287.

Alt KW, Pichler S (1991) Paläomedizinische Untersuchung einer spätägyptischen Mumie aus El Hibe (Mittelägypten). Quintessenz 42: 827–839.

Alt KW, Koçkapan C (1993) Artificial tooth-neck grooving in living and prehistoric populations. Homo 44: 5–29.

Bach A, Bach H, Ehmer U (1975) Ernährungsbiologische Aspekte der Kiefer- und Gebißbefunde beim ur- und frühgeschichtlichen Menschen. Ausgrabungen und Funde 20: 222–227.

Benninghoff A, Goerttler K (1975) Lehrbuch der Anatomie des Menschen, Bd. 1. Allgemeine Anatomie, Cytologie und Bewegungsapparat. Urban & Schwarzenberg, München, p. 503–505.

Brabant H (1960) Observations odontologiques et anthropologiques sur les ossements provenant des fouilles executees dans la cathedrale Sainte-Gertrude a Nivelles. Acta Stomatol Belg 57: 17–41.

Brasili Gualandi P (1992) Food habits and dental disease in an Iron-Age population. Anthrop Anz 50: 67–82.

Brothwell DR (1981) Digging up Bones. Oxford Univ Press, Oxford.

Carranza FA (1979) Glickman's Clinical Periodontology, 5th Ed. Saunders, Philadelphia.

Clarke NG, Carey SE, Srikandi W, Hirsch RS, Leppard PI (1986) Periodontal disease in ancient populations. Am J Phys Anthrop 71: 173–183.

Clarke NG, Hirsch RS (1991) Physiological, pulpal and periodontal factors influencing alveolar bone. In: Kelley MA, Larsen CS (eds) Advances in Dental Anthropology. Wiley, New York, p. 241–266.

Cripps S (1984) Aetiology, Periodontal Disease, Recognition, Interception and Prevention. Quintessence, New York.

Costa RL (1982) Periodontal Disease in the Prehistoric Ipiutak and Tigara Skeletal Remains from Point Hope, Alaska.

Curilovic Z, Renggli HH, Saxer UP, Germann MA (1972) Parodontalzustand bei einer Gruppe von Schweizer Rekruten. Schweiz Mschr Zahnheilk 83: 437–451.

Czarnetzki A, Uhlig C, Wolf R (1982) Menschen des frühen Mittelalters im Spiegel der Anthropologie und Medizin. Württembergisches Landesmuseum (Hrsg), Stuttgart.

Davies RM, Downer MC, Hull PS, Lennon MA (1974) Alveolar defects in human skulls. J Clin Periodont 1: 107–111.

Dobney K, Brothwell D (1986) Dental calculus: its relevance to ancient diet and oral ecology. In: Cruwys E, Foley RA (eds) Teeth and Anthropology. BAR Int Ser 291, p. 55–81.

Elliott JR, Bowers GM (1963) Alveolar dehiscence and fenestration. Periodontics 1: 245–248.

Flemmig TF (1993) Parodontologie. Ein Kompendium. Thieme, Stuttgart.

Fuhrmann R, Klein HM, Wehrbein H, Günther RW, Diedrich P (1993) Hochauflösende Computertomographie fazialer und oraler Knochendeziszenzen. Dtsch Zahnärztl Z 48: 242–246.

Goldman HM, Cohen DW (1975) Lehrbuch der Parodontologie. Medica, Stuttgart.

Greene JC, Vermillion JR (1964) The simplified oral hygiene index. J Amer Dent Ass 68: 7–13.

Grzimek D (1986) Studie von Schädeln zweier Skelettpopulationen aus dem frühen Mittelalter hinsichtlich alveolärer Dehiszenzen und Fenestrationen unter Berücksichtigung des historischen, archäologischen und anthropologischen Hintergrundes. Med Diss, Frankfurt/Main.

Guldener PHA (1982) Beziehung zwischen Pulpa- und parodontalen Erkrankungen. In: Guldener PHA, Langeland K (Hrsg) Endodontologie. Thieme, Stuttgart, p. 368–378.
Hamp SE, Nyman S, Lindhe J (1975) Periodontal treatment of multirooted teeth. Results after 5 years. J Cli Periodontol 2: 126–135.
Hanamura H, Houston F, Rylander H, Carlsson GE, Haraldson T, Nyman S (1987) Periodontal status and bruxism – a comparative study of patients with periodontal disease and occlusal parafunctions. J Periodontol 58: 173–176.
Hazen SP (1974) Indices for the measurement of gingival inflammation in clinical studies of oral hygiene and periodontal disease. J Periodont Res 9: Suppl 14, 61–77.
Hedemann M (1988) Zahn- und Kieferbefunde an Schädeln des frühmittelalterlichen Gräberfeldes Schwanenstadt/Oberösterreich. Med Diss, Marburg.
Held AJ (1989) Periodontology. Birkhäuser, Basel.
Henkel G (1961) Zur Frage der Parodontopathien bei frühgeschichtlichem Schädelmaterial. Dtsch Stomat 11: 195–204.
Herrmann B, Grupe G, Hummel S, Piepenbrink H, Schutkowski H (1990) Prähistorische Anthropologie. Springer, Berlin Heidelberg New York.
Hildebolt CF, Molnar S (1991) Measurement and description of periodontal disease in anthropological studies. In: Kelley MA, Larsen CS (eds) Advances in Dental Anthropology. Wiley, New York, p. 225–240.
Hildebolt CF, Vannier MW, Shrout MK (1992) Alveolar bone loss quantification with digital bitewing radiographs. In: Lukacs JR (ed) Culture, Ecology and Dental Anthropology. Kamla-Raj, p. 89–95.
Hillson S (1979) Diet and dental disease. World Archaeol 2: 147–162.
Hillson S (1986) Teeth. Cambridge Univ Press, Cambridge.
Karn KW, Shockett HP, Moffitt WC, Gray JL (1984) Topographic classification of deformities of the alveolar process. J Periodont 55: 336–340.
Kennedy KAR (1984) Growth, nutrition, and pathology in changing paleodemographic settings in South Asia. In: Cohen MN, Armelagos GJ (eds) Paleopathology at the Origins of Agriculture. Academic Press, New York, p. 169–192.
Krogh HW (1958) Permanent tooth mortality: A clinical study of causes of loss. J Amer Dent Ass 57: 670–675.
Lange DE (1981) Parodontologie in der täglichen Praxis. Quintessenz, Berlin.
Lange G (1983) Die menschlichen Skelettreste aus dem Oppidum von Manching. Steiner, Wiesbaden.
Larato DC (1970) Alveolar plate fenestrations and dehiscences of the human skull. Oral Surg 29: 816–819.
Lavelle CLB, Moore WJ (1969) Alveolar bone resorption in Anglosaxon and seventeenth-century mandibles. J Periodont Res 4: 70–73.
Lavigne SE, Molto JE (1995) System of measurement of the severity of periodontal disease in past population. Int J Osteoarchaeol 5: 265–273.
Lindhe J, Ericsson I (1982) The effect of elimination of jiggling forces on periodontally exposed teeth in the dog. J Periodontol 53: 562–567.
Lindhe J (1983) Textbook of Clinical Periodontology. Munksgaard, Copenhagen.
Loe H, Anerud A, Boysen H (1992) The natural history of periodontal disease in man: Prevalence, severity, and extent of gingival recession. J Periodont 63: 489–494.
Lukacs JR (1989) Dental paleopathology: Methods for reconstruction dietary patterns. In: Işcan MY, Kennedy KAR (eds) Reconstruction of Life from the Skeleton. Liss, New York, p. 261–286.
Merte K (1992) Parodontologie für die Praxis. Barth, Leipzig.
Mittermayer C (1984) Oralpathologie. Schattauer, Stuttgart.
Moore WJ, Corbett ME (1983) Dental and alveolar infection. In: Hart GD (ed) Disease in Ancient Man. Clarke Irwin, Toronto, p. 139–155.

Mutschelknauss R (1977) Die Klinik der marginalen Parodontopathien und ihre pathohistologischen Grundlagen. In: Haunfelder D, Hupfauf L, Ketterl W, Schmuth G (Hrsg) Praxis der Zahnheilkunde. Urban & Schwarzenberg, München.
Mühlemann HR, Rateitschak KH, Renggli H (1975) Parodontologie. Thieme, Stuttgart.
O'Leary T (1967) The periodontal screening examination. J Periodont 38: 617–624.
Newman HN (1974) Diet, attrition, plaque and dental disease. Br Dent J 136: 491
Newman HN (1980) Dental Plaque: the Ecology of the Flora on Human Teeth. Springfield, Illinois.
Periera CB, Katich JP, Turner CG (1994) Oral condition of three Yanomama Indian tribes of south America. Dent Anthrop Newsl 8 (3): 8–10.
Philippas GG (1952) Effects of function on healthy teeth: The evidence of ancient Athenian remains. J Amer Dent Ass 45: 443–453.
Power C (1985/6) Diet and disease: Evidence from the human dental remains in two medieval Irish populations. J Irish Archaeol 3: 49–53.
Ramfjord SP (1967) The periodontal disease index (PDI). J Periodont 38: 602–610.
Ramfjord SP, Ash MM (1979) Periodontology and Periodontitis. Saunders, Philadelphia.
Rateitschak KH, Renggli HH, Mühlemann HR (1978) Parodontologie. Thieme, Stuttgart.
Rateitschak KH, Rateitschak EM, Wolf HF (1989) Parodontologie. Thieme, Stuttgart.
Rathbun TA (1984) Skeletal pathology from the paleolithic through the metal ages in Iran and Iraq. In: Cohen MN, Armelagos GJ (eds) Paleopathology at the Origins of Agriculture. Academic Press, New York, p. 137–167.
Renggli HH (1990) Ätiologie der marginalen Parodontopathien. In: Ketterl W (Hrsg) Parodontologie, p. 33–64.
Schranz D (1962) Zahnbetterkrankungen der längst vergangenen Zeit. Z Morph Anthrop 52, 347–354
Schroeder HE (1992) Orale Strukturbiologie. Thieme, Stuttgart.
Schultz M (1988) Paläopathologische Diagnostik. In: Knußmann R (Hrsg) Anthropologie, Bd. I. Wesen und Methoden der Anthropologie. Fischer, Stuttgart, p. 480–495.
Spranger H (1980) Klinik der marginalen Parodontopathien. Hüthig, Heidelberg.
Stahl SS, Cantor M, Zwig E (1963) Fenestrations of the labial alveolar plate in human skulls. Periodontics 1: 99.
Strahan JD, Waite IM (1978) A Colour Atlas of Periodontology. Wolfe, London.
Suomi JD, Greene JC, Vermillion JR, Chang JJ, Leatherwood EC (1969) The effect of controlled oral hygiene procedures on the progression of periodontal disease in adults: Results after two years. J Periodont 40: 416–420.
Tal H (1983) Alveolar dehiscences and fenestrae in dried South African Negro mandibles. Am J Phys Anthrop 61: 173–179.
Tal H (1984) Relationship between the interproximal distance of roots and the prevalence of intrabony pockets. J Periodontol 55: 604–607.
Tallgreen A (1982) The continuing reduction of the residual alveolar ridges in complete dentures weares. J Prosthet Dent 27: 102.
Tarnow D, Fletcher P (1984) Classification of the vertical component of furcation involvement. J Periodontol 55: 283–284.
Taylor RMS (1963) Cause and effect of wear of teeth. Acta Anat 53: 97–157.
Türp J (1989) Zahnputzhölzer – Die ältesten Zahnbürsten der Welt. Quintessenz 40: 937–948.
Valentin A (1990) Stomatologische, gnathometrische und kaufunktionelle Untersuchungen an Merowingerschädeln des 5.–8. Jh. n. Chr. aus Süddeutschland. Med Diss, Frankfurt.
Volchansky A, Vieira E (1981) Alveolar dehiscence and fenestration in dried South African Negro maxillae. J Dent Assoc S Africa 36: 701–704.

Watson PJC (1986) A study of the pattern of alveolar recession. In: Cruwys E, Foley RA, Teeth and Anthropology. BAR Int Ser 291, p. 123–131.
White TF, Folkens PA (1991) Human Osteology. Academic Press, San Diego.
Zuhrt R (1956) Stomatologische Untersuchungen an spätmittelalterlichen Funden von Reckahn (12.–14. Jahrhundert). Dtsch Zahn-, Mund- und Kieferheilk 25: 16–24.

4.3 Periapical Lesions – Clinical and Anthropological Aspects

Kurt W. Alt, Jens C. Türp, and Rüdiger Wächter

Introduction

Diseases of teeth and the surrounding periodontal ligament are important topics in paleopathological investigations because they provide useful information about living conditions and dietary patterns. Since dental (pulpal) disease can cause periodontal lesions (and vice versa) (Grant et al. 1988), the two types of pathological conditions are often closely related. In early populations, the main causes of tooth loss were periodontal disease (lateral periodontitis) and pulpoalveolar disease (periapical periodontitis), both of which are attributed to oral infections.

A pulpal inflammation that spreads through the apical foramen initiates a periapical lesion. In the presence of unfavorable circumstances, such as diminished resistance of the patient, periapical lesions may even lead to a life-threatening condition as for example a spatial abscess. Yet, the differential diagnosis of alterations around the apex of a tooth is not an easy task. Accurate diagnosis requires a profound knowledge of both normal and pathological anatomy. It is the purpose of this chapter to describe the etiology, pathophysiology, and differentiation of periapical lesions from both clinical and paleopathological viewpoints. In addition, we will describe the kind, occurrence, effects, complications and importance of odontogenic infections in ancient times.

Clinical Aspects

General Description of Periapical Lesions

As a rule, periapical radiolucencies as depicted on radiographs are indications of pathological changes around the dental apex and the adjacent tissues resulting from inflammation. Although different pathological conditions may be responsible for periapical alterations (Tab. 1), periapical granulomas and radicular cysts are by far the most frequent causes (Tab. 2). According to Wood (1984), these two disease entities are responsible for approximately 95% of all periapical radiolucencies. Nonetheless, it should be borne in mind that radiographs do not allow a reliable distinction between periapical granuloma and radicular cyst (Bhaskar 1966; Wood 1984; White et al. 1994; Lee 1995). Schroeder (1991) notes that judging from a radiograph both findings are equally probable.

Table 1. Causes for periapical lesions appearing; ranked order (after Wood 1984, modified)

Periapical granuloma	most frequent
Radicular cyst	
Periapical scar (after root canal filling)	
Advanced periapical abscess	
Surgical defect	
Osteomyelitis	
Lateral periodontitis[1]	
Periapical cemental dysplasia (initial stage)	↑
Traumatic bone cysts	↓
Dentigerous cyst (in association with the crown of an unerupted tooth)	
Globulomaxillary radiolucencies[2]	
Ameloblastoma	
Cemento-ossifying fibroma (early stage)	
Malignancies (metastatic tumors; osteosarcoma; multiple myeloma; eosinophilic granuloma)	least frequent

[1] The term "lateral periodontitis" is used throughout this chapter for the condition that in most texts is referred to as "periodontitis". Newman & Challacombe (1995, 293) note that as a result of the use of the less specific latter term "it is common to find a lack of coordination between studies of changes in the periapical and lateral portions of the periodontal ligament". Therefore, a clear linguistic distinction between both disease entities seems to be warranted.
[2] Localisation between maxillary lateral incisor and maxillary canine: vertical depression of the labial alveolar plate [anatomical variation]; lateral granuloma/radicular cyst; anterior cleft; other pathologies; globulomaxillary cyst.

Some early histological investigations showed that infectious infiltration in the periapical region was more diffuse than had appeared on the radiograph (Peyer 1945). The author blamed this for the high recurrency rate after surgical treatment. The histopathologic structure of the periapical lesions in relation to the root canal of the affected teeth is therefore of particular importance (Nair 1995). Recently, the types and incidence of human periapical lesions were evaluated on the basis of 256 extracted teeth (Nair et al. 1996). The low incidence of radicular cysts (15%) found among periapical lesions is a strong argument against the widely held view that almost half of all periapical lesions are cysts.

Several studies have demonstrated that the probability for a periapical radiolucency or lesion to be a radicular cyst increases with its size (Wais 1958; Mortensen et al. 1970; Morse et al. 1973). Defects of a diameter greater 20 mm or a size exceeding 200 mm^2 are almost always considered to be cysts (Zain et al. 1989). Only rarely have lesions of this size proven to be periapical granulomas (Natkin et al. 1984). Yet, only histological differentiation allows a reliable diagnosis of periodontal defects of pulpal origin.

Table 2. Clinical evidence of periapical lesions (in %); selected studies (after Nair et al. 1996, and an own search of the literature) (N: number of periapical radiolucencies; PerGr: periapical granuloma; RadCy: radicular cyst; OthLes: other lesions)

Author	Year	N	PerGr	RadCy	OthLes	Methods
Priebe et al.	1954	101	46	54	(–)	Radiology
Baumann and Rosman	1956	121	74	26	(–)	Histology
Wais (series 1)	1958	50	64	26	10	Histology
Wais (series 2)	1958	50	84	14	2	Histology
Linenberg et al.	1964	110	62	28	10	Biopsy
Patterson et al.	1964	501	84	14	2	Biopsy
Bhaskar	1966	2308	48	42	10	Histology
Sommer et al.	1966	170	84	6	10	Biopsy
Sonnabend and Oh	1966	237	93	7	(–)	Histology
Seltzer et al.	1967	87	45	51	4	Histology
Lalonde and Luebke	1968	800	45	46	9	Histology
Mortensen et al.	1970	396	59	41	(–)	Histology
Morse et al.	1973	40	77	23	(–)	Histology/Biochemistry
Block et al.	1976	230	94	6	(–)	Biopsy
Simon	1980	35	54	17	23	Histology
Hacker and Fischbach	1985	120	27	58	15	Histology
Stockdale and Chandler	1988	1108	77	17	6	Histology
Spatafore et al.	1990	1659	52	42	6	Biopsy
Lin et al.	1991	150	(–)	19	81	Biopsy
Wayman et al.	1992	58	71	27	2	Biopsy
Nobuhara and Del Rio	1993	150	59	22	19	Biopsy
Nair et al.	1996	256	50	15	35	Histology

The histological specimens presented in this chapter were derived from newly extracted teeth by Donath's cutting-grinding technique (Donath and Breuner 1982), except for Fig. 1, which was taken from an anatomical specimen. The thin sections were stained following the descriptions made by Richardson et al. (1960), and counterstained by the method described by Laczko and Lévai (1975).

Pathogenesis of Periapical Lesions

In most cases, periapical inflammation is a reaction of the periapical tissues to an infection of the dental pulp (Schroeder 1991; Wächter et al. 1992). Due to the characteristics of the pulp cavity, which opens in the apical region, a pulpal inflammation (pulpitis, necrosis, gangrene, pulpal abscess) can spread through the apical foramen only (and, if present, through accessory foramina).

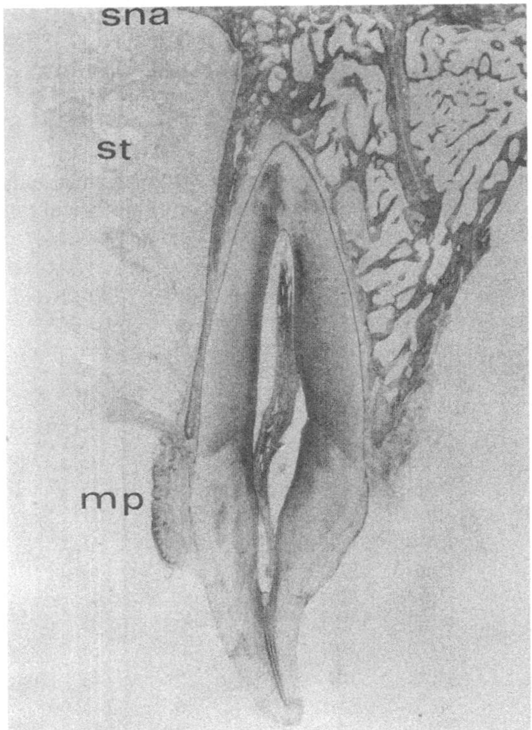

Fig. 1. Sagittal section of maxilla with tooth 11 (anatomical specimen); some shrinking of pulp due to fixation occurred (magnification: x2) (sna = spina nasalis anterior, st = soft tissue, mp = marginal periodontium).

Pulpitis

Pulpitis is an acute or chronic, focal or generalized inflammation of the pulp (Bhaskar 1986; Mittermayer 1993). It is typically accompanied by a sharp or dull, sometimes throbbing pain of moderate or severe intensity (Merskey and Bogduk 1994). Several etiologic factors for this dental disease have been identified (Tab. 3). In recent populations, the main cause for pulpitis is dental caries (Schroeder 1991). Conversely, in prehistoric populations the prevalent causes of pulpitis and subsequent periapical lesions appear to have been a dental pulp exposed by excessive attrition and caries (Taylor 1963; Alexandersen 1967; Molnar 1972; Hinton 1981; Smith 1984; Jurmain 1990; Clarke and Hirsch 1991a, b).

Two forms of pulpitis can be distinguished, namely abacterial and bacterial types. An abacterial pulpitis may lead to a pulpal necrosis, in which the pulp is dead but not (yet) infected. If a pulp necrosis becomes infected with saprophytic, protein-decomposing putrefactive bacteria a pulpal gangrene develops. A pulpal infection contains a predominantly anaerobic mixed microbial flora. Typical microorganisms are bacteroides species, peptostreptococci, streptococci, and fusobacteria (Nair and Schroeder 1983; Nair and Luder 1985; Sundquist et al. 1989;

Table 3. Etiology of pulpitis

Bacterial infection (caries)
Severe attrition with pulp perforation
Bacterial infection of pulp (pulpitis) after dental fracture or splitting
Bacterial infection of pulp after operative procedures
Spread of neighboring infections (e.g., osteomyelitis)
Spread of a lateral periodontitis (bacterial infection from deep periodontal pockets through lateral or accessory foramina)
Bacteremia/sepsis with hematogenous presence of bacteria in pulp
Physical (e.g., thermal irritation)
Chemical irritation (e.g., by dental filling material)
Tooth trauma with disturbance or interruption of blood circulation
Traumatic occlusion

Schuster 1994). According to Miller et al., "pulpitis may spread to the jawbone and result in acute localized periapical infection (abscess), chronic localized periapical infection (granuloma), apical cyst formation, or diffuse spread into bone (osteomyelitis)" (1995, 207). The chronic process may develop into a pulp abscess at any time (Schroeder 1991).

Acute Periapical Periodontitis and Periapical Abscess

The quick spread of bacteria and breakdown products of the gangrene through the apical foramen can lead to an acute inflammation of the periapical tissues and to the development of an acute periapical abscess (Bhaskar 1966; Schroeder 1991). Due to the fact that pus is being formed between the tooth and the adjacent bone, the pressure in the periapical region increases and the (non-vital) tooth is elevated in its socket. As a result of the inflammatory process, the patient experiences a severe throbbing pain, and the affected tooth is extremely tender to touch and pressure. Among the most commonly involved pathogens are bacteroides and peptostreptococcus species, *Streptococcus mitis* and *Fusobacterium nucleatum* (Williams et al. 1983). Because the abscess formation occurs rapidly, major bone resorptions cannot develop in this acute stage. Only if the abscess developed from a chronic inflammatory process do osteolytic destructions of the bone show on a radiograph (Fig. 2). In this case, well-defined round or pear-shaped periapical radiolucent areas are typically present. It is important to note that the periapical region may clinically, radiographically and histologically exhibit all signs of an inflammation even before the pulp has died completely.

Chronic Periapical Periodontitis: Periapical Granuloma

Frequently, a pulpal inflammation results in the development of a periapical granuloma. A periapical granuloma is a chronic, non-suppurative inflammation with no or only minor clinical symptoms. This condition is responsible for about 50% of

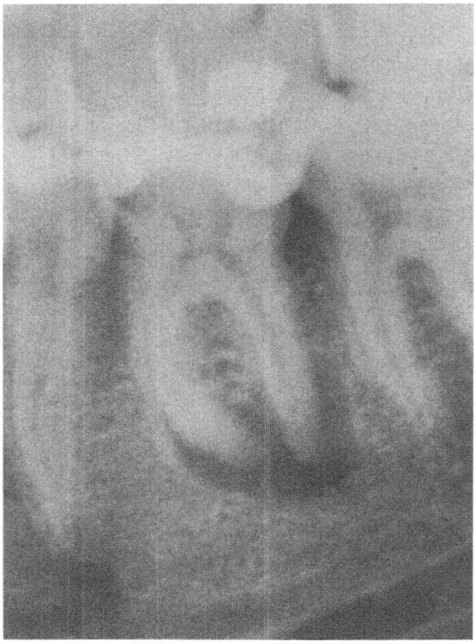

Fig. 2. Periapical granuloma at tooth 36 (panoramic radiograph/detail); there is no sharp border between lesion and bone.

all periapical radiolucencies (Shrout et al. 1993). Histologically, a periapical granuloma, which is surrounded by a capsule of connective tissue, is characterized by an infiltrate of different inflammatory cells (macrophages, lymphocytes, plasma cells, neutrophils, mast cells). The cells are located in a highly vascularized, unspecific granulation tissue which contains fibroblasts (Stern et al. 1981; Kontiainen et al. 1986). Specific antibodies are found, primarily from the IgG and IgA types, as well as components of the complement system (Nair and Schroeder 1983; Piatowska et al. 1988; Babal et al. 1989). In contrast to former beliefs, it is now recognized that bacteria are often present in periapical granulomas (Nair and Schroeder 1983). In addition, some odontogenetic epithelial cell rests (Malassez's rests) may be found. These cell rests are the pathobiological basis of the development of a radicular cyst from a periapical granuloma (Summers 1974). Nair and Schroeder (1983) report that such a transition occurs in almost 50% of all lesions.

In periapical granulomas, chronic phases usually alternate with short-lasting acute attacks in the course of which destructions occur in the periapical tissues (Bohne 1990; Wang and Stashenko 1993). Consequently, resorptions of the hard tissues (tooth and bone) are discernible in the periapical region (Fig. 3). Scanning electron microscopy (SEM) examinations clearly show that the root resorptions are more pronounced in the case of periapical granulomas than in radicular cysts (Delzangles 1989). Periapical granulomas may be asymptomatic for years even while they are increasing in size. Sometimes they heal without leaving scars (Schroeder 1991). If the balance between physical resistance and infection shifts, however, an acute exacerbation frequently ensues. Radiographically, a periapical

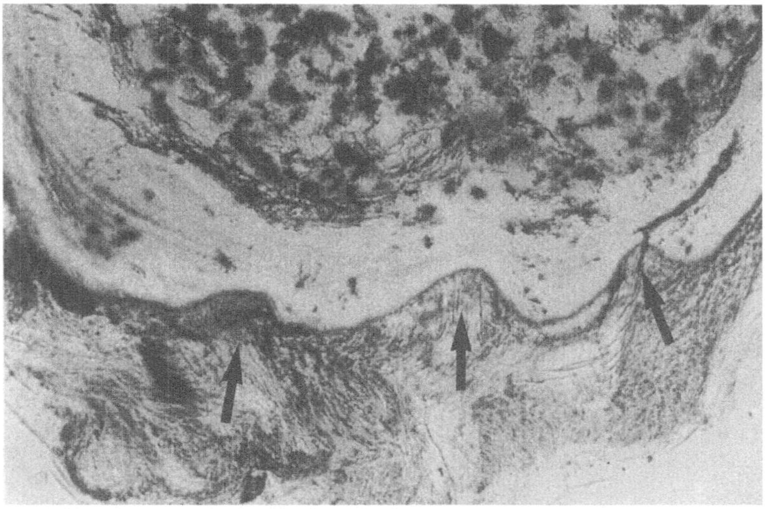

Fig. 3. Histological picture of resorption lacunae (→) in periapical granuloma (x120).

granuloma shows a well-defined radiolucency (round or oval) that is located around the tooth apex (see Fig. 2). Bone resorption in combination with root resorption may also be visible (Figs. 4 and 5).

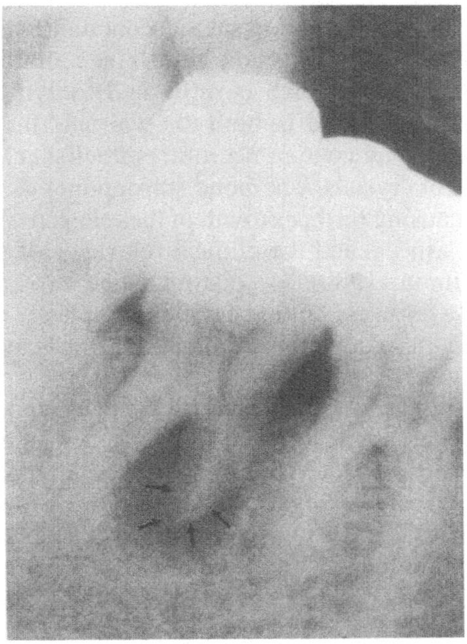

Fig. 4. Periapical granuloma at tooth 47 (panoramic radiograph/detail) showing tremendous resorption of the mesial root (→).

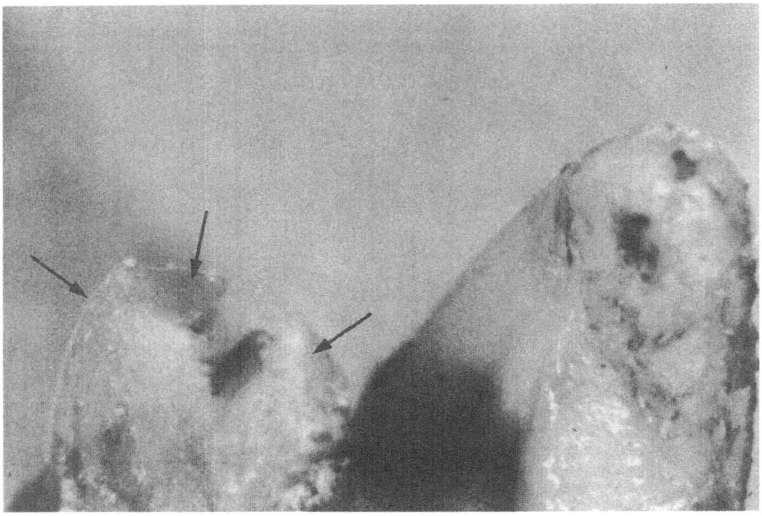

Fig. 5. Resorption of the mesial root on extracted tooth 47 (→).

Radicular Cysts

In most cases, these inflammatory odontogenic cysts develop as a consequence of a chronic periapical inflammation (periapical granuloma, secondary periapical abscess) which stimulates the epithelial cell rests of Malassez to proliferate. A characteristic feature of cysts is a soft tissue sac containing fluid, semi-fluid or gas (Lee 1995). The sac is lined by non-keratinized epithelium, overlying a vascularized connective tissue which is infiltrated by lymphocytes and plasma cells. Immunoglobulins are found in both the wall and the cavity (Smith et al. 1987). The cystic cavity may contain necrotic epithelial cells, granulocytes and, occasionally, cholesterol crystals. The bone surrounding the cyst is submitted to resorption processes, causing further growth of these lesions (Formigli et al. 1995).

The morphologic aspects and the clinical relevance of periapical cysts have been described by Simon (1980). He distinguished two distinct categories of radicular cysts, namely those containing cavities completely enclosed in epithelial lining (true cysts), and those containing epithelium-lined cavities that are open to the root canals (bay cysts).

Radicular cysts account for 60–75% of all cysts located in the jaws (Soames and Southam 1993). The vast majority of radicular cysts are located at the tooth apex. In this case, they are also referred to as apical periodontal cysts, or periapical cysts. Other forms of radicular cysts involve lateral cysts and residual cysts (Soames and Southam 1993). For residual cysts, see Oehlers (1970).

Like periapical granulomas, radicular cysts are in most cases clinically asymptomatic; yet they may cause symptoms if secondarily infected. In that case, bacteria are present in the cystic cavity (approximately 75% streptococci, 25% gram-negative germs; Rudelt 1985), and the radiolucency does not show distinct margins (Wood 1984).

Conventional root canal treatment may lead to a marked reduction of the size or to the complete disappearance of 85–90% of periapical radiolucencies (Barbakow et al. 1981; Sjögren et al. 1990; Zaldu'a 1995). On the basis of such clinical findings as well as of histopathologic studies (e.g., Bhaskar 1966; Seltzer et al. 1967), it has been suggested that most cystic lesions at the periapex heal after conventional endodontic treatment. In contrast, some oral surgeons maintain that cysts do not heal, but must be surgically removed (Winstock 1980).

Differential Diagnosis Between Periapical Granuloma and Radicular Cysts

Diseases of the periapical region often exhibit transitions from one clinical picture to another, making their differentiation difficult. On radiographs, radicular cysts appear as roundish radiolucencies which are circumscribed by a dense, at times hyperostotic (radiopaque) margin; the tooth apex projects into the radiolucency (Wood 1984; Fig. 6). Many authors have emphasized that on the basis of radiographs alone, periapical lesions cannot be reliably diagnosed as being cystic or non-cystic lesions (Baumann and Rossman 1956; Linenberg et al. 1964; Bhaskar 1966; Mortensen et al. 1970; Lentrodt 1974). In one study the diagnostic accuracy of radiographs was shown to be as high as 74% in cases of cysts, and 56% in granulomas (Linenberg et al. 1964). In a re-evaluation of 88 histologically verified periapical lesions, the validity of the diagnostic criteria applied to differentiate cysts and granulomas was reviewed by two independent investigators (Syrjänen et al.

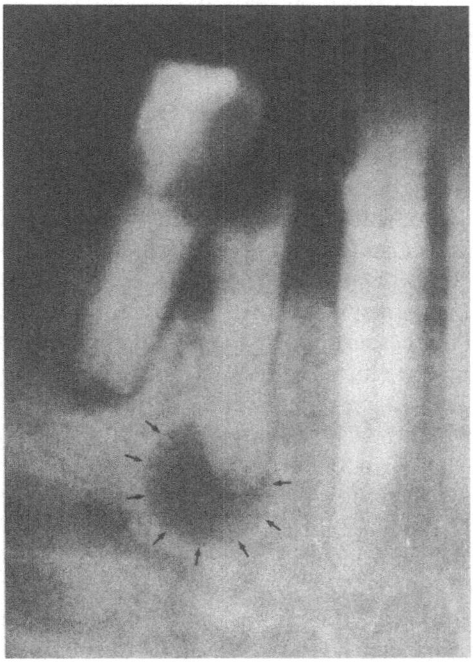

Fig. 6. Radicular cyst on root of tooth 44 showing thin sclerotic border (panoramic radiograph/detail).

1982). The authors found a high coincidence of the diagnoses based on the radiographs. The interobserver variation in the diagnoses was 9%, and the overall diagnostic accuracy was 67% for both examiners. Nevertheless, there is no doubt that a clear distinction from periapical granulomas can only be made by histological examination (Hauk et al. 1993; Nair et al. 1996) (Fig. 7 a–c). Even in cases where modern computer-assisted radiographic techniques have demonstrated an increase in the detectability of periapical lesions when compared with conventional radiography, periapical granulomas and radicular cysts cannot be distinguished (Tyndall et al. 1990). However, recently it has been reported that it is possible to differentiate between these two pathologic conditions by using a special radiometric procedure that analyzes the patterns of the gray-level distributions of digitized radiographic images (Shrout et al. 1993).

It must be taken into consideration that even after sucessful endodontic therapy the histologic status of any periapical radiolucent lesion is unknown at the time of treatment (Kerekes and Tronstad 1979; Barbakow et al. 1981; Sjögren et al. 1990). An accurate histopathologic diagnosis of periapical lesions, and of radicular cysts in particular, is possible only through serial sectioning or step-serial sectioning of the lesions after they have been removed *in toto* (Nair et al. 1996). Since a number of other processes can produce similar radiolucencies, several differential diagnoses must be taken into account (see Tab. 2). It is important to consider that anatomical variations may occasionally present a similar radiographic image (Wood 1984). One should also remember the possibility that rare malignant tumors (Gardner 1975; Wood 1984; Wannfors and Hammarström 1990) and metastases of other tumors in the periapical region (Milobsky et al. 1975) might be the reason for a periapical radiolucency.

Complications and Life-Threatening Courses of Odontogenic Infections

General Aspects – Abscess Formations

The oral cavity is physiologically colonized with microbes. In healthy individuals, the oral mucosa works as a sufficient barrier membrane that prevents penetration and infection. In the case of an odontogenic infection, there are generally two portals of entry into the organism. Starting with a deep carious lesion and following pulpitis, the infection may spread along the pulpal cavity to the periapical region involving the alveolar bone (local ostitis). In most cases, gangrenous teeth which are left untreated are the reason for this pathological process. The other possible means of entrance for bacteria is the marginal periodontal ligament. The primary localized infection – the periodontal disease (lateral periodontitis) – can spread along the neighboring soft tissues or along the alveolar process and may invade the bone. The typical clinical features of this progressive infectious disease are abscess formations in the head and neck area, and, less frequently, bony infiltrations (osteomyelitis) (Otten et al. 1998).

Three parameters influence the clinical course of the infection: The anatomic region, the patient's resistance, and the virulence of the microorganisms. Clinically, one must differentiate between acute and chronic, and between phlegmonous and abscess-forming infections. Furthermore, soft-tissue infections must be distinguished

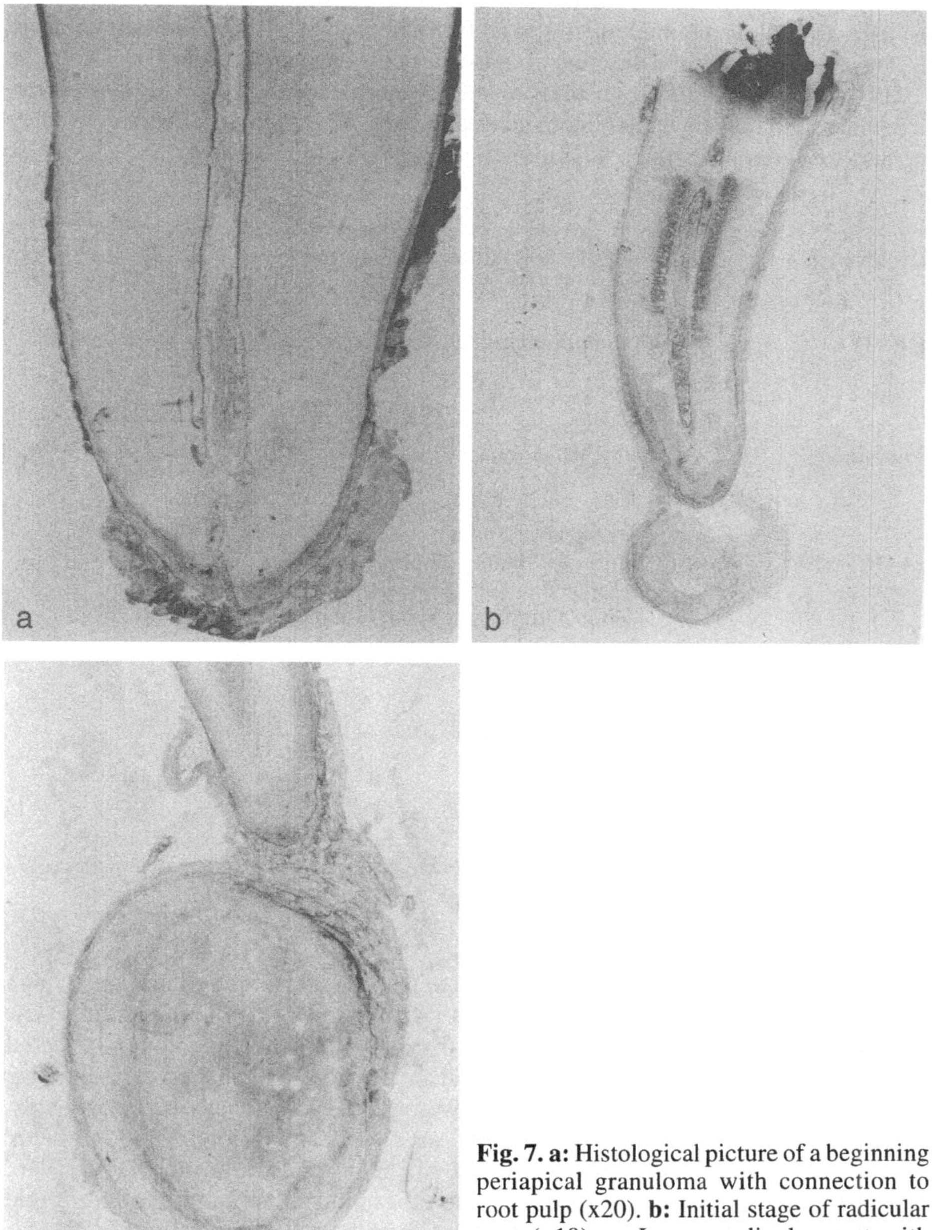

Fig. 7. a: Histological picture of a beginning periapical granuloma with connection to root pulp (x20). **b:** Initial stage of radicular cyst (x10). **c:** Larger radicular cyst with macrophages and cholesterine flakes in cystic lumen (x13).

from bony infections. The anatomical situation predisposes the extent and spread of odontogenic soft-tissue infections. Wassmund (1933) was the first researcher who precisely described the spread of the infection into the different fascial spaces. Up to now his observations are still valid. The special arrangement of the fascial spaces in the neck area allow an infection to spread from the cranium to the mediastinum (Fig. 8). These associations and the clinical consequences were later described by Dingman (1939) and by Scott (1952). In cases of soft tissue infections, cutaneous sinus tracts of dental origin may develop. Since such lesions are often diagnosed incorrectly, they are also treated ineffectively (al-Kandari et al. 1993).

CRANIUM	anterior and middle cranial fossa
⇑ ⇓	
ORBIT	ethmoid-pterygopalatine fossa – infratemporal – temporal
⇑ ⇓	
MAXILLA	maxillary sinus – canine fossa – retromaxillar
⇑ ⇓	
MANDIBLE	pterygomandibular retromandibular – cheek perimandibular – chin – sublingual submandibular – tongue – submental – parapharyngeal
⇑ ⇓	
NECK	carotid triangle – retropharyngeal
⇑ ⇓	
THORAX	mediastinum

Fig. 8. Schematic illustration of the potential spread of infection caused by dental abscesses (simplified arranged in different floors from the thorax to the cranium).

In the past, infectious diseases that started from a local tooth inflammation quite often progressed rapidly to a life-threatening situation with a lethal ending. The mortality rate of a cellulite of the floor of the mouth, which is a commonly feared condition, was about 40% before Wassmund and Axhausen published their directive surveys in the late 1920's (Axhausen 1929, Wassmund 1929). In a retrospective study, Fries (1964) reported on more than 54 cases of death following odontogenic infections occurring at the University Clinic of Vienna (Austria) from 1917 to 1946. Thanks to the introduction of penicillin and other modern antibiotics, odontogenic infections are no longer to be feared. More recently, a lethality rate of 0.2% was disclosed in a study from the University Hospital of Halle (Germany) based on a case collection from 1948 to 1975 (Schulz 1980). However, even in the era of antibiotics there are still case reports about severe or fatal courses following odontogenic infections (Quinn and Guernsey 1985; Currie and Ho 1993; Cheatham and Henry 1994; Meehan et al. 1994; Bonapart et al. 1995; Miller and Kassebaum 1995).

Odontogenic spatial abscesses may ascend to the orbital and periorbital regions (Pellegrino 1980; Harris 1988; Miller and Kassebaum 1995) or to the cranium and, on their way through the foramina, they may lead to a basal meningitis, a subdural empyema, or a brain abscess (Schulz 1980; Mathisen et al. 1984; Saal et al. 1988; Feldges et al. 1990; Kratimenos and Crockard 1991). On the other hand, the abscesses can also descend to the mediastinum causing an odontogenic mediastinitis (Murray and Finegold 1984; Musgrove and Malden 1989; Pignat et al. 1989; Petrone 1992; Bonapart et al. 1995). Further, but rare complications, such as a subphrenic abscess or a pericarditis, have been reported. The lethality rate of an odontogenic mediastinitis ranges between 40% and 60% (Garatea-Crelgo and Gay-Escoda 1991). An extremely rare situation is presented in a case report about a patient with an incurable disseminated intravascular coagulation following an acute dento-alveolar abscess with general sepsis (Currie and Ho 1993).

Between 1989 and 1995, two patients died in the Maxillofacial University Hospital of Freiburg (Germany) of sequelae related to odontogenic abscesses. Both patients suffered from insulin-dependent diabetes mellitus, whereby their resistance had considerably been reduced. In one case, a combined submandibular, parapharyngeal and retromaxillary abscess resulted in an incurable situation with a florid osteomyelitis. The second patient died as a consequence of sepsis following a submandibular abscess with mediastinitis. The abscess had spread into the para- and retropharyngeal spaces despite drainage and administration of intensive care treatment (Fig. 9 a–c).

When an infectious disease exacerbates, the walls of the veins in the neighborhood of the process may be infiltrated, which may then rapidly result in a thrombosis (thrombophlebitis). The thrombus may collapse with suppuration and may spread with the blood circulation, or may extend along the vein system. Such an expanded thrombophlebitis may reach to the internal jugular vein, or to the cavernous sinus, which was almost always fatal before the advent of antibiotics.

If an abscess formation takes place in the canine fossa, the angular vein may be involved in a thrombophlebitis. This vein passes right through the infectious area. When this happens, one must fear further expansion of the infection along the ophthalmic vein to the cavernous sinus, to which it is anastomosed. In this case, formation of a brain abscess is possible.

Infection of the Mandible and Maxilla – Osteomyelitis

An acute osteomyelitis occurs six times more frequently in the mandible than in the maxilla (Krüger 1993). The infection, which usually enters through a wound (for example after fracture treatment) or an opening in the cortical plate of bone, e.g. via the alveolar socket after extraction of a tooth, starts in the cancellous (medullary) portion of the bone. Osteomyelitis may also occur as a result of a periapical or pericoronal infection prior to surgical intervention. Still today, more than 80% of all osteomyelitis cases definitely have an odontogenic origin (Spiessl 1959; Wagner and Scheunemann 1984). The infection may be localized or it may spread diffusely through the entire medullary structure of the mandible or maxilla. It can also be preceded by septic cellulite, or it can be a sequel of an apparently

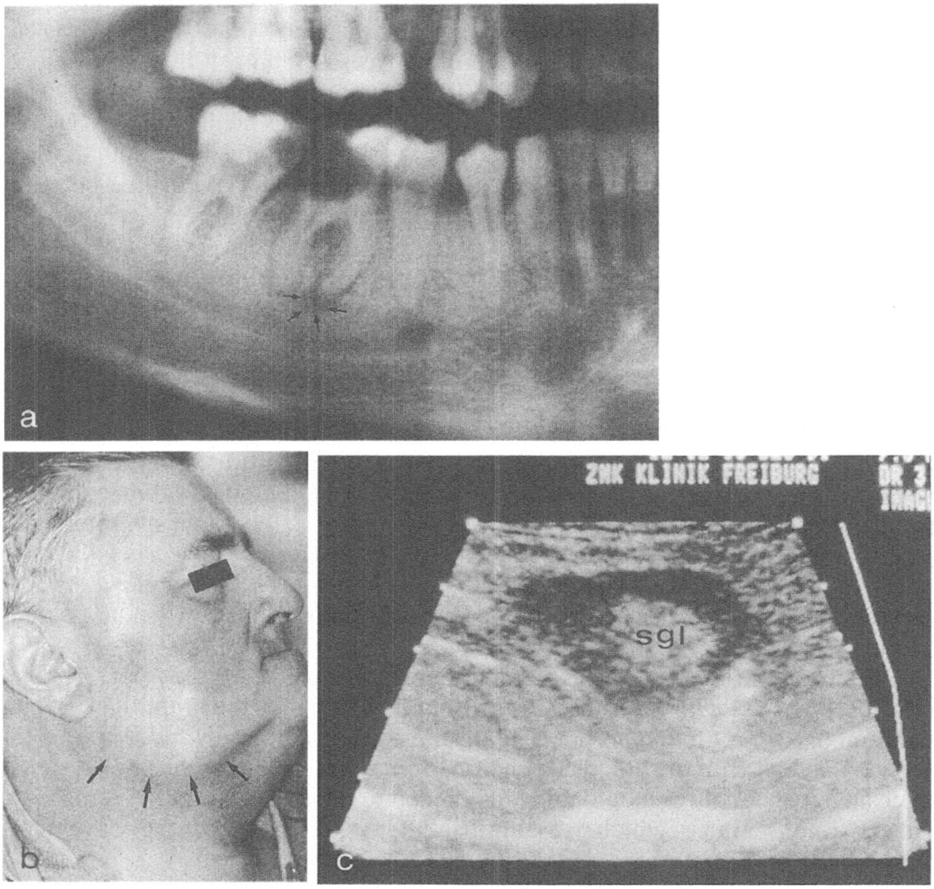

Fig. 9. a: 53 years old male with insulin-dependent diabetes mellitus (dm); periapical granuloma on tooth 46 caused by deep caries. **b:** Due to dm the infection rapidly spread to the submandibular space. **c:** Spreading of the infection to the submandibular space demonstrated in the ultrasound (B-scan) (sgl = submandibular gland); further progression of the infection led to an incurable mediastinitis with letal outcome.

simple extraction of an infected tooth (Fig. 10 a–d). A particularly severe form of osteomyelitis, commonly of hematogenous origin, can occur in the maxilla of newborns and toddlers and may even today end lethally (Mittermayer 1993).

Prior to the era of chemotherapy and antibiotics, osteomyelitic infections were not uncommon. An osteomyelitis usually does not heal. Instead, if untreated, it can spread. The factors that favor the development of an osteomyelitis are highly virulent organisms, low resistance of the patient and lack of drainage. The latter is clinically of great importance. In the case of an osteomyelitis, the pus, instead of passing through the bone along a narrow track into the soft tissues, spreads through the medulla and reaches the cortical plate at several points to lift the periosteum from the bone. It can sometimes deprive large areas of its blood supply. Eventually, the pus discharges through a sinus (Foster et al. 1992; Marasco et al.

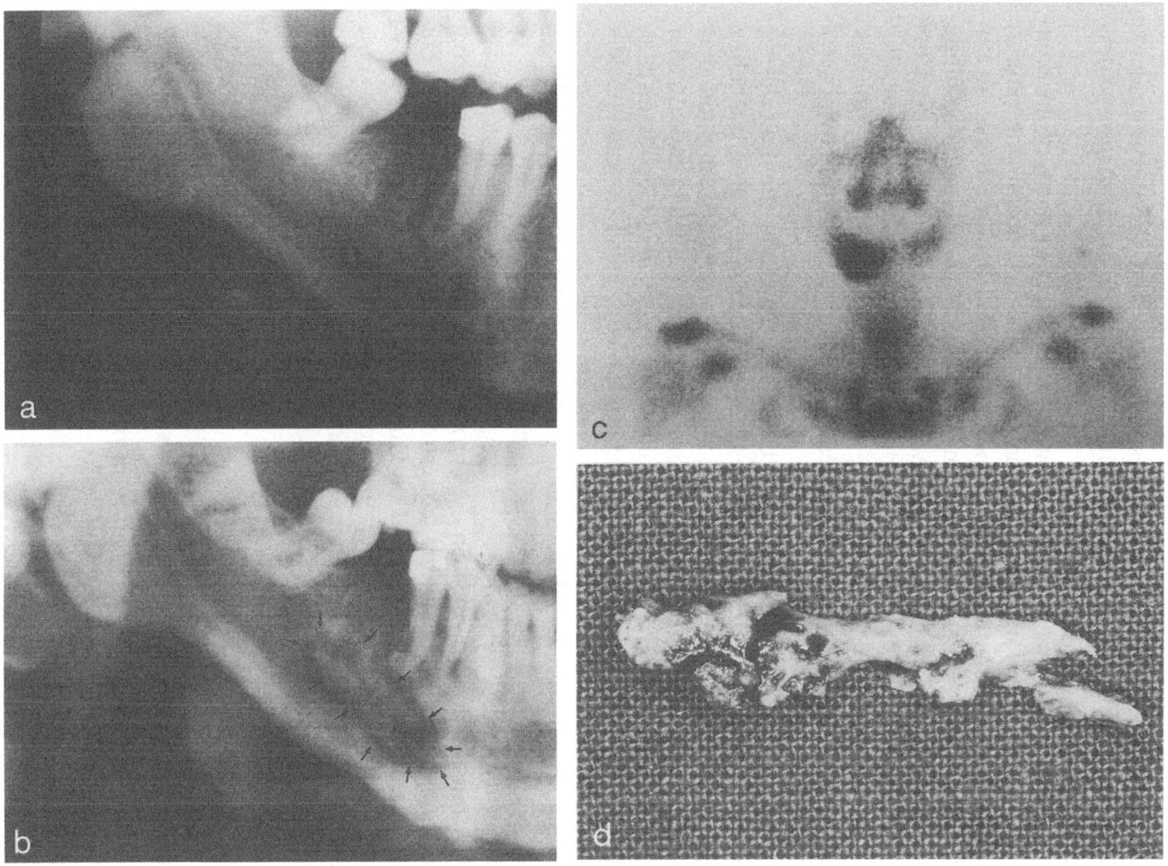

Fig. 10. a: 25 years old male: X-ray is showing the empty alveolus after extraction of the infected tooth 46. **b:** Soon after radiological signs of an osteomyelitis are visible, which are demonstrated by the sequestrum and by the surrounding radiolucent involucrum. **c:** Technetium bone scan is showing the infected area of the mandible. **d:** Removed sequestrum.

1992), and if drainage is satisfactory the condition arrives at a chronic stage (Trauner 1964; Kinnman and Lee 1968). In an attempt to limit the infection, dead bone is separated by osteoclasts and walled off by granulation tissue. Repair and support of the weakened bone is started by the osteoblasts of the periosteum (Brosch 1964). In the jaws, such a repair is seldom very marked although it does occur. Discharge continues until the sequestra are removed. Wherever drainage is inadequately slow, spread may continue indefinitely with the formation of new sequestra (Schilli 1981).

The symptoms of an acute osteomyelitis are identical to those of an acute infection, with severe pain and a raised temperature as the most prominent features. Teeth in the affected area are tender to percussion and become loose. In the mandible there is a loss of sensation in the mental nerve (Vincent's sign). Swelling of the face, if not present initially, soon follows. The swelling is caused by an edema of the overlying soft tissues, and an accompanying periostitis is also usually present. The condition may persist to a state at which the infection breaks through the cortical bone and invades the soft tissues, and induration followed by abscess formation becomes evident (Krüger 1993).

Radiographs are at first negative unless there has been a long-standing chronic infection prior to the osteomyelitis. The destructive process may progress before it can be depicted in the radiograph because the dense cortical bone may superimpose the lesion. Usually, changes are seen only approximately ten days after the start of the disease; they present as irregular radiolucent areas (Jacobsson et al. 1978). Later, sequestra show as radiopaque bodies surrounded by a radiolucent zone (Fig. 10 b).

Patients with acute osteomyelitis should be admitted to the hospital immediately, and treatment must be directed to controlling the infection with high doses of antibiotics, and to establishing drainage through an external incision (Waldvogel et al. 1970; Schilli 1981). Treatment in the chronic phase consists of administering antibiotics, maintaining drainage and sequestrectomy (Hjorting-Hansen 1970). For satisfactory removal, sequestra should be lying loose, which on a radiograph is suggested by a clear radiolucent line around the dead bone (Fig. 10 b, d). Otherwise, it is difficult to judge the line of separation during the surgical intervention, and dead bone may be left or healthy tissue needlessly traumatized.

Anthropological Aspects

Paleopathological Significance of Periapical Disease

Periapical lesions are a frequent finding in paleodontological material (Brabant 1963; Molnar and Molnar 1985; Clarke 1990; Jurmain 1990). However, paleopathological surveys seldom give detailed results on the frequency of such defects (Tab. 4) because the material as a whole is not generally submitted to radiographic examinations (Linn et al. 1987).

In skeletal remains, the distinction between periapical granuloma and radicular cyst is based mostly on the size of the lesion. This is probably the reason for the small number of radicular cysts listed in table 4. The diagnosis "radicular cyst" is only made in the case of large defects, which are less frequent than small lesions.

Table 4. Evidence of periapical lesions in past populations, as revealed on skeletal remains; selected studies (N = number of individuals; PerGr = periapical granuloma; RadCy = radicular cyst; PerLes = undifferentiated periapical lesions)

Autor	Year	N	Period	PerGr	RadCy	PerLes
Eich[1]	1939	350	500– 700 AD	(–)	(–)	52*
Zuhrt	1956	162	1100–1300 AD	125*	4*	(–)
Twisselmann and Brabant	1962	427	400–1000 AD	214*	3*	(–)
Brabant (series 1)	1963	162	0– 500 AD	(–)	(–)	103*
Brabant (series 2)	1963	116	500– 700 AD	50*	3*	(–)
Molnar and Molnar	1985	162	3000–1200 BC	(–)	(–)	35*
Petsch et al.	1982	272	1500–1800 AD	(–)	7.4+	71*
Kohl[1]	1989	122	500– 700 AD	(–)	(–)	57*
Jurmain	1990	300	500–1500 AD	(–)	(–)	31+
Baum	1990	119	4500–4000 BC	(–)	(–)	14+
Kölbel	1997	73	500– 700 AD	(–)	(–)	36+
Eckert	1998	71	500– 800 AD	90.1+	9.9+	71*

[1] Macroscopic examination only
* Number of lesions
+ In percent/individuals

Estimates of periapical inflammation in studies varied from 1.9 to 9.1% of teeth affected (Alexandersen 1967). A detailed radiographic investigation of a medieval Swedish group showed that periapical destruction was slightly more common in males than in females, and that there was a relationship to age (Swärdstedt 1966). The first molars were most often affected. It is notable that out of 283 periapical destructions detected from radiography, 230 (81.3%) resulted in a fistula that perforated the alveolar bone (Swärdstedt 1966).

The main reasons for pulpal infections in prehistoric populations are deep caries and severe attrition with exposure of the dental pulp. Both forms of tooth defects can be diagnosed macroscopically (Clarke and Hirsch 1991 a,b; Alt et al. 1992). The sequelae of this tooth decay, namely periapical granulomas and radicular cysts, are only discernible in radiographs unless a periapical abscess has led to a periapical destruction of bone with perforation of the adjacent cortical bone. If a pulpal inflammation due to caries or severe attrition is left untreated, the average time span for the development of a periapical granuloma or a radicular cyst is approximately six years (Hacker and Fischbach 1985). During this period, an exacerbation of the developing chronic periapical process is always possible.

As far as skeletal remains are concerned, reports on unspecific, supporative inflammations of the jaws caused by osteomyelitis are very rare. The main reason for this low frequency is probably related to the fact that in skeletal material jaws are commonly not routinely scrutinized for the presence of pathological alterations. In a radiographic investigation of the jaws of 272 historic individuals, Petsch et al. (1982) detected 12 cases (4,4%) with periapical inflammations indicative for a chronic osteomyelitis (periapical alterations in 3% of the teeth surveyed). Most of these findings were located in the mandible, which is not surprising since it is well known that infections have the tendency to spread more diffusely in this bone than in the maxilla (Mittermayer 1993). Since professional treatment of osteomyelitis was not possible in ancient times, we assume that severe complications due to this pathological condition were more frequent in early populations than current reports from skeletal remains suggest. It may even be speculated that a considerable number of these cases resulted in the death of the individual.

Classification of Alveolar Bone Pathology of Pulpal Origin

The integrity of the alveolar bone can be disrupted either by periodontal factors (e.g., lateral periodontitis) or by an infection of pulpal origin (primarily caused by severe attrition or caries) and its pathological consequences, such as periapical abscess and periapical lesions (Clarke and Hirsch 1991a). Chronic periapical lesions (abscess, granuloma or cyst) are usually accompanied by resorptions of the alveolar (periapical) bone and the root of the involved teeth (granuloma).

Clarke (1990) suggests the following classification of periodontal defects that are caused by pulpal pathology:

- Lateral defects: lesions below the alveolar crest, initially without connection to the marginal crest.
- Angular defects (syn.: infrabony, or vertical defects): loss of alveolar bone extending to the root surface, probably caused by chronic pulpitis.
- Furcation defects: early stage of periodontal ligament and alveolar disease of molars, caused by infection of the pulp of multi-rooted teeth.
- Apical lesions: occurrence of isolated or confluent periapical lesions (complex defects), possibly combined with an angular defect or furcation.

Not each and every acute periapical infection will necessarily develop into an alveolar abscess, however. An abscess is usually easily diagnosed in skeletal material by the presence of a discrete abscess cavity at the periapical region of the affected tooth (in most cases only discernible on radiographs). As stated previously, periapical lesions can usually be diagnosed macroscopically only in those cases where after an acute or chronic abscess a penetration of the periapical process through the alveolar bone has taken place. Abscesses almost always erupt to the labial side, a fact that Taylor explains simply and conclusively: "Warmth relieves the pain of an alveolar abscess and in the simplest ways of applying warmth the affected side is turned downwards so that hyperemia and gravitation would tend to cause the pus to track to the buccal side." (1963, 145)

Depending on the location, size, and direction of the spread of a periapical process lateral defects, angular defects or furcation defects are possible. In the most simple case, a fistula in the buccal alveolar bone of the involved tooth is present. Porosities and small perforations of the buccal bone (high vascularization of bone) are usually always associated with a chronic periapical abscess. Clarke et al. (1995) distinguish dental abscesses in periapical abscess, furcation abscess, lateral abscess, complex abscess and maxillary sinus abscess.

It seems to be rarely known that periapical lesions caused by a chronic infection of the pulp are not necessarily limited to the root apex. Only in simple cases a discrete abscess cavity is present in the periapical region. In contrast, periapical abscesses can spread retrogradely within the spongious bone from the root apex in the coronal direction, and be responsible for periodontal inflammation and destruction (Walton and Garnick 1986). A pulpal infection can also extend through additional lateral root canals causing a lateral periodontal abscess and defects in the periodontal ligament and the adjacent bone including the alveolar crest (Clarke and Hirsch 1991a; Okeson 1995).

A distinction of periodontal lesions into those of pulpal origin and of gingival origin will most probably show that, contrary to common belief, the latter may not be the main cause of tooth loss in ancient times (see also Clarke 1990). Alveolar bone defects caused by lateral periodontitis and independent defects of pulpal origin may sometimes be present at the same time, making a clear differentiation difficult. Therefore, such complex defects are often more difficult to judge with regard to their pathogenesis. In these cases, the initial cause for the destruction of the alveolar bone can most often not be determined with certainty. However, in the presence of a deep carious lesion or an open pulp caused by excessive attrition and periodontal (pathological) factors of alveolar crest loss (e.g., lateral periodontitis), sooner or later any of these pathologies will lead to alveolar bone defects.

Discrimination Between Periapical Granulomas and Radicular Cysts in Skeletal Material

In dental anthropological studies, differentiations of periapical lesions were not made because diagnostic criteria were not available. On the basis of findings gained from the radiographic analysis of clinical patients, and from macroscopic and SEM examinations of both extracted teeth and skeletal material, we present criteria that may aid paleopathologists in distinguishing periapical granulomas from radicular cysts (Tab. 5). As a rule, the more characteristics correspond to the description of one of the differential diagnoses, the more certain a discrimination of periapical granulomas and radicular cysts can be made. Illustrations were obtained from material from a Neolithic collective grave from Sorsum, northern Germany (Figs. 11 a–e, 12 a–e) and the early Medieval cemetery of Kirchheim/Ries (south Germany) (Figs. 13 a, b; 14 a, b).

Table 5. Criteria for differentiation of periapical granulomas and radicular cysts in skeletal remains (modified after Alt et al. 1992)

Periapical granuloma	*Radicular cysts*
♦ size of defect smaller 10 mm	♦ size of defect greater 10 mm
♦ in radiograph: radiolucency without radiopaque margin	♦ in radiograph: roundish radiolucency, sharply outlined, cyst wall (white line) represents a layer of cortical bone
♦ intense resorption around tip of root	♦ resorption on root minimal or absent
♦ irregularly shaped defect, surface uneven	♦ roundish, evenly shaped defect, distinct alveolus-defect border
♦ (–)	♦ bony outgrowth (buccal, lingual, maxillary sinus, etc.)
♦ in most cases distinct hypercementoses	♦ (–)
♦ cristae and spiculae covering the defect	♦ isolated cristae and spiculae
♦ bone surface of defect highly vascularized, exhibiting resorption lacunae	♦ bone surface of defect relatively even, slightly vascularized

Periapical defects of a diameter greater than 10 mm are predominantly radicular cysts; if their size exceeds 20 mm, they are almost exclusively cysts (Natkin et al. 1984). It must be considered, however, that compared to the

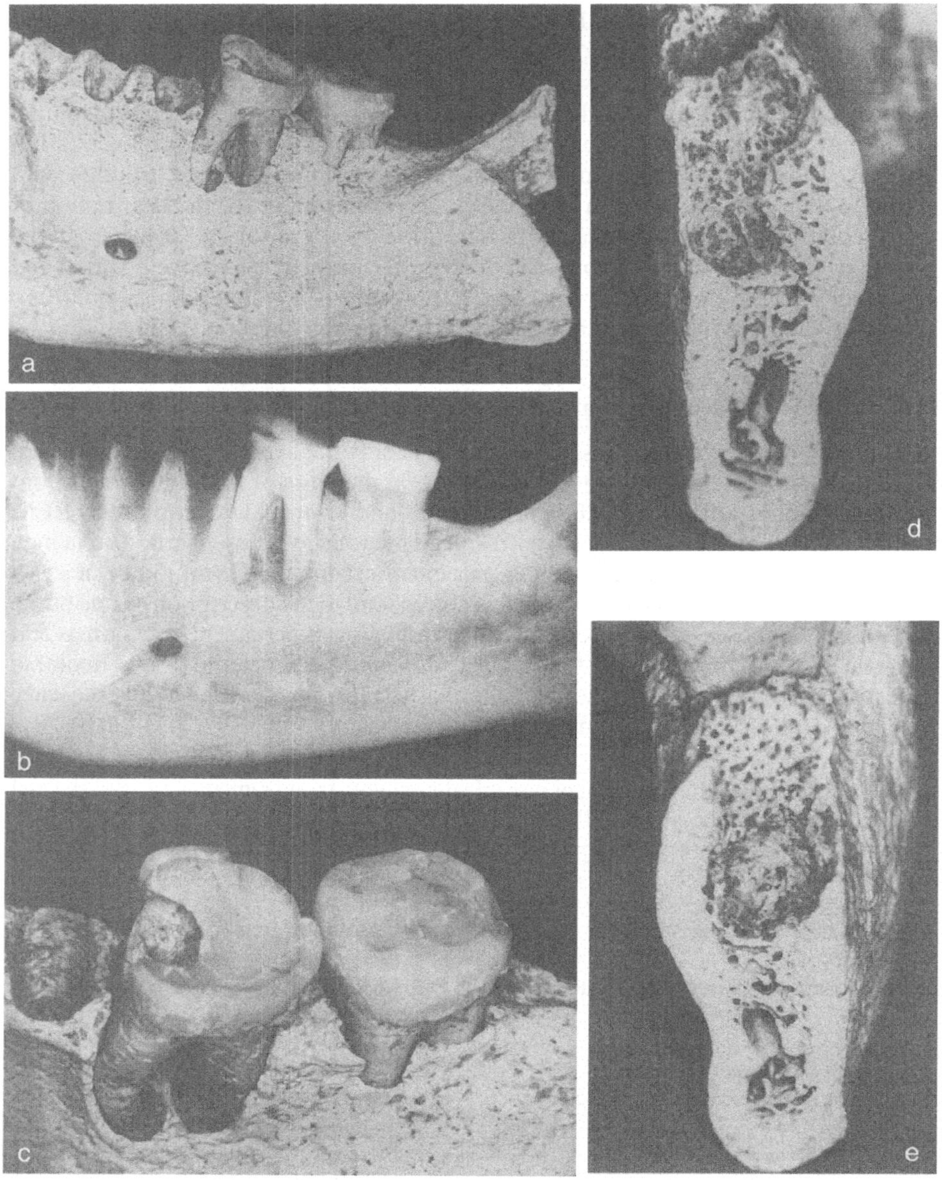

Fig. 11. a: Sorsum no. 10: lateral view of mandible with teeth 36 and 37. **b:** Panoramic radiograph of left side of mandible; periapical lesions on root-tips of tooth 36. **c:** Tooth 36; caries profunda mesial, angular bone-loss at both roots. **d:** Section at bifurcation of tooth 36; view of mesial root with resorptions by periapical granuloma. **e:** Section at bifurcation of tooth 36; view of distal root with resorptions by radicular cyst.

measured size of a radiolucency indicating a periapical defect on a radiograph the real lesion may be much larger (Peyer 1945; Barnett et al. 1984). Radicular cysts frequently exhibit a dense sclerotic border in radiographs (see Fig. 6); however, this characteristic feature may also be absent. Using apical and intracanal SEM examinations, Delzangles (1989) showed that teeth with granulomas exhibit lacunar resorption of the apical hard tissues (cementum and dentine; see Figs. 14 a and b) around the apical (main) foramen, while the hard tissues located around cysts show little resorption. The examination of the intracanal walls, on the other hand, did not reveal any significant difference in the distribution and intensity of resorption zones between periapical granulomas and radicular cysts.

In contrast to the roundish, evenly shaped cysts (Figs. 11 e, 13 a), periapical granulomas have an irregular outline (Figs. 11 d, 12 c–d). The resorptive granulation tissue of periapical granulomas causes crater-shaped, mostly slanting defects of the spongiosa, with defect and alveolus confluent (Figs. 11 d, 12 d). In contrast, the balloon-shaped osteolyses which develop from radicular cysts are clearly demarcated from the alveolus (Figs. 11 e, 13 b). There are also differences in the bony surfaces of both defects. In the case of periapical granulomas, the destruction of bone tissue caused by osteoclasts leaves semilunar hollows on the surfaces of defects (Fig. 12 e), whereas in radicular cysts the surfaces appear more spongy and porous (Fig. 13 b).

Conclusions

In ancient times, the vast majority of pulp cavities exposed by caries or severe attrition remained open. Hence, pus could flow from the tooth, and periapical abscesses probably developed less frequently. Nonetheless, large defects, infected maxillary sinuses, exposed bifurcations and alveolar fistulas found in investigations of skeletons show that such pathological processes were frequent events in early populations. For quite some time, modern paleopathology has not been limited to case reports but has tried to provide epidemiological information about frequent diseases. As yet for the differential diagnosis of periodontal disease and its consequences for the alveolar bone (lateral periodontitis, alveolar defects of pulpal origin), diagnostic criteria such as those presented by Clarke and coworkers (Clarke 1990; Clarke and Hirsch 1991 a, b; Clarke et al. 1995) are very helpful. Recent study about the discrimination between periapical granulomas and radicular cysts in skeletons could show that using the criteria listed in Table 5 makes a differential diagnosis nearly always possible (Eckert in prep.).

Chronic periapical infections are considered important foci, which sometimes may have effects in other parts of the body (Alexandersen 1967). The development of spondylitis deformans, chronic osteoarthritis, chronic endocarditis, dermatoses and pyelitis have been discussed as possible aftermaths of oral foci (Thyagarajan and Kamalam 1982; Papageorge and Kronman 1986; Meehan et al. 1994).

In some cases, acute periapical periodontitis and acute exacerbations from a chronic form may cause life-threatening conditions as a result of the

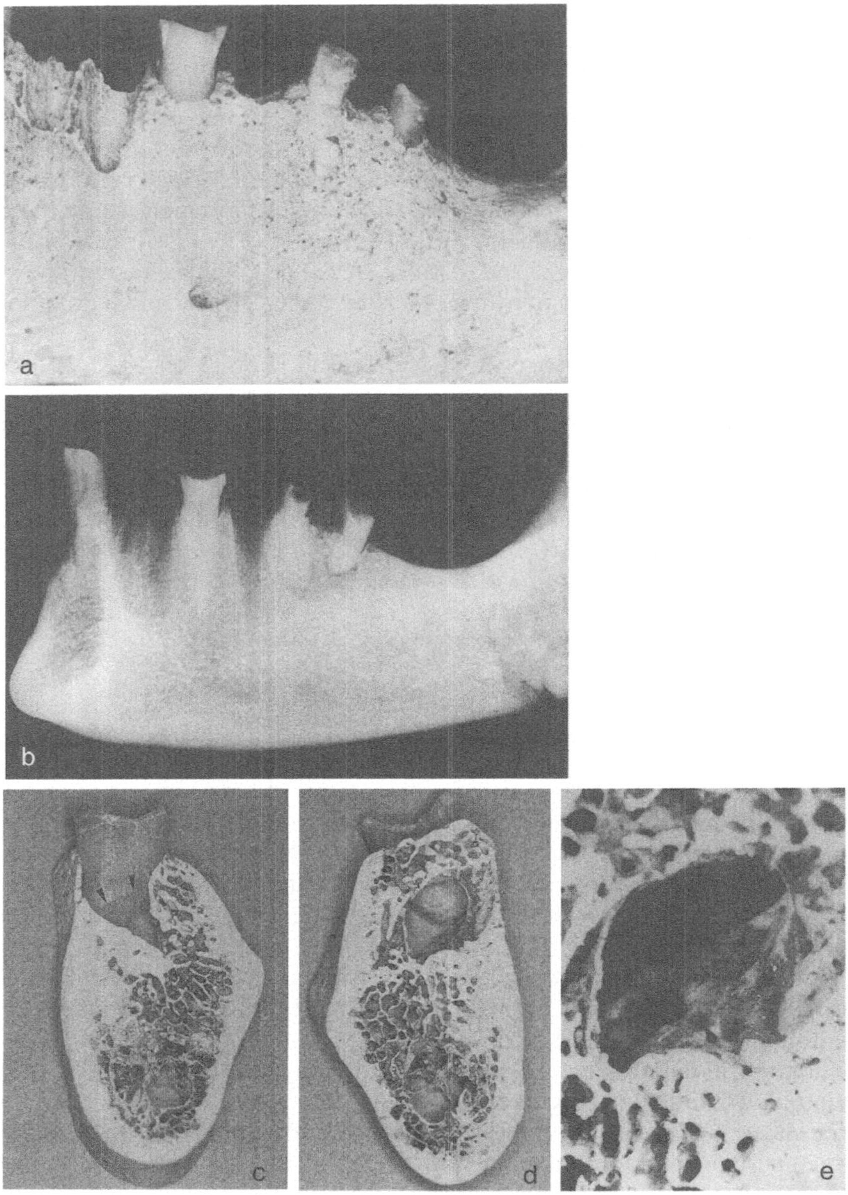

Fig. 12. a: Sorsum no. 23: lateral view of mandibula with tooth 34 and root of 36; buccal alveolar area around tooth 36 highly vascularized after inflammation (→); bone lesion at mesial root by possible abscess caused by caries or severe attrition with pulpa aperta. **b:** Panoramic radiograph of left side of mandible; apical lesions on roots of tooth 36. **c:** Separation including distal root of tooth 36; distal view with lesion slanting in lingual caudal direction and resorptions at root caused by apical granuloma (→). **d:** Separation including distal root of tooth 36; mesial view with lesion slanting in caudo-lingual direction (periapical granuloma). **e:** Detail from Fig. 7d; distinct semilunar hollows visible on surface of defect; characteristic of periapical granuloma (x10).

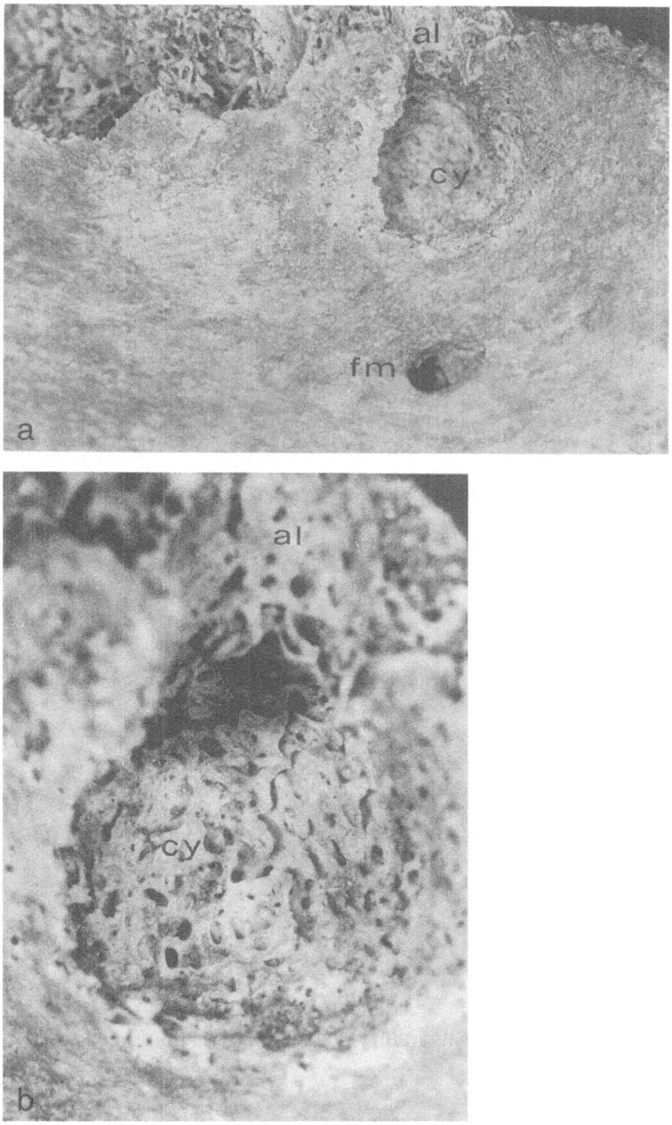

Fig. 13. a: Kirchheim/Ries no. 17: lateral view of the mandible with bone lesion by radicular cyst at tooth 36 (al = alveolus, cy = cyst, fm = foramen mentale). **b:** Detail from Fig. 8 a showing distinct, balloon-shape resorption defect (al = alveolus, cy = cyst); walls of defect more even than in periapical granuloma.

accompanying suppurations. The course of such pathologic conditions is determined by various factors, such as the anatomical situation (infection routes of the dental abscess in maxilla and mandible), the intensity of the infection, the virulence of the germs and the resistance of the affected individual. In spite of the availability of antibiotics, such dangerous situations can also occur today.

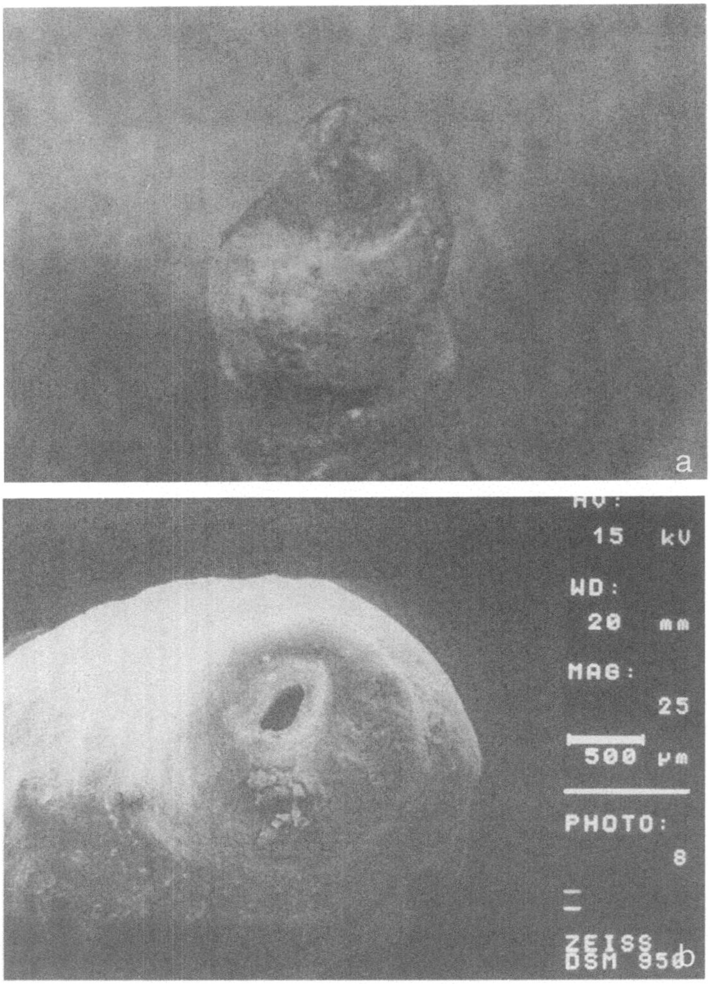

Fig. 14. a: Kirchheim/Ries no. 49: root of tooth 14 with resorptions characteristic of periapical granuloma; apical third of root with circular resorptions of hard tissue (x10). **b:** Root tip of tooth 14 (SEM).

Individuals suffering from diabetes mellitus are typical examples for patients with reduced resistance. Transferred to the situation in former times one can hypothesize that similar progressive and life-threatening infectious diseases were more frequent during periods when food was rare or nutrition was unbalanced. Because of the frequency and severity of pulpo-alveolar diseases in the ancient times it is likely that in the absence of management of the periapical diseases many cases were lethal. Numerous chronic and often multiple individual findings indicate that in the past dental diseases had a considerable high influence on the health and quality of life as well as the mortality rate of the affected individuals.

References

al-Kandari AM, al-Quoud OA, Ben-Naji A (1993) Cutaneous sinus tracts of dental origin to the chin and cheek: Case reports. Quintessence Int 24: 729–733.

Alexandersen V (1967) The pathology of the jaws and the temporomandibular joint. In: Brothwell D, Sandison AT (eds) Disease in Antiquity. Thomas, Springfield, p. 551–595.

Alt KW, Wächter R, Türp JC (1992) Pulpoalveolar disease: Etiology, incidence, and differentiation of periapical lesions. J Paleopathol 4: 163–178.

Axhausen G (1929) Zur Klassifikation der eitrigen Entzündungen am Kiefer. Chirurg 1: 459–462.

Babal P, Brozman M, Jakubovsky J, Basset F, Jany Z (1989) Cellular compostion of periapical granulomas and its function. Czech Med 12: 193–215.

Barbakow FH, Cleaton-Jones PE, Friedman D (1981) Endodontic treatment of teeth with periapical radiolucent areas in general dental practice. Oral Surg Oral Med Oral Pathol 51: 552–559.

Barnett F, Schwartzben LJ, Tronstad L (1984) Extensive periapical lesion with inconsistent radiographic findings. J Endodont 10: 26–27.

Baum N (1990) Aiterhofen-Ödmühle. Paläodontologie eines bandkeramischen Gräberfeldes in Niederbayern. Praehist Z 65: 157–203.

Baumann L, Rosman SR (1956) Clinical, roentgenologic, and histologic findings in teeth with apical radiolucent areas. Oral Surg Oral Med Oral Pathol 9: 1330–1336.

Bhaskar SN (1966) Periapical lesions – types, incidence and clinical features. Oral Surg Oral Med Oral Pathol 21: 657–671.

Bhaskar SN (1986) Synopsis of Oral Pathology. Mosby, St. Louis.

Block RM, Bushell A, Rodrigues H, Langeland K (1976) A histopathologic, histobacteriologic, and radiographic study of periapical endodontic surgical specimens. Oral Surg Oral Med Oral Pathol 42: 656–678.

Bohne W (1990) Light and ultrastructural studies of human chronic periapical lesions. J Oral Pathol Med 19: 215–220.

Bonapart IE, Stevens HP, Kerver AJ, Rietveld AP (1995) Rare complications of an odontogenic abscess: mediastinitis, thoracic empyema and cardiac tamponade. J Oral Maxillofac Surg 53: 610–613.

Brabant H (1963) Observations sur la denture humaine en France et en Belgique a l'époque Gallo-Romaine et au moyen age. Bull Group Int Rech Sci Stomatol 6: 169–296.

Brosch F (1964) Zur Entstehung des Bildes des osteomyelitischen Kiefers. Fortschr Kiefer Gesichts Chir 9: 153–156.

Cheatham BD, Henry RJ (1994) A dental complication involving Pseudomonas during chemotherapy for acute lymphoblastic leukemia. J Clin Pediatr Dent 18: 215–217.

Clarke NG (1990) Periodontal defects of pulpal origin: Evidence in early man. Am J Phys Anthrop 82: 371–376.

Clarke NG, Hirsch RS (1991a) Physiological, pulpal, and periodontal factors influencing alveolar bone. In: Kelley MA, Larsen CS (eds) Advances in Dental Anthropology. Wiley-Liss, New York, p. 241–266.

Clarke NG, Hirsch RS (1991b) Tooth dislocation: The relationship with tooth wear and dental abscesses. Am J Phys Anthrop 85: 293–298.

Clarke NG, Morrow C, Corruccini RS (1995) Periodontal status in dry skulls of Carrier Mills (Illinois, U.S.A.). In: Moggi-Cecchi J (ed) Aspects of Dental Biology: Palaeontology, Anthropology and Evolution. International Institute for the Study of Man, Firenze.

Currie WJ, Ho V (1993) An unexpected death associated with an acute dentoalveolar abscess: Report of a case. Br J Oral Maxillofac Surg 31: 296–298.

Delzangles B (1989) Scanning electron microscopic study of apical and intracanal resorption. J Endodont 15: 281–285.

Dingman RO (1939) The management of acute infections of the face and jaws. Am J Orthod 25: 780–788.
Donath K, Breuner G (1982) A method for the study of undecalcified bones and teeth with attached soft tissue. J Oral Pathol 11: 318–326.
Eckert G (1998) Differentialdiagnose osteolytischer Veränderungen dentaler Genese an Skelettmaterial. Med Diss, Freiburg (in prep.).
Eich F (1939) Pathologische Befunde an Zähnen und Kiefern von Alemannen aus den Reihengräbern der Gemarkung Mengen/Baden. Med Diss, Freiburg.
Feldges A, Heesen J, Nau HE, Schettler D (1990) Der odontogene Hirnabszess. 2 Fallberichte. Dtsch Z Mund Kiefer Gesichtschir 14: 297–300.
Formigli L, Orlandini SZ, Tonelli P, Giannelli M, Martini M, Brandi ML, Bergamini M, Orlandini GE (1995) Osteolytic processes in human radicular cysts: morphological and biochemical results. J Oral Pathol Med 24: 216–220.
Foster HK, Primack PD, Kulild JC (1992) Odontogenic cutaneous sinus tract. J Endod 18: 304–306.
Fries R (1964) Erfahrungen bei phlegmonösen Entzündungen im Kiefer-Gesichtsbereich. Fortschr Kiefer Gesichts Chir 9: 118–123.
Garatea-Crelgo J, Gay-Escoda C (1991) Mediastinitis from odontogenic infection. Int J Oral Maxillofac Surg 20: 65–68.
Gardner AF (1975) A survey of odontogenic cysts and their relationship to squamous cell carcinoma. Can Dent Assoc J 41: 161–167.
Grant DA, Stern IB, Listgarten MA (1988) Periodontics in the tradition of Gottlieb and Orban. Mosby, St. Louis, p. 911–920.
Hacker W, Fischbach H (1985) Apikale Veränderungen – Verteilung, Entwicklung und Differentialdiagnose. Dtsch Zahnärzt Z 40: 617–620.
Harris GJ (1988) Subperiosteal inflammation of the orbit. A bacteriological analysis of 17 cases. Arch Ophthalmol 106: 947–952.
Hauk KS, Raubenheimer EJ, van Heerden WFP, Buch B, Dreyer AF (1993) Differential diagnosis of dentigerous cyst-like lesions: Clinico-pathologic features of 63 cases. J Dent Assoc South Afr 48: 557–559.
Hinton RJ (1981) Form and patterning of anterior tooth wear among aboriginal human groups. Am J Phys Anthrop 54: 555–564.
Hjorting-Hansen E (1970) Decortication in treatment of osteomyelitis of the mandible. Oral Surg Oral Med Oral Pathol 29: 641–655.
Jacobsson S, Hollender L, Lindberg S, Larsson A (1978) Chronic sclerosing osteomyelitis of the mandible. Scintigraphic and radiographic findings. Oral Surg 45: 167–174.
Jurmain R (1990) Paleoepidemiology of a Central California prehistoric population from CA-Ala 329: Dental disease. Am J Phys Anthrop 81: 333–342.
Kerekes K, Tronstadt L (1979) Long-term results of endodontic treatment performed with a standardized technique. J Endod 5: 83–90.
Keudell P, Powell G, Diemer R (1982) Humoral antibodies to anaerobic bacteria isolated from patients with pulpal-periapical disease. Oral Surg 53: 194–202.
Kinnman JEG, Lee HS (1968) Chronic osteomyelitis of the mandible. Oral Surg Oral Med Oral Pathol 25: 6–11.
Kölbel J (1997) Paläopathologische und -epidemiologische Untersuchungen an frühmittelalterlichen Zahn- und Kieferresten aus Pleidelsheim, Kr. Ludwigsburg. Med Diss, Freiburg.
Kohl G (1989) Zahnmorphologische und -pathologische Untersuchungen an der frühmittelalterlichen alamannischen Population aus Hailfingen/Württemberg unter besonderer Berücksichtigung von Karies und Abrasion. Med Diss, Mainz.
Kontiainen S, Ranta H, Lautenschlager I (1986) Cells infiltrating human periapical inflammatory lesions. J Oral Pathol 15: 544–546.

Kratimenos G, Crockard HA (1991) Multiple brain abscess: A review of fourteen cases. Br J Neurosurg 5: 153–161.
Krüger E (1993) Lehrbuch der chirurgischen Zahn-, Mund- und Kieferheilkunde, Vol. 1, Quintessenz, Berlin, p. 188–211.
Laczkó J, Lévai G (1975) A simple differential staining method for semi-thin sections of ossifying cartilage and bone tissues embedded in expoxy resin. Microscopy 31: 1–4.
Lalonde ER, Luebke RG (1968) The frequency and distribution of periapical cysts and granulomas. Oral Surg Oral Med Oral Pathol 25: 861–868.
Lee KW (1995) Periodontal cysts. In: Berkovitz BKB, Moxham BJ, Newman HN (eds) The Periodontal Ligament in Health and Disease., 2nd Ed. Mosby-Wolfe, London, p. 391–407.
Lentrodt J (1974) Fehler und Irrtümer bei der Analyse und Interpretation von Röntgenbildern osteolytischer Prozesse im Kieferbereich. Dtsch Zahnärztl Z 29: 341–347.
Lin LM, Pascon EA, Skribner J, Gängler P, Langeland K (1991) Clinical, radiographic, and histologic study of endodontic treatment failures. Oral Surg Oral Med Oral Pathol 71: 603–611.
Linenberg WB, Waldron CA, DeLaune GF (1964) A clinical roentgenographic and histopathologic evaluation of periapical lesions. Oral Surg Oral Med Oral Pathol 17: 467–472.
Linn J, Srikandi W, Clarke NG, Smith T (1987) Radiographic and visual assessment of alveolar pathology of first molars in dry skulls. Am J Phys Anthrop 72: 515–521.
Marasco PV Jr, Taylor RG, Marks MW, Argenta LC (1992) Dentocutaneous fistula. Ann Plast Surg 29: 205–210.
Mathisen GE, Meyer RD, George WL, Citron DM, Finegold SM (1984) Brain abscess and cerebritis. Rev Inf Dis (Suppl 1) 6: 101–106.
Meehan S, Schmidt MC, Mitchell PF (1994) Infective endocarditis in a patient with Hodgin's lymphoma: A case report. Spec Care Dentist 14: 57–60.
Merskey H, Bogduk N (1994) Classification of Chronic Pain, 2nd Ed. IASP Press, Seattle, p. 73–74.
Miller EH, Kassebaum DK (1995) Managing periorbital space abscess. Secondary to dentoalveolar abscess. J Am Dent Assoc 126: 469–500.
Miller RL, Gould AR, Bernstein ML, Read CJ (1995) General and Oral Pathology for the Dental Hygienist. Mosby, St. Louis.
Milobsky SA, Milonsky L, Erpstein LI (1975) Metastatic renal adenocarcinoma presenting as periapical pathosis in the maxilla. Oral Surg Oral Med Oral Pathol 39: 30–33.
Mittermayer C (1993) Oralpathologie, 3nd ed. Schattauer, Stuttgart, p. 269–277.
Molnar S (1972) Tooth wear and culture: A survey of tooth function among some prehistoric populations. Curr Anthrop 13: 511–526.
Molnar S, Molnar I (1985) Observations of dental disease among prehistoric populations of Hungary. Am J Phys Anthrop 67: 51–63.
Morse DR, Patnik JW, Schacterle GR (1973) Electrophoretic differentiation of radicular cysts and granulomas. Oral Surg Oral Med Oral Pathol 35: 249–264.
Mortensen H, Winter JE, Birn H (1970) Periapical granulomas and cysts. Scand J Dent Res 78: 241–250.
Murray PM, Finegold SM (1984) Anaerobic mediastinitis. Rev Inf Dis (Suppl 1) 6: 123–127.
Musgrove BT, Malden NJ (1989) Mediastinitis and pericarditis caused by dental infection. Br J Oral Maxillofac Surg 27: 423–428.
Nair PNR (1995) Eine neue Sicht der radikulären Zysten – Sind sie heilbar? Endodontie 3: 169–180.
Nair PNR, Schroeder HE (1983) Pathogenese periapikaler Läsionen. Schweiz Monatsschr Zahnheilk 93: 935–952.
Nair PNR, Luder HU (1985) Wurzelkanal- und periapikale Flora: Eine licht- und elektronenmikroskopische Untersuchung. Schweiz Monatsschr Zahnmed 95: 992–1003.

Nair PNR, Pajarola G, Schroeder HE (1996) Types and incidence of human periapical lesions obtained with extracted teeth. Oral Surg Oral Med Oral Pathol Oral Radiol Endod 81: 93–102.

Natkin E, Oswald RJ, Carnes LI (1984) The relationship of lesion size to diagnosis, incidence, and treatment of periapical cysts and granulomas. Oral Surg Oral Med Oral Pathol 57: 82–94.

Newman HN, Challacombe SJ (1995) Infection and the periodontal ligament. In: Berkovitz BKB, Moxham BJ, Newman HN (eds) The Periodontal Ligament in Health and Disease. Mosby-Wolfe, London, p. 293–314.

Nobuhara WK, Del Rio CE (1993) Incidence of periradicular pathoses in endodontic treatment failures. J Endod 19: 315–318.

Oehlers FAC (1970) Periapical lesions and residual dental cysts. Br J Oral Surg 8: 103–113.

Okeson JP (1995) Bell's Orofacial Pains. Quintessence, Chicago, p. 235–258.

Otten JE, Drews M, Pelz K, Lauer G (1998) Odontogene Infektionen – ein systemisches Risiko? Dtsch Zahnärztl Z 53: 83–88.

Papageorge MB, Kronman JH (1986) Referral of symptoms from the trigeminal to the facial nerve as a sequel to dental infection. Oral Surg Oral Med Oral Pathol 62: 643–645.

Patterson SS, Shafer WG, Healey HJ (1964) Periapical lesions associated with endodontically treated teeth. J Am Dent Assoc 68: 191–194.

Pellegrino SV (1980) Extension of dental abscess to the orbit. J Am Dent Assoc 100: 873–875.

Petrone JA (1992) Mediastinal abscess and pneumonia of dental origin. J N J Dent Assoc 63: 19–23.

Petsch K, Krekeler G, Otten JE (1982) Pathologische Veränderungen im alveolären Bereich an Funden einer historischen Grablege. Dtsch Zahnärztl Z 37: 358–360.

Peyer E (1945) Beitrag zur Untersuchung des Alveolarknochens in der Umgebung chronischer Periodontitiden. Med Diss, Basel.

Piatowska D, Mirecka B, Pawlicka H (1988) Ultrastrukturelle Untersuchungen periapikaler Granulome. Stomatol DDR 38: 292–297.

Pignat JC, Haguenauer JP, Navailles B (1989) Les cellulites cervicales diffuses, spontanées et anaerobies. Rev Laryngol Otol Rhinol (Bord) 110: 141–144.

Priebe WA, Lazansky JP, Wuehrmann AH (1954) The value of the roentgenographic film in the differential diagnosis of periapical lesions. Oral Surg Oral Med Oral Pathol 7: 979–983.

Quinn P, Guernsey LH (1985) The presentation and complication of odontogenic septic shock. Report of a case. Oral Surg Oral Med Oral Pathol 59: 336–339.

Richardson KC, Jarett L, Finke EH (1960) Embedding in expoxy resin for ultrathin sectioning in electron microscopy. Stain Technol 35: 313–325.

Rudelt HG (1985) Das Keimspektrum der infizierten Zyste. Dtsch Zahnärztl Z 40: 590–591.

Saal CJ, Mason JC, Cheuk SL, Hill MK (1988): Brain abscess from chronic odontogenic cause: Report of case. J Am Dent Assoc 117: 453–455.

Schilli W (1981) Knocheninfektionen. In: Schwenzer N, Grimm G (eds) Zahn- Mund- Kieferheilkunde, Vol.1 Allgemeine Chirurgie, Entzündungen und Röntgenologie, Thieme, Stuttgart, p. 195–216.

Schroeder HE (1991) Pathobiologie oraler Strukturen, 2nd ed. Karger, Basel.

Schulz S (1980) Häufigkeit und Ursachen lebensbedrohlicher Ausbreitungsformen und Verläufe pyogener Weichteilinfektionen im Kiefer-Gesichts-Halsbereich an einem Krankengut von 3314 Fällen. Zahn Mund Kieferheilk 68: 20–35.

Schuster GS (1994) The microbiology of oral and maxillofacial infections. In: Topazian RG, Goldberg MH (eds) Oral and Maxillofacial Infections. Saunders, Philadelphia, p. 39–78.

Scott JH (1952) The spread of dental infection – anatomical considerations. Brit Dent J 92: 236–240.

Seltzer S, Bender LB, Smith J, Freedman I, Nazimove H (1967) Endodontic failures: An analysis based on clinical, roentgenographic, and histologic findings, Part I. Oral Surg Oral Med Oral Pathol 23: 500–516.
Shrout MK, Hall JM, Hildebolt CE (1993) Differentiation of periapical granulomas and radicular cysts by digital radiometric analysis. Oral Surg Oral Med Oral Pathol 76: 356–361.
Simon JHS (1980) Incidence of periapical cysts in relation to root canal. J Endod 6: 845–848.
Sjögren U, Hüggelund B, Sundqvist G, Wing K (1990): Factors affecting the long-term results of endodontic treatment. J Endod 16: 498–504.
Smith BH (1984) Molar wear in hunter-gatherers and agriculturists. Am J Phys Anthrop 63: 39–56.
Smith G, Matthews JB, Smith AJ, Browne RM (1987) Immunglobuline-producing cells in human odontogenic cysts. J Oral Pathol 16: 45–48.
Soames JV, Southam JC (1993) Oral Pathology, 2nd ed. Oxford Univ Press, Oxford, p. 69–111.
Sommer RF, Ostrander FD, Crowley MC (1966) Clinical Endodontics. A Manual for Scientific Endodontics, 3rd Ed. Saunders, Philadelphia, p. 409–422.
Sonnabend E, Oh CS (1966) Zur Frage des Epithels im apikalen Granulationsgewebe (Granulom) menschlicher Zähne. Dtsch Zahnärztl Z 21: 627–643.
Spatafore CM, Griffin JA, Keyes GG, Wearden S, Skidmore AE (1990) Periapical biopsy report: An analysis over a 10-year period. J Endodont 16: 239–241.
Spiessl B (1959) Osteomyelitis der Kieferknochen. In: Häupl K, Meyer W, Schuchardt K (eds) Die Zahn-, Mund- und Kieferheilkunde. Urban & Schwarzenberg, München, p. 1047.
Stern MH, Dreizen S, Mackler BF, Selbst AG, Levy BM (1981) Quantitative analysis of cellular composition of human periapical granulomas. J Endodont 7: 117–122.
Stockdale CR, Chandler NP (1988) The nature of the periapical lesion: A review of 1108 cases. J Dent 16: 123–129.
Summers L (1974) The incidence of epithelium in periapical dental granulomas and the mechanisms of cavitation in apical dental cysts in man. Arch Oral Biol 19: 1177–1180.
Sundquist G, Johansson E, Sjögren U (1989) Prevalence of black-pigmented Bacteroides species in root canal infections. J Endodont 15: 13–19.
Swärdstedt T (1966) Odontological Aspects of a Medieval Population in the Province of Jämtland/Mid-Sweden. Tiden-Barnagen, Stockholm.
Syrjänen S, Tammisalo E, Lilja R, Syrjänen K (1982) Radiological interpretation of the periapical cysts and granulomas. Dentomaxillofac Radiol 11: 89–92.
Taylor RMS (1963) Cause and effect of wear of teeth. Acta Anat 53: 97–157.
Thyagarajan K, Kamalam A (1982) Chronic urticaria due to abscessed teeth roots. Int J Dermatol 21: 606
Trauner R (1964) Die Osteomyelitis der Kiefer. Fortschr Kiefer GesichtsChir 9: 146–152.
Twiesselmann F, Brabant H (1960) Recherches sur les dents et les maxillaires d'une population ancienne de Coxyde (Belgique). Bull Group Int Rech Sci Stomatol 3: 99–171.
Tyndall DA, Kapa SF, Bagnell CP (1990) Digital subtraction radiography for detecting cortical and cancellous bone changes in the periapical region. J Endod 16: 173–178.
Wächter R, Türp JC, Alt KW (1992) Zur histomorphologischen und röntgenologischen Differenzierung von periapikalem Granulom und radikulärer Zyste – mit historischem Exkurs. Parodontologie 1: 27–42.
Wagner W, Scheunemann H (1984) Zur Therapie der chronisch rezidivierenden Osteomyelitis des Unterkiefers. Fortschr Kiefer Gesichts Chir 29: 44–48.
Wais FT (1958) Significance of findings of following biopsy and histologic study of 100 periapical lesions. Oral Surg Oral Med Oral Pathol 11: 650–653.
Waldvogel FA, Medoff G, Swartz MN (1970) Osteomyelitis: A review of clinical features, therapeutic considerations and unusual aspects (second of three parts). New Engl J Med 282: 260–266.

Walton RE, Garnick JJ (1986) The histology of periapical inflammatory lesions in permanent molars in monkeys. J Endod 12: 49–53.
Wang C, Stashenko P (1993) Characterization of bone-resorbing activity in human periapical lesions. J Endodont 19: 107–111.
Wannfors K, Hammarström L (1990) Periapical lesions of mandibular bone: Difficulties in early diagnostics. Oral Surg Oral Med Oral Pathol 70: 483–489.
Wassmund M (1929) Bemerkungen zur Arbeit von Prof. Dr. Axhausen: "Zur Klassifikation der eitrigen Entzündungen am Kiefer". Chirurg 1: 884–886.
Wassmund M (1933) Über die eitrigen Prozesse des Mundbodens und der tiefen Spalträume am Schädel. Dtsch Z Chir 240: 681–694.
Wayman BE, Murata SM, Almeida RJ, Fowler CB (1992) A bacteriological and histological evaluation of 58 periapical lesions. J Endodont 18: 152–155.
White SC, Sapp JP, Seto BG, Mankovich NJ (1994) Absence of radiometric differentiation between periapical cysts and granulomasa. Oral Surg Oral Med Oral Pathol 78: 650–654.
Williams BL, McCann GF, Schoenknecht FD (1983) Bacteriology of dental abscesses of endodontic origin. J Clin Microbiol 18: 770–774.
Winstock D (1980) Apical disease: An analysis of diagnosis and management with special reference to root lesion resection and pathology. Ann Roy Coll Surg Eng 62: 171–179.
Wood NK (1984) Periapical lesions. Dent Clin North Am 28: 725–766.
Zain RB, Roswati N, Ismail K (1989) Radiographic evaluation of lesion sizes of histologically diagnosed periapical cysts and granulomas. Am Dent J 48: 3–5, 46.
Zaldu'a CB (1995) Conservative endodontic-restorative treatment for a severely damaged and periapically compromised molar: A case report. Quintessence Int 26: 347–350.
Zuhrt R (1956) Stomatologische Untersuchungen an spätmittelalterlichen Funden von Reckahn (12.–14. Jh.). I. Die Zahnkaries und ihre Folgen. Dtsch Zahn Mund Kieferheilk 25: 1–15.

4.4 Paleopathological Evidence of Jaw Tumors

Eugen Strouhal

Introduction

Theoretically, almost all tumors, benign or malignant, can occur in the jaw regions. Some are encountered often, others occasionally. Reasons cannot be given for such differences. From the paleopathological point of view, most kinds of tumors of the jaw region can be considered extant in prehistoric or historic specimens. A lack of evidence could be due to varying preservation circumstances of the specific kinds of tumors. There are also differences in frequencies of tumors affecting jaws which can be explained by the fact that the jaws (in contrast to other bones) may cause different alterations depending on the type of tumor.

In dealing with tumors it is most important to develop a differential diagnostic distinction of tumors so that we do not confuse them with other pathological processes which may cause a similar pathomorphological appearance. As an example, we present the most ancient case described so far: A conspicuous change in the mandible from Kanam (Kenya), found by J Gitau, an assistant of LSB Leakey. The latter placed it in the Lower to Middle Paleolithic (500,000–2,000,000 BC) and attributed it to *Australopithecus* or *Homo erectus* (Leakey 1935). Later the date was assigned to Middle to Upper Pleistocene (Oakley and Tobias 1960; Tobias 1962). As the stratigraphy is uncertain, so is the absolute age (Oakley et al. 1977). The symphyseal region and the adjoining part of the left half of the mandible are thickened by a bony mass slightly protruding labially but much more lingually (maximum diameter 40.5 mm). The surface is granulated and finely roughened, but hardly any spiculae are apparent in the published pictures. Lawrence (1935), who first described this pathology, diagnosed it as a subperiosteal ossified sarcoma, which was accepted – not without hesitation – by Brothwell (1967). After reexamination, Stathopoulos (1975, 1986) concluded it to be Burkitt's lymphoma, without giving reasons for this new diagnosis. Published pictures have not been very revealing, and neither radiographs nor histological section images were published. The specimen is heavily mineralized, and these two techniques were perhaps applied without success. Clearly, a revision is needed before the case may be accepted as evidence for a malignant tumor.

Another change of opinion forms a very group: Sometimes, mandibles with protruding bony mass were considered to be tumorous. After x-ray examination they rather turned out to be excessive formations of healing callus after a fracture. As an example we present the mandible of an adult male from the late Iron Age cemetery at Varpelev (Denmark, Bennike 1985).

This review presents examples of paleopathological cases of different tumor types from different time periods of the Old World as documented in the literature and collected in field work. Only exceptionally, if other examples are lacking, cases from the New World will be presented. The accepted cases are divided into benign and malignant tumors.

Benign Tumors

The most frequently occuring benign tumors in the jaw region are fibroma, fibrotic dysplasia, chondroma, dentigerous cyst, epulis, giant cell tumor, adamantinoma and odontoma (Coley 1949). Let us review them in this order.

The nonossifying fibroma can be preserved in the paleopathological material only unusually by calcification. It can, however, be detected by evidence of areas of lytic destruction with smoothly delimited edges developed by pressure atrophy which survives even after its cause has been obliterated by postmortal changes.

A case of a nonossifying nasopharyngeal fibroma which grew from the right side of the nasal cavity into the sinus maxillaris and anterior side of the sinus sphenoidalis and subsequently caused an atrophy in the surroundings was mentioned by Brothwell (1961) in a provisionally described skull from an Anglo-Saxon cemetery in Yorkshire (England). A similar case in the skull of an adult male from the Neolithic (?) site Slagslunde, Frederiksborg (Denmark), was published by Bennike (1985: 207–209). A large cavity destroyed the face from the right side of the nasal cavity to the nasal bones, then to the left supraorbital margin, to the medial edge of the left zygomatic bone and ultimately to the upper part of the hard palate, until only the frontal part of the palatine bones remained. Intranasal structures disappeared and the left eye was undoubtedly involved in this pathology. The edges of this enormous lesion were smooth and rounded. Bennike diagnosed a benign nasopharyngeal tumor originating in the nasal cavity. A fibroma appears to be the most likely cause.

As a consequence of a fibrotic dysplasia or an ossifying fibroma a great osseous mass has formed in the region of the right maxilla of a 22–25-years-old female from Pre-Columbian Chile (Sawyer et al. 1990). The mass consists of a thick structure of sagittaly oriented spiculae firmly adhering to the remnants of the original maxillary tissue. In radiograms a heterogeneous radiopaque pitting with frayed edges is apparent. Microscopically a thick lamellar bone with a normal number of lacunae combined with immature woven bone can be observed. An invasive growth was nowhere apparent: Original bone lamellae were augmented with woven bone, but not destroyed. This differential diagnosis ruled out malignancy. The process, which took a long period to develop, must have completely corrupted the appearance of the girl, impaired her vision and ultimately – due to nutrition problems – ended her life.

Osteomas are frequently found in paleopathologically investigated skeletal populations and can thus be considered commonplace. There is, however, an interesting difference: they frequently occur in the maxilla, but they are scarce in the mandible (Coley 1949). A precise statistical study, however, is still lacking. Osteomas also occur in (rarely studied) nonhuman primates, as the two cases described in gorillas from the anatomical institutes in Frankfurt/Main and Köln attest (Schultz and Starck 1977). In one, a homogeneous and smoothly rounded bony mass grew out from the anterior aspect of the right maxilla, zygomatic bone and zygomatic process of the frontal bone. The other skull had similar outgrowths originating in the same bones, bigger on the right than the left. In the first case, not only osteoma but also osteochondroma was taken into consideration, in the second one the diagnosis of osteoma was ambiguous.

Evidence of chondromas in the jaw region, which sometimes do ossify and therefore could in principle be preserved, was not found in skeletons so far.

A dentigerous cyst develops out of the cavity around an unerupted permanent tooth. It can be found in any tooth, but it most often develops around the canines and third molars. Within the cavity the tooth crown is exposed, while its root is implanted in the wall of the cyst. The cyst, which can reach large dimensions with thin walls, is the result of an aberrant development of the dental enamel (proliferation and degeneration of the cells) before tooth eruption (Coley 1949). Several cases are listed in surveys of tumors occuring in Poland and Eastern Europe (Gladykowska-Rzeczycka 1982, 1988, 1991).

An epulis or dentoperiostal cyst develops through limited growth of the tooth periosteum of the alveolar edges, often with a narrow stem, and can expand above the gingiva and/or tooth. Being a soft tissue tumor its preservation in paleopathological material is quite unlikely.

A case of a large cavity in an extended and deformed mandibular body from a Predynastic grave at Mostagedda, Egypt (Fig. 1) was described as myeloid epulis

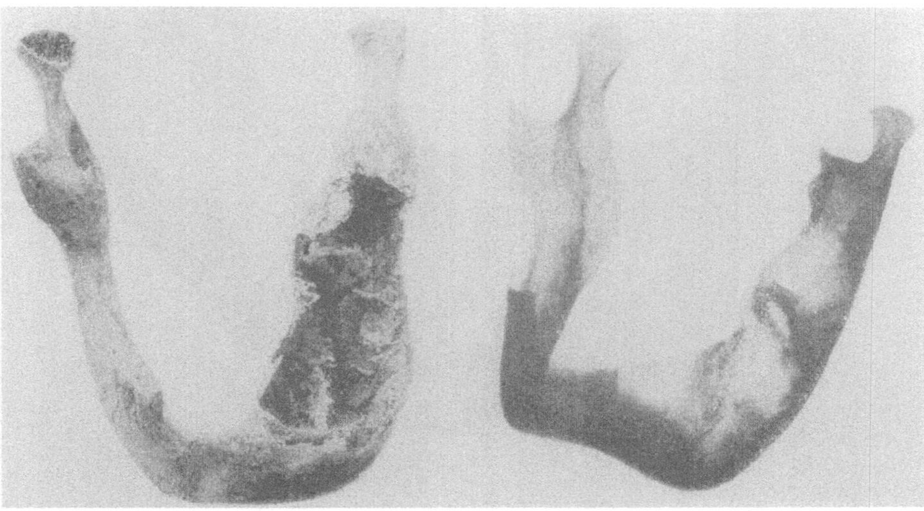

Fig. 1. Mostagedda, Egypt: A mandibular body from a Predynastic grave showing a large cavity, extension and deformation caused by a benign tumor (picture by Gardner and Urquhart 1930).

with differential diagnostic alternatives of dentigerous cyst, simple (dental) cyst or the result of fibroma (Gardner and Urquhart 1930).

Giant cell tumors develop from the central spongiotic zone of the alveolar bone and cause a slowly growing tumescence of the alveolar process. Because they do not perforate the cortical bone, they can be recognized only histologically after biopsy. Radiographically they cannot be differentiated from adamantinoma (Coley 1949). No case has yet been found in paleopathological literature.

A giant cell tumor is similar to brown tumor or osteoclastoma. It is a giant cell reaction or a restaurative granuloma consisting of giant cells in hyperparathyroidism; this has to be taken into account in differential diagnosis (Cassidy 1977).

The odontoma is a mixture of connective tissue derived from the dental papilla and the epithelium of the tooth enamel. It may stay soft, thus remaining paleopathologically undetectable. The first paleopathological evidence was found in an isolated mandible of an older (as evidenced by intravital loss of teeth) female in the labial side of the mandible under the alveolus of the right first premolar (Brothwell 1959). The specimen was found on the Island of Sokotra in the Indian Ocean and is medieval or older. The abnormal formation consists of one or two tooth-like bodies (denticuli) and parts of a less well defined dental tissue mass (Figs. 2–3). In the same mandible there was still an accessory two-rooted tooth impacted deeply within the right half of the body (Fig. 4). Three further cases were studied by Schultz (1978) who did not consider them to be tumors, but rather monstrosities. Two of these are from Pre-Columbian Mexico, the third one from Roman Germany. The first case was localized in the anterior region of the mandible, the other two in the left posterior region of the maxilla.

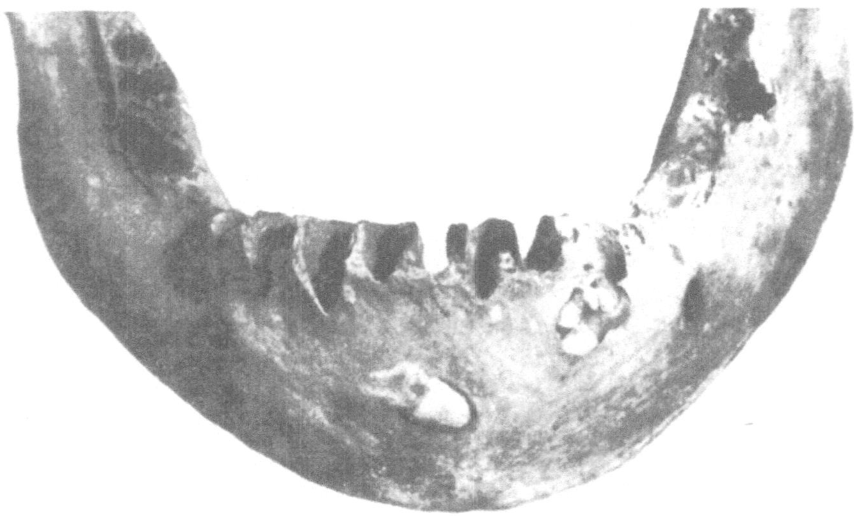

Fig. 2. Sokotra Island (Indian Ocean): Isolated mandible with an odontoma on the labial right corpus side, consisting of one or two tooth-like bodies and dental tissue mass (picture by Brothwell 1959).

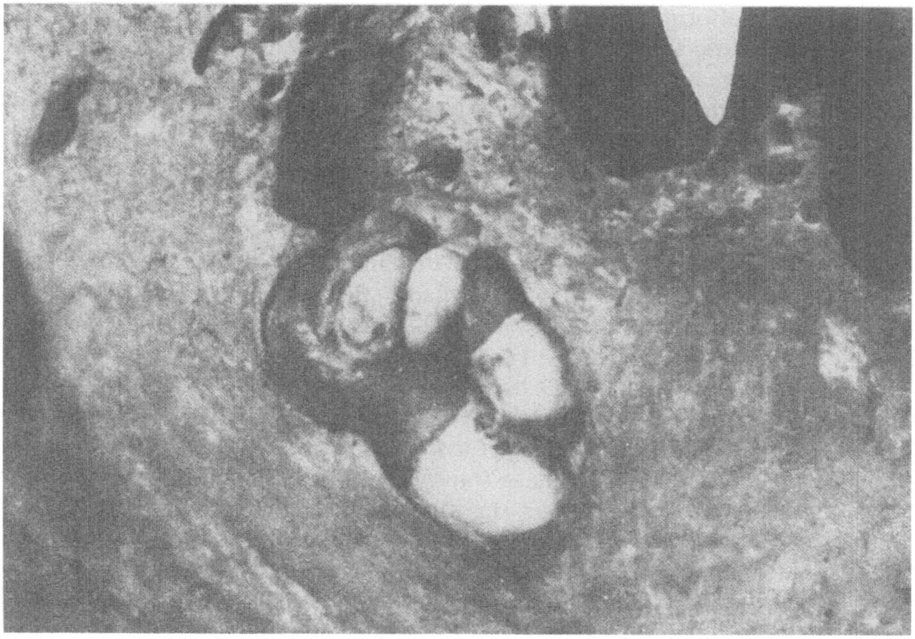

Fig. 3. Sokotra, as Fig. 2, detail (picture by Brothwell 1959).

A recently excavated case of an odontoma from medieval Canterbury, Kent, was published by Anderson and Andrews (1993). In place of the unerupted upper left third molar an anomalous mixture of two partially developed tooth crowns and

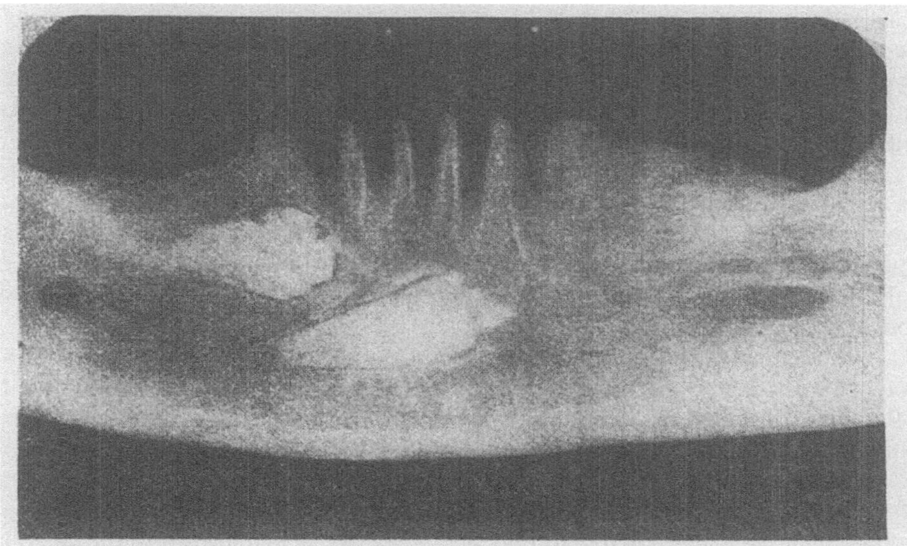

Fig. 4. Sokotra, as Fig. 2, x-ray showing an accessory two-rooted tooth on the right half of the mandibula.

remains of a malformed root were found. These were visible on the labial side, while on the palatal side the conglomeration was covered by a cement layer.

These five examples found in thousands of investigated jaws attest to the rarity of odontomas, as they are today. The real frequency in historic populations, however, must have been higher, because often the odontoma does not perforate the surface of the bone (or the gum) and is then not detectable unless x-rayed.

Adamantinoma (or ameloblastoma) develops by growth of the epithelium of the tooth enamel organ. In contrast to odontoma it causes a thickening of the jaw. It occurs more frequently in the mandible than in the maxilla. As a result of a slow growth of the tumor, lasting up to four decades, the face can become substantially deformed; the oral cavity narrows and mastication is impeded. This tumor can also become malignant, which is why it is clinically more important than odontoma. In the paleopathological literature, a case of a 40–50 year old male from a Medieval cemetery (11th–13th c.) at Czersk (Poland) was described by Gladykowska-Rzeczycka (1978). The central part of the right half of the mandibular body was smoothly elevated. In spite of the extraordinarily large bulge, the tumor did not break through the cortical bone. In the x-rays, the bony structure showed multilocular sinuous defects bordered by barely recognizable bony trabeculae. Edges of the tumor were sharply delimited and slightly sclerotic. The author preferred the diagnosis of adamantinoma instead of odontoma, fibrotic dysplasia, odontogenous or non-odontogenous cyst or even actinomycosis.

Malignant Tumors

Among the various kinds of primary osseous tumors – sarcomas – the likelihood of survival is very low for fibrosarcoma and chondrosarcoma. Despite the fragility of its structure, we can, however, expect preservation of osteosarcomas. Among the primary haematopoietic tumors, involvement of jaws can be expected in myeloma multiplex. Jaws can suffer destruction caused by invasive growth of the carcinoma into the neighbouring soft tissue. Furthermore, metastases – either lytic, osteoplastic or mixed – can involve jaws.

Let us first examine the true osseous tumor, which develops from cortical bone, i.e. the osteosarcoma. In modern practice, it occurs (together with fibrosarcoma, chondrosarcoma and reticulosarcoma) rather rarely in the jaws – in contrast to its typical location in the long bones. Only 35 cases of jaw tumors were observed prior to 1949 in the Memorial Hospital for Cancer of Cornell University, Ithaca, New York (Coley 1949).

In the paleopathological literature three cases have been described (out of a total of 10 osteosarcomas described in past populations of the Old World: Strouhal 1994) which afflicted the maxilla (while none the mandible). The oldest case was found in the skull of a 30 year old male from Giza, dated 26th Dynasty (664–525 BC; Salama and Hilmi 1950). The tumor evolved from the anterior aspect of the left half of the maxilla into the left orbita and penetrated into the fossa pterygoidea and into the nasal cavity up to the concha inferior. It reached the tips of the alveoli from the left canine to the left second molar (Figs. 5–6). In the radiographs the tumor showed a honeycomb structure with a radiating edge (Fig. 7).

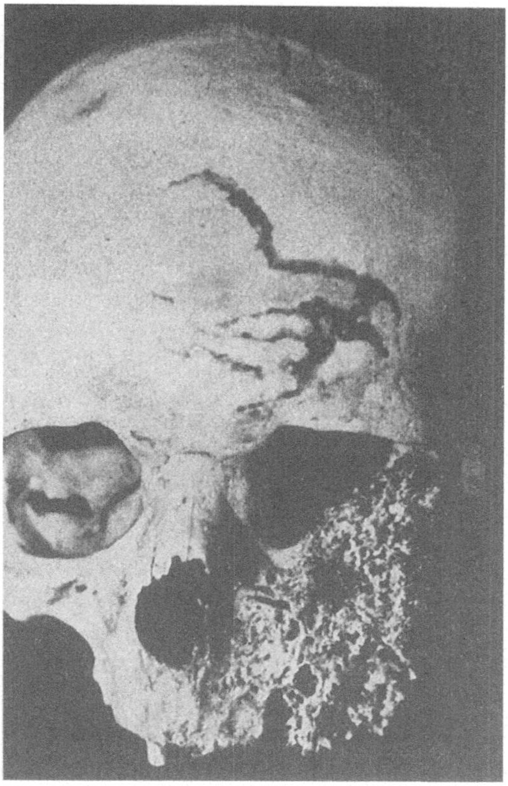

Fig. 5. Giza, Egypt: A malignant tumor evolved from the anterior part of the left maxilla into the orbita penetrating the nasal cavity and the fossa pterygoidea: frontal view (picture by Salama and Hilmi 1950).

A similar case was described in the skull of a male aged less than 40 years at death, which came from the ossuary under the St. George Chapel of the Castle of Caën (France), dated ca. 12th c. AD (Dastugue 1965). The tumor evolved from the anterior aspect of the right half of maxilla to the floor of the right orbit and the right lateral wall of the nasal cavity. Its structure was spongiotic, with irregularly ordered trabeculae and no preserved remnants of compact bone. The author's interpretation for the perforation of the right zygomatic bone and the complete lack of teeth in the maxilla is therapeutic measure, undertaken to relieve the massive pain.

Less ambiguous appears the case of a male of medium age from a New Kingdom grave in the Wadi Halfa region (Sudan, Nielsen 1970). The left maxilla was almost destroyed by invasive growth of a (most certainly) malignant tumor. The author mentions tiny spiculae and a marginal zone of osteoporosis which are not apparent in the picture (Fig. 8). Small osteolytic foci in the cranial vault are probably metastases. Nielsen prefers a diagnosis of osteosarcoma, admitting, however, the possibility of an invasive growth of a carcinoma that developed in the soft tissue.

Of the fourteen published cases of the primary haematopoietic tumor, myeloma multiplex (or plasmocytoma), detected in past populations of the Old World, in only

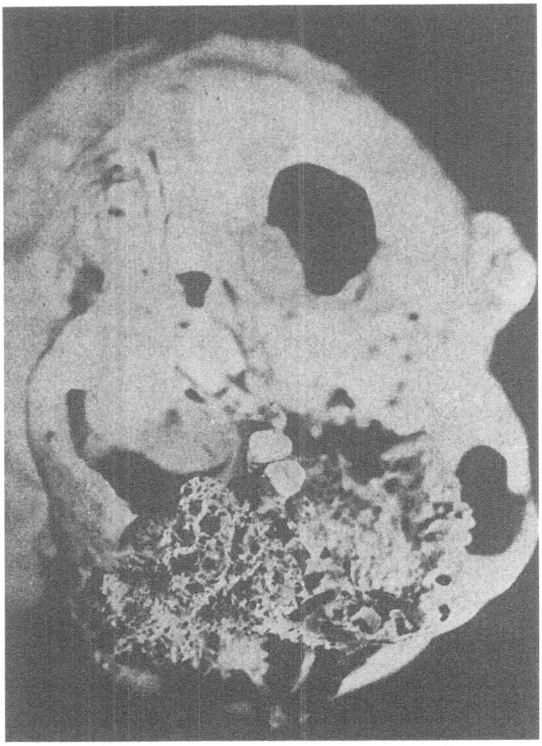

Fig. 6. Giza, as Fig. 5, basic view (picture by Salama and Hilmi 1950).

three (21%) was the mandible permeated with similarly sized small lytic foci with sharp edges (Strouhal 1994). The oldest case came from a Neolithic burial at Mauer, Vienna, Austria (Strouhal and Kritscher 1990; Figs. 9–10), followed by an Old Kingdom case from Qubbet el-Hawa, Aswan, Egypt (Rösing 1990, p. 93–95) and a Late Period (7th to 4th c. BC) case from Abusir, Egypt (Strouhal and Vyhnanek 1981, 1987). In no paleopathological case has an affliction of the maxilla by a myeloma been mentioned; this could be due to anatomical properties of that bone or to omission by the authors.

Participation of the jaw region in cases of direct invasive growth of a primary carcinoma of soft tissues of the oral cavity, lips, salivary glands, nasal cavity, paranasal sinuses and of the nasopharynx has frequently been found in paleopathological series. It should be stressed that in these cases the diagnosis of a carcinoma has been made indirectly, based on recent clinical experience. According to the localisation and shape of the defect, which – if not secondarily damaged – could represent a mold of the original soft tissue tumor, it is sometimes possible to deduct the probable place of origin.

Of thirteen Old World cases, the destruction of the maxilla was involved in eight (62%), while the mandible was afflicted only in two cases (15%; Strouhal 1994). Of the cases also involving the maxilla, four were a nasopharyngeal

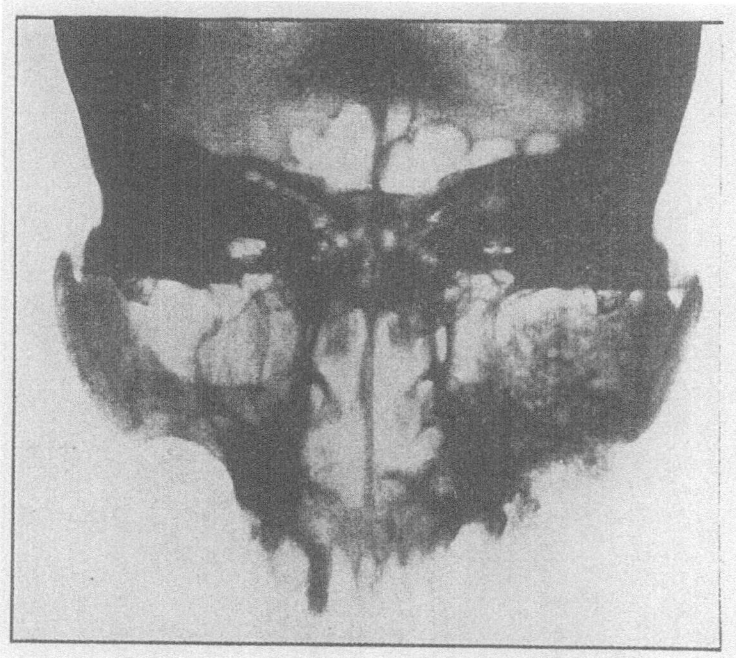

Fig. 7. Giza, as Fig. 5, x-ray, frontal view (picture by Salama and Hilmi 1950).

carcinoma, three a carcinoma of the paranasal sinuses, and one a carcinoma of the mucosa of the anterior part of the nasal cavity. Four of these cases originated in Egypt – in Qubbet al-Hawa, Old Kingdom (Rösing 1990), in an unidentified site of the 3rd to 5th Dynasty (Wells 1963), at Nag ed-Deir, 6th–12th Dynasty (Strouhal 1976, 1978, Figs. 11–12) and El-Bersha, Early Christian Period, around 500 AD (El-Rakhawy et al. 1971). Another case came from Nubia: Sayala, Christian Period, 6th-11th c. AD (Strouhal 1991, Fig. 13). The remaining cases came from Tepe Hisar in ancient Iran, 3500–300 BC (Krogman 1940), from Ferrerias on the Island of Menorca, Spain, Talayotic Culture, 1300–0 BC (Campillo 1977) while one was an Aino skull from Kitami, Hokkaido (Japan), 15th–19th c. AD (Suzuki 1984).

Among the cases involving mandibles there was a tumor of enormous dimensions which destroyed the posterior part of the cranial cavity, spread to the interior parts of the facial skeleton and eroded the inner aspect of the right mandibular ramus. It was found in a recently investigated New Kingdom tomb at Saqqara (Strouhal 1995). Another case of a carcinoma of the oral mucosa destroyed the mandible and produced a border sclerosis. It was found in a skull from Sendai in Japan, 17th to 19th c. AD (Tanaka et al. 1984).

Only rarely was the jaw region afflicted in the paleopathologically described cases of the most often occuring category of the lytic metastases of carcinoma. Of a total of thirty-three published cases from the Old World (Strouhal 1994), only six (18%) showed involvement of the mandible and for none was the maxilla mentioned as a seat of a metastasis. There were two cases involving the mandible

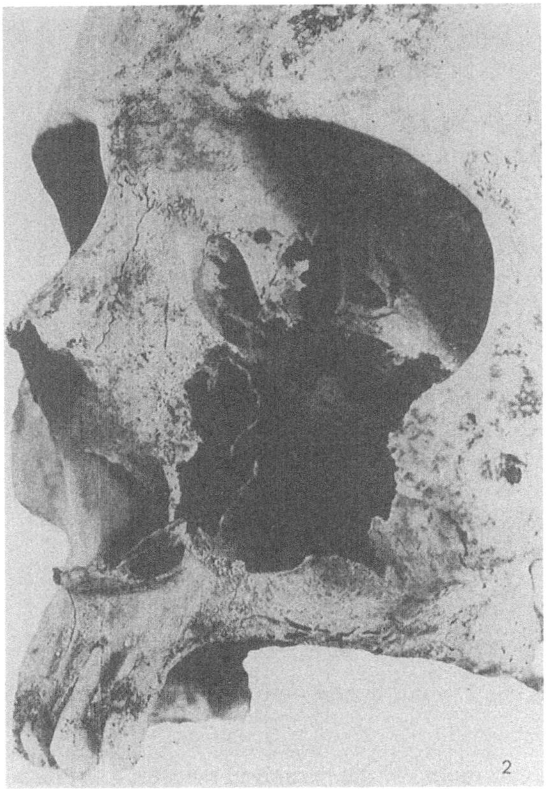

Fig. 8. Wadi Halfa, Sudan: Malignant tumor in the left maxilla showing a marginal zone of osteoporosis (picture by Nielsen 1970).

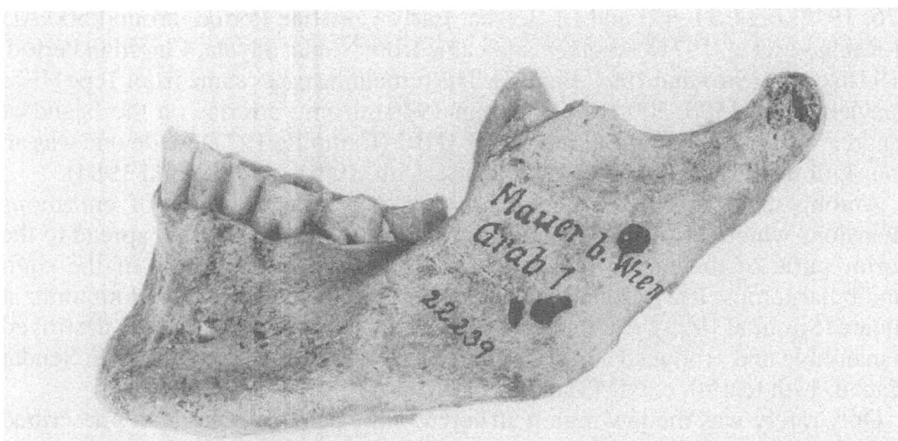

Fig. 9. Mauer, Austria: A case of myeloma multiplex (or plasmocytoma) showing the mandible permeated with small sharp edged lytic foci (picture by M. Teschler-Nicola).

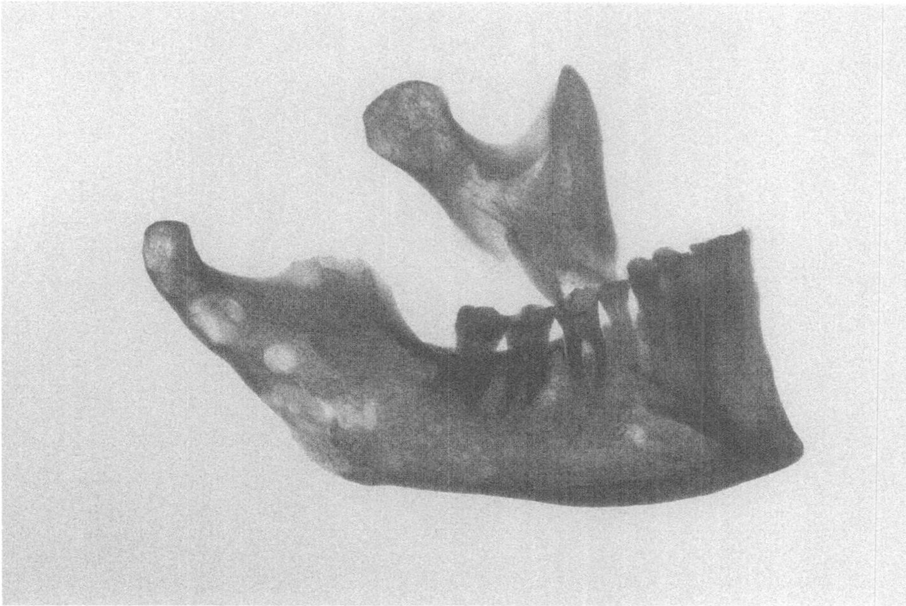

Fig. 10. Mauer, as Fig. 9, x-ray (picture by M. Teschler-Nicola).

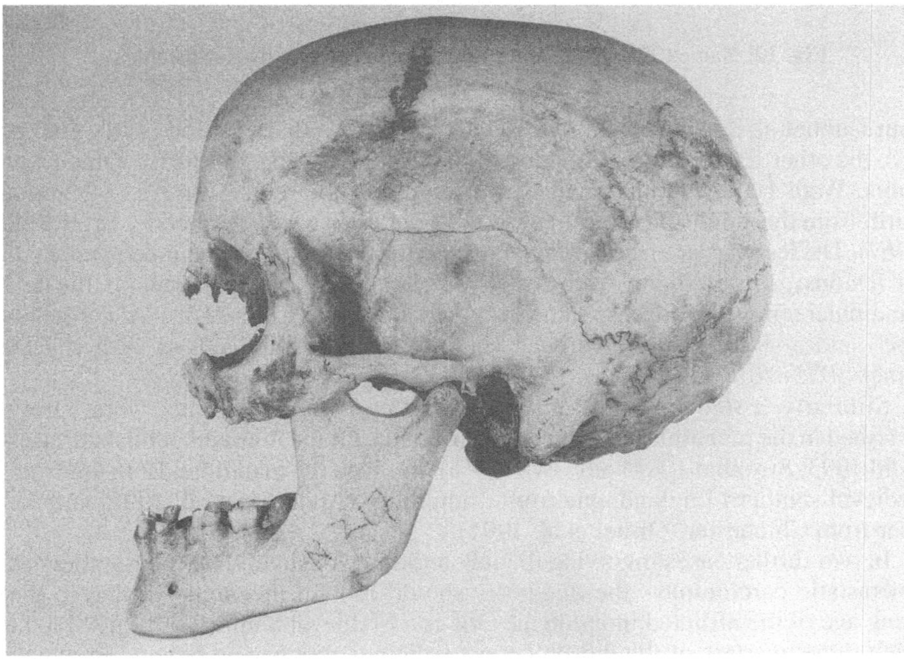

Fig. 11. Nag ed-Deir, Egypt: Malignant tumor in the maxilla, lateral view (picture by E. Strouhal).

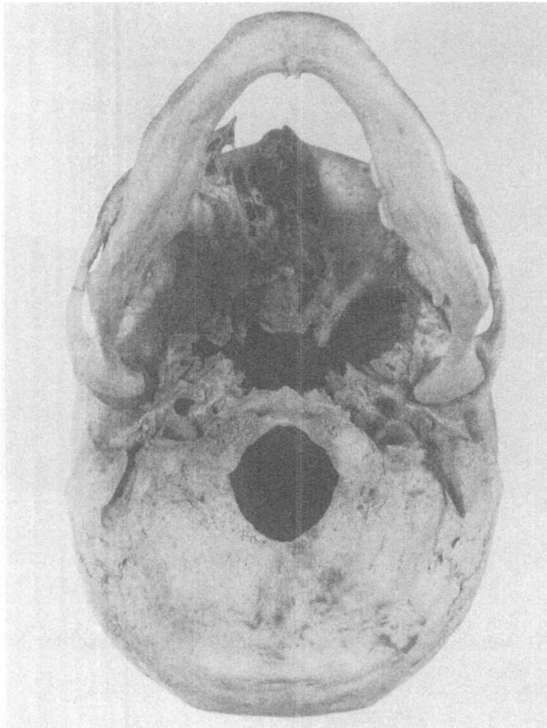

Fig. 12. Nag ed-Deir, as Fig. 11, cranial base (picture by E. Strouhal).

from Qubbet el-Hawa, Aswan; one dated 2nd Intermediary Period (17th–16th c. BC), the other Late Period (7th–4th c. BC; Rösing 1990). A third case came from Nubia, Wadi Halfa region, dated New Kingdom (Nielsen 1970; Fig. 14) and a fourth from the Anglo-Saxon cemetery of Winchester was described by Brothwell (1967). The lesions vary from irregular osteolytic destructions to rounded punched-out lesions; one of them even destroyed part of the inner aspect of the left mandibular ramus. A fifth case came from Sarkel, Russia (10th–11th c. AD, Rochlin 1965) and a sixth from Hinga near Subotica, Yugoslavia (12th to 14th c. AD, Grmek 1975–76).

Similarly, osteoplastic or mixed metastases of carcinomas were rarely described in the mandible and never in the maxilla. Of the fourteen published cases (Kühl 1993; Strouhal 1994) only two (14%) involved the mandible. Both are from Medieval southern England, one from Canterbury (Anderson et al. 1992) and the other from Chichester (Ortner et al. 1991).

In two further cases involving the jaw region – considered by their authors as a metastatic carcinoma – the diagnosis should remain inconclusive due to the young age of the afflicted individuals. One is a Medieval Sami (Lapp) maxilla, in which three quarters of the left processus palatinus was perforated and in which alveoli from the left upper canine to the second premolar were partially damaged. A further perforation was evident in the right parietal bone. In radiographs, the

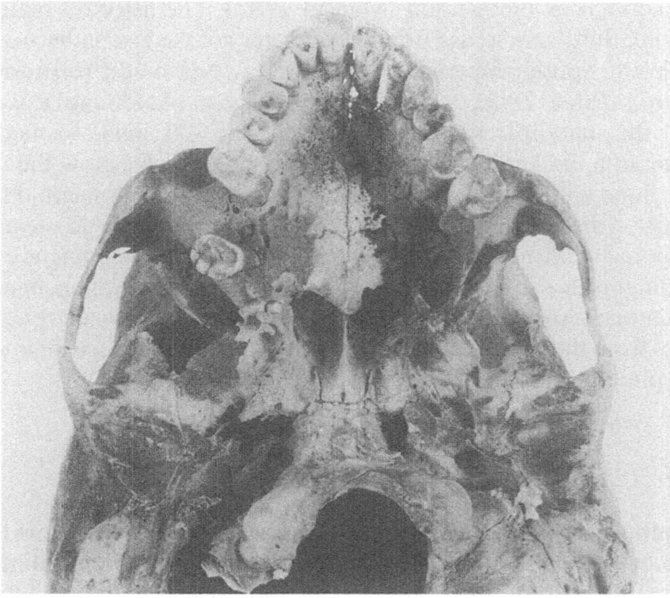

Fig. 13. Sayala, Nubia: Malignant tumor in the maxilla (picture by E. Strouhal).

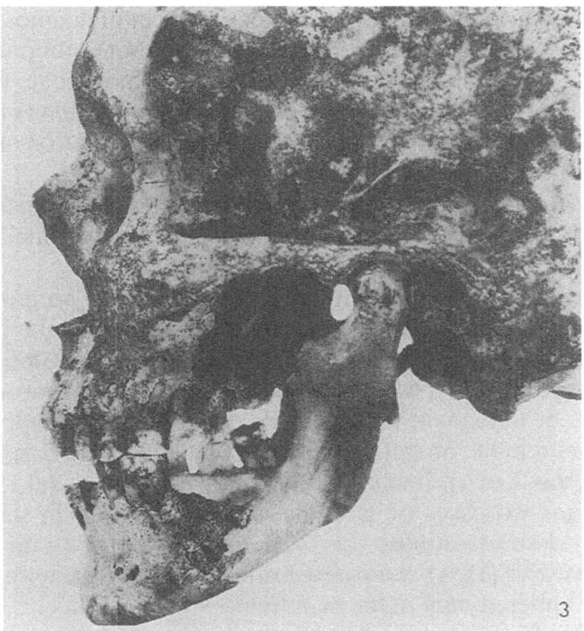

Fig. 14. Wadi Halfa, Sudan: Irregular osteolytic destructions caused by lytic metastases of a carcinoma in the left body of the mandible (picture by Nielsen 1970).

edges are described as moth-eaten (Madrid 1982). The affected male was only 21–25 years old. Similarly, a case from the cemetery of the last Sadlermiut Eskimos from Native Point, Southhampton Bay, Hudson Bay, Northwest Territories, Canada (dated transition 19th to 20th c.), concerns a 28–30 year old female (Cassidy 1977). The body of the mandible is perforated by about 300 small to medium sized openings without a thickening of the bone. The radiographs show the destruction of spongiotic bone with remnants of fine trabeculae. No reaction could be detected in the vicinity of the lesions. The author diagnosed this case as metastases of a thyroid carcinoma while also considering lip, mammary or oesophagus carcinomas. Differential diagnosis also includes brown tumor of parathyroidism, haemangioma or even haematogenous pyogenic osteomyelitis. These cases need revision, as are the few others from the Old World which involve the jaw, because their description or published photographs are inadequate for a diagnosis.

Conclusions

This paleopathological review of jaw tumors from the Old World and their classification according to the basic discernable categories is a preliminary one: First, it is not sure if all information has been collected from the widely dispersed anthropological, medical and archeological literature. Furthermore, some of the reports lack detailed information as to localization of the lesions either in the text or in the illustrations, which results in possible omissions of some cases afflicting the jaw region. In several excavated cases jaws were missing; thus the percentage of tumors afflicting jaws (as a ration of the total number of diagnosed tumors) may be too small; consequently reported percentages represent minimum numbers.

Nevertheless, based already on the present state of knowledge, we can conclude that almost all kinds of tumors known to occur in the jaw regions of recent patients which have a reasonable survival chance (because of ossification or calcifications) were present in the paleopathological evidence.

Clearly, soft tissue tumors like chondroma, epulis, giant cell tumor, fibrosarcoma, chondrosarcoma and other kinds can only be found under specific circumstances (e.g. in mummified bodies) or if they calcify.

Tumors specific to jaw regions, which develop from the primordially differentiated dental tissue like dentigerous cyst, odontoma or adamantinoma appear to be rare – then and now. Many of them, because of insufficient application of serial radiography, escape detection when they do not penetrate the compact bone surface.

Of the kinds of malignant tumor whose diagnosis, based upon the available evidence, appears reliable or sufficiently probable, the jaw region was most often afflicted by the growth of a primary carcinoma (62% maxillae and 15% mandibles). 30% of the diagnosed cases of sarcoma were found only in the maxilla. The incidence of myeloma multiplex was 21%, and it occurred only in mandibles. Rarer lytic metastases (18%) and osteoplastic or mixed metastases of carcinoma (14%) were encountered only in the mandible.

Finally, it should be stressed that tumors occurred in many parts of the Old World (Egypt, Nubia, Europe, Japan) during all eras. Thus they represent an important part of historic pathology, a piece of reality in any of these past societies.

References

Anderson T, Wakely J, Carter A (1992) Medieval example of metastatic carcinoma: A dry bone, radiological and SEM study. Am J Phys Anthrop 89: 309–323.
Anderson T, Andrews J (1993) A recently excavated odontome from medieval Canterbury, Kent. Int J Osteoarch 3: 99–104.
Bennike P. (1985) Palaeopathology of Danish Skeletons. Akademisk Forlag, Copenhagen.
Brothwell DR (1959) A rare dental anomaly in archaeological material. Brit Dent J 107: 400–401.
Brothwell DR (1961) The palaeopathology of early British Man: An essay on the problems of diagnosis and analysis. J Royal Anthrop Inst 91: 318–344.
Brothwell D (1967) The evidence of neoplasm. In: Brothwell D, Sandison AT (eds) Diseases in Antiquity. Thomas, Springfield, p. 320–345.
Campillo D (1977) Paleopatogía del cráneo en Cataluña, Valencia y Baleares. Edit. Montblanc-Martin, Barcelona.
Cassidy C (1977) Probable malignancy in a Sadlermiut Eskimo mandible. Am J Phys Anthrop 46: 291–296.
Coley BL (1949) Neoplasm of Bone and Related Conditions. PB Hoeber, New York.
Dastugue J (1965) Tumeur maxillaire sur un crâne du moyen âge. Bull Ass Franç Ét Cancer 52: 69–72.
Gardner RA, Urquhart AL (1930) Two cases of bone tumour from Ancient Egypt. Brit Med J 3631: 211.
Gladykowska-Rzeczycka J (1978) Mandibular tumour in a male skeleton from medieval burial ground in Czersk. Folia Morphol (Warszawa) 37/2: 191–196.
Gladykowska-Rzeczycka J (1982) Neoplasm from the ancient cemeteries in Poland. Anthropologie (Brno) 21: 353–364.
Gladykowska-Rzeczycka J (1988) The tumors in East and Middle Europe. Collegium Antropologicum 12 Suppl: 339.
Gladykowska-Rzeczycka J (1991) Tumors in antiquity in East and Middle Europe. In: Aufderheide M, Ortner DJ (ed) Human Paleopathology. Current Syntheses and Future Option. Smithsonian Institution Press, Washington, p. 251–256.
Grmek MD (1975–76) La paléopathologie des tumeurs osseuses malignes. Proposition d`une classification ostéo-archéologique. Hist Sci Med 1: 1–30.
Krogman WM (1940) The skeletal and dental pathology of an early Iranian Site. Bull Hist Med 8: 28–48.
Kühl I (1993) Leichenbrandanalysen, zur Information auch für Prähistoriker. Teil IV: Krebsmetastasen in einem Leichenbrand der vorrömischen Eisenzeit aus Schleswig-Holstein, Norddeutschland. Mit einem Anhang von Michael Schultz. Anthropologische Forschungen, AGW. Grasl, Wien.
Lawrence JEP (1935) Appendix A. In: Leakey LSB (ed) Stone Age Races of Kenya. Oxford University Press, London, p. 139.
Leakey LSB (1935) Stone Age Races of Kenya. Oxford University Press, London.
Madrid A (1982) Possible malignancy in a medieval Same (Lapp) maxilla. Current Anthrop 23: 200–201.
Nielsen OV (1970 Human Remains. Metrical and Non-Metrical Anatomical Variations. The Scandinavian Joint Expedition to Sudanese Nubia, Vol. 9. The Scandinavian University Books, Odense, p. 116, pl. 18.
Oakley KP Campbell BG, Molleson T (1977) Cataloque of Fossil Hominids. Part I: Africa, 2nd ed. British Museum (Natural History), London, p. 58–59.
Oakley KP, Tobias PV (1960) The Kanam jaw. Nature 185: 945–947.
Ortner DJ, Manchester K, Lee F (1991) Metastatic carcinoma in a leper skeleton from a medieval cemetery in Chichester, England. Int J Osteoarch 1: 91–98.

El-Rakhawy MT, El-Eishi HI, El-Nofely A, Gaballah MF (1971) A contribution to the pathology of Ancient Egyptian skulls. Anthropologie 9: 71–78.
Rochlin DG (1965) Bolezni drevnich ljudej. Nauka, Moskva Leningrad, p. 140–155.
Rösing FW (1990) Qubbet el Hawa und Elephantine. Zur Bevölkerungsgeschichte von Ägypten. Fischer, Stuttgart.
Salama N, Hilmi A (1950) A case of osteogenic sarcoma of the maxilla in an ancient Egyptian skull. British Dental J 88: 101–102.
Sawyer DR, Wood NK, Allison MJ (1990) An ancient tumour from pre-Columbian Chile. J. Cranio-Maxillo-Facial Surgery 18: 136–138.
Schultz M (1978) Mißbildungen odontogenen Gewebes an frühgeschichtlichem menschlichem Skelettmaterial. Dt Zahnärztl Z 33: 715–724.
Schultz M, Starck D (1977) Neue Beobachtungen und Überlegungen zur Pathologie des Primatenschädels. Folia Primatologica 28: 81–108.
Stathopoulos GP (1975) Kanam mandible's tumour. Lancet 1 (7899): 165.
Stathopoulos GP (1986) Bone tumors in antiquity. In: Retsas S (ed) Palaeo-Oncology. The Antiquity of Cancer. Ferrand Press, London, p. 13–25.
Strouhal E (1976) Tumors in the remains of Ancient Egyptians. Am J Phys Anthrop 45: 613–620.
Strouhal E (1978) Ancient Egyptian case of carcinoma. Bull N Y Acad Med 54: 290–302.
Strouhal E (1991) A case of primary carcinoma from Christian Sayala, Ancient Nubia. J Paleopath 3: 151–166.
Strouhal E (1994) Malignant tumors in the Old World. Paleopathology Newsletter 85 (Suppl): 1–6.
Strouhal E (1995) Revealing sufferings of the Ancient Egyptians. Egyptian Archaeol 6: 11–13.
Strouhal E, Kritscher H (1990) Neolithic case of a multiple myeloma from Mauer, Vienna, Austria. Anthropologie 28: 79–87.
Strouhal E, Vyhnánek L (1981) New cases of malign tumors from Late Period cemeteries at Abusir and Saqqara, Egypt. Ossa 8: 165–189.
Strouhal E, Vyhnánek L (1987) Nouveaux exemples de tumeurs osseuses malignes provenant de cimetières égyptiens de la basse époque. Bull Mém Soc Anthrop Paris Sér 14, 4: 159–170.
Suzuki T (1984) Paleopathological and paleoepidemiological study of osseous syphilis in skulls of the Edo period. Bull Univ Mus Tokyo 23.
Tanaka K, Umezu Y, Tajima T (1984) On the pathological changes in a mandibula of a king in the Sendai prefecture of Edo period. Proceedings of the 37th joint meeting of the Anthropological Society of Nippon and the Japanese Society of Ethnology. J Anthrop Soc Nippon 92: 114 ff. (Japanese).
Tobias PV (1962) A re-eximanation of the Kanam Mandible. In: Mortelmans G, Nenquin S (eds) Proceedings of the 4th Pan-African Congress of Prehistory I: 341–360.
Wells C (1963) Ancient Egyptian pathology (clinical record). J Laryngol Otol 77: 261–265.

4.5 Enamel Hypoplasias in Archaeological Skeletal Remains

Michael Schultz, Petra Carli-Thiele, Tyede H. Schmidt-Schultz, Uwe Kierdorf, Horst Kierdorf, Wolf-Rüdiger Teegen and Kerstin Kreutz

Introduction

Paleopathology enables us to understand the nature, causes and frequency of diseases of past populations. This interdisciplinary research enlarges the scope of methods and techniques which leads to a better knowledge of, in particular, the etiology and the epidemiology of ancient diseases. Furthermore, the results of paleopathological investigations on skeletal remains of prehistoric and historic populations illuminate living conditions, such as nutrition, housing and working conditions in ancient times. Enamel hypoplasias are found in many populations all around the world and from all time periods (cf. Fig. 1 a, b). Thus, anthropologists and paleopathologists are very familiar with this tooth morphology.

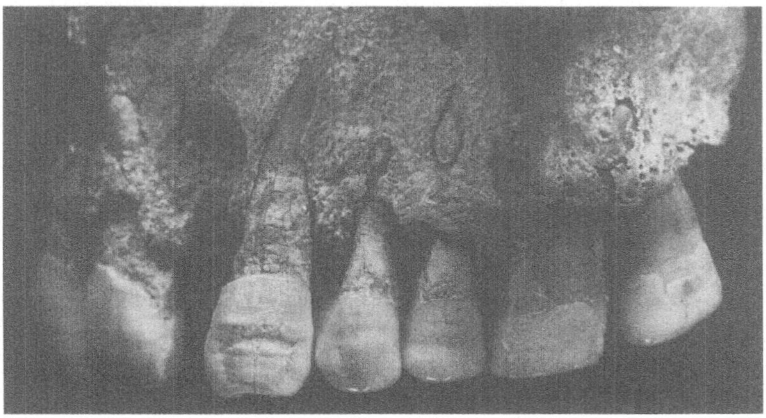

Fig. 1 a. Transverse linear enamel hypoplasias in upper teeth of an individual from Straubing (Strb-791).

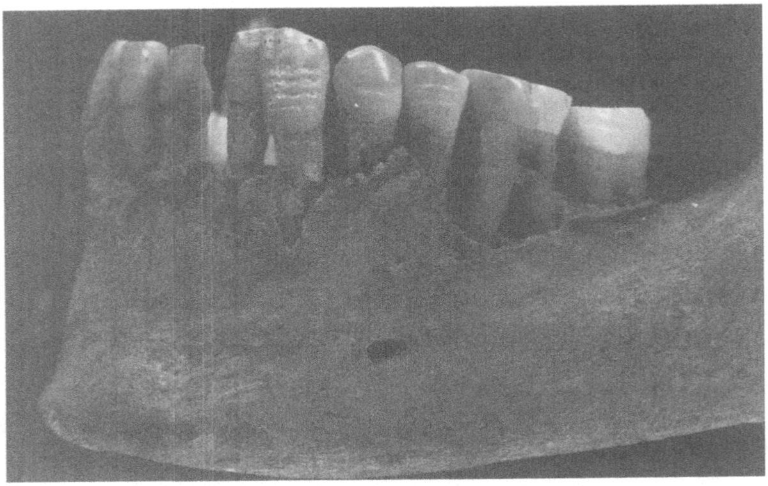

Fig. 1 b. Transverse linear enamel hypoplasias in lower teeth of an individual from Straubing (Strb-791).

The importance of enamel hypoplasias in anthropology was demonstrated by Skinner and Goodman (1992). An informative historical overview on this subject was given by Massler et al. (1941), and Sarnat and Schour (1941). The development and the structure of enamel was described by Garn et al. (1979), Yeagar (1980), Boyde (1989), Hillson (1990), and Schumacher et al. (1990). A comparison between the structure of enamel hypoplasias in human and non-human primates was given by Suckling et al. (1986), and Eckhardt et al. (1992).

This short contribution deals with the classification of enamel hypoplasias, the methods of evaluation, the etiology and the epidemiology of enamel hypoplasias in mammals, especially in humans. The frequency of transverse linear enamel hypoplasias is presented in the last chapter which deals with four pre-Columbian populations from the New World and four populations from the Old World dating to the Neolithic Age, the Iron Age, the Early and the High Middle Ages.

Classification

The Fédération Dentaire Internationale distinguishes six classes of enamel defects:

- Type 1: enamel opacities coloured white or cream
- Type 2: enamel opacities coloured yellow or brown
- Type 3: pits (in the literature also characterized as "enamel pitting")
- Type 4: horizontal grooves (also characterized as "linear enamel hypoplasia [LEH]")
- Type 5: vertical grooves
- Type 6: missing enamel.

Type 6 of this classification is responsible for the development of the Foramen caecum molare (Capasso and Di Tota 1992). Apparently, there is a genetic predisposition for the development of such a Foramen caecum (Schulze 1961). Further literature on the classification of enamel defects can be found in Clarkson (1989) and Small and Murray (1978). A classification of transverse linear enamel hypoplasias in archaeological skeletal material was made by Schultz (1988).

A large number of papers have been written on the problem of diagnosing enamel surface defects (ESD). Only a few selected examples of them will be mentioned. Condon and Rose (1992) published results on a scanning-electron microscopic investigation which is a very useful article for everyone working in this field. Marks (1992) used these techniques for the differentiation of characteristic changes of enamel surface defects into three types (I, II, and III) which are only detectable by scanning-electron microscopy.

Type I is characterized by surface changes not visible by macroscopic investigation and probably not pathological. Microscopically, however, the surface of the enamel is flat and featureless. In some cases, the prisms seem to be missing and the crests of the striae of Retzius have conglomerated.

Type II takes on many different morphological appearances during scanning-electron microscopic analysis. For this reason, it is impossible to give a detailed definition of this defect. In contrast to type I, such defects show a clear boundary between the prisms and the striae of Retzius are compressed.

Type III exhibits the most severe pattern of pathological alterations of the enamel. The defects are mostly localized at the cemento-enamel-junction and affect, as a rule, the whole circumference of the tooth. The defects are characterized by a flat and featureless enamel with compressed or twisted striae of Retzius.

Goodman et al. differentiated among five types of enamel surface defects. They gave a clear definition of a pit patch which is not frequently seen: "area of depressed enamel involving multiple hypoplastic pits and cluster bands" (1992a, 118). Their definition of cluster bands is a "series of clearly visible striae of Retzius that extend through most of the enamel thickness. These striae tend to bend and converge near the enamel surface. Cluster bands sometimes incorporate Wilson bands" (1992a, 119).

In order to arrive at these classifications, the use of a scanning-electron microscope and/or a histology laboratory is a necessity (cf. Figs. 2 a–d, 3, and 4). Such equipment is not always available, but macroscopic investigations using a magnifying glass also yield useful results. For this purpose, a classification for the external morphology of transverse linear enamel hypoplasias was developed using five degrees of enamel surface alterations (Schultz 1988).

Methods of Evaluation

Morphological Investigations

Several methods for the diagnosis of the structures caused by enamel hypoplasias are available.

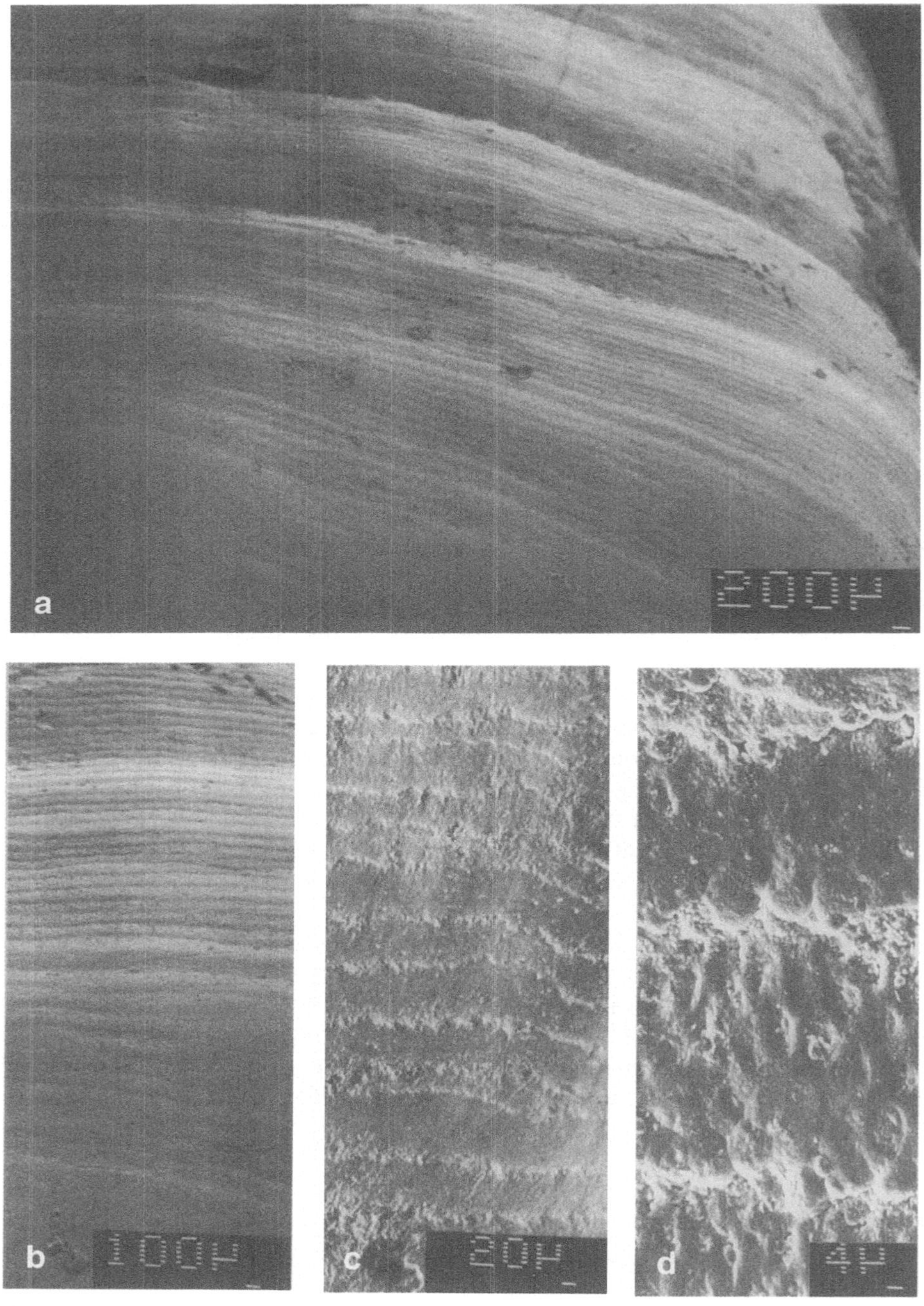

Fig. 2. Transverse linear enamel hypoplasias. Tooth No. 22 of an individual (Skh-16) from Sarai Khola. Scanning electron micrograph, different magnifications (2 a–d, see scale).

Fig. 3. Ground section through tooth No. 45 of an individual (W-2261) from Wandersleben. Microscopic view, 25fold magnification. Polarized light using an additional quartz.

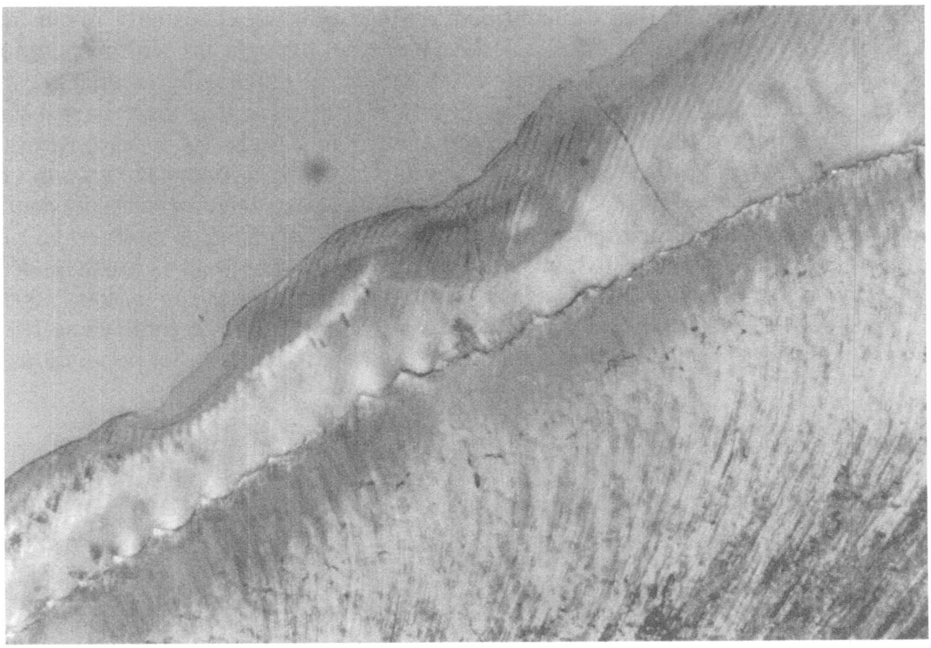

Fig. 4. Ground section through tooth No. 33 of an individual (Ai-23) from Aiterhofen. Microscopic view, 25fold magnification. Polarized light using an additional quartz.

a) Macroscopic examinations were carried out by Schour (1936), Sarnat and Schour (1941), Gustafson (1959), Wells (1967) and Schultz (1988).
b) Scanning electron microscopy was used by Whittaker and Richards (1978), Rose (1979), Paulson et al. (1984), Marks and Rose (1985), Marks (1988), Suckling et al. (1989), Marks (1992).
c) Histological investigations were carried out by Meyer (1958), Kostlan and Plackocva (1962), Marks (1988), Suga (1992), Goodman et al. (1992b).

Chemical or Biochemical Analysis

In the field of chemical or biochemical research of enamel hypoplasias, not much has been done up to now. Interesting results were presented by Gustafson and Gustafson (1967), Nikiforuk and Fraser (1981), and Warshawsky (1985).

Methods for the Evaluation of the Age of Onset of Transverse Linear Enamel Hypoplasias (LEH)

There are three common methods for the evaluation of the age of onset of transverse linear enamel hypoplasias. Goodman et al. (1980) presented a chart method, Hodges and Wilkinson (1990) a sample-specific method, while Goodman et al. (1987) describe a tooth-specific method.

A critical overview of these methods was given by Berti and Mahaney (1992). The authors point out that the confidence range (CI-index), usually given by investigators who examine transverse linear enamel hypoplasias with respect to the age of onset (cf. e.g. Moorrees et al. 1963), is much greater than these investigators reported. Berti and Mahaney (1992) confirm that there is still no reliable method to determine all growth factors (e.g. onset of amelogenesis, velocity of growth or different sizes of crown) that might influence the growth of the teeth and, thereby, the age at which the dental enamel defect was established. Sciulli (1992) introduced a new method for the determination of the age of occurrence of linear enamel hypoplasias. His method was based on a growth model for deciduous molar teeth from skeletal samples of native American infants (3000 B.P. to 400 B.P.). The author reported that his method was reliable for that specific population, but he conceded that it should be used with care for other populations.

Etiology

There are several reasons for the formation of enamel hypoplasias.

Nutrition

Jaffe et al. (1973) considered hypocalcemia to be the most important damaging agent. Duray (1990) found a strong relationship between hypocalcification and caries susceptibility in the study of the deciduous dentition of a skeletal population from the Libben Site (Ohio). Noren (1984) found that neonatal hypocalcemia correlates with the neonatal line. Brüning and Schwalbe (1913), Sweeney et al. (1971), Rose et al. (1985), and Lukacs (1989) reported on general malnutrition, but particularly rickets, as the causative factor of such enamel disturbances, while Angel (1954) and Meyer (1958) considered only rickets to be the most important cause. Probably not only severe general malnutrition of the newborn, but also birth stress in its widest sense leads to the occurrence of the neonatal line.

Bier-Katz (1980) pointed out the strong correlation between the occurrence of enamel hypoplasias and severe intestinal problems in young children. Goodman et al. (1984), Blakely and Armelagos (1985) and Rose (1977) reported that most of the enamel defects were formed between the ages of two and four. This might be a result of the physiological stress caused by the dietary shift of weaning.

Infectious Diseases

Kreshover and Clough (1953) and Meyer (1958) also discussed infectious diseases as a factor in the origin of enamel hypoplasias. Nowadays, this is not thought to be very probable. Of course, the growth of the enamel crown is retarded when an infant is seriously ill and the organism puts all its strength and energy into the healing process. Thus, a similar situation could be possible as is seen in the origin of the Harris lines. Kühl (1992) observed a co-occurrence of Harris lines and enamel hypoplasia in the young children of prehistoric cremations of Schleswig-Holstein, North Germany. However, no "one-to-one association" was found.

In the literature, several observations look relatively contradictory. Cook and Buikstra (1979) found a strong correlation between prenatal dental defects and bony evidence of anemia and infectious diseases in their study of two prehistoric populations from the Lower Illinois Valley. Also Rose et al. (1978) saw a significant correlation between hypoplastic defects and other indicators of physiological stress, such as mortality or bony evidence of infectious diseases, while Sarnat and Schour (1941) and El-Najjar et al. (1978) did not observe any correlation between the occurrence of enamel hypoplasias and infectious diseases (e.g. measles) during clinical research involving children.

Trauma

Marcsik and Kocsis (1992) described the occurrence of enamel hypoplasias in their report on the health of prehistoric and historic populations from Hungary. The authors mentioned a case of a defect which apparently originated directly from a traumatic injury. Probably, the frequency of this noxa is extremely low.

Genetic Causes or Deformities

Genetic causes of enamel hypoplasias have rarely been described (e.g. McMillan and Kashgarian 1961; Schulze 1961). As this factor has no great influence on the frequency of enamel hypoplasias in Europe, it should not be discussed in detail. An interesting study is presented by Berger (1992). He described hereditary enamel hypoplasias, probably amelogenesis imperfecta, in a Roman population from Regensburg/Harting, Germany.

Toxic Causes

Enamel deformities due to external toxins are more frequently seen than recently thought. Suckling et al. (1992) and Schumacher et al. (1990) reported on the characteristic mottled enamel caused by fluoride intoxication. However, various enamel lesions can also be due to fluorosis. In contrast to the other causes of enamel hypoplasias described briefly above, relatively little is known to anthropologists about alterations due to fluorosis. Unfortunately, anthropologists are, as a rule, not aware of these alterations. However, the morphology and the development of enamel surface hypoplasias in fluorosed dental enamel can be seen in the teeth of deer. Additionally, fluorosis seems to play a more important role in prehistory than many anthropologists are aware of.

Fluorosed dental enamel of mammals (e.g. deer) is characterized by a wide spectrum of developmental changes. These include subsurface hypomineralization of varying intensity and extent resulting from an impairment of the maturation phase of amelogenesis (Kierdorf et al. 1993, 1995). As a consequence, fluorosed enamel of deer appears opaque and is posteruptively stained. Furthermore, the teeth exhibit a loss of the enamel ridges on their occlusal or incisal surfaces, increased wear and posteruptive enamel surface lesions due to mechanical stress acting upon the hypomineralized tissue (Kierdorf 1988; Kierdorf and Kierdorf 1989; Kierdorf et al. 1993, 1996).

In addition, an enhancement of the incremental pattern due to the presence of alternating bands with highly varying mineral content is observed in more severely fluorosed teeth, denoting fluoride disturbance during the phase of enamel matrix secretion. Occurrence of enamel surface lesions of developmental origin, i.e., of enamel surface hypoplasias is also observed in the teeth (Kierdorf and Kierdorf 1989; Kierdorf et al. 1993, 1994, 1996). The shape of these hypoplasias varies considerably, both between different fluorosed teeth as well as within individual specimens (Figs. 5 and 6). Thus, rather frequently, both shallow and extended and very narrow and deep hypoplastic lesions can be observed in a single tooth (Fig. 6). Whereas the latter defects denote a severe impact on the ameloblasts during early enamel matrix formation, the former clearly result from a disturbance during a later phase of the secretory stage of amelogenesis.

In both the shallow and the deep hypoplasias, the transition zone between the lesions and the surrounding full thickness enamel is characterized by smooth, rounded walls exhibiting numerous Tomes' process pits (Figs. 6 and 7). In contrast to their margins, the bases of the hypoplasias exhibit a more irregularly structured surface (Fig. 7).

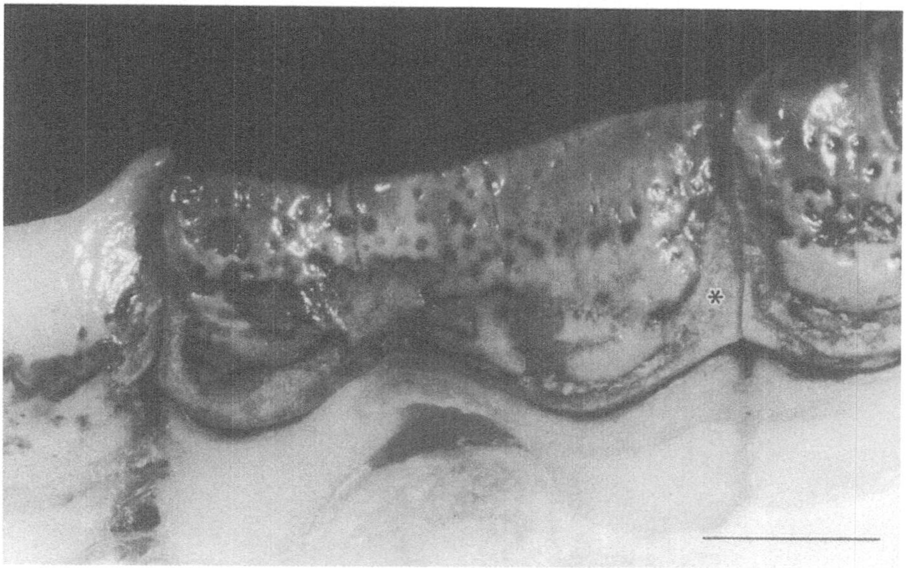

Fig. 5. Lingual enamel surfaces of P3 (right), P4 (center) and M1 (left) in the fluorosed mandible of a roe deer (*Capreolus capreolus*). Note numerous hypoplastic lesions of different size scattered over the surface of the premolars. Opaque and stained enamel in the premolars and normal enamel appearance of the M1. Asterisk = dental calculus, bar = 0.3 cm.

Surface lesions of posteruptive origin are sharply demarcated against the surrounding enamel, thus appearing as "punched out" areas with exposure of the underlying porous subsurface enamel (Kierdorf and Kierdorf 1989; Kierdorf et al. 1996). Due to these characteristic morphological differences, enamel surface lesions of fluorosed deer teeth can, using scanning-electron microscopy, with certainty be diagnosed as being either developmental or posteruptive in origin.

As is evident from Figs. 5 and 6, the spatial distribution of the enamel hypoplasias in the teeth is quite irregular. In general, however, the narrow and deep lesions are more frequently found in the cervical areas of the tooth crowns. Sometimes, these hypoplasias are aligned in more or less horizontal rows (Fig. 6), but they can also be found scattered over the entire tooth surface in an irregular pattern (Fig. 5).

Microradiographs of longitudinal sections through fluorosed enamel reveal a distinct bending of the incremental lines in accordance with the outlines of the hypoplastic lesions (Fig. 8). This again allows for the distinction of a developmental course as opposed to a posteruptive enamel surface defect, since in the latter the path of the Retzius lines does not follow the contour of the lesions (Kierdorf et al. 1996). Fig. 9 shows a low power scanning electron micrograph of an etched section through the center of a deep hypoplastic lesion in the cervical enamel of a severely fluorosed red deer premolar. It should be noted that the lesion is not symmetrical in shape, but that its cervical border (to the right) is much steeper than the coronal one. Internal to the base of the hypoplasia, a

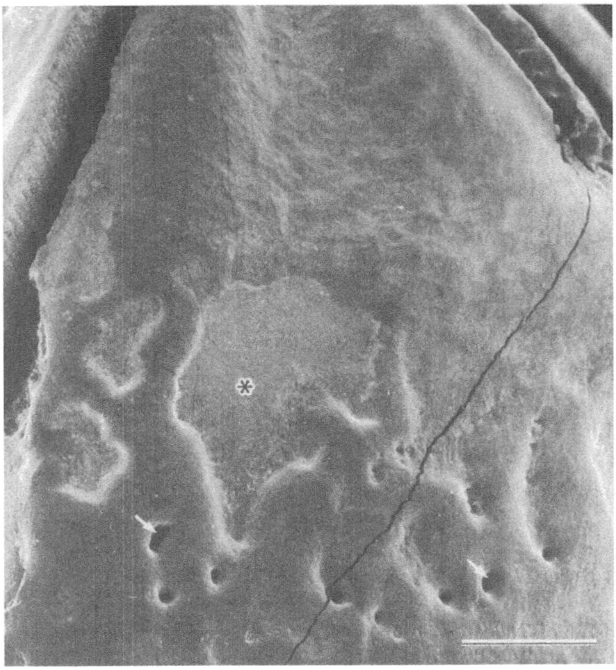

Fig. 6. Buccal enamel surface of a freshly erupted, fluorosed roe deer P4 exhibiting shallow and extended (asterisk) as well as more cervically located narrow and deep hypoplasias (arrows). Scanning micrograph, bar = 1 mm.

grossly accentuated incremental line is discernible, denoting the position of the developing enamel surface at the time when the ameloblasts were exposed to a severe (fluoride-induced) impact.

The shape of the lesion indicates that the reaction of the ameloblasts to a fluoride-induced was of varying nature and clearly related to the stage of their secretory activity. Thus, cells in an early phase of matrix production apparently reacted most intensely. Due to the very drastic reduction in secretion by these ameloblasts, only a very thin enamel layer was formed between the incremental line and the base of the hypoplastic lesion. As can be judged from the amount of enamel formed externally to the grossly enhanced incremental line, the reaction of the cells located further coronally (being in later stages of their secretory activity) was gradually less pronounced in a coronal direction until enamel of normal thickness was eventually formed coronally to the hypoplasia. The most cervically located (presecretory) ameloblasts were seemingly unaffected by the insult, whereas coronally adjacent cells, which had just entered the secretory phase were affected to a lesser extent than those in a slightly more advanced stage of matrix formation.

In addition to a reduced thickness, a conspicuous structural change is also seen in the enamel forming the walls of the deep hypoplasias. As is shown in Fig. 10, the enamel has lost its typical rod/interrod structure and instead consists of stacked

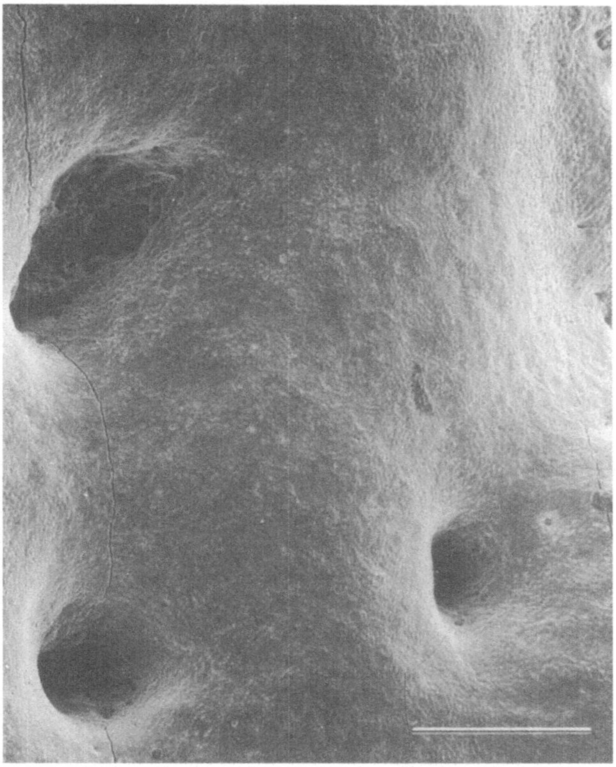

Fig. 7. Detail of the enamel surface shown in Fig. 6. Note Tomes process pits on the margins of the hypoplasias and more irregularly structured enamel at the bottom of the lesions. Scanning micrograph, bar = 200 µm.

thin layers of crystals apparently arranged in parallel. We hypothesize that this structural change is due to the fact that the secretory ameloblasts have lost the distal (rod forming) portions of their Tomes' processes (Warshawsky et al. 1981). Thus, the subsequently formed enamel crystals were all laid down with their long axes perpendicular to the now more or less flat secretory surfaces at the distal poles of the processes. This type of crystal arrangement resembles that seen in the innermost enamel secreted by the ameloblasts prior to the full establishment of the rod growth regions (Boyde 1967; Warshawsky et al. 1981; Kierdorf et al. 1991).

Systemic Diseases or Developmental Disturbances

Rarely, systemic diseases can stimulate developmental disturbances. Thus, the growth of enamel structures during early childhood can be affected (Kreshover 1960; Levine and Keen 1974; Shklar and McCarthy 1976; Noren 1983). Today, these causes are extremely rare.

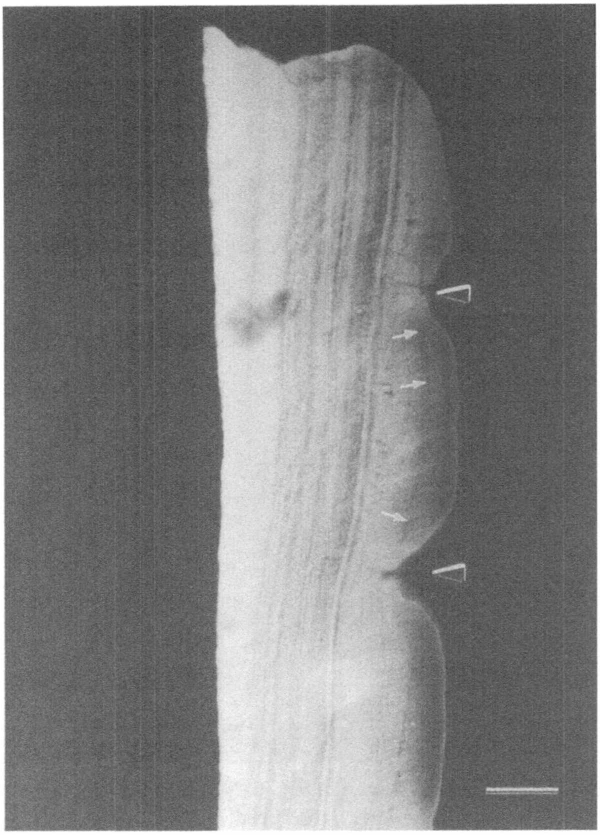

Fig. 8. Microradiograph of a longitudinal section through the buccal enamel of a fluorosed red deer (*Cervus elaphus*) M2. Note pronounced enhancement of the incremental pattern and presence of surface hypoplasias (arrowheads). The incremental lines exhibit a distinct bending (arrows) according to the outlines of the lesions, bar = 250 μm.

Epidemiology

In the New World, Storey (1992) investigated deciduous teeth of a skeletal population from the Classic Maya civilization of Copan (Honduras). The value of enamel defects (including hypoplasias and opacities) as nonspecific stress indicators for infant health was demonstrated. Enamel opacities are a result of disturbances in amelogenesis or mineralization resulting in weak and defective enamel that leads to a higher susceptibility to caries. When the caries has destroyed much of the crown, the bacteria of the oral cavity and other pathogenic microorganisms can pass through the hole and dental abscesses or bacterial infection may occur. Hematogenous dissemination and finally lethal diseases (sepsis) may result. Therefore, hypocalcification and caries are important morbidity indicators in child health. More than 70% of the individuals from this population were affected by enamel defects. Whittington (1992) found that the low status

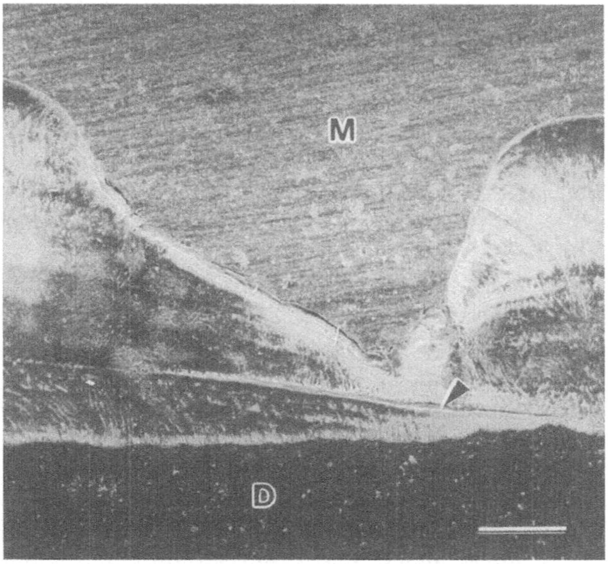

Fig. 9. Etched longitudinal section (34% phosphoric acid for 15 s) through a deep surface hypoplasia in the buccal enamel of a fluorosed red deer P4. Note grossly accentuated incremental line (arrowhead) internal to the base of the lesion. Cervical to the right, D = dentine, M = embedding medium. Scanning micrograph, bar = 200 μm.

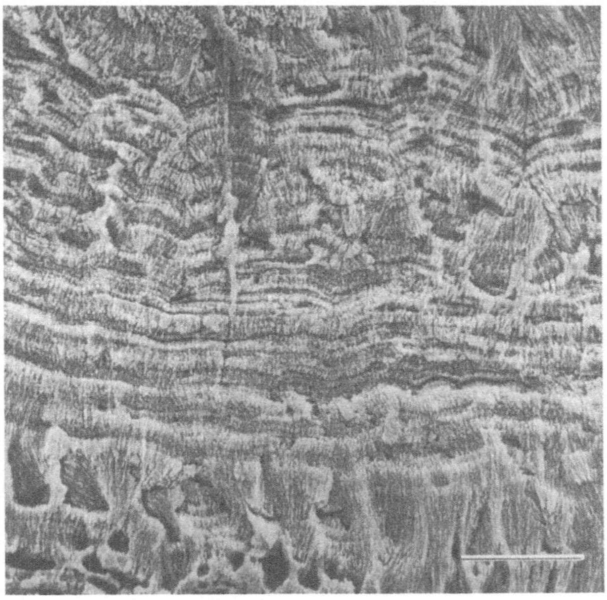

Fig. 10. Higher magnification of the enamel underlying the base of the hypoplastic lesion shown in Figure 9. Note presence of stacked thin layers of aprismatic enamel. Scanning micrograph, bar = 20 μm.

prehispanic Maya population from Copan (AD 400–700 and AD 700–800) suffered from enamel hypoplasias. All 19 individuals (= 100%) were affected by these alterations.

McHenry and Schulz (1976) stated that there is no significant correlation between enamel defects and Harris lines in their skeletal sample of prehistoric California Indians. They observed ten times more Harris lines than enamel hypoplasias. Therefore, the authors presume that mild infections may lead to the development of a Harris line, whereas only severe diseases are responsible for the formation of enamel defects.

Goodman (1991) reported on the populations from Dickson Mounds which was inhabited over a span of 400 years (AD 950–1350) and showed three different culture horizons (Late Woodland [LW], Mississippian Acculturated Late Woodland [MALW] and Middle Mississippian [MM]). There was an increase in the number of affected individuals with enamel hypoplasias over the time span. The author argues that the significant increase from 45% for the LW period to 60% for the MALW period and up to 80% for the MM period is a result of settlement (agriculture) and a higher dependence on corn. In case of a poor harvest, the people starved and especially pregnant women and young children suffered from infectious diseases and malnutrition. Goodman et al. (1992a) observed similar frequency of enamel hypoplasias in Anasazi Indians from The Black Mesa (cf. Martin et al. 1991). Almost 85% of the individuals were affected by enamel hypoplasias (one or more defects). The age of occurrence showed a peak at 2–4 years.

In the Old World, the situation with respect to the frequency of transverse linear enamel hypoplasias is a little different. Carli-Thiele (1994, 1996) examined two child populations from the Neolithic Age in Germany. There was a co-occurrence of Harris lines and transverse linear enamel hypoplasias, but, no one-to-one association was found. The number of Harris lines was, in all cases, larger than the number of the enamel defects. Eighty percent of the individuals in the population from Bavaria (Aiterhofen), and 54% of the individuals in the population from Thuringia (Wandersleben) were affected by enamel hypoplasias. The age of occurrence shows a peak at three years as do most of the Harris lines.

In the Iron Age population from Sarai Khola (Schultz et al. 1996), transverse linear enamel hypoplasias were observed in 16 out of 26 individuals (about 62%). Neglecting very slightly formed lines the frequency is only 25%. Most of the enamel defects were seen between the ages of three and five years. Probably the weaning time was relatively late in this population, which shows criteria of a social lower class. In the group of adults, 60% of males and 78% of females suffered from transverse linear enamel hypoplasias. In infants from Sarai Khola, these enamel disturbances were found in one out of seven children (about 14%).

Kreutz (1996) examined a large population of subadults from the cemetery of Straubing in Germany. The skeletons date back to the Early Middle Ages and represent a typical Bavarian group. Transverse linear enamel hypoplasias were found in 72 out of 170 individuals (about 42%). Hypoplasias were found only in the permanent teeth. Most of the enamel defects were formed between the ages of two and four (about 51%). This frequency is probably connected with the result of the physiological stress caused by the dietary shift of weaning. Pitting of enamel was seen in 48 out of 170 individuals (about 28%).

The Slav population from Starigard/Oldenburg (Northern Germany, cf. Schultz and Teegen 1996) represents an upper class population from the Early-High Middle Ages. Transverse linear enamel hypoplasias were observed in 39 out of 87 individuals (about 45%). In two out of 87 cases the lesions were also found in deciduous teeth (about 2%). Most of the enamel defects were formed between the ages of two and four years. The calculation for boys yields an age between three and five years. Perhaps boys were breast-fed longer than girls.

The examples show that we already know much about the life of prehistoric and historic populations, but, indeed, we need much more information to reconstruct ancient life conditions.

Conclusions

There are several causes for the origin of enamel hypoplasias in past populations. The most important agents are malnutrition, infectious diseases, trauma, genetic causes, toxic causes, systemic diseases, and developmental disturbances. Up to now, only a few prehistoric and historic populations have been reliably examined. With respect to the epidemiology of enamel hypoplasias, we only have some information on prehistoric populations from the New World, from peoples such as Maya from Honduras, California Indians, Anasazi, and Indians from the Mississippian Woodland. From the Old World the information is also relatively poor, but apparently somewhat better than in the New World. Attention has mainly been payed to Middle Europe, the Near and the Middle East. The results show that sometimes social factors are decisive for the occurrence of enamel hypoplasias. In most cases, enamel hypoplasias, particularly the transverse linear hypoplasias, are typical stress markers which could help in reconstructing ancient life conditions. Further investigations will hopefully fill the gap in our knowledge on the epidemiology of enamel hypoplasias.

References

Angel JL (1954) Human biology, health and history in Greece from first settlement until now. Yb Am Philos Soc: 168–174.

Berger T (1992) Amelogenesis imperfecta – Eine differentialdiagnostische Betrachtung an kaiserzeitlichen Skelettresten aus Regensburg/Harting. Med Diss, Tübingen.

Berti PR, Mahaney MC (1992) Quantification of the confidence interval of linear enamel hypoplasia chronologies. In: Capasso LL, Goodman AH (eds) Recent Contributions to the Study of Enamel Developmental Defects, J Paleopath Monogr Publ 2, Chieti (Italy), p. 19–30.

Bier-Katz G (1980) Im Schmelz spiegeln sich vielfältige Ernährungsstörungen. Zahnärztl Mitt 70, 757–759.

Blakey ML, Armelagos GJ (1985) Decidous enamel defects in prehistoric Americans from Dickson mounds: prenatal and postnatal stress. Am J Phys Anthrop 66: 371–380.

Boyde A (1967) The development of enamel structure. Proc Soc Med 60: 923–928.

Boyde A (1989) Enamel. In: Berkowitz BKB, Boyde A, Frank RM, Hohling HJ, Moxham BJ, Nalbandian J, Tonge CH (eds) Teeth. Springer, Berlin, p. 309–473.

Brüning H, Schwalbe E (1913) Handbuch der allgemeinen Pathologie und der pathologischen Anatomie des Kindesalters, Bd. 2: Erste Abtl. Bergmann, Wiesbaden.

Capasso L, Di Tota G (1992) Foramen caecum molare as a developmental defect of the enamel. In: Capasso LL, Goodman AH (eds) Recent Contributions to the Study of Enamel Developmental Defects. J Paleopath. Monogr Publ 2, Chieti (Italy), p. 91–106.
Carli-Thiele P (1994) Ursachen und Häufigkeiten von Mangelerkrankungen bei den Angehörigen früher Ackerbauernkulturen Mitteleuropas und ausgewählten präkolumbischen Populationen des nordamerikanischen Südwestens. Nat Diss, Braunschweig.
Carli-Thiele P (1996) Vestiges of deficiency diseases in stone age child skeletons. In: Schultz M (ed) Advances in Paleopathology and Osteoarchaeology, Vol. I. Goltze, Göttingen.
Clarkson J (1989) Review of terminology, classification, and indices of developmental defects of enamel. Adv Dent Res 3: 104–109.
Condon K, Rose JC (1992) Intertooth and intratooth variability in the occurrence of developmental enamel defects. In: Capasso LL, Goodman AH (eds) Recent Contributions to the Study of Enamel Developmental Defects. J Paleopath, Monogr Publ 2, Chieti (Italy), p. 61–78.
Cook DC, Buikstra JE (1979) Health and differential survival in prehistoric populations: Prenatal dental defects. Am J Phys Anthrop 51: 649–664.
Duray SM (1990) Decidous enamel defects and caries susceptibility in a prehistoric Ohio population. Am J Phys Anthrop 81: 27– 34.
Eckhardt RB, Protsch A, Protsch von Zieten RR (1992) Vertical enamel hypoplasias in a free-living population of Liberian chimpanzees: Variations in expression and frequency of incidence. In: Capasso LL, Goodman AH (eds) Recent Contributions to the Study of Enamel Developmental Defects. J Paleopath, Monogr Publ 2, Chieti (Italy), p. 107–114.
El-Najjar MY, Desanti MV, Ozebek L (1978) Prevalence and possible etiology of dental enamel hypoplasia. Am J Phys Anthrop 48: 185–192.
Fédération Dentaire Internationale (1982) An epidemiological index of developmental defects of dental enamel (DDE Index). Int Dent J 32: 159–167.
Garn SM,Osborne RH, McCabe KD (1979) The effect of prenatal factors on crown dimensions. Am J Phys Anthrop 51: 665–678.
Goodman AH (1991) Stress, adaptation, and enamel developmental defects. In: Ortner DJ, Aufderheide AC (eds) Human Paleopathology. Current Synthesis and Future Options. Smithsonian Inst Press, Washington, p. 280–287.
Goodman AH, Armelagos GJ, Rose JC (1980) Enamel hypoplasias as indicators of stress in three prehistoric populations from Illinois. Hum Biol 52: 515– 528.
Goodman AH, Armelagos GJ, Rose JC(1984) The chronological distribution of enamel hypoplasias from prehistoric Dickson mounds populations. Am J Phys Anthrop 65: 259–266.
Goodman AH, Allen LH, Hernandez GP, Amador A, Arriola LV, Chavez A, Pelto GH (1987) Prevalence and age at development of enamel hypoplasias in Mexican children. Am J Phys Anthrop 72: 7–19.
Goodman AH, Martin DL, Klein CP, Peele MS, Cruse NA, McEwen LR, Saeed A, Robinson BM (1992a) Cluster bands, Wilson bands and pit patches: Histological and enamel surface indicators of stress in the black Mesa Anasazi population. In: Capasso LL, Goodman AH (eds) Recent Contributions to the Study of Enamel Developmental Defects. J Paleopath, Monogr Publ 2, Chieti (Italy), p. 115–130.
Goodman AH, Pelto GH, Allen LH, Chavez A (1992b) Socioeconomic and anthropometric correlates of linear enamel hypoplasia in children from Solis, Mexico. In: Capasso LL, Goodman AH (eds) Recent Contributions to the Study of Enamel Developmental Defects. J Paleopath, Monogr Publ 2, Chieti (Italy), p. 373–380.
Gustafson A (1959) A morphological investigation of certain variations in the structure and mineralization of human dental enamel. Odont Tidskr 67: 361–472.
Gustafson G, Gustafson AG (1967) Microanatomy and histochemistry of enamel. In: Miles AEW (ed) Structure and Chemical Organization of Teeth. Vol. 7. Academic Press, New York, p. 75–133.

Hillson S (1990) Teeth. Cambridge Univ Press, New York.
Hodges DC,Wilkinson RG (1990) Effect of tooth size on the ageing and chronological distribution of enamel hypoplastic defects. Am J Hum Biol 2: 553–560.
Jaffe EC, Stimmler L, Osborne JA (1973) Enamel defects associated with neonatal symptomatic hypocalcemia. Proc Brit Paedodont Soc 3: 25–31.
Kierdorf U (1988) Untersuchungen zum Nachweis immissionsbedingter chronischer Fluoridintoxikation beim Reh (*Capreolus capreolus* L.). Z Jagdwiss 34: 192–204.
Kierdorf U, Kierdorf H (1989) A scanning electron microscopic study on surface lesions in fluorosed enamel of roe deer (*Capreolus capreolus* L.). Veterinary Path 26: 209–215.
Kierdorf H, Kierdorf U, Hommelsheim S (1991) Scanning electron microscopic observations on the development and structure of tooth enamel in Cervidae (*Mammalia: Ruminantia*). Anatomia, Histologia, Embryologia 20: 237–252.
Kierdorf U, Kierdorf H, Fejerskov O (1993) Fluoride induced developmental changes in enamel and dentine of European roe deer (*Capreolus capreolus* L.) as a result of environmental pollution. Arch Oral Biol 38: 1071–1081.
Kierdorf U, Schultz M, Kierdorf H (1994) Licht- und rasterelektronenmikroskopische Untersuchungen an Schmelzhypoplasien fluorotischer Rehzähne. Z Jagdwiss 40: 175–184.
Kierdorf U, Kierdorf H, Sedlacek F, Fejerskov O (1996) Structural changes in fluorosed dental enamel of red deer (*Cervus elaphus* L.) from a region with severe environmental pollution by fluorides. J Anat, 188: 183-219.
Kostlan J, Plackova A (1962) The histological investigation of the developmental hypomineralized areas of the enamel and their comparison with the carious lesion. Arch Oral Biol 7: 317–326.
Kreshover SJ (1960) Metabolic disturbances in tooth formation. Ann NY Acad Sci 85: 161–167.
Kreshover SJ, Clough O (1953) Prenatal influences on tooth development: Artificially induced fever in rats. J Dent Res 32: 565–577.
Kreutz K (1996) Ätiologie und Epidemiologie von Erkrankungen des Kindesalters bei der bajuwarischen Population von Straubing (Niederbayern). Nat Diss, Gießen.
Kühl I (1992) Six cases of enamel hypoplasia in prehistoric cremations of Schleswig-Holstein, North Germany. In: Capasso LL, Goodman AH (eds) Recent Contributions to the Study of Enamel Developmental Defects. J Paleopath, Monogr Publ 2, Chieti (Italy), p. 239–254.
Levine RS, Keen JH (1974) Neonatal enamel hypoplasia in association with symptomatic neonatal hypocalcemia. Br Dent J 137: 429–433.
Lukacs JR (1989) Dental paleopathology: Methods for reconstructing dietary patterns. In: Işcan MY, Kennedy KAR (eds) Reconstruction of Life from the Skeleton. Liss, New York, p. 261–286.
Marcsik A, Kocsis GS (1992) Occurrence of enamel hypoplasia in prehistoric and historic skeletal samples (Hungary). In: Capasso LL, Goodman AH (eds) Recent Contributions to the Study of Enamel Developmental Defects. J Paleopath, Monogr Publ 2, Chieti (Italy), p. 219–230.
Marks MK (1988) Enamel hypoplasia and hypocalcification: Correlation of defects identified by light and scanning electron microscopy. Am J Phys Anthrop 75: 246 (Abstract).
Marks MK (1992) Developmental Dental Enamel Defects: A SEM Analysis. In: Capasso LL, Goodman AH (eds) Recent Contributions to the Study of Enamel Developmental Defects. J Paleopath, Monogr Publ 2, Chieti (Italy), p. 79–90.
Marks MK, Rose JC (1985) Scanning electron microscopy of Wilson bands and enamel hypoplasias. Am J Phys Anthrop 66: 202.
Martin DL, Goodman AH, Armelagos GJ, Magennis AL (1991) Black Mesa Anasazi Health: Reconstructing Life from Patterns of Death and Disease. Southern Illinois Univ at Carbondale, Center for Archaeol Investigations, Occasional Papers 14.

Massler M, Schour I, Poncher HG (1941) Developmental pattern of the child as reflected in the calcification patterns of the teeth. Am J Dis Child 62: 33–67.
McHenry HM, Schulz PD (1976) The association between Harris lines and enamel hypoplasia in prehistoric California Indians. Am J Phys Anthrop 44: 507–511.
McMillan RS, Kashgarian M (1961) Relation of human abnormalities of structure and function to abnormalities of the dentition, I. Relation of hypoplasia of enamel to cerebral and ocular disorders. J Am Dent Assoc 63: 38–47.
Meyer W (1958) Die Zahn-, Mund- und Kieferheilkunde. Urban & Schwarzenberg, München.
Moorrees CF, Fanning EA, Hunt E (1963) Age variation of formation stages for ten permanent teeth. J Dent Res 42: 1490–1502.
Nikiforuk G, Fraser D (1981) The etiology of enamel hypoplasia: A unifying concept. J Pediatr 98: 888–893.
Noren JG (1983) Enamel structure in deciduous teeth from low birthweight infants. In: Noren JG (ed) Human Deciduous Enamel in perintal disorders. Dept of Pedodontics, Göteborg, p. 1–15.
Noren JG (1984) Microscopic study of enamel defects in deciduous teeth of infants of diabetic mothers. Acta Odont Scand 42: 153–156.
Paulson R, Schroeder K, Zacherl W, Sciulli P (1984) Scanning electron microscopy of cusptip pits in developing and mature human dentitions. Arch Oral Biol 29: 117–125.
Rose JC (1977) Defective enamel histology of prehistoric teeth from Illinois. Am J Phys Anthrop 46: 439–446.
Rose JC (1979) Morphological variations of enamel prisms within abnormal striae of Retzius. Hum Biol 51: 139–151.
Rose JC, Armelagos GJ, Lallo JW (1978) Histological enamel indicator of childhood stress in prehistoric skeletal samples. Am J Phys Anthrop 49: 511–516.
Rose JC, Condon KW, Goodman AH (1985) Diet and dentition: Developmental disturbances. In: Gilbert RI, Mielke JH (eds) The Analysis of Prehistoric Diets. Academic Press, New York, p. 281–306.
Sarnat GB, Schour I (1941) Enamel hypoplasia (chronologic enamel aplasia) in relation to systemic disease: A chronologic morphologic and etiologic classification. J Am Dent Assoc 28/29: 1989–2000/67–74.
Schour I (1936) Neonatal line in enamel and dentin of human deciduous teeth and first permanent molar. J Am Dent Assoc 23: 1946–1955.
Schultz M (1988) Paläopathologische Diagnostik. In: Knussmann R (ed) Handbuch der vergleichenden Biologie des Menschen, Band 1: Methoden. Fischer, Stuttgart, p. 480–496.
Schultz M, Lukacs JR, Schwartz P, Hemphill BE (1996) Ergebnisse paläopathologischer Untersuchungen an den eisenzeitlichen Skeleten von Sarai Khola (Pakistan). Homo 47: 85–110.
Schultz M, Teegen WR (1996) Starigard/Oldenburg. Hauptburg der Slawen in Wagrien. VI Die menschlichen Skelettreste. Offa-Bücher (im Druck).
Schulz PD, McHenry HM (1975) Age distribution of enamel hypoplasia in prehistoric California Indians. J Dent Res 54: 913.
Schulze C (1961) Anomalien, Mißbildungen und Krankheiten von Zähnen, Mund und Kiefer. In: Becker PE (Hrsg) Kurzes Lehrbuch der Humangenetik. Thieme, Stuttgart.
Schumacher GH, Schmidt H, Börnig H, Richter W (1990) Anatomie und Biochemie der Zähne. Volk und Gesundheit, Berlin.
Sciulli PW (1992) Estimating age of occurrence of enamel defects in deciduous teeth. In: Capasso LL, Goodman AH (eds) Recent Contributions to the Study of Enamel Developmental Defects. J Paleopath, Monogr Publ 2, Chieti (Italy), p. 31–40.
Shklar G, McCarthy P (1976) The Oral Manifestations of Systemic Disease. Butterworths, Boston.

Skinner MF, Goodman AH (1992) Anthropological uses of developmental defects of enamel. In: Saunders SR, Katzenberg A (eds) Skeletal Biology of Past Peoples: Advances in Research Methods. Wiley-Liss, New York, p. 157–178.
Small BW, Murray JJ (1978) Enamel opacities: Prevalence, classifications and aetiological considerations. J Dent 6: 33–42.
Storey R (1992) Patterns of susceptibility to dental enamel defects in the deciduous dentition of a Precolumbian skeletal population. In: Capasso LL, Goodman AH (eds) Recent Contributions to the Study of Enamel Developmental Defects. J Paleopath, Monogr Publ 2, Chieti (Italy), p. 171–184.
Suckling GW, Elliot DC, Thurley DC (1986) The macroscopic appearance and associated histological changes in the enamel organ of hypoplastic lesions of sheep incisor teeth resulting from induced parasitism. Arch Oral Biol 31: 427–439.
Suckling GW, Nelson DGA, Patel MJ (1989) Macroscopic and scanning electron appearances and hardness values of developmental defects in human permanent tooth enamel. Adv Dent Res 3: 219–233.
Suckling GW, Coote GE, Cutress TW (1992) Proton microprobe determination of profiles of fluoride levels in the enamel and dentine of human teeth. In: Capasso LL, Goodman AH (eds) Recent Contributions to the Study of Enamel Developmental Defects. J Paleopath, Monogr Publ 2, Chieti (Italy), p. 391–400.
Suga S (1992) Hypoplasia and hypomineralization of tooth enamel. In: Capasso LL, Goodman AH (eds) Recent Contributions to the Study of Enamel Developmental Defects. J Paleopath, Monogr Publ 2, Chieti (Italy), p. 269–292.
Sweeney EA, Saffir JA, De Leon R (1971) Linear enamel hypoplasia of deciduous incisor teeth in malnourished children. Am J Clin Nutr 24: 29–31.
Warshawsky H (1985) Ultrastructural studies on amelogenesis. In: Butler WT (ed) The Chemistry and Biology of Mineralized Teeth. Ebsco Media, Birmingham, p. 33–44.
Warshawsky H, Josephsen K, Thylstrup A, Fejerskov O (1981) The development of enamel structure in rat incisors as compared to the teeth of monkey and man. Anat Rec 200: 371–399.
Wells C (1967) A new approach to paleopathology: Harris lines. In: Brothwell DR, Sandison AT (eds) Diseases in Antiquity. Thomas, Springfield, p. 390–404.
Whittaker DK, Richards D (1978) Scanning electron microscopy of the neonatal line in human enamel. Arch Oral Biol 23: 45–50.
Whittington SL (1992) Enamel hypoplasia in the low status Maya population of prehispanic Copan, Honduras. In: Capasso LL, Goodman AH (eds) Recent Contributions to the Study of Enamel Developmental Defects. J Paleopath, Monogr Publ 2, Chieti (Italy), p. 185–206.
Yeagar JA (1980) Enamel. In: Bhaskar SN (ed) Orban's Oral Histology and Embryology. Mosby, Toronto.

5 Nutrition and Human Behaviour

5.1 Diet and Nutrition in Prehistoric Central Europe

Hermann Prossinger and Christoph Willms

Introduction

Paleonutrition is of interest in its own right. However, anthropologists have two specific reasons for their interest in this discipline: First, the results of study and analysis by archeologists further elucidate the *conditio humana*; second, theories of human evolution – anatomical, ecological, cultural, cerebral, social – may help to put paleonutritional evidence in perspective.

Various scientific disciplines assist us in comprehending the history of food and nutrition. The special approach characteristic of prehistoric research results from the uniqueness of the nature of the source material. Consequently, the concrete, visible aspects of this research tend to be stressed by archeologists. While we have been selective in this presentation, it has been with due cause: we touch upon those aspects of research which tend to be controversial and those which relate to physical anthropology.

Archeological finds are obtained primarily from systematic excavations. Research into nutrition requires interdisciplinary excavation teams; such teams enable us to enhance our fragmentary, yet gradually improving perception of the nutritional situation in prehistory. The archeologist is able to identify arrowheads as hunting tools, earthenware vessels as pots, and campfire sites as either hearths or closed ovens. Furthermore, he/she can uncover plowshare tracks, excavate storage pits, as well as recognize grinding stones.

The problem of clearly demarcating evidence from inference remains. Whereas the archeologist – at least in the traditional, restricted sense of the word – can unequivocally uncover concrete materials, theoretical archeologists have a long tradition in struggling with all the possible inferences that can be drawn from the supposedly hard evidence (Clarke 1968). In short, archeological data is meaningless without interpretation – a statement most emphatically true for the discipline paleonutrition.

A classic case is the study of taphonomy. In their study of the activity of early hominids (*Australopithecines* and *Hominines*), archeologists have often had to realize that bone assemblages may not necessarily reflect the diet of the makers of these assemblages. Indeed, there is evidence that some of the bone assemblages were not of human origin (Hill 1978, 1984)

A similar problem, closer to the topic of this article, is the localization of where the horse was domesticated. Numerous sites contain skeletal remains that seem to indicate transition forms between wild horses and the domesticated horse (Brothwell and Brothwell 1969; Bökönyi 1974; Benecke 1994).

The food itself has survived only in the rarest of circumstances (Caselitz 1986). Extensive and varied scientific investigations are often necessary before one can determine its nature from the few remaining traces (Fig. 1). Specific archeological techniques, such as sieving and elutriation, can maximize the chances of recovery. From the remains, we infer the food incorporated into the diet. We must use nontraditional techniques to infer the diet and – more importantly – attempt to assess to what extent the diet met the nutritional needs. In particular, the nutritional needs of *Homo sapiens neanderthalensis* in pleistocene Europe differ markedly from those of *Homo erectus* in the African savannah.

Methods

Archeology

In many archeological excavations, the faunal remains (bones, teeth, fish vertebrae, and calcarous shells of snails, mussels, clams and other molluscs) constitute a major portion of the assemblage. Soil conditions and excavation methods determine the extent and quality of the subsequent archeozoological analyses. Faunal species can be identified from the bone and shell remains; their relative weight can be used to assess the relative abundance, because experience has shown that bone weight most reliably expresses the weight of the consumable meat (Brothwell and Brothwell 1969; van der Merwe 1992; Benecke 1994). In the case of bone assemblages, we can determine the sex and age of the slaughtered mammals and infer their primary use (not all mammals were a source of meat); partially charred bones have been found at campfire sites (Mania 1990). But not all animals were hunted solely for food: Birds could have been killed for their feathers and some mammals for their furs.

Fish vertebrae and mussel valves inform us of the season in which the animals died. Such information is important, because it enables archeologists to determine at what time of year an archeological site was inhabited. Other seasonal indicators may be the traces left by migrating birds or hibernating animals.

Palynological analysis enables us to reconstruct paleoenvironments (van Zeist 1955; Accorsi 1985; Höpfel et al. 1992; Frenzel 1994). Of great importance for the reconstruction of diets are the partially charred or intact macrofloral remains or their imprints in pottery shards. The constituents found in a sample of cereal grains can give us information: whether we are dealing with the inclusion of threshing remains, grains, spikelets and weeds, or with the fully processed storage crop. Statistical analysis can be used to determine the variation of plant species in settlements during various time horizons and

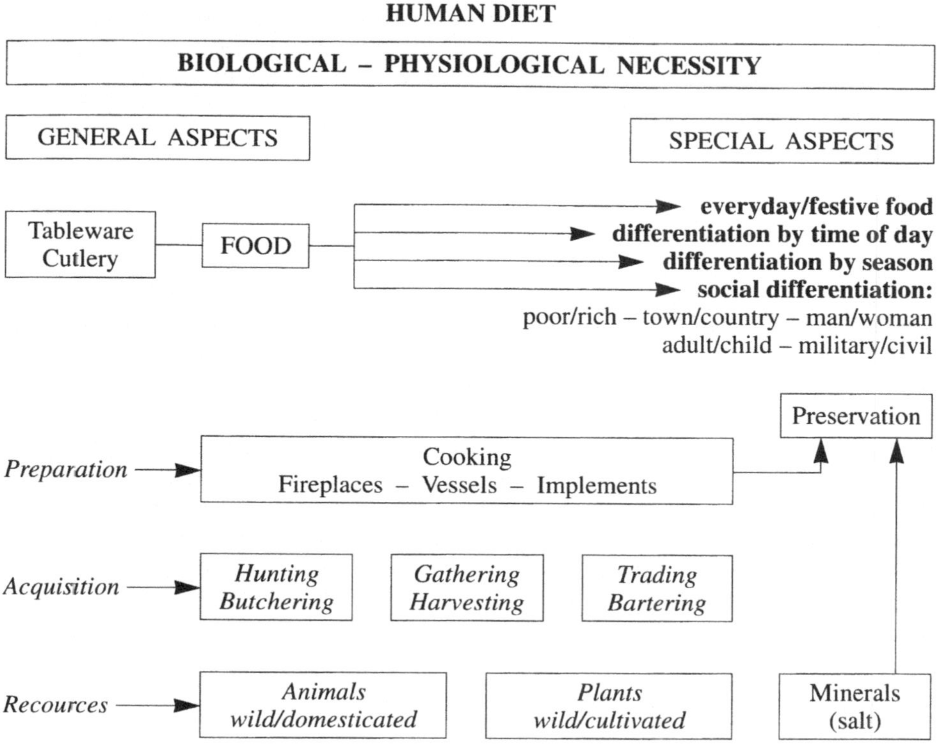

Fig. 1. The relations of research topics in paleonutrition research.

throughout various regions. For example, the transition from hunter-gatherer to agriculturalist societies is often characterized by a drastic reduction of the number of plant species: While a hunter-gatherer would incorporate up to 150 different plant species in the diet, agriculturalists could find a very few to be adequate (van der Merwe 1992). In Neolithic lakeside dwellings, on the other hand, we have evidence of up to two hundred plant taxa; of these nearly half are suitable as food or as medicine. Only in the most fortuitous cases can we furnish proof of their actual use (Willms 1991a). Usually, only shells, kernels or edible seeds have been preserved; there are no remnants of leaves or tubers, usable as legumes or salads, or mushrooms. As always, the archeological quality of the debris strongly influences the reliability of possible inferences. Intact food remains, such as dried apples, bread, etc., are very rare. Further results can be gained from the analysis of coprolites (dried faeces) and of the content of the alimentary canal of well-preserved bog bodies (Brothwell 1986). Middens are a treasure trove: some have been in use for several thousand years, permitting a statistical analysis of dietary shift. One such midden, Grotta del Moscerini in Italy, contains marine molluscs, bones and Mousterian tools and shows that the Neanderthals exploited both land and marine ecologies when given the chance (Jones et al. 1992). Not all recovered molluscan shells can be interpreted as food remains, however. A large number of sites in Eastern Austria, for example, contain shells which were used as ornaments (Simetsberger 1993). The relationship between such research and anthropology, the discipline studying the origin as well as the physical, social, and cultural development and behavior of humans is fundamental.

Paleopathology

The distinction between diet and nutrition cannot be overemphasized. The archeological faunal and floral remains supply evidence of a part of the diet. As not all foodstuffs leave traces, one must seek other approaches in order to assess the relationship between diet and nutrition.

Human skeletal remains can shed light on the nutritional status, because histological analysis allows the diagnosis of diseases and deficiency symptoms (e.g., anaemia, rickets, scurvy, and osteitis or osteomyelitis). Such deficiencies, when diagnosed, reflect the general nutritional status of the population (for examples from the Bronze Age, see Schultz and Teschler-Nicola 1989; Schultz 1990; Ziemann-Becker et al. 1991). Furthermore, there is a strong correlation between the condition of human dentition and feeding habits in a population, notably the extent of caries.

Isotopes

Human remains constitute an almost permanent repository for trace isotopes, such as carbon-13 (which is stable). The $^{13}C/^{12}C$ ratio retains the biogenic isotopic composition, especially in the structural carbonate within apatite crystals, for millions of years. From this ratio, researchers can reconstruct diets: either plants with a C3 photosynthesis pathway such as trees, shrubs, and temperate grasses or plants with a C4 photosynthesis pathway such as tropical grasses and sugar cane (Lee-Thorp and van der Merwe 1987; van der Merwe 1992). It should be noted in passing that, contrary to other reconstructions, isotope ratios suggested that *Paranthropus robustus* had a diet very similar to that of Homo (Lee-Thorp et al. 1994). Anthropologists should now review their evolutionary theory of enamel thickness relating to whether a hominoid was frugivorous, herbivorous or graminivorous (Spears and Crompton 1995).

We define

$$\delta^{13}C = (R_{sample}/R_{reference} - 1) \times 1000$$

where R is the $^{13}C/^{12}C$ ratio and $\delta^{13}C$ is expressed in per mille. A positive value for $\delta^{13}C$ means a higher $^{13}C/^{12}C$ ratio than the PDB international standard. Browsing herbivores have a $\delta^{13}C$ around –11 to –12 (Lee-Thorp et al. 1994).

The interpretation of $\delta^{13}C$ is dependent on extraneous circumstances, which can be unraveled by archeological evidence: Meat from wild game would be mixed, depending on whether browsers or grazers were hunted; gathered wild plant foods would be characterized by a C3 pathway (Lee-Thorp et al. 1989).

The ratio of nitrogen-15 to nitrogen-14 can also be used to reconstruct diet. Nitrogen-15 is enhanced at higher trophic levels. If we define

$$\delta^{15}N = (R_{sample}/R_{reference} - 1) \times 1000$$

then a large $\delta^{15}N$, therefore, implies a consumer at or near the end of the food chain.

Nonetheless, the situation is quite complicated, because legumes exhibit a decreased $\delta^{15}N$, and marine foodstuffs have an enhanced $\delta^{15}N$. A careful cross-referencing technique must be developed (Lee-Thorp et al. 1993). All other variables being equal, it can be said that enriched nitrogen-15 levels reflect a greater dependence on animal products.

Chemical Analysis

Bones are the repositories of trace elements, which permeate the environment and are acquired by living systems via the intake of foodstuffs.

Strontium in human bone can be a measure of the relative amounts of plant and animal food (Sillen and Kavanagh 1982; Grupe and Herrmann 1988). Marine environments have an enhanced strontium concentration, which consequently can be detected by consumers at higher trophic levels (Ophiel and Judd 1967). For littoral inhabitants, the effects of enhanced strontium do not have a straightforward interpretation. By comparing the Sr/Ca ratio with other element ratios (using the so-called multi-elemental analysis), one can hope to distinguish between enhanced marine food diet and floral components of a diet (Grupe 1986).

Similar analyses have also been attempted with barium, manganese, nickel, zinc and iron (Wolfsperger et al. 1993). Occasionally, one can even discover the stratification in a society, based on the variation of diet (Wolfsperger 1992). Promising as these results may sound, they are still fraught with difficulties of interpretation. It is to be hoped, however, that more extensive insights will be possible in the near future. A major difficulty is the determination of baseline controls: What constitutes a good reference sample? (Hancock et al. 1993)

Chemical examinations of archeological finds permit the assessment and evaluation of dietary situations that do not necessarily agree with the hard evidence of food remains in archeological sites. Contemporary research attempts to correlate certain vessel types with lipids (i.e., whether they are of faunal or floral fats). In the case of vegetable fats or oils, the plant species is sometimes identifiable (Rottländer 1990; Bethell et al. 1993).

Pictorial and Written Sources

The cave paintings of *Homo sapiens sapiens,* the sole member of the genus for at least the last forty thousand years, is the unique creator of cave murals. A careful analysis of the fascinating pictorial representations of beasts tells us – among other things – which animals were hunted. We have tens of thousands of such representations dating from the Paleolithic. The animals were almost always depicted as being hunted, but whether their depiction reflects them as game only is brought into question (Clottes 1989). For the anthropologist interested in paleonutrition, the interpretation of the scenes is of marginal interest, however. Many scenes show the pursuit of game; the limited stalking success of Paleolithic hunters precludes that hunting was enacted solely for ritual purposes: The flesh of the killed animals was certainly eaten, possible spiritualism and animism notwithstanding. A further example is the depiction of the practice of milking: The date of a relief from ancient Mesopotamia (Tell al'Ubaid; now in the Baghdad Museum) from the late fourth millennium BC (see illustration in Schlott and Willms 1992) must be considered a *terminus post quem*.

From the Magdalenian site of Les Trois Frères (southern France), archeologists have recovered a bone fragment which depicts a grasshopper. It is reasonable to assume that, in addition to grasshoppers, other insects were part of the diet (Brothwell and Brothwell 1969).

Since the time when humankind has acquired the skill of writing, a new type of source has been added to the archeological record. During the period of gradual transition from prehistoric to historical times, excavation materials and written sources can complement each other. Archeologists have relied on Julius Caesar's and Strabo's works as a source describing the lifestyles of Celtic and Germanic tribes – although not all passages have been found reliable. Clearly, a careful review of written sources is indispensable. All such archeological and historical evidence then allows the reconstruction of differentiated temporal systems (Fig. 2).

Analyses and Theories

Even in the most fortunate of circumstances, archeology can provide evidence for only a fraction of the human diet. We must use other methods in our attempt to assess the nutrition of the humans who left the evidence behind.

General medical considerations supply anthropologists with metabolic information. Humans need proteins, lipids and carbohydrates. There are numerous sources of these basic ingredients in the ecological niche. One aspect of human evolution gradually gaining acceptance is the realization that humans are interventionist: They alter their environment in order to change the balance of availability as well as ease of procurement. Anthropologists have long noted that the change in the masticatory apparatus changes the accessibility of necessary dietary ingredients. We thus postulate the existence of a feedback cycle: As interventionist strategies become more effective, they become less successful, necessitating ingenuity and a shift in the procurement strategies. These result in a change in the environment on which these strategies have been applied. Human nutritional needs are used as a justification for the red queen model of human evolution (Foley 1984, 1987).

Ethnography

A traditional method used to complement archeological evidence is the use of ethnographic analogies. Controversial as the results of such field research may be,

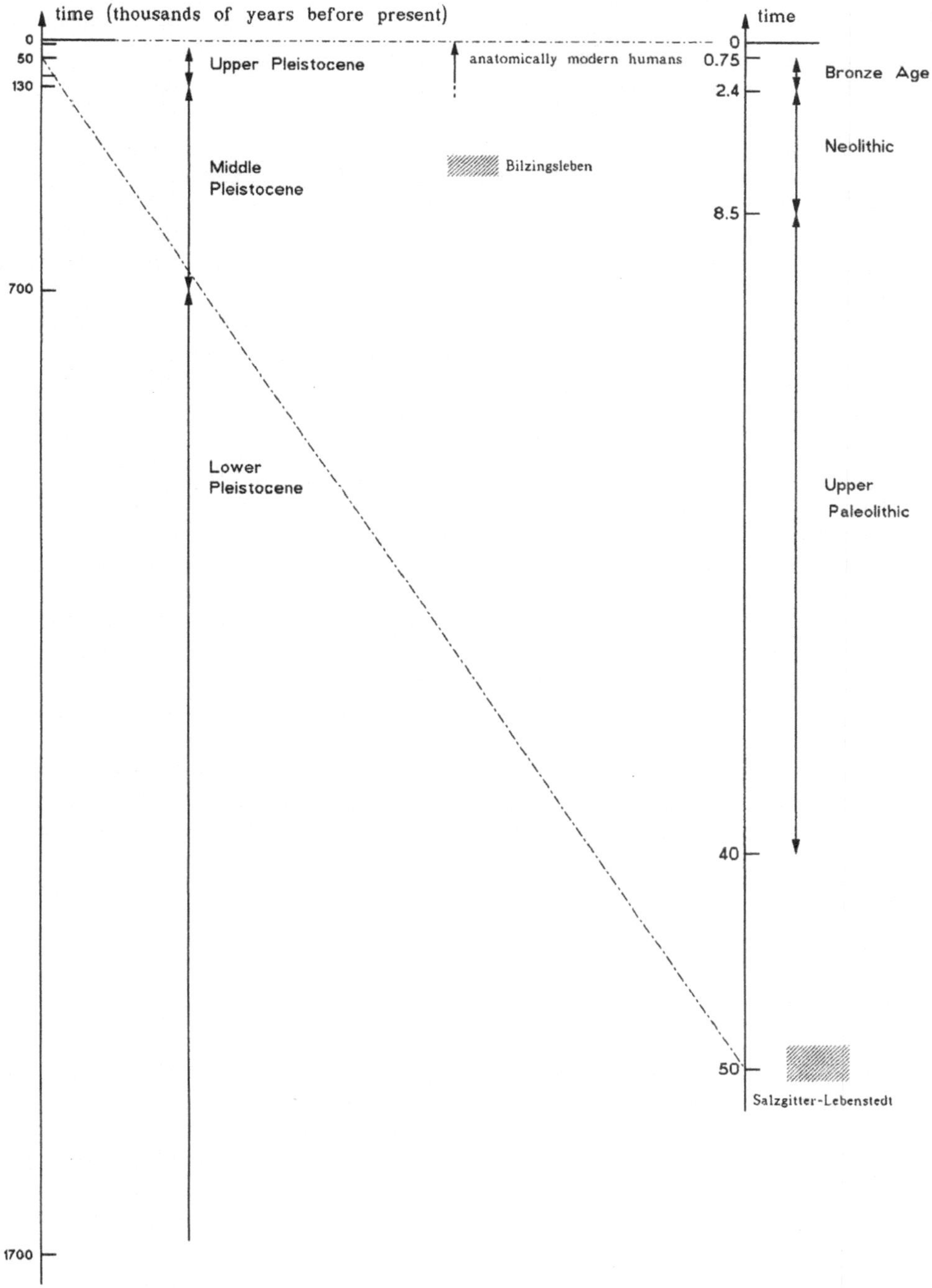

Fig. 2. The temporal structure of the Pleistocene and the subsequent prehistoric epochs as used in this article. (Cavalli-Sforza et al. 1994). Clearly, the boundaries are not the same throughout Central Europe, nor are they so precisely defined.

it is our only source of information about how hunter-gatherers could detoxify foodstuffs (Aaronson 1989) and how humans react to nutritional stress in times of scarcity. Our analysis of the hunting strategies of Inuit (Eskimos) and their shift to a protein-rich diet is of particular impact on our view of how hunters coped with the severity of arctic conditions in Ice Age Europe.

Caution is to be exercised when inferring culturally regulated food avoidances. Muslim societies forbid the consumption of pork, Hindu the consumption of beef, and Christian the consumption of dog (Harris 1985). However, we may not infer that the absence of certain food remains implies the culturally regulated avoidance of these foods. We may of course distinguish present day societies' avoidance from previous dietary acceptance if the bones are present at cooking sites.

Even more problematic is the inference, from ethnographic studies, of what combinations of foodstuffs constituted a meal in prehistoric times. Every society seems to make choices as to what to eat and what can be eaten with what (and with whom); yet it is extremely difficult to reconstruct such choices in prehistoric societies.

On the other hand, ethnographers' observations of pastoralist societies demonstrate that the interpretation of assemblages and plant remains is not straightforward. Consider the reconstruction of the transition from gatherers of wild type cereals to agriculturalists during the Neolithic: Among the many extant pastoralist societies known, some augment their meat diet by collecting seeds, tubers, etc. (Jones 1984), as doubtless also occurred during the early Neolithic. Some present-day pastoral societies exhibit both nomadic and sedentary lifestyles: the Maasai along the border between Kenya and Tanzania are known to remain sedentary long enough for millet to be sown, grown and harvested before moving on (Saitoti 1949).

Furthermore, without the studies by ethnographers, we would not know how earlier societies dealt with fungal parasites and plant toxins. Archeologists and anthropologists have become aware of the role of such substances as medicines and hallucinogens only through ethnographic field research (Aaronson 1989).

The Prehistoric Epochs

The Paleolithic

By examining the masticatory apparatus and the wear of the teeth, one can reconstruct the eating habits of the early hominids (Andrews 1992). The eating habits of the extant apes is to be contrasted with possible dietary scenarios of early hominoids (Kay 1987). We may conclude that the earliest diets consisted of roots and fruits, supplemented by eggs and insect larvae, and perhaps bone marrow. We have evidence of apes eating insects: Jane Goodall, observing chimpanzees at Gombe Stream, Tanzania, found that insects constitute up to 15% of the diet on several days per month. The insects were often eaten mixed with leaves that were picked and added to the mouthful (Goodall 1986).

There is a considerable body of evidence that supports the view that the early hominines were at the end of the food chain – they must have consumed carrion, probably predominantly the bone marrow, as evidenced by the fact that the earliest stone tools, being hammer-like, were used to crack bones (Toth 1987). The situation changed with the production of knife-edged tools, which enabled their users to dissect and dismember large animals. Indeed, one way archeologists can distinguish between bones of decayed cadavers and bones left at a butchering site is the presence of cut marks on the bone.

Systematic hunting and the control of fire resulted in hitherto uninhabitable areas becoming populated by humans. During the period from 1.5 to 0.7 million BP, major parts of the temperate zone in Europe and Asia became populated (Turner 1984). We have evidence that the proportion of meat in the diet increased. It may have even been broiled. It is reasonable to assume that, in colder climes, the campfire site became a very popular meeting point for the social interactions among the members of the various households.

The supply of floral and faunal food depended on the prevailing climatic conditions (Scott 1984). For Central Europe we can differentiate three faunal complexes (Schlott 1992):

- elephant, fallow deer and rhinoceros in a warm climate;
- wild horse, giant deer, aurochs, bison, elk and ibex in a temperate climate;
- mammoth, woolly rhinoceros and reindeer in a cold climate.

Human habitation was almost impossible in our northern latitudes during the extremes of the Ice Age. We have evidence that during the coldest periods, humans employed a foraging – as opposed to a hunting – strategy: The carcasses of the megafaunal species could be found under the snow or in the ice-covered rivers (Gamble 1987). These large carcasses were located by probing the snow or ice layer with wooden poles; such poles have often been confused with spears (Gamble 1986). During other times, however, the type of game alternated synchronously with the variation in climate. Wooden spears with fire-hardened tips were generally adequate in the successful pursuit of big game.

During the Lower Paleolithic, deer, forest elephant, and rhinoceros were the predominant game; occasionally horse and elk are included. Although the hunters were almost certainly also gatherers, it is difficult to find concrete evidence; occasionally, cherrystones and hazelnut shells can be found. An important site, dated at 350 000 BP, is Bilzingsleben (Mania 1990), probably inhabited by *Homo erectus* (Vlçek 1987).

During the Middle Paleolithic – the age of Neanderthal man – woodland elephant and rhinoceros gave way to mammoth and woolly rhinoceros. The Salzgitter-Lebenstedt site is about 50 thousand years old (Kraft 1994); there, 75% of the bones are from reindeer, while woolly rhinoceros and mammoth contribute 10% each (Busch and Schwokedissen 1991). From such data, one can conclude that most of the meat came from mammoth carcasses. Both temporally and regionally, Middle Paleolithic game varied considerably; in mountainous areas, the mighty cave bear (*Ursus spelaeus*) was hunted extensively (Schlott 1992).

The Upper Paleolithic – i.e., the period from 40 000 to 8500 BC – is well studied. The most important game in Central Europe was reindeer and wild horse. Mammoth, woolly rhinoceros and the giant deer were rarely hunted; these became extinct towards the end of this period. While some are convinced that this is due to man the hunter, we disagree with this view (Willms 1992). At the end of the Paleolithic, new weapons appear: the hurled spear, the barbed harpoon, and the bow and arrow. With the taming of the wolf, mankind produced the domestic dog (Benecke 1994). It was a cultural product that not only influenced the course of faunal evolution in Europe, but also enhanced the hunter's success.

In cooking pits, which were probably lined with animal hides, water was brought to a boil with heated stones (Dittmann 1990). In them, vegetables were probably cooked, while the meat was probably broiled over the open fire. All cooking utensils were made of either leather, wood, stone, antler or bone. One may safely assume that a proportion of the food was preserved: some by drying, some by utilizing the cold.

With the end of the Ice Age, large temperature fluctuations (Woldstedt 1954) ultimately yielded to a more temperate climate, during which the large ice sheets melted and steppe and tundra were replaced by forest. Within three thousand years, Central Europe was covered with mixed oak forest. The fauna subsequently changed: red and roe deer, wild boar, elk and aurochs roamed the forests; occasionally, elk and horse were encountered (Willms 1987). Large herds of reindeer or wild horse were no longer pursued; rather, in the period called the mesolithic, solitary game was hunted. Some hunters also seem to have occasionally turned to fishing, as evidenced by archeological sites on the banks of rivers and along the perimeters of lakes. The forests also offered a wide assortment of vegetable foodstuffs: hazelnuts, berries and mushrooms. The diet varied with the seasons.

The Neolithic

The transition from the Mesolithic to the Neolithic may have occurred where the wild type cereals, although abundant, became scarce in their natural habitats. With the reproductional success of human populations, their impact on the environment was anything but benign (Cohen 1977; Boserup 1993). Thanks to the ingenuity of some inhabitants in Asia Minor, the ability to raise crops – whose reproductive systems have been so altered by human intervention that they can no longer survive in the wild but can yield much larger harvests – led to the invention of agriculture. We have archeological evidence for such interventionist traditions evolving at the site of Tell Abu Hureyra, in Syria (Hillman 1975). By the seventh millennium BC, this agroecosystem had been introduced into Europe (Ammerman and Cavalli-Sforza 1973; Ammerman 1989; Zohary and Hopf 1993), reaching Central Europe by the middle of the sixth millennium BC.

Together with their ability to grow cultured plants, the earliest farmers acquired the skill to domesticate and breed animals (Benecke 1994). A further feature of the Neolithic is habitation in permanent villages with artifacts such as fired clay pots, being used for the storage and preparation of food, and axes with polished stone blades: they were essential for clearing forests and erecting habitation structures in woodland areas. Following V. Gordon Childe, this quantum leap in cultural evolution has been called the Neolithic Revolution (Mellars and Stringer 1989; Volkhausen 1994).

In Central Europe, the Neolithic lasted more than three thousand years; thus it is obvious that the conditions were never the same in all regions and at all times. Animal husbandry and plant cultivation reflected the specific environmental conditions, including the impact of human intervention.

While the culture of the curvilinearly decorated pottery (band ceramics culture) – the oldest permanent culture of Central Europe – was predominant in all the loess regions, numerous local variations of crops could be recognized. The main suppliers of carbohydrates were the grains emmer wheat (*Triticum dicoccum*), einkorn (*Triticum monococcum*), primitive hulled wheats, and occasionally barley. After removal of the chaff, the grains were ground between simple, functional grinding stones, which can often be found in large numbers at numerous sites. For porridges and gruels, roughly ground grain sufficed, while loaves were baked from finely ground flour. Simple ovens, located outside the dwellings, are well known in the band ceramics culture. Legumina were also planted; the peas and lentils, which contain much protein, were probably dried. Poppy seed and linseed contain oils and would be added at mealtime. These agricultural products could help minimize seasonal fluctuations in the food supply.

Throughout the following periods, water was supplied by brooks and small rivers. An extraordinarily unique situation was found at Erkelenz, near Cologne: A well was excavated, dendrochronologically dated to 5090–5050 BC (the oldest wooden construction in Europe) with wooden spades and pieces of bark (to scoop out the water) inside (Weiner 1995). An almost contemporary situation has been found in Asparn an der Zaya, Lower Austria (Windl 1996).

In the Middle Neolithic (after 4900 BC), a new kind of cereal, called bread wheat (*Triticum vulgare*), was introduced from the south. It was, as the name implies, used for making bread. The interesting story of its spread into Central Europe is gradually being written (Willms 1991b): During the Upper Neolithic (after 4400 BC) it travelled north, ultimately reaching southern Sweden, marking the appearance of the first agriculturally based cultures there. For the first time, England grew its own wheat, mainly emmer wheat and einkorn. We not only find baking ovens, but also complete loaves that have been preserved in the lake dwellings of Switzerland to this day – a proof that bread in loaf form has been in existence since the fourth millennium BC. One may add that the knowledge of baking suggests the knowledge of beer brewing (Behre 1984).

Gathered fruits, such as apples, berries, nuts and acorns were occasionally as important as cultigens. Hazelnut shells, apple remains, raspberries and blackberries have been found in archeological sites in both southern and northern Germany. The wild strawberry seems to be indigenous only south of the Main valley. Acorns were also eaten, but it was necessary to detoxify them first. Apples were halved and dried over a fire so that they could be kept for the winter months.

A unique find from the Alps is the ice mummy from the Hauslabjoch, South Tyrol, which was found in September, 1991 (Höpfel et al. 1992) dated between 3350 and 3160 BC (Bonani et al. 1992). With this mummy various grasses, grains of einkorn and maple leaves without stems were found: evidence of what a neolithic traveller would take along as food when attempting to cross the Alps. We also have documentation – judging from his bows and arrows – on how he could supplement his diet through hunting (Spindler et al. 1994).

What was the ratio of floral to faunal foodstuffs in a typical diet? The floral fraction was probably in the range of 50–75% (Schlott and Willms 1992). Of further interest is the ratio of wild to domestic animal meat. In the Lower Neolithic, game constituted only 5–10% of the meat – except in some Danube regions. In the Upper Neolithic, an increase in population necessitated an expansion of settlements into heretofore uninhabited regions. In these pioneer settlements, up to 50% of the meat was game. Red deer was hunted much more extensively than wild boar (*Sus scrofa*) or roe deer (*Capreolus capreolus*). Fish was another major source of protein, yet its proportion in the diet is difficult to assess.

The role of domestic animal breeding is quite clearly documented in the Lower Neolithic. In the southern regions, small ruminants – the caprines (sheep and goats) – predominate, while cattle husbandry can be considered a specialty of the northern ones. There, the sequence of occurrence is: cattle, caprines, then pig. In southern Germany, Switzerland, the Netherlands and southern Sweden, the pig gradually became the main domestic (and most numerous) animal (Kokabi 1987; Murray 1968). When statistically assessing the bone remains at the various sites, it is important to remember that a cow supplies up to eight times as much meat as a sheep and up to four times as much as a pig.

During the Upper Neolithic, important innovations appear. Oxen were used as draft animals for plowing and for carts; dairy farming became firmly established. Milk and cheese enhanced the variety of foodstuffs. The horse, first domesticated in the fourth millennium BC, was primarily a source of meat; only later was it used as a draft and riding animal. Horses, however, remained relatively marginal in Central Europe (in contrast to eastern steppe regions). In the Upper Neolithic, animal husbandry was an important aspect of life; breeders drank not only milk, but also the blood of their animals, as is still done by some nomadic herdsmen – Massai, Samburu, Turkana – in eastern equatorial Africa today (Amin 1981).

In the late Neolithic, porridges and stews seem to be the main meal types. Most recovered cooking vessels contain carbonised food debris, perhaps because the vessels were used repeatedly and rarely cleaned.

In Western Europe food was prepared in simple pits with hot stones. It was eaten either while squatting on the floor or sitting on low stools. Woven mats denoted the table area and bowls of wood or pottery were used as dinnerware. Wooden spoons, used to distribute gruel or soups, were also known. During festivities, consumption of food and drink must have been extensive. It remains an open question whether snails, clams and tortoises were delicacies – as they are considered nowadays – or whether they were substitute food during times of want (Schlott and Willms 1992).

Water (and other liquids) were transported and stored in a type of clay bottle. Nets, leather pouches and baskets were in widespread use.

The Bronze Age

In the Bronze Age (2200–750 BC) we observe the further stratification of society and the fact that specialization of craftsmen was on the rise and deducible from the archeological evidence in the grave fields (Mays 1987; Teschler-Nicola 1989).

Bronze Age economies continued to rely heavily on agriculture and animal husbandry; only a very few people could afford to be independent of food production in order to pursue other professions. The existence of metal – not only for jewelry or currency – permitted the accumulation of wealth and its attendant gain in power and influence.

Hulled wheat, especially emmer wheat, and barley continued to be the predominant cereals in Central Europe. Nonetheless, bread wheat, spelt wheat, and millet would alternate in importance from one region or another. In the lakeshore regions of Switzerland, one observes a decline in the production of bread wheat during the latest Upper Neolithic, decreasing further during the Bronze Age. We observe an increase in millet remains in archeological sites of the later Bronze Age. Spelt wheat was planted as winter wheat and was obviously a welcome addition to summer cereals such as emmer wheat, barley and millet.

Major cultivation areas can be identified: In a broad band from France to the Ukraine, spelt wheat, bread wheat and millet is grown. We draw the conclusion that meals in this region must have consisted mainly of gruels and porridges.

An important new cultigen was a further legumen, the broad bean (*Vicia faba*), most likely of eastern Mediterranean origin. High in protein and carbohydrate content, it spread rapidly throughout Europe and was more important than the older legumina, such as peas and lentils. This plant also plays a role in the burial rites. Some authors consider the possibility that the broad bean was used to feed horses that could not remain in pasture year round (Schlott and Willms 1992).

Whole cereal grains were stored in pits. Experimental archeology showed that the kernels could be stored this way in our latitudes, despite a rather moist winter climate (Geber 1985). Such methods were already in use in the Neolithic, especially for hulled wheat. These storage methods suggest that removing the chaff and grinding were only performed when the need arose, not exclusively at harvest time.

Animal husbandry is characterized by a difference between a western region with a predominance of pigs, an eastern and northern region with a predominance (up to 75%) of cattle, and a southern region with a larger percentage of caprines. The predominance of pigs in the western regions remained up to the Gallo-Roman period.

Horses were insignificant in numbers; their bones rarely exceed 3–5% at the archeological sites. Doubtless, horse meat was eaten; there was no taboo about eating either it or dog meat. Towards the end of the Bronze Age, a further domesticated animal makes its appearance in Central Europe: the chicken (Benecke 1994). Originally domesticated in East Asia, it arrived in Europe via Asia Minor and Greece (Fig. 3).

The remains of cult rites, which included human sacrifice (Jankuhn 1967) and possibly cannibalism (questioned by Peter-Röcher 1994; see, however, Gibbons 1997), exhibit a remarkably high number of brown hares (*Lepus europaeus*), an animal that prefers the open countryside to woodland. The percentage of small ruminants is also unusually high; their brains seem to have been highly valued for such rites. We assume that in the course of these festivities, the meat was not broiled but cooked in soups or stews.

One finds remains of unleavened bread in North German graves; in general, we observe that the baking of bread is a more widespread activity towards the end

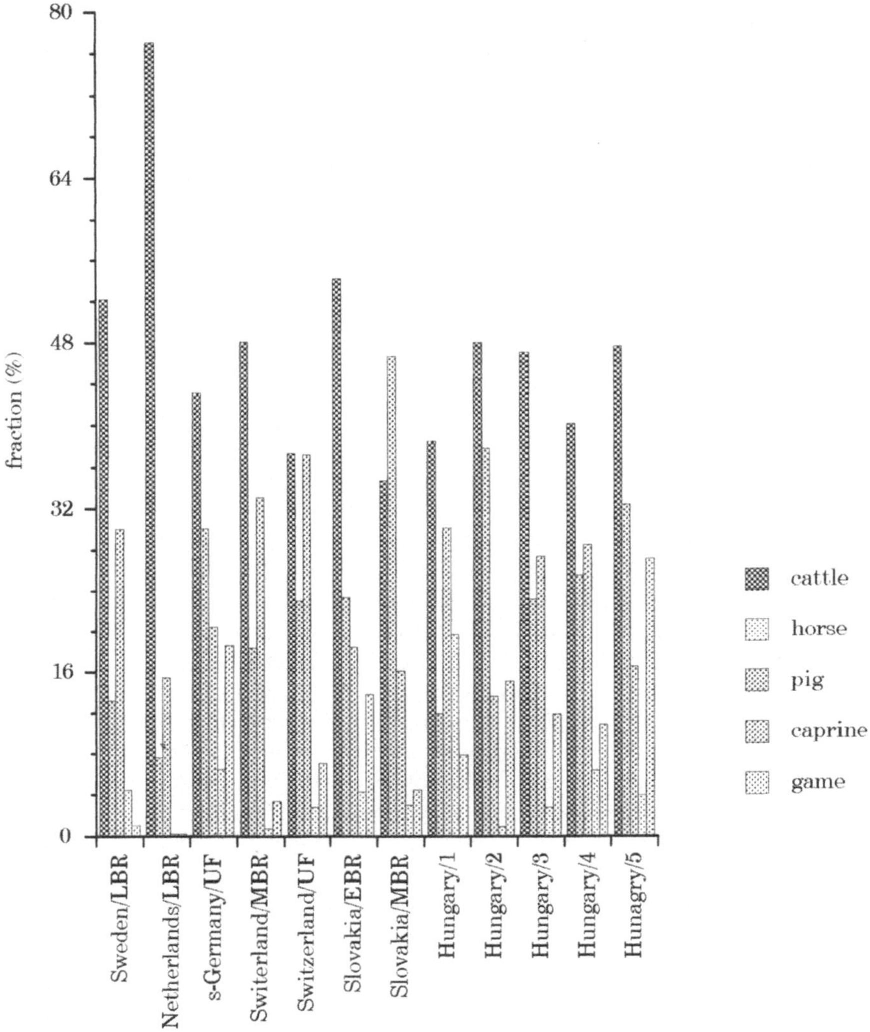

Fig. 3. The distribution of animal bones in various Bronze Age settlements (Benecke 1994). (EBR: Early Bronze Age; MBR: Middle Bronze Age; LBR: Lower Bronze Age; UF: Urnfield Culture).

of the Bronze Age. Noodle-like dough remnants can be found in sites dating from this period. During the transition to the Iron Age, we have evidence for the extraction of table salt.

Early Bronze Age graves contain not only plates but also cups and other drinking vessels; at the end of the Bronze Age, a remarkably large assemblage of containers for fluids is found in each grave, perhaps indicating elaborate drinking ceremonies during burial rites. Conspicuous is the addition of awl-like implements in bowls as extras in women's graves. These must have been small skewers which could be used as forks, since we often find with them remnants of small pieces of pierced meat. Men had pointed daggers which were apparently not only used for cutting meat, but also as skewers for dealing out small meat morsels to the women (Schlott and Willms 1992).

The Iron Age

In Europe, the Iron Age begins around 800/750 BC and ends with the appearance of the Romans in Germany. During the early Iron Age – also known as the Hallstatt culture – iron artifacts were gradually adopted for everyday use. At the onset of the later Iron Age – called Latène – we associate archeological cultures with the names of the ethnic groups that adopted the particular variants, such as Celtic, Germanic, and so forth.

In southern Germany, the cultigens of the Bronze Age were still grown and harvested. Barley remained predominant, only gradually being replaced by spelt wheat. The situation in the Rhine Valley is more diverse: At many sites, the floral remains included up to thirteen different cultigens, an indication that the nutritional base of the population was very diversified. A disastrous harvest of one cultigen could in principle be compensated by a bumper crop of another. We conclude that the meals were at least as varied as the crop cultures. Millet, especially Italian millet (*Setaria italica*), was quite common, and all the three legumina were grown, as well as poppy and flax.

Barley was predominant in northern Germany; linseed and broad beans were widespread. The cultivated food – and thus the diet – along the coast was considerably more monotonous: only barley was grown. The Rhineland was characterized by the planting of flax, carrot, lamb's lettuce and dodder seed (*Camelina sativa*); its oil is a characteristic dietary feature. During the Latène, it became widespread, then its importance gradually decreased and became virtually unknown by the Middle Ages. Millet and oats became more important towards the end of the Iron Age, while wild fruits – due to the spread of cultivated lands – became insignificant.

Agriculture acquires a new structure, which ultimately develops into what has been called the three-field system of the Middle Ages. A major innovation is the development of a heavy plow with a coulter and a curved mold board (Cunliffe 1992) that could turn over the clods. In the moist climate of northern Germany, the traditional food storage pits were gradually replaced by granaries erected some distance above the ground. Storage containers made of pottery were in use for storing grain and other foodstuffs.

It is probable that by the late Iron Age, grain cereals were kiln-dried. Particular care was given to the shape and construction of millstones in order to obtain a fine flour. Characteristic of the late Iron Age were millstones of basalt from Mayen near Koblenz; they were traded far and wide. At the end of the Iron Age, the first rotational mills made their appearance.

Only occasionally have whole ovens been excavated. They seem not to have been incorporated into the heating system of dwellings. The predominance of barley shows that bread had lost its leading role as a major dietary constituent in this period. The main meals consisted of porridges, gruels and stews – characteristically including legumes.

Cultigens with fatty fruits were assumedly not processed for their oil, but were added to the meal in seed form. It is difficult to believe that all ingredients were mixed arbitrarily. By the Iron Age, a canon of recipes had likely evolved, whereby certain products were cooked (or served) together with specific others. The reasonable diversity of foodstuffs allows us to speculate that in times of plenty the daily meals were varied: Most likely, breakfast was based on either honey or salty porridges.

Salt production played a major role with the onset of the Iron Age. Some salt was mined, often from considerable depths (e.g. in Hallstatt, Austria), some was extracted from brine (e.g., Bad Nauheim, Germany). Large quantities of salt were needed for pickling of food, mainly meat. The abundance of mines dating from the Iron Age implies that salting as a means of preservation was not yet common in the Bronze Age or the Neolithic.

The typical animals raised in northern continental Europe were cattle, in the south the small ruminants. Even the cattle, however, were small, yielding approximately 100–150 kg of meat per animal (Benecke 1994). In the late Iron Age, the raising of pigs predominates in Gaul, as Strabo reports in his *Geography.* Beginning with the late Bronze Age cattle raising becomes more intensive and stalls are erected, presumably for protection from the elements in the winter months. A type of dairy industry is well established. The cattle were not, however, only raised for milk, but also for meat; they were draft animals as well, and their hides were made into leather.

The excavation of the Celtic *oppidum* Manching (Noelle 1985) yielded the following spectrum of bones: 32% bovine, 43% swine, 18% sheep/goats, 5.5% horse, and 1.2% dog, the rest micromammalia (Hahn 1992). From such data, one must conclude that beef constituted 60–65% of the meat. An analysis of grain remains yielded the following distribution: 75.5% barley, 7.5% millet, 7.5% spelt wheat, 5.7% emmer wheat and 3.8% einkorn (Küster 1992; Fig. 4). Grazing occurred on commons within protective earthen ramparts, but one must assume that during peaceful times, the herds were certainly driven to forage outside (Köhler and Maier 1992).

Fowl were kept close to the dwellings. Although a rarity at first, the chicken becomes widespread as a domestic animal. Due to the hens' very limited egg-laying ability, eggs were perhaps used for preparing a new type of food, which would later become our pastry (maybe playing a special role in rites and ceremonies). Fowl could not be considered an important source of meat, the fraction of the diet in a typical settlement was less than one percent – 0.5% of the bone material, on average (Bökönyi 1974).

Hunting was of minor importance (less than 5% of the bones in most assemblages), and fishing surpassed hunting where conditions were favorable. Along the Baltic, we also have evidence for the collecting of clams, mussels, and other shellfish (Behre 1983; Kossack et al. 1984; Jankuhn et al. 1984).

In Bad Nauheim, a settlement where salt was precipitated from brine, archeologists have discovered, among the remains of two vessels, charred hawthorn hips (Schlott and Willms 1992). Such a discovery proves that hot beverages were made from blossoms of the local flora and were not uncommon. Safe water seems to have rarely been a problem; in some places, we find wells (Kossack et al. 1984). Water was transported in wooden barrels and wooden pails with iron handles.

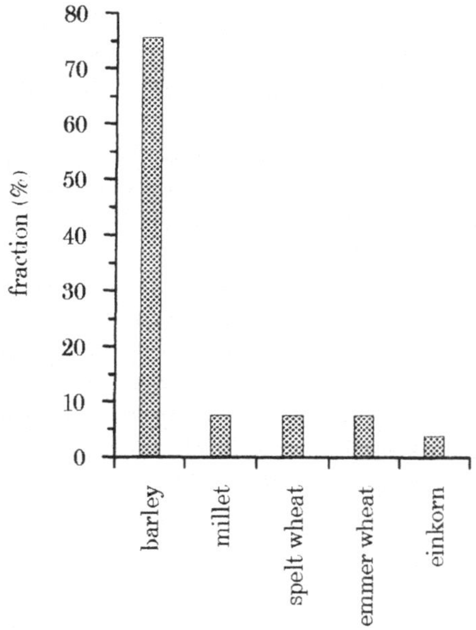

Fig. 4. The distribution of cereal types in an Iron Age settlement (Küster 1992).

With the onset of the Iron Age, we observe a further stratification of society, in which the accessibility of foodstuffs reflects social differences. Diversity, quantity and quality of the available food correlated with social rank. The most conspicuous example is the availability of wine from the Mediterranean; we can reconstruct its trade routes and consumption frequency from the recovered characteristic transport vessels, the amphorae. For the less wealthy, mead was more important. In the grave of a prince in Hochdorf, a large kettle, drinking horns, bronze plates, pans, carving knives and skewers were found (Körber-Grohne 1985; Biel 1985). Such lavishness demonstrates the importance of drinking feasts and banquets in the lives of the upper social class.

Other graves contain gridirons, three-legged kettles with suspension chains, meat forks, meat hooks, cutting and carving knives. Skewers are found either lying close to meat offerings or sticking in them. In most graves from southern Germany the meat was pork; usually head and limbs of the dressed carcass are missing. We conclude that the meat must have been roasted on a spit. A change in funeral rites must have occurred in the late Iron Age: We find animal bones with meat portions of inferior quality left on them. It seems that the good meat was eaten during the funeral festivities (Schlott and Willms 1992).

Conclusions

Archeologists have discovered sites from all epochs of human existence in Central Europe since Bilzingsleben (the time of *Homo erectus*). Early sites are rare, but site frequency increases the more recent the epochs. Many sites are extensive enough to permit glimpses of the paleodiet. We can observe the change in meat consumption, the hunting success, the types of food gathered and the shift to animal husbandry and agriculture, with attendant shifts in food preparation and storage technology.

Revealing as such uncovered evidence may be, only a part of the diet can be inferred from the uncovered archeological artifacts. Trace element analysis and isotope ratio analysis give us clues of foodstuffs that were consumed yet did not leave recoverable remnants in the assemblages.

We must not forget, however, that the reconstruction of the diet of ancient populations is not to be equated with their nutritional status, or even the availability of a balanced diet. The results of paleopathology and paleodemography clearly show that a balanced diet existed only in the rarest of cases.

The adage you are what you eat is also applicable to individuals of ancient populations: Anthropologists know who *Homo* was by knowing what *Homo* actually ate, through the concerted effort with archeologists, chemists, physicists and paleopathologists.

References

Aaronson S (1989) Fungal parasites of grasses and cereals: Their role as food or medicine, now and then. Antiquity 63: 247–257.

Accorsi C (1985) The contribution of palynology in the reconstruction of the environment. In: Peretto C (ed) Homo: Journey to the Origins of Man's History. Four Million Years of Evidence. Cataloghi Marsilio, Venezia, p. 192–202.

Ammerman AJ (1989) On the Neolithic transition in Europe. A comment on Zvelebil and Zvelebil. Antiquity 63: 162–165.

Ammerman AJ, Cavalli-Sforza L (1973) A population model for the diffusion of early farming in Europe. In: Renfrew C (ed) The Explanation of Cultural Change. Duckworth, London, p. 343–357.

Amin A (1981) Cradle of Mankind. Chatoo & Windus, London.

Andrews P (1992) Evolution and environment of the Hominoidaea. Nature 360: 641–646.

Behre K-E (1983) Ernährung und Umwelt der wikingerzeitlichen Siedlung Haithabu. Die Ergebnisse der Untersuchungen der Pflanzenreste. Wachholtz, Neumünster.

Behre K-E (1984) Zur Geschichte der Bierwürzen nach Fruchtfunden und schriftlichen Quellen. In: van Zeist W, Casparie WA (eds) Plants and Ancient Man. Studies in Palaeoethnobotany. Balkema, Rotterdam, p. 115–122.

Benecke N (1994) Archäozoologische Studien zur Entwicklung der Haustierhaltung in Mitteleuropa und Südskandinavien von den Anfängen bis zum ausgehenden Mittelalter. Springer, Berlin.

Bethell PH, Evershed RP, Goad LJ (1993) The investigation of lipids in organic residues by gas chromatography/mass spectrometry. Applications to palaeodietary studies. In: Lambert JB, Grupe G (eds) Prehistoric Human Bone. Springer, Berlin Heidelberg New York, 229–256.

Biel J (1985) Die Ausstattung des Toten, Reichtum im Grabe, Spiegel seiner Macht. In: Planck D, Biel J, Süsskind G, Wais A (ed) Der Keltenfürst von Hochdorf. Landesdenkmalamt Baden-Württemberg/K. Theiss, Stuttgart, p. 78–105.

Bonani S, Ivy S, Niklaus T, Suter M, Housley R, Bronk C, Klinken van G, Hedges R (1992) Altersbestimmung von Milligrammproben der Ötztaler Gletscherleiche mit der Beschleunigermassenspektroskopie-Methode. In: Höpfel F, Platzer W, Spindler K (ed) Der Mann in Eis, Band 1. Eigenverlag Univ Innsbruck.

Boserup E (1993) The Conditions of Agricultural Growth. The Economies of Agrarian Change under Population Pressure. Aldine, London.
Bökönyi S (1974) The History of Domestic Mammals in Central and Eastern Europe. Akadémiai Kiadó, Budapest.
Brothwell D (1986) The Bog Man and the Archaeology of People. British Museum Press, London.
Brothwell DR, Brothwell P (1969) Food in Antiquity. Thames & Hudson, London.
Busch R, Schwokedissen H (1991) Der altsteinzeitliche Fundplatz Salzgitter-Lebenstedt. Naturwissenschaftliche Untersuchungen, Band 2. Böhlau, Wien.
Caselitz P (1986) Ernährungsmöglichkeiten und Ernährungsgewohnheiten prähistorischer Bevölkerungen. Dissertation Hamburg. British Archaeol Reports, Int Series 314, Oxford.
Cavalli-Sforza L, Menozzi P, Piazza A (1994) The History and Geography of Human Genes. Princeton Univ Press, Princeton.
Clarke DL (1968) Analytical Archaeology. Methuen, London.
Clottes J (1989) A question of identity. In: Murphy H (ed) Animals in Art. Unwin Hyman, London.
Cohen MN (1977) The Food Crisis in Prehistory. Overpopulation and Origins of Agriculture. Yale Univ Press, New Haven.
Cunliffe B (1992) The Celtic World. Constable, London.
Dittmann A (1990) Das Kochen mit Steinen. Marburger Studien zur Volkskunde, 7. Reimer, Berlin.
Foley R (1984) Early man and the red queen. Tropical African community evolution and hominid adaptation. In: Foley R (ed) Hominid Evolution and Community Ecology: Prehistoric Human Adaptation in Biological Perspective. Academic Press, London, p. 85–110.
Foley R (1987) Another Unique Species. Patterns in Human Ecology. Longman, Harlow.
Frenzel B (ed) (1994) Evaluation of Land Surfaces Cleared from Forests in the Roman Iron Age and the Time of Migrating Germanic Tribes Based on Regional Pollen Diagrams. Fischer, Stuttgart.
Gamble C (1986) The Palaeolithic Settlement of Europe. Cambridge Univ Press, Cambridge.
Gamble CS (1987) Man the shoveller. Alternative models for middle Pleistocene colonization and occupation in northern latitudes. In: Soffer O (ed) The Pleistocene Old World. Regional Perspectives. Plenum, New York.
Geber W (1985) Jungbronzezeitliche und eisenzeitliche Getreidevorratshaltung in Rullsdorf, Ldkr. Lüneburg – ein Zeugnis urgeschichtlicher Vorratswirtschaft. In: Wilhelmi K (ed) Ausgrabungen in Niedersachsen 1979–1985. Theiss, Stuttgart, p. 146–150.
Gibbons A (1997) Archaeologists rediscover cannibals. Science 277: 635–637.
Goodall J (1986) The Chimpanzees of Gombe. Patterns of Behavior. Cambridge Univ Press, Cambridge.
Grupe G (1986) Multielementanalyse. Ein neuer Weg für die Paläodemographie. Materialien zur Bevölkerungswissenschaft, Sonderh 7. Bundesinst. für Bevölkerungswiss., Wiesbaden.
Grupe G, Herrmann B (eds) (1988) Trace Elements in Environmental History. Springer, Berlin Heidelberg New York.
Hahn E (1992) Die Tierknochenfunde. In: Maier F, Geilenbrögge U, Hahn E, Köhler H-J, Sievers S (ed) Ergebnisse der Ausgrabungen 1984-1987 in Manching. Fischer, Stuttgart, p. 235–269.
Hancock RGV, Grynpas MD, Fkesson K, Obrant KB, Turnquist J, Kessler MJ (1993) Baselines and variabilities of major trace elements in bone. In: Lambert JB, Grupe G (eds) Prehistoric Human Bone. Springer, Berlin Heidelberg New York, p. 189–202.
Harris M (1985) Good to Eat. Riddles of Food and Culture. Simon & Schuster, New York.

Hill A (1978) Hyaenas, bones and fossil man. Kenya Past and Present 9: 8–14.
Hill A (1984) Hyaenas and hominids. Taphonomy and hypothesis testing. In: Foley R (ed) Hominid Evolution and Community Ecology. Prehistoric Human Adaptation in Biological Perspective. Academic Press, London, p. 111–128.
Hillman G (1975) The plant remains from Tell Abu Hureyra. A preliminary report. Proc Prehist Soc 41: 70–77.
Höpfel F, Platzer W, Spindler K (1992) Der Mann im Eis. Bericht über das Internationale Symposium 1992 in Innsbruck, Band 1. Eigenverlag Univ Innsbruck.
Jankuhn H (1967) Archäologische Beobachtungen zu Tier- und Menschenopfern bei den Germanen in der römischen Kaiserzeit. Ann. Göttingen.
Jankuhn H, Schietzel K, Reichstein H (ed) (1984) Archäologische und naturwissenschaftliche Untersuchungen an ländlichen und frühstädtischen Siedlungen im deutschen Küstengebiet vom 5. Jahrhundert v. Chr. bis zum 11. Jahrhundert n. Chr. Acta humaniora, Weinheim.
Jones DK (1984) Shepherds of the Desert. Elmtree Books, London.
Jones S, Martin R, Pilbeam D, Bunney S (eds) (1992) The Cambridge Encyclopedia of Human Evolution. Cambridge Univ Press, Cambridge.
Kay RF (1987) The nut-crackers. A new theory of the adaptations of the Ramapithecinae. In: Ciochon RL, Fleagle JG (eds) Primate Evolution and Human Origins. Aldine-de Gruyter, New York, p. 230–237.
Kokabi M (1987) Die Tierknochenfunde aus dem neolithischen Seeufersiedlungen am Bodensee. Archäologische Nachrichten aus Baden 38/39: 80–An.
Köhler H-J, Maier F. (1992) Der nördliche Wall. In: Maier F, Geilenbrügge U, Hahn E, Köhler, H-J, Sievers S (eds) Ergebnisse der Ausgrabungen 1984–1987 in Manching. Fischer, Stuttgart, p. 340–356.
Körber-Grohne U (1985) Pflanzliche und tierische Reste aus dem Fürstengrab von Hochdorf. Die Biologie als Hilfswissenschaft der Archäologie. In: Planck D, Biel J, Süsskind G, Wais A (eds) Der Keltenfürst von Hochdorf. Methoden und Ergebnisse der Landesarchäologie. Landesdenkmalamt Baden-Württemberg/Theiss, Stuttgart, p. 116–123.
Kraft I (1994) Studien zur Kultur und Umwelt im Mittelpleistozän Europas. Univ'forsch zur prähist Archäol 24. Habelt, Bonn.
Kossack G, Behre K-E, Schmid (ed) (1984) Archäologische und naturwissenschaftliche Untersuchungen an ländlichen und frühstädtischen Siedlungen im deutschen Küstengebiet vom 5. Jahrhundert v. Chr. bis zum 11. Jahrhundert n. Chr. Acta humaniora, Weinheim.
Küster H (1992) Vegetationsgeschichtliche Untersuchungen. In: Maier F, Geilenbrügge U, Hahn E, Köhler H-J, Sievers S (eds) Ergebnisse der Ausgrabungen 1984–1987 in Manching. Steiner, Stuttgart, p. 433–476.
Lee-Thorp JA, van der Merwe NJ (1987) Carbon isotope analysis of food bone apatite. S Afr J Sci 83: 712–715.
Lee-Thorp JA, van der Merwe NJ, Brain CK (1994) Diet of Australopithecus robustus at Swartkrans from stable carbon isotopic analysis. J Hum Evol 27: 361–372.
Lee-Thorp JA, Sealy JC, van der Merwe NJ (1989) Stable carbon isotope ratio differences between bone collagen and bone apatite and their relationship to diet. J Archaeol Sci 16: 585–599.
Lee-Thorp JA, Sealy JC, Morris AG (1993) Isotopic evidence for diets of prehistoric farmers in South Africa. In: Lambert JB, Grupe G (eds) Prehistoric Human Bone. Springer, Berlin Heidelberg New York, p. 99–120.
Mania D (1990) Auf den Spuren des Urmenschen. Die Funde von Bilzingsleben. Theiss, Stuttgart.
Mays S (1987) Social Organization and Social Change in the Early and Middle Bronze Age of Central Europe. A Study Using Human Skeletal Remains. Thesis Southampton.

Mellars P, Stringer C (eds) (1989) The Human Revolution. Behavioural and Biological Perspectives on the Origins of Modern Humans. Edinburgh Univ Press, Edinburgh.

Merwe NJ van der (1992) Reconstructing prehistoric diet. In: Jones S, Martin R, Pilbeam D, Bunney S (eds) The Cambridge Encyclopedia of Human Evolution. Cambridge Univ. Press, Cambridge, p. 369–372.

Murray J (1968) Some aspects of ovicaprid and pig breeding in Neolithic Europe. In: Coles J, Simpson DDA (eds) Studies in Ancient Europe. Studies Presented to Stuart Piggott. Leicester Univ Press, Leicester.

Noelle H (1985) Die Kelten und ihre Stadt Manching. Noelle, Pfaffenhofen.

Ophiel IL, Judd JM (1967) Skeletal distribution of strontium and calcium and strontium/calcium ratios in several species of fish. In: Lenihan JMA, Loutit JF, Martin JH (eds) Strontium Metabolism. Academic Press, London, p. 103–110.

Peter-Röcher H (1994) Kannibalismus in der prähistorischen Forschung. Studien zu einer paradigmatischen Deutung und ihren Grundlagen. Univ'forsch zur prähist Archäol 20. Habelt, Bonn.

Rottländer RCA (1990) Die Resultate der modernen Fettanalytik und ihre Anwendung auf die prähistorische Forschung. Archaeo-Physik 12: 1–354.

Saitoti TO (1949) Maasai. Reprint 1980, Abrams, New York.

Schlott C (1992) Lange Zeit der Jäger und Sammlerinnen. In: Pfeiffer A (ed) Vom Mammutfleisch bis zur Kartoffel. Heilbronner Museumsheft 15. Terra Incognita, Frankfurt, p. 8–73.

Schlott C, Willms C (1992) Nach der Vertreibung aus dem Paradies. In: Pfeiffer A (ed) Vom Mammutfleisch bis zur Kartoffel. Heilbronner Museumsheft 15. Terra Incognita, Frankfurt, p. 74–157.

Schultz M (1990) Erkrankungen des Kindesalters bei der frühbronzezeitlichen Population von Ikiztepe (Türkei). In: Andraschko FM, Teegen W-R (eds) Gedenkschrift für Jürgen Driehaus. Zabern, Mainz, p. 83–90, Tf 1: 14–16.

Schultz M, Teschler-Nicola M (1989) Osteologische Untersuchungen an bronzezeitlichen Kinderskeletten aus Franzhausen, Niederösterreich. Verh Anat Ges Leipzig 82: 407–409.

Scott K (1984) Hunter-Gatherers and Large Mammals in Glacial Britain. In: Foley R (ed) Hominid Evolution and Community Ecology. Academic Press, London, p. 219–236.

Sillen AG, Kavanagh M (1982) Strontium and paleodietary research. A review. Yb Phys Anthrop 15: 67–90.

Simetsberger B (1993) Die fossilen Molluskenreste aus jungpaläolithischen und neolithischen Fundstellen in Niederösterreich und Burgenland Österreich. Diplomarbeit, Universität Wien.

Spears I, Crompton RH (1995) Mechanisms of breakdown induced in food by hominoid molars. In: Radlanski RJ, Renz H (eds) Proceedings of the 10th International Symposium on Dental Morphology, Berlin, Sept. 6–10, 1995. Brünne, Berlin, p. 387–390.

Spindler K, Rastbichler-Zissernig E, Wilfing H, zur Nedden D, Nothdurfter H (1994) Der Mann im Eis. Neue Funde und Ergebnisse, Band 2. Springer, Wien.

Strabo: Meineke A (ed) (1851/52) Strabo, Geografia. Teubner, Leipzig. English translation in: Jones, HL (1960–69) The Geography of Strabo. Loeb Classical Library, Cambridge, MA.

Teschler-Nicola M (1989) Soziale und biologische Differenzierung in der frühen Bronzezeit am Beispiel des Gräberfeldes F von Gemeinlebarn, Niederösterreich. Ann Naturhist Mus Wien 90: 135–145.

Toth N (1987) The first technology. Sci Am 256: 112–121.

Turner A (1984) Hominids and fellow travellers: Human migration into high latitudes as part of a large mammal community. In: Foley R (ed) Hominid Evolution and Community Ecology. Academic Press, London, p. 193–213.

Vlcek E (1987) Les anténéandertaliens en Europe central et leur comparaison avec l'homme de Tautavel. L'Anthropologie 90: 503–514.
Volkhausen B (1994) Ethnographische Parallelen und Vergleiche zum Prozeß der Neolithisierung. Dissertation Hamburg. Europäische Hochschulschriften, Frankfurt/Main.
Weiner J (1995) Wasserversorgung in der Jungsteinzeit. Archäol in Deutschland 4: 6–11.
Willms C (1987) Der Elch (Alces alces L.) im nacheiszeitlichen Mitteleuropa. Eine paläozoogeographische Untersuchung auf quantitativer Ebene. Archeologia Polska 32, 249–291.
Willms C (1991a) Mensch und Natur in der Jungsteinzeit. Über die Eingriffe des Menschen in seine Umwelt in der Zeit 5600–2400 v. Chr., Band 1. Museum für Ur- und Frühgesch., Frankfurt.
Willms C (1991b) Getreide im europäischen Frühneolithikum. Saeculum 42: 104–110.
Willms C (1992) Homo sapiens sapiens. Ein ausrottendes Lebewesen von Beginn an? Zum Mensch-Umwelt-Verhältnis in vorchristlicher Zeit. Plattform 1: 43–53.
Windl H (1996) Archäologie einer Katastrophe und deren Vorgeschichte. Katalog des NÖ. Landesmuseums, N.F. 393, Asparn/Zaya.
Woldstedt P (1954) Das Eiszeitalter. Grundlinien einer Geologie des Quartärs. Enke, Stuttgart.
Wolfsperger M (1992) Rekonstruktion sozialer und diachroner Differenzierungen des Nahrungsverhaltens mittelalterlicher und frühneuzeitlicher österreichischer Populationen anhand der Sr- und Zn-Gehalte im Skelett. Wissenschaftliche Zeitschrift der Humboldt-Universität zu Berlin, R. Medizin 41: 154–158.
Wolfsperger M, Wilfling H, Matiasek K, Teschler-Nicola M (1993) Trace elements in ancient Peruvian mummy hair. A preliminary report. Int J Anthrop 8: 27–33.
Zeist W van (1955) Pollen Analytical Investigations in the Northern Netherlands. With Special Reference to Archaeology. North Holland, Amsterdam.
Zieman-Becker B, Pirsig W, Teschler-Nicola M (1991) Schwerhörigkeit durch eine Stapesankylose vor 4000 Jahren. Ein Fund aus der frühen Bronzezeit in Franzhausen-Niederösterreich. Abstract zur 22. Tagung der Ges Anthrop und Humangenetik, Bochum, p. 48.
Zohary D, Hopf M (1993) Domestication of Plants in the Old World. The Origin and Spread of Cultivated Plants in West Asia, Europe and the Nile Valley. Clarendon, Oxford.

5.2 "Archives of Childhood" – The Research Potential of Trace Element Analyses of Ancient Human Dental Enamel

Gisela Grupe

Introduction

Trace element analysis is an archaeometric approach to addressing a variety of questions aimed at reconstructing past human life-style and behaviour. Of these aims, two are of major interest: First, trace element uptake and the resulting elemental composition of a consumer's tissue yields clues to ancient dietary habits, which in turn permit the reconstruction of subsistence strategies in connection with the development of anthropogenic landscapes. The second aspect concerns environmental history in the sense that body stores of heavy elements indicate environmental pollution resulting from human activities. Both have had, and still have, a major impact on human population development, and a great number of trace element analyses have been carried out by numerous research groups in different parts of the world (Grupe and Herrmann 1988; Lambert and Grupe 1993; Sandford 1992, 1993). Much work has been done with appreciable success, however an equal part possibly does not meet expectations. Therefore, the "trace element boom" of the last decade has been followed by a waning optimism in recent years.

Common agreement now exists as to why the interpretation of trace element patterns in human bodily relics is much more difficult than previously expected. Major faults in the past are due to either too few samples (trace element profiles can be interpreted properly on a group level only), a too small sample size, and thus sample inhomogeneity, insufficient consideration of diagenesis, or all of the above (cf. e.g. Radosevich 1993). Since the technique and the necessary laboratory procedures are both time- and money-consuming, a reflexion on how to pose the question and what exactly is indicated from concentrations of trace elements must take place prior to starting such an invasive technique.

This paper is not meant to present a comprehensive review of trace element studies of dental enamel, but rather to point out the informations hidden in trace element patterns of this material, the advantages of this tissue compared to bone, and the special problems associated with it.

Dental Enamel as a Substrate for Trace Element Analysis

Advantages of Analysing Enamel

Looking through all the trace element work which has been carried out on ancient hard tissue remains, including animal finds, dental enamel obviously had a high priority. This is of course due to the fact that teeth are more frequently preserved than bones, and is in turn due to the unique properties of enamel. It is both the hardest and most heavily mineralized part of the body (Schroeder 1987; Lowenstam and Weiner 1989), the composition of which is dramatically different from bone. The difference does not concern the mineralized matrix as such, which in both cases is basically a biological apatite, but the size and packing of the crystals, the water content and the amount and nature of organic components.

While mineral crystals in compact human bone average 50 nm in length, 30 nm in width and 2–3 nm in thickness (Lowenstam and Weiner 1989), the crystals of mature human enamel are much larger: 30–90 nm wide and 20–60 nm thick, and can grow to the enormous length of 100 µm or more (Schroeder 1987). Crystal growth is under the strict control by ameloblasts and their cell products and results in very dense prism layers.

Maturation of dental enamel is accompanied by a loss of organic components. As a cell-free structure, it does not represent a tissue *sensu strictu*, but rather the mineralized end-product of specialised cells. After its formation and maturation, dental enamel is not remodelled. Mature enamel consists of no more than 1–2 vol % organic matrix components which are free of hydroxylated amino acids. These components are acidic, non-collagenous proteins with a high affinity for the mineral (Lowenstam and Weiner 1989).

Being both hard and poor in organic matter renders dental enamel much less susceptible to diagenesis than bone. It is not very attractive to heterotrophic microorganisms, which are responsible for the initial stages of decomposition and therefore for the further fate of dead tissue. This latter aspect is of major importance, since, in the case of bone, soil fungi and bacteria are capable of actively transporting soil cations into the bone structure. Bacterial cell-walls, especially in case of gram-negative bacteria, act as a loosely packed ion-exchange column by binding cations. Biogenic contamination of a sample is thus a big problem for bones which frequently show markers of microbial invasion, but a much smaller problem for teeth (Grupe and Piepenbrink 1989; Grupe and Dreses-Werringloer 1993).

In sum, dental enamel is less susceptible to changed trace element profiles as a result of diagenesis than bone because of its lesser porosity, the small specific surface and the absence of substantial amounts of organic matter. This does not mean that enamel does not experience decomposition at all, and a decontamination must be carried out prior to trace element analysis.

Choice of Trace Elements and Sample Processing

Since the vast majority of trace element analyses still focus on the reconstruction of dietary habits, one must consider first which elements are suitable for addressing this task. Some dietary indicators such as copper (Cu) and zinc (Zn) are essential elements which have a vital function in the organism, either as part of metalloenzymes, metalloproteins or enzyme activators. Such elements are under homeostatic control with the result that rising uptake is not necessarily followed by enhanced absorption. Their concentrations in a consumer's tissue thus varies within quite narrow limits. Next are so-called candidates for essential elements like vanadium (V) which means that their essentiality is most probable, but not yet proven. The meaning of such elements in ancient hard tissues thus remains largely obscure. Best for archaeometric work are the non-essential elements, or better those elements for which no essentiality is yet evident. These elements, such as strontium (Sr) and barium (Ba) do indeed reflect uptake by daily diet (Grupe 1986).

With the exception of radiogenic ^{90}Sr, little literature is available on the metabolism of non-essential trace elements, due first to the lack of clinical relevance, and secondly to the fact that they are well recognised by the organism and subsequently excreted. It was therefore mostly up to the physical anthropologists to investigate by either controlled feeding experiments or analysis of free-ranging omnivores the most important aspects of dietary uptake, absorption, transport, distribution, storage and excretion of diet-indicating trace elements (Schoeninger 1979; Price et al. 1985; Grupe and Krüger 1990; Lambert and Weydert-Homeyer 1993). The most important results of these investigations are first that an oversimplified estimation of the amounts of gross dietary components such as meat, milk or vegetables is not justified, but that trace element concentrations in biological apatite are rather correlated with certain dietary constituents such as fiber- and mineral-content (Lambert and Weydert-Homeyer 1993; Burton and Wright 1995). Second, it is not possible to simply put the omnivore humans somewhere between herbivorous and carnivorous mammals and to estimate the amount of consumed animal products by differential trace element concentrations. An estimation of this type is only possible when certain prerequisites are met, e.g. for well known predator/prey relationships (Sillen and LeGeros 1991). Also, one must be aware that large species-specific differences exist with regard to tolerance and toxicity of trace element uptake.

Since the mineral matrices of bone and enamel are rather similar, information obtained from bone can be readily used for the interpretation of trace element patterns in dental enamel. In addition, the necessary decontamination protocol is similar for both (laboratory sample processing is described in detail elsewhere, cf. Grupe 1992). Contaminating elements and diagenetic recrystallization products within the sample are solubilized by etching, which at the same time removes the sample surfaces. As far as enamel is concerned, these surfaces are not only the site of element adsorption after death, but additionally some trace element uptake takes place *in vivo*. This holds for both the dentino/enamel-junction and for the enamel surface exposed to the oral cavity (Molleson 1988). Etching enamel with 2% nitric acid for 5–10 minutes ultrasonically removes on average 0.2 mm of the surface. Removal of thicker layers does not lead to significant changes in trace element detections in the remaining sample.

Special Informations from Trace Element Concentrations in Human Enamel

Since dental enamel is not remodelled after its maturation, with the exception of adsorption at its inner and outer surface, its composition reflects trace element uptake during formation. Since the mineralization of permanent teeth ranges from late fetal times up to juvenile age, trace element concentrations in enamel of adult individuals should give clues to elemental intake at these ontogenetic stages. Therefore, for one individual which has survived its childhood, ontogenetic changes in trace element uptake could be reconstructed.

Enamel formation starts at the tip of the crown, and the tooth is built up by subsequent enamel layers which are produced during certain limited periods according to tooth type. Since it is impossible to sample enamel layers formed at a certain age, we sampled the complete enamel of the teeth. Mineralization periods for the crown of permanent teeth are listed up in Table 1.

Table 1. Mineralization period of the enamel of mandibular permanent teeth. fm = fetal month, m = months *post partum*, y = years of age. Average periods according to Schroeder (1987)

Tooth type	Onset of mineralization maxilla/mandibula	End of mineralization maxilla/mandibula
I-1	3 – 4 m	3.3 – 4.1 / 3.4 – 5.4 y
I-2	10 – 12 / 3 – 4 m	4.4 – 4.9 / 3.1 – 5.9 y
C	4 – 5 m	4.5 – 5.8 / 4.0 – 4.7 y
P-1	18 – 24 m	6.3 – 7.0 / 5.0 – 6.0 y
P-2	24 – 30 m	6.6 – 7.2 / 6.1 – 7.1 y
M-1	7 – 8 fm	2.1 – 3.5 / 2.1 – 3.6 y
M-2	30 – 36 m	6.9 – 7.4 / 6.2 – 7.4 y
M-3	7 – 9 / 8 – 10 y	12.8 – 13.2 / 12.0 – 13.7 y

Sampling the complete enamel of a certain tooth type, the trace element content will reflect the total elemental uptake during the crown's formation. For example, the first permanent molar represents trace element uptake from the last months in utero until early childhood. Differing elemental profiles in the enamel of the first molar and of compact bone in an adult individual (the latter tissue reflecting elemental uptake during the last five to ten years prior to death) therefore reflects changing elemental absorptions from childhood to adult age. Since Ehlken (1991) was able to show that variability of trace element concentrations within all four teeth of one type (e.g. first premolar) is less than the variability of element contents between teeth of different types, we conclude that the trace element concentrations in human enamel can be used as "archives of childhood" of individuals who survived this developmental period. Some examples illustrating the resulting perspectives for physical anthropology are given in the following sections.

Weaning Age

The elements Sr and Ba discriminate well between vegetable food and milk and milk products respectively, since they are positively correlated with the fiber content of food. Both elements have thus been frequently used to detect the weaning age in prehistoric people (Sillen and Smith 1984; Grupe and Bach 1993) by means of Sr analyses of childrens' bones. A major objective which cannot be met completely by analysing skeletal remains of small children results from the fact that all these children had died early. They must have suffered from something, or their mothers must have been ill and not capable of nursing their babies properly. If it would be possible to estimate weaning age by use of dental enamel of adult people who survived this period of dietary change, one could strongly support implications about childcare in prehistory up to this point obtained only from bone analyses.

To determine the validity of this approach, Sr/Ca ratios in the different types of permanent teeth were determined, taking into account age-specific discrimination factors against Sr, favouring Ca. The resulting relative Sr/Ca ratios in enamel of the various tooth types were arranged in order of their mineralization (Tab. 1). Analyzing more than 170 samples of dental enamel from medieval burials showed that for the majority of individuals tested, relative Sr/Ca ratios followed two trends (Fig. 1). The first trend is a constantly rising Sr/Ca ratio from the first permanent molar to the first premolar, whereas in the second trend, Sr/Ca ratios drop after an

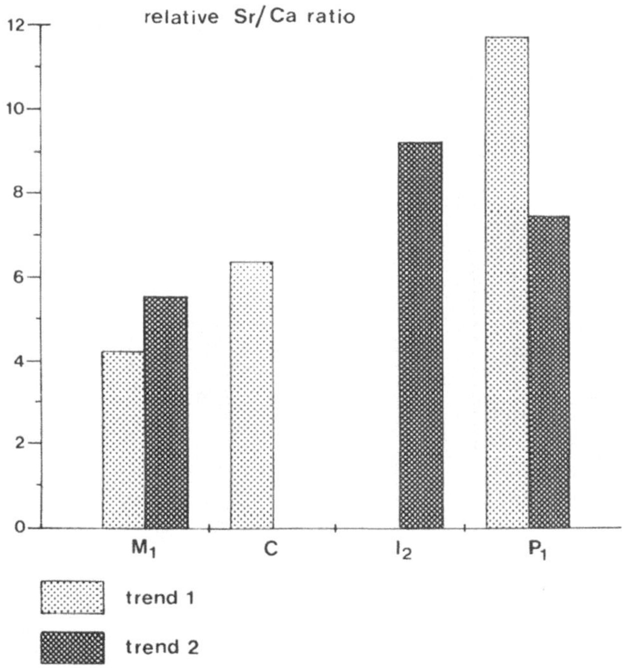

Fig. 1. Major trends of relative Sr/Ca ratios in the enamel of mandibular permanent teeth reflecting varying Sr uptake at different ontogenetic stages, shown by two representative adult medieval skeletons. Data from Ehlken (1991).

initial rise. Translated into an individual's life history, trend 1 means a rise of Sr uptake until approximately four years of age, and trend 2 means an increasing uptake until the second or third year with a following decline. One must be aware that these trends do not say much as long as nothing is known about the average composition of the adults' diet. Therefore, we strongly suggest the analysis of compact bone of the individuals tested in addition to the enamel analyses. In case of an adult's diet wich is poor in Sr (such as from a diet poor in fiber with an appreciable amount of meat and milk/milk products), the second trend indicates an earlier weaning age than does trend 1.

Another trace element study including enamel from the first permanent molar revealed hints of differential child care in a socially stratified society (Siegert 1993). Different hard tissue samples of 18 individuals from the early medieval burial site at Altenerding (upper Bavaria) were analysed for a number of trace elements, including Sr and Ba. Depending upon whether the burials were poor, average or rich in grave goods, the individuals were grouped into three social categories. Fig. 2 shows that both Sr and Ba contents in the enamel of the first molar discriminate between the "poor" and the other two categories, respectively, in as much as Sr/Ca and Ba/Ca

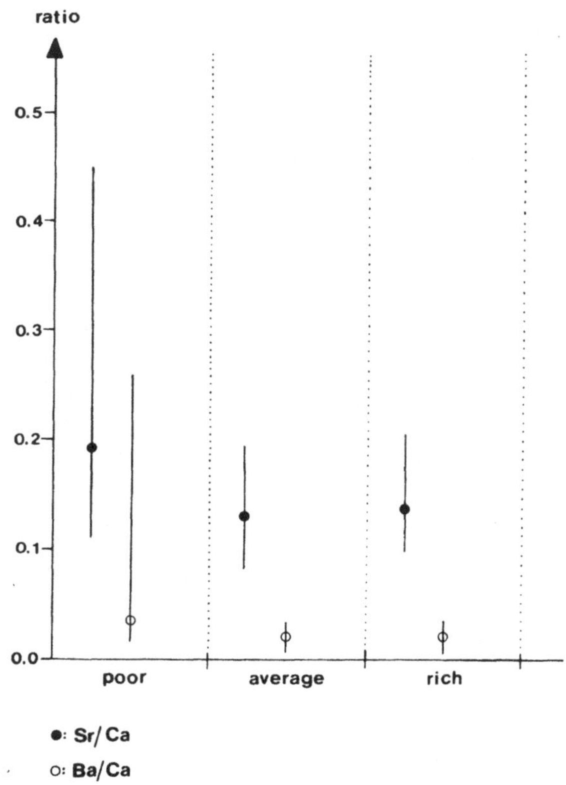

Fig. 2. Variability and median (dots) of Sr/Ca and Ba/Ca ratios respectively, for 18 individuals from a socially stratified early medieval population (see text). Each social group consists of six individuals only. Data from Siegert (1993).

ratios in the underprivileged people were higher on average and had a markedly higher variability. It is necessary to stress that the sample size is much to small in this case (six individuals in each categorie) because the reconstruction of socially diverging dietary habits was not the ultimate aim of the study. However, this by-product of the investigation is at least worth mentioning, since it points toward a direction of possible future research. If such a trend was supported by a sufficient amount of data, its interpretation would be that mothers of the lower social class weaned their babies earlier than did mothers of the privileged classes. One possible explanation could be that a higher work load for the mothers necessitated early weaning and supervision of the babies by elder siblings or other relatives.

Heavy Metal Burden

Analogous to the ontogenetic trends of Sr uptake, heavy metal exposure during childhood may be reconstructed by e.g. lead (Pb) analyses of dental enamel. Relating the Pb/Ca ratios of dental enamel to the time it takes for a complete mineralization of the respective tooth (Tab. 1), annual Pb/Ca ratios can be estimated. Among the more than 170 tooth samples analysed for Sr, including deciduous teeth in this case, two major trends of Pb/Ca intake per year could be detected (Ehlken 1991; Fig. 3).

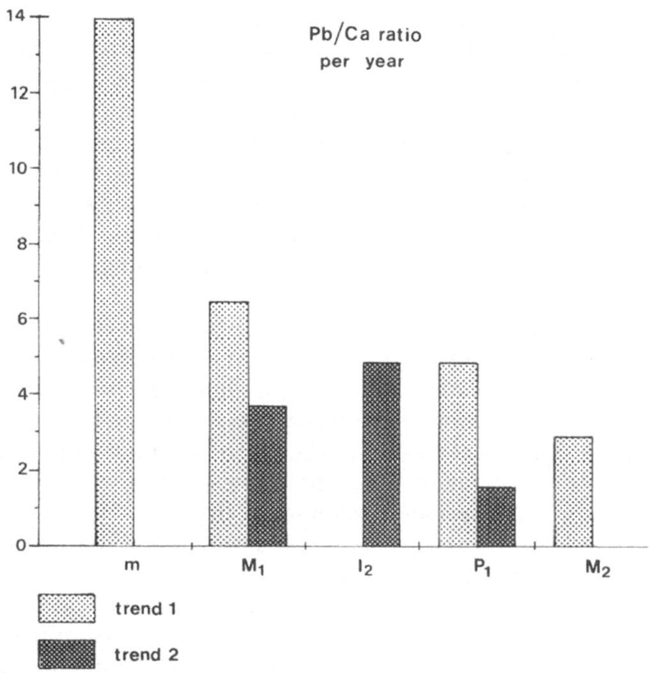

Fig. 3. Major trends of annual Pb/Ca input into the enamel of deciduous (m = deciduous molar) and mandibular permanent teeth reflecting different ontogenetic lead burdens, shown by two representative adult medieval skeletons. Data from Ehlken (1991).

The first trend is characterized by a very high lead burden reflected in the deciduous molars. This means that the childrens' mothers must have alredy been exposed to considerable lead pollution, since this Pb uptake could only have occured during the late fetal and following nursing period. The second trend shows a rising Pb uptake with toddlers' age which is easily explained by small childrens' behaviour. Frequent hand-to-mouth contacts at this age leads to unintended soil ingestion which is still a significant Pb source in todays' developing countries (Thornton 1988). Pb analyses of dental enamel thus permits the reconstruction of time and mode of heavy metal exposure in prehistoric people.

Migration

In addition to trace element content in dental enamel and bone, the determination of isotopic composition of elements yields a new perspective for physical anthropology. Since Sr is a good indicator for ontogenetic trends in dietary habits, stable Sr isotopes are determinants of an individual's residential patterns, i.e. migration (Price et al. 1994).

Sr exists as stable isotopes ^{88}Sr, ^{87}Sr, ^{86}Sr and ^{84}Sr. ^{87}Sr is a product of radiogenic ^{87}Rb (rubidium) and therefore occurs in rocks depending from the initial ^{87}Rb content and geological age. Geologists characterize rocks by the $^{87}Sr/^{86}Sr$ ratio, two Sr isotopes which occur in comparable amounts (about 10% of total Sr each). The local Sr isotopy is site-specific since weathering of rocks and Sr transport into soil and groundwater and subsequently into the food chain does not change Sr isotope ratios. The mass differences of this quite heavy element are so low that transport processes are not accompanied by fractionation factors, in contrast to the well-documented isotopic fractionation of lighter elements such as carbon and nitrogen (Graustein 1989). As a consequence, Sr concentrations in hard tissues reflect the consumer's diet, whereas Sr isotopic ratios reflect the geology of the consumer's habitat. For a non-mobile, permanently settling population, the subsistence strategy of which is the hunting and gathering of local game and plants, Sr isotopy of all tissues reflect the geology of the area.

For a mobile population, the isotopic ratios of ingested Sr change with the place of temporary settlement (Ericson 1985). Since Sr in the enamel of e.g. the first permanent premolar is taken up during early childhood and is not remodelled, and the Sr content of compact femoral bone represents the Sr uptake of the last 5 to 10 years prior to death, significant differences between Sr isotope ratios in enamel and bone should indicate that the individual under investigation spent its childhood in a location different from its residence prior to death.

The prerequisites for using this approach are manyfold but are usually met: The local geology must be variable with regard to Sr isotope ratios, the place of burial must be known, and the population should not have had access to the open sea, since Sr isotopy differs significantly in marine and terrestrial habitats.

A pilot study was carried out to test whether the archaeometric approach for the reconstruction of migration could help solve the controversy about the Bell Beaker folk. The rapid appearance of beakers across Central and Western Europe around 2500 BC has been thought to reflect the migration of new peoples into this area. Earlier interpretations of Bell Beaker involved a distinct group of people who were invaders. Although this latter view is no longer completely accepted, the true nature of this people is still unclear.

Eight morphologically healthy adult individuals from two Bavarian Bell Beaker sites (Augsburg and Weichering, respectively) were chosen and Sr isotope ratios were determined in compact femoral bone and enamel of the first permanent molar (Price et al. 1994). $^{87}Sr/^{86}Sr$ ratios are listed up in Table 2:

Table 2. Sr isotope ratios in compact bone and dental enamel of adult Bell Beaker skeletons

Sample no	$^{87}Sr/^{86}Sr_{bone}$	$^{87}Sr/^{86}Sr_{1st\ molar}$
Weichering WA	0.708554	0.709861
Weichering WD	0.708847	0.712821
Weichering W3	0.708641	0.709321
Weichering W5	0.708640	0.708821
Weichering W17	0.708509	0.709524
Augsburg A4	0.708151	0.708551
Augsburg A5	0.708152	0.708600
Augsburg A10	0.708191	0.716319

The differences in Sr isotope ratios are seemingly very small. Sr isotopy of rocks range from > 0.730 (rocks with high Rb/Sr ratios such as shale or igneous rock) to < 0.704 (such as basalt) only. These variations are, however, exceptionally large from a geological standpoint, and far in excess of the analytical error (± 0.00001 to ± 0.00003).

Sr isotope ratios around 0.709 are typical for marine sediments and thus reflect the Sr isotopy of the northern Alpine region which was formed by the elevation of calcareous (shell) rocks. Sr isotope ratios greater than 0.71 are in general typical for the regions north-east from the Danube river. It is clear from the data in table 2 that the individuals A10 and WD, possibly also WA, had different Sr isotope ratios in tooth enamel compared to bone, which strongly suggests migration (Price et al. 1994).

However, a recent reconsideration of the stratigraphy at the Weichering site revealed that the individuals WA and WD were much younger and rather date into the LaTène period. This could open up one more perspective for archaeometry: If prehistoric people had settled at a single site for a considerable time, phases of different mobility could be distinguished and in turn related to changes of e.g. environment and/or subsistence patterns. The preliminary results on the Bell Beaker skeletons indicate that the isotopic technique applied to dental enamel may adequately answer questions concerning mobility and migration. It should be noted that diagenesis is not likely to cause substantial problems: Even in cases where some diagenetic Sr remains in the bone, contamination can add nothing but a local component; Sr isotope ratios will be like those intrinsic to bone, whereas the dental enamel should in general be more resistant to diagenetic additions of Sr. The routine cleaning procedures applied prior to trace element analysis also proved valuable for the analysis of Sr isotope ratios.

Research Potential of Trace Element Analyses of Human Dental Enamel

Trace element patterns in human dental enamel function as an „archive of childhood" in as much as certain aspects of ontogenetic importance can be detected. These include parental investment and child care such as nursing. Not only on a population level, but also within one population, differences in weaning practice can be related to the general circumstances of life. Since babies were at a high risk of falling ill or even die at weaning age especially in preindustrial societies, the estimation of weaning age in individuals who definitely survived this period until adulthood is of special interest.

Children are highly susceptible to heavy metal intoxication. Lead analysis of dental enamel permits the identification of ontogenetic stage and source of the heavy metal exposure during childhood.

The direct reconstruction of migration from an individuals' bodily relic is far superior to the traditional morphological approach. Our pilot study has shown that it should also be possible to tell mobile from non-mobile people who buried their dead at the same place. Possibly, a differential mobility of males and females might give clues to other behavioural aspects such as endogamy versus exogamy.

We conclude that dental enamel is most suitable for trace element analyses due to its matrix properties, and in most aspects poses fewer problems than does bone. Its outstanding advantage is the preservation of patterns of life-style of the investigated individual.

Acknowledgement

Ongoing reserach on stable Sr isotopes of Bell Beaker skeletons was financially supported by the Deutsche Forschungsgemeinschaft. A considerable mobility of Bell Beaker people of southern Bavaria was verified, and exogamy was indicated (Grupe et al. 1997).

References

Burton JH, Wright LE (1995) Nonlinearity in the relationship between bone Sr/Ca and diet: Paleodietary implications. Am J Phys Anthrop 96: 273–282.
Ehlken B (1991) Ontogenetische Trends des Strontium- und Bleieintrages in den Zahnschmelz mittelalterlicher Individuen. Dipl Arb, Göttingen.
Ericson JE (1985) Stontium isotope characterization in the study of prehistoric human ecology. J Hum Evol 14: 503–514.
Graustein WC (1989) $^{87}Sr/^{86}Sr$ ratios measure the sources and flow of strontium in terrestrial ecosystems. In: Rundel PW, Ehleringer JR, Nagy KA (eds) Stable Isotopes in Ecological Research. Springer, New York, p. 491–512.
Grupe G (1986) Multielementanalyse: Ein neuer Weg für die Paläodemographie. Bundesinstitut für Bevölkerungsforschung (ed) Materialien zur Bevölkerungswissenschaft, Sonderheft 7, Wiesbaden.
Grupe G (1992) Analytisch-chemische Methoden in der prähistorischen Anthropologie: Spurenelemente und stabile Isotope. In: R. Knußmann (ed) Anthropologie. Handbuch der vergleichenden Biologie des Menschen, Band I, 2. Teil. Fischer, Stuttgart, p. 66–73.
Grupe G, Bach H (1993) Life style, subsistence and mortality in the Slavonic village at Espenfeld (Kr. Arnstadt, FRG). A trace element study. Anthrop Anz 51: 317–332.

Grupe G, Dreses-Werringloer U (1993) Desomposition phenomena in thin sections of excavated human bone. In: Grupe G, Garland AN (eds) Palaeohistology. Methods and Diagnosis. Springer, Berlin Heidelberg New York, p. 27–36.
Grupe G, Herrmann B (eds) (1988) Trace Elements in Environmental History. Springer, Berlin Heidelberg New York.
Grupe G, Krüger HH (1990) Feeding ecology of the stone and pine marten revealed by element analysis of their skeletons. Sci Tot Envir 90: 227–240.
Grupe G, Piepenbrink H (1989) Impact of microbial activity on trace element concentrations in excavated bones. Appl Geochem 4: 293–298.
Grupe G, Price TD, Schröter P, Söllner F, Johnson CM, Beard BL (1997) Mobility of Bell Beaker people revealed by stable strontium isotope ratios of tooth and bone. A study of southern Bavarian skeletal remains. Appl Geochem 12: 517–525.
Lambert JB, Grupe G (eds) (1993) Prehistoric Human Bone – Archaeology at the Molecular Level. Springer, Berlin Heidelberg New York.
Lambert JB, Weydert-Homeyer JM (1993) Dietary inferences from element analyses of bone. In: Lambert JB, Grupe G (eds) Prehistoric Human Bone – Archaeology at the Molecular Level. Springer, Berlin Heidelberg New York, p. 217–228.
Lowenstam HA, Weiner S (1989) On Biomineralization. Oxford Univ Press, New York.
Molleson T (1988) Trace elements in human teeth. In: Grupe G, Herrmann B (eds) Trace Elements in Environmental History. Springer, Berlin Heidelberg New York, p. 67–82.
Price TD, Connor M, Parsen JD (1985) Bone chemistry and the reconstruction of diet: Strontium discrimination in white-tailed deer. J Archaeol Sci 12: 419–442.
Price TD, Grupe G, Schröter P (1994) Reconstruction of migration patterns in the Bell Beaker period by stable strontium isotope analysis. Appl Geochem 9: 413–417.
Radosevich SC (1993) The six deadly sins of trace element analysis: A case of wishful thinking in science. In: Sandford MK (ed) Investigations of Ancient Human Tissues. Chemical Analyses in Anthropology. Gordon & Breach, Langhorne, p. 269–332.
Sandford MK (1992) A reconsideration of trace element analysis in prehistoric bone. In: Saunders SR, Katzenberg MA (eds) Skeletal Biology of Past Peoples: Research Methods. Wiley-Liss, New York, p. 79–103.
Sandford MK (ed) (1993) Investigations of Ancient Human Tissue. Chemical Analyses in Anthropology. Gordon & Breach, Langhorne.
Schoeninger MJ (1979) Diet and status at Chalcatzingo: Some empirical and technical aspects of strontium analysis. Am J Phys Anthrop 51: 295–310.
Schroeder HE (1987) Orale Strukturbiologie. Entwicklungsgeschichte, Struktur und Funktion normaler Hart- und Weichgewebe der Mundhöhle und des Kiefergelenkes. Thieme, Stuttgart.
Siegert G (1993) Ontogenetische Marker für den Spurenelementeintrag in das menschliche Skelett. Dipl Arb, München.
Sillen A, LeGeros R (1991) Solubility profiles of synthetic apatites and of modern and fossil bones. J Archaeol Sci 18: 385–397.
Sillen A, Smith P (1984) Weaning patterns are reflected in strontium-calcium ratios of juvenile skeletons. J Archaeol Sci 11: 237–245.
Thornton I (1988) Soil features and human health. In: Grupe G, Herrmann B (eds) Trace Elements in Environmental History. Springer, Berlin Heidelberg New York, p. 135–144.

5.3 Gross Dental Wear and Dental Microwear in Historical Perspective

Jerome C. Rose and Peter S. Ungar

Introduction

The wearing away of the tooth surface during the chewing of food is a natural process to which the teeth have continuously adapted since even before they were used by Devonian fish. Since then, teeth have been altered in growth pattern, size, morphology, and structural integration of dentine and enamel to accommodate the various diets exploited over time. The processes of adapting to tooth wear during mastication are in themselves works of art and wonder to be appreciated and marveled over. We take just one example from modern human teeth. As the occlusal enamel and underlying dentine are worn away by the tough gritty foods the pulp or vital living portion of the tooth is in danger of being exposed to the oral environment. Once this happens the pulp becomes infected and abscessed which eventually leads to tooth loss. Not only will the tooth's owner experience excruciating pain and the health dangers of an active infection, but once the tooth is lost, reduced ability to chew and eat. All of this potentially contributes to premature death. But, this usually does not happen and countless teeth are found which have been worn right down to the gums without pulp exposure. As the occlusal surface is worn away, the odontoblasts lining the pulp chamber deposit secondary dentine along the pulp chamber, which provides more tooth substance between the pulp and approaching chewing surface. If the rate of secondary dentine formation can match the rate of wear then there will always be dentine to form the occlusal surface and protect the pulp. Frequent pulp exposures are rare in ancient human groups and only seem to appear at times of initial dietary transitions such as the Mesolithic of northeast Africa (Armelagos et al. 1984; Rose et al. 1993). It is not long before wear rates are reduced and pulp exposures become infrequent. How and why cultures make this adjustment in their wear rates is just one of the many research questions that can be addressed by dental wear researchers.

Gross dental wear is the loss of dental enamel and eventually dentine from occlusal surfaces of the teeth. This chapter begins by reviewing the history of gross dental wear analysis. We show how first the use of wear as an aging technique and then for dietary reconstruction advanced the scoring and analysis of the wear process in ancient human teeth. The strengths and weaknesses of various methods are discussed and recommendations for analysis are made. The gross wear literature is vast and thus only publications making significant or widely adopted analytical advances are included.

Dental wear is produced by tooth-tooth contact (attrition) and tooth-food interaction (abrasion) during mastication. Examining the tooth surface with both light and scanning electron microscopes reveals the scratches, pits, and other features (i.e., microwear) produced during the wear process. Here again we review the history of this new and exciting chapter in dental wear analysis. Because this field of research is new and not widely known, we cover the literature in far greater detail. Again, we explore the strengths and weaknesses of the various methodological and interpretive contributions to microwear research. During the discussion, and in the conclusion, we attempt to envision the directions in which this research effort might be headed. In particular, we believe that the eventual marriage of gross tooth wear and microwear analyses will make a significant contribution to dietary reconstruction of ancient humans and the fossil primates.

Gross Dental Wear

Early osteologists and archaeologists waxed eloquently upon the extensive dental wear observed on the ancient skeletons that they were excavating or examining in their laboratories. From these simple observations in the latter half of the nineteenth century to today, osteologists, dental anthropologists and paleontologists have been examining the causes of dental wear, developing new methods for recording and analyzing the data, and employing these methods to interpret the past. In his review of dental wear research Molnar (1972) identified three pervasive themes: cultural modifications, determining age at death, and dietary reconstruction. Cultural modifications include altering the teeth for aesthetic or ideological reasons, using the teeth as tools in the production of crafts or in food preparation, and incidental behavioral activities that cause tooth wear such as the localized wear from pipe smoking. This aspect of dental wear is discussed in Chapter 5.3 by Alt and Pichler. Because there were few methods for determining age at death from skeletal remains, tooth wear became a major aging technique for osteologists (Brothwell 1963). Although dental wear aging techniques are critically evaluated in Chapter 6.2 by Rösing and Kvaal, we will revisit this topic here because we cannot consider dietary implications of dental wear without an understanding of the effects of age on wear. Osteologists have studied dental wear for more than a century, still, with the advent of the "New Archeology", the "New Physical Anthropology", and Bioarcheology during the past three decades, dental wear research has flowered in methodology, interpretation, and the number of studies. It is this later theme that is the focus of this chapter's first section.

We begin with the earliest phases of research in the late nineteenth and early twentieth centuries where age determination dominated research (although some attention was paid to dietary reconstruction). The temporal approach is used to document the development of dental wear recording schemes and place them within their historical contexts. The earliest improvements in dental wear data collection were made in this context of age determination. Furthermore, all interpretation of dental wear for dietary reconstruction requires that wear rates (i.e. dental wear over time) be calculated and thus, age determination and dietary reconstruction research are inherently interwoven. We then turn to the development of dental wear methods for dietary reconstruction. Here the focus is upon diet, but the relationship of foods and masticatory processes cannot be ignored. The concluding section makes methodological recommendations and hints at future research directions.

Beginning of Dental Wear Research

Mummery (1870) was the first to demonstrate convincingly that dental decay was positively associated with increasingly sophisticated food processing technology in ancient Britain, and, as needs to be the case, he incorporated changes in dental wear within this discussion. Subsequent researchers have all had to elucidate the interaction of wear and caries in diet reconstruction research (see Powell 1985). His results and conclusions have withstood the test of time, including numerous recent analyses of new data from ancient British dentitions. He singled out ancient Egyptian molars for special comment and was the first to attribute the extensive wear to food contamination by the ubiquitous desert sand (Mummery 1870). Since Mummery, interest in the extensive wear on ancient Egyptian molars has continued unabated. There are at least 61 publications which examine or extensively discuss tooth wear among the ancient Egyptians and Nubians (see Rose et al. 1996). Most researchers have attributed Egyptian/Nubian dental wear to two factors: first, living on the edge of the desert and, second, food preparation technology which introduces numerous large particles into the bread (e.g., Smith and Jones 1910; Ruffer 1919, 1920, 1921; Smith 1932; Leek 1972).

Despite the long discussions and association of tooth wear with a coarse diet, tooth wear was never really used in dietary reconstruction of dynastic Egyptians, probably because the authors assumed that they knew the diet from the archeological and historical records. After all, the actual loaves of bread are in museums where the particles are easily observed (Samuel 1993). This widely read literature seemed to eliminate any interest in using dental wear for dietary reconstruction. Additionally, in the early years, osteology was devoted almost exclusively to the attribution of skulls to various racial and subracial groups and paleopathology was confined to identification and description, thoughts of dietary reconstruction were far in the future.

Broca (1879) was the first, or at least is the most extensively cited, to have designed an ordinal dental wear scoring system for molars. This five level system (scoring 0 to 4) was frequently used by early osteologists. For example, Aleš Hrdlička, an early American osteologist trained in Europe, used this system in his osteological analyses of ancient American skeletons which were published as appendices to archeological excavation reports. Unfortunately for those wishing to use his data, he often reported the scores for dental attrition without defining the scale employed (e.g. Hrdlička 1908, 1909). It is clear, however, that he was using the Broca scale for scoring tooth wear although he does not cite its source even in his laboratory manual (Hrdlička 1939) where the scale was defined.

Determination of Age at Death

In the first half of this century there were few methods for determining age at death of human skeletons and molar wear scores were of major importance. In Hrdlicka's (1939) *Practical Anthropometry*, a molar wear scoring system was one of his five methods for determining age at death. He presented a verbal description of five stages with a specific age range for each stage. For example, Stage 2 is with the cusps of the molars worn off and is assigned to the age range of 26 to 33 years. He recognized that both diet and age determine dental wear so the reader was warned that this system is only applicable to "American aborigines" and that the system must be recalibrated for use with other groups. Molar wear scores collected with this system were widely reported in the early literature and were only replaced by other methods in the 1960's. Smith (1984) said that more than a dozen molar wear scoring systems were reported in the literature following this early work, but she did not name or cite them, while Lunt (1978) cited three modifications.

The use of molar wear scores for age determination is not confined to ancient human skeletons. Field zoologists have long used dental wear for determining ages at death and zooarchaeologists have borrowed and modified these wear scoring systems to determine age at death of various species found in archaeological sites (e.g. Chaplin 1971; Hillson 1986). For example, faunal mortality profiles derived from these ages have been used to distinguish patterned hunting from scavenging activities (e.g. Klein 1982). Patterns of tooth wear have also been used to determine the age at death of extant primates (e.g. Bramblett 1969; Meikle 1977).

The age-diet interdependence requires that when molar wear scores are used for either age determination or diet reconstruction that the other factor is held constant. This is achieved by developing methods for calculating wear rates (wear by time) rather than simply comparing wear score averages. In one of the first attempts at wear rate estimation (holding diet constant) Rainer Zuhrt (1955) used the age intervals between the eruption of the first, second and third molars (approximately 6 years between each) to calculate the rate of wear and, thus, age his historic German skeletal series. He continued using this method in his study of Neolithic Nubian skeletons (Zuhrt 1967). Zuhrt's method seems to have attracted little attention from osteologists, except a citation by Brothwell (1963).

Miles (1958) was the next to employ the years between successive molar eruptions to ascertain the yearly rate of attrition using an ordinal scoring system. Subsequent publication of this system (Miles 1963) in Brothwell's widely read *Dental Anthropology* provided wide exposure and use of this system by some osteologists. Miles (1978) employed individuals who had been aged by dental and epiphyseal development to be younger than 20 years to calculate the wear rates (wear scores per units of time). He matched wear on the second and third molars with that on the first and used the years between successive eruptions (approximately six years) to determine how many years were required for the posterior molars to reach the same stages of wear as the first. In other words, he calculated a gradient of 6.0:6.5:7.0, which means that it took the second molar 6.5 years and the third molar 7.0 years to reach the same stage of wear as the first molar did in six years (Miles 1978). This gradient was employed by others in their tooth wear aging systems (e.g. Lovejoy 1985). Despite the obvious advantage of using a diet independent system, most osteologists still preferred the simple system of pictorial molar wear stages with an age range attached to them for aging skeletons and recording dental wear.

Brothwell (1963) provided such a simple system in his osteology textbook, *Digging Up Bones,* which has gone through numerous editions and printings. He provided a 12 stage scoring system which included drawings of molars showing successive cusp wear, exposure of dentine, and eventual wearing away of the entire crown. For age determination there were four age groups with each group having drawings of all three molars showing three successive stages of dentine exposure. He noted that wear rates do not change significantly between the Neolithic and Medieval periods and, thus, the system can be used to age most ancient British skeletons. Brothwell cautioned (following Hrdlička) that this molar wear aging system applies only to British skeletons and that it must be recalibrated for use elsewhere where diets and other factors differ.

Brothwell's wear/age chart was also reproduced in Bass' (1995) *Human Osteology* manual, which has also gone through numerous editions and printings. Ubelaker (who wrote the chapter on teeth in the most recent edition) clearly states that this system is culture/diet specific and cannot be used on different populations. But how many archeologists and beginning osteologists (including the numerous physicians turned osteologist) have carefully read or remembered the "warning labels" in these two very popular osteology field manuals? The total number of skeletal appendices throughout the world which cite either Brothwell's or Bass' manuals as their osteological methods are legion. The fact that these manuals are often the only methods cited implies that this dental wear chart may have made a major contribution to recording molar wear on many thousands of skeletons throughout the world. The Brothwell system has mostly replaced the Broca/Hrdlicka system for the routine analysis of skeletons throughout the world, although Broca's scheme can still be found in use today (e.g. Mucič and Đurič-Srejič 1996).

Dietary Reconstruction and Determination Rates of Wear

While most researchers were employing dental wear for age determination, others were exploring the relationship between diet and wear (for a more detailed discussion see Molnar 1972; Powell 1985). Campbell (1925), in his study of Native Australians, pointed out the numerous features of the environment and diet which contributed to advanced tooth wear, while Waugh (1933) discussed the influence of dietary composition on the teeth of Alaskan Eskimos. Buxton (1920) noted that subsaharan African pastoralists exhibited softer diets and less dental wear than agricultural groups. Both Leigh (1925) and Moodie (1929) relate diet and dental wear in their studies of North American skeletal samples and like most North

American researchers attributed the variation in dental wear to whether or not stone grinders were used in food preparation. Molnar (1972) in his review article said that the literature abounds with assertions that tooth wear can be used in dietary reconstruction. Tooth wear (as well as size and morphology) has from the beginning been used in the reconstruction of early fossil hominid diets (e.g. Robinson 1956; Grine 1981, among hundreds of others).

In two publications, Murphy (1959a, b) introduced the first system to replace the Broca/Hrdlička scores in dental wear research (though most osteological analyses found in site report appendices continued to rely on Brothwell). Murphy provided a graded system of ordinal scores to describe the gradual exposure of dentine using drawings of the occlusal surfaces of all tooth classes and textual descriptions of each grade. The molar wear scoring system is provided here:

- Dentine is exposed on one cusp only;
- Dentine is exposed on two cusps;
- Dentine is exposed on three cusps;
- Dentine is exposed on all four (or more) cusps;
- Two areas of exposed dentine have coalesced, leaving two discrete areas;
- Three areas of exposed dentine have coalesced, leaving one discrete area;
- Areas of exposed dentine on all the cusps have coalesced, leaving an island or peninsula of enamel still occupying a portion of the center of the occlusal surface; and
- Dentine is exposed over the complete occlusal surface, save for a rim of enamel partially or completely surrounding the surface (Murphy 1959b).

He then proposed that the difference in wear scores between the successively erupting molars provided a wear gradient calculation that estimated the magnitude of molar wear independent of age (see also Miles 1958, 1963, 1978).

This independence from age is the critical concept which permits dental wear to be used for dietary reconstruction, so the calculation of wear rates (or other statistical methods to control for age) became the methodological focus for all subsequent research. Murphy recognized that it is not the total amount of wear (mean wear scores) observed in a skeletal sample that is important, it is the rate of wear that differentiates diets. If skeletons could be precisely aged and placed into groups with small intervals of years between them, then wear rates could be calculated. Even differences in wear between same aged groups would permit differences in wear rates to be measured. Because this is not possible, the wear gradient score attempts to estimate rates using the differences in exposure to the diet of the first, second and third molars produced by the differences in eruption times between them (approximately 6 years). This is, in theory, a simple process whereby the wear scores of one molar (e.g. the second) is subtracted from the other (e.g. the first) and the difference is the wear gradient which estimates the wear rate independently of age. Powell (1985) provided an example of how the Murphy gradient is calculated and used in dietary reconstruction. However, this is not really so simple and subsequent innovations focused upon more detailed and comprehensive scoring systems and the application of various statistical procedures to increase sensitivity of the data analysis.

The 1960's and 1970's witnessed a theoretical and methodological revolution in American archeology. Archaeologists advocated the study of cultural process, ecological relationships, subsistence pattern reconstructions, and the development of field and analytical methods to accomplish these goals (Willey and Sabloff 1974). These changes in archeology provided motivation for enhancing the integration of archeology and the study of human skeletons, in addition to producing the archeological data necessary to conduct this integrated research. The commingling of the "New Archeology" (Willey and Sabloff 1974) with the "New Physical Anthropology" (Washburn 1951) produced the field known in the U.S. as Bioarcheology (see Blakely 1977; Buikstra 1991). These events were the primary stimulants for generating interest in and developing new methods for the reconstruction of ancient diets using tooth wear, among many other skeletal features such as bone biodynamics, trace element and stable isotope analyses to name just a few (see Larsen 1987). A new era of wear studies began where archeological subsistence reconstructions were tested with skeletal data. Molnar (1972, 515) stated: "the study of the dynamics of tooth wear, its relation to the oral-facial complex, and the interaction between certain aspects of culture and teeth is only beginning."

An early study reflecting this association of archeology and dental wear analysis was by Patricia Smith (1972) who examined the physical consistency of the Natufian diet. Natufians are protohorticulturalists occupying the Jordan River region of the Middle East and this study tested the dietary reconstruction produced by archeological interpretation. Recognizing the high age-tooth wear correlation and the inability to place skeletons within small age brackets, she advocated using wear gradients rather than mean wear scores. She employed a five point scale similar to the Broca/Hrdlička system, but added the refinement of scoring each molar cusp separately, which presaged the Scott system to be discussed later. The rankings of score differences between the first, second and third molars were compared using Spearman's rank correlation coefficient. The slopes (angle of increase) of the regression lines showed that two sites were very similar with high wear gradients and different from the third with a lower gradient. The high wear was attributed to high grain consumption consistent with the interpretation of Natufian subsistence, while the third was associated with hunter and gatherer archeological assemblages. In a subsequent comparison of Neanderthal groups she emphasized that it is the pattern, not the severity of wear, that leads to understanding masticatory processes and diet (Smith 1976). Regression analyses of wear scores of first and second molar, and canine and first molar pairs was employed to demonstrate a homogeneity of all Mousterian fossil groups.

Lunt (1978) tested the proposition that wear decreases with a more refined diet by contrasting medieval Danish with prehistoric skeletons. The Murphy system was selected over Broca's (and its modifications) because it had a larger number of grades and is "entirely objective." The mathematical difference in successive molar wear scores (or wear gradients) was used in the analysis rather than the wear scores themselves. Her procedural contribution was emphasizing that wear scores, which are ordinal data, require nonparametric statistics for analysis. Interpretation of those differences in wear gradients found to be statistically significant simply employed wear gradient frequency distributions plotted against first molar wear scores to control for age (Lunt 1978). Lunt selected this method because, she argued, analysis of mean gradients (as done by Murphy 1959b and others) or even use of the Mann-Whitney U test are statistically unsound.

Beyond Wear Scores: Occlusal Surface Shapes and Angles

A number of researchers noted that various factors including diet and food preparation technology not only produced differing rates of wear, but also differences in occlusal surface shapes (Murphy 1959a; Brace 1962; Taylor 1963). For example, Leek (1972) reported three patterns for ancient Egyptian molar occlusal surfaces: horizontal, oblique, and concave. Molnar (1971) built upon these ideas and proposed a comprehensive methodology for relating tooth wear to diet and food preparation technology. He criticized earlier recording schemes for only scoring molar occlusal surface loss and ignoring the remaining dentition along with the complexities of the wear pattern, and was the first to provide a comprehensive scoring system for all functional tooth classes which recorded degree of wear, angle of wear, and occlusal surface shape. Molnar presented drawings and textual descriptions for recording degrees of occlusal wear. He further developed a technique and scoring scheme for measuring the angle of the wear plane, and a classification system for surface shape such as flat, cupped, and rounded. This system was tested by analyzing the wear patterns of three North American skeletal series, the results of which showed excellent concordance with established dietary and food preparation reconstructions. This system was comprehensive and methodologically sound, but, unfortunately, it was sufficiently complex and time consuming that it has not often been adopted for routine analyses of skeletal collections, although it has been used in a number of specialized studies of dental wear.

Interstitial Wear

In addition to occlusal surfaces, teeth also experience interstitial (interproximal) wear – the wear that occurs between teeth where they touch as they move up and down during mastication. Increase in interstitial wear is associated with both increased masticatory forces and time spent chewing. It is directly related to the overall toughness of the food and development of the muscles of mastication. The amount of enamel lost at the contact areas between teeth can be dramatic. This process has been of greatest concern to those studying fossil groups (see Wolpoff 1971 for a review), but is seldom measured when studying modern humans, with some exceptions (e.g. Reith 1990). One interstitial research domain is Begg's (1954) notion that interstitial wear reduced the mesio-distal length of the tooth and prevented malocclusion in ancient peoples and those modern groups still subsisting on an unrefined diet. Excellent reviews and refutations of this concept have recently been published (Corruccini 1990, Kaul and Corruccini 1990).

Wear and the Origins of Agriculture

If one theme characterizes the bioarchaeological/osteological research of the 1980's, it is the origins of agriculture: when, where, and how did it occur, and what were its health consequences? The literature abounds with articles, commentaries, and edited symposia. The origin of agriculture quest also had its impact upon dental wear research and, in turn, this research has made significant contributions to our understanding of this phenomenon in both the New (e.g. Powell 1985) and Old Worlds (e.g. Smith 1984). The role of dental plane angles, discussed by so many over the decades and championed by Molnar (1971), was brought to bear on these questions.

Hinton (1981) took Molnar's advice and examined the changes in pattern of anterior tooth wear between hunter-gather and agricultural samples from several parts of the world. He scrutinized both the rate of wear and changes in the occlusal surfaces (i.e. flat, rounded, cupped). Not being able to calculate wear gradients he used molar wear scores as an estimate of functional age. The differences in wear on different teeth erupting at the same ages were compared to molar wear scores as the constant in place of age. For example, mean wear scores (i.e., distribution centroids) of mandibular central incisors and first molars were plotted as a function of second molar wear scores. Although the results were in the predicted direction, the analysis ignores Lunt's statistical cautions.

B. Holly Smith (1984) contended that the concept of flat molar wear being associated with "tough–fibrous" foods could be used to identify the earliest stages of the food producing revolution. The contention was that the angle of molar wear should change independently of changes in rates of wear. Using a modified protractor and customized Murphy wear scoring system, Smith measured the bucco-lingual inclinations of molar occlusal surfaces on 15 skeletal samples representing hunting and gathering subsistence strategies and 17 samples of food producers from North America, Europe, and the Middle East. Age variation within and between the samples was controlled by calculating least squares regressions of occlusal angles on the eight wear stages of the first molars. The slopes of the regression lines differentiated the subsistence strategies. A more gradual slope indicated a slower increase in occlusal angle with advancing wear. The hunter-gatherers maintained flatter wear although all groups showed increasing occlusal angles as wear progresses. The consistent trend was that both subsistence types started off with similar occlusal angles and began to separate when the molars were at the third of eight wear stages. By the time individuals at stage 5 wear were compared, the two subsistence groups have quite distinct occlusal angles and the distances between the regression lines increased as individuals with higher wear scores were compared (i.e. the regression lines differ in slope). Smith (1984) concluded that wear plane angles were sensitive to differences in food and food preparation technology independent of variation in the rate of wear. Thus, in examining skeletal sequences over time, a change in the wear angle within any given wear stage should identify the earliest phases of this subsistence revolution. Smith's (1984) methodological contributions were also significant. The modified protractor is easy to construct and use, and good replicability ±1.5° was consistently achieved. Various wear scoring systems were tested and the Murphy system was found to be the most useful. A modified wear scoring system with 8 categories for each of the three functional classes of teeth (incisors and canines, premolars, and molars) was presented in the traditional manner of a graded series of occlusal surface drawings of dentine exposure accompanied by textual descriptions. This methodology has since been effectively employed by others (e.g. Schürmann 1986).

Tooth wear analysis continues to play an important role in understanding the biological consequences of subsistence change, but it must be kept in mind that it is just one component of this complex analysis. To emphasize this point one example, from the many available, is provided here. Northern Sudanese Mesolithic (circa 12,000 B.P.) peoples are characterized by large teeth with complex morphology, large faces, bun-shaped skulls, massive brow ridges, sloping foreheads and heavy musculature (see Armelagos et al. 1984; Rose et al. 1993 for summaries of the numerous osteological and dental studies concerned with this topic). Craniometric analysis demonstrated cranial changes from the Mesolithic to the Christian period (550–1100 A.D.) that reflected changes from a hunter-gatherer subsistence to agriculture: smaller teeth with simpler morphology, elevated forehead, rounding of the posterior cranium, and decreased musculature. They postulated that selective pressures changed with the transformation from a coarse diet with extensive tooth wear to a softer high caries diet (Armelagos et al. 1984). Smith (1984) used her regression technique to demonstrate a reduction in tooth wear consistent with this model. Other studies have shown a parallel decrease in tooth size and an increase in caries rates within this temporal sequence of skeletons (for a review see Rose et al. 1993). In this case, derived from more than a dozen publications, tooth wear analysis played a crucial role, but it was only one of many methodological elements.

Continued Advances in Scoring Wear

Scott (1979a, b) offered the next innovation in refining the ordinal molar wear scoring systems by increasing the number of potential scores per molar from 8 to 36. Scott divided each of the molars into four quadrants, as did Smith (1972), and scored each quadrant (simplifying to a four cusp molar and ignoring the fifth) independently using a 10 point system:

0. No information or tooth missing or unerupted.
1. Wear facets invisible or very small.
2. Wear facets large, but large cusps still present and surface features very evident. It is possible to have pinprick size dentine exposures or dots which should be ignored. This is a quadrant with much enamel.
3. Any cusp in the quadrant area is rounded rather than being clearly defined as in 2. The cusp is becoming obliterated but is not yet worn flat.
4. Quadrant area is worn flat (horizontal) but there is no dentine exposure other than a possible pinprick sized dot.
5. Quadrant is flat, with dentine exposure one-fourth of quadrant or less.
6. Dentine exposure greater: more than one-fourth of quadrant area is involved, but there is still much enamel present. If the quadrant is visualized as having

three sides the dentine patch is still surrounded on all three sides by a ring of enamel.

7. Enamel is found on only two sides of the quadrant.
8. Enamel on only one side (usually outer rim) but the enamel is thick to medium on this edge.
9. Enamel on only one side as in 8, but the enamel is very thin – just a strip. Part of the edge may be worn through at one or more places.
10. No enamel on any part of the quadrant – dentine exposure complete. Wear is extended below the cervicoenamel junction into the root.

The four quadrant scores are then summed to produce a wear score ranging between 4 and 40 for scorable teeth. Scott contended that this system was reliable and produced lower variances than the Murphy or Molnar system. Scott's (1979a) ultimate goal was to provide a method for using molar wear to distinguish dietary patterns and, thus, the effect of age must be controlled. She used a principal axis statistical procedure to provide the comparisons of molar wear rates independent of age. This procedure entailed calculating the principal axis of an ellipsoid produced by wear score plots of pairs of teeth. The slopes of the axes were used to compare the different data sets and establish the changes in wear rates with the adoption of agriculture among her North American skeletal samples. Scott compared the results obtained from her system with those derived from Molnar's scoring method. The principal axis technique produced better results when using her method rather than Molnar's because the confidence regions intevals were smaller.

The Scott system was tested by Cross and coauthors (1986, 103) and found to be "straightforward and easy to use." They tested the system using two observers on 33 adult dentitions. They found the two sets of scores to be highly correlated and found only significant differences when scoring the earliest stages of wear. They conclude that the system is objective and repeatable, and therefore recommend that it be used in attrition studies.

All wear studies discussed to this point employed ordinal wear scores and many researchers have recognized that the intervals between sequential scores do not represent a linear or equal loss of tooth substance (i.e. an interval/linear measurement). Lunt (1978) suggested that these ordinal wear data present a statistical challenge. He argued that because the wear scores are not normally distributed the use of regression techniques is suspect and, further, that the Pearson product moment correlation coefficient (r) does not measure the amount of wear between molars as Smith (1972) assumed (Benfer and Edwards 1991).

We would add that ordinal data may not present the challenge that previous workers envisioned. Although such data are not parametric, Conover and Iman (1980) have demonstrated that data ranks can be used in conventional tests such as anovas, linear regressions and Pearson's correlation coefficients to form a bridge between parametric and nonparametric procedures. Although tests on rank-transformed data are less powerful than comparable tests on normally distributed data, they may allow comparisons of ordinal wear values between groups using conventional and readily available statistical procedures.

Further, Scott (1979a) and later Benfer and Edwards (1991) pointed out that the Scott four quadrant scoring system with its large number of gradations actually does approach a normal distribution. Despite this fact, they advocated the principal axis method, which does not require the assumption of a normal distribution as it is a distribution free calculation (Benfer and Edwards 1991). While Scott (1979a) used the slope of the axis to differentiate wear rates among skeletal samples, Benfer and Edwards (1991) contended that the axis intercept is a measure of six years of wear on the first molar. Thus, the intercept value can be divided by six to produce a measure of the rate of wear per year permitting calculation of wear rates without having to use numerous adolescent dentitions (the rarest of skeletons due to low mortality rates during these years) as required by the Miles (1978) method. The authors implied that this statistical procedure overcomes the limits of an ordinal scale and approaches the results obtainable if tooth wear could be measured on an interval scale.

Walker (1978) recognized the limits of ordinal wear scores and offered a direct measure of dentine exposure areas as an interval-ratio technique. Photographs were taken perpendicular to the occlusal surfaces along with a scale and the negatives were projected at 70 magnifications for measurement. The outlines of the occlusal surface and all dentine exposures were traced with a compensating polar planimeter and the surface areas were calculated (accuracy = 0.003 mm^2). The differences between areas of exposed dentine on successive teeth (M1 minus M2 and M2 minus M3) were added together to provide the area of dentine exposed over a 12 year period (wear rate). These indices of the "abrasive quality of the diet" were standardized for body size to control for caloric intake (i.e. volume of food consumed). The advantage of this technique is that it meets the parametric statistic requirements, but it is a time consuming method and the rate of dentine exposure (or enamel removal) is not truly linear.

Walker and coworkers (1991) next proposed using crown heights and occlusal surface angles to measure wear rates and to provide an aging technique. Molar crown heights were measured from the cemento-enamel junction to the occlusal surface on all four quadrants and summed for analysis, while occlusal angles were measured with a modified protractor similar to Smith's (1984). These data were then used to develop an aging method using dental development and pubic symphysis criteria to determine the "known" ages at death. Various predictive models were developed and their predictive accuracy was tested using a "jackknife" technique and the sum of squares of the errors rather than the usual correlation coefficient. Interestingly, normalization of crown height data for tooth size (bucco-lingual and mesiodistal diameters) did not improve age prediction, which suggests that this factor may not have to be considered in dietary reconstruction, at least when comparing skeletal series which are not genetically distinct. However, incorporating occlusal angles into the models did significantly improve accuracy, which suggests that occlusal angles need to be included in dietary reconstruction methods, although Benfer and Edwards (1991) stated that scoring the four tooth quadrants for wear should incorporate this factor. Walker et al. (1991) demonstrate that dental wear is nonlinear with the rate of crown height loss decreasing with age (i.e. as wear approaches the cemento-enamel junction) and approximating a negative exponential curve. The rate of permanent molar crown height reduction was significantly different between those in the first and those in later decades of life. This forced them to use the square of summed crown heights in their predictive model. This technique does meet the parametric statistical requirement of an interval-ratio scale, but it still suffers from the same problems as the ordinal scoring schemes such as nonlinearity of wear. Benfer and Edwards (1991) also used crown height measurements and demonstrated that they are no better for calculating wear rates than the Scott scoring system and principal axis analysis.

Where Are We now and Where Are We Going?

Having provided a condensed literature review, we summarize where we are now, make recommendations, and hazard some possibilities of where we might go with dental wear research. The preceding discussion placed dental wear research into two interrelated venues: determination of age at death and dietary reconstruction. There have been and still are many osteologists working with archeological expeditions who use dental wear to age the skeletons excavated from sites large and small throughout the world. Here the Brothwell system still holds almost exclusive domain despite the fact that during the past four decades dental anthropologists and osteologists have proven beyond any doubt that both diet and age determine the degree of wear observed on ancient teeth. Considerable effort has been expended developing a variety of techniques for determining rates of dental wear that can then be used in age determination, while controlling for dietary variation. Despite the potential limitations of dental wear aging (see Rösing and Kvaal, this volume) these efforts are motivated by the simple fact that, while teeth are always the best preserved portion of the skeleton, the preferred aging features (e.g. pubic symphysis, auricular surface) are the most prone to being rendered useless. It is simply a fact that the vast majority of skeletons are poorly preserved and fragmentary and thus dental wear will be used as an aging technique and technical innovations will continue (e.g. Lucy et al. 1995). However, despite documented success, it is rare to find a routine skeletal analysis (an osteological chapter or appendix in a site report) that has employed one of the appropriate techniques for aging ancient skeletons (e.g. Lovejoy 1985; Benfer and Edwards 1991; Walker et al. 1991). There is a great need for one or more of these techniques to be incorporated into an osteology manual which will be widely used. We suspect that a technique such as Lovejoy's will be more readily embraced as an aversion for extensive statistical manipulation militates against those offered by Benfer and Walker.

Studies included in this synthesis, and numerous others not mentioned but just as significant, have demonstrated that gross dental wear analysis has and can successfully contribute to dietary reconstruction. The transition to agriculture has been repeatedly documented using dental wear data. Increasing refinement in food preparation technology is reflected in reduced tooth wear rates, while even more subtle changes in foods and food sources have been detected. Dental wear has proven to be a valuable tool in the weaponry of those engaged in dietary reconstruction. It is interesting to note that studies of tooth wear employing the simplest recording schemes, such as Broca's, to the most sophisticated, such as Molnar's and Scott's, have all successfully documented changes in diets and food preparation technology. If this is true, what is the reason for continuous development of more sophisticated methods? On the surface and clearly stated in the introductions of most publications it is searching for greater sensitivity in detecting the subtlest of dietary changes.

Underlying this obvious motive for methods development is the real reason. It is changes in rates of dental wear which must be compared, not the total amount of wear, and to do this it is necessary to know the precise ages at death of the skeletons. If this were possible, which it is not, then the crudest ordinal dental wear scoring system would be sufficient. Thus, greater refinement in scoring dental wear along with evaluation of appropriate statistical methods have been necessary to approximate yearly rates of wear. The use of the Scott system and principal axis analysis seems to have achieved this goal, although the analysis of crown heights appears to be as successful, but possibly not worth the additional time in data collection. Considerable development effort has also gone into various schemes to record occlusal angles and changing occlusal surface shape (e.g. cupped, rounded). These techniques have proven to offer greater resolution in detecting dietary change, but the increased effort in data collection seems to have discouraged their widespread adoption.

In addition to those conducting specialized dental research there are the bioarcheologists dealing with the collection and analysis of all osteological data from large series of skeletons from one or more archeological sites and for whom dental wear is just one of a large number of data sets that must be collected. For this group, standardization of data collection has been a major concern for some decades (e.g., see the recommendations of the Workshop of European Anthropologists for determining sex and age of skeletons; Ferembach et al. 1980). Most recently in the U.S., passage of federal and state repatriation laws (e.g. Native American Graves Protection and Repatriation Act, Rose et al. 1996) has forced federal and state agencies, museums, universities, and individual archeologists and osteologists to seek standardization in their hurried attempts to collect skeletal data prior to repatriation and possible reburial of tens of thousands of skeletons. In the early stages of this process the Field Museum of Natural History in Chicago, faced with immediate skeletal repatriation, obtained financial support from the National Science Foundation to convene a workshop on osteological data standardization (see Buikstra and Ubelaker 1994). Ten osteologists and other specialists were assembled to develop a standardized system for collecting and recording osteological data from the entire skeleton. Discussions of which methods to employ for each data category included consideration of what could be learned, the time it takes to collect the data, the cost of specialized equipment for data collection, the limited financial resources of most institutions, and acceptability to the broader profession of osteology. Each data collection protocol was debated by the participants keeping all of these limitations in mind.

The final recommendations for dental wear analysis include the recording system of Smith (1984) for the anterior teeth and premolars because it is easy to use and has been profitably applied to a variety of dental research projects (Buikstra and Ubelaker 1994). The molar wear scoring system of Scott (1979b) was selected because of its enhanced discrimination when molar wear is slight to moderate. It was thought that when these systems are employed along with caries data on the many thousands of skeletons which would be studied using this system that dietary reconstructions would be possible.

What might the future hold? First, advances and decreasing costs in computer technology and imaging techniques may very well permit the more complex previously discussed methods to be routinely employed. Rather than four caliper measurements of crown heights, occlusal areas, total volume of remaining crown, as well as occlusal angles may be measurable using semiautomated image analysis routines and affordable photographic/video equipment. Further, recent advances in morphometrics techniques may allow us to get an even better handle on gross tooth wear. We can now generate intricate three dimensional representations of teeth using landmark data. Ungar et al. (1994) recently demonstrated that a 3-space (Polhemus Corp.) electromagnetic digitizer can be used to map the occlusal surfaces of the molars of humans and non-human primates. Other techniques, such as convergence videogrammetry (see Spencer and Spencer 1993) hold similar promise for characterizing tooth form and wear. We can then use coordinate-free shape statistical tests, such as procrustes analyses and euclidean distance matrix analyses (Bookstein 1991; Rohlf 1991; Lele and Richtsmeier 1995) to describe and localize form changes in teeth through a given wear sequence.

Most studies reviewed in this chapter used skeletal series from various regions to test the techniques; what is now needed are numerous routine analyses and comparisons of data between temporally sequential skeletal series from the same cultures and ecological regions. This will eventually come to be if standardized recording schemes would be routinely used in all skeletal analyses permitting the aggregation of data from numerous studies. Secondly, and probably foremost, advances in dental microwear research (to be discussed below) show promise of being able to characterize the physical properties of diets and in some cases specific foods and food processing techniques. Dietary reconstruction using gross dental wear will come into its own when it is combined with the results of microwear analysis. Wear rates will characterize total mastication effort and grit load, while microwear will provide the fine detail of diet content. Together the two methods will make the next leap in dietary reconstruction.

Dental Microwear Research

Researchers have also recognized that microscopic traces of dental wear might be used to infer aspects of diet and anterior tooth use in humans and other mammals. The study of microscopic wear on tooth enamel has become the subject of a relatively new branch of dental anthropology: dental microwear research. Practitioners of dental microwear studies have recognized that patterns of tooth wear can tell us something of ingestive behaviors and the sorts of foods eaten by an individual. This approach has clear implications for both bioarcheology and paleoanthropology – the reconstruction of diet and tooth use in past peoples and fossil primates. In this section we take an historical approach to review this relatively recent addition to the field of dental anthropology. What is dental microwear and why is it important? What can dental microwear tell us, and what is in store for this burgeoning approach as we head into the 21st century? Recent work underscores the potential of such research, and provides a glimpse at things to come.

The history of dental microwear research can be divided into four phases: 1) embryonic stage (1950s–1960s); 2) infancy (1970s); 3) paradigm shift (early to mid 1980s); 4) rebirth (mid 1980s); and 5) microwear today. We will discuss each stage in turn, focusing principally on anthropological applications of this important technique.

The Embryonic Stage (1950s and 1960s)

In 1933, Simpson proposed a correlation between jaw movement in mammals and principles of molar occlusion. He suggested that teeth met in alternation, opposition, shearing and grinding. Teeth were said to serve as guides for specific types of jaw movements, which reflect particular diets. The earliest writings on microscopic tooth wear set out to look for evidence of such jaw movements. For example, Butler (1952) argued that since molar facets are produced by contact or close approximation of upper and lower teeth, they record a part of jaw movement. He and colleague Mills suggested that the direction of jaw movement was reflected in the orientations of microscopic, subparallel striations on molar teeth (Butler 1952; Mills 1955, 1963, 1967; Butler and Mills 1959). These and subsequent studies (e.g., Kay and Hiiemae 1974) focused more on jaw movements and the mechanics of chewing *per se* than on direct associations between foods eaten and patterns of microscopic wear on teeth.

Some early researchers also examined the microscopic wear caused by food abrasives for more direct evidence of diet. Among the first studies was that of Baker, Jones and Wardrop (1959) on New Zealand sheep molars. They suggested that individual wear features resulted from chewing, and were caused by opal phytoliths in grass blades and quartz in the soil because these minerals are harder than tooth enamel. This idea was corroborated by the presence of fractured phytoliths and angular quartz fragments in the sheep feces.

Such studies did not enter the anthropological literature until 1963 when Dahlberg and Kinzey examined several human teeth with a light microscope. They proposed that the nature of microscopic scratches on tooth enamel might relate to properties of the foods that caused those scratches. They further suggested that studies of variation in microwear within and among species would allow the inference of concomitant differences in diet.

Infancy Stage (The 1970s)

There was little microwear research in the decade that followed Dahlberg and Kinzey's publication. Then, in the late 1970s, three critical papers on living mammals followed. First, Philip Walker (1976) published an analysis that used a light microscope to examine the incisors of a series of living non-human primates *(Colobus, Macaca, Papio, Presbytis).* This study suggested that terrestrial monkeys possess more striated dentine surfaces than arboreal forms because of feeding substrate, siliceous material in foods eaten, and the mechanical demands of food breakdown. Also, striation orientation differences between colobines and cercopithecines led Walker to speculate that folivores preferentially strip leaves laterally across the incisors.

Then in 1978, two further studies introduced anthropologists to the use of the scanning electron microscope (SEM) in microwear analyses – though neither project focused on humans or non-human primates. John Rensberger (1978) used the SEM to document molar microwear differences among six genera of rodents. He defined several discrete types of microwear, and related differences to a number of factors, including enamel structure and diet. At the same time, Alan Walker and colleagues (1978) used the SEM to document molar microwear patterns in hyraxes. This study showed that seasonal changes in the diet of *Procavia johnstoni* are associated with microwear changes such that browsing produces pitted surfaces whereas grazing leads to a preponderance of parallel striations. Further, fecal analyses of *Procavia* and *Heterohyrax* (another hyrax species) suggested that opal phytoliths were the most likely cause of these microwear features. These and related pioneering studies (Walker 1980, 1981) suggested that microwear might be of considerable significance to studies of palaeobiology.

Fossil Studies

At about the same time, attempts were being made to use dental microwear to infer the diets of fossil hominids (Albertini and Puech 1977; Grine 1977, 1981; Puech 1979; Puech and Prone 1979; Ryan 1980; Puech et. al. 1981; Walker 1981). These early studies were largely qualitative, though comparative in nature. For example, Puech and coauthors attempted to experimentally create a microwear "template" with which to compare various early hominids. They examined the appearances of enamel surfaces on the molars and incisors of *Homo heidelbergensis*, *Homo habilis*, *Paranthropus boisei* and *Australopithecus afarensis*. These authors suggested that australopithecines had a varied but grit-laden vegetarian diet, that *Homo habilis* was a frugivore/insectivore and that later hominids were essentially human-like hunter-gatherers.

In contrast, Alan Walker (1981) compared microwear on East African early hominid molars to that of living mammals. This work contributed much to the rejection of some of the most popular dietary scenarios of the day and showed that early hominids were not grass seed eaters, nor were they bone crushing scavengers. In contrast, Walker argued that East African early hominid dental microwear looked like that of living frugivores such as mandrills, chimpanzees and orangutans.

Ryan (1980, see also Ryan and Johanson 1989) took a similar approach to the study of anterior tooth microwear. He examined the incisors of *Australopithecus afarensis*, *Homo neanderthalensis* and an assortment of living primates, including *Pan troglodytes, Gorilla gorilla, Papio hamadryas* and human Inuit, Amerinds, Micronesians and Europeans. Much like Walker, Ryan used associations between diet and microwear in living primates to infer diet from microwear in fossil forms. He suggested for example, that *Australopithecus afarensis* had a mosaic of gorilla and baboon-like microwear features indicating the use of incisors to strip gritty plant parts including seeds, roots and rhizomes. In addition, he argued that Neandertal incisor microwear resembles that of some modern humans, reflecting power grasping unrelated to diet (e.g., hide-processing).

Finally, Grine (1977, 1981) compared deciduous molar enamel microwear patterns between South African early hominid species. He noted in 1981 for example, that *Australopithecus africanus* had more striated facets, whereas *Paranthropus robustus* showed more heavily pitted molar surfaces. He took this to suggest the consumption of small-hard food objects by the "robust" australopithecines, and softer foods, such as fruits and immature leaves by the "gracile" forms.

Paradigm Shift (The Early to Mid 1980s)

At the close of the 1970s, the future of dental microwear studies looked bright indeed. Researchers documented patterns of microwear, compared them between and among living and fossil taxa, and optimistically related what they observed directly to diet. Then, between 1981 and 1983, a series of papers burst the bubble of optimism and forced a complete reevaluation and eventually a paradigm shift in dental microwear research. First, Covert and Kay (1981) conducted a three month experimental study on three test groups of American opossum, *Didelphis marsupialis.* One group was fed cat food with plant fiber (soybean hulls), another cat food with insect chitin, and a third control group cat food only. Covert and Kay were not able to distinguish the three groups by microwear patterns, and so concluded that microwear analyses could not distinguish herbivory from insectivory in fossil primates.

Further, Peters (1982) experimentally produced microwear on a human tooth using chert, bone, mongongo nuts, *Grewia* berries, carob beans and wild-onion bulbs. Peters suggested that while puncturing and crushing bones and hard legumes produce distinctive patterns of microwear, scratches caused by exogenous grit on foods are rather similar to those evinced by grass phytoliths in foods. He further argued that some primate foods, such as dicotyledonous seeds, do not by themselves even produce microwear.

Covert and Kay's study provoked an acrimonious response by Gordon and Alan Walker (1983) in a paper entitled *Playing 'Possum: A Microwear Experiment.* Gordon and Walker questioned the conclusions of the *Didelphis* study, arguing that its results reflected methodological limitations of their experiment rather than fundamental limits to what dental microwear can tell us. Kay and Covert (1983) rebutted in a paper entitled *True Grit: A Microwear Experiment* by reiterating that grit and plant opals leave similar microwear, and by repudiating microwear studies on samples that lack dietary control. This interchange, more than any other, forced a reassessment of the potentials and limits of dental microwear research. Indeed, much of the research that has followed, continuing to this day, has been designed to assess the potentials, and to push the limits of dental microwear research.

Rebirth (The Mid to Late 1980s) – Quantification and Standardization

Much of the microwear research that followed in the mid to late 1980s set out to begin to quantify and control variables that might effect microwear. Gordon's (1982,

1984) pioneering work provides an excellent example. She conducted quantitative analyses on *Pan troglodytes* molars taken at fixed magnifications of 120x – 130x. Her investigations found that many factors affect microwear patterning, including the part of the tooth examined, and the position of that tooth in the row. Feature lengths and scratch frequencies decreased from mesial to distal molars, and microwear patterns differed markedly on shearing as compared with crushing facets. In sum, Gordon's study made clear the importance of maintaining control over the sites sampled for microwear features (i.e., homologous facets and molar positions). It further established that microwear technique refinements, such as quantification and subsequent statistical analyses could reveal microwear differences not obvious in previous, qualitative studies that depended on the individual observer to describe the patterns they saw. In subsequent work, Gordon (1988) noted that other factors, such as magnifications chosen and SEM instrumentation settings can also affect microwear study results. This stressed the need for standardization, both within and among projects to allow for comparability of results.

Early attempts at standardization involved efforts to automate microwear quantification. Kay (1987; Grine and Kay 1987) for example, suggested that analysis of power spectra obtained from numerical Fourier transformation of digitized micrograph images might reveal differences that can be used to distinguish primates with dissimilar diets. Further, Phillip Walker and coauthors (1987) suggested that thresholding techniques could be used to isolate microwear features from the rest of an image, thereby allowing the computer to count and measure those features automatically. While automated techniques have not been adopted by most microwear researchers because of limitations in pattern recognition technology, they do reiterate the importance of developing standards in microwear quantification to facilitate comparisons across studies.

Enamel Structure and Microwear

Researchers also began to consider the effects of enamel structure on tooth wear in the mid 1980s. These workers (Boyde 1984 a, b; Walker 1984; Fortelius 1985; Boyde and Fortelius 1986; Maas 1988) noted that because prismatic dental enamel is anisotropic, wear resistance differs depending on the direction of abrasion with respect to the underlying structure of the enamel prisms and the orientation of the hydroxyapatite crystals of which they are made. For example, Maas (1988, 1991, 1994) conducted abrasion experiments with tooth enamel and silicon carbide grits of different sizes. She found that under shearing loads, while striation breadth increased with particle size for nonprismatic enamel, variation in crystallite orientation swamped possible striation breadth differences in prismatic enamel. Under compressive loads, however, abrasive particle size was the primary determinant of microwear feature size. These studies further emphasized the need to understand and control for non-diet related factors that might effect microwear formation.

At about this time, Teaford and Walker (1983a, b) began to explore the mechanics of microwear feature formation. Among the most fundamental questions they addressed was "are microwear features caused by contact between opposing teeth (attrition) or by contact between teeth and foods eaten (abrasion)"? They reasoned that because permanent teeth of guinea pigs (*Cavia porcellus*) are erupted and worn *in utero*, the presence or absence of microwear striations on still-born guinea pigs would determine whether such scratches formed as the result of tooth-to-tooth contact or if dietary abrasives were needed. Teaford and Walker found that still-borns lacked microwear features typically found on adult guinea pig molars and, therefore, that abrasives in foods rather than attrition produce microwear.

Quantitative Microwear and Living Primates

Teaford and Walker (1984) then examined molar microwear in a series of wild-caught museum specimens representing several species of extant primates with known differences in diet. They carefully selected homologous facets on homologous teeth for sampling microwear, used consistent magnifications, and quantified feature dimensions and densities. This study clearly demonstrated an association between patterns of dental microwear and diet. For example, they found that the frugivores *Cercocebus*, *Cebus*, *Pongo*, and *Pan* show a higher ratio of pits to striations (which they distinguished by a length to width ratio threshold of 10:1) than do more folivorous *Colobus*, *Gorilla*, and *Alouatta* (subsequent studies have employed a length-to-width ratio of 4:1 to distinguish pits from striations, see Teaford 1988). Further, among frugivores, hard-object specialists (*Cercocebus*, *Cebus apella*) have the highest relative frequencies of pits.

Encouraged by these results, Teaford spent much of the rest of the 1980s testing for microwear differences between primates with subtler differences in diet. He found that he could distinguish molar microwear patterns between closely related *Colobus guereza* and *Procolobus badius* (Teaford 1986), and even among congeners *Cebus capucinus*, *C. olivaceus* and *C. apella* (Teaford 1985). Differences in microwear pit percentages again mirrored differences in diet such that more folivorous taxa had relatively more scratches than frugivores, and among frugivores, hard-object specialists had relatively more pits. Once interspecific differences within genera had been demonstrated, the next step was to look for differences within species. Teaford and Robinson (1989) examined a series of *Cebus olivaceus* specimens with known dates and locations of capture collected by the Smithsonian Venezuelan project. Their results indicated that while intraspecific differences were not of a magnitude that would obscure differences between species, molar microwear patterns did reflect both seasonal and ecological zone differences within the capuchin sample.

Paleobiological Applications

Microwear also began to play an increasingly important role in paleobiology and bioarcheology during the mid to late 1980s. The database of groups studied increased, and some workers incorporated the new, quantitative methods into their research projects. Kelley (1986) for example examined incisor microwear in a variety of living primates and Miocene catarrhines. He proposed an association between labial face feature density and degree of incisal preparation of foods. While his methods were qualitative, he suggested differences among taxa such that some (*Proconsul nyanzae, Ouranopithecus macedoniensis, Dryopithecus laietanus*) had little wear and were presumably folivorous whereas others (*Proconsul heseloni, P. major, Rangwapithecus gordoni, Sivapithecus indicus*) showed more incisor striations and were presumably frugivorous.

Grine (1986, 1987a, b) quantified molar microwear in fossil hominids, and found significant differences between South African "robust" and "gracile" australopithecines such that *Paranthropus robustus* had greater feature densities, higher pit frequencies, greater striation orientation heterogeneity and shorter/wider scratches than did *Australopithecus africanus*. He compared these data directly with that collected on extant primates by Teaford and Alan Walker (see Teaford 1988) and showed that *Paranthropus*' pit percentages fell between the hard-object feeders *Cercocebus albigena* and *Cebus apella* whereas *Australopithecus* values fell between the soft-fruit eaters *Pan troglodytes* and *Pongo pygmaeus*. Grine's study demonstrated that microwear could both reveal differences in diet among closely related fossil forms, and suggest some properties of actual foods eaten by those forms with the help of an extant baseline series.

Bioarcheological Applications

Bioarcheologists began to apply the results of these dental microwear studies to prehistoric humans to infer diets and subsistence practices. The work of Rose and colleagues (Blaeuer and Rose 1982; Rose et al. 1983; Rose 1984; Rose and Marks 1985; Marks et al. 1985; Harmon and Rose 1988) on prehistoric North American Caddo and Mississippi Valley cultures provides an example. They proposed the following associations: 1) high incidences of compression fractures and hickory nut consumption; 2) a dominance of microwear striae with sharp edges and the use of stone grinders, and 3) striae with rounded edges or microscopic tooth polish and a fibrous diet. They used these associations, in conjunction with archeological evidence, to suggest seasonality of individual site occupation and the transition to maize agriculture in the Red River and Lower Mississippi Valleys. While this and other early work on prehistoric human microwear (e.g., Puech et. al. 1983; Hojo 1989) were largely qualitative, it was instrumental in introducing microwear analyses to bioarcheology.

Experimental Studies

In the latter part of 1980s, research began to focus on learning more about the processes of microwear formation and on improving methods of data collection and analysis. A number of experimental laboratory studies set out to learn more about the actual effects of specific foods and tooth enamel structure on wear patterns. For example, Teaford and Oyen (1989 a–c) published a series of papers on a long-term *in vitro* study of 15 vervet monkeys, *Chlorocebus* (*Cercopithecus*) *aethiops,* obtained from a wild-ranging colony on St. Kitts Island in the Caribbean.

These authors tackled several issues with these studies, including 1) reliable methods for replicating occlusal surfaces of live primates without harming the animals; 2) determining how long individual microwear features remain on a tooth's surface or the rate of feature turnover; and 3) assessing the actual effects of differences in diet on dental microwear in a controlled environment. These studies demonstrated a direct and causal relationship, such that those primates fed dry, hard monkey chow showed more microwear on crushing facets than did those fed moistened, soft chow.

Teaford and Oyen (1989 b, c) also considered the important issue of microwear turnover rates. As described above, Alan Walker and colleagues (1978) first demonstrated that microwear patterns can change over time due to seasonal variation in diet. One subsequent major criticism of microwear research was that items eaten shortly before death might have a disproportionate effect on wear patterns, a notion now commonly referred to as the "Last Supper" phenomenon (Grine 1986). Teaford and Oyen (1989 b, c) indeed demonstrated that individual microwear features can be formed and obliterated quite rapidly – in some cases within 24 hours. Even given minimum reported rates of cusp height reduction (about 70 mm per year), it is clear that the average microwear striation recorded at 500x magnification (< 1 μm) will be worn away less than one week after formation. These studies demonstrated both the need for large samples (to get a representative cross-section of a group or species to infer dietary proclivities), and the potential of microwear to discern seasonal and short-term variation in diet, as also noted by the bioarcheologists working on ancient human teeth.

Microwear Today (The 1990s and Beyond)

Microwear investigations have grown exponentially in the 1990s in both number and in scope. These most recent studies have gone in many different directions, including: 1) an expansion of the extant comparative database of non-human primate microwear; 2) a proliferation of palaeobiological studies to document dental microwear patterns of fossil primates; 3) quantitative analyses of bioarcheological samples; 4) the development of new methods to facilitate microwear quantification; and 5) attempts to better understand microwear formation processes through field studies.

Expansion of the Extant Comparative Database

Researchers have continued to expand the extant primate microwear baseline database for the interpretation of dental microwear on fossil teeth. Teaford (Teaford and Runestad 1992; Teaford 1993, personal communication) has now examined molar microwear on hundreds of wild-caught anthropoids representing 24 genera: *Alouatta, Aotus, Ateles, Cebus, Cercocebus, Chlorocebus, Cercopithecus, Chiropotes, Colobus, Gorilla, Homo, Hylobates, Macaca, Nasalis, Pan, Papio, Pithecia, Pongo, Presbytis, Procolobus, Rhinopithecus, Saimiri, Theropithecus,* and *Trachypithecus* (Strait 1991, 1993) has further expanded the baseline series to include molars of the prosimian grade

Tarsius, and the strepsirhines *Arctocebus, Eulemur, Galagoides* and *Propithecus*. In addition, Ungar (1990, 1992, 1994a, in progress) has documented dental microwear on the incisors of primates representing seven extant genera, including *Alouatta, Cebus, Homo, Hylobates, Macaca, Pongo*, and *Presbytis*. Such studies continue to demonstrate associations between differences in microwear and differences in diet and ingestive behaviors.

Paleobiological Use of Dental Microwear

Work has also progressed on the paleobiological front, as researchers began the immense task of documenting occlusal microwear patterns of fossil primates. Paleoprimatologists have now examined dental microwear on species from every epoch from which true primates are known. Strait (1991), for example, has examined dental microwear of several small-bodied Eocene Omomyids. Her results, in connection with studies of molar form, suggest a range of diets from soft to hard-object faunivory (the consumption of animals, including insects, other invertebrates and in this case, perhaps small vertebrates).

Teaford and coauthors (1997) examined the molar microwear of Oligocene Fayum catarrhines including *Aegyptopithecus, Parapithecus*, and *Apidium*. This evidence suggested that these primates were predominantly frugivorous, though there were some significant differences among these forms.

A number of researchers have also examined dental microwear in a variety of Miocene forms including the Asian genera *Gigantopithecus* (Daegling and Grine 1994) and *Sivapithecus* (Ungar et al. 1996), the African forms *Dendropithecus, Micropithecus, Proconsul, Prohylobates* and *Rangwapithecus* (Walker et al. 1994; Ungar and Teaford 1996; Ungar et al. 1996), and the European forms *Anapithecus, Dryopithecus, Oreopithecus, Ouranopithecus* and *Pliopithecus* (Ungar 1996). Molar pit percentages suggest a broader variety of food preferences than seen in living apes, ranging from hard-object feeding to soft-fruit frugivory to folivory.

Researchers have also studied microwear patterns found on the molars of Plio-Pleistocene primates. Some researchers have examined molar microwear in cercopithecoids such as *Cercocebus, Cercopithecoides, Colobus, Papio, Paracolobus, Rhinopithecus* and *Theropithecus* (Teaford and Leakey 1992; Lucas and Teaford 1994; Ungar and Teaford 1996). Many of these primates show patterns comparable to those of their modern analogs, but not all. For example, the colobine *Cercopithecoides* shows microwear consistent with terrestrial feeding, a scenario supported by its locomotor adaptations. In addition, the first quantitative study of incisor microwear on South African australopithecines was completed in 1991. Ungar and Grine (1991) found notable differences between *Paranthropus* and *Australopithecus* such that the former had less microwear on its incisors than the latter, but also showed higher incidences of enamel prism exposure. This suggested to these authors that the "robust" australopithecines used their front teeth less in ingestion than did the "gracile" forms, but that the former ate foods more effective at etching enamel than did the latter.

Finally, researchers have even begun to examine dental microwear on subfossil Malagasy strepsirhines such as *Archaeolemur* and *Megaladapis* (Rafferty and

Teaford 1992; Rafferty et al. in prep). *Megaladapis* was found to possess a more restricted diet of foliage, whereas *Archaeolemur* evidently incorporated more fruit in its diet.

Bioarcheological Research

Bioarcheological microwear studies have also continued to expand in the 1990s as dental anthropologists began to embrace the quantitative approach in attempts to infer diets, and especially changes in subsistence practices over time. The results of these studies clearly refined the interpretations derived from previous wear analyses. Bullington (1991) found age-related differences in molar occlusal microwear in children from Middle Woodland horticulturalist and Mississippian agriculturalist groups from the lower Illinois River valley in the United States. She also showed that the two groups differed significantly in that the horticulturalists had harder, more varied diets than agriculturalists. Teaford (1991) examined the microwear of ancient native Americans from Georgia-Florida coast archeological sites, and documented increased pitting and broadening of striations between teeth from post-contact and pre-contact times. He suggested that these results are consistent with a shift from hunting and gathering marine resources towards an increased dependence on maize agriculture. Further, Pastor (1992, 1993, 1994) examined dental microwear of Mesolithic, Neolithic, Chalcolithic and Bronze Age peoples of the Indian Subcontinent. His multivariate analyses were easily able to distinguish the groups. He also associated specific patterns of microwear with subsistence changes such that, for example, increasingly coarse microwear features accompanied agricultural intensification in the greater Indus Valley. In addition, Molleson and coauthors (1993) found associations between dental microwear patterning and changes in diet and food preparation technology (e.g., the addition of domesticated cereals to the diet, the introduction of pottery, etc) during the Neolithic in Syria. Finally, Lalueza and Pérez-Pérez (Lalueza et al. 1993, Pérez-Pérez et al. 1994) have advocated the analysis of dental microwear on molar buccal surfaces in archeological samples. For example, they found age-related variability in number and average length of striations that evidently reflect changes in diet from weaning to adulthood in a medieval agriculturalist population from Palencia, Spain.

New Methods of Quantification

As more researchers became active in microwear investigations, and more work was done to expand the extant and fossil primate microwear databases, workers have become increasingly concerned about comparability of results across studies. While efforts continue (Boyde and Fortelius 1991; Ungar in progress), pattern recognition technology has not yet provided us with a commonly used automated method of microwear quantification. The problem stems from the fact that digitized microwear photomicrographs differ in grey level (light intensity) over a single image and vary in the intensity of transition to neighboring pixels. Therefore, it is exceedingly difficult for a computer to automatically isolate features and define their boundaries, even though most humans are able to do this with little difficulty. Ungar and colleagues (Ungar et. al. 1991; Ungar 1995a) have therefore developed a "semi-automated" image analysis procedure, which depends on the user to identify individual features on a computer screen using a mouse-driven pointer. The user marks feature edges, and the computer tallies and measures dimensions (i.e., major and minor axis lengths) and orientations. While this technique still has an element of subjectivity (it depends on human recognition of feature boundaries), it does allow several researchers to quantify features using a single protocol, and it facilitates microwear image sharing and distribution using venues such as the World Wide Web (e.g., see http://comp.uark.edu/~pungar/). This software is now being used by microwear workers in North America, Europe and Africa, and is offered as the first "industry standard" in dental microwear research. This system uses a personal computer and flatbed scanner, and is cost-effective because almost everyone has access to the required hardware. This is especially critical if microwear is to become a routine part of bioarchaeology.

Field Studies

The final direction that dental microwear research has taken in the 1990s is attempting to determine causes of wear by examining primates in their natural habitats. Such studies have been conducted by Ungar (1990, 1992, 1994a, b, 1995b, 1996a, Ungar et al. 1995) on platyrrhines in Venezuela and anthropoids in Indonesia, by Teaford and colleagues on howler monkeys in Costa Rica (Burnell et al. 1994; Pastor et al. 1995; Pokempner et al. 1995; Teaford and Glander 1991, in press; Teaford et al. 1994; Ungar et al. 1995), and by Strait and Overdorff (1994) on strepsirhines on Madagascar.

Ungar's investigations have combined analyses of incisor microwear with field studies to document ingestive behaviors that cause anterior dental microwear, and evidence of the abrasives responsible for that microwear. The first such study involved a comparison of sympatric red howler monkeys (*Alouatta seniculus*) and wedge-capped capuchins (*Cebus olivaceus*) from the Orinoco River valley in the Llanos of Venezuela. Ungar (1990, 1992) noted that the capuchins use their incisors more during ingestion than do sympatric howler monkeys, even when the two species consume the same foods. He also found that capuchins had higher densities of microwear on their incisors than did howlers, and therefore related incisor microwear density to degree of anterior tooth use on abrasive foods. Ungar (1992, 1994a, b, 1995b, 1996b) then conducted a one-year study of *Hylobates lar, Macaca fascicularis, Pongo pygmaeus,* and *Presbytis thomasi* in the Gunung Leuser National Park on Sumatra, Indonesia. During this study, he examined food preferences, ingestive behaviors, feeding heights, grit in the soil and the canopy, and other factors likely to affect incisor microwear. Results of this study, in connection with a study conducted on museum specimens collected mostly in this part of northern Sumatra, suggest that striation density related to degree incisor use in food processing, that scratch orientations related to the directions that food items are passed across the teeth, and that striation breadths related to the types of abrasives contacting the front teeth (perhaps the ratios of exogenous grit to phytoliths contributing to dietary abrasives).

Further, Strait and Overdorff (1994) conducted a study of strepsirrhines in the Ranomafana National Park on Madagascar. First, they observed *Propithecus diadema*, *Eulemur fulvus* and *Eulemur rubiventer* in the wild, and examined material properties of foods eaten by these primates. Then they caught the individuals observed, and collected dental impressions for microwear analysis. While their published results are still preliminary, they do already suggest an association between food item hardness and ratio of pits to scratches on molar surfaces.

Finally, Teaford, Glander and colleagues (Burnell et. al. 1994; Pastor et. al. 1995; Teaford and Glander 1991, in press; Ungar et al. 1995) have been conducting an investigation of microwear formation in mantled howler monkeys (*Alouatta palliata*) at Hacienda La Pacifica in Guanacaste Province, Costa Rica. These researchers are in the process of examining relationships between material properties of foods, abrasives in and on foods, feeding behaviors and dental microwear of the actual individuals studied in the field. While results are preliminary, Teaford and colleagues have already detected microwear differences related to age, sex, season and microhabitat preferences.

Where Do We Go from Here?

So, where do we go from here? We see anthropological microwear studies continuing in four principle directions: 1) development of more automated methods of analysis stressing comparability across studies; 2) carrying on the work to unravel microwear formation processes; 3) expansion of the comparative database; 4) persisting in the documentation of wear patterns of fossil primates and prehistoric humans.

More Automated Methods and Comparability of Studies

First, we need to develop more efficient ways of "doing" microwear research. Current techniques are time-consuming and prohibitively expensive to many. Specimens must be cleaned, molded, cast, mounted on SEM stubs, coated, imaged, photographed and scanned into computer files even before one can begin collecting microwear data! Indeed, it typically takes several hours to process a single specimen. Even with current "semiautomated" techniques, the researcher must manually identify up to hundreds of microwear features on a single micrograph image. Further, SEM time, photomicrographic film, and the supplies to prepare casts for analysis can be rather expensive – even a modest project requiring little travel can cost thousands of dollars.

We need more rapid, inexpensive ways of quantifying microwear without a significant reduction in the quality of data output. A first step may be to look for alternatives to the conventional SEM. While the SEM does have remarkable depth of focus, allowing researchers to effectively image even curved tooth surfaces, it is not the ideal tool for quantitative microwear analyses. Beside the time and expense of preparing specimens for analysis, we must remember that scanning electron microscopy is fundamentally different from optical microscopy. The image formed on an SEM screen is not a magnification of an actual tooth's surface, but is rather a representation formed by the interaction of that surface and an electron beam. The settings we choose on an SEM can therefore greatly affect microwear images (see Gordon 1988). Collector-specimen geometry, the type of electrons used in creating the image (e.g., backscattered *versus* secondary), voltage, working distance and other parameters all have different effects that can lead to differing images. While most researchers control for these differences by maintaining consistency within studies and reporting parameters used, it would still be best to have a "true" image of the surface being examined. Researchers are now beginning to consider other possible imaging tools such as the profilometer and the atomic force microscope (e.g., Walker and Hagen 1994). Such studies should continue into the next century.

In addition, we need more automated methods of quantifying dental microwear, regardless of whether surfaces are imaged using a SEM or another technique. Repeatability of results in dental microwear quantification is a problem. Because some micrographs present a seemingly infinite number of overlapping scratches and pits, two researchers can look at the same image and get different feature counts and boundary measurements. An automated procedure that gives the same numbers every time an image is analysed by every researcher is clearly a priority. There are two ways that this has been approached: 1) attempts to have the computer automatically identify individual microwear features (much like the researcher does now), and 2) attempts to have the computer identify patterns of grey level distributions that characterize and distinguish micrographs without isolating individual features. Early attempts on both fronts have met with only limited success.

SEM microwear images are just too complicated for simple thresholding followed by pattern recognition (*sensu* Walker et al. 1987). Because features vary in grey level over a single image, it is difficult to isolate them and to define their borders using thresholding. Further, given SEM collector geometry, photomicrographs are rarely if ever evenly "illuminated" and thus thresholding does a better job identifying surface shading than anything else. If this approach is to progress, we must develop a procedure that deals with the complex nature of microwear images. Perhaps a multi-step routine that first controls for the effects of background shading using appropriate filtering (e.g., Boyde and Fortelius 1991), and then isolates features using edge detection and pattern recognition technology is our best bet.

Early attempts to look for patterning in micrographs without isolating individual features have met with some success. Kay (1987) suggested that analysing power spectra obtained from numerical Fourier transformation of microwear images could differentiate primates with dissimilar diets. Unfortunately, this approach does not present direct information concerning feature density and dimensions, and therefore may not be able to separate taxa as well as manual procedures can (see Grine and Kay 1987). This approach does have promise, but the application of several different techniques each examining slightly different aspects of microwear would clearly better differentiate surfaces. For example, histograms of edge orientations detected using Sobel edge extraction routines can distinguish linearity, and determine its degree and preferred orientation on a microwear surface (Ungar, ongoing research). As hardware and software technology develops, these and other routines will no doubt come to play an increasingly important role in dental microwear studies, whether researchers use SEM imaging or another approach.

Microwear Formation Processes

Work also continues on unraveling the processes of microwear formation. What are the roles of exogenous grit and opal phytoliths in microwear formation? How do different sorts of tooth-use behaviors and the material properties of foods eaten affect wear patterns? How do enamel microstructure, overall tooth morphology and gross dental wear affect microwear patterns? Ongoing research promises to help answer these and other questions. For example, a multidisciplinary team including Mark Teaford, Ken Glander, Dolores Piperno, Mary Maas, Peter Lucas and Peter Ungar are now preparing to converge on Hacienda La Pacifica in Costa Rica to determine the causes of wear in howler monkeys. These researchers and accompanying students will document feeding behaviors of the primates through behavioral observations, and they will analyse stomach contents and feces, document phytoliths and other food abrasives, examine the physical properties of foods and exogenous dust in the canopy, and study enamel structure, occlusal morphology, gross wear and microwear of howler monkey teeth. Such multidisciplinary studies need to continue, and will hopefully lead us to the causal agents responsible for the dental microwear patterns we see on primate teeth.

Expansion of the Extant Comparative Database

Microwear researchers have now described dental microwear for the incisors or molars of species representing nearly thirty genera of primates. Still, with approximately 200 extant primate species, each with a unique feeding adaptation, many taxa remain to be considered for microwear investigation. As habitats shrink and more of these species become threatened with extinction, it behooves us to continue to document microwear in relation to diet and feeding behavior in a naturalistic setting. Because microwear research employs the comparative method, its ultimate resolution for the inference of diet and feeding behaviors of fossil primates is only as good as the extant baseline series used to interpret patterns of wear. The same applies for studies of human dental microwear. As museums begin the task of repatriating their human skeletal collections, comparative samples are dwindling. In sum, we must continue to expand our extant microwear database, both for humans and non-human primates.

In addition, we must begin to include other teeth in our analyses. Recent quantitative studies have focused on central incisors and second molars. Further consideration of other teeth, especially canines and premolars, are critical to document "whole row" patterns of dental microwear that may yield more complete pictures of diet and feeding behaviors. In addition, dental microwear work is needed to better understand non-diet or "cultural" uses of the teeth through examination of ethnographically well-documented human populations. Issues concerning the use of "teeth as tools" are important both to studies of prehistoric modern humans and fossil hominids (e.g., Alt and Pichler this volume, Bermudez de Castro et al. 1988, Lalueza et al. 1993, Spencer and Ungar 1997, Ungar et al. 1997).

Documentation of Wear Patterns of Fossil Primates and Prehistoric Humans

Finally, there are many projects now in progress using dental microwear to reconstruct diets and feeding behaviors of fossil primates and prehistoric humans. Such work must continue. Some would argue that we need a better understanding of microwear in living groups before we can apply what we have learned to the fossil record. We disagree. Expanded comparative databases in the past few years and new analytical techniques are allowing us to discriminate finer differences in diet than ever before. Patterns of microwear on fossil primate and prehistoric human teeth can yield insights into important issues in archeology and paleoanthropology today, and hopefully those patterns already documented will take on increased meaning as we become better able to discriminate diet and tooth use in the future. Further, this identification of interesting patterns of wear in the fossil and archeological records will help direct the studies of extant groups and microwear formation processes to best explain those patterns. Some researchers have spent most of the last two decades building a comparative database and assessing the potentials and limitations of dental microwear research. It is now time to look back at the fossil and archeological records using the knowledge we have gained as this, after all, is the ultimate goal of anthropological dental microwear research.

Conclusions

Gross tooth wear analysis has firmly established its ability to identify changes in diet and food processing technology during the study of ancient human skeletal remains. The greatest present need in bioarcheology and paleoanthropology is for standardization in data collection protocols. We do not need another study of how a new and improved technique can establish a change in dental wear with the adoption of agriculture, what we need are comprehensive regionally based studies of dental wear variation over time, and between cultures and fossil groups. This goal can only be accomplished if all researchers use the same data collection protocol and the data from numerous researchers and skeletal collections can be pooled. It is never going to be possible for one researcher to collect all of the data necessary to develop a comprehensive understanding of regionally specific dietary change. With the imminent lose of skeletal collections in the United States, Israel, and Australia we will not be given the luxury of being able to study skeletal collections twice. At present the scoring schemes of Scott (1979b) for the molars and Smith (1984) for the anterior tooth classes is recommended. In addition to wear rates, the analysis of occlusal angles (Smith 1984) has also been shown, by a number of studies, to be critical for documenting dietary change. If technological advances are to come, they will be in the application of computer assisted morphometrics techniques to the collection and analysis of wear data which will enhance efficiency, cost effectiveness, and standardization.

Dental microwear analysis has demonstrated its utility for identifying many of the dietary components that cause the tooth abrasion so readily seen in the gross wear of teeth. Our historical overview shows that this young field is expanding rapidly. Increasing numbers of dental anthropologists are incorporating microwear analyses into their research, and the number of graduate students pursuing dental microwear dissertations is growing exponentially. As we head into the new millennium, the future looks bright indeed. Still, microwear research has far from reached its ultimate potential; it will certainly continue to grow in several directions in the coming years. The last decade has seen significant technical advances, but we still need cheaper and faster methods for recording microwear. We need to continue to expand our library of microwear patterns associated with known diets and feeding behaviors. At the same time the standardization of data collection is critical for making use of this comparative data base. We also need to continue research to improve our understanding of the processes that produce microwear. Realistic diet interpretations cannot be made unless microwear is understood within the context of foods (and other materials introduced into the oral environment) interacting with enamel, tooth morphology, and the biomechanics of mastication.

Future progress in dietary reconstruction will depend upon the eventual analytical integration of gross wear and microwear data. We not only need to know the variation in dental wear rates, but also be able to identify the specific dietary components which cause the loss of tooth substance. We must also realize that tooth wear is not a panacea to answer all questions regarding feeding behaviors and subsistence practices. While tooth wear can tell us much about the material properties of foods, the way those foods pass between the teeth during ingestion and mastication, and aspects of food preparation technology, it will never provide all the details of diet important to understanding paleoecology. Therefore, we should interpret dental wear data within the context of a large body of evidence. We must continue to incorporate other indicators of diet, including bony evidence of nutrition stress, isotopes, craniofacial biomechanics, and contextual or archaeological evidence into our reconstructions for a better picture of what past peoples, and our more distant simian ancestors ate.

Acknowledgements

We thank the editors, K. W. Alt, F. W. Rösing and M. Teschler-Nicola, for the opportunity to prepare this chapter. J. C. Rose gratefully acknowleges the assistance of F. W. Rösing, the American Center of Oriental Research, Amman, Jordan, and the librarians of the Institute of Archaeology and Anthropology, Yarmouk University, for their assistance in aquiring the necessary literature.

References

Armelagos GJ, Van Gerven DP, Martin DL, Huss-Ashmore R (1984) Effects of nutritional change on the skeletal biology of northeast African (Sudanese Nubian) populations. In: Clark JD, Brandt SA (eds) From Hunters to Farmers: The Causes and Consequences of Food Production in Africa. Univ Press, Berkeley CA, p. 132–46.

Baker G, Jones LHP, Wardrop ID (1959) Cause of wear in sheep's teeth. Nature 184: 1583–1584.

Bass W M (1995) Human Osteology: A Laboratory and Field Manual. Missouri Archaeological Society, Columbia MO.

Begg PR (1954) Stone age man's dentition. Am J Orthodont 40: 298–312, 373-383, 462–475, 517–531.

Benfer RA, Edwards DS (1991) The principal axis method for measuring rate and amount of dental attrition: Estimating juvenile or adult tooth wear from unaged adult teeth. In: Kelley MA, Larsen CS (eds) Advances in Dental Anthropology. Wiley-Liss, New York, p. 325–340.

Bermúdez de Castro JM, Bromage TG, Fernández-Jalvo Y (1988) Buccal striations on fossil human anterior teeth: Evidence of handedness in the middle and early Upper Pleistocene. J Hum Evol 17: 403–412.

Blakely RL (ed) (1977) Biocultural Adaptations in Prehistoric America. Univ of Georgia Press, Athens GA.

Blaeuer MW, Rose JC (1982) Bioarcheology of the Powell Canal site. In: House JH (ed) Powell Canal. Fayetteville, Arkansas, Arkansas Archeological Survey 19: 72–84.

Bookstein FL (1991) Morphometric Tools for Landmark Data. Cambridge Univ Press, London.

Boyde A (1984a) Airpolishing effects on enamel, dentine, cement and bone. Brit Dent J 156: 287–291.

Boyde A (1984b) Dependence of rate of physical erosion on orientation and density of mineralized tissues. Acta Embryol 170: 57–62.

Boyde A, Fortelius M (1986) Development, structure and function of Rhinoceros enamel. Zool J Linnean Soc 87: 182–214.

Boyde A, Fortelius M (1991) New confocal LM method for studying local relative microrelief with special references to wear studies. Scanning 13: 429–430.

Brace CL (1962) Cultural factors in the evolution of the human dentition. In: Montagu MFA (ed) Culture and the Evolution of Man. Oxford Univ Press, New York, p. 343–354.

Bramblett CA (1969) Non-metric skeletal age changes in the Darajani Baboon. Am J Phys Anthrop 30: 161–172.

Broca P (1879) Instructions relative à l'étude anthropologique du système dentaire. Bull Soc Anthrop Paris 2: 128–152.

Brothwell DR (1963) Digging Up Bones. British Museum, London.

Buikstra JE (1991) Out of the appendix and into the dirt: Comments on thirteen years of bioarchaeological research. In: Powell ML, Bridges PS, Mires AMW (eds) What Mean These Bones? University of Alabama Press, Tuscaloosa AL, p. 172–188.

Buikstra JE, Ubelaker DH (eds) (1994) Standards for Data Collection from Human Skeletal Remains. Arkansas Archeological Survey, Fayetteville AR.

Bullington J (1991) Dental microwear of prehistoric juveniles from the lower Illinois River Valley. Am J Phys Anthrop. 84: 59–73.

Burnell CL, Teaford MF, Glander KE (1994) Dental microwear differs by capture site in live-caught *Alouatta* from Costa Rica. Am J Phys Anthrop Suppl 18: 62.

Butler PM (1952) The milk molars of perissodactyla with remarks on molar occlusion. Proc Zool Soc Lond 121: 777–817.

Butler PM, Mills JRE (1959) A contribution to the odontology of *Oreopithecus*. Bull Brit Mus Natl Hist Geol 4: 1–26.

Buxton LHD (1920) The teeth and jaws of savage man. Transact Brit Soc Stud Orthodont 1916–1920: 79–88.
Campbell TD (1925) Dentition and Palate of the Australian Aboriginal. Hassell Press, Adelaide.
Chaplin RE (1971) The Study of Animal Bones from Archaeological Sites. Seminar Press, London.
Corruccini RS (1990) Australian aboriginal tooth succession interproximal attrition and Begg's theory. Am J Orthodont Dentofac Orthoped 97: 349–357.
Conover WJ, Iman RL (1981) Rank transformations as a bridge between parametric and nonparametric statistics. Am Statist 35: 124–129.
Covert HH, Kay RF (1981) Dental microwear and diet: Implications for determining the feeding behavior of extinct primates, with a comment on the dietary pattern of *Sivapithecus*. Am J Phys Anthrop 55: 331–336.
Cross JF, Kerr NW, Bruce MF (1986) An evaluation of Scott's method for scoring dental wear. In: Cruwys E, Foley RA (eds) Teeth and Anthropology. BAR International, Oxford, p. 101–108.
Daegling DJ, Grine FE (1994) Bamboo feeding, dental microwear, and diet of the Pleistocene ape *Gigantopithecus blacki*. S Afr J Sci 90: 527–532.
Dahlberg AA, Kinzey W (1962) Étude microscopique de l'abrasion et de l'attrition sur la surface des dents. Bull Group Int Rech Sc Stomat 5: 242–251.
Ferembach D, Schwidetzky I, Stloukal M (1980) Recommendations for age and sex diagnosis of skeletons. J Hum Evol 9: 517–549.
Fortelius M (1985) Ungulate cheek teeth: Developmental, functional and evolutionary interrelations. Acta Zool Fennica 180: 1–76.
Gordon KD (1982) A study of microwear on chimpanzee molars: Implications of dental microwear analysis. Am J Phys Anthrop 59: 195–215.
Gordon KD (1984a) Hominoid dental microwear: Complications in the use of microwear analysis to detect diet. J Dent Res 63: 1043–1046.
Gordon KD (1984b) Orientation of occlusal contacts in the chimpanzee, *Pan troglodytes verus*, deduced from scanning electron microscopic analysis of dental microwear patterns. Arch Oral Biol 29: 783–787.
Gordon KD (1988) A review of methodology and quantification in dental microwear analysis. Scan Microsc 2: 1139–1147.
Gordon KD, Walker A (1983) Playing 'possom: A microwear experiment. Am J Phys Anthrop 60: 109–112.
Grine FE (1977) Analysis of early hominid deciduous molar wear by scanning electron microscopy: A preliminary report. Proc Elect Microsc Soc S Afr 7: 157–158.
Grine FE (1981) Trophic differences between "gracile" and "robust" australopithecines: A scanning electron microscope analysis of occlusal events. S Afr J Sci 77: 203–230.
Grine FE (1986) Dental evidence for dietary differences in *Australopithecus* and *Paranthropus*. J Hum Evol 15: 783–822.
Grine FE (1987) Quantitative analysis of occlusal microwear in *Australopithecus* and *Paranthropus*. Scan Microsc 1: 647–656.
Grine FE, Kay RF (1987) Early hominid diets from quantitative image analysis of dental microwear. Nature 333: 765–768.
Harmon AM, Rose JC (1988) The role of dental microwear analysis in the reconstruction of prehistoric diet. In: Kennedy BV, Le Moine GM (eds) Diet and Subsistence: Current Archaeological Perspectives. Archeol Assoc Univ of Calgary, Calgary, Alberta, p. 267–272.
Hillson S (1986) Teeth. Cambridge Univ Press, Cambridge.
Hinton RJ (1981) Form and patterning of anterior tooth wear among aboriginal human groups. Am J Phys Anthrop 54: 555–564.

Hrdlička A (1908) Report on a collection of crania from Arkansas. J Acad Nat Sci Philadelphia 13: 558–563.
Hrdlička A (1909) Report on an additional collection of crania from Arkansas and Louisiana. J Acad Nat Sci Philadelphia 14: 171–249.
Hrdlička A (1939) Practical Anthropometry. Wistar Institute of Anatomy & Biology, Philadelphia.
Kaul SS, Corruccini RS (1990) Interproximal attrition and dental arch length reduction in modern Australian Aborigines. Indian J Phys Anthrop Hum Gen 16: 9–16.
Hojo T (1989) Dietary differences and microwear on the teeth of late Stone Age and early modern people from Western Japan. Scan Microsc 3: 623–628.
Kay RF (1987) Analysis of primate dental microwear using image processing techniques. Scan Microsc 1: 657–662.
Kay RF, Covert HH (1983) True grit: A microwear experiment. Am J Phys Anthrop 91: 33–38.
Kay RF, Hiiemae KM (1974) Jaw movement and tooth use in recent and fossil primates. Am J Phys Anthrop 40: 227–256.
Kelley J (1986) Paleobiology of Miocene Hominoids. PhD Dissertation, Yale University.
Klein RG (1982) Age (mortality) profiles as a means of distinguishing hunted species from scavenged ones in Stone Age archaeological sites. Paleobiol 8: 151–158.
Lalueza C, Pérez-Pérez A, Turbón D (1993) Microscopic study of the Banyoles mandible (Girona, Spain): Diet, cultural activity and toothpick use. J Hum Evol 24: 281–300.
Larsen CS (1987) Bioarchaeological interpretations of subsistence economy and behavior from human skeletal remains. In: Schiffer MB (ed) Advances in Archaeological Method and Theory 10. Academic Press, New York, p. 339–445.
Leek FF (1972) Bite attrition and associated oral conditions as seen in ancient Egyptian skulls. J Hum Evol 1: 289–295.
Leigh RW (1925) Dental pathology of Indian tribes of varied environmental and food conditions. Am J Phys Anthrop 8: 179–199.
Grine FE (1981) Trophic differences between "gracile" and "robust" australopithecines: A scanning electron microscope analysis of occlusal events. S Afr J Sci 77: 203–230.
Lele S, Richtsmeier JT (1995) Euclidean distance matrix analysis: Confidence intervals for form and growth differences. Am J Phys Anthrop 98: 73–86.
Lovejoy CO (1985) Dental wear in the Libben population: Its functional pattern and role in the determination of adult skeletal age at death. Am J Phys Anthrop 68: 47–56.
Lucy D, Pollard AM, Roberts CA (1995) A comparison of three dental techniques for estimating age at death in humans. J Archeol Sci 22: 417–428.
Lunt DA (1978) Molar attrition in medieval Danes. In: Butler PM, Joysey KA (eds) Development, Function and Evolution of Teeth. Academic Press, New York, p. 465–482.
Lucas PW, Teaford MF (1995) Significance of silica in leaves eaten by long-tailed macaques (*Macaca fascicularis*). Folia Primatol 60: 30–36.
Maas MC (1988) The Relationship of Enamel Microstructure and Microwear. PhD Diss, Univ New York, Stony Brook.
Maas MC (1991) Enamel structure and microwear: An experimental study of the response of enamel to shearing forces. Am J Phys Anthrop 85: 31–50.
Maas MC (1994) A scanning electron-microscopic study of in vitro abrasion of mammalian tooth enamel under compressive loads. Arch Oral Biol 39: 1–11.
Marks MK, Rose JC, Buie EL (1985) Bioarcheology of the Seminole Sink. In: Turpin SA (ed) Seminole Sink: Excavation of a Vertical Shaft Tomb, Val Verde County, Texas. Univ of Texas, Austin, Texas Archeol Surv Res Report 93: 9–152.
Meikle WE (1977) Molar wear stages in *Theropithecus gelada*. Kroeber Anthrop Soc Papers 50: 21–25.
Miles AEW (1958) The assessment of age from the dentition. Proc Roy Soc Med 51: 1057–1060.

Miles AEW (1963) Dentition in the assessment of individual age in skeletal material. In: Brothwell DR (ed) Dental Anthropology. Pergamon Press, Oxford, p. 191–209.
Miles AEW (1978) Teeth as an indicator of age in man. In: Butler PM, Joysey KA (eds) Development, Function and Evolution of Teeth. Academic Press, New York, p. 455–464.
Mills JRE (1955) Ideal dental occlusion in the primates. Dent Pract 6: 47–61.
Mills JRE (1963) Occlusion and malocclusion in the teeth of primates. In: Brothwell DR (ed) Dental Anthropology. Oxford Univ Press, Oxford, p. 29–52.
Mills JRE (1967) A comparison of lateral jaw movement in some mammals from wear facets on the teeth. Arch Oral Biol 12: 645–661.
Molleson T, Jones K, Jones S (1993) Dietary change and the effects of food preparation on microwear patterns in the Late Neolithic of Abu Hureyra, northern Syria. J Hum Evol 24: 455–468.
Molnar S (1971) Human tooth wear, tooth function, and cultural variability. Am J Phys Anthrop 34: 175–190.
Molnar S (1972) Tooth wear and culture: A survey of tooth functions among some prehistoric populations. Curr Anthrop 13: 511–515.
Moodie RL (1929) Studies in paleodontology, XI: Dental attrition and its results among certain ancient Indians from southern California. Pacific Dent Gazette 37: 217–227.
Mucič D, Đurič-Srejič M (1996) Abrasion of teeth in population groups from historical periods in the region of former Yugoslavia. Dent Anthrop Newsl 11: 7–10.
Mummery JR (1870) On the relations which dental caries – As discovered amongst the ancient inhabitants of Britain and amongst existing aboriginal races – May be supposed to hold to their food and social condition. Transact Odontol Soc Great Britain NS 2: 27–102.
Murphy T (1959a) The changing pattern of dentine exposure in human tooth attrition. Am J Phys Anthrop 17: 167–178.
Murphy T (1959b) Gradients of dentine exposure in human molar tooth attrition. Am J Phys Anthrop 17: 179–186.
Pastor RF (1992) Dietary adaptations and dental microwear in Mesolithic and Chalcolithic South Asia. J Hum Ecol Spec Issue 2: 215–228.
Pastor RF (1993) Dental Microwear among Prehistoric Inhabitants of the Indian Subcontinent: A Quantitative and Comparative Analysis. PhD Diss, Univ Oregon.
Pastor RF, Teaford MF, Glander KE (1995) Methods for collecting and analyzing airborne abrasive particles from neotropical forests. Am J Phys Anthrop Suppl 20: 168.
Pastor RF (1994) A multivariate dental microwear analysis of prehistoric groups form the Indian subcontinent. Am J Phys Anthrop Suppl 18: 158–159.
Peters CR (1982) Electron-optical microscopic study of incipient dental microdamage from experimental seed and bone crushing. Am J Phys Anthrop 57: 283–301.
Pérez-Peréz A, Lalueza C, Turbón D (1994) Intraindividual and intragroup variability of buccal tooth striation pattern. Am J Phys Anthrop 94: 175–188.
Powell ML (1985) The analysis of dental wear and caries for dietary reconstruction. In: Gilbert RI Jr, Mielke JH (eds) The Analysis of Prehistoric Diets. Academic Press, Orlando FL, p. 307–338.
Puech PF (1979) The diet of early man: Evidence from abrasion of teeth and tools. Curr Anthrop 20: 590–592.
Puech PF, Prone A (1979) Reproduction expérimentale des processus d'usure dentaire par abrasion: implications paléoécologiques chez l'homme fossile. CR Acad Sci Paris D 289: 895–898.
Puech PF, Prone A, Albertini H (1981) Reproduction expérimentale des processus d'altération de la surface dentaire par friction non abrasive et non adhésive: Application à l'étude de l'alimentation de l'Homme fossile. CR Acad Sci Paris II 293: 729–734.
Puech PF, Serratrice C, Leek FF (1983) Tooth wear as observed in ancient Egyptian skulls. J Hum Evol 12: 617–629.

Rafferty K, Teaford MF (1992) Diet and dental microwear in Malagasy subfossil lemurs. Am J Phys Anthrop Suppl 14: 134.
Reith T (1990) Untersuchungen zur Altersbestimmung der Individuen zweier frühneolithischer Skelettserien aus Baden-Württemberg anhand verschiedener Abrasionsphänomene. Med Diss, Tübingen.
Rensberger JM (1978) Scanning electron microscopy of wear and occlusal events in some small herbivores. In: Butler PM, Joysey KA (eds) Development, Function and Evolution of Teeth. Academic Press, New York, p. 415–438.
Robinson JT (1956) The Dentition of the Australopithecinae. Transvaal Museum, Pretoria.
Rose JC, Armelagos GJ, Perry LS (1993) Dental anthropology of the Nile Valley. In: Davies WV, Walker R (eds) Biological Anthropology and the Study of Ancient Egypt. Brit Mus Press, London, p. 61–74.
Rohlf J (1991) Morphometrics. Ann Rev Ecol Syst 21: 299–316.
Rose JC, Green TJ, Green VD (1996) NAGPRA is forever: Osteology and the repatriation of skeletons. Ann Rev Anthrop 25: 81–103.
Rose JC (1984) Bioarcheology of the Cedar Grove Site. In: Trubowitz N (ed) Cedar Grove. Fayetteville, Arkansas. Arkansas Archeol Survey 23: 255–261.
Rose JC, Marks MK (1985) Bioarcheology of the Alexander Site. In: Hemmings ET, House JH (eds) The Alexander Site. Fayetteville, Arkansas. Arkansas Archeol Survey 24: 76–98.
Rose JC, Marks MK, Riddick EB (1983) Bioarcheology of the Bug Hill Site. In: Altshul JH (ed) Bug Hill: Excavation of a Multicomponent Midden Mound in the Jackfork Valley, Southeast Oklahoma. Pollock, Louisiana. New World Res Report of Investigation No. 81: 241–278.
Rose JC, Tucker TL, Lovell N, Filer J (1996) Bioarchaeology of Ancient Egypt and Nubia: A Bibliography. Brit Mus, London.
Ruffer MA (1919) Food in Egypt. Mémoires de l'Institut d'Égypte 1: 1–88.
Ruffer MA (1920) Study of abnormalities and pathology of ancient Egyptian teeth. Am J Phys Anthrop 3: 335–382.
Ruffer MA (1921) Studies in the Palaeopathology of Egypt. Univ Press, Chicago.
Ryan AS (1980) Anterior Dental Microwear in Hominoid Evolution: Comparisons with Human and Nonhuman Primates. PhD Diss, Univ of Michigan.
Ryan AS, Johanson DC (1989) Anterior dental microwear in *Australopithecus afarensis*. J Hum Evol 18: 235–268.
Samuel D (1993) Ancient Egyptian bread and beer: An interdisciplinary approach. In: Davies WV, Walker R (eds) Biological Anthropology and the Study of Ancient Egypt. Brit Mus Press, London, p. 156–164.
Schürmann R (1986) Soziale Differenzierung nach frühgeschichtlichen Ernährungsmustern am Beispiel von Zahnabrasionen in merowingerzeitlichen Bevölkerungen. Diplomarbeit Biologie, Ulm.
Scott EC (1979a) Principal axis analysis of dental attrition data. Am J Phys Anthrop 51: 203–212.
Scott EC (1979b) Dental wear scoring technique. Am J Phys Anthrop 51: 213–218.
Simpson GG (1933) Paleobiology of Jurassic mammals. Paleobiol 5: 127–158.
Smith BH (1984) Patterns of molar wear in hunter-gatherers and agriculturalists. Am J Phys Anthrop 63: 39–56.
Smith GE (1932) The incidence of dental diseases in ancient Egypt. Brit Med J 2: 760.
Smith GE, Jones FW (1910) Archaeological Survey of Nubia 1907/1908, Vol II: Report on the Human Remains. Ministry of Finance Survey Department, Cairo, Egypt.
Smith P (1972) Diet and attrition in the Natufians. Am J Phys Anthrop 37: 233–238.
Smith P (1976) Dental pathology in fossil Hominids: What did Neanderthals do with their teeth? Curr Anthrop 17: 149–151.

Spencer MA, Ungar PS (1997) Craniofacial biomechanics of five modern human populations. Am J Phys Anthrop Suppl 24: 215.
Strait SG (1991) Dietary Reconstruction in Small-Bodied Fossil Primates. PhD Diss, Univ New York, Stony Brook.
Strait SG (1993) Molar microwear in extant small-bodied faunivorous mammals: An analysis of feature density and pit frequency. Am J Phys Anthrop 92: 63–79.
Strait SG, Overdorff DJ (1994) A preliminary examination of molar microwear in strepsirhine primates. Am J Phys Anthrop Suppl 18: 190.
Taylor R (1963) Cause and effect of wear on teeth. Further non-metrical studies on the teeth and palate in Moriori and Maori skulls. Acta Anat 53: 97–157.
Teaford MF (1985) Molar microwear and diet in the genus *Cebus*. Am J Phys Anthrop 66: 363–370.
Teaford MF (1986) Dental microwear and diet in two species of *Colobus*. In: Else J, Lee P (eds) Proc 10th Ann Int Primatol Conf. Vol II: Primate Ecology and Conservation. Cambridge Univ Press, Cambridge, p. 63–66.
Teaford MF (1988) A review of dental microwear and diet in modern mammals. Scan Microsc 2: 1149–1166.
Teaford MF (1991) Dental microwear: What can it tell us about diet and dental function? In: Kelley MA, Larsen CS (eds) Advances in Dental Anthropology. Liss, New York, p. 341–356.
Teaford MF (1993) Dental microwear and diet in extant and extinct *Theropithecus*: Preliminary analyses. In: Jablonski NG (ed) *Theropithecus*: The Life and Death of a Primate Genus. Cambridge Univ Press, Cambridge, p. 331–349.
Teaford MF, Glander KE (1991) Dental microwear in live, wild-trapped *Alouatta palliata* from Costa Rica. Am J Phys Anthrop 85: 313–319.
Teaford MF, Glander KE (1997) Dental microwear and diet in a wild population of mantled howlers (*Alouatta palliata*). In: Norconk M, Rosenberger A, Garber P (eds) Adaptive Radiations of Neotropical Primates. Plenum Press, New York (in press).
Teaford MF, Leakey MG (1992) Dental microwear and diet in Plio-Pleistocene cercopithecoids from Kenya. Am J Phys Anthrop Suppl 14: 160.
Teaford MF, Maas MC, Simons EL (1997) Dental microwear and microstructure in Early Oligocene primates from the Fayum, Egypt: Implications for diet. Am J Phys Anthrop (in press).
Teaford MF, Oyen OJ (1989a) Differences in rate of molar wear between monkeys raised on different diets. J Dent Res 68: 1513–1518.
Teaford MF, Oyen OJ (1989b) In vivo and in vitro turnover in dental microwear. Am J Phys Anthrop 80: 447–460.
Teaford MF, Oyen OJ (1989c) Live primates and dental replication: New problems and new techniques. Am J Phys Anthrop 80: 73–81.
Teaford M F, Robinson J G (1989) Seasonal or ecological zone differences in diet and molar microwear in *Cebus nigrivittatus*. Am J Phys Anthrop 80: 391–401.
Teaford MF, Runestad JA (1992) Dental microwear and diet in Venezuelan primates. Am J Phys Anthrop 88: 347–364.
Teaford MF, Ungar PS, Kay RF, Leakey MG (1996) The evolution of diet in Old World Monkeys. Am J Phys Anthrop Suppl 22: 227.
Teaford MF, Walker A (1983a) Prenatal jaw movements in the guinea pig, *Cavia porcellus*: Evidence from patterns of tooth wear. J Mammal 64: 534–536.
Teaford MF, Walker A (1983b) Dental microwear in adult and still-born guinea pigs. Arch Oral Biol 28: 1077–1081.
Teaford MF, Walker A (1984) Quantitative differences in dental microwear between primate species with different diets and a comment on the presumed diet of *Sivapithecus*. Am J Phys Anthrop 64: 191–200.

Ungar PS (1990) Incisor microwear and feeding behavior in *Alouatta seniculus* and *Cebus olivaceus*. Am J Primatol 20: 43–50.
Ungar PS (1992) Incisor Microwear and Feeding Behavior of Four Sumatran Anthropoids. PhD Diss, Univ New York, Stony Brook.
Ungar PS (1994a) Incisor microwear of Sumatran anthropoid primates. Am J Phys Anthrop 94: 339–363.
Ungar PS (1995a) A semiautomated image analysis procedure for the quantification of dental microwear II. Scanning 17: 57–59.
Ungar PS (1995b) Fruit preferences of four sympatric primate species at Ketambe, northern Sumatra, Indonesia. Int J Primatol 16: 221–245.
Ungar PS (1996a) Dental microwear of European Miocene catarrhines: Evidence for diets and tooth use. J Hum Evol 30: 335–366.
Ungar PS (1996) Feeding height and niche separation of four sympatric primate species at Ketambe, northern Sumatra, Indonesia. Int J Primatol (in press).
Ungar PS, Fennell KJ, Gordon K, Trinkaus E (1997) Neandertal incisor bevelling. J Hum Evol 32: 407–421.
Ungar PS, Grine FE (1991) Incisor size and wear in *Australopithecus africanus* and *Paranthropus robustus*. J Hum Evol 20: 313–340.
Ungar PS, Kay RF, Teaford MF, Walker A (1996) Dental evidence for diet in Miocene apes. Am J Phys Anthrop Suppl 22: 232–233.
Ungar PS, Simons JC, Cooper JW (1991) A semiautomated image analysis procedure for the quantification of dental microwear. Scanning 13: 31–36.
Ungar PS, Teaford MF (1996) A preliminary examination of non-occlusal dental microwear in anthropoids: Implications for the study of fossil primates. Am J Phys Anthrop 100: 101–113.
Ungar PS, Teaford MF, Glander KE, Pastor RF (1995) Dust accumulation in the canopy: A potential cause of dental microwear in primates. Am J Phys Anthrop 97: 93–99.
Ungar PS, Walker A, Coffing K (1994) Reanalysis of the Lukeino Molar (KNM-LU 335). Am J Phys Anthrop 94: 165–173.
Walker A (1980) Functional anatomy and taphonomy. In: Behrensmeyer AK, Hill AP (eds) Fossils in the Making. Chicago Univ Press, Chicago, p. 182–196.
Walker A (1981) Diet and teeth: Dietary hypothesis and human evolution. Phil Trans Roy Soc Lond 292: 57–64.
Walker A (1984) Mechanisms of honing in the male baboon canine. Am J Phys Anthrop 65: 47–60.
Walker AC, Hoeck HN, Perez L (1978) Microwear of mammalian teeth as an indicator of diet. Science 201: 808–810.
Walker A, Teaford MF, Ungar PS (1994) Enamel microwear differences between species of *Proconsul* from the early Miocene of Kenya. Am J Phys Anthrop Suppl 18: 202–203.
Walker PL (1976) Wear striations on the incisors of cercopithecoid monkeys as an index of diet and habitat preference. Am J Phys Anthrop 45: 299–308.
Walker PL (1978) A quantitative analysis of dental attrition rates in the Santa Barbara Channel area. Am J Phys Anthrop 48: 101–106.
Walker PL, Bernstein SA, Gordon KD (1987) An image processing system for the quantitative analysis of dental microwear. Am J Phys Anthrop 72: 267.
Walker PL, Dean G, Shapiro P (1991) Estimating age from tooth wear. In: Kelley M, Larsen CS (eds) Advances in Dental Anthropology. Wiley-Liss, New York, p. 77–119.
Walker PL, Hagen EH (1994) A topographical approach to dental microwear analyses. Am J Phys Anthrop Suppl 18: 200.
Washburn SL (1951) The new physical anthropology. Transact New York Acad Sci 13: 298–304.
Waugh LM (1933) Influence of diet upon the jaws and face of the American Eskimo. J Dent Res 13: 149–151.

Willey GR, Sabloff JA (1974) A History of American Archaeology. Freeman, San Francisco CA.
Wolpoff MH (1971) Interstitial wear. Am J Phys Anthrop 34: 205–228.
Zuhrt R (1955) Stomatologische Untersuchungen an spätmittelalterlichen Funden von Reckkahn (12.–14. Jh.) I. Die Zahnkaries und ihre Folgen. Dtsch Zahn Mund Kieferheilk 25: 1–15.
Zuhrt R (1967) Gebißbefunde bei Khartoumneolithikern vom Jebel Shaqadud (Republik Sudan). Z Morph Anthrop 59: 36–39.

5.4 Artificial Modifications of Human Teeth

Kurt W. Alt and Sandra L. Pichler

Introduction

Artifical modifications of the human body – often referred to as *exotica* – are of a common interest within the cultural and human sciences. The present paper deals with artificial modifications relevant to the field of dental anthropology, i.e. masticatory and non-masticatory uses of teeth. Its primary aim is the systematic description of artificial changes as well as a characterisation of their variety with the help of selected examples.

After eruption, dental hard tissues are subject to various aging processes. These are natural occurrences if there is normal use and occlusal relationships are correct. Mechanical wear of occlusal and interproximal tooth surfaces is well known (Pindborg 1970). They are mainly characterised by two parameters: first by the quality of dental hard tissues, which is determined by heredity, and secondly by varying exogenous influences reflecting lifestyle and habits (dietary composition, bruxism etc.). The degree of wear is inevitably determined by the length of use. However, considerable individual variation exists.

Aside from natural wear, artificial modifications of teeth may reflect individual or collective habits. They are specific in character both regionally and diachronically and may indicate historical developments. Unusual wear patterns must be viewed in the context of the preparation and consumption of food as well as of parafunctions. The characteristic appearance of wear patterns and their frequently generalised occurrence in the jaws make modifications of this type easy to recognize. In contrast, isolated or generalised defects of the hard-tissues originating in cultural practises, occupational or individual habits are more difficult to classify. Additional modifications include dental mutilations determined by socio-cultural parameters, and from therapeutic measures for curative or cosmetic purposes.

The literature on artificial modifications in the oral region is extensive but mostly deals with the recurring topics of abrasion and tooth mutilations. In contrast, occupational and habitual modifications or signs of therapeutic measures are presented casuistically. Depending on the emphasis of the researchers, papers are either cultural, historical, or medical in orientation. In accordance with the type of source material used (prehistoric skeletons, ethnographic field studies, extant groups), the findings differ in the form and frequency of their occurrence as well as in their expression. Up to now, classifications of dental modifications have been rather preliminary (Milner and Larson 1991; Scott 1991; Cruwys et al. 1992). We have developed a new and comprehensive system of classification which covers all areas of artificial dental modifications (Tab. 1).

Classification of Artificial Dental Modifications

Previous subdivisions of artificial dental modifications were based upon the concept of "teeth as artifacts of human behaviour" (Milner and Larson 1991, 357), which Scott states "indirectly reflect four classes of human behaviour: (1) dietary; (2) implemental; (3) incidental cultural; and (4) intentional cultural" (1991, 798). In contrast, our classification distinguishes two classes of modifications: active (or intentional) changes vs. passive (or unintentional) changes (Tab. 1). Passive modifications of the oral region are side effects of certain activities (accidental dental modifications), whereas active modifications are caused intentionally and are thus forced or wanted (intentional dental modifications). Accidental changes are also characterised by generally not being caused by a single event, whereas intentional changes serve a given purpose and are executed at a certain time.

Accidental modifications can be differentiated into three groups. The first group is made up of changes caused by the selection, preparation and consumption of food, which are associated primarily with the process of urbanisation and socio-economic status. Davies (1963, 213) states that "these factors determine the physical character of the food, and this affects the vigor and duration of mastication". However, a distinction should be made between natural, physiological attrition (e.g. by tooth contact in mastication) and unusual patterns of abrasion (e.g. by abrasive elements in the diet). The second group includes all modifications caused by activities where teeth are used as tools or as a third hand. Changes caused by habitual activities which inadvertantly affect the teeth are included in the third group.

Intentional modifications are divided into two groups. Oral or dental mutilations which have a ritual or socio-cultural background in the broadest sense belong to the first group. These are recognised as the classical artificial modifications of the oral region, and are represented by great variety and wide distribution chronologically, spatially, and in frequency. The second group of active modifications consists of all measures relating to the art of healing practiced for both therapeutic and cosmetic reasons.

Accidental Dental Modifications

The wear of the hard dental tissues, enamel and dentine, by mechanical use is referred to as attrition and abrasion. Whereas German authors generally use the term *abrasion* for the loss of hard tissues by mechanical wear (cf. Horn 1967), English speakers differentiate between *attrition*, *abrasion* and *erosion* (Smith 1975; Eccles 1982), but do not do so uniformly. Furthermore, definitions and nomen-

Table 1. Classification of artificial modifications of human teeth

ARTIFICIAL DENTAL MODIFICATIONS				
Accidental modifications			**Intentional modifications**	
Dietary / parafunctional	**Occupational / traumatic**	**Habitual**	**Dental (oral) mutilations**	**Dental therapy**
Abrasion	**Teeth as tool activities**	**Hygienic habits**	**Tooth mutilations**	**Prosthetics**
♦ occlusal wear (mechanical wear from mastication; effect of subsistence strategy) ♦ interproximal wear (mastication) ♦ microwear patterns (characteristic of specific diet)	♦ hard tissue defects in the anterior teeth (e.g. leather working; LSAMAT) ♦ hard tissue defects in isolated or few teeth (indicator of specific activities: basket maker, sinew stripping etc.; also modern professional activities: shoemaker, musician)	♦ extensive hard tissue defects (wedge-shaped defects from tooth brush or chewing stick) ♦ isolated hard tissue defects (e.g. interproximal grooves from toothpick)	♦ filing ♦ chipping ♦ decoration ♦ ablation ♦ bleaching ♦ dyeing ♦ change of position ♦ amputation ♦ germectomy	♦ artificial dentures ♦ bridgework ♦ crown-work **Conservative dentistry** ♦ metal fillings ♦ non-metal fillings
Attrition	**Dental trauma**	**Individual dental marks**		**Oral surgery**
♦ contact wear (physiological attrition from contact friction) ♦ bruxism (unphysiological attrition caused by stress factors etc.)	♦ craze lines ♦ cracked teeth ♦ fractured cusps ♦ split teeth ♦ vertical root fractures ♦ chipped teeth ♦ fractured teeth ♦ traumatic loss of teeth	♦ specific hard tissue defects in anterior teeth (e.g. pipe notches, wear facets from lip plug, disk, labret or "botoru") ♦ unspecific hard tissue defects (e.g. defects from seed cracking, wear facets from pebble sucking)		♦ tooth extraction ♦ tooth trepanation ♦ tooth replantation ♦ fracture treatment
Dental Erosion	**Dental Erosion**	**Dental Erosion**	**Oral mutilations**	**Periodontal therapy**
♦ erosive foodstuffs (e.g. acidic fruits) ♦ erosive drinks (e.g. lemon juice)	♦ hard tissue defects affecting complete dentition (miners, stone masons, quarry and chemistry workers etc.)	♦ infant care practices (feeding divices, pacifiers etc.) ♦ chronic vomiting	♦ modern body piercing	♦ gold ligatures

clature of abrasive processes on the hard tissues differ between anthropology and dentistry.

Three processes are usually summarised under the term *abrasion* (Tab. 1): first, tooth wear caused by chewing (demastication; Fig. 1 a), which is intensified by abrasive elements (e.g. silicates) and by foreign materials (e.g. sand) in food, and by the use of teeth for the production and preparation of food (Eccles 1982; Brothwell 1989); second, wear caused by direct contact between occlusal (incisal) and approximal tooth surfaces (contact friction) during talking, swallowing, chewing etc. (physiological attrition; Fig. 1 b; Graf and Zander 1964; Russel and Grant 1983); and, third, wear caused by pathological jaw movements (unphysiological attrition) during grinding and pressing of teeth (parafunctions; eg. bruxism or excentric occlusal/incisal positions; Fig. 1 c; Gebrande 1973; Xhonga 1977). Interproximal (interstitial) attrition as a consequence of slight movements of adjacent teeth is a further, little noted type of dental attrition (Wolpoff 1971; Hinton 1982). Demasticative processes are characterised by rounded abrasion surfaces, whereas attrition and parafunctions result in the formation of opposing sharp edges which match perfectly in occlusion. The extent of these abrasive processes differs in each individual and increases with age (Barrett 1977; Brothwell 1989).

Specific loss of hard dental tissues is caused by erosive processes (Fig. 2). Loss of this type can be due to dietary habits, e.g. the consumption of acidic fruit juices (Eccles 1979), psycho-social disorders such as anorexia nervosa, where gastric acid acts as an aggressive agent upon the teeth (Smith and Knight 1984), or to a person's occupation (Ten Brugen Cate 1968; Alt and Pichler 1995).

From a clinical point of view, a differentiation between abrasion, attrition and erosion is necessary for therapeutic reasons. This differentiation is unneccessary in findings on prehistoric skeletal material or in recent traditional societies, since in these cases all abrasive processes overlap and can hardly be aetiologically differentiated.

Aside from the above-mentioned processes resulting in the accidental loss of enamel and dentine, individual, cultural or occupational habits may damage the hard tissues of the tooth. In these cases, the term *usur* is commonly used (Alt and Koçkapan 1993).

Crown wear studies usually describe position, degree, and shape of abrasions (cf. Miles 1963; Perizonius and Pot 1981; Brothwell 1989) as well as the angle of the abrasion surface to the occlusal plane (Neff 1993). A large number of quantitative, qualitative, and functional indices and schemata have been developed for the registration and evaluation of dental abrasions. To this day, however, standardisation in the registration of tooth wear is lacking, and comparative population studies remain difficult (see Rose and Ungar this volume).

In the past, the interpretation of abrasive processes has been used primarily for the estimation of age. In recent times, attrition phenomenona have also been evaluated from a socio-ecological and socio-cultural perspective. These studies aim to differentiate subsistence (e.g. hunter-gatherers vs agriculturalists), dietary habits, and environmental factors, to identify specific attritional differences between neighboring populations with regard to gender and social status as well as diachronic trends (Molnar 1972; Cybulski 1974; Lavelle 1979; Smith 1984; Schürmann 1986).

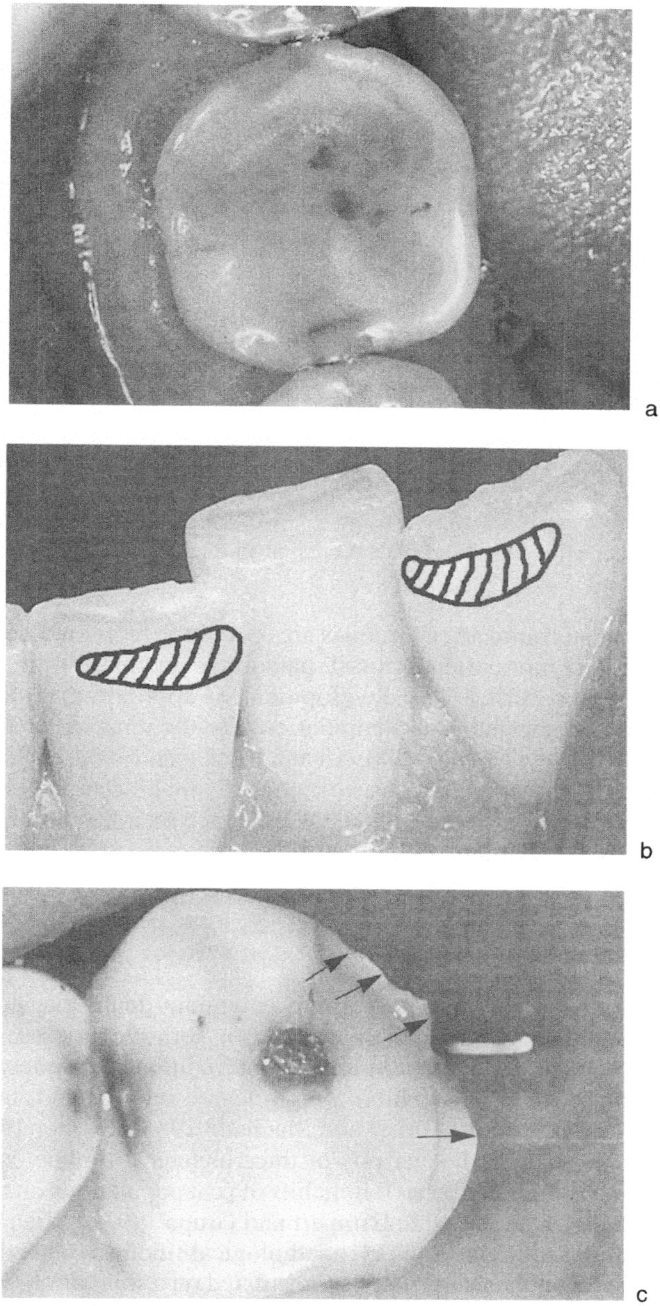

Fig. 1 a–c. Tooth wear in clinical patients; **a:** Abrasion by demastication – well-rounded occlusal surfaces (right mandibular molar); **b:** Physiological attrition by contact wear – regular, circumscribed wear-facets by antagonist contact (mandibular incisors); **c:** Unphysiological attrition by parafunction – jagged, irregular wear-facets (maxillary molar) (from Hickel 1989).

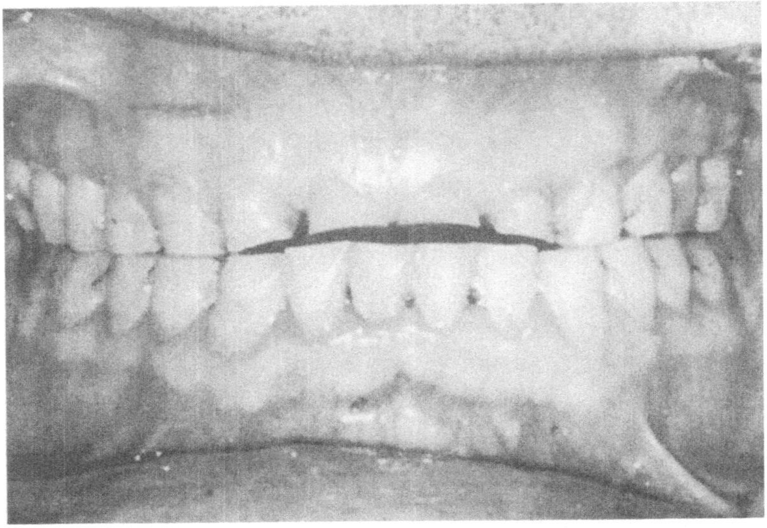

Fig. 2. Acid-induced occlusal and buccal erosion in chemistry-worker (from Kreter and Pantke 1979).

Approximal and subvertical grooves are specific defects on the interproximal surfaces diagnosed more often in fossil, paleolithic and mesolithic material than in younger skeletal series. The development of approximal grooves may be attributed to active mechanical manipulations for the removal of food particles (N.N. 1988; Brown and Molnar 1990; Alt and Koçkapan 1993; see below). Certain characteristics of subvertical grooves of interproximal facets in Neandertal man suggest that erosion-abrasion mechanisms influence their formation (Kaidonis et al. 1992; Villa and Giacobini 1995; Fig. 3).

Dietary Modifications

The dietary behaviour of any given group is without doubt the most important factor in accidental dental modifications (Tab. 1). Abrasive processes are relevant primarily for prehistoric populations and recent traditional societies. The teeth of individuals from both groups exhibit a high degree of wear and, in most cases, specific patterns of abrasion (Turner and Machado 1983; Heymer 1986). Modern research methods such as the analysis of trace elements in bone contribute to a reconstruction of the dietary behaviour habits of past populations on the molecular level (Grupe and Herrmann 1988; Lampert and Grupe 1993). When the results of these investigations are related to paleopathological findings on teeth, the dietary behaviour of past populations can be reconstructed on a solid basis (Hillson 1979; Gilbert and Mielke 1985; Lukacs 1989).

The type of foods consumed, the degree of their preparation, their composition and possible foreign elements contained in them (e.g. as a result of primitive grinding techniques) are responsible for the mechanical wear of teeth. Since hunter-gatherers and agriculturalists use different resources, different factors influence

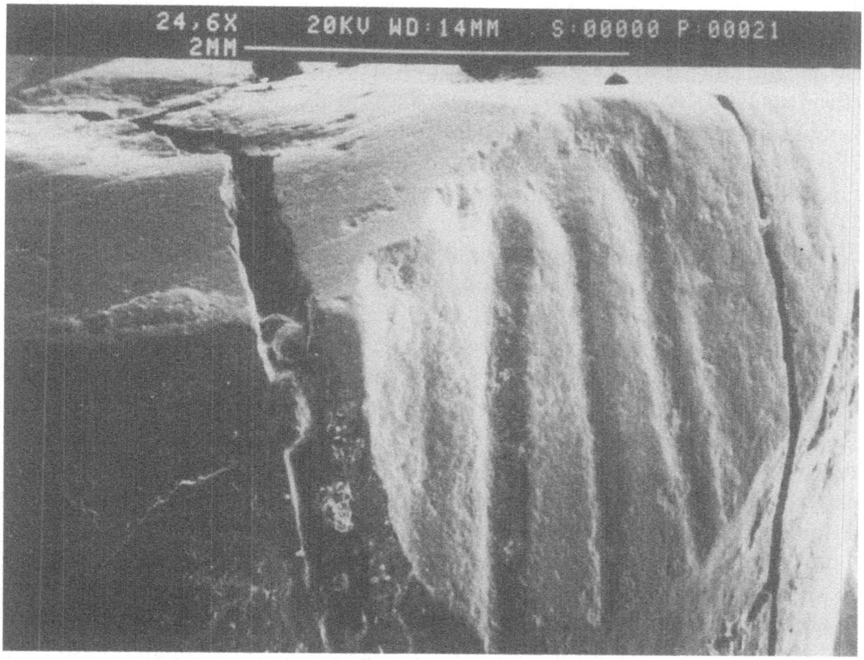

Fig. 3. Subervertical grooves in Neandertal man (Genay, France). SEM image of mesial interproximal facet of first lower right premolar; grooves perpendicular to occlusal surface (after Villa and Giacobini 1995).

dental abrasion and leave characteristic marks on occlusal surfaces (Smith 1984; Fig. 4 a,b). The employment of the teeth in the production and preparation of food represents a further abrasive process. At the same time, genetic factors determining thickness and durability of enamel also influence the degree of contact wear.

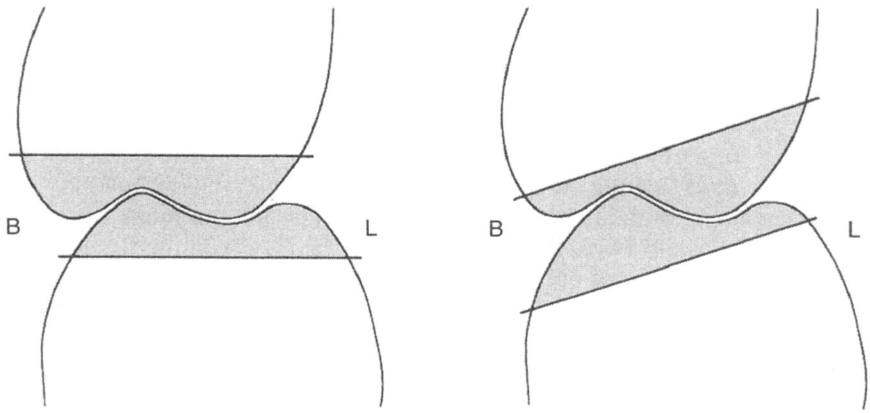

Fig. 4 a, b. Subsistence specific wear patterns reflected by different angles of occlusal plane; **a:** Hunter-gatherer – flat molar wear; **b:** Agriculturalist – oblique molar wear.

Crown wear due to specific diets was a natural process in prehistoric times. It is of special interest only in regard to comparative studies and specific patterns of attrition and abrasion. In extant populations, especially in industrial nations, attrition caused by dietary behaviour plays only a minor role, whereas unphysiological attrition as a result of parafunctions occurs quite frequently (Xhonga 1977). The abundance of anthropological literature on the subject makes a further discussion of these problems unneccessary (cf. Miles 1963; Perizonius and Pot 1981; Brothwell 1989).

Since the 1980s, the systematic analysis of dental microwear has been a pivotal source of information on the eco-ethology of recent, (pre)historic, and fossil species (e.g. Puech et al. 1986; Teaford 1991; Fig. 5 a, b; see also above Rose and Ungar this volume). If one assumes that the observed variation in microwear pattern among primates is due to differences in diets, one would expect that consumers of similar foods show similar tooth wear patterns, whereas those preferring different foods should exhibit divergent patterns. Hence, a close correlation between diets and patterns of dental microwear should exist (Henke and Rothe 1996). However, research on dental microwear by various investigators has yielded ambiguous results (Gordon 1982; Grine 1986; Walker and Teaford 1989). For example, taxa such as *Pan* und *Pongo*, which, according to current knowledge, hardly differ with regard to food preferences, are characterized by divergent wear patterns (Walker and Teaford 1984). It therefore remains controversial to what extent dental microwear is able to provide reliable information regarding the diet of recent and fossil species. Most probably, biomechanical (masticatory) aspects act as significant, yet often neglected modifications (Gordon 1982). Thus, as far as patterns of dental microwear are concerned, we concur with Teaford's conclusion that "much work remains to be done, but one thing is clear: if we proceed cautiously, we can provide new insights into the evolution of diet and tooth use" (1994, 17).

Occupational Modifications

In connection with a person's occupation, specific accidental modifications of teeth may develop which are described as *occupational modifications* (Tab. 1). In recent populations, modifications of this type can be found among seamstresses/dressmakers (thread), shoemakers/carpenters (nails), butchers (string), glassblowers and musicians (mouthpieces), office workers (pens; Fig. 6), and also among jugglers and trapeze artists. In these cases, the assignment of characteristic modifications to specific occupations poses few problems.

In prehistoric skeletal material, such findings are far more difficult to interpret (Blakely and Beck 1984; Fig. 7 a, b). The observed modifications are regarded as an indicator of occupational activities often paraphrased as *teeth as a tool* or *teeth as third hand* use. Through comparisons with similar defects in ethnographic contexts, in some cases conclusions can be drawn as to the occupational activities represented by such modifications (Larson 1985; Lous 1970; Merbs 1983). The occlusal grooves of basket makers are a frequently cited example (Fig. 8; Milner and Larsen 1991). Stripping of sinew (Brown and Molnar 1990), shaping of wood (Hylander 1973), chewing of leather (Lous 1970;

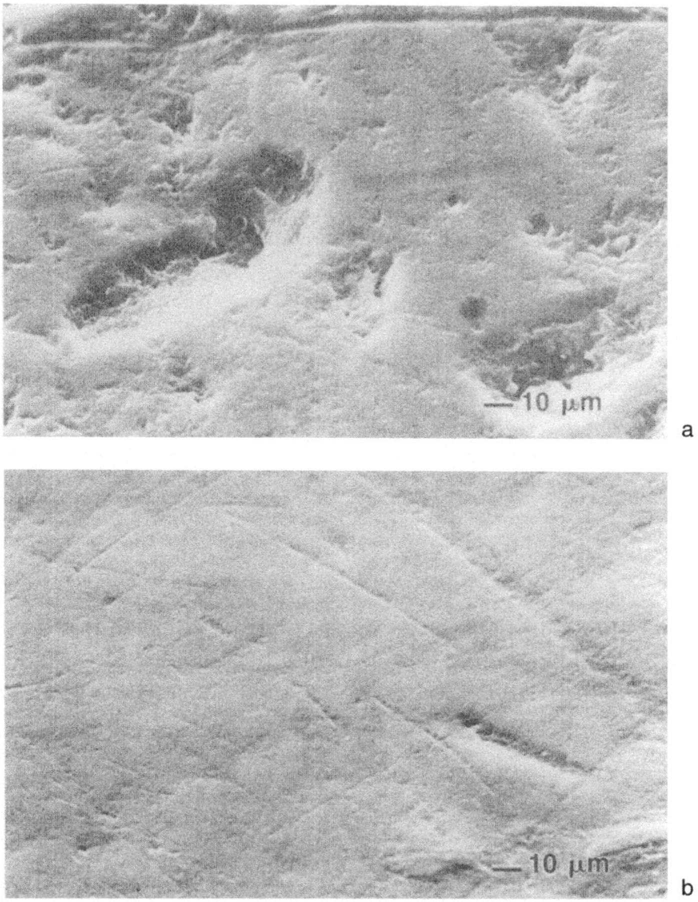

Fig. 5 a, b. Dental microwear. SEM of wear-facets on maxillary first molars of Native American individuals from the Georgia coast dating before and after European contact; **a:** Individual from precontact site (Mary's mound); **b:** Individual from late contact site (SCDG-SM36B). Increased enamel pitting and wider scratches on the teeth of the precontact individuals suggests that these hunter-gatherers were ingesting more hard objects as compared to the postcontact groups. These findings suggest that a gradual dietary shift occured within the time-frame documented by the samples (from Teaford 1991).

Fig. 9 a, b), and, among prehistoric mineworkers, holding of pine torches with the teeth (Iron age Hallstatt salt mine; pers. com. M. Singer) are further examples of such *third hand* activities.

In modern times, excessive abrasion of tooth crowns is found among individuals employed in jobs with high dust pollution. Occupational exposure to mineral, metal or vegetational dust affects quarrymen and mineworkers (Fig. 10), stone masons, millworkers and so on (Dechaume 1938; Ring 1984; Enbom et al. 1986; Hickel 1989). Erosive modifications on tooth surfaces can be found among persons working in acid-enriched environments, such as in chemistry plants (Ten

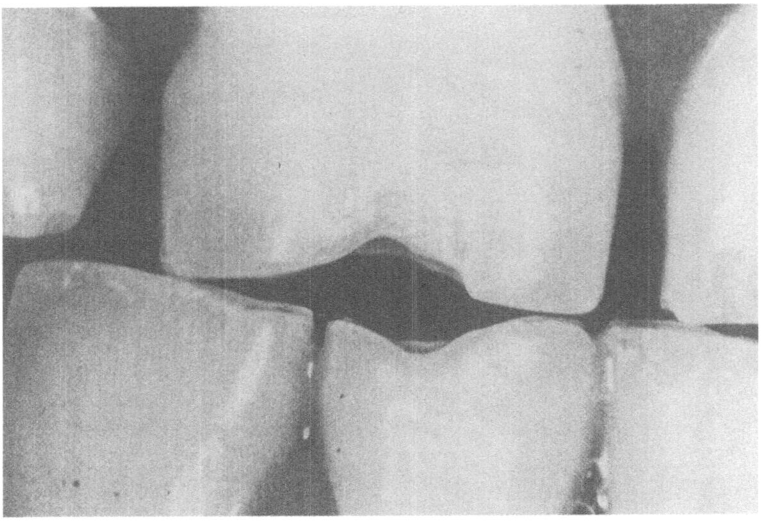

Fig. 6. "Usur" by habitual biting of pen; 28-year old clinical patient (from Hickel 1989).

Bruggen Cate 1968). In the interest of the persons affected, it is imperative that epidemiological studies on the loss of dental hard tissues of occupational origin are conducted and that the effects are acknowledged as occupational diseases (Hickel 1989).

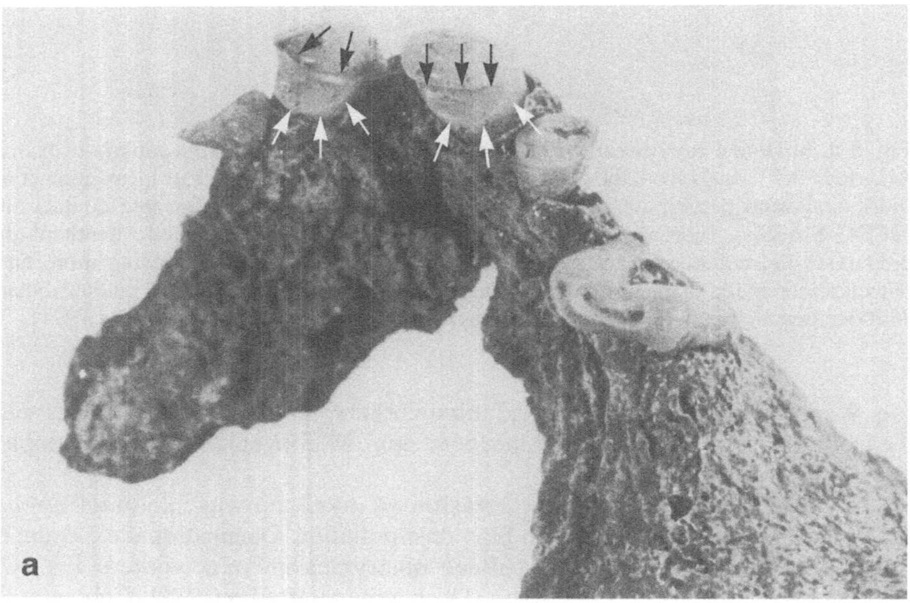

Fig. 7 a. Teeth-as-tool use in Neolithic individual (Wandersleben, Thuringia); Lingual grooves on central maxillary incisors (Photo: A. Bach).

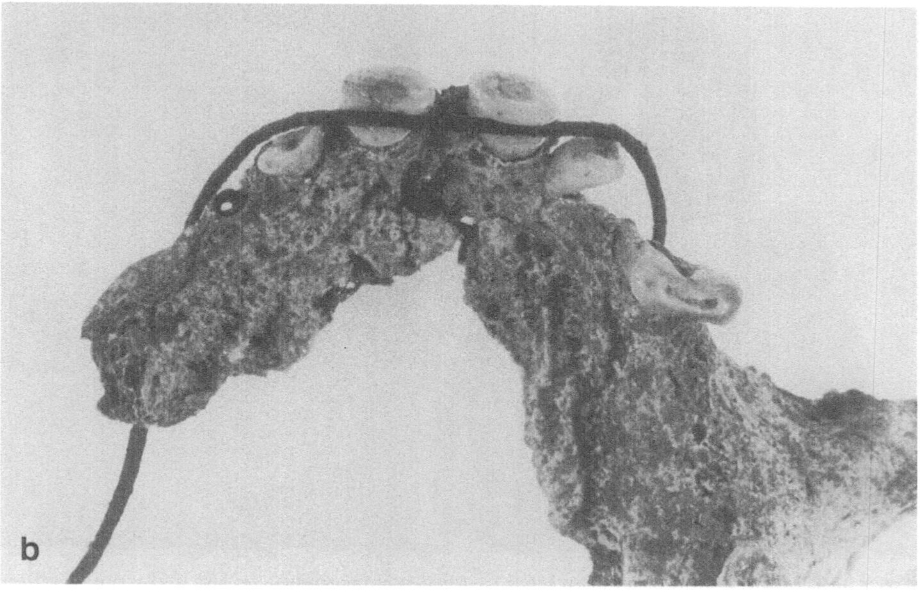

Fig. 7 b. Teeth-as-tool use in Neolithic individual (Wandersleben, Thuringia); Thread inserted in grooves to show hypothetical activity – origin of modifications unknown (Photo: A. Bach).

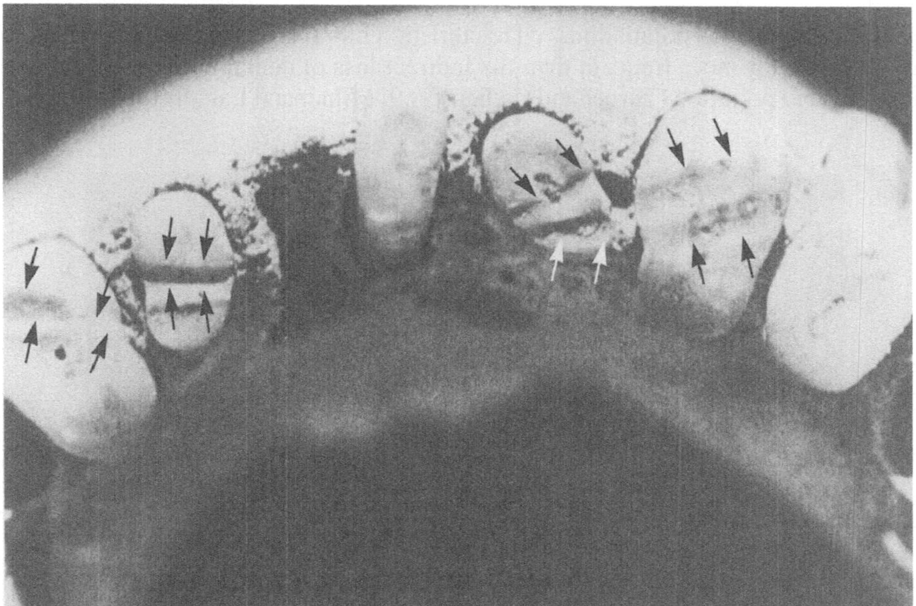

Fig. 8. Task-related wear in Great Basin Indian. Transversal grooves on occlusal surfaces of mandibular anterior teeth by processing of plant fibers for basketry or cordage (from Larsen 1985).

Fig. 9 a, b. Modification of anterior teeth by third-hand activities; **a:** Eskimo woman softens skin by incisal mastication (photo: Lous 1970); **b:** First right incisor of an early medieval mature male individual (grave no. 353) from Kirchheim/Ries, southwest Germany; curved wear plain and cupped crown margin probably caused by processing of soft material (leather?).

In (pre-)historic populations, characteristic chipping of the enamel and tooth fractures are far more frequent than the indirect loss of dental hard tissue through dust and acid pollution (Turner and Cadien 1969; Milner and Larson 1991; Fig. 11).

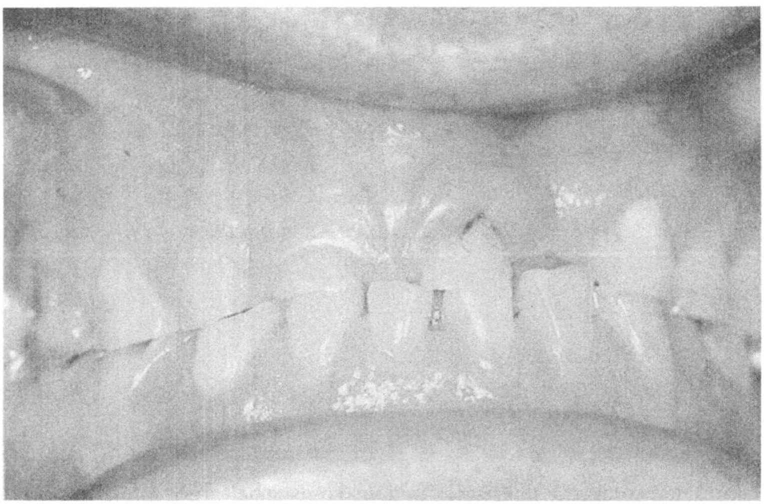

Fig. 10. Severe abrasion of anterior maxillary teeth due to occupational exposure to granite dust in quarry-worker (from Hickel 1989).

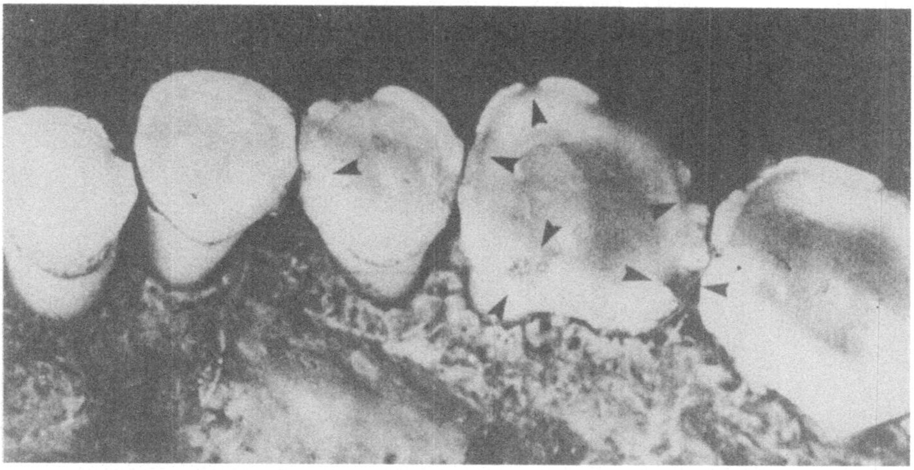

Fig. 11. Chipped teeth – traumatized tooth surfaces characterized by marginal fractured enamel (from Hansen et al. 1985).

In most cases, these findings are probably associated with the use of the *teeth as tools*, but their direct causes, whether they be in the context of subsistence or occupational in nature, can rarely be differentiated. Walton (1996) divided the subject of recent traumatic dental modifications into five major categories: craze lines, cracked teeth, fractured cusps, split teeth, and vertical root fractures. However, it is useful to look at dental trauma in modern population in the broader context of human behavior, such as circumstances that produce trauma.

Turner and Machada (1983) were the first to report on a form of abrasion known as *lingual surface attrition of the maxillary anterior teeth* (LSAMAT), which is defined as the result of progressive local wearing of upper anterior lingual tooth surfaces without corresponding wear on the lower teeth (see also Irish and Turner 1987). They postulated a connection between these findings in archaic south and central American Indian populations and the processing (peeling and chewing) of manioc, the main component of their diet. LSAMAT patterns can also be found among central European early medieval populations (Alt and Pichler 1994; Fig. 12 a, b; see also Fig. 9 a, b), but in lower frequencies than reported by Turner and co-authors. As there was no correlation with accompanying archeological finds, there is of yet no precise determination of the causes. The origin of these artificial modifications in European populations can only be determined from interdisciplinary investigations.

Individual Habitual Modifications

Dental modifications or unusual patterns of abrasion caused by personal habits are generally very individual in character (Davies 1963; Tab. 1). They usually occur unilaterally and are most frequently found on the front teeth, such as the oval defects characteristic of pipe-smoking (Morris 1988; Kvaal and Derry 1996; see Fig. 13 a, b) or the chewing of pens (Hickel 1989; see Fig. 6). A habit common to the Mediterranean area, the cracking and chewing of dried seeds (sunflower-, pumpkin seed etc.) also

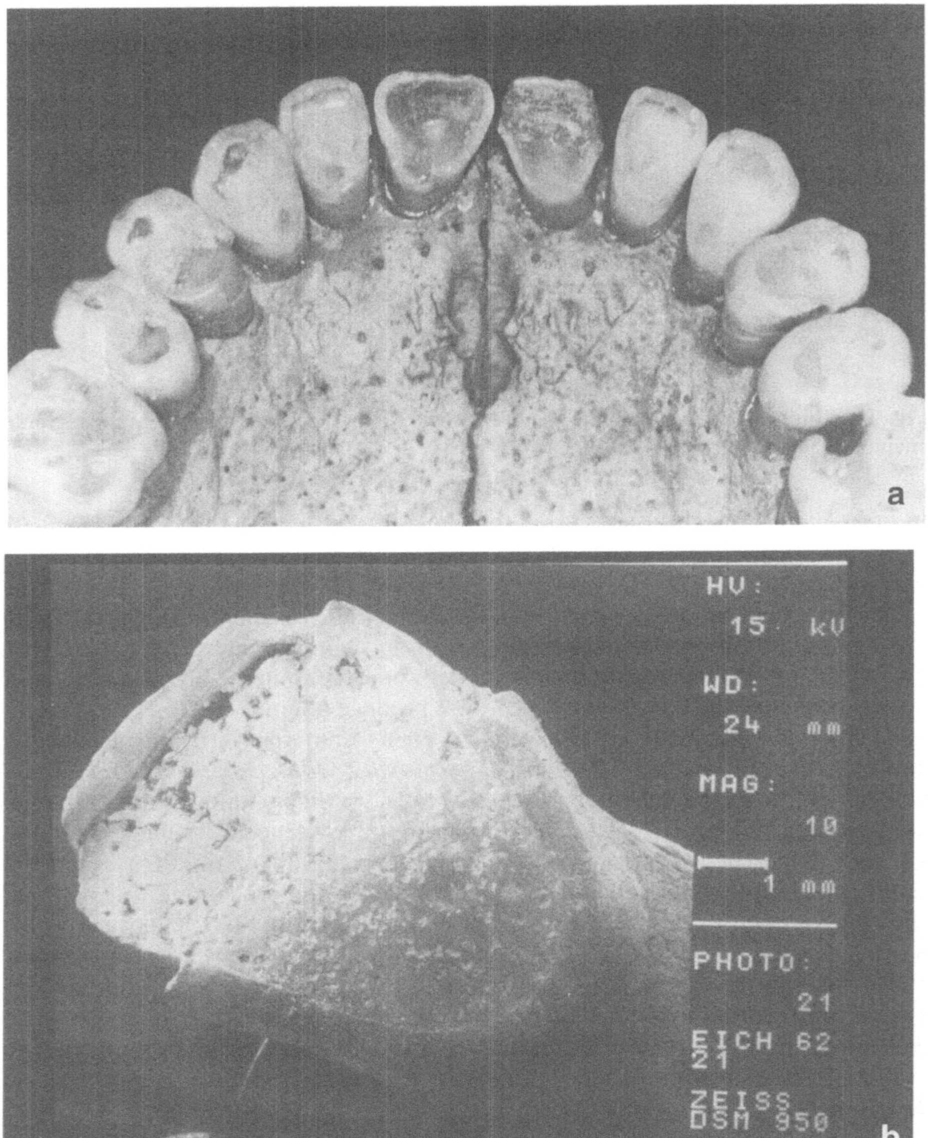

Fig. 12 a, b. LSAMAT phenomena in an early medieval skeleton (grave no. 62) from Eichstetten, southwest Germany; **a:** Maxillary anterior teeth of an adult male with advanced lingual wear of first incisors, lingual facets less developed on second incisors and canines; postcanine teeth show occlusal wear pattern typical of this period. **b:** SEM image of first left upper incisor illustrating flat wear plain and distinct crown margin of lingual surface. Wear pattern probably due to teeth-as-tool activites/occupational modifications of unknown genesis (compare superficially similar, yet distinctly different crown margin in Fig. 9).

causes characteristic defects (Popoviciu 1969). The wide distribution of such defects is demonstrated by the large number of extant and ethnographic findings (Tab. 1).

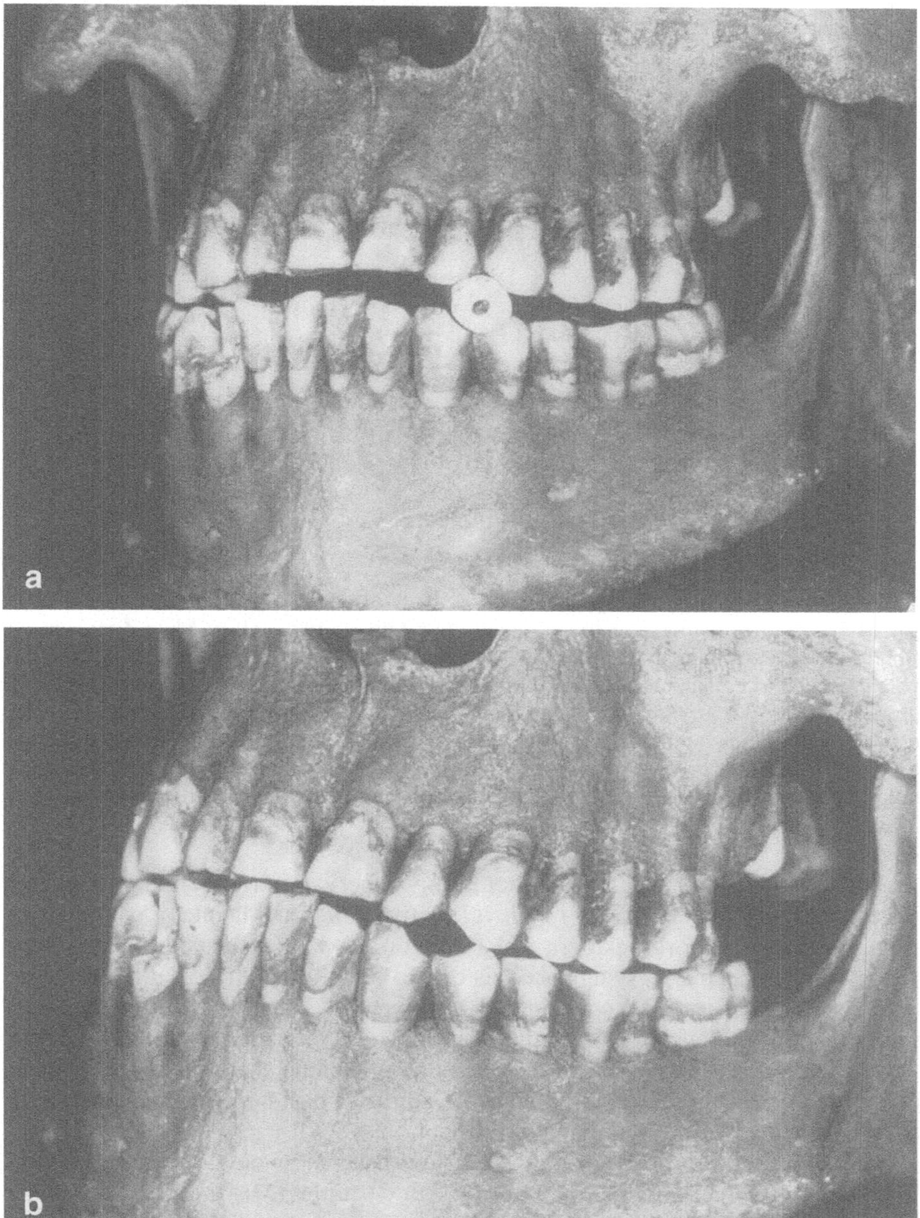

Fig. 13 a, b. Dentition of a female skull from the House of Correction in Oslo with distinctive habitual dental marks; **a:** The close fit of a fractured pipe-stem to the aperture; **b:** The aperture normally formed by four teeth: two teeth from the upper jaw and two from the lower (Photos: SI Kvaal).

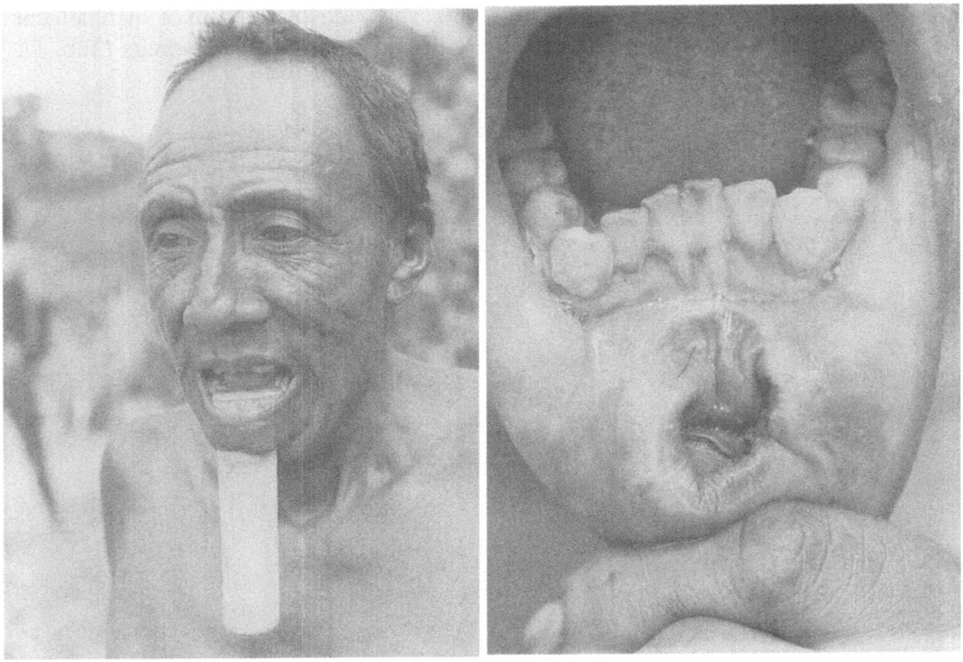

a b

Fig. 14 a, b. Deformation of teeth and mandible by body ornament; **a:** Male Zoé Amazon Indian from Rio Cuminapanema wearing a "boteru", a round labret of the lower lip, ca. 4 cm in diameter and 15 cm in length. **b:** 25 year-old Zoé Indian exhibiting severe "boteru" induced deformations. Perforation for labret clearly visible in pulled-down lower lip. Anterior teeth pressed inward hemispherically conforming to shape and size of "boteru"; anterior section of jaw-bone deformed correspondingly. Labial wear-facets extending to canines; first incisors liable to loss as consequence of dehiscence on roots and periodontal insufficiency (Photos: R. Garve).

Williams and Curzon report interesting observations on dietary behavior concerning the primary teeth of medieval British children: "Dental erosion/caries of a previously unreported pattern was found on the palatal surfaces of maxillary primary incisors. This was suggested as due to feeding methods with cariogenic food such as pap, oatmeal, honey and milk with bread. The use of feeding devices, made of pottery, cow horn or leather, may have been a factor akin to modern nursing bottles" (1986, 210). Studies of human skeletal material reveal a wide spectrum of dietary practices in infants and adults of past populations which may cause a variety of accidental dental modifications.

Various manipulations in the oral region can cause accidental dental modifications. One may cite the labrets of Eskimo peoples (Merbs 1983), the *boteru* (a labret of the lower lip), worn by Brazilian Amazon Indians (Garve 1995), or the lip discs of African tribes, all of which require that the lips be pierced in order to fix the ornaments (Menzel 1957; Hoffmann and Treide 1976). When worn, such objects come in contact with the labial surfaces of the opposing teeth and result in the formation of characteristic, smoothly polished wear facets. The *boteru* even causes

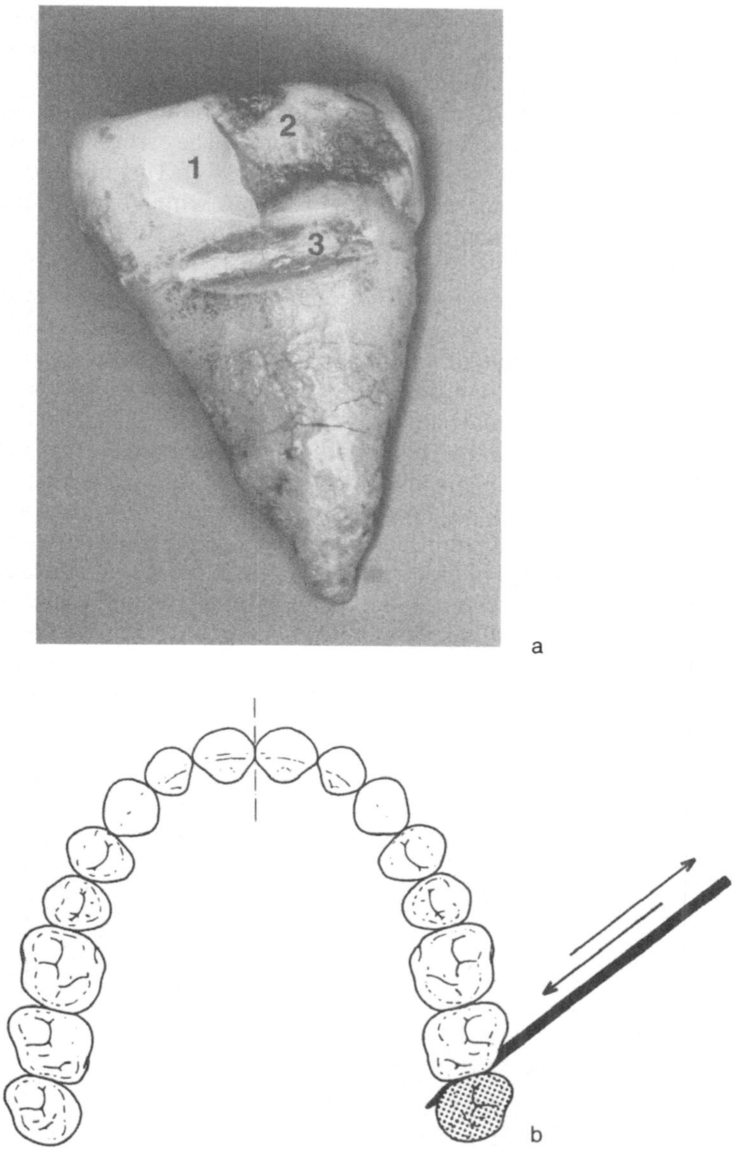

Fig. 15 a, b. Approximal groove in an upper paleolithic tooth from Kohlerhöhle near Brislach, Switzerland; **a:** Mesial surface of isolated left upper third molar with interproximal attrition (1), postmortem defect (2) and approximal groove (3). The concave groove is located at the mesial enamel-cementum line. It is approximately 1 mm wide and 4 mm long. SEM-studies show an oval outline with a round concavity that ends abruptly in the lingual direction. The shape of the defect suggests that the interproximal embrasure was repeatedly manipulated with a roundish, relatively hard object; **b:** Reconstruction of manipulation with "tooth-pick" assumed to have caused the observed approximal groove on the third molar. Instruments such as bone needles suited to remove food particles from in-between teeth were found in corresponding strata of the cave.

semi-symmetrical tooth malpositioning (Fig. 14 a, b). As the dental modifications caused by the wearing of such types of ornaments are unintentional, they are subsumed under the category of accidental dental modifications.

Individual habits of oral hygiene can also result in dental modifications and are also subsumed under accidental modifications. The earliest findings of this type date far back into prehistory (Freyer and Russel 1987). The repeated use of tooth-picks made of wood or bone, or cleaning processes with sinew or strands of grass can cause the formation of approximal grooves (Turner 1988; Brown and Molnar 1990; Alt and Koçkapan 1993; Willey and Hofman 1994; Fig. 15 a, b). Diagnosis of such defects is comparatively easy. The cleaning processes were probably carried out to remove food particles stuck in-between teeth, which disturbed the individuals' well being. Individual habitual modifications originating from acts of oral hygiene are proof of prophylactic, curative and palliative measures in the past.

Damage to the dental hard tissue through false cleaning also belong to this group of artificial modifications. This applies to modern tooth brushes (Slop 1986) as well as to traditional instruments of dental care (e.g., chewing sticks; Türp 1990). Tooth brushes can cause wedge-shaped defects of the tooth-necks (Lee and Eakle 1984; Alt and Koçkapan 1993; Klimm and Graehn 1993; Fig. 16). In Africa and Asia, chewing sticks are used instead of tooth brushes (Betram 1981; Butt and Dunning 1986). The ends of the sticks are chewed to loosen the fibre structure so that a kind of brush is formed. Incorrect handling of this instrument causes extensive flat or through-shaped erosions on the labial surfaces which are difficult to distinguish from defects resulting from acid erosion by drinks or chronic vomiting (Eccles 1979; Robb et al. 1991; Cruwys and Duhling 1993).

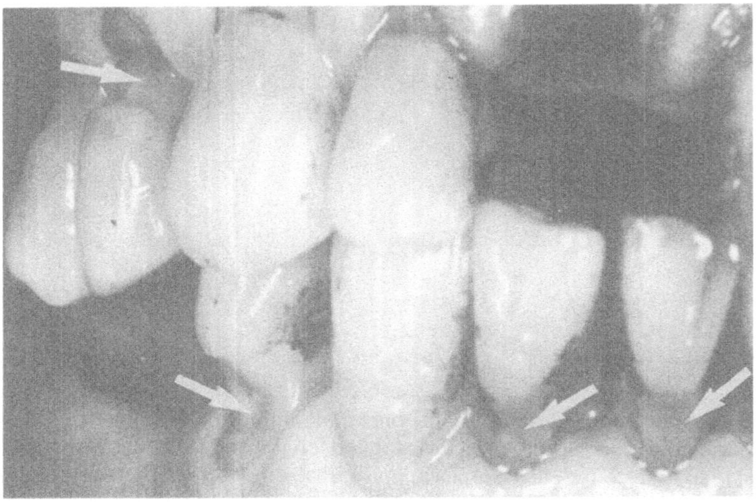

Fig. 16. Clinical patient with distinctive, wedge-shaped defects (arrows) on maxillary and mandibular teeth. The highly polished, notch-like incisions in the vestibular region of the tooth necks are characterized by a chronic, progressive loss of hard substance in the shape of an inverted "L". Etiologic factors such as gingiva recession, tooth-brush traumata, dentifrice abrasiveness, faulty mineralization of the cervical crown, malocclusion, occlusal stress etc. are discussed.

Determining the causes of unusual facets or patterns of wear often proves to be difficult. Extraordinary, unusually located defects, i.e. on the buccal surfaces of both maxillary and mandibular molars, were detected in three upper paleolithic burials (Vlček 1986). The lesions are presumed to have been caused by pebble sucking to produce saliva. Although frequently encountered in nomads, in this case findings and interpretation are difficult to bring into accord because the defects occur unilaterally in all individuals, and because saliva flow from the sublingual glands is more easily provoked by retaining objects in the mouth cavity itself. So in this as in many other cases from prehistory, the causes for the unusual dental modifications are generally unknown.

Intentional Dental Modifications

Intentional modifications of human teeth are subdivided into two groups: one comprises the field of dental mutilations, the other includes all effects of therapeutic measures on the teeth (Tab. 1). Tooth mutilations belong to the artificial modifications of the human body, such as trepanations, artificially deformed skulls, tattoos, and deforming operations on nose, lips, ears, and genitals (Kunter 1971; Hoffmann and Treide 1976). In modern times, dental mutilations are uncommon, whereas tattoos and the piercing of nostrils, eyebrows, lips, ears etc. are increasingly encountered (Ottermann 1979; Herion 1985). In contrast, therapeutic measures on teeth are widespread throughout the population (at least in the industrial nations). In earlier times, therapeutic measures were a privilege of the upper social classes. Intentional modifications on human teeth other than those of therapeutic origin are of little relevance in the present, except in the forensic sciences.

Dental (Oral) Mutilations

Dental mutilations belong to the field of *body decoration*, which includes all attempts to improve (or flatter) appearance (Tab. 1). In the oral region, one differentiates between deformations of the teeth, the lips (tattoos, attachment of decorations) and the tongue (oral body art). Because tattoos and scarification from prehistoric times are rarely preserved (e.g. on mummies), and because lip or tongue decorations can only be assessed indirectly through archaeological grave goods (labrets etc.) or specific wear facets on corresponding tooth surfaces, ethnographic observations have always been the most important source for studies of body decoration (Hoffmann and Treide 1976; see Fig. 14 b).

Methods to alter or deform the human teeth include filing, colouring and bleaching, inlaying with ornaments, ablation, amputation, germectomy, and the changing of the position of teeth (Tab. 1; cf. Jahnke 1970). Tooth deformations are a world-wide phenomenon, but characteristic forms are specific to continental or regional centres (Verger-Pratoucy 1970).

The most striking areas for tooth mutilations are Africa (with the exception of the north; Gould et al. 1984), Central and South America (Fastlicht 1976), and the

Philippines and the Malayian Archipelago in East Asia (Galang 1941). Filing is among the most common forms of tooth mutilation. It shows the greatest variety even when compared to the numerous forms of inlaying in pre-Columbian American cultures (Maya, Aztec, Inca; Fastlicht 1976; Linné 1940; Fig. 17). In Africa, tooth-filing as well as other forms of dental mutilations are practised to this day (Briedenhann and Van Reenen 1985; Heymer 1986; Jones 1992); the situation in east Asia is similar (Headland 1977). Under the growing influence of western culture, these traditional customs are on the decline (Van Reenen and Briedenhann 1986). The original motives for dental mutilations can rarely be traced as they are not securely handed down and are partly superimposed by secondary processes. The extensive literature on tooth deformations among historic and extant traditional communities supplements reports on prehistoric cases of dental mutilations (Jackson 1915; Fastlicht 1976; Alt et al. 1990).

Damaging after-effects (opening of pulp-cavity, tooth-loss) are irrelevant in the practice of dental mutilations (usually at the beginning of puberty), as the filing of teeth drastically demonstrates. The instruments used (stones, iron files) often damage the soft tissues surrounding the "worked" tooth. Due to the loss of contact points, tooth movements and position changes may occur as an after-effect of tooth-filing. Surprisingly, findings on prehistoric teeth and ethnological

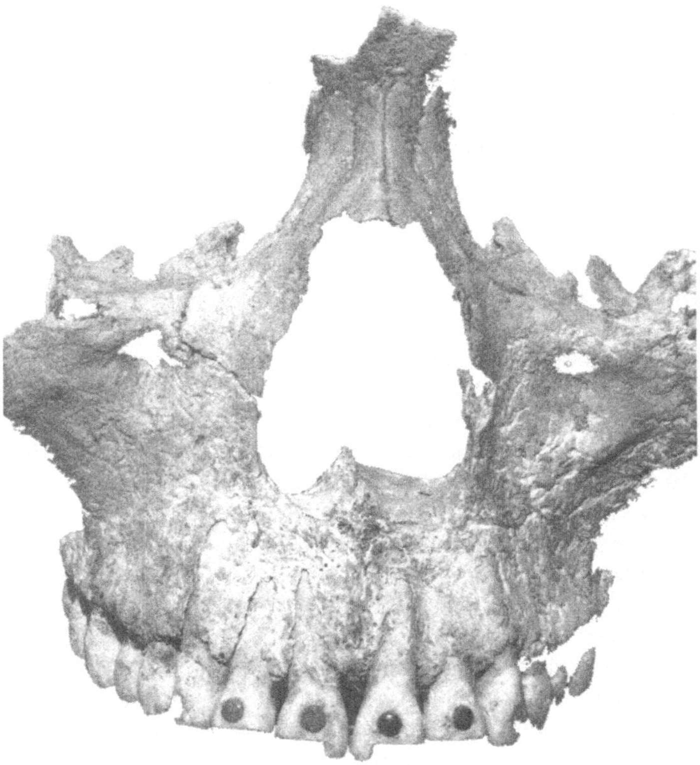

Fig. 17. Mutilations of maxillary anterior teeth in a skull from Tlaxala, Mexico, 500–900 AD. Dental inlays of pyrit and jade; filing of incisors (from Lässing and Müller 1983).

reports demonstrate that damage to the dental pulp is rare (Fitting 1989; Fig. 18). One must therefore assume that the executing individuals have some anatomical knowledge. Germectomy in newborn babies has a tremendous influence on facial growth, dentition and occlusion, and also involves great health risks (Abusinna 1979).

The acceptance of both unpleasant side effects during the performance of dental mutilations and of possible after-effects give us an idea of the importance of this ritual. Individual, group specific or ritual motives have priority over possible disturbances of the individual's well-being and negative after-effects of the process (cf. Alt et al. 1990). Body decoration can help ethnologists, archaeologists and anthropologists to establish the ethnic, cultural, social and religious affiliation of individuals. Present-day fashions of body decoration are of interest mainly with regard to sociological and forensic aspects (for modern *oral body art* see Scully and Chen 1994).

Dental Therapy

Healing and nursing developed from instinctive and empirically learned therapeutic measures and mark an important step in the cultural development of the human race. Acts of healing and nursing from the early stages of humankind can only be detected by inference from both archaeological findings and the evaluation of bio-historical sources (Tab. 1). Due to the fact that the teeth are usually better preserved than other parts of the skeleton, palaeodontological investigations frequently yield indications of former treatments of teeth and jaws. Of all the disciplines of dentistry, prosthetics is best represented historically. Artificial dentures and parodontal

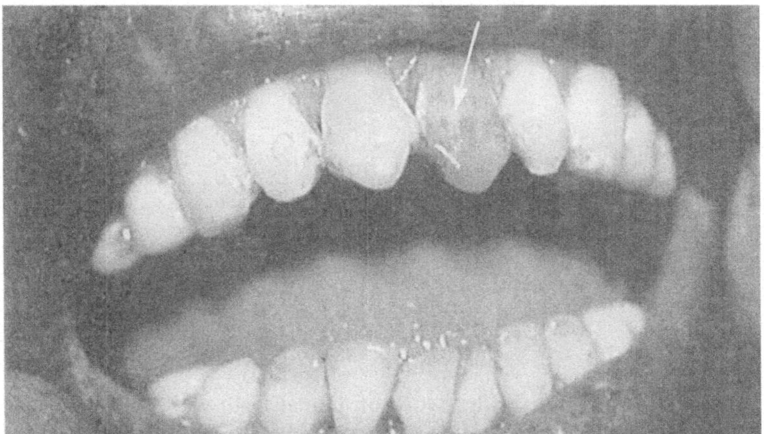

Fig. 18. Modern tooth-filing in Africa (Photo: W. Fitting). Pulp gangrene on the first left maxillary incisor developed as result of filing of the anterior maxillary teeth. Further side-effects may be apical periodontitis, periapical abscess, periapical granuloma, radicular cyst, and tooth-loss.

splinting to preserve loosened teeth date back to the middle of the 1st millenium BC as demonstrated by Phoenician, Etruskan and Attic finds (Hoffmann-Axthelm 1981; Lässig and Müller 1993; Alt 1994; Becker 1994).

In central Europe, proof of artificial dentures dates to the 12th century AD. An individual from the Slavic grave field of Sanzkow (Germany) exhibits a mandibular denture for which the individual's anterior teeth had probably been used after extraction (Ulrich 1973). This case seems to be unique, since no other artificial dentures appear until the 17th and 18th centuries (Thierfelder et al 1987; Czarnetzki and Alt 1991). Around 1750 AD, burials from churchs in Geneva (Switzerland) and Göppingen (Germany) show the splinting of parodontally insufficient anterior teeth with gold ligatures, bridge work carved from ruminant teeth, and artificial dentures and bridge-works made of ivory (Alt 1993; Alt and Newesley 1994; Figs. 19, 20). Whittaker and Hargreaves (1991) found nine artificial dentures among 987 burials from an 18th-century London spitalfield which were made of gold and ivory. Other materials used for artificial dentures include bone, hippopotamus teeth and walrus ivory.

In prosthetics, there was little progress from antiquity until the middle of the 18th century (Czarnetzki and Alt 1991; Alt 1994). Most finds come to light in

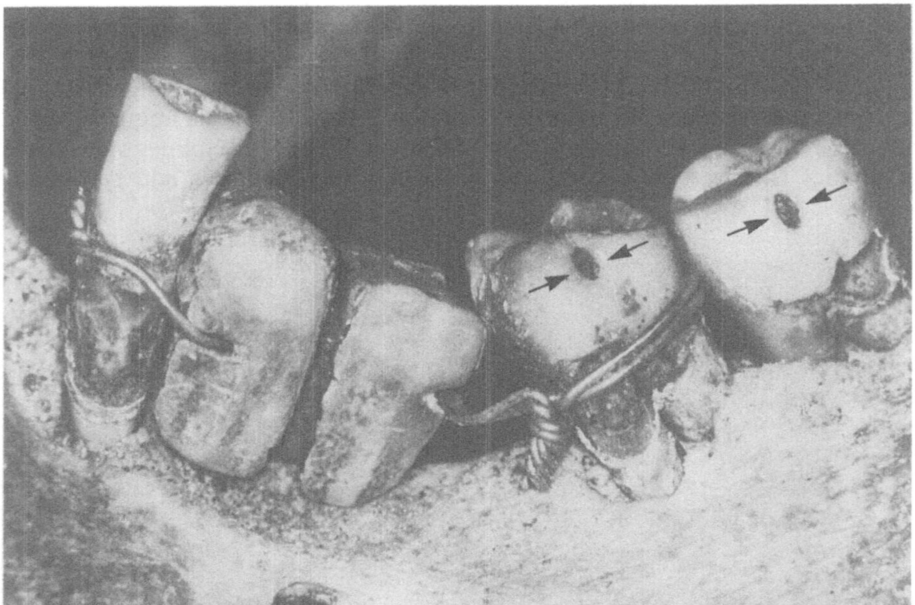

Fig. 19. Historical prosthetics. Bridge-work of animal teeth with gold wire from Saint-Hippolyte Church (grave no. 37), Geneva, Switzerland. In the left mandible, the bridge replaced three postcanine teeth. The bridge is perforated for its attachment with the gold wire on both the mesial and distal ends. Tin fillings on the buccal surfaces of the left mandibular second and third molars. The individual was identified as François-Bénigne du Trousset d'Hericourt, deacon, originally from Paris, France, who died in Geneva in 1761 at the age of 58.

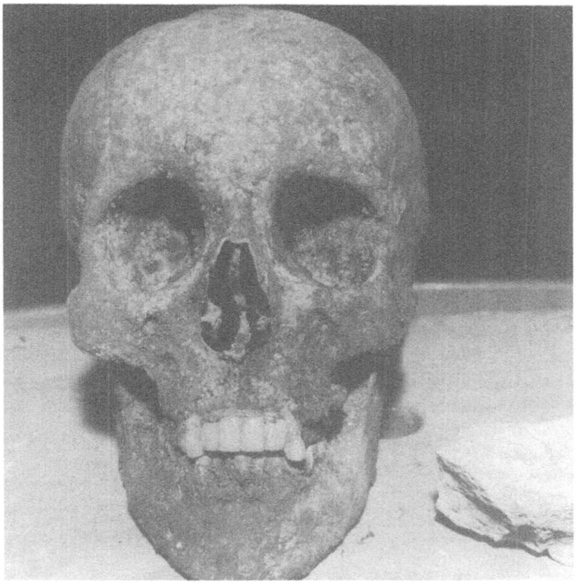

Fig. 20. Historical prosthetics (16./17. century). Ivory bridge-work from Oberhofener Church, Göppingen, Germany in situ (after excavation). The ivory bridge for anterior area of maxilla is perforated lingually for its attachment with the gold wire to the maxillary canines.

churches, indicating that only the upper social classes could afford such luxury. With the help of grave inscriptions and written records it was possible to establish the identity of some indivduals with dentures and periodontal therapy (Whittaker and Hargreaves 1991; Alt 1993; Whittaker 1993).

Restorative therapy appears to begin at a later time, since no archaeological findings predate the 16th century, and only rare indicators appear up to the 19th century. Again, this may be interpreted as representing a social privilege as in the case of aritificial dentures. Early amalgam and gold fillings were found on Princess Anna Ursula von Braunschweig and Lüneburg who died 1601 (Czarnetzki and Ehrhardt 1990). Much later, there are tin and gold foil fillings in the already mentioned burials from Geneva (Alt 1993; Fig. 21 a, b), as well as silver and gold fillings from 18th century English graveyards (Harvey 1968; Whittaker and Hargreaves 1991; Whittaker 1993). Empty cavities suggest that the number of fillings must be greater than reported (Alt 1994).

Medicinal plants have been used in palliative or therapeutic treatment of dental diseases including treatment of toothaches since prehistoric times (Elvin-Lewis 1979). Among Plains Indians, for example "a number of plants were used to treat oral, throat, and tooth disorders" (Willey and Hofman 1994, 152).

Surgical findings are rare except for tooth extractions, which occur worldwide. In a neolithic burial, Bennike and Fredebo (1986) diagnosed a trepanation carried out intra-vitam to relieve a tooth abscess. More recently, Schwartz et al. (1995) published a case of tooth drilling associated with a root apex abscess of an incisor

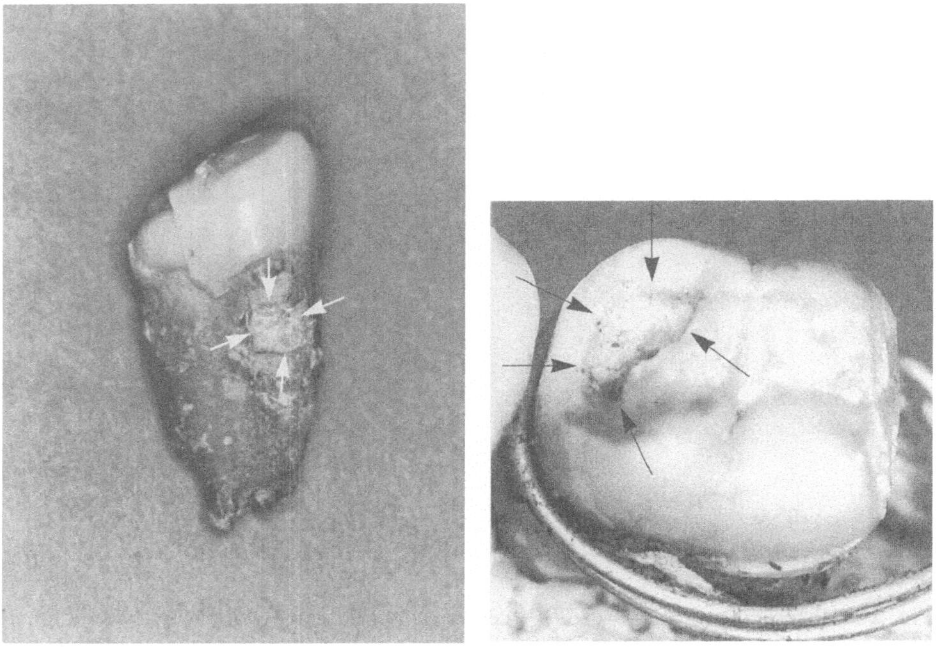

a b

Fig. 21 a, b. Historical restorative therapy. Examples of metal fillings observed in François-Bénigne du Trousset d'Hericourt (see Fig. 19); **a:** Gold foil on neck of mandibular right third molar; **b:** (Tin?) filling probably lost post portem from prepared disto-occlusal cavity of mandibular left second molar (note also the small tin fillings on the buccal surfaces of the left mandibular molars, Fig. 19).

in an Alaskan Arctic group (Point Hope, 1300–1700 A.D.). A 2,5 mm long bronze pin was found in the root canal of a maxillary incisor in an individual from a mass grave in the Negev desert dating around 200 BC (Zias and Numeroff 1987). The reasons for this unusual measure are unknown. The only possible explanation may be a connection with the belief in "tooth worms", which was quite common in earlier times (Alt 1989).

The cases last described here are isolated findings and are consequently not representative for their time. Nevertheless, they do show that exceptional dental and therapeutic findings may be encountered at any time during archaeological excavations and anthropological investigations. Chronologically, the examples reach far back into the history of mankind. Some findings can be traced in unbroken traditions up to the modern age and document the important role of tradition in the artificial modifications of human teeth.

Conclusions

Aside from physiological ageing and wear, teeth can be subject to artificial modifications in their period of use. These reflect human behaviour from different areas of life and may be classed in two groups: accidental and intentional dental modifications. Natural and specific abrasive processes dependant on dietary behaviour as well as modifications caused by occupational and personal habits are among the accidental changes. Intentional dental modifications include different forms of tooth mutilations and all therapeutic measures (Tab. 1).

The classification of artificial dental modifications we have developed meets various reqirements. It permits the easy assignment of all artificial modifications of teeth to specific sub-classes, irrespective of diachronic, etiological and functional aspects. With the help of a standardised recording – if need be by numerical codification – it will be possible to clearly and unequivocally describe prehistoric and ethnographic findings as well as observations in present day populations and thereby ensure the comparability of data in population studies for the first time.

Acknowledgment

We wish to thank the following researchers and publishers for their permission to reproduce the figures indicated in the text: A. Bach, Jena, R. Garve, Geesthacht, R. Hickel, München, F. Kreter, Frankfurt, H. E. Lässig, München, R. A. Müller, München, H. Pantke, Giessen, Germany; W. Fitting, Buchillon, Switzerland; G. Villa and G. Giacobini, both Turin, Italy; M. Teaford, New York, C. S. Larsen, West Lafayette, USA; I. Lous and J. P. H. Hansen, both Copenhagen, Denmark; J. Wiley & Sons, New York, USA; International Institute for the Study of Man, Florence, Italy; Quintessenz, Berlin, and DuMont, Köln, Germany. We are expecially grateful to O. Temme, Duesseldorf, for his kindly assistance in computer graphics.

References

Abusinna IM (1979) Lugbara teeth germectomy of canines for the newborn babies. A magico-religious phenomena in some African tribes. Egyptian Dent J 25: 209–214.

Alt KW (1989) Archäologische Befunde aus Archäologie und Anthropologie. Zahnärztl Mitt 79: 785–796.

Alt KW (1993) Praktische Zahnmedizin im 18. Jahrhundert. Schweiz Monatschr Zahnmed 103: 1147–1154.

Alt KW (1994) Prosthetics, periodontal therapy, and conservative dentistry in the eighteeth century – Archeological findings from Grand-Saconnex, Geneva, Switzerland. Bull Hist Dent 42: 39–43.

Alt KW, Koçkapan C (1993) Artificial tooth-neck grooving in living and prehistoric population. Homo 44: 5–29.

Alt KW, Newesely H (1994) Ein Anfangskapitel der modernen Prothetik. Zahnärztl Mitt 84: 66–68.

Alt KW, Pichler S (1995) Unusual tooth wear pattern – abrasive food, chronic vomiting or teeth as tool? In: Jacob B, Bonte W (eds) Forensic sciences in 1993. Proc 13th Meeting Int Assoc Forensic Sciences, Düsseldorf 1993, p. 268-271.

Alt KW, Parsche F, Pahl WM, Ziegelmayer G (1990) Gebißdeformation als "Körperschmuck" – Verbreitung, Motive und Hintergründe. Zahnärztl Mitt 80: 2448–2456.

Barrett MJ (1977) Masticatory and non-masticatory uses of teeth. In: Wright RVS (ed) Stone Tools as Markers. Austral Inst Aboriginal Studies, Canberra, p. 18–23.

Becker MJ (1994) Etruscan gold dental appliances: Origins and functions as indicated by an example from Orvieto, Italy, in the Danish national Museum. Dent Anthrop Newsletter 8 (3): 2–8.

Bennike P, Fredebo L (1986) Dental treatment in the stone age. Bull Hist Dent 34: 81–87.
Betram K (1981) Comparative efficacy of oral hygiene procedures practiced in Sri-Lanka. Odontostomatol Trop 4: 35–42.
Blakely RL, Beck LA (1984) Tooth-tool use versus dental mutilations: A case study from the prehistoric South east. Midcont J Archaeol 9: 269–277.
Briedenhann SJ, Van Reenen JF (1985) Tooth extraction and tooth mutilating practices amongst the Herero peoples of South West Africa (Namibia). J Dent Assoc S Afr 40: 531–536.
Brothwell DR (1989) The relationship of tooth wear to aging. In: Iscan MY (ed) Age Markers in the Human Skeleton. Thomas, Springfield IL, p. 303–316.
Brown T, Molnar S (1990) Interproximal grooving and task activity in Australia. Am J Phys Anthrop 81: 545–553.
Butt MH, Dunning JM (1986) Reimplementation of chewing stick oral hygiene in Kenya. Odontostomatol Trop 9: 35–37.
Cruwys E, Duhig C (1993) A possible case of dental erosion in a skull from the French Congo (People's Republic of the Congo). J Palaeopath 5: 47–52.
Cruwys E, Robb ND, Smith BGN (1992) Anterior tooth notches: An Anglo-Saxon case study. J Paleopath 4: 211–220.
Cybulski JS (1974) Tooth wear and material culture: Precontact patterns in the Tsimshian area, Brithish Columbia. Syesis 7: 31–35.
Czarnetzki A, Alt KW (1991) Eine Frontzahnbrücke aus Flußpferdzahn – Deutschlands älteste Prothese? Zahnärztl Mitt 81: 216–219.
Czarnetzki A, Ehrhardt S (1990) Re-dating the Chinese amalgam-filling of teeth in Europe. Int J Anthrop 5: 325–332.
Davies GN (1963) Social customs and habits and their effect on oral disease. J Dent Res 42: 209–232.
Dechaume M (1938) Lésions buccales, dentaires et maxillaires dans les maladies professionnelles. Arch Mal Prof Hyg Toxicol Ind 1: 200–220.
Eccles JD (1979) Dental erosion of non-industrial origin. A clinical survey and classification. J Prosth Dent 42: 649–656.
Eccles JD (1982) Tooth surface loss from abrasion, attrition and erosion. Dent Update 9: 373–378.
Elvin-Lewis M (1979) Empirical rationale for teeth cleaning plant selection. Medical Anthrop 3: 431–458.
Enbom L, Magnusson T, Wall G (1986) Occlusal wear in miners. Swed Dent J 10: 165–170.
Fastlicht S (1976) Tooth mutilation and Dentistry in Pre-Columbian Mexiko. Quintessenz, Berlin.
Fitting W (1989) Les mutilations dentaires dans le cadre des mutilations rituelles. Actualités Odonto-Stomat 166: 191–203.
Frayer DW, Russel MD (1987) Artificial grooves on the Krapina Neanderthal teeth. Am J Phys Anthrop 74: 393–405.
Galang RE (1941) Filing and blackening of teeth among some Philippine ethnic groups. Philippine J Science 75: 425–431.
Garve R (1995) Indianer am Amazonas. Tanner, Adliswil.
Gebrande R (1973) Exzessive Schliffacetten, exzentrische Okklusion und ihr Einfluß auf das Bewegungsmuster des Unterkiefers. Med Diss, Tübingen.
Gilbert RI, Mielke JH (1985) The analysis of prehistoric diets. Academic Press, Orlando.
Gordon KD (1982) A study of microwear on chimpanzee molars: Implications for dental microwear analysis. Am J Phys Anthrop 59: 195–215.
Gould AR, Forman AG, Corbitt D (1984) Mutilations of the dentition in Africa: A review with personal observations. Quintessenz Int 15: 89–94.

Graf H, Zander HA (1964) Okklusale Zahnkontakte und Muskelaktivität beim Kauen und Schlucken. Schweiz Mschr Zahnheilk 74: 495–502.
Grine FE (1986) Dental evidence for dietary differences in *Australopithecus* und *Paranthropus*; a quantitative analysis of permanent molar microwear. J Hum Evol 15: 783–822.
Grupe G, Herrmann B (eds) (1988) Trace elements in environmental history. Springer, Berlin Heidelberg New York.
Harvey W (1968) Some dental and social conditions of 1692–1852 connected with St. Bride's Church, Fleet Street, London. Med Hist 12: 62–75.
Headland TN (1977) Teeth mutilations among the Casiguran Dumagat. Phillip Quart Cult Soc 5: 54–64.
Henke W, Rothe H (1996) Zahnphylogenese der Hominiden. In: Alt KW, Türp JC (ed) Die Evolution der Zähne. Phylogenie, Ontogenie, Variation. Quintessenz, Berlin, p. 273–354.
Herion P (1985) Schmücken, ein Urtrieb der Völker. Königsbach.
Heymer A (1986) L'abrasion dentaire chez les Pygmées Bayaka en fonction des conditions écologiques, alimentaires et d'une mastication accrue. Homo 37: 160–188.
Hickel R (1989) Zahnabrasion und beruflich bedingte Einflüsse bei Steinbrucharbeitern. Hauptverband der gewerblichen Berufsgenossenschaften (ed), St. Augustin.
Hillson SW (1979) Diet and dental disease. World Archeol 11: 147–162.
Hinton RJ (1982) Difference in interproximal and occlusal tooth wear among prehistoric Tennessee Indians: Implications for masticatory function. Am J Phys Anthrop 57: 103–115.
Hoffmann E,. Treide B (1976) Schmuck früher Kulturen und ferner Völker. Stuttgart.
Hoffmann-Axthelm W (1981) History of Dentistry. Quintessenz Publ. Co. Inc., Chicago.
Horn H (1967) Untersuchungen über die Abrasio dentis des alternden Gebisses. Med Diss, Berlin.
Hylander WL (1973) The adaptive significance of Eskimo craniofacial morphology. In: Dahlberg AA, Graber TM (eds) Orofacial Growth and Development. Mouton, The Hague.
Irish JD, Turner CG (1987) More lingual surface attrition of the maxillary anterior teeth in American Indians: Prehistoric Panamanians. Am J Phys Anthrop 73: 209–213.
Jackson JW (1915) Dental mutilations in neolithic human remains. J Anat Physiol 49: 72–79.
Jahnke E (1970) Die Zahnverstümmelungen bei Natürvölkern unter besonderer Beachtung der Motive und Hintergründe. Med. Diss. Köln.
Jones A (1992) Tooth mutilation in Angola. Br Dent J 173: 177–179.
Kaidonis JA, Townsend GC, Richards LC (1992) Interproximal tooth wear: A new observation. Am J Phys Anthrop 88: 105–107.
Klimm W, Graehn G (1993) Der keilförmige Defekt. Quintessenz, Berlin.
Kreter F, Pantke H (1979) Einführung in die Zahnheilkunde mit Grenzinformationen. Quintessenz, Berlin.
Kunter M (1971) Zur Geschichte der Tatauierung und Körperbemalung in Europa. Paideuma 17: 1–20.
Kvaal SI, Derry TK (1996) Tell-tale teeth: Abrasion from the traditional clay pipe. Endeavour NS 20: 28-30.
Lambert JB, Grupe G (eds) (1993) Prehistoric human bone – archaeology at the molecular level. Springer, Berlin Heidelberg New York.
Lässig H, Müller R (1983) Die Zahnheilkunde in Kunst- und Kulturgeschichte. DuMont, Köln.
Larsen CS (1985) Dental modifications and tool use in the western Great Basin. Am J Phys Anthrop 67: 393–402.
Lavelle CLB (1979) Population contrasts based upon teeth with varying degrees of attrition. J Det 7: 126–132.

Lee WC, Eakle WS (1984) Possible role of tensile stress in the etiology of cervical erosive lesions of teeth. J Prosthet Dent 52: 374–379.
Linné S (1940) Dental decoration in aboriginal America. Ethnos 5: 2–34.
Lous I (1970) The masticatory system used as a tool. Dent Abstr 15: 457–458.
Loux F (1981) L'ogreet la dent. Berger-Levrault, Paris.
Lukacs JR (1989) Dental paleopathology: Methods for reconstructing dietary patterns. In: Iscan MY, Kennedy KAR (eds) Reconstruction of Life from the Skeleton. Liss Inc., New York, p. 262–286.
Lukacs JR, Pastor RF (1988) Activity-induced patterns of dental abrasion in prehistoric Pakistan: Evidence from Mehrgarh and Harappa. Am J Phys Anthrop 76: 377–398.
Menzel B (1957) Deformierender Gesichtsschmuck südamerikanischer Naturvölker. Baessler Archiv NF 5: 1–120.
Merbs CF (1983) Patterns of activity – Induced pathology in a Canadian Inuit population. Nat. Mus. of Man Mercury Series 118. Nat. Mus. of Canada, Ottawa.
Miles AEW (1963) The dentition in the assessment of individual age in skeletal material. In: Brothwell DR (ed) Dental Anthropology. Pergamon, London, p. 191–209.
Milner GR, Larsen CS (1991) Teeth as artifacts of human behavior: Intentional mutilation and accidental modification. In: Kelley MA, Larsen CS (eds) Advances in Dental Anthropology. Wiley-Liss, New York, p. 357–378.
Molnar S (1972) Tooth wear and culture: A survey of tooth functions among some prehistoric populations. Curr Anthropol 13: 511–526.
Morris AG (1988) Archaeological evidence of "pipe-smoker's wear". Off J Dent Ass S Afr 43: 361–364.
Nagel A (1989) Zähne mit keilförmigen Defekten im Tierreich. Med Dipl, Berlin.
Neff UV (1993) Untersuchungen zur Altersbetimmung der Individuen zweier spätmittel-alterlicher Skelettserein aus Baden-Württemberg anhand verschiedener Abrasions-phänomene. Med Diss, Tübingen.
N.N. (1988) Interproximal grooving of teeth: Additional evidence and interpreatation. Current Anthrop 29: 663–671.
Oettermann S (1979) Zeichen auf der Haut. Die Geschichte der Tätowierung in Europa. Syndikat, Frankfurt.
Perizonius WRK, Pot TJ (1981) Diachronic dental research on human skeletal remains excavated in the Netherlands, I: Dorestad's cemetery on "the Heul". Berichten van de Rijksdienst voor het Oudheidkundig Bodemonderzoek 31: 369–413.
Popoviciu AM (1969) Verstümmelung der Frontzähne durch Sonnenblumenkernknacken. Dtsch Zahnärztl Z 16: 1018–1022.
Puech P, Cianfarani F, Albertini H (1986) Dental microwear features as an indicator for plant food in early hominids: A preliminary study of enamel. Hum Evol 6: 507–515.
Ring A (1984) Zur Frage berufsbedingter Abrasionsschäden bei Steinmetzen und Steinhauern. Dtsch Zahnärztl Z 39: 36–42.
Robb ND, Cruwys E, Smith BGN (1991) Is "Lingual surface attrition of the maxiallary anterior teeth (LSAMAT)" caused by dental erosion? Am J Phys Anthrop 85: 345–351.
Russel MD, Grant AA (1983) The relationship of occlusal wear to occlusal contact area. J Oral Rehabil 10: 383–391.
Schürmann R (1986) Soziale Differenzierung nach frühgeschichtlichen Ernährungsmustern am Beispiel von Zahnabrasionen in merowingerzeitlichen Bevölkerungen. Dipl Arb. Ulm.
Schwartz JH, Brauer J, Gordon-Larsen P (1995) Brief communication: Tigaran (Point Hope, Alaska) tooth drilling. Am J Phys Anthrop 97: 77–82.
Scott R (1991) Dental anthropology. In: Dulbecco R (ed) Encyclopedia of Human Biology, Vol 2. Academic Press, San Diego.

Scully C, Chen M (1994) Tongue piercing (oral body art). Br J Oral Maxillofac Surg 32: 37–38.
Slop D (1986) Abrasion of enamel by toothbrushing. Med Diss, Groningen.
Smith B (1975) Dental erosion, attrition and abrasion. Practioner 214: 341–347.
Smith BH (1984) Patterns of molar wear in hunter-gatherers and agriculturalists. Am J Phys Anthrop 63: 39–56.
Smith BGN, Knight JKA (1984) A comparison of patterns ot tooth wear and aetiological factors. Br Dent J 157: 16–19.
Teaford MF (1991) Dental microwear: What can it tell us about diet and dental function? In: Kelley MA, Larsen CS (eds) Advances in Dental Anthropology. Wiley-Liss, New York, p. 341–356.
Teaford MF (1994) Dental microwear and dental function. Evol Anthrop 3: 17–30.
Ten Bruggen Cate HJ (1968) Dental erosion in industry. Brit J Ind Dent 25: 249–253.
Thierfelder C, Hesse H, Schott L, Sommer K (1987) Zahnersatz im alten Berlin. Med Aktuell 13: 585.
Türp JC (1990) Zahnfegen, Zahnpinsel, Zahnputzhölzer: Zur Aktualität traditioneller Formen der Mund- und Zahnhygiene. Curare 13: 75–81.
Turner CG (1988) Interproximal grooving of teeth: Additional evidence and interpretation. Curr Anthropol 29: 664–665.
Turner CG, Cadien JD (1969) Dental chipping in Aleuts, Eskimos and Indians. Am J Phys Anthrop 31: 303–310.
Turner CG, Machado LMC (1983) A new dental wear pattern and evidence for high carbohydrate consumption in a Brazilian archaic skeletal population. Am J Phys Anthrop 61: 125–130.
Ullrich H (1973) Behandlung von Krankheiten in frühgeschichtlicher Zeit. In: Ber II. Int. Kongr Slaw Archäol, Berlin, Bd. 3, p. 475–481.
Van Reenen JF, Briedenhann SJ (1986) Further observations on the tooth mutilating practices of Vassekela and Kung Bushmen. Tydskr Tandheelkd Ver S Afr. 41: 557–562.
Verger-Pratoucy IC (1970) Recherches sur les mutilations maxillo-dentaires préhistoriques. Bull Group Int Rech Sc Stomat 13: 133–310.
Villa G, Giacobini G (1995) Subvertical grooves of interproximal facets in Neandertal posterior teeth. Am J Phys Anthrop 96: 51–62.
Vlček E (1986) Die Mammutjäger von Dolní Vestonice. Amt für Museum und Archäologie, Liestal.
Walker A, Teaford MF (1989) Inferences from quantitative analysis of dental microwear. Folia Primatol 53: 177–189.
Walton RE (1996) Cracked tooth and vertical fracture. In: Walton RE, Torabinejad M (eds) Principles and practice of endodontics. Saunders, Philadelphia, p. 474–492.
Whittaker DK (1993) Oral health. In: Molleson T, Cox M (eds) The Spitalfields Project, Vol 2. The Anthropology. CBA Res Rep 86, York, p. 49–65.
Whittaker DK, Hargreaves AS (1991) Dental restorations and artificial teeth in a Georgian population. Br Dent J 171: 371–376.
Willey P, Hofman JL (1994) Interproximal grooves, toothaches, and purple coneflowers. In: Owsley W, Jantz RL (eds) Skeletal Biology in the Great Plains. Smithsonian Institution Press, Washington, p. 147–157.
Williams SA, Curzon MEJ (1986) Observations on dental caries in primary teeth in some medieval British skull material. In: Cruwys E, Foley RA (eds) Teeth and Anthropology. BAR Int Ser 291, p. 201–212.
Wolpoff MH (1971) Interstitial wear. Am J Phys Anthrop 34: 205–228.
Xhonga FA (1977) Bruxism and ist effect on the teeth. J Oral Rehabil 4: 65–76.
Zias J, Numeroff K (1987) Operative dentistry in the second century. J Am Dent Ass 114: 665–666.

6 Age and Sex Estimation

6.1 Dental Age Estimation of Non-Adults. A Review of Methods and Principles

Helen M. Liversidge, Berthold Herdeg and Friedrich W. Rösing

Introduction

Tooth formation is widely used as a growth marker, allowing assessments and comparisons between individuals and populations in dentistry, pediatrics and anthropology. Human ecology, archeology, and forensic sciences also rely on standards of dental formation. Of all the growth systems tooth formation has the highest time stability and as such provides the most accurate way of calibrating growth and development. Dental growth standards define which stages of tooth formation occur on average when; for example the age at which teeth begin to mineralise and when they are fully mineralised. Age standards that document the sequence and timing of events during somatic and dental growth are a powerful screening device to (1) assess the maturation level of an individual, (2) assess the ecological situation of a population, (3) plan for medical interception in the case of abnormal development, (4) measure the response to treatment of abnormal growth, and (5) estimate age at death of an immature skeleton or corpse. In addition, investigations of the control systems and dynamics of tooth formation lead to a better understanding of the biology of growth in general.

Historically, dental eruption was first used as a measure of maturity in 1837 by Saunders with respect to the age of factory children in England. Since then several hundred other studies of dental growth focused on eruption, often not differentiating between the initial emergence of a tooth through the gum and its final functional position in occlusion. The number of teeth in the mouth was first used to assess age by Cattell (1928), but this coarse approach is limited to the time of emergence of teeth. More differentiated information was provided by dissection data between 1873 (Legros and Magitot) and 1941 (Massler et al.). With a substantial overlap radiography was used next, starting in the twenties of this century (Bustin et al. 1929, 1930). Other studies combined the two data types (Logan and Kronfeld 1933). Much of the early work on somatic growth was descriptive, but a new era of study was heralded by D'Arcy Thompson (1917) who combined the philosophy and geometry of biological growth so allowing the prediction of growth from a mathematical model. Eighty years on, renewed anthropological and comparative anatomical research into incremental growth of enamel and dentine is revealing more about the mechanisms of growth of hominids. During the last decade the understanding of human tooth growth has been placed more firmly into an evolutionary context through advances in our knowledge of tooth formation and a better understanding of variation in ape and human patterns or sequences of tooth growth (see Macho and Wood 1995). The growth rate of both enamel and dentine is now known to vary considerably during different tooth formation stages as measured by perikymata spacing, root cone angle or incremental markers like tetracycline (Dean 1985, 1987; Dean et al. 1993a).

The objective of this chapter is to review mineralisation standards of human deciduous and permanent teeth, including only a brief mention of tooth emergence. The different methods of age estimation and their problems are discussed, comparable methods pooled and recommendations given.

Growth Variation Factors

The most notable characteristic of dental development is the time stability: the same growth phenomenon occurs at a similar time in the life of different children. This is due to a high heritability of dental development, lying around 0.8, higher than for other growth parameters. On closer inspection, however, there is considerable fluctuation around a time mean. Probably the most important factor responsible for this variation is chance. All more complex biological characteristics follow the Gaussian curve of stochastic distribution, even after elimination of systemic influences that cause dispersion.

Depending on the general scientific position genetic differences between individuals is the next important or even more important factor for variation.

The third important factor of variation is sex. This can be well quantified because all studies except the early ones subdivide their samples as to sex ("gender"). It has an importance for all traits, and girls reach most stages ahead of boys. This difference amounts to roughly 0.3 years with a tendency to increase with age (see below), reaching a maximum of 1.7 years (Haaviko 1970).

Like any other human biological characteristic, dental formation has a regional or large-scaled geographic variation which is sometimes inappropriately called ethnic; ethnicity is the result of common tradition, language and sometimes personal decision, whereas regional variation in biological traits is an effect of genetic group differentiation due to climate, migration, history etc. One study found larger differences of up to one year between regional groups (Harris and McKee 1990). Other regional results are difficult to interpret as information is often lacking in sampling or methods (Loevy 1983; Mappes et al. 1992). Recent studies on Swedish children using several types of Scandinavian standards showed noticeable differences within one population; this evidently reflects methodology and age make up and not real differences between the groups of children being studied (Mörnstad et al. 1994, 1995). Comparisons of dental maturation of normal children using the method of Demirjian et al. (1973) suggest that French-Canadian children show a slower maturation than children in other parts of the world (Prahl-Andersen et al. 1979; Proy et al. 1981; Loevy 1983; Nichols et al. 1983; Nyström et al. 1986, 1988; Kataja et al. 1989; Ocholla 1990; Davis and Hägg 1994). From these few examples it is unclear if mineralisation has similar geographic gradients to eruption and other developmental changes, particularly menarche. Moreover, it is not clear to what extent these gradients are genetic or environmental (see below ecology).

Fifth, there is a secular trend which influences variation. A study investigating this correlated the year of study and developmental stage (Herdeg 1992); e.g. for the mineralised crown between 10 and 21 samples could be used. Of the 28 correlation coefficients (7 teeth, 2 jaws, 2 sexes) for this stage only one was significantly different from zero (5% level), which can be explained by chance alone. In total 26 of 140 were significant; this is not impressive either because the corresponding regressions yielded 15 positive slopes meaning retardation, 5 negative slopes meaning acceleration, and 4 with a slope of zero. Retardation was also found by Kahl and Schwarze (1988). This means that the developmental acceleration of the last century which is so well documented for puberty and stature is not detectable in tooth development. This might be explained by the larger genetic control of teeth – or by uncontrolled other factors including differing methods. For eruption, however, acceleration has been shown several times (Adler-Hradecky 1959; Schützmannsky 1957; Adler 1958; Grosch and Joksch 1960; Henke 1960; Valöík and Fábryková 1964; Oster 1960; Weise and Bruntsch 1965; Grivu et al. 1967; Seichter et al. 1980; Kromeyer and Wurschi 1996). A difference of almost one standard deviation has been noted between the standards of Moorrees (1963b) and a recent investigation which was attributed to a secular trend (Ito et al. 1993), but might also be attributed to method differences (see below). These unclear and contradictory results clearly indicate that a new, modern, critical and mathematically oriented reexamination is needed. Judging from the large human biological knowledge of growth differentials, there must be a secular trend in tooth mineralisation, too, but at the moment nothing substantial can be said.

Sixth, growth is also influenced by nutrition, hygiene, health, education and income. This package of factors represents human ecology, a field that has recently emerged, particularly in eastern Europe, also to monitor the national status of modernisation (Wolanski 1983; Verenich 1984; Prokopec et al. 1981; Wolanski et al. 1991). This effect is also illustrated by epidemiological studies, which for example demonstrated that dentition is delayed in low-birthweight African-American children (Harris et al. 1993).

Seventh, growth is also influenced by climate. Although this probably affects clinical emergence, no such effect has been demonstrated with respect to mineralisation with its higher general stability.

Finally, eighth, time has another influence on growth variation: there is a steady increase in variance with age. This not only applies to teeth but to all ontogenetic phenomena. Processes before birth are maximally controlled by genetics; moreover, in this phase the growing organism is relatively well protected from environmental influences. Postnatally, environmental effects increase as the genetic influence decreases, so that eventually only few percent of the variation of time at death is controlled biologically. This has a direct and important meaning: age diagnosis and developmental assessment using teeth in infants can be far more precise than in adolescents.

As other growth parameters suggest, many of these variation factors should be connected like sex and region or ecology and secular trend – but again little is known for dental formation. However, an understanding and appreciation of the many biological factors that result in variation is important not only to explain differences between specific methods but also to select appropriate methods for application.

Developmental Traits

Eruption may be defined as the progressive movement of a tooth towards occlusion with its antagonist of the opposite jaw. Clinical emergence into the oral cavity is considered to be only one stage in this process of eruption. Many hundreds of modern population studies exist, but these standards of emergence are inappropriate for determining age of skeletal material and also suboptimal for corpses. However, one study (Haaviko 1970) provides useful information about the time between alveolar eruption (when a tooth emerges beyond the crest of the alveolar bone) and clinical emergence (when a tooth penetrates the gingiva).

Table 1. Ages of alveolar (alv) and clinical (clin) emergence of permanent teeth. SD – standard deviation. Data from cross sectional study by Haaviko (1970)

		Boys				Girls			
		Maxillary median ± SD		Mandibular median ± SD		Maxillary median ± SD		Mandibular median ± SD	
Tooth	stage	median	SD	median	SD	median	SD	median	SD
I1	alv	6.2	0.86	5.9	0.74	6.1	0.35	5.8	0.43
	clin	6.9	0.86	6.3	0.70	6.7	0.66	6.2	0.55
I2	alv	7.3	1.29	6.9	0.78	7.0	0.90	6.5	0.55
	clin	8.3	1.25	7.3	0.70	7.8	0.86	6.8	0.70
C	alv	11.2	1.21	9.8	1.09	9.3	1.25	8.8	0.63
	clin	12.1	1.41	10.4	1.17	10.6	1.45	9.2	1.06
P1	alv	9.8	1.41	9.6	1.29	9.0	1.09	9.1	0.90
	clin	10.2	1.41	10.3	1.80	9.6	1.37	9.6	1.48
P2	alv	11.1	1.60	10.3	1.72	9.5	1.37	9.2	1.64
	clin	11.4	1.48	11.1	1.72	10.2	1.60	10.1	0.67
M1	alv	5.3	0.74	5.3	0.35	5.3	0.47	5.0	0.39
	clin	6.4	0.63	6.3	0.55	6.4	0.55	6.3	0.55
M2	alv	11.4	1.09	10.8	1.02	10.3	0.90	9.9	1.06
	clin	12.8	1.25	12.2	1.41	12.4	1.17	11.4	1.41
M3	alv	17.7	1.52	18.1	2.15	17.2	2.46	17.7	2.34

Mineralisation of the crown and root of a tooth can usually be seen on radiographs and this allows stages of growth to be assessed. The stages most clearly identified are initial mineralisation, the end of crown growth, furcation of the roots of posterior teeth, and the end of root growth with closure of the apex. Resorption of the root in deciduous teeth has also been used. Other stages of formation include fractions of the crown and root length such as half root completed. Without knowing the total length of the root at completion such fractional stages can only be a rough guess and should be avoided for age assessments of importance.

It should be noted that the radiographic stage does not always correspond to the actual stage as seen by direct observation in a dissection or histological investigation. Some stages of formation are especially difficult to assess radiographically. These include crown completion and initial root furcation in premolars and molars. The radiographic assessment of crown completion of permanent incisors and canines is difficult due to the enamel contour that extends on the buccal and lingual surfaces 2 to 4 mm more towards the root than on the mesial and distal sides (Black 1902). Initial root formation occurs on the approximal surfaces of the tooth before the last enamel is formed on the buccal and lingual surfaces which can take up to 2 years (Beynon 1992). The margin in these last areas of enamel formation is not clearly visible from radiographs and thus age of true enamel completion of these teeth is later than radiographic age. Most drawings of stages depict radiographic stage of crown completion on the mesial and distal aspects of a tooth before initial root formation has actually begun buccally or lingually.

Instead of subjectively estimating fractions of a potential length the existing length of a tooth may be measured (Stack 1960; Liversidge 1993). This parameter does not rely on speculation of a possible final length. This final length does have an influence, but it becomes part of the dispersion.

Histological studies contribute to an understanding of individual variation and the pattern of growth processes and influencing factors. Several types of incremental markings occur in enamel and dentine that are observed histologically from ground sections of teeth. Enamel incremental growth markers include daily cross striations, striae of Retzius, and their associated perikymata that are visible on the crown surface. An exaggerated Retzius line is sometimes visible that is formed during the first few weeks after birth and is known as the neonatal line. The number of cross striations from the neonatal line to the forming enamel front represents the number of days of enamel formation postnatally. If death occurs before enamel completion, the duration of amelogenesis (in number of days) can be determined by counting the cross striations from a ground section to an accuracy of about 10% (Boyde 1963). This method of calculating age from incremental growth has also been applied in forensic odontology (Skinner and Anderson 1991; Dean et al. 1993a), archeology (Dean and Beynon 1991; Huda and Bowman 1995; FitzGerald et al. 1996) and ontogeny of fossil tooth times and of hominids themselves (Bromage and Dean 1985; Beynon and Wood 1987; Dean et al. 1993b). Further studies might increase accuracy. On the other hand, it is often held that some factors of variation have not yet been quantified, which might decrease accuracy.

Another histological variable is cement annulation which has been used for adults only (see next chapter by Rösing and Kvaal), but should be applicable for non-adults just as well. The same applies to racemisation, which is clearly the most precise age estimation method at all, but is only applicable for fresh corpses.

The applicability of growth standards depends largely on consistent and clear definitions of tooth formation stages. Very few studies define or describe mineralisation stages clearly in words. One exception to this general situation provides both a radiograph, a line drawing of the radiograph and a clearly worded descriptive criteria of the stages adopted (Demirjian et al. 1973).

Techniques

There are two different types of technical data extraction: anatomical and radiographic. The chronology and variation in human tooth formation from fetus to early childhood is based largely on small samples of autopsy material.

Anatomical Studies

Early calcification and development of the prenatal (e.g. Krauss and Jordan 1965; Sunderland et al. 1987) and postnatal deciduous dentition have been well documented (see e.g. Lunt and Law 1974). Postnatal growth data are largely derived from histological studies (Legros and Magitot 1873; Black 1883; 1908; Peirce 1884; Röse 1891, 1909; Zuckerkandl 1891; Berten 1985; Paltauf 1902; Broomell and Fischelis 1923; Logan and Kronfeld 1933, 1935). The cardinal problem of these investigations is a small sample, below 30. Despite this, most results are comparable to larger studies (Tab. 2 and 3).

Table 2. Development of deciduous teeth at birth: crown stage: fractions of total crown height; Coc: occlusal outline complete; Cco: coalescence of cusp tips; Crown height in mm: along long axis, from incisal tip to developing edge 5. Data from Liversidge (1993)

	Crown stage	Crown height at birth
i1	4/5	5.4 ± 1.04
i2	3/5	4.5 ± 0.82
c	1/3	3.4 ± 0.94
m1	Coc	4.0 ± 0.38
m2	Cco	3.1 ± 0.55

Table 3. Age of formation stages of deciduous teeth. Age in years, Cc – crown complete, Rc – root complete, Ac = apex closed. Data from Moorrees et al. (1963a), Fass (1969) and Liversidge (1995b)

		Cc	Rc	Ac
i^1		0.1	1.5	2.6
i_1		0.1	1.1	2.3
i2		0.2	1.5	2.4
c	girls	0.68 ± 0.14	2.04 ± 0.29	3.00 ± 0.38
	boys	0.67 ± 0.14	1.92 ± 0.27	3.10 ± 0.38
m1	girls	0.33 ± 0.12	1.25 ± 0.20	1.80 ± 0.26
	boys	0.42 ± 0.12	1.30 ± 0.20	1.95 ± 0.27
m2	girls	0.68 ± 0.14	1.98 ± 0.28	2.85 ± 0.36
	boys	0.69 ± 0.14	2.08 ± 0.28	3.08 ± 0.38

Radiographic Studies

Dental growth can be more easily assessed radiographically using carefully defined stages of tooth growth and maturation (Tab. 4, 5 and 6). Data from about 70 radiographic studies detail the development of some deciduous and most permanent teeth of predominantly North American children of European origin (see review below).

Table 4. Age of formation stages of first permanent molar. "Cc" – radiographic crown complete. Data from longitudinal study by Gleiser and Hunt (1955)

	"Cc" Mean ± SD		Rc mean ± SD		Ac mean ± SD	
Boys	3.5	0.47	8.4	0.64	8.9	0.62
Girls	3.3	0.35	7.8	0.68	8.5	0.50

Table 5. Age of radiographic formation stages of permanent teeth. Ci – initial mineralisation, Cc – crown complete, Rc – root length complete, SD – standard deviation. Data from cross sectional study by Haaviko (1970)

		Boys				Girls			
Tooth	stage	Maxillary median ± SD		Mandibular median ± SD		Maxillary median ± SD		Mandibular median ± SD	
I1	Cc	3.3	–	–	–	3.3	–	–	–
	Rc	8.7	0.90	7.2	0.66	8.2	0.66	6.8	0.51
I2	Cc	4.6	0.66	3.3	0.55	4.4	0.90	–	–
	Rc	9.6	0.66	8.1	0.82	8.5	0.51	7.1	0.74
C	Cc	4.6	0.51	4.3	0.63	4.5	0.82	4.1	0.63
	Rc	12.3	1.02	11.6	1.48	11.2	0.98	10.3	1.02
P1	Cc	6.8	0.78	5.9	0.59	6.3	0.47	5.4	0.55
	Rc	11.5	1.25	11.8	1.05	10.9	1.05	11.1	1.13
P2	Cc	7.1	0.78	7.0	0.98	6.6	0.74	6.4	0.59
	Rc	12.0	1.29	12.1	1.41	11.3	1.17	11.5	1.09
M1	Cc	3.6	0.59	3.5	0.35	3.5	0.31	3.5	0.31
	Rc	8.1	0.74	7.3	0.74	7.5	0.55	6.9	0.78
M2	Ci	3.7	0.55	3.9	0.82	3.8	0.63	3.9	0.74
	Cc	7.3	0.55	7.4	0.59	6.9	0.86	7.0	0.66
	Rc	13.6	1.13	13.4	0.98	12.5	1.37	12.5	1.21
M3	Ci	9.0	1.60	9.8	2.42	9.4	1.56	9.6	1.68
	Cc	13.2	2.07	13.7	1.95	12.8	1.84	13.3	1.68
	Rc	18.1	1.25	18.4	1.25	18.1	1.05	18.7	1.48

Table 6. Age of radiographic formation stages of permanent teeth in girls. Stages from Demirjian et al. 1976, medians and standard deviations from Kahl and Schwarze (1988). Stages: A – beginning of calcification at the superior level of the crypt in the form of inverted cones. B – fusion of calcified points to a regularly outlined occlusal surface. C – complete enamel formation at the occlusal surface. D – crown formation complete to the cemento-enamel junction. E – uniradicular teeth: pulp chamber walls form straight lines, larger pulp horn than in D; molars: initial radicular bifurcation. F – uniradicular teeth: pulp chamber walls form a isosceles triangle; molars: semi-lunar bifurcation. G – parallel root canal walls but open apical end. H – apex completely closed. For a more complete description and type radiographs see source

Tooth	stage	Maxillary median	±SD	Mandibular median	±SD	Tooth	stage	Maxillary median	±SD	Mandibular median	±SD
						I2	C	6.7	0.6	–	
I1	D	6.7	0.6	–			D	7.1	0.8	–	
	E	7.4	0.8	–			E	8.0	1.0	6.7	1.0
	F	8.4	1.1	7.3	1.6		F	9.1	0.8	7.4	1.2
	G	9.8	1.3	9.3	1.2		G	10.2	1.4	9.5	1.1
	H	13.3	3.6	12.7	3.6		H	13.6	3.7	12.8	3.6
C	D	7.4	0.8	7.4	0.9	**P1**	D	7.8	1.1	7.3	0.9
	E	8.4	1.1	8.0	1.1		E	9.6	1.1	8.7	1.2
	F	10.0	1.2	9.7	1.1		F	10.3	1.4	9.8	1.2
	G	12.4	1.6	11.5	1.6		G	12.6	1.4	11.7	1.5
	H	16.0	3.6	15.3	3.6		H	15.8	3.6	15.2	3.5
P2	C	7.6	2.1	7.6	3.1	**M1**	E	5.4	–	–	
	D	8.3	1.3	7.6	0.9		F	7.4	1.1	6.3	1.1
	E	9.8	1.3	9.5	1.2		G	9.7	1.3	9.2	1.3
	F	10.4	1.4	10.3	1.6		H	13.3	3.6	12.5	3.6
	G	12.7	1.5	12.8	1.3						
	H	15.8	3.5	15.8	3.6	**M3**	A	10.0	1.4	9.8	1.4
							B	9.5	1.4	9.8	1.6
M2	C	7.1	1.6	7.1	0.7		C	10.4	1.7	11.3	1.7
	D	8.5	1.2	8.2	1.0		D	12.7	1.8	13.1	1.9
	E	9.7	1.2	9.7	1.2		E	15.6	1.2	15.3	1.9
	F	11.0	1.4	10.3	1.4		F	16.7	3.5	16.5	3.4
	G	13.3	1.5	13.0	1.6		G	20.7	3.7	20.0	3.5
	H	17.4	3.4	17.4	3.4		H	23.0	1.5	23.0	1.4

Cross-Sectional and Longitudinal Studies

Radiograhic studies can be either cross-sectional where many children are investigated only once (first: Bustin et al. 1929, 1930) or longitudinal where fewer individuals are investigated several times (first: Gleiser and Hunt 1955). A mixed type are the studies including both types of data (first: Broadbent 1941). Only the longitudinal design gives a sufficient access to the individual dynamics of growth. Because of the wide variation in growth, standards from any cross-sectional study give little indication of individual growth rate over time. On the other hand, the state of growth patterns in a population can only be assessed by combining children who mature early, average and late. In longitudinal studies the time between examinations should be noted as this influences accuracy. Moreover, the films should always be scored by the same observer in order to reduce errors.

Dahlberg and Menegaz-Bock (1958) highlighted the problems comparing cross sectional and longitudinal studies, suggesting medians from cross sectional data are better compared to means from longitudinal data. The attainment of a stage of tooth development is a fleeting event and in a longitudinal study with a long interval between examinations the child has often passed the stages to be assessed. In this case the actual age of attainment was some time previous to the last examination. If the child is radiographed near his or her birthday annually, and a stage of tooth formation is scored as being present, it could have reached that stage any time during the previous year, i.e. between 1 and 51 weeks earlier. This explains why in the few cases of comparability the age of attainment of any one stage from longitudinal studies is generally later than the age from cross sectional studies, if no allowance is made for the recall interval. For example tooth formation standards of Sardinian children from a cross-sectional study were compared to the longitudinal data of Moorrees et al. (1963b) and found to be between 1 and 2 years delayed (Diaz 1993). These differences may be due to inappropriate comparisons between the methods. It is notable that the standards of Moorrees et al. (1963a, b) are considerably earlier than many other studies. A difference of almost one standard deviation has been noted (Ito et al. 1993).

Sampling and Strategies

Sample Size

Early studies described single cases (e.g. Zuckerkandl 1891). As in any type of quantitative research it is important to have a sufficient sample size in growth studies. The general recommendations from statistics (30 as a minimum or 100 as a low comfortable size) cannot be applied in this study type because age is a continuous variable and data are needed for more than one phenomenon, therefore theoretically larger samples are needed.

The sample size required is connected with the technique: dissection data are less easily gained than radiographic data and this leads to smaller groups. The number of individuals in cross sectional studies is much higher than in longitudinal studies; sample size in longitudinal studies, however, is the number of recorded events and not the number of individuals.

Even if these differenciations are taken into account, it is clear that a total sample size of 30 dissected children is not sufficient. This is the number of one of the most frequently used studies (Logan and Kronfeld 1933; data used again in Kronfeld 1935, 1954; Schour and Massler 1940, 1941). Most of these children had died from infectious diseases.

Even Age Structure

Methods of growth analysis are sensitive to the age structure of the sample and require an even age distribution (similar number of subjects in each age interval), otherwise the normal distribution of a growth event might be cut off. For example, the first radiographic study of crown and root formation investigated growth of the permanent first molar in 50 children between the ages of birth and 10 years (Gleiser and Hunt 1955). This longitudinal study reported the age of crown completion as 3.5 ± 0.47years (mean ± standard deviation) for boys and 3.3 ± 0.35 years for girls. It follows that studies based on children where the youngest is 3 years will include only unusually late maturing children and results for mean values of crown completion will be considerably higher than the true mean age of crown completion. Two cross sectional studies of similar design to Gleiser and Hunt report marked differences in the mean age of crown completed stage of the first permanent molar (Finn 3.5 years, sample 6 children; Japanese 2.7 years, sample 148 children). These differences may partly reflect the number of children in the youngest age group between 2 and 3 years (Haaviko 1970, Diato et al. 1989).

Age Classes

Account should also be taken of how children are grouped into age intervals in growth studies. If children are ranked in age categories of 1 year, then those children aged one day over 5 years are effectively judged equal in age to those aged one day under 6 years. Larger age classes than one year are not appropriate.

Cumulation, Median, and Mean

The knowledge of the method of univariate data grouping used to compute results is useful. Cumulative frequency distribution is considered to be appropriate in assessing attainment of a growth stage (reviewed by Smith 1991). This is done by plotting a cumulative curve of a developmental stage, where the proportion of children who have attained the stage is plotted against age. Thus different percentiles may be quickly derived from the figure. The distribution center is the median. Other studies calculated means. In a normal distribution these are very similar, but not if the distribution is skewed.

Error, Accuracy, and Precision

Accuracy is the closeness of a computed value to its true value. The accuracy of a method of age estimation defines how well chronological age can be estimated and can be expressed as the difference between dental age and chronological or actual age. Recent analysis of accuracy and precision of dental standards have compared various methods in early childhood from small samples of skeletal remains where age at death is known (Saunders et al. 1993, Liversidge 1994) as well as larger samples of radiographs of living children (Crossner and Mansfeld 1983; Hägg and Matsson 1985; Pöyry et al. 1986; Staaf et al. 1991; Mörnstad et al. 1995). Some previous reports on accuracy have calculated the correlation coefficient between dental and actual age. However, some have discounted this as a measure of accuracy since it is greatly influenced by the age range of the sample under examination (Thorson and Hägg 1991).

Accuracy may be calculated in different ways: from the reference sample or from a new test sample (theoretically the latter gives higher values); the parameter may be a normal mean or the mean disregarding signs (often called bias); the dispersion measure may be the standard deviation or any percentage of the confidence interval. It is crucial to know which of the alternatives have been chosen, but many papers are mute in this respect, particularly during the first half of this century. Unfortunately the test for regression towards the mean is rarely used: in many biological phenomena an estimation in high values gives results that are low and vice versa. Naturally this is of great importance in forensic cases.

Precision (the closeness of repeated measurements of the same quantity) is a measure of intra-observer variability and methods using well defined criteria have high precision (Levesque and Demirjian 1980; Gat et al. 1984; Hägg and Matsson 1985).

Statistically descriptive growth curves are not suitable for prediction, and ideally age should be the independent variable. This concept has been noted by Smith (1991) who adapted the data of Moorrees et al. (1963b) for prediction and this is more accurate than using the original age of attainment data (Liversidge 1994).

Two Scandinavian studies have tested the accuracy and precision using data from Haaviko (1970, 1974). Staaf et al. (1991) examined radiographs of children 5.5 to 14.5 years and found higher accuracy for children younger than 10 years (0.4 ± 0.67 year) compared to those older (0.5 ± 0.92 year). Mörnstad et al. (1995) found this method to be very accurate for boys aged 5 years (0.0 ± 0.50) while accuracy of girls for this age was 0.2 ± 0.67; at age 12 years the accuracy was 0.2 ± 0.75 for boys 0.3 ± 0.50 for girls.

Maturity scores of seven mandibular teeth by Liliequist and Lundberg (1971) gives ages in half-year intervals. The accuracy of this method between 3.5 and 12.5 years was from 0.4 ± 0.50 for the younger children to 0.6 ± 1.00 for the older children (Hägg and Matsson 1985). Another study found this method very accurate for girls (0 ± 0.75) but less so for boys (0.6 ± 0.97 year) for the ages 5.5 to 14.5 years (Staaf et al. 1991). Similar results report higher accuracy for 5 year old girls 0.0 ± 0.67 than boys 0.6 ± 0.67 at this age (Mörnstad et al. 1995).

The maturity scale of dental formation (Demirjian et al. 1973, 1976) has also been tested as a method of age estimation although its use is limited with fragmentary skeletal remains. Important population differences in dental maturation have been noted (Swedish, French and Chinese children are about one year ahead of French-Canadians); therefore this method may over-estimate age by one year (Hägg and Matsson 1985, Staaf et al. 1991; Davis and Hägg 1994; Mörnstad et al. 1995). However, accuracy for young children aged 3.5 to 6.5 was 0.1 ± 0.66 for boys and 0 ± 0.63 for girls (Hägg and Matsson 1985) and this method may be suitable for young children. In a study by Staaf et al. (1991) this method consistently over-estimated age by about 0.8 ± 1.00 year for both sexes between 5.5 and14.5 years. Two other small studies testing this method found similar results with Kenyan (Nichols et al. 1983) and Mexican-American children (Ocholla 1990).

The longitudinal data of Moorrees et al. (1963a, b) have been tested on the Spitalfields children (Liversidge 1994). Deciduous tooth standards were accurate to 0.5 ± 0.62 years for the groups of children from 0 to 4 years. Permanent tooth standards were accurate to 0.6 ± 0.42 years between 0 and 5.4 years. These standards have been modified specifically for prediction (see Smith 1991) and appear to be more accurate (0.3 ± 0.39) than using unmodified data in early childhood (Liversidge 1994).

The variability of tooth formation increases with chronological age as environmental influences affect the rate of growth (Garn et al. 1959; Haavikko 1970; Anderson et al. 1976; Haavikko 1974) and greater accuracy can be obtained by choosing early developing teeth that show the least variability (Haavikko 1974). Deciduous teeth are not only smaller than permanent teeth but also develop far more quickly. This means that in early childhood many indicators of short duration are available and greater accuracy is possible during the fetal and early postnatal period when growth is rapid (Deutsch et al. 1985; Stack 1964; Luke et al. 1978). In contrast, accuracy of age prediction is poor during early adulthood and maturity (Johanson 1971; Levesque et al. 1981; Engström et al. 1983; Nortjé 1983).

Age, Stage, and Maturity Score

In general growth standards can be expressed as development at a particular age or the average age of a child who has reached a particular stage of tooth formation and some knowledge of the different methods of calculation is necessary as they differ fundamentally depending upon the purpose and design of each study (reviewed by Smith 1991).

Development at a particular age indicates the average development (or mean stage) at a given age. This method is used to estimate age by comparing either the overall dental formation of the jaws or for each individual tooth from radiographs. The best known is a modified version of the Schour and Massler atlas (1941). The atlas is an easy method to use but difficulties arise where no clear differentiation of stages are apparent between adjacent diagrams or when some teeth are ahead of others. Another type of chart combining histological, radiographic and emergence data from 20 sources was drawn up by Gustafson and Koch (1974). Each tooth is represented as a line against age, with a number of triangles indicating several stages of formation. It does not allow for sex and is limited to three mineralisation stages and emergence for each maxillary and mandibular tooth (excluding the third molars) for both deciduous and permanent dentitions. This chart is difficult to read but does give a measure of variability. Some radiographic studies present data as the average stage per age (Nanda and Chawla 1966; Nolla 1960; Nyström et al. 1977), but computing averages of ranked stages in two of these studies introduces some degree of error.

Data can be expressed as the average age of a child who has reached a particular stage. This type of data (mean age per stage with standard deviation or percentile) provides age of attainment of stages calculated by cumulative distribution functions. This is done by plotting a cumulative curve of a developmental stage, where the proportion of children who have attained the stage is plotted against age. The mean age of attainment is equivalent to the age reached by 50% of the sample. Data of this type have been provided by Garn et al. (1958, 1959), Fanning (1961), Moorrees et al. (1963a, b), Demirjian and Levesque (1980). This method offers several advantages over alternate types of data. Dispersion for each stage of individual forming teeth can be ascertained for both boys and girls. The longitudinal studies (Moorrees et al. 1963a, b) are of a large sample with many children from each age class with frequent radiographs that begins from birth and data are given for both dentitions.

A maturity scale combines aspects of both of these types of data presentation. This method assesses maturity of children of known age by means of a weighted score (reviewed by Demirjian 1986) and is based on the principle of skeletal age by assessing maturity from wrist ossification status by Tanner (1962). Dental age is determined by scoring stages of formation of several individual developing teeth (Demirjian et al. 1973; Demirjian and Levesque 1980). This allows comparison of individual children both normal and with growth disturbances or diseases (Myllärniemi et al. 1978, Oliver and Nixon 1995, Seow 1995a, b) as well as comparing maturity between population groups, although their use is limited where specimens are fragmentary.

Among populations clear differences in the timing of initial mineralisation of one permanent tooth to another may be found (Fanning and Moorrees 1969; Owsley and Jantz 1983; Tompkins 1996). This kind of relative developmental difference is masked when a maturity score for several developing teeth is used (as done by Wolanski 1966; or Demirjian et al. 1973). The maturity score is hardly comparable to other methods. Several studies overcome this problem by presenting results that allow easy comparison (Kahl and Schwarze 1988). Another approach is to compare dental maturation at a particular skeletal age rather than chronological age using Demirjian's criteria for stage assessment (Chertkow 1979, 1980).

Representation of Developmental Data

Age can be determined from developing teeth in several ways. The first is a comparison with an atlas showing the mean stage of development for individual teeth or the whole dentition at particular ages (e.g. Schour and Massler 1940 or Gustafson and Koch 1974).

A second method of age estimation from developing teeth is to use age and sex specific standards or maturity scales from radiographic growth studies which rely on defined stages of crown and root development although few are designed for prediction (see Smith 1991).

Another method used to estimate age from developing teeth is to use regression equations of quantitative data derived from individuals of known age (Tables 7 and 8). This comprises a comparison of tooth length and weight data against age from anatomical studies during fetal and early childhood (Stack 1960, 1963, 1964, 1967, 1971; Luke et al. 1978; Deutsch et al. 1981, 1984, 1985; Liversidge et al. 1993) and tooth length, measured from radiographs (Ledley et al. 1971; Israel and Lewis 1971; Carels et al. 1991; Inoue and Suzuki 1992). Accuracy of quantitative methods using deciduous tooth length and weight up to the age of one year is reported as 0.1 ± 0.15 (Liversidge 1994). Tooth length can be measured from good quality radiographs allowing for distortion and magnification. Isolated developing teeth pose special problems. Direct vision allows true stage assessment particularly where tooth tissue is thin and "burned out" by x-rays. The age of true crown completion is later than age from radiographic studies (Liversidge 1995a). Quantitative methods such as those that measure tooth length during growth are suited to archaeological or forensic studies of isolated developing teeth where individuals are young, i.e. fetal to early childhood.

Table 7. Equations of increasing deciduous tooth length. Age in years; h: tooth length in mm (only appropriate if the root is incomplete, i.e. the tooth is still growing). SD: Standard deviation of regressive estimate. Tooth length in mm along long axis, from incisal tip to developing edge. Data from Liversidge et al. (1993)

Tooth	regression	SD
i1	age = –0.653 + 0.144h	± 0.19
i2	age = –0.581 + 0.153h	± 0.17
c	age = –0.648 + 0.209h	± 0.22
m1	age = –0.814 + 0.222h	± 0.25
m2	age = –0.904 + 0.292h	± 0.26

Table 8. Equations of permanent tooth length and age. Age in years; h: tooth height in mm, appropriate for tooth length less than the maximum; SD: standard deviation of estimate; max t/l: maximum tooth lenght on which data is based. Data from Liversidge (1993)

Tooth	regression	SD	max t/l
I1	age = $0.237 - 0.018h + 0.042h^2$	± 0.21	< 11.3 mm
I^2	age = $-0.173 + 0.538h + 0.003h^2$	± 0.14	< 9.9
I_2	age = $0.921 - 0.281h + 0.075h^2$	± 0.12	< 9.8
C	age = $-0.163 + 0.294h + 0.028h^2$	± 0.25	< 9.8
M1	age = $-0.942 + 0.441h + 0.010h^2$	± 0.25	< 11.5

Review of Methods

The first large group of methods for the ageing of non-adults is the use of dissection data (Legros and Magitot 1873; Black 1883; Peirce 1884; Röse 1891; Zuckerkandl 1891; Berten 1895; Paltauf 1902, Broomell and Fischelis 1923; Eidmann 1923;

Brady 1924; Logan and Kronfeld 1933). Their cardinal element is the restricted group size of up to 30 individuals, which by modern standards is insufficient. Despite the additional draw-back of a high disease load the wide-spread method by Logan and Kronfeld is of a good accuracy and in congruence with others (Brauer and Bahador 1942; Miles 1958; Herdeg 1992; Liversidge 1994); possibly this was the effect of "smoothing" along the lines which have been set by the previous studies (Herdeg 1992).

The Schour and Massler scheme (1940, 1941) has also been updated (to include data from other radiographic studies) which appear in several dentistry, osteology, or anthropology textbooks. Occasionally the canine and third molar are excluded because of their high variability. One version includes the "lower" range of a number of cross-sectional and longitudinal studies supposedly suitable for archaeological material though originally adapted for American Indians (Ubelaker 1978; Ferembach et al. 1978, 1980).

This type of expert pooling has been presented several times (Berten 1895; Black 1908; Kronfeld 1935, 1954; Neureiter et al. 1940; Hohmann 1954; Lewis and Garn 1960; Legoux 1962; Falkner 1966; Gustafson 1966; Krogman 1968; Acsádi and Nemeskéri 1970; Johanson 1971; Cameron and Sims 1974; Lunt and Law 1974; Hotz 1976; Hunger and Leopold 1978; van der Linden 1978; Demirjian 1986; Zimmermann and Angel 1986). For an analysis of these and other secondary presentations and the primary sources they used see Herdeg (1992).

Another example is the chart of Gustafson and Koch (1974), which is not better or worse than the others, but which can be assessed as to accuracy. Age is read in 2 monthly intervals up to 1 year, thereafter in one year intervals. Hägg and Matsson (1985) tested the method and found it more accurate for boys. For the age group 3.5 to 6.5 years and 9.5 to 12.5 years accuracy for boys was 0 ± 0.58 and 0.1 ± 0.98; corresponding values for girls was 0.6 ± 0.79 and 0.3 ± 0.83. Similarly, a higher accuracy for boys was noted by Mörnstad et al. (1995); 0.1 ± 0.75 year for 5 year olds to 0.2 ± 1.00 for 12 year olds compared to girls (0.6 ± 0.67 for 5 year olds to 0.7 ± 1.00 for 12 year olds). Reported overall accuracy between birth and 5 years of age is 0.1 ± 0.37 years (Liversidge 1994).

Quantitative data collection for development of the prenatal deciduous dentition was pioneered by Stack (1960, 1963, 1964, 1967, 1971) who measured developing tooth length and weight for age. Further quantitative studies extend this type of data up to the end of the first postnatal year (Deutsch et al. 1981, 1984, 1985). Data for tooth length, where age is known, for the permanent dentition is based on cross sectional data from the Spitalfields collection for early growth (Liversidge et al. 1993) and also from radiographs (Ledley et al. 1971; Israel and Lewis 1971; Carels et al. 1991; Inoue and Suzuki 1992; Mörnstad at al. 1994; Kullman et al. 1995). Age can be predicted from regression lines of data for tooth length for age, giving mean age of the sample from these cross sectional studies. However, this mean value and the growth velocity, implied in the slope of the curve, represents the group as a whole and individual patterns of changing growth rate are masked. Growth velocity or mean rate of growth of an individual can only be determined from longitudinal studies and data of this type is meagre (Carlson 1944; Inoue and Suzuki 1992; Dean et al. 1993b). Nevertheless, data from both types of studies are in close agreement.

The next type of original study is the cross-sectional investigation of radiographies (Bustin et al. 1929, 1930; Hess et al. 1932; Andreas 1956; Grøn 1962; Haataja 1965; Nanda and Chawla 1966; Haavikko 1966, 1970; Fass 1969; Schopf 1970; Liliequist and Lundberg 1971; Leinonen et al. 1972; Trodden 1982; Crossner and Mansfield 1983; Kahl and Schwarze 1988).

Finally the historically third and last group are longitudinal or semi-longitudinal studies of radiographs (Broadbent 1941; Broadbent et al. 1975; Gleiser and Hunt 1955; Garn et al. 1959; Nolla 1960; Bradley 1961; Moorrees et al. 1963a, b; Haataja 1965; Fanning and Brown 1971; Sapoka and Demirjian 1971; Anderson et al. 1976).

A critical comparisons of all these methods (Herdeg 1992) reveals that within each group the quality of the studies is well comparable. Nevertheless there are very clear preferences, which may be explained by "school" and national or language limits.

Application

First some general recommendations for application should be developed. Theoretically growth standards should be appropriate for the sex, the regional group, the historic time, and the subsistence pattern of the individual or group being assessed. Most of these variation factors (except sex), however, are not sufficiently known (see above). At the moment this specifity demand is more an ideal which in many applications cannot lead to selection strategies.

If an individual is to be aged, it is important to use as many teeth as possible, and to use them separately. Next a mean of the single values is calculated. The use of scores is not recommended because it ignores a part of the variablity. Moreover, all available traits should be used. This means not only qualitative stages of development but also the use of quantitative measures such as tooth length.

The result of an age estimation is an extrapolation by a probability procedure. Therefore it is impossible to give a precise date; the only appropriate way is to give an age interval for a specific confidence, moreover to state which interval is meant (e.g. 1 SD). This in turn allows a selection of methods: those which do not give dispersion parameters are unsuited.

Concerning single stages, crown completion is a difficult stage to assess accurately, particularly on permanent incisors and canines. Another difficulty is the subjective assessment of crown and root fractions in the absence of longitudinal radiographs; it should be avoided and replaced by tooth lengths. Inter- and intra-observer variability should be kept to a minimum by a common training programme for workers as the interpretation and accurate rating of formation stages is crucial in determining precision (Roche 1980).

The Gustafson and Koch chart (1974) is easy to use and appropriate during early childhood. Radiographic stage assessment of individual teeth as a method of age prediction for older children, however, has several advantages. Each tooth is staged from clear drawings of formation stages such that dental age can be determined for each developing tooth. Recent results suggest that selecting the least variable permanent teeth (Haaviko 1974) provides a more accurate and precise prediction of age compared to a dental maturity scale (Staaf et al. 1991; Mörnstad et al. 1995). The least variable mandibular permanent teeth in children under the age of ten are the first and second molars, first premolar and central incisor. After this age, the mandibular second molar, first premolar and lower canines are least variable (Haaviko 1974).

The traditional Schour and Massler atlas (1941) may still be used for infancy and early childhood for reasons of accuracy and ease. The recommended choice of radiological stage assessment in older children is that of Demirjian et al. (1976). It has the advantage of being very clear with radiographs, drawings and written criteria. A recent study by Kahl and Schwarze (1988) presents the mean, standard deviation and median age for stages defined by Demirjian, as well as updated drawings which represent development. More recent standards of tooth length (Mörnstad et al. 1994) remain to be tested.

Despite the many factors of variation and the problems of methodology, developing teeth are likely to remain a widely used method to assess age and maturity. Recommended standards (including those reproduced in our tables) do not all give dispersions and lack continuity as well as details of population description. It is hoped that future research will focus on the gaps and in so doing provide better assessment of ontogeny.

References

Acsádi G, Nemeskéri J (1970) History of Human Life Span and Mortality. Akadémiai Kiadó, Budapest.

Adler P (1958) Die Akzeleration des Zahnwechsels. Archiv Kinderheilk 157: 23–33.

Adler-Hradecky C (1959) Die Bestimmung des individuellen Zahnalters. Z Kinderheilk 82: 16–22.

Anderson DL, Thompson GW, Popovich F (1976) Age of attainment of mineralisation stages of the permanent dentition. J Forens Sci 21: 191–200.

Andreas A (1956) Röntgenologische Darstellung der physiologischen Entwicklung des zweiten unteren Prämolaren. Med Diss, Mainz.

Berten J (1895) Hypoplasie des Schmelzes. Teile I–IV. Dtsch Mschr Zahnheilk 13: 483–497.

Beynon AD (1992) Circaseptum rhythms in enamel development in modern humans and plio-pleistocene hominids. In: Smith P, Chernov E (eds) Structural Function and Evolution of Teeth. Freund, London, p. 295–309.

Beynon AD, Wood BA (1987) Patterns and rates of enamel growth in the molar teeth of early hominids. Nature 326: 493–496.

Black GV (1883) Tables of lines of contemporaneous calcification. Ann J Illinois State Dental Soc 1883: 1.

Black GV (1902) Descriptive Anatomy of the Human Teeth, 4th Ed. White, Philadelphia, p. 16–19.

Black GV (1908) A Work on Operative Dentistry. Medico-Dental Publ, Chicago.

Boyde A (1963) Estimation of age at death of young human skeletal remains from incremental lines in the dental enamel. In: Proceedings of Third International Meeting of Forensic Immunology, Medicine, Pathology and Toxicology, London. Reprinted in DeRousseau C (ed) (1990) Primate Life History and Evolution. Wiley-Liss, New York, p. 260–267.

Bradley RE (1961) The relationship between eruption, calcification, and crowding of certain mandibular teeth. Angle Orthod 31: 230–236.

Brady WJ (1924) Chart of Average Time of Development, Eruption, and Absorption of Teeth. Private Print, Kansas City. Quoted after Sicher and Tandler 1928.

Brauer JC, Bahador MA (1942) Variation in calcification and eruption of the deciduous and the permanent teeth. J Am Dent Ass 29: 1373–1387.

Broadbent BH (1941) Ontogenetic development of occlusion. Angle Orthod 11: 223–241.

Broadbent BH Sr, Broadbent BH Jr, Goldon WH (1975) Bolton Standards of Dentofacial Development Growth. Mosby, St Louis.

Bromage TG, Dean MC (1985) Re-evaluation of the age at death of of immature fossil hominids. Nature 317: 525–527.

Broomell I, Fischelis P (1923) Anatomy and Histology of the Mouth and Teeth. Blakiston, Philadelphia.

Bustin E, Leist M, Priesel R (1929; 1930) Röntgenologische Studie am kindlichen Gebiß. Fortschr Geb Röntgenstr 40: 80–88; 41: 49–53.

Cameron JM, Sims BG (1974) Forensic Dentistry. Churchill Livingstone, London.

Carlson H (1944) Studies on the rate and amount of eruption of certain human teeth. Am J Orth Oral Surg 30: 575–588.

Carels CEL, Kuijpers-Jagtman AM, van der Linden FPGM, van't Hof MA (1991) Age reference charts of tooth length in Dutch children. J Biol Bucc 19: 297–303.

Cattell P (1928) The eruption and growth of the permanent teeth. J Dent Res 8: 279–287.

Chertkow S (1979) The relationship between tooth mineralisation and early radiographic evidence of the ulnar sesamoid. Angle Orthod 49: 282–288.

Chertkow S (1980) Tooth mineralisation as an indicator of the pubertal growth spurt. Am J Orthod 77: 79–91.

Crossner C, Mansfeld L (1983) Determination of dental age in adopted non-European children. Swed Dent J 7: 1–10.

D'Arcy Thompson W (1917) On Growth and Form. Cambridge Univ Press, Cambridge.

Dahlberg AA, Menegaz-Bock RM (1958) Emergence of the permanent teeth in Pima Indian children. J Dent Res 37: 1123–1140.

Davis PJ, Hägg U (1994) The accuracy and precision of the "Demirjian System" when used for age determination in Chinese children. Swed Dent J 18: 113–116.

Dean MC (1985) Variation in the developing root cone angle of modern man and certain fossil hominids. Am J Phys Anthrop 68: 233–238.

Dean MC (1987) The dental developmental status of six East African juvenile fossil hominids. J Hum Evol 16: 197–213.

Dean MC, Beynon AD (1991) Histological reconstruction of crown formation times and initial root formation in a modern human child. Am J Phys Anthrop 86: 215–228.

Dean MC, Beynon AD, Reid DJ and Whittaker DK (1993a) A longitudinal study of tooth growth in a single individual based on long and short period incremental markings in dentine and enamel. Int J Osteoarch 3: 249–264.

Dean MC, Beynon AD, Thackeray JF, Macho GA (1993b) Histological reconstruction of dental development and age at death of a juvenile *Paranthropus robustus* specimen, SK63 from Swartkrans, South Africa. Am J Phys Anthrop 91: 401–419.

Demirjian A (1986) Dentition. In: Falkner F, Tanner JM (eds) Human Growth. A Comprehensive Treatise, Vol II. Plenum, New York, p. 269–298.

Demirjian A, Goldstein H (1976) New systems for dental maturity based on seven and four teeth. Ann Hum Biol 3: 411–421.

Demirjian A, Goldstein H, Tanner JM (1973) A new system of dental age assessment. Hum Biol 45: 211–227.

Demirjian A, Levesque GY (1980) Sexual differences in dental development and prediction of emergence. J Dent Res 59: 1110–1122.

Deutsch D, Goultschin J, Anteby S (1981) Determination of human fetal age from the length of femur, mandible and maxillary incisor. Growth 45: 232–238.

Deutsch D, Pe'er E, Gedalia I (1984) Changes in size, morphology and weight of human anterior teeth during the fetal period. Growth 48: 74–85.
Deutsch D, Tam O, Stack MV (1985) Postnatal changes in size, morphology and weight of developing postnatal deciduous anterior teeth. Growth 49: 202–217.
Diato M, Kawahara S, Tanaka M, Imai G, Nishihara G, Hieda T (1989) Calcification of the permanent first molars observed in panoramic radiographs. J Osaka Dent Univ 23: 45–55.
Diaz G, Maccioni P, Zedda P, Cabitza F, Cortis IM (1993) Dental development in Sardinian children. J Craniof Genet Dev Biol 13: 109–116.
Engström C, Engström H, Sagne S (1983) Lower third molar development in relation to skeletal and chronological age. Angle Orthod 53: 97–106.
Falkner F (1966) Human Development. Saunders, Philadelphia.
Fanning EA (1961) A longitudinal study of tooth formation and root resorption. NZ Dent J 57: 202–217.
Fanning EA, Brown T (1971) Primary and permanent tooth development. Australian Dental J 16: 41–43.
Fanning EA, Moorrees CFA (1969) A comparison of permanent mandibular molar formation in Australian aborigines and caucasoids. Arch Oral Biol 14: 999–1006.
Fass EN (1969) A chronology of growth of the human dentition. J Dent Child 36: 391–401.
Ferembach D, Schwidetzky I, Stloukal M (1980) Recommendations for age and sex diagnoses of skeletons. J Hum Evol 9: 517–549.
FitzGerald C, Foley RA, Dean MC (1996) Variation of circaseptan cross striations in the tooth enamel of three modern human populations. Am J Phys Anthrop Suppl 22: 104.
Garn SM, Lewis AB, Koski K, Polacheck D (1958) The sex difference in tooth calcification. J Dent Res 37: 561–567.
Garn SM, Lewis AB, Polacheck DL (1959) Variability of tooth formation. J Dent Res 39: 135–148.
Gat H, Sarnat H, Bjorvatn K and Dayan D (1984) Dental age evaluation. A new six developmental stage method. Clin Prev Dent 6: 18–22.
Gleiser I, Hunt EE (1955) The permanent mandibular first molar: its calcification, eruption and decay. Am J Phys Anthrop NS 13: 253–283.
Grivu O, Theiss E, Ghibu-Petcú C, Mecher E, Patrut G (1967) Beitrag zum Studium des Zahnwechsels. Dtsch Zahnärztl Z 22: 1120.
Grøn AM (1962) Prediction of tooth emergence. J Dental Res 41: 573–585.
Grosch FC, Joksch HC (1960) Statistische Untersuchungen der Durchbruchszeiten bleibender Zähne. Dtsch Zahnärztebl 14: 336–342.
Gustafson G (1966) Forensic Odontology. Staples, London.
Gustafson G, Koch G (1974) Age estimation up to 16 years based on dental development. Odont Revy 25: 297–306.
Haataja J (1965) Development of the mandibular permanent teeth of Helsinki children. Suom Hammaslääk Toim/Proc Finn Dent Soc 61: 45–63.
Haavikko K (1970) The formation and alveolar and clinical eruption of the permanent teeth, an orthopantomograph study. Proc Finn Dent Soc 66: 104–170.
Haavikko K (1974) Tooth formation age estimated on a few selected teeth. A simple method for clinical use. Proc Finn Dent Soc 70: 15–19.
Hägg U, Matsson L (1985) Dental maturity as an indicator of chronological age: The accuracy and precision of three methods. Eur J Orthod 7: 25–34.
Harris EF, Barcroft BD, Haydar S (1993) Delayed tooth formation in low birthweight African-American children. Pediatr Dent 15: 30–35.
Harris EF, McKee JH (1990) Tooth mineralisation standards for Blacks and Whites from the Middle Southern United States. J For Sci 35: 859–872.

Henke W (1960) Untersuchungen über Dentitionsverschiebungen im bleibenden Gebiß. Dtsch Zahnärztebl 14: 699–706.
Herdeg B (1992) Die Zahnentwicklung beim Menschen. Kritische Analyse der bisherigen Zahlen und Zitierreihen. Med Diss, Ulm.
Hess AF, Lewis JM, Roman B (1932) A radiographic study of the teeth from birth to adolescence. Dental Cosmos 74: 1053–1061.
Hohmann A (1954) Untersuchungen über die Beziehungen zwischen Mineralisationsgrad von Zahndentin und Lebensalter als Beitrag zu den forensischen Methoden zur Altersbestimmung des Menschen. Med Diss, Marburg.
Hotz R (1976) Zahnmedizin bei Kindern und Jugendlichen. Thieme, Stuttgart.
Huda TFJ, Bowman JS (1995) Age determination from dental microstructure in juveniles. Am J Phys Anthrop 97: 135–150.
Hunger H, Leopold D (1978) Identifikation. Springer, Berlin.
Inoue Y, Suzuki Y (1992) Study on eruption of maxillary permanent incisors using panoramic tomography. Pediatr Dent J 2: 83–92.
Israel H, Lewis AB (1971) Radiographically determined linear permanent tooth growth from age 6 years. J Dent Res 50: 334–342.
Ito RK, Vig KWL, Garn SM, Hopwood NJ, Loos PJ, Spalding PM, Deputy BS, Hoard BC (1993) The Influence of growth hormone (rhGH) therapy on tooth formation in idiopathic short statured children. Am J Orthod Dentofacial Orthop 103: 358–364.
Johanson G (1971) Age determination from human teeth. Odont Revy Suppl 22: 1–126. [incorrectly printed as suppl 21]
Kahl B, Schwarze CW (1988) Aktualisierung der Dentitionstabelle von I Schour und M Massler von 1941. Fortschr Kieferorthop 49: 432–443.
Kataja M, Nyström M, Aine L (1989) Dental maturity standards in southern Finland. Proc Finn Dent Soc 85:187–197.
Krauss BS, Jordan RE (1965) The Human Dentition Before Birth. Lea and Febinger, Philadelphia.
Krogman WM (1968) Biological timing and the dento-facial complex. J Dent Child 35: 175–185, 328–341, 377–381.
Kromeyer K, Wurschi F (1996) Zahneruption bei Jenaer Kindern in der ersten Phase des Wechselgebisses. Anthrop Anz 54: 57–70.
Kronfeld R (1935) Development and calcification of the human deciduous and permanent dentition. The Bur 15: 18–25.
Kronfeld R (1954) Development and calcification of the human deciduous and permanent dentition. In: Steward TD, Trotter M (eds) Basic Readings on the Identification of Human Skeletons: Estimation of Age. Wenner-Gren, New York.
Kullman L, Martinsson T, Zimmerman M, Welander U (1995) Computerized measurements of the lower third molar related to chronological age in young adults. Acta Odontol Scand 53: 211–216.
Leinonen A, Wasz-Höckert B, Vuorinen P (1972) Usefulness of the dental age obtained by orthopantomography as an indicator of the physical age. Suom Hammaslääk Toim/Proc Finn Dent Soc 68: 235–242.
Ledley RS, Huang HK, Pence RG (1971) Quantitative study of normal growth and eruption of teeth. Comp Biol Med 1: 231–241.
Legros C, Magitot E (1873) Origine et formation du folliculaire dentaire chez les mammifères. J Anat Physiol 5: 449–503.
Legoux P (1964) Détermination de l'âge dentaire de quelques fossiles de la lignée humaine. Rev Fr Odonto-Stom 11: 375–392.
Levesque GY, Demirjian A (1980) The inter-observer variation in rating dental formation from radiographs. J Dent Res 59: 1123–1126.

Levesque GY, Demirjian A, Tanguay R (1981) Sexual dimorphism in the development, emergence and agenesis of the mandibular third molar. J Dent Res 60: 1735–1741.
Lewis AB, Garn SM (1960) The relationship between tooth formation and other maturational factors. Angle Orthod 30: 70–77.
Liliequist B, Lundberg M (1971) Skeletal and tooth development. A methodologic investigation. Acta Radiol 11: 97–112.
Linden FPGM van der (1978) Gebißentwicklung. Univ Nijmegen.
Liversidge HM (1993) Human Tooth Development in an Archaeological Population of Known Age. PhD Thesis, London.
Liversidge HM (1994) Accuracy of age estimation from developing teeth of a population of known age (0–5.4 years). Int J Osteoarch 4: 37–45.
Liversidge H (1995a) Crown formation of the permanent dentition in humans. In: Moggi-Cecchi J (ed) Aspects of Dental Biology, Palaeontology, Anthropology and Evolution. International Institute for the Study of Man, University of Florence, Italy, p. 267–275.
Liversidge HM (1995b) Growth standards of human deciduous teeth. In: Radlanski RJ, Renz H (eds) Proceedings of 10th International Symposium on Dental Morphology. Brünne, Berlin, p. 184–189.
Liversidge HM, Dean MC, Molleson T (1993) Increasing human tooth length between birth and 5.4 years. Am J Phys Anthrop 90: 307–313.
Loevy H (1983) Maturation of permanent teeth in Black and Latino children. Acta Odont Pediatr 4: 59–62.
Logan WHG, Kronfeld R (1933) Development of the human jaws and surrounding structures from birth to the age of fifteen years. J Am Dent Ass 20: 379–427.
Luke DA, Stack MV, Hey EN (1978) A comparison of morphological and gravimetric methods of estimating human fetal age from the dentition. In: Butler PM, Joysey KA (eds) Development, Function and Evolution of Teeth. Academic Press, London, p. 511–518.
Lunt RC, Law DB (1974) A review of the chronology of calcification of deciduous teeth. J Am Dent Ass 89: 599–606.
Macho GA, Wood BA (1995) The role of time and timing in Hominid dental evolution. Evol Anthrop 4: 17–31.
Mappes MS Harris EF, Behrents RG (1992) An example of regional variation in the tempos of tooth mineralisation and hand-wrist ossification. Am J Orthod Dentofacial Orthop 101: 145–151.
Massler M, Schour I, Poncher G (1941) Developmental pattern of the child as reflected in the calcification of the teeth. Am J Disease Children 41: 33–67.
Miles AEW (1958) The assessment of age from the dentition. Proc Roy Soc Med 51: 1057–1060.
Moorrees CFA, Fanning EA, Hunt EE (1963a) Formation and resorption of three deciduous teeth in children. Am J Phys Anthrop 19: 99–108.
Moorrees CFA, Fanning EA, Hunt EE (1963b) Age variation of formation stages for ten permanent teeth. J Dent Res 42: 1490–1502.
Mörnstad H, Staaf V, Welander U (1994) Age estimation with the aid of tooth development, a new method based on objective measurements. Scand J Dent Res 102: 137–143.
Mörnstad H, Reventlid M, Teivens A (1995) The validity of four methods for age determination by teeth in Swedish children. Swed Dent J 19: 121–130.
Myllärniemi S, Lenko HL, Perheentupa J (1978) Dental maturity in hypopituitarism, and dental response to substitution treatment. Scand J Dent Res 86: 307–312.
Nanda RS, Chawla TN (1966) Growth and development of dentition in Indian children, I. Development of permanent teeth. Am J Orthod 52: 837–853.
Neureiter Fv, Pietrusky F, Schütt E (1940) Handwörterbuch der gerichtlichen Medizin und naturwissenschaftichen Kriminalistik. Springer, Berlin.

Nichols R, Townsend E, Malina R (1983) Development of permanent dentition in Mexican-American children (Abstract). Am J Phys Anthrop 60: 232.
Nolla CM (1960) The development of the permanent teeth. J Dent Child 27: 254–266
Nortjé CJ (1983) The permanent mandibular third molar. Its value in age determination. J For Odonto-Stom 1: 27–31.
Nyström M, Haataja J, Kataja M, Evälahti M, Peck L, Kleemola-Kujala E (1986) Dental maturity in Finnish children, estimated from the development of seven permanent mandibular teeth. Acta Odont Scand 44: 193–198.
Nyström M, Kilpinen E, Kleemola-Kujala E (1977) A radiographic study of the formation of some teeth from 0.5 to 3.0 years of age. Proc Finn Dent Soc 73: 167–172.
Nyström M, Ranta R, Kataja M, Silvola H (1988) Comparisons of dental maturity between the rural community of Kuhmo in north-eastern Finland and the city of Helsinki. Community Dent Oral Epidemiol 16: 215–217.
Ocholla TJO (1990) The accuracy and reliability of radiographically determined dental age observations from two groups of 4–16 year olds. J Dent Res 69: 967
Oliver JA, Nixon F (1995) Do premature, light-for-date, and premature/light-for-date infants have delayed dental formation? Br J Orthod 22: 397.
Oster H (1960) Liegen die Durchbruchszeiten der bleibenden Zähne heute anders als früher, und haben wir es dabei mit einem Symptom der Akzeleration zu tun? Kinderärztl Praxis 2: 80.
Owsley DW, Jantz RL (1983) Formation of the permanent dentition in Arikara Indians. Timing differences that affect dental age assessments. Am J Phys Anthrop 61: 467–471.
Paltauf H (1902) Der Zahn als Maßstab der Entwicklung des Menschen. In: Scheff: Handbuch der Zahnheilkunde II, 2.
Peirce CN (1884) Calcification and development of mandibular teeth. Dental Cosmos 26: 449–455.
Pöyry M, Nyström M, Ranta R (1986) Comparison of two tooth formation rating methods. Proc Finn Dent Soc 82: 127–133.
Prahl-Anderson B, Kowalski CJ, Heydendael PHJM (1979) A Mixed-Longitudinal Interdisciplinary Study of Growth and Development. Academic Press, New York.
Prokopec M, Dutkova L, Vignerova J (1988) Dospivani ceskych divek. (Summary: Maturation of Czech girls). Cs Pediatrie 43: 690–691.
Proy E, Sempé M, Ajacques JC (1981) Étude comparée des maturations dentaire et squelettique chez des enfants et adolescents français. Rev d'Ortho Dento-fac 15: 309–320.
Roche AF (1980) The measurement of skeletal maturation. In: Johnston FE, Roche AF, Susanne C (eds) Human Physical Growth and Maturation. Methodologies and Factors. Plenum Press, New York, p. 61–82.
Röse C (1891) Über die Entwicklung der Zähne des Menschen. Arch Mikr Anat 38: 447–491.
Röse C (1909) Über die mittlere Durchbruchszeit der bleibenden Zähne des Menschen. Dtsch Mschr Zahnheilk 27: 553–570.
Sapoka AM, Demirjian A (1971) Dental development of the French Canadian child. J Can Dental Ass 37: 100–104.
Saunders E (1837) The Teeth a Test of Age, Considered with Reference to the Factory Children, Addressed to the Members of Both Houses of Parliament. Renshaw, London.
Saunders S, DeVito C, Herring A, Southern R, Hoppa R (1993) Accuracy tests of tooth formation age estimations for human skeletal remains. Am J Phys Anthrop 92: 173–188.
Schopf PM (1970) Wurzelmineralisation und Zahndurchbruch im Wechselgebiß. Fortschr Kieferorthop 31: 39–56.
Schour I, Massler M (1940) Studies in tooth development. J Am Dental Ass 27: 1778–1793, 1918–1931.

Schour I, Massler M (1941) The development of the human dentition. J Am Dent Ass 28: 1153–1160.
Schützmannsky G (1957) Akzeleration und Zahndurchbruch. Dt Stomat 7: 404–409.
Seichter U, Lange W, Pfahr E, Schübel F (1980) Untersuchungen an 5899 Düsseldorfer Schulkindern über den Ablauf der zweiten Dentition – eine statistische Erhebung. Dtsch Zahnärztl Z 35: 291.
Seow WK (1995a) Dental development in amelogenesis imperfecta. A controlled study. Pediatr Dent 17: 26–30.
Seow WK (1995b) Dental development and molar root length in children with cleidocranial dysplasia. Pediatr Dent 17: 101–105.
Sicher H, Tandler J (1928) Anatomie für Zahnärzte. Julius Springer, Wien.
Skinner M, Anderson GS (1991) Individualization and enamel histology, a case report in forensic anthropology. J For Sci 36: 939–948.
Smith BH (1991) Standards of human tooth formation and dental age assessment. In: Kelley M, Larsen CS (eds) Advances in Dental Anthropology. Alan R. Liss, New York, p. 143–168.
Staaf V, Mörnstad H, Welander U (1991) Age estimation based on tooth development, a test of reliability and validity. Scand J Dent Res 99: 281–286.
Stack MV (1960) Forensic estimation of age in infancy by gravimetric observations on the developing dentition. J For Sci 1: 49–59.
Stack MV (1963) Retardation of foetal dental growth in relation to pathology. Arch Dis Child 38: 443–446.
Stack MV (1964) A gravimetric study of crown growth rate of the human deciduous dentition. Biol Neonat 6: 197–224.
Stack MV (1967) Vertical growth rates of the deciduous teeth. J Dent Res 46: 879–882.
Stack MV (1971) Relative rates of weight gain in human deciduous teeth. In: Dahlberg AA (ed) Dental Morphology and Evolution. Chicago Univ Press, Chicago IL, p. 59–62.
Sunderland EP, Smith CJ, Sunderland R (1987) A histological study of the chronology of initial mineralisation in the human deciduous dentition. Arch Oral Biol 32: 167–174.
Tanner JM (1962) Growth at Adolescence. Blackwell, Oxford.
Thorson J, Hägg U (1991) The accuracy and precision of the third mandibular molar as an indicator of chronological age. Swed Dent J 15: 15–22.
Tompkins RL (1996) Human population variability in relative dental development. Am J Phys Anthropol 99: 79–102.
Trodden B (1982) A radiographic study of the calcification and eruption of the permanent teeth in Inuit and Indian children. Archeol Survey Canada 112: 1–136.
Ubelaker DH (1978) Human Skeletal Remains. Excavation, Analysis, and Interpretation. Aldine, Chicago.
Valöík JA, Fábryková E (1964) Einige Beobachtungen über die Eruption der bleibenden Zähne in der Nordslowakei. Dtsch Stomatol 64: 263–274.
Verenich GI (1992) Health status of rural school children from Byelorussian Polesye: Twenty years shifts. Studies in Human Ecol 10: 155–163.
Weise W, Bruntsch E (1965) Röntgenologische Untersuchungen zum Nachweis und zur Entwicklung des Weisheitszahnes. Zahnärztl Rundschau 74: 205.
Wolanski N (1966) A new method for the evaluation of tooth formation. Acta Genet 16: 186–197.
Wolanski N (1983) Zmiany srodowiskowe a rozwój biologiczny czlowieka. Ossolineum, Wroclaw.
Wolanski N, Tomonari K, Chung S, Harada S, Januszko L, Liocheva V, Seiwa H, Tsushima S (1991) Comparative study on socio-economic and biological properties of families from Bulgaria, Japan, South Korea and Poland. Studies in Human Ecol 9: 151–166.

Zimmermann MR, Angel JL (1986) Dating and Age Determination of Biological Material. Billing, Worcester.
Zuckerkandl E (1891) Anatomie der Mundhöhle unter besonderer Berücksichtigung der Zähne. Hölder, Wien.

6.2 Dental Age in Adults – A Review of Estimation Methods

Friedrich W. Rösing and Sigrid I. Kvaal

Introduction

Time is a wicked thing. It has no tangible substance, and yet it has an extremely solid autonomy, advancing ever so constantly. And, most wickedly, its correlation to other things is often lamentably weak. This weak interdependence between time and many other natural phenomena is the cardinal reason for the difficulties which exist in human biology in assessing the two variables historic dating of human remains and age at death. In particular the problem of age estimation has been approached by many methods developed in the last decades – which illustrates that the existing techniques were always insufficient.

The objective of this contribution is to provide a survey of the possibilities for reconstructing the age at death of adult humans, including an evaluation of the sufficiency of the techniques available today. In keeping with the theme of this book emphasis is placed on teeth, but other tissues will occasionally be mentioned as well. There are several earlier surveys which deserve mention (e.g. Euler 1940; de Jonge 1950; Sognnaes 1960, 1977; Johanson 1971; Holm-Pedersen 1978; Altini 1983; Mjör 1986; Bang 1989; Solheim 1993a; Kvaal 1995); the present contribution lays emphasis first on the evaluation of the many methods for practical application and then on a complete compilation of the sources. The goal of a complete coverage of the literature was reached to the extent that at the end of the logistics of papers, just thirteen (7%) were not available; naturally this count cannot take into consideration hidden papers or closed circles of quotation from isolated schools, i.e. such work as does not appear in normally accessible literature, bibliographies, and literature bases.

The Array of Changes and Techniques

Teeth undergo the most conspicuous time-related change in children, in whom they are initially formed in spectacular events such as eruption, but in adults, too, temporal variability is considerable. At least eleven different change features may be distinguished: 1. number, 2. color and fluorescence, 3. attrition, 4. periodontal recession, 5. cementum apposition to the root, 6. root resorption, 7. secondary dentine in the pulp, 8. root translucency, 9. peritubular dentine, 10. racemisation, and 11. cementum annulation. These are the phenomena of ageing which have been quantified sufficiently and which do exhibit a significant and high time gradient. There are other phenomena which do not fulfill these criteria and which are therefore hardly suitable for application: changes in chemical composition (Bhussry and Emmel 1955; Brudevold 1957; Schranz 1960; Kósa et al. 1990), pulp nerve sheath changes (Armenio et al. 1956), decrease of vascular pulp structures (Bennett et al. 1965), formation of pulp stones (Saunders and Röckert 1967), pulp nerve calcification (Bernick 1967; see also Morse 1991), decrease in crown length (López-Nicolás et al. 1990, 1991), continuing "eruption" of mandibular side teeth (Whittaker 1992), resistance to etching (Yonan and Fosdick 1964), increase in hardness (Beust 1931; Kiesel 1973), specific weight (Kühns 1895; Förster and Happel 1959), density (Sutor 1937; Heuschkel et al. 1979; Dufkova and Branik 1983; Dufkova 1984a, b), and loss of water (Toto et al. 1971).

Most of the features represent biological ageing processes and, to a lesser degree, responses to tooth use (cf. Bödecker 1925; Euler 1940; Sognnaes 1960; Dalitz 1962; Miles 1963a, b; Port 1965; Gustafson 1966; Altini 1983; Ketterl 1983; Bang 1989; Whittaker 1992; Xu et al. 1992). The features are embedded in a complex living, working, and regulated system. The last two features (nos. 10–11), however, are of a different character, in that they seem to be autonomous chronobiological and chemical processes which are independent of tooth use. Several features are often combined because this normally results in higher accuracy (Maat 1987); this applies to the first nine only, the autonomous features 10 and 11 are new and have not yet been combined with others. The attempt to do so might pose difficulties, as no dispersion homogeneity is to be expected between them and the conventional ones. Only in one recent study (Kvaal et al. 1995) were the parameters to be combined selected by calculation (principal component analysis) rather than by decision.

Some of the age-related features have been known for a long time. Occasionally, however, attempts have been made to escape from the confinement of old knowledge by the formation of new features which are nevertheless of a classical character. An example is provided by Ito (1972, 1975). He developed the idea that attrition and secondary pulp chamber dentine apposition might be combined directly in one feature, and not only later during data processing. In sections of molars the enamel and the pulp areas were added (because both decrease during time) and standardised by the area of the unaffected crown dentine. The result was a rather low correlation of 0.63. Another very thorough attempt was a combination of all possible measurements of the pulp in order to arrive at a higher accuracy (Kvaal et al. 1995). Again the success was limited: The best single correlations were –0.77 and –0.73, and for the complete combination the correlations amounted to –0.83. Such results should be regarded as an indication that a new success for age diagnosis would have to come from outside the "classic" inventory, as it indeed happened with the introduction of the autonomous processes.

Little is known about the differentiation of these changes, except in one respect: In some of the newer presentations sex differences have been tested and were found to be negligible (e.g. Drusini 1995). Attrition is different, males tend to wear their teeth faster. None of the other usual anthropological differentiations (race, region, genetics; physique; life style, diet) have been tested directly.

There are several techniques for the assessment or quantification of these age features: 1. examination with the naked eye, 2. microscope, 3. digitalisation, 4. weighing, 5. sectioning a thin slice or halving the tooth, 6. non-destructive penetration like x-rays or ultrasound, and 7. chemical analysis. Some techniques may be combined with each other, e.g. count or digitalisation after section or radiograph. Digitalisation has been applied occasionally (Ricco et al. 1984; López-Nicolás et al. 1990).

Processing Age-Dependent Features

Although some of the studies which will be cited here clearly give the impression of art for art's sake, the only real goal of method quantification is the application in biohistory and in forensic anthropology and odontology. For that purpose the value of the methods must be assessed. This is best done via the standard error (m or s(y.x) or S.E.) of the regression of a variable to time. It represents the dispersion of the individual observations around the regression line. The single error encompasses 68% of the observations in the reference sample, the double error roughly 95%, and the triple error 99.72%. These percentages represent the theoretical correctness expectation for the application to an individual diagnosis. The quality of a method may also be assessed by its application to a test sample. Such external tests are regrettably rare (most recent exception: Kemkes-Grottenthaler 1996).

Besides the error m, the correlation coefficient r also indicates the value of a method, but in a more general and theoretical sense. Generally, more studies give the correlation than the error (see table below). Moreover, the error varies much more, probably due to different calculation methods. Often this cannot be controlled because the equation used is not indicated. For these practical reasons r is the better assessment variable and not the theoretically superior m.

Some papers, particularly those of older date, do not give the standard error m but the percentage of reliability for a certain year span, e.g. 48.8% for ± 5 y (Pilin 1981). By means of the variable z of the normal distribution (see the respective table in most statistical handbooks) this may be converted to m = 7.63.

The statistical confidence interval has an eminently practical meaning: If a method reveals an error of 10 years (which is an average value) and if an application diagnosis yields an age of 50 years (which is also an average assumption), then the single range is 40 to 60 years. The probability of being correct is only 68%. Particularly in forensics much higher precision is needed, namely the triple error range; this is the consensus in many fields (e.g. paternity diagnostics or skeletal identification). In this example it means that only a span of 20 to 80 years has a satisfactory correctness probability of above 99% – and this is the complete normal adult life span which has been given to humans by God and genetics.

A less unfavourable example, in fact one of the most favourable ones for the present state of the art of age estimation: Let the age by the cementum annulation method (see below method 11) of an unknown female body be 28, the standard error 3 years, all epiphyses closed. Then the correctness probability of 99.72 holds for an age range of 23 to 37 years.

This allows of an assessment: Methods with a standard error of the regression of more than five or seven years are not suited for routine forensic application, and they ought to be avoided in biohistory, too. Their use is justified only when better methods are not at hand, e.g. when equipment or time is unavailable or when the necessary tissues are not preserved.

Besides mean statistical estimation errors, there are phenomena of bias, i.e. systematic error. A first bias may originate in the lack of a distinction by sex: Many biological features have a dimorphism which causes every estimate to have an additional error. For example attrition (see below, method 3): Males grind down their teeth faster than females. When an equation disregarding sex is applied, then an age diagnosis for a male will tend to be too high and for a female too low. A second source of bias is the general tendency for the age distribution in an application group to imitate the reference group distribution (Bocquet-Appel and Masset 1977). If the reference group follows the modern age distribution with a clear dominance of the older ages, then historic groups will tend to give the same, though false, picture. A third source of bias is the regression towards the mean which is a wide-spread biological phenomenon: The ages of young persons tend to be overestimated and those of old ones underestimated. This has been shown for combined methods (Seifert 1958; Solheim and Sundnes 1980; Nkhumeleni et al. 1989) and for a comparison of secondary dentine deposition (method 7), root translucency (method 8), and osteological criteria (Kvaal et al. 1994b). A variant of this is the constant increase of the dispersion around the regression line with advancing age (most clearly shown in Helfman and Bada 1976), and another variant is an undescribed non-linear regression line (to be found in Dalitz 1962). A fourth source of bias is the astonishingly frequent application of suboptimal feature definitions (e.g. the use of age or trait classes or Gustafson's score system instead of a complete quantification), incorrect statistics (e.g. May 1952 first calculated class means and then the regression), and the frequent calculation errors. Of course, if a paper does not contain either of the bivariate parameters correlation and regression, then this is not a source of bias for an application but it excludes any application (see also below, limitations). A fifth source of bias is the disregard of the exact type of tooth tested; some authors think this has no influence on results (e.g. Johanson 1971 or Drusini 1995), but those tests which differentiate according to tooth type do find substantial differences (e.g. Dalitz 1962; Kvaal et al. 1995). Finally, a sixth bias is caused by the disregard of the state of health of the teeth investigated. Most teeth are extracted because they are involved in a disease process which usually influences the normal development – but it is unknown to what extent and for how long (Kvaal et al 1994a; Kvaal and Solheim 1995). As a general conclusion it must be stated that the statistical error is an underestimation; because of bias the empirical error, i.e. the only one which counts for a specific application case, is normally greater, but it is rarely known to what extent. At any rate, in a presentation of an age diagnosis before a court of law this discrepancy must be taken into account and must be openly stated.

It is only in a few investigations of a rather recent date that the linearity of the regression was checked (e.g. Bang and Ramm 1970; Drusini et al. 1989; Rosenberg and Rösing, see Fig. 1 below). In these investigations non-linear regressions described the data better than linear ones.

The processing of the data in a regression analysis is the next most important tool after univariate parameters. Without the regression equation or line a test sample cannot be applied to new cases. There is, however, a superior method: factor analysis or its derivative, principal component analysis. In cases of larger studies using many tooth features it gives objective indications of the differential value of the features. Thus features which are useless as receiving no significant residual influence from age may be excluded, or others combined (Kvaal et al. 1995).

Most authors of specific methods give only the correlation coefficient r between their chosen variable and age. For its conversion into the crucial standard regression an estimation equation may be used (Bocquet-Appel and Masset 1977).

Limitations of the Sources

There are hundreds of papers dealing with age and teeth, but many of them have some limiting properties. The most bewildering limitation is a lack of statistical processing. Dental materials have been prepared, measurements taken or observations made – but then at best univariate parameters are calculated, yet none of the bivariate parameters which are indispensable for quality assessment and application (Brudevold 1957; Seifert 1958; Bennett et al. 1965; Bernick 1967; Bang and Monsen 1968; Čechova and Titlbachova 1971; Čechova 1972; Dufkova and Branik 1983; Dufkova 1984b). It is equally insufficient to classify age and then compare the means of the classes (as done e.g. by Henry and Weinmann 1951). Statistical insufficiency has not only occurred in some distant dark ages, but also more recently (e.g. Philippas and Applebaum 1967, 1968), and it means that the results cannot be used for application or quality assessment. Therefore this survey excludes them.

Another limitation is the testing of new preparation methods which then turned out to yield weak results (Miller et al. 1988; Kvaal and Solheim 1995, see below method 11). This problem, however, should not lead to exclusion, but should be discussed and the result finally included in recommendations on how to proceed.

From a statistical viewpoint recommendations for individual numbers may be given: If data collection is difficult (as sometimes in clinical studies or with fossils), the minimum group size should be 30 or 50; a good size is 100 when univariate or bivariate parameters are intended to be given. A sample size of 1000, on the other hand, is sheer luxury if no subdivision is made. If a multivariate analysis is intended, then the number of individuals should be at least ten times larger than the number of effects aimed at or factors to be extracted. However, these are only vague rules, and in the framework of this chapter there is an important exception: If tests are possible such as checking for a significant difference of a correlation from zero, then the number of individuals is included in the test and does not need to be regarded separately. Moreover, the major problem of insufficient group size, large chance variation of the resulting parameters, does not play too large a role in dental age diagnostics, because for every important phenomenon there are several results which balance the few meagre studies. For these reasons group size was not used as an exclusion criterion.

Of course, a limiting factor for the inclusion of studies is the knowledge of the chronological age of the teeth used. However, in odontology this is no problem because identified teeth are easily available. One study with historic, non-identified material has nevertheless been included (Mays et al. 1995), because it used juvenile tooth age for evaluating attrition; naturally, this adds a considerable statistical error. Another problem is that group size is the number of individuals and not the number of teeth; it is not correct to count several teeth from an individual dentition as independent cases. This has been done by some of the classical authors (e.g. Gustafson 1947 or Bang and Ramm 1970). In a larger number of investigations this may be suspected but cannot be checked, because the author only gives an insufficient description of his sample.

1. Number of Teeth in the Dentition

This variable deserves first mention for its simplicity and its age: Since time immemorial it has been used for mammals, particularly horses ("Don't look a gift horse in the mouth"). In their case it is the first choice for fast and simple visual evidence of age, whereas in humans facial skin alteration is first choice. In the sixties of this century tooth counting was rediscovered for humans and by science. There are more than two dozen studies quantifying this (summarized in Endris 1979). The largest, most recent, and most illustrative investigations are by Hoefig (1964), Lindemaier et al. (1989), and Rosenberg and Rösing (hitherto unpublished). From the data of the last-named study a smoothed non-linear regression was calculated and graphed (Fig. 1). The standard error of 10 or 12 years is normal as compared to other classic and simple estimators, but excessive for a wide application. A good example for forensic application is the isolated edentate mandible, for which an age estimate of 72 ± 12 years (one standard error) can be given.

Closely related to the number of teeth is the amount of caries, because a large majority of teeth are lost through caries. The problem is not so much the uneven age distribution or the individual, geographic, and above all secular variation (as stated e.g. in Bang 1989), but the fact that there has been no quantitative development of this feature yet. In the life of a dentition a carious tooth is just a short transitional state, the final state being the loss of the tooth. So for the purpose of quantification carious and lost teeth would have to be taken together. This feature should then have a correlation to time comparable to the number of teeth alone.

2. Tooth Color and Fluorescence

This too is an ancient variable: It has long been generally known that the teeth gradually become more yellowish and brownish towards old age (e.g. Rheinwald 1966; Tsuchiya 1973; Endris 1979; Burchett and Christensen 1988). The reasons are degradation of organic material in enamel and cementum, deposition of external substances in the tooth tissues, mineralisation of dentine, and apposition of

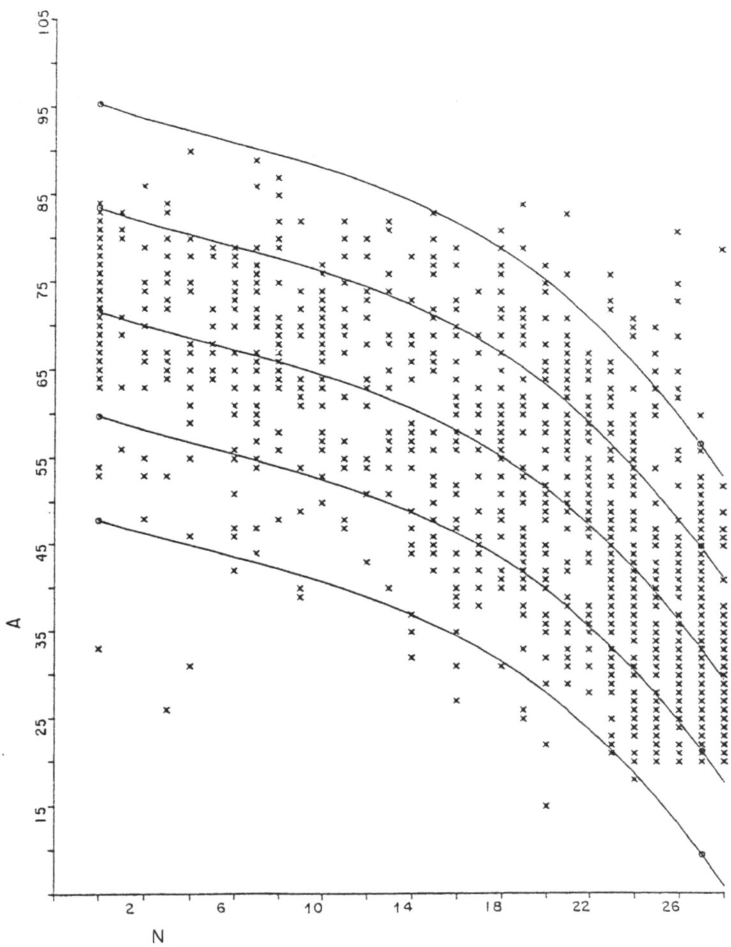

Fig. 1. Number of teeth (N) in a dentition and age (A). Non-linear regression line (polynomial of 3rd degree), n = 1120 dentitions from an Ulm dental practice, r = 0.567, m = 12.0.

cementum (see short review in Bang 1989). The correlation with age is high, fully comparable to other features. For the root there is an applicable scientific quantification by Ten Cate et al. (1977), who gave a correlation of r = 0.9. The best elaboration so far is by Solheim (1988a), who used one of the many comparative dental shade guides for scoring. In that study, color was found to be sufficiently correlated with age (0.59 – 0.84); it forms a part of a combined method (Solheim 1993b, see respective subheading below). Application is more or less limited to forensics; in archaeological material discoloration, especially soil darkening, overrules the age gradient.

A newly discovered property is the age-related increase of red fluorescence of the halved tooth under green light (Kvaal and Solheim 1989). The correlations (in dentine 0.73, in cementum 0.77) are highly significant but low in comparison to other estimators, though comparable to color.

3. Attrition

Attrition, the gradual loss of tooth substance from the functional contact zones between antagonistic teeth during chewing, is the necessary consequence of normal mastication, because no diet is really and completely soft and because even the extreme hardness of enamel is limited. The process of grinding off tooth surfaces is variable, depending on several physical factors, such as the material properties of diet and teeth and the dynamic properties of chewing – but it is also clearly dependent on time length of use. This is why attrition may be used for an age assessment as in the first scientific proposals of the last century (Broca 1879; Baume 1882).

A classic elaboration of tooth attrition for the age estimation of juveniles and adults is by Miles (1963a, b). It is based on a simple and elegant principle: In a pre-modern population, i.e. without secular acceleration, the first permanent molar erupts around an age of six years, the second around fourteen. This difference leads to an additional eight years' worth of attrition in the first molar: The exact attrition difference between the molars indicates the individual's speed of attrition; a large difference means a high speed. The amount of attrition of the second molar then allows the estimation of age.

This principle does not work well in reality, often the results grossly deviate from others (see also Santini et al. 1990); particularly, under-estimations seem to occur easily. One reason might be the insufficient quantification. The core of Miles' work was a pictorial gradation of molar attrition (redrawn several times by secondary authors). This meant the fixation of the dynamic principle to a certain medium state. The main problem, however, is the considerable natural and cultural variation of the hard components in diets in interaction with the properties of the dentition, and this applies to the subsistence and population levels as well as to the individual level and even to the lower level of limited time periods in the ontogenesis of a single dentition (Euler 1939; Molnar 1971; Johnson 1976; Kieser et al. 1983; Smith 1984; Schürmann 1986; Walker et al. 1991). Miles' method did not sufficiently quantify this complex variability.

Other elaborations did quantify. Sometimes attrition was used only in combination with other features, which does not allow an assessment of accuracy (Gustafson 1947; Matsikidis 1981; Matsikidis and Schultz 1982). The available figures average out at a single error of 10.7 years, thus suggesting that attrition should only be used as a rough guide and not in cases or groups of minor importance. Specific error sources which account for the low accuracy are variable diet composition, bruxism and other personal habits, occlusion anomalies, and the use of teeth for other functions. Moreover, the third molar is much more variable than the others (Tomenchuk and Mayhall 1979; Richards and Miller 1991; Mays et al. 1995), it should not be used.

In the case of Lovejoy's method (1985), although developed only on the basis of a prehistoric group of unknown ages, the external accuracy has been tested by application to two other prehistoric populations (Kemkes-Grottenthaler 1996): In comparison with several other bone methods attrition strongly underestimates age and yields significantly different distributions for the sexes and the jaws. These results mandate high caution.

Attrition may also be measured indirectly by crown length (axial direction). Of course the resulting correlations are poor, varying around 0.4, because no information on the original crown length was included. Ito (1972, 1975) created a new and more complex concept: Attrition does not stand alone, but inseparably occurs together with the constriction of the pulp cavity. Consequently he formed an index of the coronal pulp area to the coronal dentine area in radiographs. Others followed this idea (Takei 1984; Wei and Feng 1984; Feng 1985; Drusini 1993), sometimes with modifications. When the components attrition and pulp constriction are treated separately, constriction has the major age gradient and not attrition (Solheim, personal communication).

4. Periodontal Recession

The natural fixation of teeth to jaws, the periodontal attachment, is dependent on age: There is a gradual recession of gingival borders and consequently of alveolar bone. Dentists regard this to be the consequence of inflammation or malfunction – but it occurs also in their absence as a function of time (Bödecker 1925), possibly as a compensation for dental tissue loss due to attrition or abrasion (Osborn and Ten Cate 1983). This retraction of the periodontal attachment has been used, normally in conjunction with other parameters (Gustafson 1947; Matsikidis 1981; López-Nicolás et al. 1991; Lamendin et al. 1992; Solheim 1993b). Only one source (Solheim 1992b) gives a separate correlation (the mean of four surface positions and ten teeth: $r = 0.31$).

5. Cementum Apposition

In old individuals and in strongly reduced dentitions the root has often more of a club shape than the cone shape of the young. This is the result of a continuous slight apposition of cementum, which in the end may lead to a highly conspicuous hypercementosis. The effect was discovered by Magitot (1878; see also Zander and Hürzeler 1958). In extreme periodontitis with a highly reduced alveolar crest, apposition accelerates in a physiological adaptive attempt to anchor the tooth in the jaw, despite an ever decreasing contact area.

Again this is a variable which is normally employed in combined methods (Gustafson 1947; Matsikidis 1981; Popov et al. 1992; Solheim 1993b). In three sufficiently large studies the continuous cementum apposition was studied alone. The correlations to age vary around 0.5, giving the impressive standard error of about 12 years.

6. Root Resorption

In some teeth the surface of the root is resorbed in minute areas, producing a rough appearance. This seems to be increasingly the case in older teeth. The first expert to use this feature was Gustafson (1947) in his combined method. Later it was held that the correlation is low (Henry and Weinmann 1951; Dalitz 1962; Saunders 1965; Miles 1976; Matsikidis 1981; Solheim and Kvaal 1993) or even negative (Maples 1978). Although some studies found a significant difference from zero, the contribution of resorption to age variance is unimportant. To reverse the argument: There must be other and more important factors than age which steer resorption. The most recent study (Solheim and Kvaal 1993) demonstrated, however, that despite this poor age correlation, resorption does contribute usefully to a multiple regression method of estimating age, but to a lower degree than do other features.

7. Secondary Dentine

The living tissues of teeth change and grow: Not only does the cementum layer increase (see above, 5 and below, 11) or become irregular (6), but also the dentine walls of the pulp form appositional tissue (Bödecker 1925). The main biological formative factor should be normal use: mechanical stress during mastication and abrupt temperature changes; yet this has also been doubted (Philippas 1961). The apposition is strongest in the coronal roof of the pulp, close to the stress source. Moreover, dentine is formed more rapidly when attrition approaches the pulp. The nature of these processes implies a high individual variation, i.e. the correlation between apposition and age should not be too high.

It is difficult to differentiate primary from secondary dentine. The principal method of assessing secondary dentine is an indirect method: the size reduction of the pulp cavity. This reduction may be quantified by linear measurements on the thin slice or on the halved tooth (Gustafson 1947; Dalitz 1962; Johanson 1971; Kiesel 1973; Solheim 1992; Drusini 1993). Variants are the pulp area relative to crown area (1) on sections (Ito 1972, 1975; Wei et al. 1983, López-Nicolás et al. 1990) or (2) on radiographs (Matsikidis 1981; Popov et al. 1992; Drusini 1993), (3) the ratio between width and length of the pulp on radiographs (Wei and Feng 1984; Kósa and Antal 1989), or (4) pulp width or length in relation to tooth width or length on radiographs (Kvaal and Solheim 1994). A digitalisation of area parameters in principle reduces accuracy (but improves speed and observer stability). Consequently the attempt of López-Nicolás et al. (1991) led to a high error (10.2 y).

Correlations range from 0.21 to 0.89, averaging out at almost 0.6. Frontal teeth tend to give higher correlations than lateral teeth, and coronal measurements tend to give higher correlations than apical measurements.

8. Root Translucency

During the life of a tooth the dentine becomes more and more calcified. Starting from the apex, a sclerosis of the dentine tubules occurs (Beust 1931; Pilz 1959; Bang and Monsen 1968; Hess 1970), resulting in more homogeneous dentine which makes the root translucent like milk glass (often called transparent). It is most evident in a thin section of a tooth but also visible in the intact root; it disappears in a decalcified tooth. Despite former assumptions (e.g. Vlček and

Mrklas 1975), it may be visible in historic excavated teeth (Kvaal et al. 1994b; Stermer et al. 1994). It has also been observed in a deciduous upper first incisor which persisted to the age of eleven (Lamendin 1972). The length of the translucent zone covaries with the chronological age of the individual; the correlation does not increase when tooth age is used (Bang and Ramm 1970). Although discovered already in the middle of the last century (Bang 1989), the nature and cause of the process is not sufficiently clear: E.g. there is only a loose connection between sclerosis and translucency (Brinkmann and Hartmann 1979; Vasiliadis et al. 1983). Among the classical variables translucency is rather autonomous, it is little influenced by diseases and environment; only one observation suggests an influence from advanced periodontitis (Pilz 1959). Therefore it is supposed to be useful as a single age criterion (Bang 1972, 1989), which is supported by rather good correlations of about 0.7. This is challenged, however, by some low or mediocre correlations (e.g. Johnson 1968: 0.28, not significantly different from zero) and impressive standard errors (e.g. Wegener and Albrecht 1980: 15.3 y).

Most studies have been performed after tooth section. However, the feature is so simple and conspicuous that it can be assessed by the unassisted eye. The measurement criterion for the length of the transparent zone is then the mean height of the borderline. Since this borderline is not straight, Sognnaes et al. (1985) proposed to digitalise the transparent portion; although at that time only two-dimensional digitalisation was accessible, the proposed method arrived at a certain 3D measure by rotation of the tooth and repeated area determination. This complicated method has never been applied to a series of age-identified teeth, as has only normal 2D digitalisation (López-Nicolás et al. 1990, 1991; Drusini et al. 1991).

9. Peritubular Dentine

A factor which might contribute to the increasing translucency of root apices is the gradual filling of the dentine tubules or canals, which may be observed in the scanning electron microscope. The sclerosed tubules appear to have a different texture and consistency from that of the intertubular dentine, but the deposited crystals are the same (Nalbandian et al. 1960). For this phenomenon early studies found weak (Dreyfuss et al. 1964) or no correlations to age (Lamendin 1972). Endris (1979) seems to have been the first to propose the actual measurement of the tubule diameter. A significant difference in tubular diameter between two age-groups has been demonstrated (Traub et al. 1988), but the thickness of the intertubular dentine was found to be a stronger parameter (Kvaal et al. 1994a). The quantification gave a correlation (0.69) which is too low for general applicability. This might be different if the variable were included in a combined method.

10. Racemisation

Some of the stereo-isomeric biochemical compounds (e.g. aspartic acid) have an optical property: They turn polarised light towards the left. This L version is the only natural one. After formation there is a gradual transition to D versions, which turn the light towards the right. The final result after prolonged periods of time is a racemic mixture where both versions occur equally often. This effect has been widely used as a dating method for the historic time scale – and it is known to be a problematic one, because racemisation as a chemical process is heavily dependent on temperature and in most samples temperature conditions are not well known for the whole time since death.

The temperature problem is less pressing when the racemisation effect is used for the diagnosis of age at death in forensic sciences: The temperature conditions of teeth in living persons are known and rather stable (Helfman and Bada 1975, 1976). When applied without delay or in the living, the error of the method is low: on average 2.6 years, see Table 1 below. The normal correlation to age in most reference samples lies above 0.99. The correlation decreases and the error increases with increasing time since death (see the extreme but rather untypical single case by Masters (Helfman) 1986). So the method is unsuitable for historic specimens and for forensic cases where the corpse has been exposed to raised temperatures. In forensic applications dentine turned out to be more accurate than enamel. When using enamel with its more unstable temperature conditions, the correlation decreased from 0.979 to 0.921 (Helfman and Bada 1975) or from 0.995 to 0.928 (Ohtani and Yamamoto 1992). The initial studies were performed with amino acid analysers or gas chromatography – expensive procedures. Mörnstad et al. (1994) demonstrated that the cheaper HPLC (high pressure liquid chromatography) suits just as well.

Forensic sciences needs age diagnosis not only for dead bodies but also for the living, namely when decisions are to be made on whether a suspect is to be treated by juvenile or by adult procedure and when a person claims to have a higher age than documented and would therefore be entitled to start a pension. In these cases the extraction of a tooth for the racemisation investigation is at least problematic. A solution might be trepan biopsy of a dentine cube of 1 x 1 mm (Ritz et al. 1995).

11. Cementum Annulation

Mammals appose a thin layer of cementum, consisting of both a darker and a lighter zone, around their roots every year, comparable to the growth dynamics of temperate zone trees. The biology of this growth, i.e. the dynamics and reasons of formation, is virtually unknown, and there is little knowledge of the physical and none of the chemical properties of the two or possibly more different tissue fractions. As a phenomenon, however, the rings are well known and applicable to age diagnosis, and it seems that they represent a chronobiological process, i.e. an endogenous rhythm matched to the annual cycle.

This cementum annulation (lat. annulus = ring) was discovered in game, and also the first applications to humans were proposed and performed by game biologists (Stott et al. 1981). Later it was confirmed several times (see list below), and Großkopf (1990) demonstrated that annulation is also present in very old cremated teeth. There seems, however, to exist the biological irregularity that some humans disobediently appose two layers every year. If this is so, then it is necessary to apply one of the other age diagnosis methods, in order to detect those doublers. Moreover, in older humans layers might be skipped irregularly.

The technique is of a histological character: A thin slice of root is sawn or ground, usually horizontally through the root, but a longitudinal section is also possible. Obviously the section thickness is crucial; Miller et al. (1988) did not find a sufficient correlation between ring number and known age, probably because their sections were much thicker ("200 to 400 µm") than the optimal 100 µm (cf. Lorton 1988; Renz and Radlanski 1995). The section is inspected by light microscopy. Besides this wide-spread preparation technique others may be applied: Several of the numerous available staining techniques give good visibility after decalcification (Kvaal et al. 1996). When the decalcification is applied to a sample, however, the resulting correlation of rings to age is considerably lower than with the other techniques (Kvaal and Solheim 1995). Probably also the magnification, technical properties of the microscope, and the adjustment are important. Lipsinic et al. (1986) produced an increasing underestimation with advancing age, and Kvaal and Solheim (1995) found an observer sensitivity.

All rings consisting of a dark and a light zone are counted, including the last dark ring at the surface. An age is obtained by adding the developmental age of the exact section position. Example: If a first upper premolar of a male was cut directly at the crown, an age of 6.6 years should be added (Herdeg 1992, see also the chapter by Liversidge et al. on subadult tooth age in this volume); if the same tooth was cut in the middle of the root, an age of 8.5 years should be added. No sample published so far was adjusted according to this ontogenetic method; instead fixed values for eruption age or for age at root completion were added. This discrepancy might explain some low correlations.

The accuracy of the cementum ring count method can be assessed well because most authors calculated the mean empirical error or the standard error of regression. They range between 1.3 and 10.8 years and average at 5.1; the average correlation is 0.85. Possibly undetected doublers, an unsuitable eruption age adjustment, counting technique and other method variants raised some of the errors. So at the present state of incomplete knowledge it might be hoped that the real error will amount to less than 3 years. More might be known when some present research projects will yield their results.

Combined Methods

Even the most elegant and precise methods produce stochastic errors. Therefore it is advisable to combine several age-dependent processes into one method. The first and best-known elaboration of this kind is by Gustafson (1947, 1950, 1955). In fact this was the first combined age diagnosis method ever. Six variables are determined in a longitudinally sectioned tooth and assessed according to a scoring system. This is a suboptimal processing method (Metzger et al. 1980); when multiple regressions are calculated instead, the correlation to age increases from 0.805 to 0.934 in the same sample (Burns and Maples 1976). Moreover, recalculation of Gustafson's data revealed that his regression line was wrong

(Maples and Rice 1979, Nkhumeleni et al. 1989). Gustafson's idea has been quantified again several times (see list below).

In a direct comparison (Solheim and Sundnes 1980) between combined methods (by Dalitz 1962 and Johanson 1971) and single variable methods (Miles 1963a; Bang and Ramm 1970) differences in reliability appear – but they are astonishingly small. The best procedure (Johanson's) has a mean empirical error in the 100 tested teeth of slightly above +1 year, but the absolute error disregarding signs amounts to 8.3 years, suggesting high caution in any single diagnosis case of importance. All methods produce an impressive regression towards the mean: Ages below 40 are overestimated, above 50 underestimated. In general the goal of an error reduction has been reached by variable combinations: In 22 reference samples the mean error amounted to only 8.0 years, clearly below that for any of the single classic parameters. Nevertheless the combined tooth feature methods have rarely been applied, in forensic and more particularly in historic material, even in times when more precise methods were clearly not available.

Evaluation of the Methods

As stated in the introduction, a direct evaluation of the applicability of methods may be derived from the confidence interval m of the reference sample. A similar but less illustrative measure is the correlation coefficient r. Moreover, it is important that a method be based on a large individual number n in the reference sample. So these three parameters will be summarized in the following table, in which only those methods have been included for which either m or r is known. When several values have been reported, the mean was calculated for Table 1 (C – variable used only in a combined method).

Table 1. Methods of age estimation

Variable	Technique	Author	n	r	m
1 Number of teeth	Sight	Lindemaier et al. 1989	11374	–	10
		Rosenberg and Rösing unpubl.	1120	–	12
2a Tooth color	Densitometry	Ten Cate et al. 1977	36	0.90	
	Comparison	Ten Cate et al. 1977	26	–	11
	Table	Solheim 1988a, 1993a, b	758	0.82	
2b Fluorescence	Measurement	Kvaal and Solheim 1989	100	0.75	
3a Attrition	Sight	Philippas 1961	93	0.63	
		Takei 1970: 1 tooth	200	0.88	
		Johanson 1971	165	0.49	10.8
		Takei et al. 1981: maxilla	200	0.89	
		Takei et al. 1981: mandible	200	0.87	
		Takei et al. 1981: one side	200	0.90	
		Takei et al. 1981: all 28 teeth	200	0.93	
		Tomenchuk a. Mayhall 1979: M1	85	0.84	9.2
		Tomenchuk a. Mayhall 1979: M2	85	0.79	10.9
		Tomenchuk a. Mayhall 1979: M3	85	0.54	23.4

Table 1 (continued)

Variable	Technique	Author	n	r	m
		Takei 1984	1000	0.81	6.2
		Wei and Feng 1984	97	0.26	14.3
		Zhang and Ji 1988	262	0.91	
		Song and Jia 1989	880	0.71	
		Richards and Miller 1991	119	0.80	10.0
3b Attrition	Section	Tomaru et al. 1993	83	0.61	
		Solheim 1988b	758	0.55	
3c Attrition	X-ray	Matsikidis 1981	40	C	
3d Crown height	X-ray	Ito 1972, 1975	302	–0.59	7.3
		Kambe et al. 1991	147	–0.55	
		Kvaal et al. 1995	100	–0.17	
3e Crown height	Section	López-Nicolás et al. 1990	173	–0.20	
3f Crown height	Whole tooth	Mays et al. 1995	87	–0.56	
4a Periodontal recession	Section	Johanson 1971	165	0.49	11.9
		López-Nicolás et al. 1990	173	0.18	
		Solheim 1992b	758	0.52	
4b Periodontal recession	Whole tooth	Kvaal and Solheim 1994	452	0.28	
4c Periodontal recession	X-ray	Matsikidis 1981	40	C	
5a Cementum apposition	Section	Zander and Hürzeler 1958	133	–	15.5
		Dalitz 1962	146	0.33	
		Johanson 1971	165	0.55	11.5
		Solheim 1990 (mean multiple r)	758	0.67	
5b Cementum apposition	X-ray	Matsikidis 1981	40	C	
6a Root resorption	Section	Dalitz 1962	146	0.40	
		Johanson 1971	165	0.24	13.3
		Solheim and Kvaal 1993	758	0.56	
6b Root resorption	X-ray	Matsikidis 1981	40	C	
7a Secondary dentine	Section	Philippas 1961: pulp roof	93	0.21	
		Philippas 1961: pulp floor	93	0.68	
		Philippas 1961: pulp wall	93	0.51	
		Dalitz 1962	146	0.61	
		Moore 1970	200	0.62	
		Johanson 1971	165	0.66	10.3
		Lamendin 1988	23	0.64	
		López-Nicolás et al. 1990	173	0.19	
		Solheim 1992a	758	0.76	
7b Secondary dentine	X-ray	Philippas 1961: pulp roof	93	0.21	
		Philippas 1961: pulp floor	93	0.68	
7c Pulp width	Section	Moore 1970	200	–0.62	
		López-Nicolás et al. 1990	173	–0.24	
7d Pulp width	X-ray	Philippas 1961	93	–0.41	
7e Pulp height	X-ray	Philippas 1961	93	–0.75	
		Kvaal et al. 1995	100	–0.59	
7f Pulp area	Section	Ito 1972, 1975	302	–0.59	7.3
		Haertig and Durignon 1978	69	–	15.0
		Wei et al. 1983	97	–0.62	12.5
		Feng 1985	91	–0.63	12.3

Table 1 (continued)

Variable	Technique	Author	n	r	m
		Lamendin 1988	23	–0.64	
		López-Nicolás et al. 1990	173	–0.17	
		Drusini 1995	166	–0.89	
7g Pulp width index	X-ray	Ito 1972, 1975	302	–0.11	
		Kósa and Antal 1989	65	–0.78	
		Kvaal and Solheim 1994: upper	452	–0.61	
		Kvaal and Solheim 1994: mid	452	–0.46	
		Kvaal et al. 1995: upper	100	–0.69	
		Kvaal et al. 1995: mid	100	-0.63	
		Kvaal et al. 1995: lower	100	–0.59	
7h Pulp height index	X-ray	Kvaal and Solheim 1994	452	–0.44	
		Kvaal et al. 1995	100	–0.64	
7i Dentine thickness	X-ray	Kiesel 1973	142	0.71	
		Philippas 1961: mesial	93	0.51	
		Philippas 1961: distal	93	0.37	
8a Root translucency	Section or half tooth	Miles 1963a, b	118	0.73	12.0
		Johnson 1968	27	0.28	
		Bang and Ramm 1970	265	0.76	10.8
		Lamendin 1972	217	0.45	
		Falter 1974	43	0.93	
		Hiemer 1975	103	0.83	
		Solheim and Sundnes 1980	68	–	9.9
		Wegener and Albrecht 1980	50	0.67	15.3
		Kósa et al. 1983	93	0.86	
		Colonna et al. 1984	65	0.92	7.1
		Lamendin 1988	65	0.55	
		Solheim 1989, 1993b	758	0.79	
		Drusini 1989: linear regression	311	0.76	7.7
		Drusini 1989: non-linear regression	311	0.81	6.8
		López-Nicolás et al. 1990	173	0.35	
		Drusini et al. 1990	46	0.55	7.3
		Drusini et al. 1991	134	0.83	12.8
		Drusini 1995	46	0.52	7.4
8b Root translucency	Whole tooth	Bang and Ramm 1970	265	0.77	10.9
		Lamendin 1972, 1973, 1978	217	0.45	
		Drusini 1991	295	0.85	10.5
		Drusini et al. 1991	36	0.84	10.7
		Kvaal and Solheim 1994	452	0.45	
		Drusini 1995	295	0.77	10.2
9. Peritubular dentine	Half tooth	Kvaal et al. 1994	50	0.69	
10. Racemisation	Chemistry	Helfman and Bada 1975	19	0.92	
		Helfman and Bada 1976	20	0.98	
		Ogino et al. 1985	74	0.99	4.0
		Ohtani and Yamamoto 1987	221	0.99	1.9
		Ohtani and Yamamoto 1991	27	1.00	1.0
		Ohtani and Yamamoto 1992	16	1.00	1.4
		Yamamoto 1992	14	0.99	

Table 1 (continued)

Variable	Technique	Author	n	r	m
		Ritz et al. 1993	70	0.99	2.3
		Mörnstad et al. 1994	32	0.97	6.2
		Ritz et al. 1995	62	0.99	2.9
11. Cementum annulation	Section	Stott et al. 1981	3	–	1.3
		Condon et al. 1986	80	0.78	6.9
		Lipsinic et al. 1986	31	0.84	8.0
		Charles et al. 1986, 1989	55	0.86	7.4
		Großkopf 1990	36	–	3.2
		Jacobshagen and Kunter 1992	7	–	4.1
		Stein and Corcoran 1994	42	0.93	
		Kvaal and Solheim 1995	95	0.84	10.8
		Kvaal and Solheim 1995: healthy teeth	39	0.88	
Combinations:					
3b+4a+5a+6a+7a+8a		Gustafson 1947	41	0.91	3.6
3b+4a+5a+6a+7a+8a		Nalbandian 1959	40	0.87	6.0
3b+4a+5a+6a+7a+8a		Dechaume et al. 1960	100	–	3.9
3b+4a+5a+6a+7a+8a		Burns and Maples 1976, Maples 1978	168	0.81	9.3
3b+4a+5a+6a+7a+8a		Haertig and Durignon 1978	69	–	12.0
3b+4a+5a+6a+7a+8a		Kilian 1986, Kilian and Vlček 1989	116	–	2.6
3b+4a+5a+6a+7a+8a		Lamendin et al. 1992	39	–	14.2
3b+4a+7a+8a		Dalitz 1962	146	0.87	6.0
3b+4a+7a+8a		Dalitz 1962: control group	29	0.88	
3b+4a+7a+8a		Solheim and Sundnes 1980	68	–	9.1
3a+4b+5a+6a+7a+8b		Johanson 1971	165	0.92	5.2
3a+4b+5a+6a+7a+8b		Solheim and Sundnes 1980	68	–	8.2
3b+5a+6a+7a		Vlček and Mrklas 1975: 1 tooth	41	–	8.0
3b+5a+6a+7a		Vlček and Mrklas 1975: 5 teeth	41	–	3.6
7e+8a		Schwarz et al. 1978	151	0.94	
7e+8b		Schwarz et al. 1980	68	0.70	
3c+4c+5b+6b+7b		Matsikidis 1981	40	–	5.6
1+3a+4b+8b		Pilin 1981	81	–	7.6
3b+7f		Wei and Feng 1984	97	0.86	7.7
3d+4a+7c+7f+8a		Lopez et al. 1990	126	0.43	
3d+2x7d+7e+7i		Popov et al. 1992	420	0.93	6.2
4b+8b		Lamendin et al. 1992	208	0.57	8.9
2+3b+4a+5a+6a+7a+8a		Solheim 1993b	758	0.86	9.3
4c+7d+2x7e+8b		Kvaal and Solheim 1994	452	0.90	8.5
3c+7e+3x7g+7h		Kvaal et al. 1995: 1 tooth	100	0.41	
3c+7e+3x7g+7h		Kvaal et al. 1995: 6 teeth	100	0.87	8.6

Application

When the age of a skeleton is to be diagnosed, the most important criterion for selecting a method is its reliability. Then come the operational criteria of time consumption and application field. These criteria will be used for a set of recommendations for practice in forensic osteology as well as historic and prehistoric anthropology. Since teeth are no island, the best known and most applicable bone methods are included in the following recommendations:

(1) Racemisation (10) has the highest accuracy, the average error m amounts to less than three years, but it has serious restrictions in applicability. It is best suited for fresh corpses. If the temperature conditions since the time of death are not known, or if incineration or even cremation is involved, then this method cannot be applied because any chemical reaction speeds up under heat. In case of a shorter time since death corrections should be made for the lower temperatures. For historic and prehistoric teeth racemisation is not suitable.

(2) A close competitor is cementum annulation (11) for reliability and general applicability. It can even be used in professionally cremated remains. Reliability at present is only slightly better in comparison to the other methods, but there are concrete indications for future improvements. Then the error of now 5 years might develop towards 3 years.

(3) The next choice are the combined histological methods for teeth (Johanson 1971 or Solheim 1993b) and for bones (Kerley 1965 and successive improvements, e.g. Drusini and Businaro 1990). In historic material the non-destructive methods should be preferred (Kvaal and Solheim 1994; Kvaal et al. 1995). The average error is 8 years. – This third position is a certain Rubicon: Other methods are too unreliable or are restricted to certain fields.

(4) Then and only then follow the combined bone methods like Nemeskéri's (Acsádi and Nemeskéri 1970) or Lovejoy's et al. (1985) with an error of roughly 11 years. In the case of Nemeskéri's method the error was empirically estimated later, the monograph sufficiently states that their value of 2.5 years in the case of all four parameters is a guess and not a calculation.

(5) If a first quick glance is the only technique available, then attrition, number of teeth, and skull suture closure might be used in conjunction. However, the user should be aware that this is a makeshift method only and that the applicability of such home brews is entirely unknown. When one of these parameters is used on its own, the average standard error lies between 14 and 20 years, sometimes even higher.

In the case of the histological and classical methods (recommendations 2 to 5) the experienced researcher will often modify the result of the method application, if there is additional, controversial, or untypical information. Thus in series of those forensic cases, where the success can be checked later, the empirical error tends to be very low, lower than the theoretical standard error derived from the reference sample.

Conclusion

The general conclusion of this review is that the modern techniques for age estimation in adults are no longer insufficient. In particular the comparably new processes of cementum annulation and racemisation, but also the development of new histological and combined methods have strongly increased the accuracy in applications.

Now it may also be stressed that teeth are better suited for age estimation than bones. Therefore, both the European custom of diagnosing historic skeletons by means of the combined Nemeskéri method only as well as the surviving American fashion of relying on one of the pubic symphysis methods should be regarded as transitional practices.

Acknowledgements

For reading the manuscript we are grateful to Andrea Drusini (Padova), Birgit Großkopf (Göttingen), Ariane Kemkes-Grottenthaler (Mainz), Maria Teschler-Nicola (Wien), and especially Tore Solheim (Oslo). Kurt W. Alt (Freiburg), as so often, supplied invaluable moral support. Sigmar Schnutenhaus (Ulm) started the literature logistics. Both Thomas Kingston Derry (Oslo) and Heidi Asmus (Ulm) have improved the English language. May our work serve the preservation of wisdom and not only the increase of knowledge.

References

Acsádi G, Nemeskéri J (1970) History of Human Life Span and Mortality. Académiai Kiadó, Budapest.

Altini M (1983) Age determination from the teeth – a review. J Dent Ass S Afr 38: 275–279.

Armenio G, Terio B, Laforgia PD (1956) Aspetti della degenerazione senile delle fibre nervose della polpa dentale. Riv Clin Odont 11: 1–15.

Bang G (1972) Factors of importance in dental identification: Five case reports. Int For Sci J 1: 91–102.

Bang G (1989) Age changes in teeth: Developmental and regressive. In: Işcan MY (ed) Age Markers in the Human Skeleton. Thomas, Springfield, p. 211–235.

Bang G, Monsen R (1968) Möglichkeiten der Altersbestimmung durch Untersuchungen an Zähnen. Dtsch Zahnärztl Z 23: 934–36.

Bang G, Ramm E (1970) Determination of age in humans from root dentin transparency. Acta Odont Scand 28: 3–35.

Baume R (1882) Die Defekte der harten Zahnsubstanzen. Felix, Leipzig.

Bennett CG, Kelln EE, Biddington WR (1965) Age changes of the vascular pattern of the human pulp. Arch Oral Biol 10: 995–98.

Bernick S (1967) Effect of age on the nerve supply to human teeth. J Dent Res 46: 694–99.

Beust TB (1931) Physiologic changes in dentine. J Dent Res 11: 267–275.

Bhussry BR, Emmel V (1955) Changes in the nitrogen content of enamel with age. J Dent Res 34: 672.

Bocquet-Appel JP, Masset C (1977) Estimateurs en paléodémographie. L'Homme 17: 65–90.

Bocquet-Appel JP, Masset C (1982) Farewell to paleodemography. J Hum Evol 11: 321–333.

Bödecker CF (1925) A consideration of the changes in the teeth from young to old age. Dent Cosmos 67: 543–549.

Brinkmann B, Hartmann C (1979) Determination of the mineral content and the transparency of the root dentine of human teeth. For Sci Int 15: 99–101.

Broca P (1879) Instructions relatives à l'étude anthropologique du système dentaire. Bull Mém Soc Anthrop Paris 2: 128–152.

Brudevold F (1957) Aldersforandringer i tannemaljen. Norske Tannlaegefor Tids 67: 451–458.

Burchett PJ, Christensen LC (1988) Estimating age and sex by using color, form, and alignment of anterior teeth. J Prosth Dent 59: 175–179.

Burns KR, Maples WR (1976) Estimation of age from individual teeth. J For Sci 21: 343–356.
Charles DK, Condon K, Cheverud JM, Buikstra JE (1986) Cementum annulation and age determination in Homo sapiens, I. Tooth variability and observer error. Am J Phys Anthrop 71: 311–320.
Charles DK, Condon K, Cheverud JM, Buikstra JE (1989) Estimating age at death from growth layer groups in cementum. In: Işcan MY (ed) Age Markers in the Human Skeleton. Thomas, Springfield, p. 277–301.
Čechová L (1972) Abrase zubů u současné pražské populace. (In Czech, summary.) Čs Stomatol 72: 27–34.
Čechová L, Titlbachová S (1971) Attrition of teeth. In: Anthropological Congress Dedicated to Aleš Hrdlička. Academia, Praha, p. 395–404.
Colonna M, Introna F, Favia G, Pesce Delfino V (1984) Valutazione della trasparenza della dentina per la determinazione dell'età: revisione metodologica e analisi di un campione. In: de Fazio F, Vernole B (eds) La laurea in odontoiatria e protesi dentaria. I problemi medico-legali in odontostomatologia. CIC Edizioni Int, p. 357–368.
Condon K, Charles DK, Cheverud JM, Buikstra JE (1986) Cementum annulation and age determination in Homo sapiens. II. Estimates and accuracy. Am J Phys Anthrop 71: 321–330.
Dalitz GD (1962) Age determination of adult human remains by teeth examination. J For Sci Soc 3: 11–21.
Dechaume M, Dérobert L, Payen J (1960) De la valeur de la détermination de l'âge par l'examen des dents en coupes minces. Ann Méd Lég 40: 165–167.
de Jonge TE (1950) Das Altern des Gebisses. Parodontol 4: 83–98.
Dreyfuss F, Frank RM, Gutmann B (1964) La sclérose dentinaire. Bull Group Int Rech Sci Stomat 7: 207–229.
Drusini AG (1991) Age-related changes in root transparency of teeth in males and females. Am J Hum Biol 3: 629–637.
Drusini AG (1993) Age estimation from teeth using soft X-ray findings. Anthrop Anz 51: 41–46.
Drusini AG (1995) Age-dependent dental variables in human adults. In: Moggi-Cecchi J (ed) Aspects of Dental Biology: Palaeontology, Anthropology and Evolution. Int Inst Study Man, Florence, p. 87–92.
Drusini A, Businaro F (1990) Skeletal age determination by mandibular histomorphometry. Int J Anthrop 5: 235–243.
Drusini AG, Businaro F, Volpe A (1989) Age determination of root dentine transparency of intact human teeth. Cah d'Anthrop Biométr Hum 7: 109–127.
Drusini A, Calliari I, Volpe A (1991) Root dentine transparency: Age determination of human teeth using computerized densitometric analysis. Am J Phys Anthrop 85: 25–30.
Dufkova J, Branik E (1983) Vergleichende Untersuchungen zur Dichte der Zähne unter gerichtsmedizinischen Gesichtspunkten. Beitr Gerichtl Med 41: 377–382.
Dufkova J (1984a) Untersuchungen zur Dichte der Zähne. Beitr Gerichtl Med 42: 177–185.
Dufkova J (1984b) Beitrag zur Identifizierung von Leichen/Skeletteilen unter Berücksichtigung von Brandleichen. Beitr Gerichtl Med 42: 233–248.
Endris R (1979) Forensische Odonto-Stomatologie. Kriminalistik-Verlag, Heidelberg.
Euler H (1939) Die Anomalien, Fehlbildungen und Verstümmelungen der menschlichen Zähne. Lehmann, München.
Euler H (1940) Über das Altern des menschlichen Zahnsystems. Dtsch Zahnärztl Wschr 43: 393–403.
Falter G (1974) Altersbestimmung an Zähnen zur Identifizierung unbekannter Toter. Med Diss, München.

Feng J (1985) Age determination from structure of teeth (Chinese; Summary). Renleixue-xuebao/Acta Anthrop Sinica 4: 380–384.

Förster A, Happel G (1959) Untersuchungen zur Altersbestimmung des Menschen auf Grund des Mineralisationsgrades des Zahndentins. Dtsch Zschr Gesamte Gerichtl Med 48: 195–201.

Großkopf B (1990) Individualaltersbestimmung mit Hilfe von Zuwachsringen im Zement bodengelagerter menschlicher Zähne. Z Rechtsmed 103: 351–59.

Gustafson G (1947) Åldersbestämningar på tänder. Odont Tidskr 54: 470–79; (1950 in English: Age determination of teeth. J Am Dent Assoc 41: 45–54); (1955 auf Deutsch: Altersbestimmungen an Zähnen. Dtsch Zahnärztl Zschr 10: 1763–68).

Gustafson G (1966) Forensic Odontology. Elsevier, New York.

Haertig A, Durignon M (1978) Identification dentaire. Comparaison des méthodes de Gustafson et de Shiro-Ito (modifiée). A propos de 156 cas. Bull Méd Lég Toxicol 21: 603–608.

Helfman P Masters, Bada JL (1975) Aspartic acid racemisation in tooth enamel from living humans. Proc Nat Acad Sci USA 72: 2891–2894.

Helfman P Masters, Bada JL (1976) Aspartic acid racemisation in dentine as a measure of ageing. Nature 262/5566: 279–281.

Henry JL, Weinmann JP (1951) The pattern of resorption and repair of human cementum. J Am Dent Assoc 42: 270–90.

Herdeg B (1992) Die Zahnentwicklung beim Menschen. Kritische Analyse der bisherigen Zahlen und Zitierreihen. Med Diss, Ulm.

Hess JC (1970) Les problèmes de l'endodontie en fonction de l'âge mûr. Cahiers de l'Odonto-Stomatol 1: 17–23.

Heuschkel HJ, Koch R, Süssmann H (1979) Ermittlung der Gesamtzahndichte permanenter menschlicher Zähne und ihr Aussagewert zum Lebensalter des Zahnträgers. Zahn-, Mund- Kieferheilkd 67: 583–590.

Hiemer HR (1975) Untersuchungen zur Altersbestimmung an Zähnen. Med Diss, München.

Hoefig W (1974) Zahnzahl alter Menschen. Dtsch Zahnärztl Z 29: 660

Holm-Pedersen P (1978) Ageing in Oral Tissues. The Royal Dental College, Århus.

Ito S (1972) Research on age estimation based on teeth (Japanese; Summary). Nihon-hoigaku-zasshi (Jap J Legal Med) 26: 31–41.

Ito S (1975) Age estimation based on tooth crowns. Int J Forensic Dent 7: 9–14.

Jacobshagen B, Kunter M (1992) Präzisierte Altersschätzungen nach dem Zementzonen-verfahren anhand der Zähne einer mittelneolithischen Skelettpopulation aus Trebur/ Südhessen. Vergleich mit den Ergebnissen der kombinierten Methode. Ms Gießen.

Johanson G (1971) Age determination from human teeth. Odontologisk Revy 22 (Suppl 21): 1–126.

Johnson CC (1968) Transparent dentine in age estimation. Oral Surg Med Path 25: 834–838.

Johnson JS (1976) A comparison of age estimation using discriminant function analysis with some other age estimations of unknown skulls. J Anat 121: 475–484.

Kambe T, Yonemitsu K, Kibayashi K, Tsunenari S (1991) Application of a computer assisted image analyzer to the assessment of area and number of sites of dental attrition and its use for age estimation. Forensic Sci Int 50: 97–109.

Kemkes-Grottenthaler A (1996) Critical evaluation of osteomorphognostic methods to estimate adult age at death: A test of the "complex method". Homo 46: 280–292.

Kerley ER (1965) The microscopic determination of age in human bone. Am J Phys Anthrop 23: 149–164.

Ketterl W (1983) Age-induced changes in the teeth and their attachment apparatus. Int Dent J 33: 262–271.

Kiesel W (1973) Untersuchungen zur Altersbestimmung des Menschen auf Grund von Messungen der Wandstärke und Härte des Dentins. Med Diss, Mainz.

Kieser JA, Preston CB, Evans WG (1983) Skeletal age at death: An evaluation of the Miles method of ageing. J Archaeol Sci 10: 9–12.
Kilian J (1986) Určováni věku dospělých osob podle chrupu. Med Diss, Pilsen.
Kilian J, Vlček E (1989) Age determination from teeth in the adult. In: Işcan MY (ed) Age Markers in the Human Skeleton. Thomas, Springfield, p. 255–275.
Kósa F, Szendrényi J, Tóth V (1983) Fogak transzparenciájának viszgálata az életkor megállapítására. (Summary: Dental transparency and its role in the determination of the age.) Morphologiai és Igazságügyi Orvosi Szemle 23: 286–291.
Kósa F, Antal A (1989) Metrical and structural study of the dental canal to determine the individual age. Acta Med Leg Soc 39: 349–364.
Kósa F, Antal A, Farkas I (1990) Electron probe microanalysis of human teeth for the determination of individual age. Med Sci Law 30: 109–114.
Krause N, Reimann N (1978) Altersbestimmung durch Messung der Wurzeltransparenz an intakten Zähnen. Vortrag auf der 5. Jt Gerichtsmed Ges DDR in Halle. Cit in Hopp, Blick 1980.
Kühns C (1895) Untersuchungen über die chemische Zusammensetzung der harten Zahnsubstanzen des Menschen in verschiedenen Altersstufen. Dtsch Mschr Zahnheilk 13: Heft 8.
Kvaal SI (1995) Age-Related Changes in Teeth. Thesis, Oslo.
Kvaal SI, Kolltveit KM, Thomsen IO, Solheim T (1995) Age estimation of adults from dental radiographs. For Sci Int and in Kvaal SI: Age-Related Changes in Teeth. Thesis, Oslo.
Kvaal SI, Koppang HS, Solheim T (1994a) Relationship between age and deposit of peritubular dentine. Gerontology 11: 93–98.
Kvaal SI, Sellevold BJ, Solheim T (1994b) A comparison of different non-destructive methods of age estimation in skeletal material. Int J Osteoarchaeol 4: 363–370.
Kvaal SI, Solheim T (1989) Fluorescence from dentin and cementum in human mandibular second premolars and its relation to age. Scand J Dental Res 97: 131–138.
Kvaal SI, Solheim T (1994) A non-destructive dental method for age estimation. J For Odontostomatol 12: 6–11.
Kvaal SI, Solheim T (1995) Incremental lines in human dental cementum in relation to age. Eur J Oral Sci 103: 225–230.
Kvaal SI, Solheim T, Bjerketvedt D (1996) Evaluation of different preparation, staining and microscopic techniques for counting incremental lines in human teeth. Biotechnic Histochem (in print) and in Kvaal SI 1995: Age-Related Changes in Teeth. Thesis, Oslo.
Lamendin H (1972) Dentine radiculaire et estimation d'âge. L'Information Dentaire 17: 1647–1662.
Lamendin H (1973) Observations on tooth roots in the estimation of age. Int J For Dent 1: 4–7.
Lamendin H (1978) Critères dentinaires pour appréciation d'âge: Etudes de la translucidité et des canalicules. Intérêt en odonto-stomatologie légale. Rev d'Od-Stom 7, 111–119.
Lamendin H, Baccino E, Humbert JF, Tavernier JC, Nossintchouk RM (1992) A simple technique for age estimation in adult corpses: The two criteria dental method. J For Sci 37: 1373–1379.
Lindemaier G, Schuller E, Müller K, Grabmann S (1989) Lebensalter und Zahnzahl – ein Hilfsmittel zur Identifizierung unbekannter Leichen. Beitr Gerichtl Med 47: 515–518.
Lipsinic FE, Paunovich E, Houston GD, Robison SF (1986) Correlation of age and incremental lines in the cementum of human teeth. J For Sci 31: 982–989.
López-Nicolás M, Canteras M, Luna A (1990) Age estimation by IBAS image analysis of teeth. For Sci Int 45: 143–150.
López-Nicolás M, Luna A (1991) Application of automatic image analysis (IBAS system) to age calculation. Efficiency in the analysis of several teeth from a single subject. For Sci Int 50: 195–202.

Lorton L (1988) Discussion of "Failure of use of cemental annulations in teeth to determine the age of humans". J For Sci 33: 1305–1306.

Lovejoy CO (1985) Dental wear in the Libben population. Its functional pattern and role in the determination of adult skeletal age at death. Am J Phys Anthrop 68: 47–56.

Lovejoy CO, Meindel RS, Mensforth RP, Barton TJ (1985) Multifactorial determination of skeletal age at death: A method and blind test of its accuracy. Am J Phys Anthrop 68: 1–14.

Maat GJR (1987) Practising methods of age determination. Comments on methods combining multiple age indicators. Int J Anthrop 2: 293–299.

Magitot ME (1878) Treatise on Dental Caries. Houghton, Osgood & Co, Boston.

Maples WR (1978) An improved technique using dental histology for estimation of adult age. J For Sci 23: 764–70.

Maples WR, Rice PM (1979) Some difficulties in the Gustafson dental age estimations. J For Sci 24: 168–172.

Masters PM (1986) Age at death determinations for autopsied remains based on aspartic acid racemisation in tooth dentin: Importance of postmortem conditions. For Sci Int 32: 179–184.

Matsikidis G (1981) Altersbestimmung aus Zahnfilmen. Med Diss, Heidelberg.

Matsikidis G, Schulz P (1982) Altersbestimmung nach dem Gebiß mit Hilfe des Zahnfilms. Zahnärztl Mitt 72: 2527–2528.

May G (1952) Untersuchungen über die Möglichkeit einer objektiven Altersbestimmung durch Messung des Aschengehaltes der Zähne. Stoma 5: 67–73.

Mays S, de la Rua C, Molleson T (1995) Molar crown height as a means of evaluating existing dental wear scales for estimating age at death in human skeletal remains. J Archaeol Sci 22: 659–670.

Metzger Z, Buchmer A, Gorski M (1980) Gustafson's method for age determination from teeth. A modification for the use of dentists in identification teams. J For Sci 25: 742–749.

Miles AEW (1963a) The dentition in the assessment of individual age in skeletal material. In: Brothwell DR (ed) Dental Anthropology. Pergamon, Oxford, p. 191–209.

Miles AEW (1963b) Dentition in the estimation of age. J Dent Res 42: 255–263.

Miles AEW (1976) Age changes in dental tissues. In: Cohen, Kramer (eds) Scientific Foundations of Dentistry. Heinemann, London, p. 37–38.

Miller CS, Dove SB, Cottone JA (1988) Failure of use of cemental annulations in teeth to determine the age of humans. J For Sci 33: 137–143.

Mjör IA (1986) Age changes in teeth. In: Holm-Pedersen P, Löe H (eds) Geriatric Dentistry. A Textbook of Oral Gerontology. Munksgaard, Copenhagen, p. 94–101.

Molnar S (1971) Human tooth wear, tooth function, and cultural variability. Am J Phys Anthrop 34: 175–189.

Moore GE (1970) Age changes occurring in the teeth. J For Sci Soc 10: 179–180.

Mörnstad H, Pfeiffer H, Teivens A (1994) Estimation of dental age using HPLC-technique to determine the degree of aspartic acid racemisation. J For Sci 39: 1425–1431.

Morse DR (1991) Age-related changes in the dental pulp complex and their relationship to systemic aging. Oral Surg Med Path 72: 721–745.

Nalbandian J (1959) Age changes in human teeth. J Dental Res 38: 681–682.

Nalbandian J, Gonzales F, Sognnaes RF (1960) Sclerotic age changes in root dentin of human teeth as observed by optical, electron, and x-ray microscopy. J Dent Res 39: 598–607.

Nkhumeleni FS, Raubenheimer EJ, Monteith BD (1989) Gustafson's method for age determination revised. J For Odonto-Stomatol 7: 13–16.

Ogino T, Ogino H, Nagy B (1985) Application of aspartic acid racemisation to forensic odontology: post-mortem designation of age at death. For Sci Int 29: 259–267.

Ohtani S, Yamamoto K (1987) Age estimation using the racemisation of aspartic acid on human dentin. Jap J Legal Med 41: 181–190.
Ohtani S, Yamamoto K (1991) Age estimation using the racemisation of amino acid in human dentin. J For Sci 36: 792–800.
Ohtani S, Yamamoto K (1992) Estimation of age from a tooth by means of racemisation of an amino acid, especially aspartic acid – comparison of enamel and dentin. J For Sci 37: 1061–1067.
Osborn JW, Ten Cate AR (1983) Age changes in the dental tissues. In: Advanced Dental Histology. Wright, Bristol, p. 198–203.
Philippas GG (1961) Influence of occlusal wear and age on formation of dentin and size of pulp chamber. J Dent Res 40: 1186–1198.
Philippas GG, Applebaum E (1966) Age factor in secondary dentin formation. J Dent Res 45: 778–789.
Philippas GG, Applebaum E (1967) Age changes in the permanent upper lateral incisor. J Dent Res 46: 1002–1009.
Philippas GG, Applebaum E (1968) Age change in the permanent upper canine tooth. J Dent Res 47: 411–417.
Pilin A (1985) Identification of unknown bodies from the point of view of forensic dentistry. Acta Med Leg Soc 35: 310–312.
Pilz W (1959) Das Transparenzphänomen der Zahnwurzel als ein Ausdruck der Biomorphose des menschlichen Dentins. Z Alternsforsch 13: 139–52.
Popov VL, Dudarev AL, Prokopchuk NI (1992) Kompleksnaia metodika opredeleniia vozrasta po sostoianiiu zubov s ispol'zovaniem analogovykh EVM. (Summary: A complex method for determining age by tooth status using analog computers.) Sudebno-Meditsinskaia Ekspertiza 35: 6–10.
Port TA von (1965) Altersvorgänge der Zähne und der Kiefer. Dtsch Zahnärztebl 19: 801–880.
Renz H, Radlanski RJ (1995) The appearence of cemental annulations in light and transmission electron microscopy. In: Radlanski RJ, Renz H (eds) Proc 10th Int Symp on Dental Morphology. Brünne, Berlin, p. 195–201.
Rheinwald U (1966) Über das Altern der Zähne. Dt Zahn Mund Kieferheilkde 46: 177–209.
Richards LC, Miller SLJ (1991) Relationships between age and and dental attrition in Australian aborigines. Am J Phys Anthrop 84: 159–164.
Ricco R, Colonna M, Vacca E, Introna F, Bufo P, Pesce Delfino V (1984) Modificazioni della trasparenza della dentina: Analisi densitometrica computerizzata. Boll Soc Ital Biol Sper 12: 2215–2221.
Ritz S, Schütz HW, Schwarzer B (1990) The extent of aspartic acid racemisation in dentin: A possible method for a more accurate determination of age at death? Z Rechtsmed 103: 457–462.
Ritz S, Schütz HW, Peper C (1993) Postmortem estimation of age at death based on aspartic acid racemisation in dentin: Its applicability for root dentin. Int J Legal Med 105: 289–293.
Ritz S, Stock R, Schütz HW, Kaatsch HJ (1995) Age estimation in biopsy specimens of dentin. Int J Legal Med 108: 135–139.
Santini A, Land M, Raab GM (1990) The accuracy of simple ordinal scoring of tooth attrition in age assessment. For Sci Int 48: 175–184.
Saunders M (1965) Dental factors in age determination. Med Sci Law 5: 34–37.
Saunders RL, Röckert HOE (1967) Vascular supply of dental tissues, including lymphatics. In: Miles EAW (ed) Structural and Chemical Organisation of Teeth. Academic Press, London, p. 245–254.
Schranz D (1960) Die quantitativen Veränderungen einzelner Bestandteile der Zähne und deren Brüchigkeit bei Extraktionen in fortgeschrittenem Alter. Dtsch Zahnärztl Z 15: 614–619.

Schürmann R (1986) Soziale Differenzierung nach frühgeschichtlichen Ernährungsmustern am Beispiel von Zahnabrasionen in merowingerzeitlichen Bevölkerungen. Biol Dipl Arb, Ulm.

Schwarz HR, Blick U, Liebhardt E (1978) Untersuchungen zur Altersbestimmung an Zähnen. Beitr Ger Med 36: 369–372.

Schwarz HR, Blick U, Liebhardt E, Pohl H (1980) Eine rationelle Methode zur Altersbestimmung an Zähnen. Beitr Gerichtl Med 38: 289–291.

Seifert E (1958) Über die Brauchbarkeit von Altersbestimmungen an Zähnen (Methode Gustafson). Med Diss, Leipzig.

Smith BH (1984) Patterns of molar wear in hunter-gatherers and agriculturalists. Am J Phys Anthrop 63: 39–56.

Sognnaes RF (ed) (1960) Oral Aspects of Aging. Washington DC: Am Ass Adv Sci Publ 65.

Sognnaes RF (1977) Forensic stomatology. (In three parts.) New England J Med 296: 79–85, 149–153, 197–203.

Sognnaes RF Gratt BM, Papin PJ (1985) Biomedical image processing for age measurements of intact teeth. J For Sci 30: 1082–1089.

Solheim T (1984) Dental age estimation. An alternative technique for tooth sectioning. Am J For Med Path 5: 181–184.

Solheim T (1988a) Dental color as an indicator of age. Gerodontics 4: 114–118.

Solheim T (1988b) Dental attrition as an indicator of age. Gerodontics 4: 299–304.

Solheim T (1989) Dental root translucency as an indicator of age. Scand J Dent Res 97: 189–197.

Solheim T (1990) Dental cementum apposition as an indicator of age. Scand J Dent Res 98: 510–519.

Solheim T (1992a) Amount of secondary dentin as an indicator of age. Scand J Dent Res 100: 193–199.

Solheim T (1992b) Recession of periodontal ligament as an indicator of age. J For Odonto-Stomatol 10: 32–42.

Solheim T (1993a) Dental Age-Related Regressive Changes and a new Method for Calculating the Age of an Individual. Thesis, Oslo.

Solheim T (1993b) A new method for dental age estimation in adults. For Sci Int 59: 137–147.

Solheim T, Kvaal S (1993) Dental root surface structure as an indicator of age. J For Odonto-Stomatol 11: 9–21.

Solheim T, Sundnes PK (1980) Dental age estimation of Norwegian adults – a comparison of different methods. Forensic Sci Int 16: 7–17.

Song H, Jia J (1989) The estimation of tooth age from attrition of the occlusal surface. Med Sci Law 29: 69–73.

Stein TJ, Corcoran JF (1994) Pararadicular cementum deposition as a criterion for age estimation in human beings. Oral Surg Med Pathol 77: 266–270.

Stermer Beyer Olsen EM, Bang G, Sellevold BJ (1994) Dental root dentine translucency used in age determination of medieval Norwegians from Trondheim. Int J Osteoarcheol 4: 305–310.

Stott GG, Sis RF, Levy BM (1981) Cemental annulation as an age criterion in forensic dentistry. J Dent Res 61: 814–817.

Sutor M (1937) Eine korrelationstheoretische Berechnung zwischen Lebensalter der Zähne und deren zunehmender Verkalkung. Korrespbl Zahnärzte 61: 10–14.

Takei T (1970) The use of tooth attritions in age estimation (in Japanese; Summary). Nihon-hoigaku-zasshi 24: 4–17.

Takei T (1984) Age estimation from dental attrition and state of dental treatment. J Nihon Univ School Dent 26: 119–132.

Takei T, Maruyama T, Takahashi T, Zaha H, Yonashiro T, Nemoto M (1981) The use of tooth attrition in age estimation (part 2) (in Japanese; Summary). Nihon-hoigaku-zasshi 35: 400–410.

Ten Cate AR, Thompson GW, Dickinson JB, Hunter HA (1977) The estimation of age of skeletal remains from the color of roots of teeth. J Can Dent Ass 43: 83–86.
Tomaru Y, Uchiyama Y, Kobayashi K, Kudo Y, Mikami H, Endo M, Tsukamoto T, Terezawa K (1993) Age estimation from tooth attritions of lower incisors – Discussion of the "Amano's method". Nippon Hoigaku Zasshi/Jap J Legal Med 47: 13–17.
Tomenchuk J, Mayhall JT (1979) A correlation of tooth wear and age among modern Igloolik Eskimos. Am J Phys Anthrop 51: 67–78.
Toto PD, Kastelic EF, Duyvejonck KJ, Rapp GW (1971) Effect of age on water content in human teeth. J Dent Res 50: 1284–1285.
Traub HR, Altini M, Hille JJ (1988) A comparison of radicular dentinal tubule size in two different age groups. J For Odonto-Stomatol 6: 43–54.
Tsuchiya K (1973) A calorimetric study of anterior teeth (in Japanese; Summary). Shikwa Gakuho 73: 87–120.
Vasiliadis L, Darling AI, Levers BGH (1983) The histology of sclerotic human root dentine. Arch Oral Biol 28: 693–700.
Vlček E, Mrklas L (1975) Modification of Gustafson method of determination of age according to teeth on prehistorical and historical osteological material. Scripta medica 48: 203–208.
Walker P, Dean G, Shapiro P (1991) Estimating age from tooth wear in archaeological populations. In: Kelly MA, Larsen CS (eds) Advances in Dental Anthropology. Wiley-Liss, New York, p. 169–178.
Wegener R, Albrecht H (1980) Zur Schätzung des Alters anhand der Zahnwurzeltransparenz. Z Rechtsmed 86: 29–34.
Wei B, Feng J (1984) A trinial regression formula of estimation age with both the degree of tooth attrition and the tooth index (in Chinese; Summary). Renleixue-xuebao/Acta Anthrop Sinica 3: 270–276.
Wei B, Feng J, Fang Z (1983) The relationship between the construction of maxillary first molar and age (in Chinese; Summary). Renleixue-xuebao/Acta Anthrop Sinica 2: 72–79.
Whittaker D (1992) Quantitative studies on age changes in the teeth and surrounding structures in archaeological material: A review. J Royal Soc Med 85: 97–101.
Xu X, Philipsen HP, Jablonski NG, Pang KM, Zhu J (1992) Age estimation from the structure of adult human teeth: Review of the literature. For Sci Int 54: 23–28.
Yamamoto K (1992) Molecular biological studies on teeth inquests (in Japanese; Summary). Nihon-hoigaku-zasshi 46: 349–355.
Yonan T, Fosdick LS (1964) The degree of etching as a function of the age of the teeth. J Dent Res 43: 629–630.
Zander HA, Hürzeler B (1958) Continuous cementum apposition. J Dent Res 37: 1035–1044.
Zhang J, Ji Y (1988) Tooth attrition and age estimation of Chinese males (in Chinese; Summary). Renleixue-xuebao/Acta Anthrop Sinica 7: 230–234.

6.3 Degeneration in Dental Hard Tissues and Pulp

Tore Solheim

Introduction

The estimation of age is essential for the identification of corpses with unknown identity and skeletal remains in forensic medicine, and an important factor for demographic reconstructions in archeology (Işcan 1989). Dental traits are particularly suited for age estimation. In adults, a number of classic and new procedures are used; the best results may be reached by a combination of methods (Alt 1995). Degeneration indicates changes in dental hard tissues which occur as a result of disturbances in nutrition or metabolism. The extent of such alterations increases with age and is sometimes termed regressive change. These terms are more commonly applied to soft tissues than to hard tissues such as teeth, but age-related changes in the teeth are often described as degenerative changes (Kvaal 1995). A general description of such changes together with the normal histology can be found in a number of standard textbooks on the microstructure of the teeth (Scott and Symons 1977; Bhaskar 1980; Hillson 1986; Berkovitz et al. 1989; Schroeder 1991).

Having studied such changes in 1000 extracted teeth, excluding molars, I present a description of age-related changes in enamel, dentine and cementum and include some of my findings. Based on these findings statistical calculation formulae for age calculations were developed (Solheim 1993). A number of methods exists for calculating the age of an individual from teeth. Using only the length of the translucent zone in the root as the indicator of age, Bang developed a single formula for each tooth (Bang and Ramm 1970). In contrast, Johanson used a standard set of scores for age indicators and presented a single formula for all types of teeth (Johanson 1971). Using a multiple regression method, Maples included correction factors for tooth type (Maples 1978).

Enamel

Enamel is the hardest tissue in the human body. As it does not contain cells, only small changes occur with age and these are due to physiochemical processes

(Schroeder 1991). The hardness of the enamel enables it to resist the abrasive effects of chewing. Nevertheless, a small loss of substance in the incisal/occlusal part of the tooth occurs which has been called attrition. The use of toothbrushes or tooth picks, especially when used together with abrasive toothpaste, results in a loss of tooth substance called abrasion. Abrasion is found mostly in the cervical areas of teeth and is dependent on the frequency, intensity and forcefulness of the cleaning process.

Attrition may vary from one population to another and in different periods of time (Molnar 1971; Brothwell 1989). Archaeological populations appear to show a more regular increase with age than do modern populations, but the real age of individuals is usually unknown and the degree of attrition may have influenced the estimation of age. Also the degree of attrition is more pronounced in prehistoric populations (Molnar 1972). Present-day diet is very variable, even within the same population. Individual parafunctions are frequently found, and the correlation in modern populations between attrition and age is rather weak.

In this work, the strongest correlation of attrition with age was found in premolars, and the weakest in canines (Solheim 1988b). Pearson's correlations varied from $r = 0.22$ to $r = 0.68$ for the various types of teeth. Men tended to have higher attrition scores than women, and teeth with severe attrition were found only in men.

Increased translucency of enamel is seen to accompany advancing age and influences tooth color by allowing darker dentin to become more visible (Solheim 1988a). This observation, however, only partly explains the darkening color of teeth with increasing age. Other reasons may include the loss of water by enamel, and the increase the amounts of nitrogen, fluoride, and other trace elements with age.

The permeability of enamel in less in older than in younger people and the enamel more brittle. Since elasticity is less than that found in dentin, enamel tends to crack or fracture, as a result of loading pressure and trauma; thus cracks and lost enamel prisms are often seen in the elderly. Especially when attrition extends far into the dentin, the harder enamel remains as a thin wall, and clefts from lost prisms may be seen in this occlusal rim of enamel. Organic material may collect in the enamel cracks, thus increasing the amount of such material in teeth from older individuals. As this material is usually dark it may contribute to color change in older teeth.

Dentin

Dentin is formed by odontoblasts during odontogenesis (Schroeder 1991). Cytoplasmic processes are left behind in the dentin to form dentinal tubuli and odontoblasts remain in the pulp near the predentin wall. After odontogenesis is complete odontoblasts continue to form a more irregular, or secondary, dentin at a lower rate. Dentin becomes irregular possibly as a result of crowding of the odontoblasts as pulpal volume is reduced. Volume reduction may also result in degeneration and apoptotic cell death of a number of odontoblasts, thereby ending their dentin production.

When odontoblastic processes are irritated by caries, cavity preparation or exposed dentin such as in the cervical area of the tooth, the formation rate of secondary dentin increases. This type of dentin may be particularly irregular, and is sometimes called tertiary or reparative dentin.

In our investigation, the amount of secondary dentin was measured as a function of the area of the coronal part of the pulp. At the root, it was measured at 4 levels (Fig. 1), and the ratio between root and pulp measurements calculated (Solheim 1992). Amounts of secondary dentin were scored according to three different systems. Correlation with age was found to be from $r = 0.60$ to $r = 0.80$ for the various types of measurements and teeth. Only small differences in correlation with age resulted among the different types of teeth, with premolars being weakest. More secondary dentin was found in the lingual part of the pulps of maxillary incisors and canines than in the vestibular part.

Chewing normally causes stress and attrition on the lingual sides of maxillary teeth, so from an evolutionary point of view it would be advantageous to form additional secondary dentin on the lingual sides of maxillary teeth. In addition, more secondary dentin was observed in the teeth of men than in those of women. Furthermore a reduced rate of secondary dentin formation was found in older individuals. Controlling for the effect of age, no correlation was found between the amount of secondary dentin and the reduction in periodontal attachment or increased attrition as has also been found by others (Philippas 1961).

Tiny tubuli containing cytoplasmic processes from odontoblasts penetrate the dentin. Dentin between tubuli is called intertubular dentin. The density of tubuli is higher near the pulp than towards the surface. With increasing age calcification of in tubuli occurs from the periphery, thus creating what is called peritubular dentin.

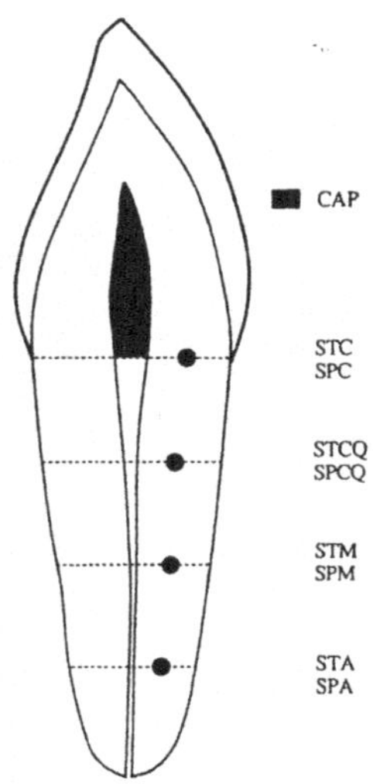

Fig. 1. Drawing of a tooth showing where the various measurements were made (CAP = area of crown pulp; STC = total tooth width at cemento-enamel junction; SPC = pulp width at the cemento-enamel junction; STCQ = total width with a coronal $^1/_4$ of root; SPCQ = pulp width at coronal $^1/_4$ of root; STM = total tooth width at midroot area; SPM = pulp width at midroot area; STA = total tooth width at apical $^1/_4$ of root; SPA = pulp width at apical $^1/_4$ of root.

The diameter of the tubules is in this way reduced with advancing age. The lumen is eventually totally obliterated by peritubular dentin, so that the number of open dentinal tubuli is reduced (Fig. 2).

Obliterated or nearly obliterated tubuli change the refractive index of dentin, rendering it more translucent (Nalbandian et al. 1960). This process starts at the apex and proceeds in the coronal direction, while from the surface of the root translucency

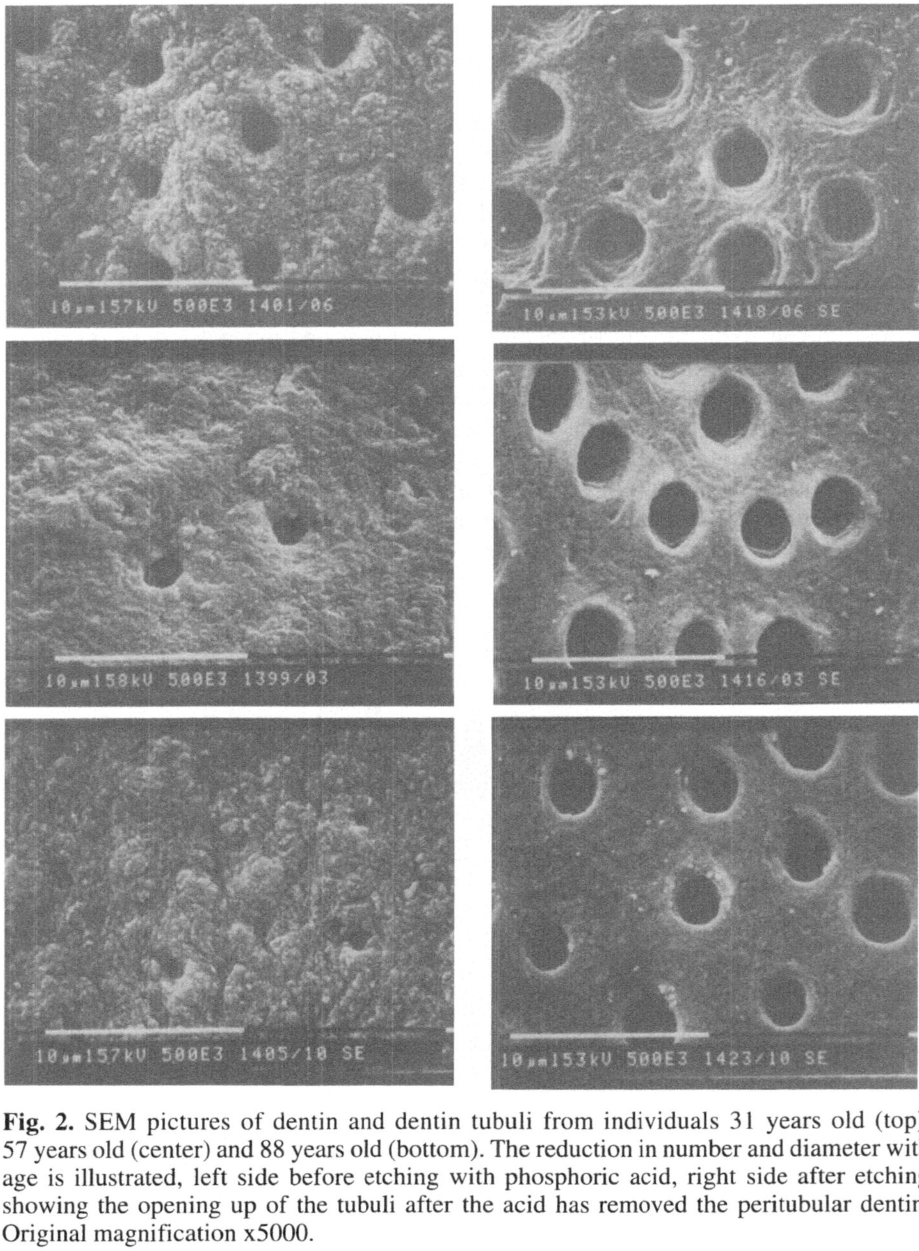

Fig. 2. SEM pictures of dentin and dentin tubuli from individuals 31 years old (top), 57 years old (center) and 88 years old (bottom). The reduction in number and diameter with age is illustrated, left side before etching with phosphoric acid, right side after etching showing the opening up of the tubuli after the acid has removed the peritubular dentin. Original magnification x5000.

spreads in the pulpal direction. The border between translucent and opaque dentin cannot always be clearly distinguished, making it more difficult to assess the exact length of this zone. The use of the ratio of the area of the translucent zone to the total area of the root might appear to be more rational, but investigations have shown that this ratio is not related more closely to age than are simple measurements of the zone length. The translucent zone is, however, closely related to age and in our investigations varied from $r = 0.65$ to $r = 0.85$ for different types of teeth (Solheim 1989). The length of this zone increased linearly with age, so that tooth roots of individuals above 70 years of age were almost completely translucent.

The translucent zone tends to be longer in males than in females. Controlling for the effect of age, it was found that the extent of the translucent zone was positively correlated with color and with cementum thickness, but not with attrition or the retraction of the periodontal ligament.

Tooth Color

Teeth become more brownish-yellow with increasing age. In the crown this may be a result of increased translucency of the enamel allowing the darker dentin to show through. That is, however, not the whole explanation. We have, after removing the cementum, measured root dentin color which was increasingly darker with age (Solheim 1988a). This disclosed a strong correlation with age, varying from $r = 0.60$ to $r = 0.80$ in the different tooth types. Teeth of males were darker than those of females and the rate of color change decreased with advancing age.

Dental color is difficult to measure. In dental practices dentists usually make a subjective assessment of color after comparison with a shade guide (Sproull 1973). Spectrophotometric measurement is considered to be the most objective (Goodkind et al. 1985), but does not measure what is seen by the eye. Measurements may be corrected according to a standard observer, namely the CIE-system (Commission Internationale d'Éclairage) (MacEntee and Lakowski 1981). In this work assessments in a subjective five-graded scale correlated more strongly with age than did spectrophotometric measurements.

The reasons for color darkening with age are not known. We found that red fluorescence in dentin increased with age and correlated strongly with increased darkness of tooth color (Kvaal and Solheim 1989). The increase in fluorescence may be caused by deposition of externally derived products in the dentin. Possibly extravasated erythrocytes decompose and porphyrins with a red fluorescence slowly diffuse into dentin over the years. Other fluorophores such as tryptophane and hydroxypyridium have also been identified in dentin, but these substances have a blue fluorescence. Lipofuscin is another fluorizing substance known to increase with age in many tissues.

Cementum

Cementum is the dental hard tissue that is most similar to bone. Cells are not included in slowly formed cementum, but when formed more rapidly, cementum contains cells called cementocytes. Cementum is continuously deposited on the root surface throughout life and triples in quantity from 20 to 70 years (Zander and Hürzeler 1958). The thickness at 1/3 of the root length from the apex strongly correlates with age (Solheim 1990). Using a stereomicroscope, measurements were taken of surfaces which had been ground from one side to the mid-pulpal area (Fig. 3). From logarithmic transformations of cementum thickness Pearson correlation with age varied from 0.40 to 0.65. The formation rate was reduced with age, and males had thicker cementum than females.

For some teeth pathological changes appeared to influence cementum formation. Thickness at the apex was found to be two to three times that at one-third of the root length from the apex, and was not strongly correlated with age. For some teeth cementum thickness was positively correlated with the degree of darkened color, extent of root translucency and with topography of the root surface.

In reindeer and moose, biologists count annulation rings in the cementum to determine the animal's age. One or more rings are formed per year in these animals, but the reason for this regularity is not known. Possibly nutrition and the influence of light play a roles. Annulation rings sometimes seen in the dental cementum of humans possibly form at the rate of one ring every year. In humans

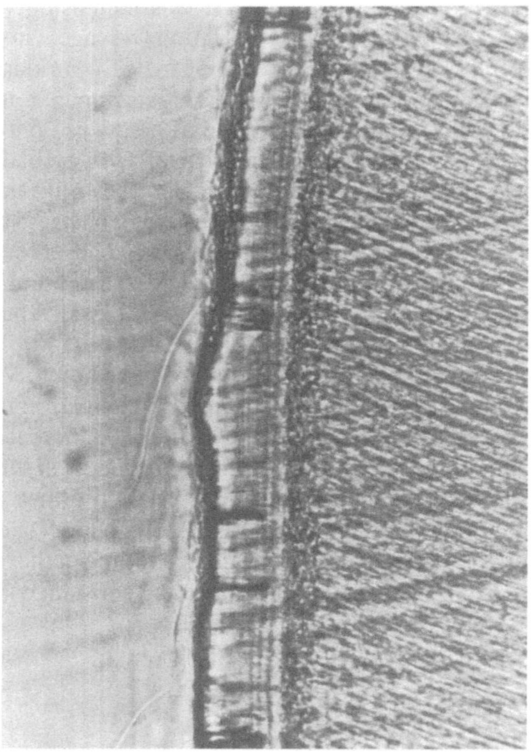

Fig. 3. Microscopic visualization of the cementum layer in a tooth 35 ground from one surface to the mid-pulpal area.

the rings are more difficult to count, especially in older individuals where a large number of closely packed rings are present (Condon et al. 1986). Lines are easily overlooked; counting is more feasible in animals which do not live more than 5–6 years.

With increasing age, pathological conditions and trauma to the tooth may contribute to resorptions of the root surface. Root resorption may be an age-related change and has been used as a method of age estimation. According to our studies it is only weakly related to age. The preparation technique may have contributed to this; where resorptions occur damage may be rapidly repaired by cellular cementum, which is not easily detected in a stereomicroscope.

Study of its general structure has revealed a cementum surface which becomes steadily more irregular with increasing age (Solheim and Kvaal 1993). The reason may be that cementum apposition does not take place in a regular fashion and that calcification occurs along the collagen bundles of the periodontium. A grading system from 1 to 4 resulted in a correlation with age from $r = 0.50$ to $r = 0.70$ for the different types of teeth. The reproducibility of the estimate for scoring of the surface roughness of the root was rather low, but the factor was included in some formulae for age calculations. Scores of this roughness were correlated with the amount of secondary dentin, but no explanation of this originating from external influences could be put forward.

Cementocytes which are located in lacunae in the cementum may necrotize and the lacuna will then be empty. The effect is difficult to estimate, but necrotic cementum may possibly function as well as acellular cementum. It has been suggested, however, that cementum necrosis may be a cause of periodontal retraction.

Cementum apposition at the apex may result in a narrowing of the apical pulp opening, which may lead to circulatory disturbances and degeneration in the pulp tissue. In extreme cases venous draining may be so hampered that pulp tissue necrotizes.

Pulp

The pulp contains a fine fibrillar specialized connective tissue with odontoblasts bordering on the dentin surface. The odontoblasts are highly specialized cells which are easily disturbed. Low-grade irritation stimulates the cells to increase the formation of secondary dentin. A higher degreee of irritation such as inflammation or circulatory disturbance results in degeneration of the odontoblasts, and the formation of secondary dentin may come to a stop (Siskin 1973; Seltzer and Bender 1984; Schroeder 1991).

Another type of degeneration in the pulp is fibrosis. Fibrosis can also be found in bone tissue during inflammation and a number of other disease processes and indicates an increased amount of coarse collagen fibres, which may be packed so closely together that a hyalinized appearance (hyalinization) results. It can usually be seen in the apical area of the pulp and may be considered to be an almost normal finding in this location. However, when fibrosis progresses in a coronal direction and is found in the pulp chamber it may be considered to be degeneration. Fibrosis may be found in quite young individuals, but occurs with increased frequency in older ones. Fibrotic tissue has a reduced number of blood vessels resulting in reduced nutrition. As fibrous tissue tolerates these conditions better than does normal pulp tissue, fibrosis tends to increase.

Calcification is often seen in the pulp tissue even of young people, but its frequency increases with age. These calcifications are irregular structures without cells. They are often found along vessels and calcium salts are deposited in degenerated connective tissue, usually hyalinized. This process is called dystrophic calcification and may be found in many types of tissue.

Dentin may be formed beginning with nidi of degeneration or dystrophic calcification. The result may be a more or less round stone in the pulp, a true denticle. Denticles may grow and fill almost the entire pulp and become attached to the dentin in the pulp wall. The hard structure in the dystrophic calcifications or denticles approximating nerves or blood vessel may result in pain or circulatory disturbances and degenerations. The latter could result in necrosis of the pulp tissue.

Pulp necrosis is usually the result of inflammation, but can also be caused by trauma or exposure of the pulp as a result of attrition. Degenerative changes can in some cases lead to necrosis. The outcome of pulp necrosis is the necrosis of the whole pulp channel and inflammation in the periapical bone tissue, called apical periodontitis, which may eventually result in the formation of an apical cyst. Granulation tissue which forms in apical periodontitis requires space, and results in bone resorption visible on radiographs of the jaws; when this process penetrates the cortical bone it may be seen macroscopically in archeological materials. Cysts usually require even more space, and a large, round, well defined radiolucent area at the root apex is a typical radiographic finding.

Chronic pulpitis may cause internal resorptions, starting at the pulpal dentin wall. Such resorptions may penetrate the entire dentin thickness and result in communication with the periodontium. Reparative processes often occur where new hard tissue formation may be found in resorption lacunae. Reparative tissue of this type is a cellular cementum rather than dentin.

Conclusions

Fifty years ago, Gustafson developed the first systematic dental method for the estimation of age (Gustafson 1947). Age estimation from dental development (children) as well as regressive dental changes in the teeth and surrounding structures (adults) may be important for the identification of unknown corpses and skeletons. Today, many methods and techniques exist to determine age in both the outer part (e.g. attrition, color changes, cementum annulation, root resorption) and the inner part of the tooth (e.g. dentin hardness, root transparency, secondary dentin deposition) in adults (Bang 1989). It is the role of forensic odontologists or anthropologists to extract this information and use it for the identification of bodies and skeletons. This chapter presents some of the changes currently used for this purpose.

References

Alt KW (1995) Dental anthropology in forensic odontology and archaeology. A review. In: Jacob B, Bonte W, Alt KW, Pieper P (eds) Advances in Forensic Sciences, Vol 7. Forensic Odontology and Anthropology. Köster, Berlin, p. 255–262.
Bang G, Ramm E (1970) Determination of age in humans from root dentin transparency. Acta Odontol Scand 28: 3–35.
Bang G (1989) Age changes in teeth: Developmental and regressive. In: Iscan MY (ed) Age Markers in the Human Skeleton. Thomas, Springfield, p. 211–235.
Berkovitz B, Boyde A, Frank R, Höhling H, Moxham BNJ, Tonge C (1989) Teeth. Springer, Berlin Heidelberg New York.
Bhaskar S (1980) Orban's Oral Histology and Embryology. Mosby, St Louis.
Brothwell DR (1989) The relationship of tooth wear to aging. In: Iscan MY (ed) Age Markers in the Human Skeleton. Thomas, Springfield, p. 303–316.
Condon K, Charles DK, Cheverud JM, Buikstra JE (1986) Cementum annulation and age determination in Homo sapiens, II. Estimates and accuracy. Am J Phys Anthrop 71: 321–330.
Goodkind RJ, Keenan KM, Schwabacker WB (1985) A comparison of chromascan and spectrophotometric color measurements of 100 natural teeth. J Prosth Dent 53: 105–109.
Gustafson G (1947) Åldersbestämningar på tänder. Odontol Tidskr 55: 556–568.
Hillson S (1986) Teeth. Univ Press, Cambridge.
Işcan MY (ed) (1989) Age Markers in the Human Skeleton. Thomas, Springfield.
Johanson G (1971) Age determination from teeth. Odont Revy 22 (Suppl 21): 1–126.
Kvaal S, Solheim T (1989) Fluorescence from dentin and cementum in human mandibular second premolars and its relation to age. Scand J Dent Res 97: 131–138.
Kvaal S (1995) Age-Related Changes in Teeth. A Microscopic and Radiographic Investigation of the Human Permanent Dentition. Thesis, Oslo.
MacEntee M, Lakowski R (1981) Instrumental color measurement of vital and extracted human teeth. J Oral Rehabil 8: 203–208.
Maples WR (1978) An improved technique using dental histology for estimation of adult age. J For Sci 23: 764–770.
Molnar S (1971) Human tooth wear, tooth function and cultural variability. Am J Phys Anthrop 34: 175–190.
Molnar S (1972) Tooth wear and culture: A survey of tooth functions among some perhistoric populations. Curr Anthrop 13: 511–526.
Nalbandian J, Gonzales F, Sognnaes RF (1960) Sclerotic age changes in root dentin of human teeth as observed by optical, electron, and X-ray microscopy. J Dent Res 39: 598–607.
Philippas GG (1961) Influence of occlusal wear and age on formation of dentin and size of pulp chamber. J Dent Res 40: 1186–1198.
Schroeder HE (1991) Oral Structural Biology. Thieme, New York.
Scott J, Symons N (1977) Introduction to Dental Anatomy. Livingstone, Edinburgh.
Seltzer S, Bender IB (1984) The Dental Pulp. Biologic Considerations in Dental Procedures. Lippincott, Philadelphia.
Siskin M (1973) The Biology of the Human Dental Pulp. Mosby, St. Louis.
Solheim T (1988a) Dental color as an indicator of age. Gerodontics 4: 114–118.
Solheim T (1988b) Dental attrition as an indicator of age. Gerodontics 4: 299–304.
Solheim T (1989) Dental root translucency as an indicator of age. Scand J Dent Res 97: 189–197.
Solheim T (1990) Dental cementum apposition as an indicator of age. Scand J Dent Res 98: 510–519.
Solheim T (1992) Amount of secondary dentin as an indicator of age. Scand J Dent Res 100: 193–199.

Solheim T (1993) A new method for dental age estimation in adults. For Sci Int 59: 137–147.
Solheim T, Kvaal S (1993) Dental root surface structure as an indicator of age. J For Odontostomatol 11: 9–21.
Sproull RC (1973) Color matching in dentistry, Part II. Practical application of the organization of color. J Prosth Dent 29: 556–566.
Zander H, Hürzeler B (1958) Continuous cementum apposition. J Dent Res 37: 1035–1044.

6.4 Sex Determination Using Tooth Dimensions

Maria Teschler-Nicola and Hermann Prossinger

Introduction

Among the many parameters used in studies of prehistoric skeletal series, two are fundamental prerequisites for any statistical analysis: the age at death and the sex of an individual. Various demographic, biological or pathological features can be studied when these parameters are known. Biological age at death assessments are needed for estimating mortality rates, which subsequently have implications for the possible conclusions to be drawn about living conditions, nutritional status, epidemiology, social stratification etc. in prehistoric societies.

A broad spectrum of qualitative and quantifiable features – with variable reliability – are at the investigator's disposal when determining the sex of adult individuals (Sjøvold 1988; Hoyme and İşcan 1989). Often, however, the knowledge of the sex of immatures is of high importance: Goble and Konopka (1973), for example, present a sex-specific disposition for various types of infectious diseases with a greater incidence among boys. This phenomenon has also been observed in historic skeletal series (e.g. Schutkowski 1990; Sperl 1990; Teschler-Nicola et al. 1994). In skeletal series, however, the sexual diagnosis of young – i.e., immature – individuals is usually tentative; consequently, the conclusions risk being much more precarious. The (questionable) reliance on extrapolations from the sparse data matrix is due primarily to the incompleteness of most immature skeletons. Nonetheless, the challenge posed by unfortunate circumstances must be dealt with. Sex has (among other things) implications for the determination of biological age: there is a sexually determined variability in ossification, mineralization and growth rates. Endocrine functions exercise a biologically controlled sexual differentiation from an early stage of the uterine development onwards up to the end of puberty (Mazess and Cameron 1972; Ounsted et al. 1981; Stini 1985; Specker et al. 1987; Saunders 1992).

First attempts at sexual diagnosis of subadult individuals were based on the intra-individual comparison of tooth mineralization and the developmental status of the postcranial skeleton (Hunt and Gleiser 1955). Assuming that these methods could be applied at all (accuracy levels between 73 and 81% have been obtained), they have been rarely used in practice, most obviously due to the inadequate preservation status of the carpal bones of immature skeletons in prehistoric grave fields. Also, the sex-specific ossification (Hunger and Leopold 1978) and the sex-specific tooth mineralisation (Bailit and Hunt 1964) are unsuitable for sex determination of immatures, because they themselves depend on biological age.

Problems are also encountered when using the method suggested by Choi and Trotter (1970) for diagnosing the sex of fetal skeletons by relying on the ratio of long bone length to mass. Despite the fact that the analyses by Boucher (1955, 1957), Fazekas and Kósa (1969, 1978), Weaver (1980, 1986) and Schutkowski (1987, 1989, 1990) have demonstrated that the observations of nonmetric and metric traits are reproducible when applied to the postcranial part of the fetal or neonatal skeleton (most of them could be obtained for the pelvis, since it shows the highest dimorphism in the adult) they, too, have rarely been used for prehistoric skeletal series.

Novel techniques such as sex chromatin determination also in decomposed bones (Duffy et al. 1991) or molecular analysis by means of PCR (Pääbo et al. 1989; Mullis et al. 1994) have yielded remarkable results in test cases and promise to be applicable in forensic practice. A newly developed sexing method using morphological and metric features (Schutkowski 1990) is feasible, too, but not easily tranferable (Majo et al. 1993) and is often not applicable due to the inadequate preservation state of the skeletons. Teeth are more often extant, so their analysis remains the method of choice.

As most of the available measurements used for sex determination are the mesiodistal and buccolingual tooth crown diameters of the permanent dentition (Moorrees 1957; Moorrees et al. 1957; Moorrees 1959; Schranz and Bartha 1964; Garn et al. 1964, 1966, 1977; Maudrich 1977; for further data see Kieser 1990; Schnutenhaus 1994), their sexual differences have appeared to be the most attractive to investigate (see Figs. 1–2). The crowns of the permanent teeth develop early and remain invariant during the subsequent biological changes of the individual – unless there are extraneous functional, intentional, pathological or nutritional circumstances. Consequently, one infers that any mathematical combinations (linear or otherwise) – as assumed in methods such as discriminant analysis – of crown dimensions would offer an implementable solution to the problem of sex diagnosis of immature individuals in prehistoric populations. Investigations of the dimensions of permanent dentition in recent (Lundström 1943, 1944; Moorrees 1957; Hanihara and Koizumi 1979; Garn et al. 1979; Potter et al. 1981; Kieser et al. 1985a, 1985b; Paul 1990; Rösing et al. 1995) and ancient populations (Lunt 1969; Ditch and Rose 1972; Grimm and Hildebrand 1972; Rottstock 1975; Sciulli et al. 1977; Frayer 1980; Rösing 1983; Langenscheidt 1983; Owsley and Webb 1983; Teschler 1992a; Rieger 1993; Alt et al. 1995; Stermer and Alexandersen 1995) seem to confirm this, although not all of these analyses have been carried out for the specific problem of sexing immatures.

While published data for permanent teeth show that metric averages are sex-discriminatory, this seemingly favorable situation has rarely been implemented in practice, mainly because the set of variables used for calculating the discriminant functions differ between different investigations, problably because of variability in tooth size between populations and sexual dimorphism (Dahlberg 1963; Subow

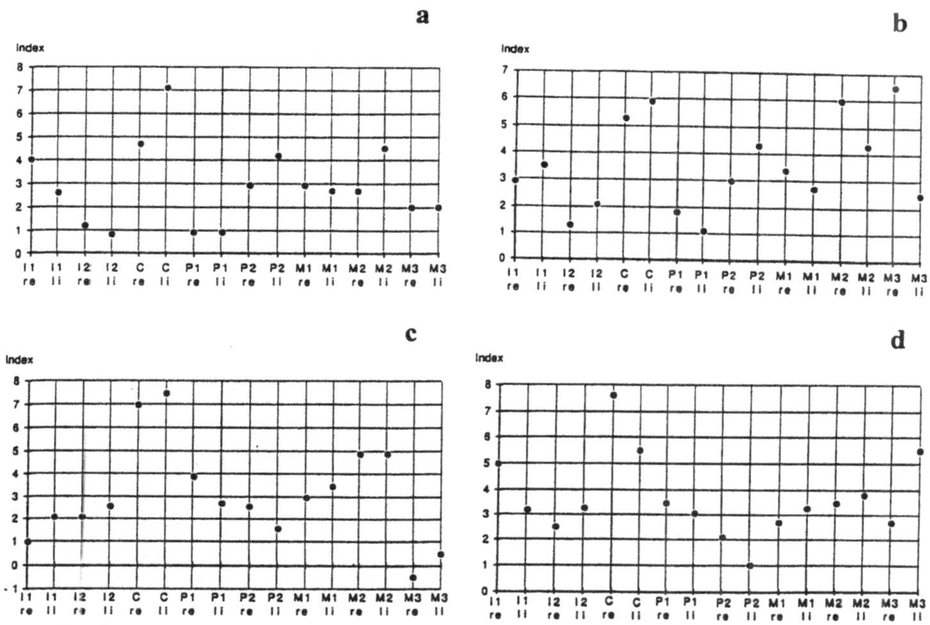

Fig. 1. Sexual dimorphism (BL and MD tooth crown dimensions) of the permanant dentition of an early Bronze age population (Franzhausen I, Teschler 1992a). *Legend*: a = mesiodistal dimension of the maxillary tooth; b = buccolingual dimension of the maxillary tooth; c = mesiodistal dimensions of the mandibular tooth; d = bucolingual dimensions of the mandibular tooth.

1965; Lavelle 1972, 1984; Kieser 1990; Schnutenhaus and Rösing, this volume). This seems to be a rather random variation, geography rarely influences tooth size to a systematic degree (Rösing et al. 1995). Such variations would imply that the multivariate functions developed for one population may rarely be applied to another population. The majority of the published data sets and discriminant analyses are based on the permanent dentition (Tab. 1; Fig. 3), few use the deciduous dentition (Moorrees et al. 1957; Black 1978; DeVito and Saunders 1990; Teschler 1992a).

Ditch and Rose (1972) were the first to demonstrate that stepwise discriminant analysis carried out with tooth diameters can be successfully used in the determination of the sex of fragmentary historic human archeological remains. All the skeletons included in their analysis were sexed with reference to the postcranial remains. Due to the small but sensible diagnosis error this procedure is inferior to using series of identified skeletons or of living persons. The percentage of correctly classified individuals lies between 89 and 96% (Tab. 1). Rösing (1983) then raised the argument that non-adults should be sexed in general, and that teeth are best suited for this. The resulting accuracy rates of 80–97% cover a wider range, probably because a total of sixty-six analyses were performed; a problem is the inclusion of crown height, which is subject to attrition. As a reaction to his small sample, Langenscheidt (1983) tried to reduce possible spurious sampling effects

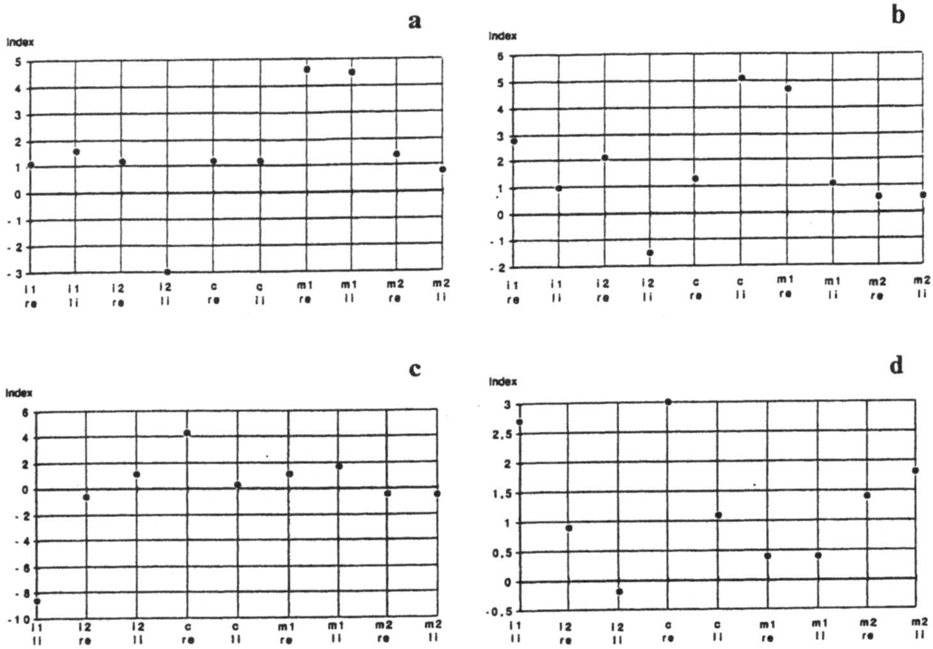

Fig. 2. Sexual dimorphism (BL and MD tooth crown dimension) of the deciduous dentition of Franzhausen. *Legend*: a = mesiodistal dimension of the maxillary tooth; b = buccolingual dimension of the maxillary tooth; c = mesiodistal dimensions of the mandibular tooth; d = bucolingual dimensions of the mandibular tooth.

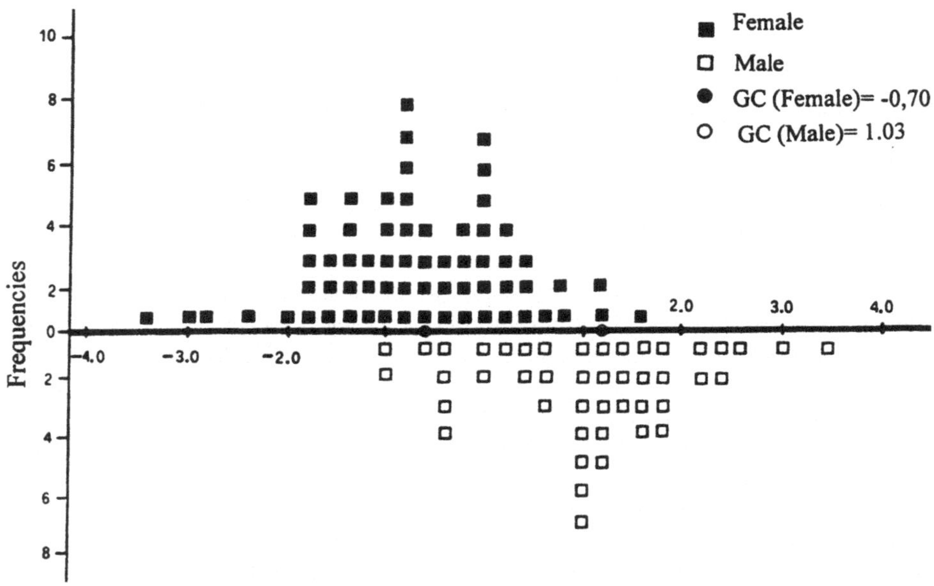

Fig. 3. The result of applying discriminant function 1 of Franzhausen: Group centroids of males (□) and females (■) in the discriminant axis.

Table 1. Stepwise discriminant functions for sexing human skeletal remains using tooth measurements of deciduous and permanent dentitions. *Abbreviations*: U = upper jaw, L = lower jaw, MD = mesiodistal diameter, BL = buccolingual diameter, CH = crown height, RH = root height, R = right; L = left; 1–8 = permanent tooth; i1, m2, c, . . . = deciduous tooth

1a) Ditch & Rose 1972, function A: UBL1 x 0.721 + UMD3 x (–0.939) + UBL3 x 1.030 (88% correct classifications in reference sample)
1b) Ditch & Rose 1972, function F: UMD3 x (–0.970) + UBL3 x 0.837 + UBL5 x (–1.302) + UBL6 x 1.094 + LBL3 x 1.526 (96% correct classifications in reference sample)
2) Rösing 1983, function 64: LMD2 x (–0.0716) + LBL2 x (–0.0090) + LCH2 x (–0.0415) + LMD3 + 0.0837 x LBL3 0.0445 + LCH3 x 0.0935 – 10.2913 (95% correct classifications in reference sample)
3a) Langenscheidt 1983, function 5: LMD1 x (–1.654) + LBL1 x 1.565 + LBL3 x 1.847 + (–14.845) (85% correct classifications in reference sample)
3b) Langenscheidt 1983, function 6: LMD3 x 0.759 + LBL3 x 1.743 + (–18.285) (78% correct classifications in reference sample)
4a) Paul 1990, function I: UMD3 x (–1.1565) + UMD5 x 1.6123 + LMD3 x (–1.4536) + LMD6 x (–1.04373) + 19.8881 (83% correct classifications in reference sample)
4b) Paul 1990, function II: UMD6 x (–0.7057) + UBL5 x 0.8204 + LMD3 x (–2.1896) + LMD6 x (–0.315) + 21.9204 (82% correct classifications in reference sample)
5a) De Vito & Saunders 1990, function D (three maxillary and one mandibular variables): UBLi1R x 1.899 + UBLm2L x 1.174 + UBLcL x (–1.750) + LMDcR x 1.653 + (–20.138) (74% correct classifications in reference sample)
5b) De Vito & Saunders 1990, function B (four maxillary and one mandibular variables): UBLi1R x 1.380 + UBLcR x 0.896 + UBLm2L x 0.357 + UBLcL x (–1.474) + LMDcR x 2.266 + (–19.736) (80% correct classifications in reference sample)
6a) Teschler 1992a, function 5: UBL3 x 1.53 + UMD3 x 1.89 + UBL6 x (–1.03) + (–15.36) (80% correct classifications in reference sample)
6b) Teschler 1992a, function 2: UBL3 x 1.54 + UMD3 x 1.69 + UBL5 x (–0.49) + UMD5 x 1.43 + UBL6 x (–0.98) + UMD7 x (–037) + (–16.07) (81% correct classifications in reference sample)

by introducing the jackknife variant of discriminant analysis. Consequently her correctness rate was lower: 73–81%. Simultaneously and independently Owsley and Webb (1983) used the jackknife technique; moreover, the validation techniques involved sample resubstitution and the use of a holdout sample. Their classifications were correct to 65–81%, but the discriminant functions were not published.

The first discriminant analysis application to both deciduous and permanent teeth was performed by DeVito and Saunders (1990), again using a recent population. Their rate of correct attributions was in the range of 76–90% for deciduous teeth, and of 83–85% for combinations of deciduous and permanent teeth. Paul (1990) applied discriminant analysis to a large recent population (219 immatures), achieving a correct sex determination in 77–83% of all cases. Moreover, he developed a method for estimating individual sex assignment probabilities, which is particularly appropriate for forensics. Teschler (1992a) used permanent teeth; her sample is not only large (172 immature individuals), but it is also drawn from a prehistoric population. Furthermore, in order to perform discriminant analysis which requires complete data sets for every sampled individual, she introduced a linear regression imputation method to reconstruct some missing tooth crown diameters. She achieved a correct classification level between 75 and 81%. The most recent method presentation so far revisited the problem of classification correctness (Alt et al. 1995); using strict criteria, the success rates amounted to 50–78% for single teeth. They, too, however, do not publish functions for our Tab. 1.

The application of such functions is fraught with difficulties. The functions are usually more descriptive (reflecting the sex differenciation of the population) than predictive (they are rarely used to estimate the sex of individuals from sexually unknown skeletal series; published exception: Rösing 1983 applied in 1990). The difficulties are many: (1) When discriminant functions are to be used for diagnosis, the sex in the reference sample should be known, not only diagnosed by e.g. the pelvis. (2) Many systematic variation factors of tooth size are known in general but not in specific cases, so the applicabiliy of specific functions is rarely clear. (3) Often, the systematic measurement errors due to interproximal attrition is disregarded. (4) Many investigators apply functions without mathematically incorporating uncertainties such as the possibility of measurement variabilities. Teschler (1992a) could document a statistically significant age-dependent attrition trend of the MD diameters of deciduous molars; she suggested that changes due to attrition influence the small deciduous teeth more than the – larger – permanent teeth. Consequently, sexually specific differences could be blurred by differences due to abrasion. (5) The variable combinations extracted by the discriminant functions limit the applicability to only a few individuals from large (incomplete) data sets.

With these difficulties in mind, we investigate limitations of the possibility of using metric dental parameters for sexing immatures in this article. We first analyze an Early Bronze Age population with respect to variability and sexual dimorphism of both deciduous and permanent tooth crown diameters. Then, after introducing various mathematical methods, we apply the results of our analyses to a more recent population. We then look at possible applications to and implications for demography.

Materials and Data Acquisition

Our investigation is based on two populations: Early Bronze Age Franzhausen I in the Traisen Valley, Lower Austria (Neugebauer 1987; Neugebauer-Maresch and Neugebauer 1988; Neugebauer and Gattringer 1988) and Early Mediaeval Gars/Thunau (Friesinger and Friesinger 1975; Teschler et al. 1994). These populations are temporally distant but geographically close to each other.

The deposition rites of the Early Bronze Age population to the south of the Danube (Unterwölbling culture) is sex-specific: males lie on their left side, heads oriented northwards; the females lie on their right, their heads southwards. The consistency of this burial rite has been shown by comparing the anthropologically determined sex with the archeological evidence (Teschler 1992b). We take advantage of this fortunate circumstance: thus the sex of almost all immature individuals is known. Only few graves have been so severely disturbed that the original body orientation could not be reconstructed. Thus the sex is known archeologically for 172 juveniles (85 male and 87 female).

In the Mediaeval population from Gars/Thunau, we cannot assign the sex to any individual, because the archeologists could not find any sex-specific clues. We have some implications that the high infant and juvenile mortality rate observed in this – second – population is due to nutritional deficiencies and/or infectious diseases. One reason for critically evaluating the methodology presented here is our interest in knowing whether there are any epidemiological or pathological phenomena that could be sex specific. As a first step in attempting to answer this question, we use the known sex of the Franzhausen immatures in order to mathematically assign a sex to the immature individuals of Gars/Thunau.

For the sex determination, we first measured the mesiodistal and buccolingual tooth crown diameters of all extant deciduous and permanent teeth of immatures (younger than 20 years at death) in both populations. We did not include adults in this data set, because their teeth exhibit interstitial abrasion and extensive crown reduction. The crown diameters were measured according to standard procedures (Remane 1927; Bräuer 1988) using a digital caliper. Due to the bad preservation state of the jaws, most measurements were taken from isolated teeth or from the dental germs, if these had at least developed beyond the dental neck.

Methodologies Using Tooth Dimensions

All data matrices from both grave fields are very sparse. In fact, there are more blanks than entries. In the published literature dealing with the problem of sex determination, several solutions have been suggested: one method uses imputation (a filling-in approach) by either estimating, reconstructing or guessing missing elements in the data matrices, while another attempts to devise a direct statistical approach (Vach 1994). We will not discuss the latter method here, because (1) it appears to be effective only for large sets of extant values and (2) a somewhat detailed comparison of merits and drawbacks will have to be postponed to a later publication, as investigations are not complete.

In the imputation approach, one can either simply replace a missing value by the average of the sample of corresponding extant values (mean imputation method, Little and Rubin 1987), or one laterally mirrors missing left/right values by complementary right/left values. The mean imputation method can rarely be justified mathematically. Furthermore the sample mean is undefined for a population whose individuals' sex we are attempting to determine, as the two sexes are pooled! The straightforward mirroring strategy is not always justified, but there exists a mathematically intriguing and challenging computational method – we propose to call it the *enhanced mirroring technique* –, in which one linearly interpolates assuming uncertainties in both variables with a Laplace error distribution (see Prossinger, this volume) which can be recommended, as it has survived several tests (see below). We are currently developing more sophisticated mathematical reconstruction techniques which should shed light on the applicability of the currently practised replacement strategies.

In this article, we compare a simple method which does not depend on the sparseness of the data matrix with some published discriminant methods, with and without the lateral replacement strategy. Some of the discriminant analysis methods use reconstructed tooth dimensions (Teschler 1992a). Conventionally – and unfortunately –, the (exclusively linear) interpolations are performed using the pooled sample of the reference population, rather than interpolating the extant crown dimensions of the specific individual whose missing dimensions are to be reconstructed. We use the results of the latter method, although with some trepidation.

A much simpler method was originally applied by Breitinger to the dentition of four immatures found in an urn field grave at Stillfried (Breitinger 1980). The method is based on the observation that the average male permanent tooth dimension is larger for every tooth crown, both buccolingually and mesiodistally. The first step is to compare every measured tooth dimension from an individual's dentition with the average values (male and female) from modern reference populations (Moorrees 1959; Saunders 1992). If the majority of tooth dimensions of an individual is closer to – or greater than – the male average, then the sex of the individual is assumed to be male. We suspect that the apparent success of this method when applied to the Stillfried population has made it tempting ever since. However, we would like to mention three caveats: (1) the method was applied to a population of only four individuals, (2) the averages used were from a reference population that was *both* temporally and geographically remote and (3) the method does not consider the fact that there are two standard deviations of the tooth dimensions for the reference populations. As we have no other sex determination method that can be applied to the *four* immature individuals from Stillfried, we cannot ascertain the reliability of the sex determination; all the same, we have our doubts concerning the appropriateness of the method. A further indication of the problems associated with this procedure can be seen in Saunders (1992 Fig. 2): archeological cases do not always lie between the male and female averages and a sizable fraction can quite often be smaller than the female average of the reference population.

We have applied the Breitinger method using the Early Bronze Age population from Franzhausen as our reference population. We discovered that the standard deviations overlap quite often. For example, Fig. 4 shows the case for the maxillar deciduous tooth dimensions. In order to investigate results derived from this simple method the standard deviations must be included; we therefore introduce the average $\langle x_k \rangle$ and standard deviation σ_k of the male ($k = 1$) and the female ($k = 2$) subsets of each tooth diameter from the Franzhausen sample (in which the sex is known), and we use the four values – $\langle x_{\text{male}} \rangle$, σ_{male}, $\langle x_{\text{female}} \rangle$ and σ_{female} – obtained for each tooth diameter to form a weighted boundary (abbreviated WB) viz.

$$WB = \langle x_{\text{female}} \rangle + \frac{\sigma_{\text{female}} \cdot (\langle x_{\text{male}} \rangle - \langle x_{\text{female}} \rangle)}{\sigma_{\text{male}} + \sigma_{\text{female}}} \tag{1}$$

This formula ensures that the weighted boundary always lies between the male and female averages of the reference population. (Thus we overcome the difficulty

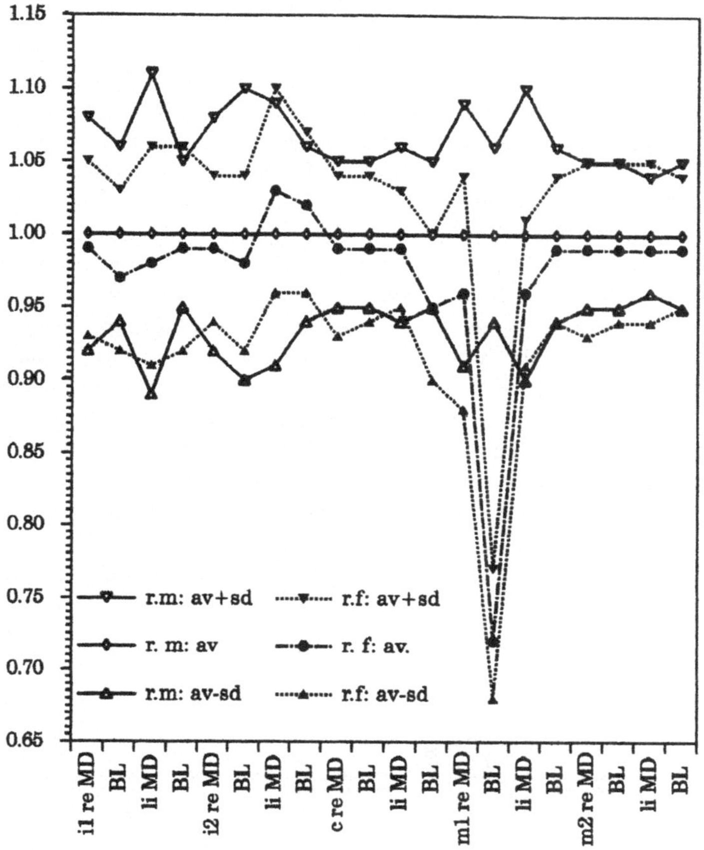

Fig. 4. Deciduous tooth diameters for Franzhausen. All dimensions have been graphed relative to the (normalized) male dimensions. Some female tooth dimensions are larger than in males. Standard deviations vary considerably, and the male standard deviation intervals almost always overlap the female ones. *Symbols*: r.m: av – male average (normalized to 1.00); r.m: av + sd – male average (normalized to 1.00) + standard deviation (relative to the male average); r.m: av –sd – male average (normalized to 1.00) –standard deviation (relative to the male average); r.f: av – relative female average (i.e., relative to the male average); r.f: av + sd – relative female average + standard deviation (relative to the male average); r.f: av –sd – relative female average –standard deviation (relative to the male average).

evident in Fig. 2 of Saunders (1992); note the overlap of the standard deviation intervals). The weighted boundary is used to determine whether a measured dimension from an individual of unknown sex is on the male or female side. A weighted boundary has the advantage of being sensitive to the spread in diameter values of each of the sex specific data subsets. The possible differences of the male standard deviation from that of females also ensures that the boundary is not always midway between the two averages.

The sexual diagnosis is the result of determining the sites of each individual's extant tooth dimensions relative to the boundary calculated for the pooled data of the male and female reference samples for that dimension. Breitinger's method suggests how to calculate the probability of an individual's sex very simply: the probability of assignment to a specific sex is estimated according to the fraction of dimensions that fall on one side of the boundaries. We will subsequently call this method of sex determination the weighted boundary method.

The following four graphs (Figs. 5–8) show how the boundary varies between the male and female average values for the deciduous and permanent dentitions of the maxilla and the mandibula. (For clarity, the male averages have been normalized to about 1.0, the female averages and the boundaries to the same scale.) One can observe that some boundaries are much better discriminators than others – a fact that is often used to choose the variables for discriminant function analysis methods. Also note that the boundaries are usually closer to the female average, which can be considered a justification for using the weighted boundary method.

Fig. 9 shows an actual case of a set of extant tooth diameters from one immature individual (grave 47) from Gars/Thunau and their relation with the boundary values from the Early Bronze Age reference population. The open circles are the boundary values for those tooth dimensions that are extant in the dentition of the individual. The upward pointing triangles are those dimensions that are less than the boundary values – and thus indicating a female attribute –, the downward pointing triangles

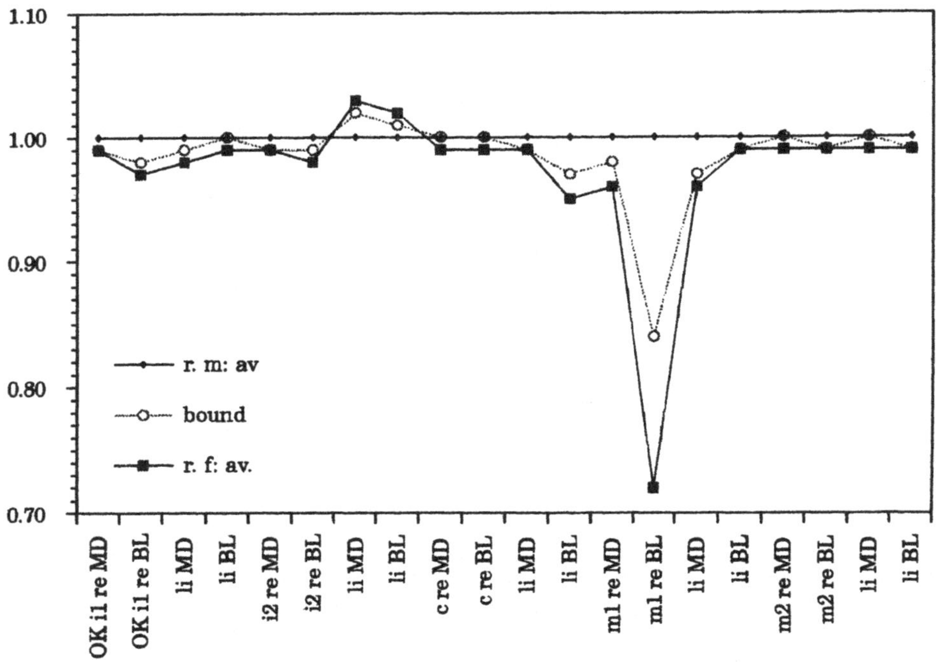

Fig. 5. Deciduous upper jaw tooth diameters for Franzhausen. Both male (r.m: av) and female (r.f: av) dimensions, as well as the boundary (labelled bound), graphed relative to the (normalized) male dimensions. The weighted boundary varies considerably and is not always closer to the female average. The first right molar female average is much smaller than its male counterpart. (For an interpretation, see Teschler 1992a).

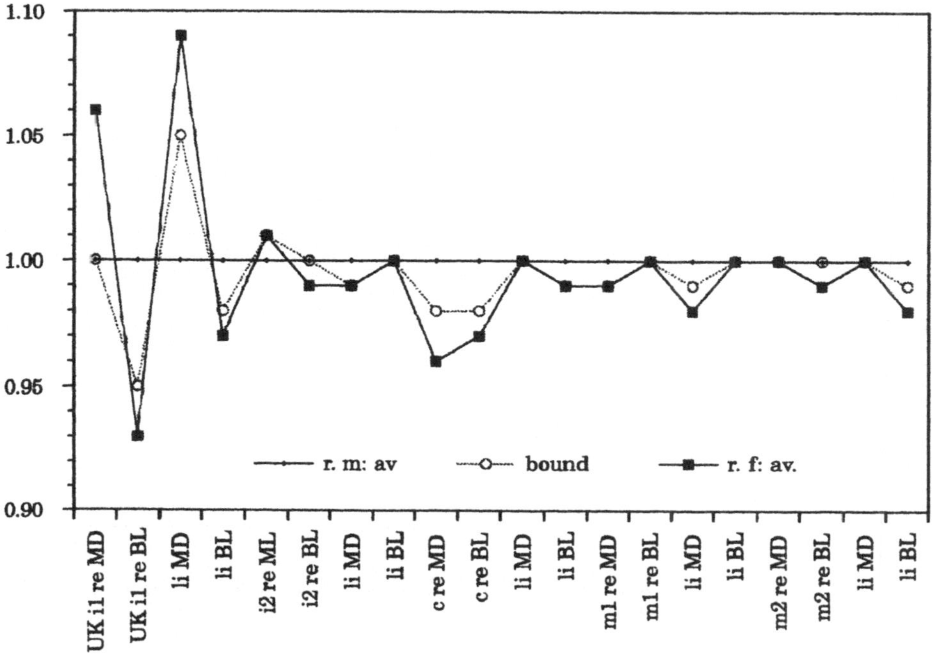

Fig. 6. Deciduous lower jaw tooth diameters in Franzhausen. The first right and left incisor mesiodistal diameter average for females is larger than for males, as is the mesiodistal diameter of the second right incisor.

are the ones corresponding to the male attributes. Of the 36 dimensions in this sample, 25 (69%) can be considered male attributes, 11 (31%) female attributes. If we assume the weighted boundary method for sex determination, we conclude that this immature individual is male.

Comparisons

In a further step of the analysis we would like to know how good a sex determination can be achieved with this method, basing our hopes on the observation that we could thereby obtain a larger sample for demographic analyses. We did in fact use this method for the sex determination of the individuals of Gars/Thunau, but we could not test the result directly, as we have no other way of determining the sex of these immature individuals.

However, we can apply the test on the reference population directly: we know whether an immature individual is male or female in the Franzhausen sample. The weighted boundary method should determine the sex to be the same as the archeologically determined one.

The test becomes quite involved, because we have many parameters at our disposal. As a first test, we simply sample all individuals that have any tooth diameters (i.e., deciduous and/or permanent) present. For males we obtain a correct

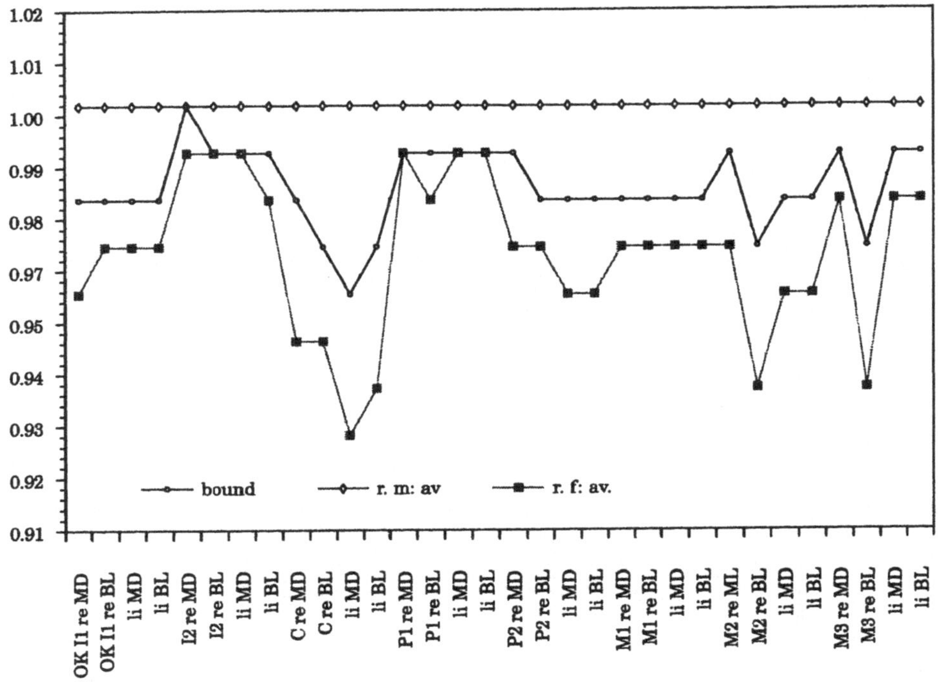

Fig. 7. Permanent upper jaw tooth diameters in Franzhausen. All female averages (r.f: av.) are less than the male (r. m: av) ones: the boundary (labelled bound) fluctuates considerably, but is usually closer to the female averages. Note the discriminatory potential of the right and left canine. These left female average diameters are smaller than their right counterparts by about 2%.

classification of 54 out of 82 individuals (66%), for the females we obtain 51 out of 84 individuals (61% correctly classified). We note that the percentage of correct classification is not much greater than a random guess probability, even though the number of individuals in the sample is quite large. One suspects that the reason for the low correct classification probability is due to the erratic boundary behavior of the deciduous tooth dimensions.

If one tests for correct classification results for permanent tooth diameters only, one obtains: 53 of 76 males (70% correct classification), 51 of 84 females (61% correct classification). Although a slightly higher correct classification has been achieved, one must still consider the fraction relatively low.

In order to assess the weighted boundary method further, we compared the results with a discriminant function obtained by Teschler (1992a). The variables of the discriminant function are: canine BL and MD, 2nd premolar BL and MD, 1st molar BL and 2nd molar MD – all left maxilla. Clearly the sample size decreases considerably, because not all dentitions contain these tooth dimensions (this is the major problem of using the functions presented in the literature). Using the weighted boundary method, we obtain 39 of 59 males (66% correctly classified) and 42 of 65 females (65% correctly classified). One observes that the correct classification fraction has increased, but not significantly. A good reference is the

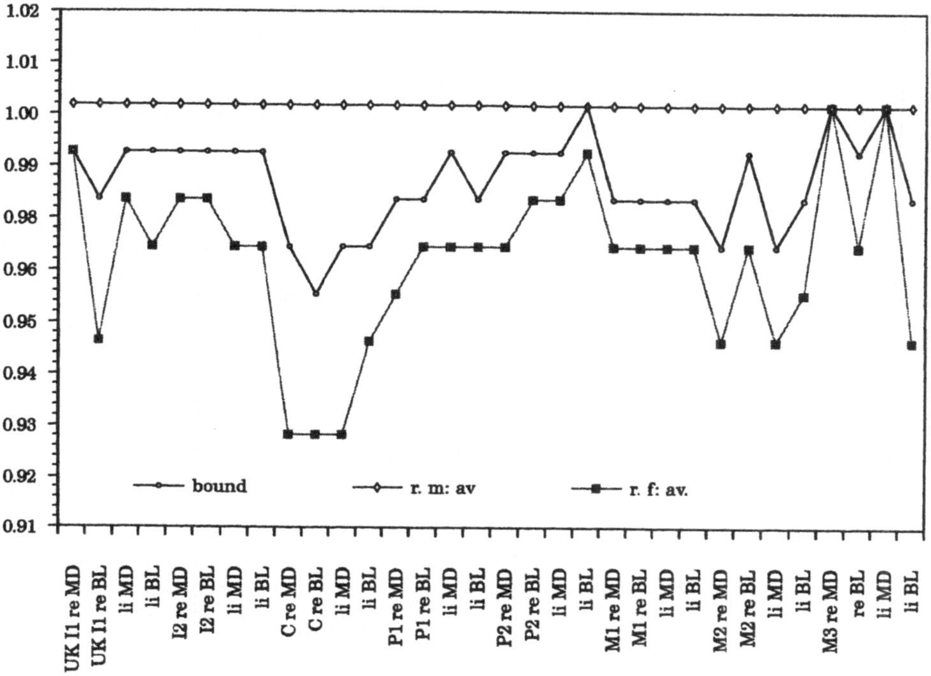

Fig. 8. Permanent lower jaw tooth diameters in Franzhausen. All female averages are less than the male ones: the boundary fluctuates considerably, but is predominantly closer to the female averages. Note the discriminatory potential of the right and left canine. The difference between right and left canine averages are not as pronounced as in those of the upper jaw.

observation that the discriminant function with these six variables yields 19 of 21 correctly classified males (91%) and 10 of 15 correctly classified females (67%). Thus, a correct classification of 65% for the weighted boundary method is to be compared with a 81% correct classification achieved with the discriminant analysis method.

Why not reject the weighted boundary method outright? Before we recommend doing so, we add the following observations: the weighted boundary method samples more individuals (124) than does the discriminant function method (36), because the discriminant function method necessitates all six dimensions to be nonzero for each individual, while the weighted boundary method can still be applied, even with up to five missing values.

If we limit ourselves to those individuals with six extant (i.e., nonzero) crown diameters, we can ask: how well do the sex determinations found by the weighted boundary method agree with those found by the discriminant function method? Of the 21 male individuals that were used in the discriminant method, 12 had the same prediction as the weighted boundary method (57% agreement), but of the correctly classified ones, 11 agreed (92% agreement). For the 15 females, 12 had the same prediction (80% agreement) and 9 (i.e., 75%) predicted the sex correctly. Thus, we conclude that only in the most fortunate circumstances can the prediction

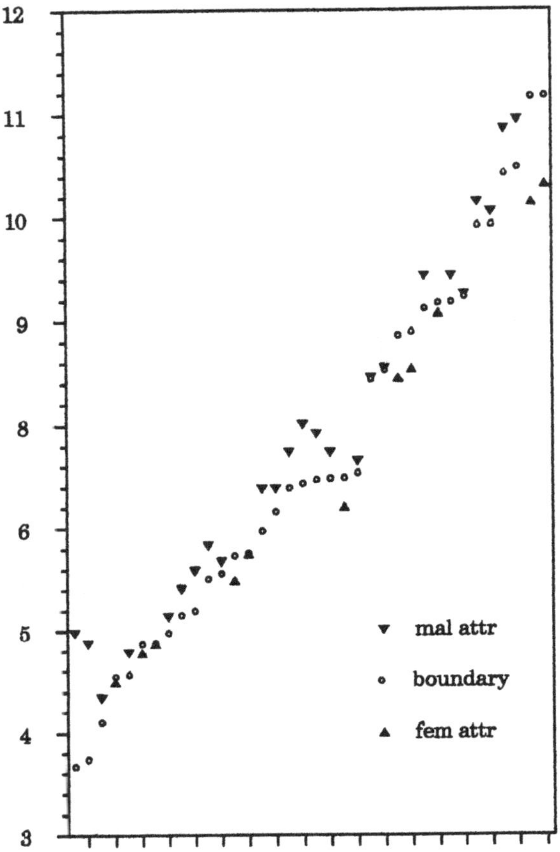

Fig. 9. The application of the weighted boundary method in order to determine the sex of individual Nr. 47 of the medieval population from Gars/Thunau. The open circles represent the boundary values for Franzhausen, calculated with formula (1). The downward-pointing triangles labelled "mal attr" are male attributes, as their values are larger than the boundaries; "fem attr" are the labels of the upward-pointing triangles that represent female attributes. This individual has 69.5% male attributes and is thus determined to be male – by this method.

probability using the weighted boundary method be comparable with a discriminant analysis method, but never surpass it.

Because the samples used for testing the predictions are so small, we have applied the enhanced mirroring technique in order to reconstruct some missing dimensions. By doing so, we pursue a double purpose: we would like to (1) know whether prediction probabilities change significantly (and thus also test the applicability of the enhanced mirroring technique) and (2) use – if the prediction probabilities remain comparable – the enlarged sample for demographic investigations.

After enlarging the sample by including dimensions obtained with the enhanced mirroring technique, we found the following: Of the 33 individuals that were used in the discriminant method, 22 had the same prediction as the weighted boundary method (67% agreement), but of the correctly classified ones, 16 agreed (73% agreement).

For the 28 females, 23 had the same prediction (82% agreement) and 16 (i.e., 70%) predicted the sex correctly. Thus, we conclude that the reliability of sex prediction is somewhat better; using the enhanced mirroring technique can be recommended.

Mortality Rates

Franzhausen

Fig. 10 compares the mortality curves for 85 male and 87 female individuals from the Early Bronze Age grave field in Franzhausen. We can observe a sex specificity: about twice as many females die between 10 and 15 years of age as do their male counterparts. In the 16 to 20 year age bracket, the situation seems reversed. Such straightforward interpretations are not possible in the age brackets of the very young. One explanatory factor may be the difficult repository conditions in the riverine deposits of the Traisen valley, which rarely allow the preservation of the graves of the very young (Neugebauer and Gattringer 1987).

Gars/Thunau

In this sample, the age at death of 164 immature individuals has been determined (labelled "all" in the graph). Of these, 134 had extant tooth dimensions. Fig. 11

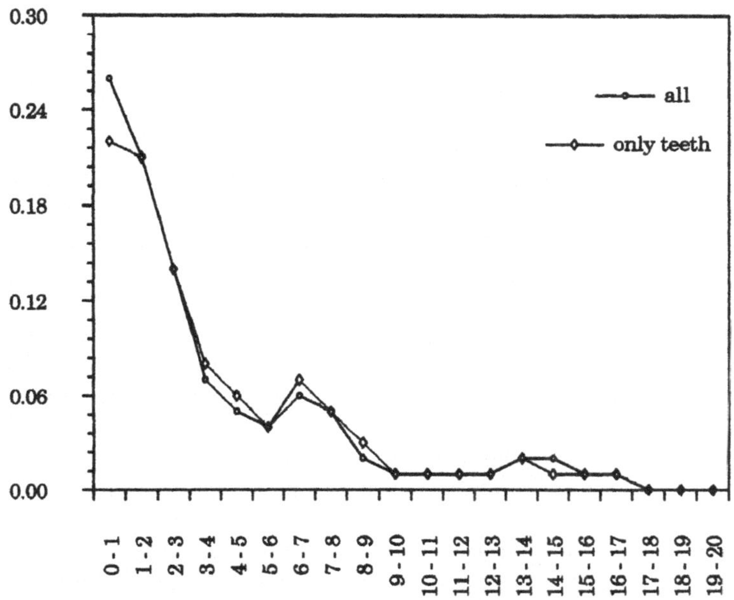

Fig. 10. The mortality curves for the 85 male and 87 female immature individuals from Franzhausen. Note that there is an observable difference in mortality. There are very few individuals in the first age bracket.

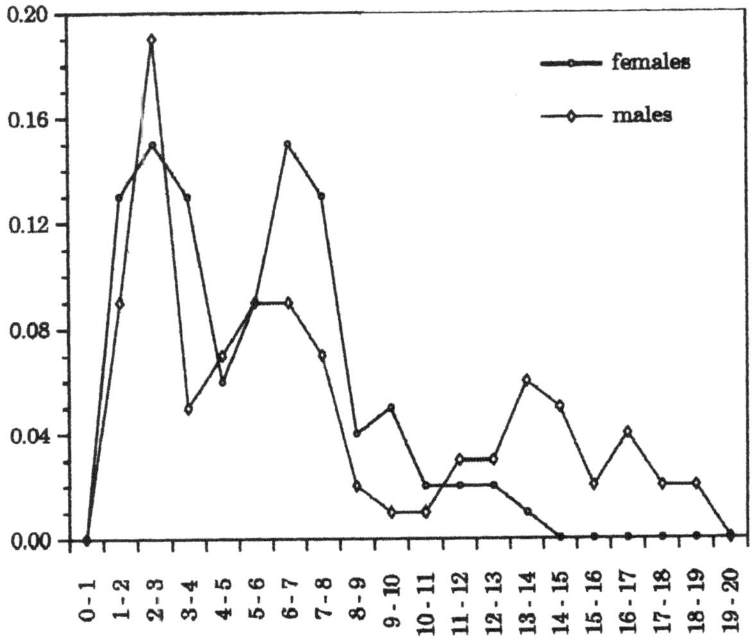

Fig. 11. The mortality curve for the individuals of the medieval population from Gars/Thunau compared with a subsample of those individuals with extant teeth. The differences are slight. Note that the number of infants in the first age brackets is very large.

compares the mortality curve of all the immatures with the mortality curve of the immatures with teeth. We observe that the (normalized) distribution curves differ insignificantly between the two samples.

In order to make some demographic comparisons, the following procedure was adopted: from the enlarged sample of immatures with extant tooth dimensions (i.e., including dimensions reconstructed with the enhanced mirroring technique), we create two subsamples by applying the weighted boundary technique in order to determine the sex. One subsample contains all the males we find this way (a total of 16), the other all the females (a total of 26).

Fig. 12 shows a slight difference between males and females, which varies by age group. However, we caution against observing a difference in mortality rate: the differences are smaller than the reliability of sex determination by the weighted boundary technique.

A graph like this would be meaningless if the enhanced mirroring technique had not been used, as the sample size would have been much smaller, and the probability of correct sex determination as well. Using the discriminant function method is ruled out – despite its much higher sex determination reliability – for practical reasons: only three (!) immature individuals have all six tooth dimensions (either extant or reconstructed using the enhanced mirroring technique). One reason why the very young individuals are not present in the sex specific mortality curves is the absence of permanent teeth. We must thus look for methods of sex

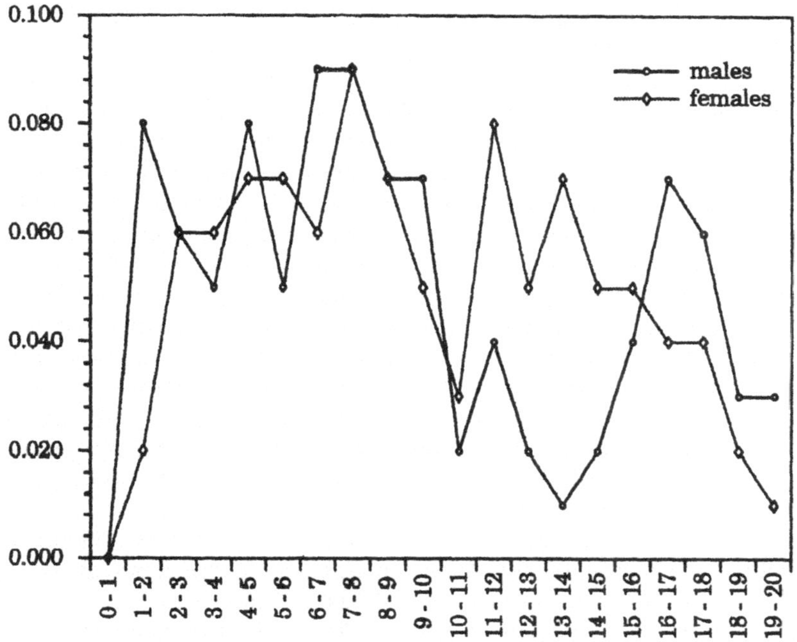

Fig. 12. The mortality curves for the 16 male and 26 female immature individuals from Gars/Thunau whose sex has been determined by enhanced mirroring technique together with the weighted boundary method (see text). There are very few individuals in the first age bracket.

determination based on deciduous teeth. One proposal to remedy the situation is to attempt to reconstruct mandibular tooth dimensions from maxillary ones (and vice versa) and permanent tooth dimensions from deciduous ones; the possibility of success has not been ruled out (Prossinger and Teschler 1995).

Conclusions

We can use the (relativly simple) weighted boundary technique, where the average male and female permanent tooth dimensions are derived from a reference population, in order to reliably determine the sex of immature individuals, by using the enhanced mirroring technique. Nonetheless, one cannot achieve reliabilities comparable with those obtained with the discriminant functions of adult dentitions.There are several reasons for presenting such disappointing results.

First, our rejection of the weighted boundary method, *despite* its singularly attractive feature (namely: that only very few extant tooth dimensions need be used) is based on the outcomes of two numerical tests. Obviously, one can test the prediction performance on the same population which was used to derive the boundaries. This self-consistency test thus rules out possibilities of population-dependent effects. The other test of consistency was to see how well the sex determined by discriminant analysis agrees with the one defined by the weighted boundary method. Neither test was overwhelmingly encouraging.

Second, we suspect that if results presented at workshops, conferences or in published papers which use this method (or some variant) seem to justify it, the situation may be fortuitous; justification should pass a more severe test. The population sizes used in our investigations are quite large (134 immatures with extant teeth in Gars/Thunau and 172 in Franzhausen), so we can feel confident that our conclusions carry some weight. By using only archeologically retrieved material, rather than relying on modern reference populations, we introduce bias into our samples – namely, the bias that the survival of evidence is determined by geological, climatic and agricultural circumstances that existed then and during the intervening repository period. This bias, on the other hand, is important, as we aim to draw conclusions about prehistoric populations, and we must draw these very conclusions concerning sex specific death risks from similarly biased samples. While, at first sight, this bias may seem to be a culprit, we actually welcome its existance: paleoanthropology tacitly assumes biased data. Anthropologists would like to draw conclusions concerning sex specific death risks, but we (and presumably others) presently cannot do it in a succsessful manner, due to methodological limitations – as explained in our extensive rejection analysis of the currently employed sex determination methods for immatures.

The third reason is epistemological. As Amos (1996), Dunthorn (1996) and West (1996) have shown, the empirical sciences are susceptible to a statistically driven feedback cycle that may enhance the respectability of erroneous work. If an investigation showing an acceptable phenomenon or an anticipated result is published, then it determines what future research results are considered publishable, irrespective of their validity. The reason is that the empirical sciences, especially the inductive ones (in particular, anthropology) are susceptible to looking for results that confirm methods previously published. Indeed, the result of a primary study will establish the validity of future published work with a probability greater than 0.5 at better than 5% confidence level (Dunthorn 1996). This selection process can be modelled by a classical combinatorial probability game. Scientific methods have, however, some incorporated safeguards: if a further study of some phenomenon is at extreme odds with a first published study, then the probability can just as well be reversed to some other, stable probability value. Presenting disappointing results (especially if they have included consistency tests) may, in fact, be ocassionally necessary.

Fourth, we use the results of these investigations not only as a justification for rejecting them, but also for justifying why we are pursuing other, more promising – yet somewhat more ambitious methods (Prossinger and Teschler-Nicola 1995).

References

Alt KW, Rieger S, Vach W, Krekeler G (1995) Odontometrische Geschlechtsbestimmung. Evaluierung frühmittelalterlicher Bestattungen. (Summary: Sex determination by odontometric parameters. Evaluation of early medieval burials.) Rechtsmedizin 5: 82–87.

Amos B (1996) Nature 379: 484.

Bailit H, Hunt E (1964) The sexing of children's skeletons from teeth alone and its genetic implications. Am J Phys Anthrop 22: 171–174.
Black TK (1978) Sexual dimorphism in tooth crown diameters of the deciduous teeth. Am J Phys Anthrop 48: 77–82.
Boucher BJ (1955) Sex differences in the foetal sciatic notch. J Forensic Med 2: 51–54.
Boucher BJ (1957) Sex differences in the foetal pelvis. Am J Phys Anthrop 15: 581–600.
Bräuer G (1988) Osteometrie. In: Knussmann R (ed) Anthropologie, Band I/1. Fischer, Stuttgart, p. 160–232.
Breitinger E (1980) Skelette aus einer späturnenfelderzeitlichen Speichergrube in der Wallburg von Stillfried an der March, NÖ. Forschungen in Stillfried 4, Veröff der österr Arbeitsgem für Ur- und Frühgesch 23/24: 45–106.
Choi S, Trotter M (1970) A statistical study of the multivariate structure and race-sex differences of American White and Negro fetal skeletons. Am J Phys Anthrop 33: 307–312.
Dahlberg AA (1963) Analysis of the American Indian dentition. In: Brothwell DR (ed) Dental Anthropology. Pergamon, Oxford, p. 149–177.
De Vito C, Saunders S (1990) A discriminant function analysis of deciduous teeth to determine sex. J For Sci 35: 845–858.
Ditch L, Rose J (1972) A multivariate dental sexing technique. Am J Phys Anthrop 37: 61–64.
Duffy JB, Waterfield JD, Skinner MF (1991) Isolation of tooth pulp cells for sex chromatin studies in experimental dehydrated and cremated remains. For Sci Int 49: 127–141.
Dunthorn D (1996) Nature 380: 477.
Fazekas I, Kósa F (1969) Geschlechtsbestimmung bei Feten auf Grund der Hüftknochenmaße. Arch Kriminol 143: 49–57.
Fazekas IG, Kósa F (1978) Forensic Fetal Osteology. Akadémiai Kiadó, Budapest.
Frayer DW (1980) Sexual dimorphism and cultural evolution in the Late Pleistocene and Holocene of Europe. J Hum Evol 9: 399–415.
Friesinger H, Friesinger I (1975) Die Befestigungsanlagen von Thunau. 5000 Jahre Siedlung im Garser Raum. Katalogreihe des Krahuletzmuseums Eggenburg 3.
Garn SM, Lewis A, Kerewsky R (1964) Sex difference in tooth size. J Dent Res 43: 306.
Garn SM, Lewis A, Kerewsky R (1966) Sexual dimorphism in the buccolingual tooth diameter. J Dent Res 45: 1819.
Garn SM, Cole PE, Wainwright R, Guire K (1977) Sex discriminatory effectiveness using combinations of permanent teeth. J Dent Res 56: 697.
Garn SM, Cole, PE, van Alstine WL (1979) Sex discriminatory effectiveness using combinations of root length and crown diameters. Am J Phys Anthrop 50: 115–118.
Goble F, Konopka E (1973) Sex as a factor in infectious disease. Transa New York Acad Sci, 2nd Ser 35: 325–346.
Grimm H, Hildebrand H (1972) Odontometrischer Beitrag zur Einordnung der Neolithiker vom Djebel Shanadud (Sudan). Anthropologie (Prag) 10: 17–20.
Hanihara K, Koizumi K (1979) Sexing from crown diameters in the permanent teeth by discriminant function method. J Anthrop Soc Nippon 87: 445–456 (in Japanese; Summary).
Hoyme LE, İşcan MY (1989) Determination of sex and race: accuracy and assumptions. In: İşcan MY, Kennedy KAR (eds) Reconstruction of Life from the Skeleton. Liss, New York, p. 53–93.
Hunger H, Leopold D (1978) Identifikation. Springer, Berlin.
Hunt E, Gleiser I (1955) The estimation of age and sex of preadolescent children from bones and teeth. Am J Phys Anthrop 13: 479–487.
Kieser JA (1990) Human Adult Odontometrics. The Study of Variation in Adult Tooth Size. Cambridge Univ Press, Cambridge.
Kieser JA, Groeneveld HT, Preston CB (1985a) An odontometric analysis of the Lengua Indians dentition. Hum Biol 57: 611–620.

Kieser JA, Groeneveld HT, Preston CB (1985b) A metrical analysis of the South African caucasoid dentition. J Dent Ass S Afr 40: 121–125.
Langenscheidt F (1983) Diskriminanzanalytische Geschlechtsbestimmung an Hand von Zahnmaßen unter Verwendung von Verfahren zur angenähert unverzerrten Schätzung der Trennstärke. Homo 34: 22–27.
Lavelle CLB (1972) Maxillary and mandibular tooth size in different racial groups and in different occlusal categories. Am J Phys Orthod 62: 29–37.
Lavelle CLB (1984) A metrical comparison of maxillary first premolar form. Am J Phys Anthrop 63: 397–403.
Little RJA, Rubin DA (1987) Statistical Analysis with Missing Data. Wiley, New York.
Lundström A (1943) Intermaxillära tandbreddsförhfllanden och tandställningen. Svensk Tandl Tidskrift 36: 575.
Lundström A (1944) Förtidiga mjölktandsförluster och tandställningen. Svensk Tandl Tidskrift 37: 698.
Lunt D (1969) An odontometric study of medieval Danes. Acta Odontol Scand 27 (Suppl. 55): 1–173.
Majo T, Tillier A-M, Bruzek J (1993) Test des fonctions discriminantes de Schutkowski impliquant l'ilium pour la détermination du sexe dans des séries d'enfants de sexe et d'âge connus. Bull Mém Soc d'Anthrop Paris ns 5: 61–68.
Maudrich U (1977) Odontometrische Untersuchungen: Die Zahngröße als ein Merkmal zur Geschlechtsbestimmung. Diplomarb Med Univ, Berlin.
Mazess R, Cameron J (1972) Growth of bone in school children: Comparison of radiographic morphometry and photon absorptiometry. Growth 36: 77–92.
Moorrees C (1957) The Aleut Dentition. A Correlative Study of Dental Characteristics in an Eskimoid People. Harvard Univ Press, Cambridge, MA.
Moorrees C (1959) The Dentition of the Growing Child. Harvard Univ Press, Cambridge, MA.
Moorrees C, Thomsen SO, Jensen E, Yen PKJ (1957) Mesiodistal crown diameters of the deciduous and permanent teeth in individuals. J Dent Res 36: 39–47.
Mullis K, Gibbs R, Ferre F (eds) (1994) The Polymerase Chain Reaction. Wiley, New York.
Neugebauer J-W (1987) Die Bronzezeit im Osten Österreichs. Forschungsber zur Ur- und Frühgesch 13. Niederöst Pressehaus, St. Pölten.
Neugebauer-Maresch Ch, Neugebauer J-W (1988) Das frühbronzezeitliche Hockergräberfeld Franzhausen I in urgeschichtlicher Sicht. In: Windl H, Neugebauer J-W, Teschler-Nicola M, Neugebauer-Maresch C (eds) Mensch und Kultur der Bronzezeit. Katalog des Niederösterr Landesmus NF 208: 16–36.
Neugebauer J-W, Gattringer A (1988) Rettungsgrabungen im Unteren Traisental im Jahre 1987. Fundber Österr 26: 35–72.
Ounsted M, Scott A, Moar V (1981) Proportionality and gender in small-for-dates and large-for-dates babies. Early Hum Devel 5: 289–298.
Owsley DW, Webb RS (1983) Misclassification probability of dental discriminant functions for sex determination. J For Sci 28: 181–185.
Paul G (1990) Geschlechtsbestimmung von Skeletten mit Diskriminanzfunktionen für bleibende Zähne. Dental Diss, Univ Ulm.
Pääbo S, Higuchi RG, Wilson AC (1989) Ancient DNA and the polymerase chain reaction. J Biol Chem 264: 9709–9712.
Potter RHY, Alcazaren AB, Herbosa FM, Tomaneng J (1981) Dimensional characteristics of the Filipino dentition. Am J Phys Anthrop 55: 33–42.
Prossinger H, Teschler-Nicola M (1995) Metrication meets morphology when attempting to reconstruct both permanent from deciduous and maxilla from mandibular tooth dimensions. In: Radlanski R, Renz H (eds) Proceedings of the 10th International Symposium on Dental Morphology. Brune, Berlin, p. 343–346.

Remane A (1927) Zur Meßtechnik der Primatenzähne. In: Abderhalden E (ed): Handbuch der biologischen Arbeitsmethoden 7, Teil 1, p. 609–675.
Rieger S (1993) Untersuchung zur Geschlechtsbestimmung der Individuen einer merowingerzeitlichen Skelettserie aus Baden-Württemberg anhand der Zahnmaße. Med Diss, Freiburg/Br.
Rottstock F (1975) Odontometrische Untersuchungen an afrikanischen Schädeln und ihre Anwendung auf Kieferreste aus dem Khartoum-Neolithikum. Med Diss, Humboldt-Universität Berlin.
Rösing FW (1983) Sexing immature human skeletons. J Hum Evol 12: 149–155.
Rösing FW (1990) Qubbet el Hawa und Elephantine. Zur Bevölkerungsgeschichte von Ägypten. Fischer, Stuttgart.
Rösing FW, Paul G, Schnutenhaus S (1995) Sexing skeletons by tooth size. In: Radlanski RJ, Renz H (eds) Proceedings of the 10th International Symposium on Dental Morphology. Brune, Berlin, p. 373–376.
Saunders S (1992) Subadult skeletons and growth related studies. In: Saunders S, Katzenberg A (eds) Skeletal Biology of Past Peoples: Research Methods. Wiley-Liss, New York, p. 1–20.
Schnutenhaus S (1994) Weltverteilung der Zahngrößen. Überprüfung der Eignung von Zahnmaßen für eine Taxonomie des Menschen. Dental Diss, Ulm.
Sciulli PW, Williams JA, Gugelchuk GM (1977) Canine size: An aid in sexing prehistoric Amerindians. J Dent Res 56: 1424.
Schranz D, Bartha M (1964) Geschlechtsbestimmung an Zähnen. Z gerichtl Med 54: 10–15.
Schutkowski H (1987) Sex determination of fetal and neonate skeletons by means of discriminant analysis. Int J Anthrop 2: 347–352.
Schutkowski H (1989) Beitrag zur Alters- und Geschlechtsdiagnose am Skelett nichterwachsener Individuen. Anthrop Anz 47: 1–9.
Schutkowski H. (1990) Zur Geschlechtsdiagnose von Kinderskeletten. Morphologische, metrische und diskriminanzalytische Untersuchungen. Nat Diss Göttingen.
Sjøvold T (1988) Geschlechtsdiagnose am Skelett. In: Knußmann R (ed) Anthropologie, Band I/1. Fischer, Stuttgart, p. 444–480.
Specker B, Brazerol W, Tsang R, Levin R, Searcy J, Steichen J (1987) Bone mineral content in children 1 to 6 years of age. Am J Dis Child 141: 343–344.
Sperl J (1990) Harris-Linien bei den frühmittelalterlichen Alemannen von Kirchheim/Teck. Med Diss Ulm.
Stermer Beyer-Olsen E, Alexandersen V (1995) Sex assessment of medieval Norwegian skeletons based on permanent tooth crown size. Int J Osteoarchaeol 5: 274–281.
Stini W (1985) Growth rates and sexual dimorphism in evolutionary perspective. In: Gilbert R, Mielke J (eds) The Analysis of Prehistoric Diets. Academic Press, Orlando, FL, p. 191–226.
Subow A (1965) Anthropological odontology and the historical sciences (russ.). Sov Etnografia 1: 4–12.
Teschler-Nicola M (1992a) Sexualdimorphismus der Zahnkronendurchmesser: Ein Beitrag zur Geschlechtsdiagnose subadulter Individuen anhand des frühbronzezeitlichen Gräberfeldes von Franzhausen I, Niederösterreich. Anthrop Anz 50: 51-65.
Teschler-Nicola (1992b) Untersuchungen zur Bevölkerungsbiologie der Bronzezeit in Ostösterreich: Phänetische Analyse koninuierlicher und nichtkontinuierlicher Skelettmerkmale. Habilitationsschrift, Univ Wien.
Teschler-Nicola, M, Wiltschke-Schrotta K, Prossinger H, Berner M (1994) The epidemiology of an early Mediaeval population from Gars/Thunau, Lower Austria. Homo 45 (Suppl): 131.
Vach W (1994) Logistic Regression with Missing Values in the Covariates. Springer, Berlin.

Weaver D (1980) Sex differences of the ilia of a known sex and age sample of fetal and infant skeletons. Am J Phys Anthrop 52: 191–195.
Weaver D (1986) Forensic aspects of fetal and neonatal skeletons. In: Reichs KJ (ed) Forensic Osteology. Thomas, Springfield, p. 90–100.
West R (1996) Nature 380: 477.

6.5 The Reconstruction of Missing Tooth Dimensions as a Prerequisite for Sex Determination

Hermann Prossinger

Introduction

Prehistoric anthropologists investigating the remains of individuals in cemeteries are interested in, among other things, the sex-linked frequencies of pathologies, epidemiological traits and further demographic data (Teschler-Nicola et al. 1994). We are investigating very large grave fields with a considerable number of immature skeletons: Franzhausen I (Teschler-Nicola 1992) and Gars am Kamp (Friesinger and Friesinger 1975; Teschler-Nicola and Prossinger, this volume). We believe to have developed a method that enhances our ability to determine the sex of these individuals. Although developed for immatures, this method improves sexing of adults just as well.

The multivariate techniques usually used by anthropologists require a complete data set (Dillon and Goldstein 1984), which is often lacking with prehistoric populations. Various strategies have been proposed to complete the data set, often called imputation techniques (Little and Rubin 1987). These techniques can be broadly classified as hot deck imputation (a missing value is replaced by the nearest similar value preceding it), mean imputation (missing values are replaced by arithmetic means of the extant values) and regression imputation (missing values are estimated by some regression model of the extant values). We used a linear regression model in this analysis.

Our linear regression model generates functions that reconstruct missing tooth diameters and we use both these generated functions and the reconstructed tooth dimensions to find sex-differentiating properties in the data set.

Description of Data

The method presented here was developed using two collections of immature individuals: One collection consists of 207 individuals from an Early Bronze Age cemetery (Franzhausen I) in Lower Austria, the other 135 individuals from a medieval cemetery in Gars, Lower Austria. The mesiodistal and buccolingual diameters of all extant deciduous and permanent teeth of all immature individuals from both grave fields were measured.

Due to the sex-specific burial rites in the Early Bronze Age, we know the sex of the buried immature individual (male: head orientated north, body on its side facing west; female: head orientated south, body on its side facing east; for further details see: Neugebauer and Gattringer 1988; Teschler-Nicola 1992). The sex of the individuals from the medieval cemetery is unknown to us and the primary objective of the presented method is to find functional relations between values in the data matrices that enable the determination of sex.

Description of Variables

We generate a data matrix whose rows are the buccolingual and mesiodistal diameters of deciduous and permanent teeth of an immature individual from the grave field. Each column, therefore, is the set of all measured diameters of a specific type of a specific tooth in the grave field. Even in an optimal case, there are missing elements in this matrix. (An individual cannot have all deciduous and permanent teeth simultaneously.) In the grave fields we study, the data matrix is very sparse – most values are missing. Missing values are usually set to zero. The matrix always has 104 columns, but the number of rows depends on the number of individuals in the grave field.

$$
\begin{array}{c}
 \\
i = 1 \ldots N \\
\downarrow \textit{individuals}
\end{array}
\quad
\begin{array}{c}
j = 1 \ldots M \longrightarrow \textit{tooth diameters} \\
\begin{pmatrix}
X_{11} & X_{12} & \cdot & \cdot & X_{1j} & \cdot & \cdot & X_{1M} \\
X_{21} & \cdot & \cdot & \cdot & \cdot & & & X_{2M} \\
\cdot & & & & \cdot & & & \cdot \\
\cdot & & & & \cdot & & & \cdot \\
X_{i1} & \cdot & \cdot & \cdot & X_{ij} & \cdot & \cdot & X_{iM} \\
\cdot & & & & \cdot & & & \cdot \\
\cdot & & & & \cdot & & & \cdot \\
X_{N1} & \cdot & \cdot & \cdot & X_{Nj} & \cdot & \cdot & X_{NM}
\end{pmatrix}
\end{array}
\qquad (1)
$$

Multivariate methods require a compact matrix. One way of meeting this requirement is by striking rows and/or columns with zero values in order to obtain a compact submatrix (i.e., a matrix with only nonzero values) and use it for subsequent multivariate analysis. However, it is our experience that for most grave fields such submatrices do not exist, except for trivial 1×1 matrices. We use the imputation method of replacing values in the matrix with dimensions obtained from nonzero values in the data matrix by a linear regression model that discards the notion of an explanatory value and controls the influence of outliers.

The Conventional LSQ Model and Its Implementation

We look for linear dependencies between various elements in the data matrix. Such linearities can be represented as functions; function values are then used to replace nonzero values in the data matrix. For example, Fig. 1 and Fig. 2 show two possible linear relations between corresponding left and right tooth diameters. There are two options: One possibility is to find out how all measured left diameters (of males) relate to all measured right diameters (of these males) in a grave field (in this case: Franzhausen I). Another possibility is to find out how all measured left diameters of one immature individual (a female from Franzhausen I) relate to her measured right tooth diameters. In either case, we look for a linear function relating the pairs (X_i, Y_i) of the form

$$Y(X_i) = a + bX_i \tag{2}$$

that best fits the data.

We write

$$deviation_i = Y_i - Y(X_i) = Y_i - a - bX_i \tag{3}$$

and expect the sum of the squares of all the deviations, normalized by their standard deviations, to be a minimum (called the method of least squares – LSQ), viz.

$$\chi^2 = \sum_{i=1}^{N} \frac{(Y_i - Y(X_i))^2}{\sigma_i^2} \longrightarrow \text{minimum.} \tag{4}$$

If there are no repeated measurements of the pairs X_i and Y_i (as is usually the case in anthropology), then $\sigma_i \equiv 1\ \forall i$.

This leads to the best fit condition:

$$\chi^2 = \sum_{i=1}^{N} (Y_i - a - bX_i)^2 \longrightarrow \text{minimum.} \tag{5}$$

Setting the first derivatives to zero as a condition for the extremum, we obtain

$$\frac{\partial \chi^2}{\partial a} = 0 = -2\sum_{i=1}^{N} (Y_i - a - bX_i) \tag{6}$$

and

$$\frac{\partial \chi^2}{\partial b} = 0 = -2\sum_{i=1}^{N} X_i(Y_i - a - bX_i) \tag{7}$$

with the solutions

$$b = \frac{N\sum_{i=1}^{N} X_i Y_i - \sum_{i=1}^{N} X_i \sum_{i=1}^{N} Y_i}{N\sum_{i=1}^{N} X_i^2 - (\sum_{i=1}^{N} X_i)^2} \tag{8}$$

and

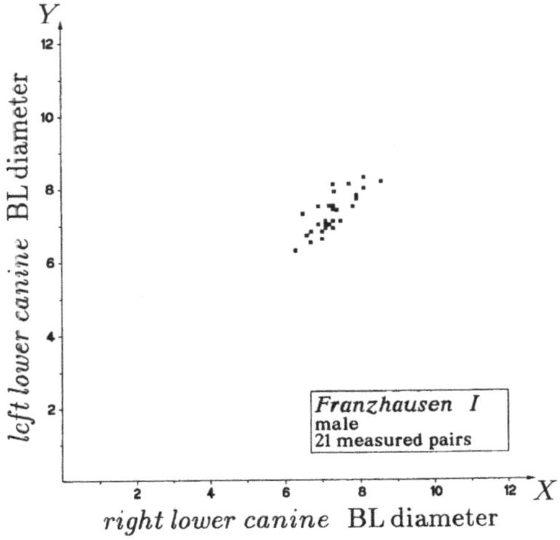

Fig. 1. Measured tooth dimensions of one tooth from a population.

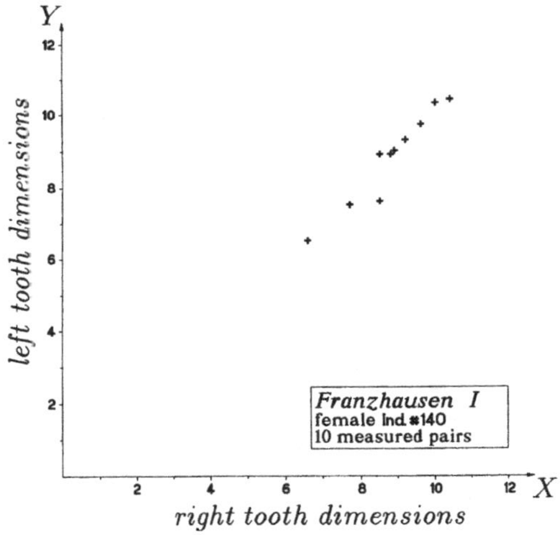

Fig. 2. Measured tooth diameters of teeth from one individual.

$$a = \frac{1}{N}\left(\sum_{i=1}^{N} Y_i - b\sum_{i=1}^{N} X_i \right) \tag{9}$$

Note that the forms (8) and (9) of the solution are not numerically stable; if solutions are to be obtained using digital computers, then various algebraic transformations of (8) and (9) are necessary.

If a tooth dimension Y_k is missing, and its corresponding dimension X_k is known, then one can reconstruct the missing dimension by substitution in the obtained linear function

$$Y_k = a + bX_k \tag{10}$$

If a tooth dimension Y_k is known, and its dimension X_k is missing, then the substitution

$$X_k = \frac{1}{b}(Y_k - a) \tag{11}$$

is *not permissible*, because the deviations have been minimized along the ordinate. This error is frequently encountered in the literature.

An advantage of this linearization approach – one that we want to retain – must be stressed: any choice of a function *F*(*X*) may be used as long as the interpolation function *Y*(*F*(*X*)) is linear in *a* and *b* (Bunke and Bunke 1989; Seber and Wild 1989). In the statistical literature, one distinguishes between the space of all possible solutions $(a, b) \in \mathrm{R}^2$ and the most likely value. As we are not investigating uncertainties in the estimations (in this article), we will not make such a distinction.

Deficiencies of the Conventional LSQ Model

Explanatory or Control Variables

As noted, the interpolation is not symmetric in *Y* and *X*. One would assume that a straightforward remedy of this situation would be: if values X_k are present, use the nonzero pairs (X_k, Y_k) and interpolate *Y*(*X*) to replace missing Y_k values as shown in (10). If values Y_k are present, use the nonzero pairs (Y_k, X_k) and interpolate *X*(*Y*) (with solutions a' and b') to replace missing X_k values with

$$X_k = a' + b'_k \tag{12}$$

We most emphatically stress that the slopes obtained by the two interpolations *X*(*Y*) and *Y*(*X*) are not reciprocals of each other (as will be shown below) and this nonidentity is an indication of a methodological shortcoming in the approach (5).

Fig. 3 shows that in the conventional LSQ model the deviations are determined by the ordinate values only – in effect assuming that the abscissa values are sufficiently exact. The requirement that abscissa values are sufficiently exact so that they can be considered control variables or explanatory variables and that the ordinate variables are dependent variables is often encountered in the physical sciences. Data collected by anthropologists never satisfies this requirement (Orear 1982; Fuller 1987; Reed 1989). Rather, the uncertainties in both abscissa and ordinate values are comparable and the model must be refined to account for this property of the collected data (MacDonald 1975; Jeffreys 1980; Eichhorn and Standish 1981; Jeffreys 1981; Fuller 1987). After all, the observer measured the dimensions of teeth without knowledge of the possible absence of its laterally symmetric counterpart.

Furthermore, a model cannot be considered satisfactory if the linear interpolation of left vs. right is different from the interpolation right vs. left. Neither (tooth) dimension can have preferential status. We deny the implication (which

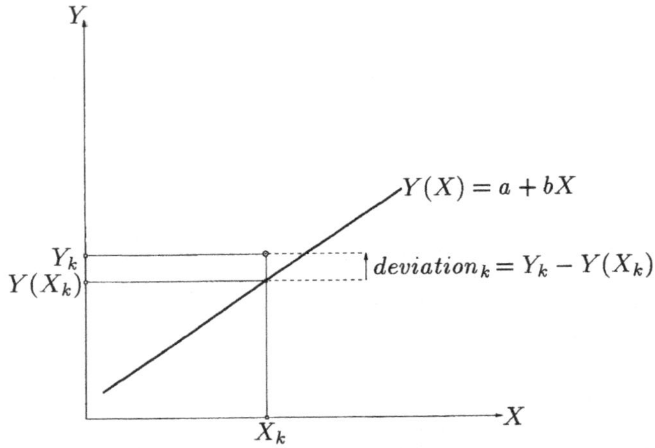

Fig. 3. Conventional LSQ linear interpolation with dependent and independent (control or explanatory) variable.

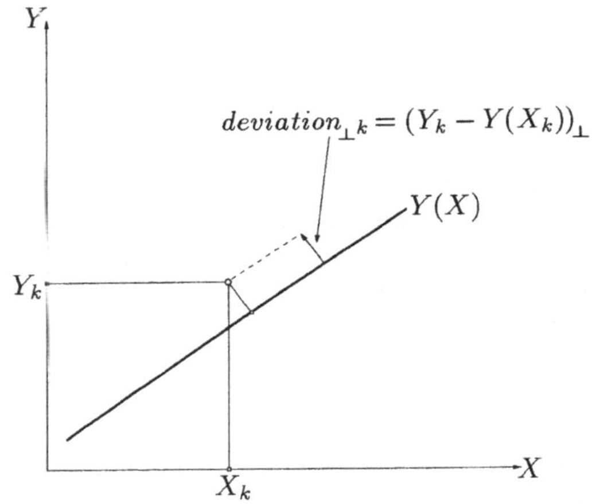

Fig. 4. Linear interpolation assuming uncertainties in both variables.

follows from the assumptions of the conventional LSQ model) that a right tooth dimension explains a left tooth dimension (or vice versa).

If the data has measurement uncertainties in both ordinate and abscissa values, then the deviation must be perpendicular to the interpolated line. We must therefore replace the condition

$$\sum_{i=1}^{N} deviation_i^2 \longrightarrow \text{minimum} \tag{13}$$

with the condition

$$\sum_{i=1}^{N} d_i^2 = \sum_{i=1}^{N} \frac{deviations_i^2}{1 + slope^2} \longrightarrow \text{minimum} \tag{14}$$

If we assume a linear function relating the variables in the form

$$Y = \alpha + \beta X \tag{15}$$

we obtain the *least squares fit with uncertainties in both dimensions*

$$\sum_{i=1}^{N} d_i^2 = \sum_{i=1}^{N} \frac{(Y_i - \alpha - \beta X_i)^2}{1+\beta^2} \longrightarrow \text{minimum}. \tag{16}$$

The denominator in (16) makes the first derivative

$$\frac{\partial}{\partial \beta} \sum_{i=1}^{N} d_i^2 = \frac{\partial}{\partial \beta} \sum_{i=1}^{N} \frac{(Y_i - \alpha - \beta X_i)^2}{1+\beta^2} \tag{17}$$

rather unwieldy. In the next section, we present a method for solving the (nonanalytic) minimization condition (16) which can be applied to similar, more computationally challenging conditions.

Outliers

A further limitation of the conventional LSQ model needs to be dealt with: the problem of outliers.

Both the assumptions (5) and (16) imply that the deviations and the perpendicular distances are normally distributed around the most likely value $Y(X_k)$. Working with such a distribution is rarely justified. The formalism of the conventional LSQ model (minimization of least squares of deviations along the ordinate) is often implemented because of its algorithmic convenience. For example, the derivatives of (5) leads to the system of linear equations (6) and (7). Assuming any other distribution of deviations introduces formidable computational challenges.

In the normal distribution, the values exhibit the most central tendency possible and values far from the mean are rare. If the number of data points is small, even a few large departures from the mean will dominate the solution $Y(X)$ because of the squares in the terms in (5) and (16). In statistics, this problem is well-known; the LSQ regression and the normal distribution are nonrobust (Rousseeuw and Leroy 1987).

If we are faced with the problem that values far from the mean are not so rare, then we are dealing with the outlier problem. We call data points outliers if they have large deviations from the assumed regression model. Outliers often occur in anthropological data sets. In the case of tooth diameters, measurements may generate outliers because of pathologies, genetic anomalies, unusual wear, etc. In other words, we may not assume that outliers are due to measurement errors or faulty data recording, but rather that they express biological phenomena. Detection of outliers is, therefore, of interest to biologists in general and anthropologists in particular.

It is desirable to constrain the outliers' influence in any regression model, linear or nonlinear. Since they cannot be discarded from the data set – the matrix is sparse already –, the strategy of implementing some sort of robust technique (Box 1953) is required. There is a gradually growing body of literature dealing with robust statistics and robust techniques (Rousseeuw and Leroy 1987; Bunke and Bunke 1989).

In order to introduce the various strategies that are being developed, some notational conveniences are appropriate.

We can describe the LSQ condition (5) as a special case of

$$\chi = \sum_{i=1}^{N} \rho(z_i) \longrightarrow \text{minimum} \tag{18}$$

where $\rho(z)$ is not necessarily a quadratic function of the deviations z_i. (In the mathematical statistics literature, the deviations z_i are called residuals, as they are considered to be the unexplained part of a distribution model. We prefer to call them deviations, because this name expresses the deviation from the regularity of some kind of biological description.)

In general, for a linear regression model (assuming $\sigma_i \equiv 1 \ \forall i$)

$$z_i = Y_i - Y(X_i;\alpha) = Y_i - Y(X_i;\alpha,\beta)$$

with

$$Y(X_i;\alpha,\beta) = \alpha + \beta X_i \tag{19}$$

The condition (18) implies

$$\begin{aligned} 0 &= \sum_{i=1}^{N} \psi(z) \frac{\partial Y}{\partial \alpha} \\ 0 &= \sum_{i=1}^{N} \psi(z) \frac{\partial Y}{\partial \beta} \end{aligned} \tag{20}$$

where $\psi(z)$ is the derivative of $\rho(z)$ with respect to z:

$$\psi(z) = \frac{\partial \rho(z)}{\partial z}$$

The solutions to these two equations are called the maximum likelihood estimators (Seber and Wild 1989). Note that the two equations (20) are only linear in a and b if $Y(X; a, b)$ is linear in a and b (as in (19)) and $\rho(z)$ is quadratic in z.

The function $\rho(z)$ determines the type of distribution of the deviations. Here is a sample of various distributions, together with the various names used in the literature:

1) Gauss (normal distribution) (Plackett 1972)

$$\rho(z) = cz^2, \qquad \psi(z) \sim z$$

2) Laplace (double-sided exponential) (Johnson and Klotz 1970; Press et al. 1992)

$$\rho(z) = c|z|, \qquad \psi(z) \sim \operatorname{sgn}(z)$$

3) Cauchy-Laurentian (Huber 1981; Press et al. 1992)

$$\rho(z) = \ln\left(1 + \frac{z^2}{2}\right), \qquad \psi(z) \sim \frac{z}{1 + \left(z^2/2\right)}$$

4) Andrew's sine (Hogg 1979; Press et al. 1992)

$$\rho(z) = c\left(1 - \cos\frac{z}{c}\right), \qquad \psi(z) \sim c \sin\frac{z}{c}$$

5) Tukey's biweight (Hogg 1979; Press et al. 1992)

$$\rho(z) = -\frac{c^2}{6}\left(1 - \frac{z^2}{c^2}\right)^3, \qquad \psi(z) \sim z\left(1 - \frac{z^2}{c^2}\right)^2$$

In Fig. 5, the probability densities for a few of these distributions are graphed about the mean $z = 1$. From the curves, one can see that the Gaussian distribution

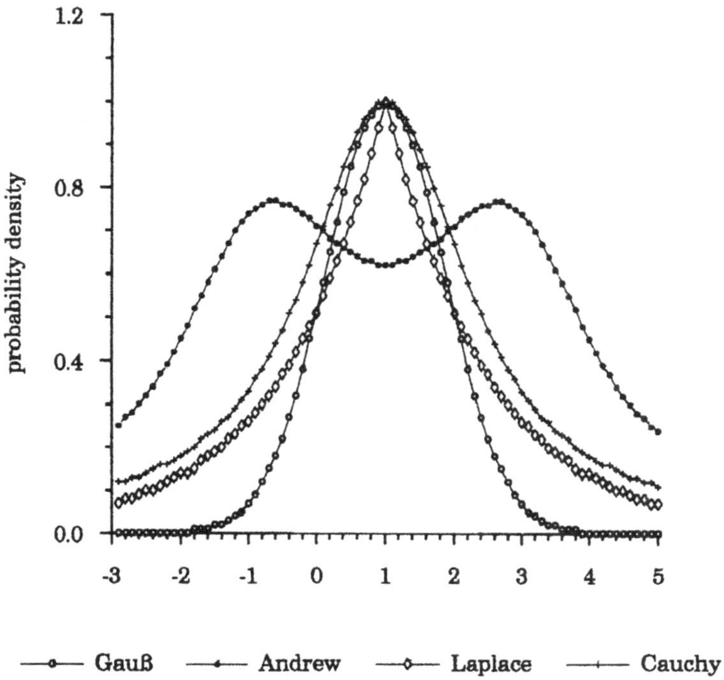

Fig. 5. Various distribution functions that can be used to constrain the influence of outliers.

converges fastest to zero. This extremely rapid convergence makes the LSQ method of determining the maximum likelihood estimators so sensitive to outliers. The other distributions converge more slowly towards zero and, consequently, offer the advantage of being more robust for regression models.

When comparing conventional LSQ methods with more robust methods, Bunke and Bunke (1989, p.135) observed that "The classical procedures are highly sensitive to the gross errors (i.e., to the outliers and long-tailed distributions): 10% of the outliers with standard deviation 3*s* contribute a variance equal to that of the remaining 90% of the cases with standard deviation 1*s*. The outliers can double or triple the variance, so that cutting out their effect could only increase the precision."

In this article, we investigate the performance of outlier control by assuming that the deviations are distributed as a *double* or *two-sided exponential*, with a probability distribution

$$P = \prod_{i=1}^{N} \exp\left[-\left|\frac{Y_i - Y(X_i)}{\sigma_i}\right|\right] \tag{21}$$

The condition of best fit (again assuming $\sigma_i \equiv 1\ \forall i$) is

$$\chi = \sum_{i=1}^{N} \left|Y_i - Y(X_i)\right| = \sum_{i=1}^{N} \left|Y_i - \alpha - \beta X_i\right| \longrightarrow \text{minimum} \tag{22}$$

which is the case if the deviations follow a Laplace distribution (Johnson and Kotz 1970).

The analytic properties of the Laplace distribution are intricate and cumbersome (Johnson and Kotz 1970). The difficulty (and, we suspect, the reason for the unpopularity of this distribution) is the analytic discontinuity in the condition for the extremum when calculating the derivative of χ with respect to β and setting the result equal to zero:

$$\frac{\partial \chi}{\partial \beta} = 0 = \sum_{i=1}^{N} X_i \operatorname{sgn}(Y_i - \alpha - \beta X_i) \tag{23}$$

which contains the derivative of the absolute value function (as does the derivative of χ with respect to *a*), namely, the discontinuous function sgn(x), which is defined as

$$\operatorname{sgn}(x) = \begin{cases} 1, & x > 0 \\ 0, & x = 0 \\ -1, & x < 0 \end{cases} \tag{24}$$

A Generalized Linear Regression Model

"Anthropologists, as far as they use the mathematical multivariate statistical approaches at all, tend to rely on standard statistical packages. As a consequence, the advantages that specifically designed approaches might offer are insufficiently recognized." (van Vark and Schaafsma 1992) Combining the control of outliers (by postulating a Laplace distribution of deviations) with the fact that there are uncertainties in both the ordinate and abscissa values leads to the condition

$$\chi = \sum_{i=1}^{N} \left| \frac{Y_i - Y(X_i)}{\sqrt{1+\beta^2}} \right| = \sum_{i=1}^{N} \frac{|Y_i - \alpha - \beta X_i|}{\sqrt{1+\beta^2}} \longrightarrow \text{minimum.} \tag{25}$$

This condition is both analytically and computationally even more challenging than (17), because the derivative with respect to β is

$$\frac{\partial \chi}{\partial \beta} = \sum_{i=1}^{N} \frac{X_i \sqrt{1+\beta^2}\, \mathrm{sgn}(Y_i - \alpha - \beta X_i) - |Y_i - \alpha - \beta X_i| (\partial / \partial \beta) \sqrt{1+\beta^2}}{1+\beta^2}. \tag{26}$$

This derivative is extremely unwieldy and, furthermore, possesses discontinuities.

However, we are dealing with large data matrices and we need not look for solutions for α and β analytically. Nelder and Mead (1965) have presented an algorithmic method – called the simplex method – which can be used to minimize non-analytic functions of many variables. We use an implementation of this general algorithm – aptly described as the AMOEBA – from Numerical Recipes in FORTRAN (Press et al. 1992) to minimize the function $\chi = \chi(\alpha, \beta)$. We find implementing this procedure preferable to other techniques of solving (26), because the AMOEBA routine is extraordinarily robust (in the computational sense of being insensitive to rounding errors) and is not restricted to numerical solutions of linear models. We can generalize the model – via error bars in both dimensions and a Laplace distribution of deviations – and will present such results at a later date.

The function in two variables $\chi = \chi(\alpha, \beta)$ which is to be minimized can be represented as a two-dimensional surface in three-dimensional space. Starting with three points that bracket the minimum, the AMOEBA procedure gradually moves downhill towards the lowest point within the bracketed points – thus a local minimum is found. The choice of the initial bracketing points is crucial for a successful completion of a computational run. Programming care must be exercised to ensure that the procedure of moving downhill does not skip over the bracketed minimum. Numerical constraints specified by the programmer define how close to the minimum the procedure terminates.

The minimum of $\chi = \chi(\alpha, \beta)$ satisfying the condition (25) is the solution $Y = \alpha + \beta X$ that linearizes the nonzero pairs (Y_k, X_k); such a solution is necessarily symmetric in Y_k and X_k; in fact, we use this symmetry as a control when finding solutions during the AMOEBA iteration steps. Any missing values on either the ordinate or the abscissa can now be reconstructed by direct substitution into this solution:

$$\begin{aligned} \text{if } (X_k, 0) &\Rightarrow Y_k = \alpha + \beta X_k \\ \text{if } (0, Y_k) &\Rightarrow X_k = \frac{1}{\beta}(Y_k - \alpha) \end{aligned} \tag{27}$$

Using the Model in the Strategy of Sex Determination

The general strategy can be described as a series of steps:

1) Start with the raw data matrix as defined by (1).

2) Choose corresponding pairs of nonzero data points in a row, corresponding to a single individual. We advise against using pairs within a column, because this would mask sex-specific relations between data pairs. However, the interpolations within a row using the method described above will retain sex-specific traits of the individual. The anthropologist's intuition and expertise are necessary guidelines for finding anthropologically meaningful pairs.

3) Develop a linear model with the condition (25). The solution α, β is used in the substitution (27) to reconstruct missing data points. Both abscissa and ordinate reconstructions will replace nonzero values in each row of the matrix.

4) Repeat steps 2) and 3) for each row. Note that the number of interpolations will be

$$\begin{matrix} \textit{number of} \\ \textit{combinations} \\ \textit{of pairs} \end{matrix} \times \begin{matrix} \textit{number of} \\ \textit{individuals} \\ \textit{per matrix} \end{matrix} \qquad (28)$$

i.e., a very large number of minimization computations. In this present investigation, there are 87 immature females and 85 immature males in Franzhausen I, and 135 immature individuals of indeterminate sex in Gars am Kamp. If interpolations reconstruct dimensions via only 2 pairs per individual, then this project requires 614 interpolations using AMOEBA. In a later section, a discussion of the overall computational burden and computational costs per reconstructed tooth dimension will be presented.

5) After some implementations of steps 2) to 4), the matrix will be considerably less sparse. Then devise a programmable algorithm that determines a compact submatrix by striking rows and columns containing zero values.

6) We have two grave fields, and consequently two data matrices. This method can be applied to any matrix. The data matrix in one of the grave fields (Franzhausen I) can be subdivided according to sex: one matrix for males, one matrix for females. Apply discriminant analysis (Dillon and Goldstein 1984; Backhaus et al. 1989) to find two functions of the matrix elements that discriminate maximally between the two sexes, i. e., the two submatrices.

7) Use the discriminant functions found in 6) to determine the sex of each individual of the second grave field (in our case: Gars am Kamp) – or any other grave field. The necessary discussion as to whether populations can be compared is not the topic of this article. We stress that the interpolation *per se* – relying only on the data from one individual at a time – is not population dependent.

Results

This article describes the method of reconstructing missing tooth dimensions in order to differentiate immature individuals in grave fields according to sex, not the sex distribution found in the grave fields using this method.

First, we look at results of the linear least squares interpolation (the conventional LSQ model).

Fig. 6 shows how the least squares interpolation differs when minimizing along the ordinate versus minimizing along the abscissa. Notice another reason why interpolating a sample of same tooth dimensions in a population (even when the sex is known) is misguided: the tooth dimensions are clustered so closely together that any slight measurement error along one dimension axis can throw the interpolation way off. For a detailed discussion of such effects, see Rousseeuw and Leroy (1987).

Fig. 7 shows the result of linear least squares interpolation of the tooth dimensions of an individual (in this case, an immature female from Franzhausen I). Although the

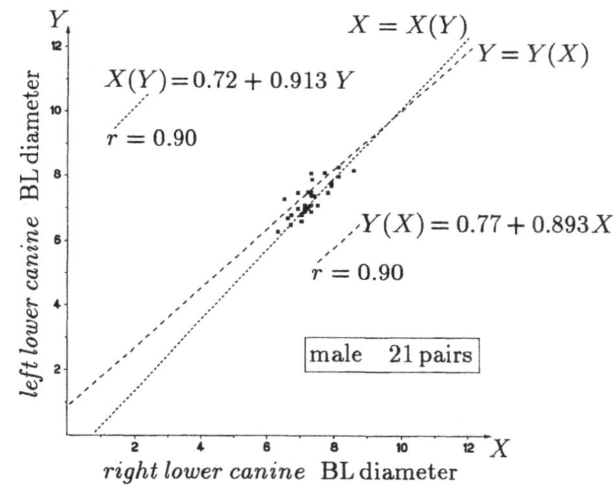

Fig. 6. Conventional least squares (Gauss distribution) unsymmetric linear fit: pooled data rom the population.

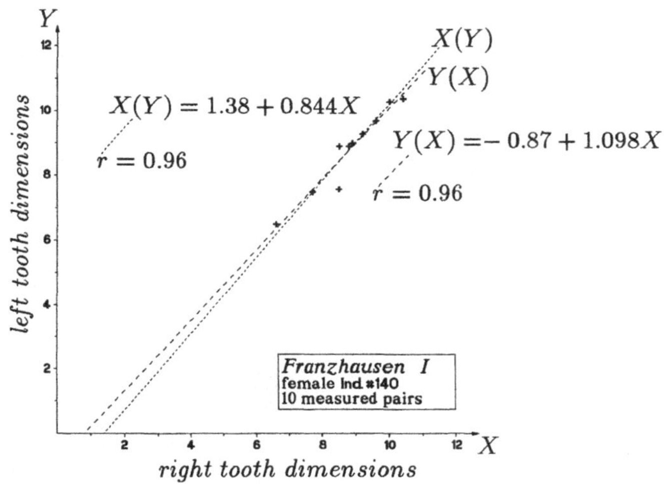

Fig. 7. Conventional least squares (Gauss distribution) unsymmetric linear fit: single individual.

two interpolation lines are closer than in the previous case, reconstruction again depends on whether the sum of the ordinate deviations or sum of the abscissa deviations has been minimized.

The next two graphs show the performance of a linear interpolation if no preference is given to ordinate or abscissa values. Fig. 8 shows a standard least squares interpolation, but with the modification of taking uncertainties in both ordinate and abscissa measurements into account. Notice how the outliers influence the solution.

If, however, we do not assume a Gaussian distribution of uncertainties in both coordinate directions, but rather a two-tailed exponential probability distribution, we obtain the result shown in Fig. 9. It demonstrates a successful linearization: the Laplace distribution does constrain the influence of the outliers, the assumption of

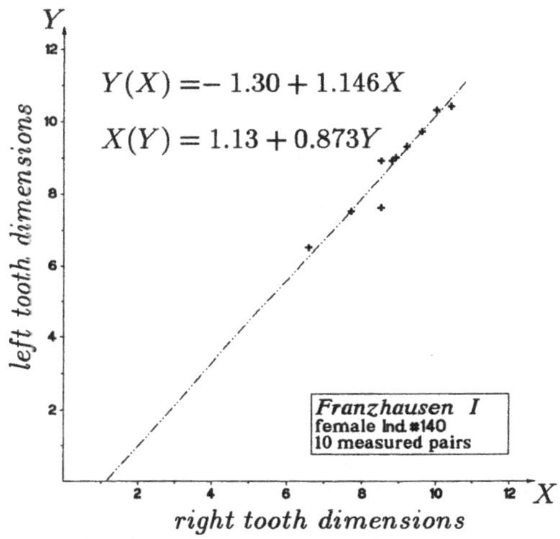

Fig. 8. Symmetric linear fit – Gauss distribution of uncertainties.

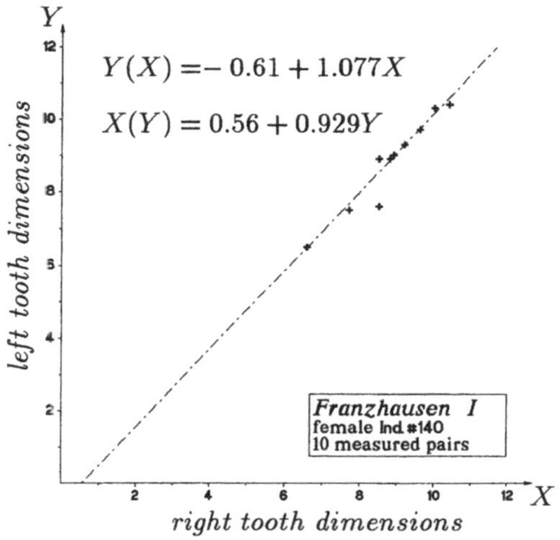

Fig. 9. Symmetric linear fit – Laplace distribution of uncertainties.

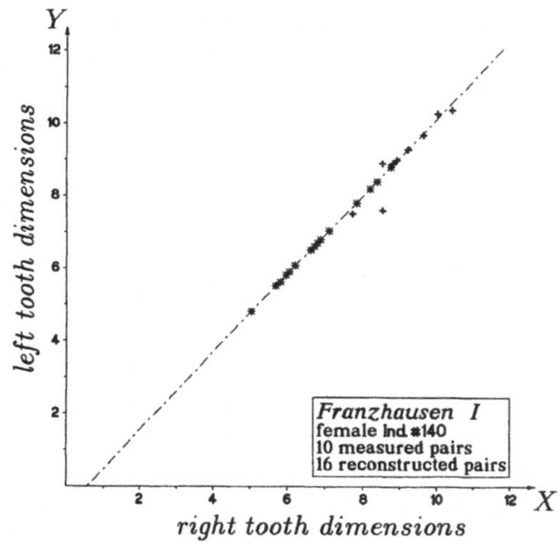

Fig. 10. Symmetric linear fit (Laplace distribution) and reconstructed pairs.

measurement uncertainties in both ordinate and abscissa values generates a symmetric solution of the condition (25). In this particular case, 86 iterations of AMOEBA were required. The iterative routine was stopped when the solutions for the slope and the intercept remained stable in the fourth decimal place – ensuring adequate precision to avoid spurious rounding error effects.

The solution $Y(X)$ shown in Fig. 9 has been used to reconstruct 16 missing dimensions, as shown in Fig. 10: the crosses represent measured data, the stars represent the reconstructed data. Because the solution is symmetric in $X \leftrightarrow Y$, only one linear function is necessary. The outliers will be retained for further analysis, but they have not overly determined the values of the reconstructed solutions.

Computational Burden

An overall tally of the computational demands is difficult to assess. A few numbers may elucidate the technical aspects of this project.

1) Presently, we assume a Laplacian distribution of deviations and assume uncertainty in measurements in both ordinate and abscissa. We use the AMOEBA routine to find the minimum condition (25). This routine is iterative and is terminated when the merit function $\chi(\alpha, \beta)$ does not vary in the fourth decimal place. This requires at least 50 iterations. The number of iterations depends on a good choice of initial values (i.e., the quality of the bracket). In some cases, however, more than 250000 (!) iterations – even with a good choice of an initial bracket – will not detect a minimum. This happens when the two dimensional surface is very flat around the minimum (due to many outliers) and AMOEBA crawls around the minimum region, failing to decide where it actually is to be found. Experience shows, however, that such a flat minimum region is still characterized by a pair of values for α and β that are a (numerically) satisfactory solution of the minimization condition (25).

2) To obtain a good initial bracketing of a possible solution, a conventional least squares linear interpolation is performed first. For the present project, this requires more than 600 least squares interpolations.

3) The AMOEBA must update the terms in (25) several times per iteration – anywhere from 3 to 30.

4) For each reconstructed tooth dimension, a substitution in the linear equation (27) is necessary.

5) The number of reconstructed data points depends on the 'quality' of the data: if few teeth have been lost, the number of computations will be high, and the number of reconstructed tooth dimensions will be small, but these reconstructed dimensions can be expected to be very realistic. If many individuals have few teeth, then the number of computations decreases dramatically, and the number of reconstructions can be quite high, but the reconstructed dimensions might not be so reliable.

6) In general, a computer program replaces zero values in the data matrix with reconstructed dimensions. Whether these reconstructed dimensions are used in a further interpolation for some other unmatched pair is determined by the reliability of the interpolation functions. At least one indicator of reliability that can be translated into a control flag in the evaluation program is, for every interpolation, the root mean square perpendicular distance of the measured data to the interpolated function. Furthermore, it is possible to estimate the uncertainty of the slopes and intercepts with a type of Monte Carlo method. Presently, it is not possible to include a complete theory of cutoff guidelines because the Laplace distribution is so unwieldy, which reflects the difficulty of outliers in palaeoanthropological data. We emphasize that we cannot discard outliers, because doing so would make the data matrix unacceptably sparse (and outliers do contain information for anthropologists). There are various options that are being developed to further assess the quality of reconstructed values in the data matrix (1).

7) In both grave fields, an obvious first choice of pairs (left versus right) usually doubles the number of nonzero data elements in the matrix with – in the cases under study – very acceptable reconstructions. Another choice of pairs increases the number of nonzero data elements by about 30% – depending on the choice of pairs and the acceptability of uncertainties. We stress that these percentages depend on the extent and degree of preservation of teeth in the grave fields.

Discussion

We consider the use of a LSQ approach in a linear regression model numerically sensitive to outliers and biologically unjustified, because we cannot assume one tooth dimension to be more precisely known than the other (or one tooth dimension to be explanatory for the other). Furthermore, we have used an errors-in-both-variables (linear) regression model and have assumed that the outliers' influence can be constrained by assuming a Laplace distribution.

In this article, we have not demonstrated the overall performance of this model for all individuals in both grave fields (more than 300 dentitions) for both deciduous and permanent dentition. Such details are available directly from the author.

Furthermore, we have not compared the performance of a Laplace distribution with any of the other similarly robust distributions due to the limitations on the length of this article. Such a comparison will be published elsewhere. Sufficient to say: The performance of the Laplace distribution can indeed be considered satisfactory.

We also intend to investigate the ramifications when using discriminant analysis with data reconstructed by this model. We would like to point out, however, that the assumptions about outliers in this model do not imply that the distribution of tooth dimensions used in discriminant analysis must also be non-normal. One important advantage of this model is that its reconstruction technique is individual by individual. All the same, the corresponding tooth dimensions of the individuals in a grave field may nonetheless be normal. Outliers of tooth dimensions in one dentition do not imply outliers for the population ensemble.

Conclusion

We have adopted a model construction approach in dealing with missing values in a data matrix. The formalism presented here has been used to model inter-dependencies in tooth dimensions. While the statistical description of data is valid and necessary, the results of the calculations indicate that the number of possible interdependencies is high enough for a model description (preliminary to a statistical analysis) to be considered meaningful.

The computational burden for each reconstructed datum can be high and the formalism necessary is unusually intricate, but very promising. In fact, the experience gained from analysis of such models indicates how to estimate uncertainties in reconstructions, the subsequent implications of sex determination reliability and the possibility of even more powerful generalizations (Prossinger and Teschler-Nicola 1995).

Acknowledgements

Many thanks to M. Teschler-Nicola for her help and patience in discussing and clarifying anthropological standards of measurement quality. Thanks also to her team of anthropologists who prepared, measured, collated and entered all the data into electronic storage. The computations were completed on a powerful computer mainframe at the University of Vienna Computing Center. The generous allocation of CPU time, the supply of support facilities and the advice by the staff is gratefully acknowledged. In particular, many discussions with the late R. Wytek were very constructive. Some of the routines which have been implemented in this project are based on suggestions made by H. Störi from the Institut für Allgemeine Physik, Technische Universität Wien.

References

Backhaus K, Erichson B, Plinke W, Schuchard-Fischer C, Weiber R (1989) Multivariate Analysemethoden. Springer, Berlin.

Box GEP (1953) Non-normality and tests on variance. Biometrika 40: 318-335.

Bunke H, Bunke O (1989) Nonlinear Regression. Functional Relations and Robust Methods. Volume II: Statistical Methods of Model Building. Wiley, New York.

Dillon W, Goldstein M (1984) Multivariate Analysis: Methods and Applications. Wiley, New York.
Eichhorn H, Standish Jr M (1981) Remarks on nonstandard least-squares problems. Astron J 86(1): 156-159.
Friesinger H, Friesinger I (1975) Die Befestigungsanlagen von Thunau, 5000 Jahre Siedlung im Garser Raum. Katalogreihe des Krahuletzmuseums Eggenburg 3.
Fuller W (1987) Measurement Error Models. Wiley, New York.
Hogg RV (1979) An introduction to robust estimation. In Launer RL, Wilkinson GN (eds): Robustness in Statistics. Academic Press, New York, p. 1–18.
Huber PJ (1981) Robust Statistics. Wiley, New York.
Jeffreys W (1980) On the method of least squares. Astronom J 85: 177–181.
Jeffreys W (1981) On the method of least squares II. Astronom J 86: 149–155.
Johnson N, Kotz S (1970) Distributions in Statistics: Continuous Univariate Distributions. Wiley, New York.
Little RJA, Rubin DB (1987) Statistical Analysis with Missing Data. Wiley, New York.
MacDonald J (1975) Comment on "Simple method for fitting data when both variables have uncertainties". Am J Phys 43: 372–374.
Nelder J, Mead R (1965) A simplex method for function minimization. Computer J 7: 308–313.
Neugebauer J-W, Gattringer A (1988) Rettungsgrabungen im Unteren Traisental im Jahre 1987. Fundber Österr 26: 35–72.
Orear J (1982) Least squares when both variables have uncertainties. Am J Phys 50: 912–916.
Plackett RL (1972) Studies in the history of probability and statistics, XXIX: The discovery of the method of least squares. Biometrika 59: 239–251.
Press W, Teukolsky S, Vetterling W, Flannery WB (1992) Numerical Recipes in FORTRAN. The Art of Scientific Computing. Cambridge Univ Press, Cambridge.
Prossinger H, Teschler-Nicola M (1995) Metrication meets morphology when attempting to reconstruct both permanent from deciduous and maxilla and mandibular tooth dimensions. In: Radlanski R, Renz H (eds) Proceedings of the 10th International Symposium on Dental Morphology, Sept. 6–10, 1995. Brünne, Berlin, p. 343–346.
Reed B (1989) Linear least-squares fits with errors in both coordinates. Am J Phys 57: 642–646.
Rousseeuw PJ, Leroy AM (1987) Robust Regression and Outlier Detection. Wiley, New York.
Seber GAF, Wild CJ (1989) Nonlinear Regression. Wiley, New York.
Teschler-Nicola M (1992) Sexualdimorphismus der Zahnkronendurchmesser. Ein Beitrag zur Geschlechtsbestimmung subadulter Individuen anhand des frühbronzezeitlichen Gräberfeldes von Franzhausen I, Niederösterreich. Anthrop Anz 50: 51–65.
Teschler-Nicola M, Wiltschke-Schrotta K, Prossinger H, Berner M (1994) The epidemiology of an early medieval population from Gars/Thunau, Lower Austria. Homo 45 (Suppl): 131.
Vark GN van, Schaafsma W (1992) Advances in the quantitative analysis of skeletal morphology. In: Saunders S, Katzenberg A (ed) Skeletal Biology of Past Peoples: Research Methods. Wiley-Liss, New York, p. 225-257.

7 Geographical and Familial Tooth Variation

7.1 World Variation of Tooth Size

Sigmar Schnutenhaus and Friedrich W. Rösing

Introduction

During the many lively decades of odontology the size of teeth has been measured very often, thus creating the opportunity to develop a taxonomy. This is the objective of the present contribution: the collection of literature data for a differential and hierarchical assessment of affinities between human groups in time and space. Taxonomies allow the possibility to infer ancient migrations, adaptations, chance genetic fluctuations, and other population genetic processes which lead to racial, regional, and local differentiation.

There is a series of advantages for which teeth should be well suited for taxonomy:

- As teeth are not covered by soft tissues the same structure can be measured in the living as well as in the skeleton. Problematic conversions are not needed, and the number of groups increases. On the other hand, the missing cover means constant wear. Such traces of wear, however, are clearly seen and can thus be taken into account.
- After formation, a tooth does not continuously change its shape and composition as bones do with their physiological turn-over. This means a higher ontogenetic stability of traits.
- Due to their high density and hardness teeth usually survive for longer historic periods after the death of the bearer than do bones. This extends and enlarges analyses.
- On the average the heritability of tooth traits is higher than that of other traits, and this property of a dominant genetic determination is useful for taxonomy. On the other hand, heritability is not complete as in biochemical polymorphisms, which impedes taxonomic interpretation.

The idea to use tooth size for taxonomy is not new (see e.g. Wajeman and Levy 1979; Lavelle 1984; Rottstock 1985; Harris and Rathbun 1991). Former investigations, however, did not make sufficient use of the rich literature data or suffered from inferior or lack of numeric taxonomy methods.

Materials

From the literature 160 groups with metric data on the permanent dentition were collected (Tab. 1). One of them was measured in our unit of population history (No 119, Paul 1990). The distribution to continents: Africa 16, America 40, Asia 49, Australia and Oceania 34, Europe 21 populations. One group was non-recent (No 230 Pecos). Moreover, 31 groups were described by deciduous dentition diameters. They are mentioned here spuriously only, for reasons of reduced differentiation. The uneven distribution is obvious; it is the consequence, for example, of wealth and science development, in whose shadow we find Africa; and it is the consequence of innovation in science, in whose shadow we find the vast territories and large populations of the Soviet Union which never really discovered dental anthropology before her end. – The numerical minimum for the inclusion of a group was an average n =10 for all measurements.

The list below reviews the material by the following specifications:

No – group number.
Group – name of population.
Source – author and year of publication.
m – classification and number of measurements:
- A – all 32 diameters measured, namely BL and MD of all teeth of one side,
- B – 30 diameters, BL of tooth 3 not measured,
- C – 28 diameters, tooth 3 not measured,
- D – 28 diameters, tooth 8 not measured,
- E – 28 diameters, BL of teeth 1 and 2 not measured,
- F – 26 diameters, BL of teeth 1, 2, and 3 not measured,
- G – 24 diameters, teeth 1 and 2 not measured,
- H – 24 diameters, teeth 7 and 8 not measured,
- I – 24 diameters, BL of teeth 1 and 2 and tooth 3 not measured,
- J – 24 diameters, BL of teeth 1 and 2 and tooth 8 not measured,
- K – 22 diameters, BL of teeth 1, 2, and 3 and tooth 8 not measured,
- L – 20 diameters, teeth 1, 2, and 3 not measured,
- M – 18 diameters, BL of teeth 1 to 5 and 7 to 8 not measured,
- N – 16 diameters, only MD of all teeth measured,
- O – 16 diameters, only upper jaw measured,
- P – 16 diameters, only teeth 5 to 8 measured,
- Q – 14 diameters, only MD of teeth 1 to 7 measured,
- R – 12 diameters, only MD of teeth 1 to 6 measured.

n – minimal number of individuals excluding the third molar,
- () – no distribution parameter given (sd, se, or cv).

Sex, containing the classes
- MF – separate parameters of males and females,
- (MF) – the group size of one sex is less than half the size of the other sex,
- M – only males,
- bar – sexes pooled.

A complete documentation of all available measurement means, standard deviations, and individual numbers is given in Schnutenhaus (1994), including populations with deciduous dentition measurements and derived parameters for taxonomy.

Table 1. Groups with metric data on the permanent dentition

No	Group	Source	m	n	sex
	Europe				
101	Icelanders	Axelsson and Kirveskari 1983	C 28	196	MF
102	Fins	Alvesalo et alii 1976	C 28	(84)	MF
103	Fins	Hjelmman 1928 in Pedersen 1949	G 24	(?)	MF
104	Skolt Lapps, Fins	Kirveskari 1978	A 32	122	MF
105	Lapps Norway	Selmer-Olsen 1949	A 32	158	MF
106	Norwegians	Alvesalo 1970 in Frayer 1978	D 28	84	MF
107	Swedes	Ebeling et alii 1973	A 32	(26)	M
108	Swedes	Seipel 1946 in Moorrees 1957	N 16	220	MF
109	Swedes	Lysell and Myrberg 1982	Q 14	835	MF
110	Swedes	Forsberg 1988	Q 14	74	MF
111	Swedes	Lundström '44 in Selmer-Olsen 1949	R 12	103	MF
112	English, Midlands	Lavelle 1973	D 28	300	MF
113	English	Lavelle 1973	R 12	40	MF
114	English	Lavelle 1973	D 28	80	M
115	English	Radnzic 1987	R 12	120	M
116	Irish, Aran	Dockrell 1956	R 12	35	MF
117	Germans, Freiburg	Jonas and Schienle 1984	D 28	46	MF
118	Germans, Erfurt	Eismann 1983	H 24	32	MF
119	Germans, Ulm	Paul 1990	K 22	63	MF
120	Swiss, Basel	Jonas and Schienle 1984	D 28	24	MF
121	Hungarians	Boboc 1965	N 16	58	MF
	America				
201	North Am., Whites	Moorrees 1957	Q 14	106	MF
202	North Am., Whites	Keene 1971	Q 14	382	M
203	North Am., New England	Keene 1971	Q 14	(34)	M
204	North Am., South Atlantic	Keene 1971	Q 14	(64)	M
205	North Am., Michigan	Keene 1971	Q 14	(17)	M
206	North Am., Indiana	Keene 1971	Q 14	(32)	M
207	North Am., Ohio	Keene 1971	Q 14	(41)	M
208	North Am., Illinois	Keene 1971	Q 14	(22)	M
209	North Am., East South Central	Keene 1971	Q 14	(19)	M
210	North Am., West North Central	Keene 1971	Q 14	(31)	M
211	North Am., West South Central	Keene 1971	Q 14	(45)	M
212	North Am., Mountain, Pacific	Keene 1971	Q 14	(33)	M
213	North Am., Whites	Keene 1971	Q 14	(44)	M
214	North Am., Whites	Black 1902	A 32	(?)	–
215	North Am., Chicago	Hanihara 1976	Q 14	57	(MF)
216	North Am., Iowa	Bishara et alii 1986	R 12	57	MF
217	North Am., Ohio	Garn et alii 1964	Q 14	(150)	MF
218	North Am., Ohio	Garn et alii 1967	Q 14	294	MF
219	North Am., Ohio	Garn et alii 1968	Q 14	359	MF
220	North Am., Chicago	Dahlberg 1960	A 32	53	M

Table 1 (continued)

No	Group	Source	m	n	sex
221	North Am., Blacks	Keene 1979	Q 14	52	M
222	North Am., Washington DC	Hanihara 1976	Q 14	22	MF
223	North Am., St Louis	Corruccini et alii 1982	G 24	?	–
224	North Am., Blacks	Macko et alii 1979	Q 14	44	MF
225	North Am., Blacks	Richardson et alii 1975	Q 14	80	MF
226	Aleuts	Moorrees 1957	E 28	27	MF
227	Eskimo, Alaska	Doyle and Johnston 1977	L 20	70	M
228	Eskimo, Canada	Ritchie 1923 in Pedersen 1949	G 24	(?)	M
229	Pima, Chicago	Hanihara 1976	Q 14	120	MF
230	Pecos Pueblo	Nelson 1938	A 32	141	–
231	Indians, Pecos AZ	Doyle and Jahnston 1977	L 20	62	M
232	Mexico, Chihuahua	Bishara et alii 1986	R 12	60	MF
233	Mexico, Tlaxcala	O'Rourke and Crawford 1980	H 24	42	(MF)
234	Mexico, Saltillo	O'Rourke and Crawford 1980	H 24	88	(MF)
235	Mexico, San Pablo	O'Rourke and Crawford 1980	H 24	108	(MF)
236	Mexico, Cuanalan	O'Rourke and Crawford 1980	H 24	77	MF
237	Columbia, Ticuna Ind.	Harris and Nweeia 1980	D 28	41	MF
238	Paraguay, Lengua Ind.	Kieser et alii 1985b	F 26	103	MF
239	Peru, Indios	Goaz and Miller 1966	I 24	135	–
240	Brazil, Xingu Indios	DeSmet 1966	A 32	(?)	–
	Australia, Oceania				
301	Aborigines	Campbell 1925	A 32	(43)	–
302	Aborigines, Swanport	Smith 1982	O 16	7	M
303	Aborigines, Yuendumu	Smith 1982	O 16	154	M
304	Aborigines, Kalumburu	Smith 1982	O 16	7	M
305	Aborigines, Broadbeach	Smith 1982	O 16	7	M
306	Aborigines, Roonka	Smith 1982	O 16	7	M
307	Aborigines, Broadbeach	Smith et alii 1981	B 30	8	(MF)
308	Aborigines, Murray Basin	Brace 1979	A 32	(150)	–
309	Aborigines, Tasmania	Brace 1979	A 32	(210)	–
310	Aborigines, Walbiri	Brace 1979	A 32	(7)	–
311	Aborigines, Walbiri	Barret et alii 1963, 1964	A 32	77	MF
312	Aborigines, Adelaide	Hanihara 1976	Q 14	102	MF
313	Whites, Adelaide	Townsend 1983	D 28	88	MF
314	Nasioi, Melanesia	Sofaer et alii 1971	D 28	305	–
315	Nasioi, Melanesia	Bailit et alii 1966	A 32	82	MF
316	West Nakanai, Melan.	Turner and Swindler 1978	E 28	174	M
317	Neu–Pommern, Melan.	Janzer 1927	A 32	(74)	MF
318	Javanese Jogjakarta	Brace 1976	A 32	48	–
319	Javanese	Mijsberg 1931	A 32	164	(MF)
320	Awa, New Guinea	Boyd 1972	M 18	84	MF
321	Auyana, New Guinea	Boyd 1972	M 18	40	(MF)
322	Tairora, New Guinea	Boyd 1972	M 18	129	(MF)
323	Gadsup, New Guinea	Boyd 1972	M 18	138	(MF)
324	Sasaura, New Guinea	Boyd 1972	M 18	139	M
325	Akuna, New Guinea	Boyd 1972	M 18	68	MF

Table 1 (continued)

No	Group	Source	m	n	sex
326	Onteau, New Guinea	Boyd 1972	M 18	31	MF
327	Baieanabuta, New G.	Boyd 1972	M 18	57	M
328	Abiera, New Guinea	Boyd 1972	M 18	19	M
329	Babaraai, New Guinea	Boyd 1972	M 18	23	M
330	Batainabura, New G.	Boyd 1972	M 18	22	M
331	Tauna, New Guinea	Boyd 1972	M 18	25	M
332	Ilakia, New Guinea	Boyd 1972	M 18	20	M
333	Lufa, West Irian	Doran and Freedman 1974	A 32	21	M
334	Goroka, West Irian	Doran and Freedman 1974	A 32	27	M
	Africa				
401	North Africans	Chamla 1980	D 16	73	(MF)
402	North Africa, Jews	Koyoumdjisky-Kaye et alii 1978	I 24	43	MF
403	Teso, Uganda	Barnes 1969	A 32	17	–
404	San (Bush), Sotho	Haeussler et alii 1989	O 16	67	MF
405	Bantu, Central Sotho	Haeussler et alii 1989	O 16	91	MF
406	South Africa, Whites	Kieser et alii 1985a	K 22	99	MF
407	South Africa, Whites	Kieser et alii 1987	D 28	99	MF
408	Bamum, South Africa	Abel 1933	A 32	(21)	–
409	Khoi Coloured, S.A.	Abel 1933	A 32	(42)	–
410	Bantu, South Africa	Abel 1933	A 32	(33)	–
411	San, South Africa	Abel 1933	A 32	(24)	–
412	San Coloured, S. A.	Abel 1933	A 32	(53)	–
413	Hutu, Ruanda	Brabant 1963	O 16	(3)	–
414	San, South Africa	Drennan 1929	A 32	(47)	–
415	Bantu, South Africa	Shaw 1931	A 32	(124)	–
416	Tristan da Cunha	Thomsen 1955	E 28	97	MF
	Asia				
501	Chinese, Beijing	Brace et alii 1984	A 32	41	(MF)
502	Chinese, Harbin	Brace et alii 1984	A 32	13	MF
503	Chinese, Nanjing	Brace et alii 1984	A 32	5	MF
504	Chinese, Yunnan	Brace et alii 1984	A 32	7	MF
505	Chinese, Shanghai	Brace et alii 1984	A 32	26	MF
506	Chinese	Hosaka 1936	E 28	70	M
507	Chinese, Hong Kong	Brace 1976	A 32	27	–
508	Chinese, Liverpool	Goose 1976	D 28	99	MF
509	Chinese	Brace in Frayer 1978	A 32	25	(MF)
510	Ami, Taiwan	Liu 1977	I 24	184	MF
511	Atayal, Taiwan	Liu 1977	I 24	200	MF
512	Ainu, Japan	Brace and Nagai 1982	A 32	7	MF
513	Ainu, Japan	Hanihara 1976	Q 14	34	MF
514	Japanese, Tokyo	Ozaki et alii 1987	A 32	154	(MF)
515	Japanese, Tokyo	Gonda 1959	A 32	(268)	MF
516	Japanese	Yamada 1932 in Moorrees 1957	E 28	186	(MF)
517	Jap., Tokyo, Hokkaido	Hanihara 1976	Q 14	100	MF
518	Japanese	Miyibara 1916	A 32	(106)	MF

Table 1 (continued)

No	Group	Source	m	n	sex
519	Japanese, Kyoto	Brace and Nagai 1982	A 32	33	MF
520	Koreans	Brace and Nagai 1982	A 32	14	(MF)
521	Philippinos, Manila	Potter et alii 1981	D 28	152	MF
522	Thai	Brace 1976	A 32	92	–
523	Bhutaner	Prakash et alii 1979	J 24	70	M
524	Bodhs, North India	Bhasin et alii 1985	D 28	29	M
525	Khatris, Punjab, India	Bhasin et alii 1985	D 28	74	M
526	Jats, India	Bhasin et alii 1985	D 28	132	M
527	Ahirs, India	Bhasin et alii 1985	D 28	32	M
528	Brahmans, India	Bhasin et alii 1985	D 28	34	M
529	Dangis, India	Bhasin et alii 1985	D 28	86	M
530	Kunbis, India	Bhasin et alii 1985	D 28	60	M
531	Varlis, India	Bhasin et alii 1985	D 28	54	M
532	Brahmans, Bengal	Bhasin et alii 1985	D 28	74	M
533	Brahmans,Tamil Nadu	Bhasin et alii 1985	D 28	72	M
534	Jats, India	Kaul and Prakash 1984	K 22	284	MF
535	India	Biviji et alii 1967, 1971	A 32	(153)	–
536	Panjabis, India	Sharma and Kaul 1977	F 26	145	M
537	Tibetans, India	Sharma 1983	F 26	47	MF
538	Chochin, Jews, Israel	Koyoumdjisky-Kaye et alii 1978	J 28	51	MF
539	Chochin, Jews, Israel	Rosenzweig and Zilberman 1967	K 22	57	MF
540	Egyptians, Cairo	Eid and El-Nemarwy 1984	R 12	95	MF
541	Habani, Jews, Israel	Rosenzweig and Smith 1971	K 22	68	MF
542	Yemenite Jews, Israel	Rosenzweig and Zilberman 1967	K 22	42	MF
543	Beduins, Israel	Rosenzweig and Zilberman 1969	K 22	97	MF
544	Circassians, Israel	Koyoumdjisky-Kaye et alii 1977	J 24	55	MF
545	Druses, Israel	Koyoumdjisky-Kaye et alii 1977	J 24	50	MF
546	Kurdic Jews, Israel	Koyoumdjisky-Kaye et alii 1976	J 24	68	MF
547	Yemenite Jews, Israel	Koyoumdjisky-Kaye et alii 1976	J 24	43	MF
548	Samaritans, Israel	Rosenzweig et alii 1969	K 22	134	MF
549	Iraqis	Ghose and Baghdady 1979	R 12	161	MF

Methods

The dental parameters of this study are the mesio-distal (MD) and bucco-lingual (BL) diameters of the crown, i.e. measurements Nos 81 and 81(1) in Bräuer (1988), the central standard for any morphological measurement technique in skeletons and the living. Other measurements were not regarded because they have been taken far less frequently, because they sometimes compete (for originating in different schools or countries), or because they are imprecise in themselves (e.g. crown height). In the case of every group retrieved from the literature the standard congruence of measurement technique was checked – usually without success because most authors gave no or insufficient descriptions of what they did. Dental measurements seem to be very easy, but it is well known that there are still numerous sources of error and bias. Since this cannot be checked directly in the sources it will be part of the discussion below.

For the creation of order among populations one of the prerequisites is an independence of the describing parameters. If the parameters are only available as population means, then the distance by Penrose (1954) has been employed very often (latest and largest example Schwidetzky and Rösing 1992). This method has a makeshift property in its manner of reaching that independence: an average intercorrelation between the traits. This does not make sufficient use of the highly variable entries in the covariance matrix. To improve this we have adopted a different procedure: In a first step the traits were subjected to a factor analysis. For factor extraction the principal component procedure was used. The derived factor loadings were subjected to a varimax rotation for an easier interpretation (see also Hanihara 1976, Lin 1976, Rottstock et al. 1983, Harris and Rathbun 1991). The dental measurements were standardized in order to reduce the influence of size. As commonly done, eigenwerte of less than one were disregarded.

After the factor analytical reduction of the variable number (in the largest procedure from 24 to 4) the factor scores were subjected to a multivariate distance calculation. The one by Mahalanobis could not be used because it demands individual measurements, and the one by Penrose was rejected because it has the makeshift element that the covariance matrix is ignored by using a mean intercorrelation; clearly this does not account for the frequent and large correlation variation between the groups. Therefore the Euclidian distance was chosen (Sokal and Sneath 1963, Creel 1968; Cormack 1971; Steinhausen and Langer 1977; Deichsel and Trampisch 1985).

In the third and last step the large triangular distance matrix is reduced to the taxonomic content by a clustering technique. From at least nine different procedures (Gordon 1987), for this study the hierarchical dual sequential or single linkage method was selected. The resulting trees represent the morphological dental shape similarity structure among the collected groups. Only for the Penrose distance a significance evaluation has been developed (Rahman 1963, details and application see Rösing 1990), not for the Euclidian distance employed here. Therefore only qualitative linkage levels can be observed and described.

This stepwise procedure was run 30 times for different measurement sets (BL always without incisors) and different group numbers in males/females/sexes pooled:

1. All teeth except M3, MD and BL, 24 measurements, 50/32/47 groups
2. All teeth except M3, I1, and I2, MD and BL, 20 measurements, 53/36/54 groups.
3. All teeth except M3, MD, 14 measurements, 104/60/74 groups.
4. All teeth except M2 and M3, MD, 12 measurements, 121/73/99 groups.
5. Upper jaw except M3, MD and BL, 12 measurements, 55/34/50 groups.
6. Lower jaw except M3, MD and BL, 12 measurements, 52/33/51 groups.
7. Distal teeth P2 to M2, MD and BL, 10 measurements, 79/48/68 groups.
8. Distal teeth P1 to M2, BL, 8 measurements, 70/58/70 groups.

Deciduous dentition:

9. All teeth, md and bl, 16 measurements, 11/11/15 groups.
10. All teeth, md, 10 measurements, 26/26/30 groups.

Results

The correlations between the measurements are very high: For the permanent dentition there is an average r = 0.594 for males and 0.634 for females and in the deciduous dentition r = 0.653 for males and 0.582 for females. Measurements of antagonists lie clearly above this average, r = 0.807 for both sexes and both diameters pooled. Also BL vary more closely among each other, thus indicating differentiated genetic control. On the other hand there are several low correlations, e.g. of the MD of the first lower incisors; these are dispersed throughout the matrix. Practically all correlations are significantly different from zero; with the chosen procedure of this paper, i.e. without individual data, the border value for a 1% significance is r = 0.393. When individual data are available, the border is much lower, moreover then there is more inter-group variation (Rottstock 1985).

As in most other metric analyses the number of extracted factors is very high: between 16 and 24. Of them the first explains on the average 70% of the total variation; this factor is caused by size in general, but has a slightly higher loading on MDs (see also Hanihara 1976). The second factor explains 11%, the third 6%, and the fourth 5%; with them the BLs, frontal teeth, and side teeth are loaded. There is no indication of a differentiation between the jaws. The limitation to an eigenwert of more than one means that on the average 81% of the total variation was explained.

The communality, i.e. the part of trait variation which is explained by the factors, was an average h = 0.84. Communalities were higher for BL and lower for first incisors; the latter indicates a rather strong independence of this tooth. The morphogenetic field theory (Butler 1939) could not be confirmed: There is no increased communality for premolars and molars, contrary to other studies which processed MDs and BLs separately (Lombardi 1975; Townsend and Brown 1979: Kolakowski and Bailit 1981).

Fig. 1 gives the dendrogram for Procedure 4 with 12 measurements of the permanent dentition of males, selected because this makes use of the maximal group number of 121. First of all it displays a gradual affinity between the groups, typical for stochastic effects, and not a clear separation of clusters. Moreover, geographic, genetic, or historic differentiation seems to govern individual positions only slightly. Of the rather close clusters 12 may be defined visually, which is marked in the dendrogram. Cluster 1 is dominated by East Asia, clusters 2 to 8 have a slight dominance of Europid groups. Among them cluster 6 is interesting as it comprises most groups by Keene, whereas other North American groups are rather dispersed; this is a typical case of a personal method artefact. Cluster 9 for once is a bit more homogeneous, comprising most morphologically archaic groups of Australia and Oceania, along with the two groups of North American Negrids (Nos 222 and 225). Although these groups occupy neighbouring positions, their similarity is often poor, see e.g. No 620 Australids Walbiri. After some rather chaotic groupings the end of the dendrogram is formed by an explosion of distances for most of the Indian groups, most of them by Bhasin (1985). Normally author clusters are particularly homogeneous, this one is particularly heterogeneous. On a racial level there is a certain lumping of Europids in the first part of the dendrogram, and of Mongolids and

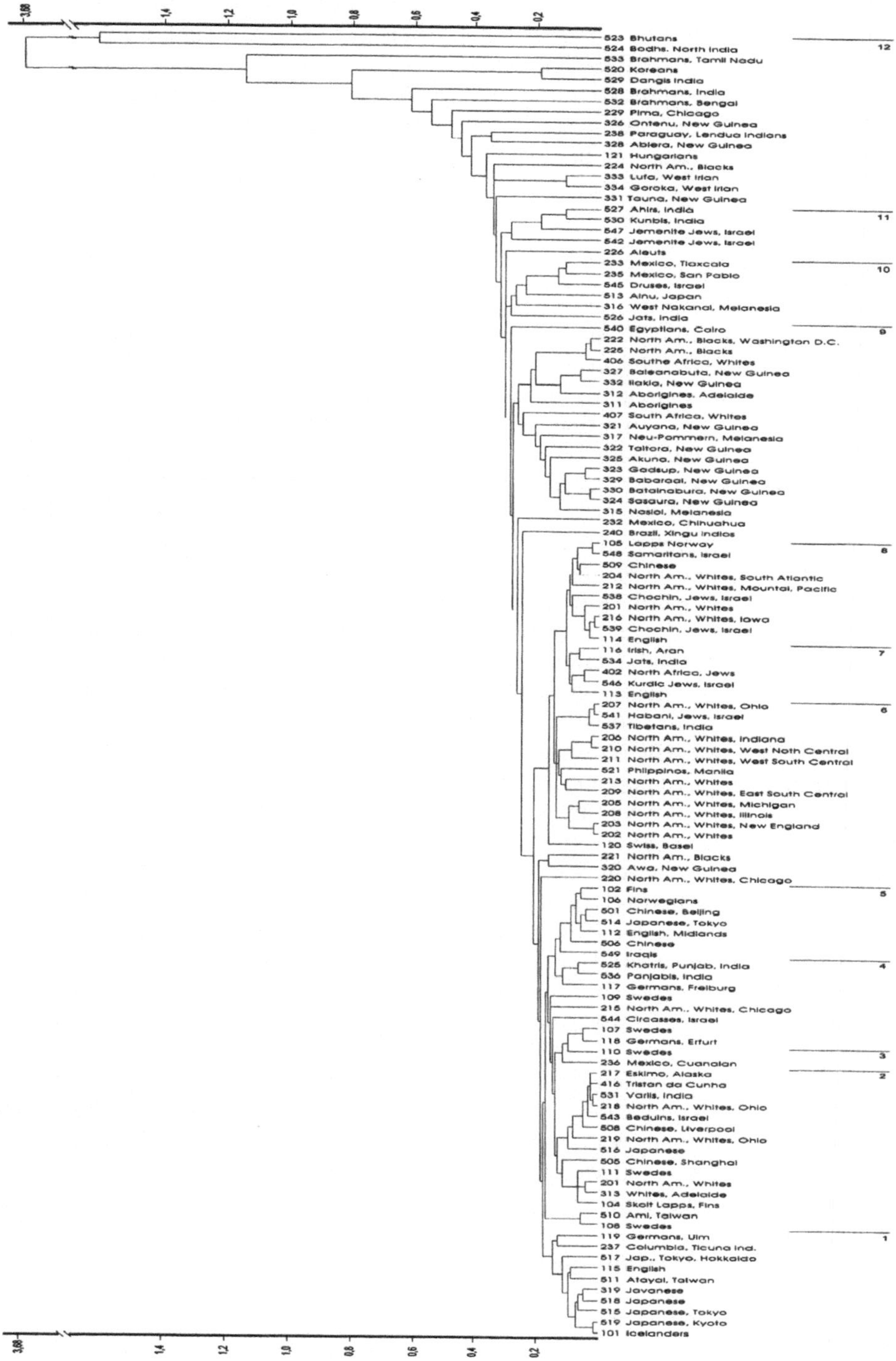

Fig. 1. Dendrogram for procedure 4, 12 measurements permanent dentition of males.

Australids/Melanesids in the second part. The clearest impression, however, is a quite limited order. This is corroborated when specific groups are followed. Only rarely are they in an expected neighbourhood.

The 30 dendrograms (10 different measurement sets; males, females, and sexes pooled) do not differ too much from each other. The single groups are often in a different specific position, but the general picture is always very much the same. Also the partition in clusters and the neighbourhood distribution are similar.

Discussion

A surprise is the emergence of two very clear author clusters. One of them (Keene 1971) is characterized by an order which obviously does not exist. Reasons for such artefacts may be: data and procedure manipulations in order to arrive at "nice", "clean" results. This ambition violates scientific standards and yields nonsense results – but it occurs quite often. The other author cluster (Bhasin et al. 1985) is characterized by a disorder which obviously does not exist either. Reasons for this are less clear. Were data randomly chosen? Were the groups measured by different observers? Was there no standardisation of techniques? Whatever the reasons, they, too, yielded nonsense results.

The influence of different techniques may be seen when comparing different tooth taxonomies, ideally performed by the same author with the same groups. There is one such investigation (Hanihara 1976), which makes use of the distance measures by Mahalanobis and by Penrose. The results are highly different.

These two characteristics – high author sensitivity and strong method dependence – clearly indicate that tooth measurements are not suited for a taxonomy of modern populations. The absence of order is obviously not the result of unknown influence variables, maybe history or minute ecological adaptation, but the result of "convergence", i.e. chance variation which results in chance similarity. Other characteristics like head or postcephal body measurements, pigmentation, dermatoglyphis, or blood and tissue traits have a much better discrimination power. A different picture emerges only when the evolution of tooth sizes is studied (newest and largest example: Wei 1996); in the course of time much stronger differences emerge which then facilitate the reconstruction of order. Moreover teeth have an assessable use for sex diagnosis (see respective chapter in this book).

When we look at the racial level, there is a certain order: Groups with large dentitions from Oceania are more or less clearly separated. Moreover, when factor analyses are performed with individual data, then large groupings like Europids, Melanesids, Negrids, and Mongolids differ (Rottstock 1985). These results suggest that there might be useful taxonomic information in tooth sizes, too, which should emerge when all technical problems are handled with care and when these traits are analysed together with others.

Acknowledgement

We thank Heidi Asmus for correction reading and language revision of the manuscript.

References

Abel W (1933) Zähne und Kiefer in ihren Wechselbeziehungen bei Buschmännern, Hottentotten, Negern und deren Bastarden. Z Morph Anthrop 31: 314–361.

Alvesalo L (1970) The Influence of Sex-Chromosome Genes on Tooth Size in Man. PhD Diss, Univ Turku, Finland. Cit. in Frayer DW (1978).

Alvesalo L, Osborne RH, Kari M (1976) The 47, XYY-Male, Y-Chromosome and tooth size: A preliminary communication. In: Dahlberg AA, Graber TM (eds) Orofacial Growth and Development. Aldine, Chicago, p. 109–118.

Axelsson G, Kirveskari P (1983) Crown size of permanent teeth in Icelanders. Acta Odontol Scand 41: 181–186.

Bailit HL, DeWitt SJ, Leigh RA (1966) The size and morphology of the Nasioi dentition. Am J Phys Anthrop 28: 271–288.

Barnes DS (1969) Tooth morphology and other aspects of the Teso dentition. Am J Phys Anthrop 30: 183–194.

Barret MJ, Brown T, MacDonald MR (1963) Dental observations on Australian aborigines: Mesiodistal crown diameters of permanent teeth. Austral Dent J 8: 150–155.

Barrett MJ, Brown T, Arato G, Ozols IV (1964) Dental observations on Australian aborigines: Buccolingual crown diameters of deciduous and permanent teeth. Austr Dent J 9: 280– 285.

Bhasin MK, Sharma A, Singh JP, Walter H (1985) Morphological and metric dental study on Indians. Z Morph Anthrop 76: 77–90.

Bishara SE, Garcia AF, Jakobsen JR, Fahl JA (1986) Mesiodistal crown dimensions in Mexico and the United States. Angle Orthod 65: 315–323.

Biviji AT, Bhatt A (1971) Odontometric data of Indian teeth: II. Posterior teeth. J Indian Dent Ass 43: 250–252.

Biviji AT, Shourie KL, Singh B, Menezes CG (1967) Odontometric data of Indian teeth: I. Anterior teeth. J Anat Soc India 16: 128–134.

Black GV (1902) Descriptive Anatomy of the Human Teeth. SS White Dental Manufac, Philadelphia.

Boboc G (1965) Studiu statistic si citeva consideratii asupra latimii dintilor temporari si definitivi. Stomatologia 12: 345–354.

Boyd RC (1972) An odontometric and observational assessment of the dentition. In: Littlewood RA (ed) Physical Anthropology of the Eastern Highlands of New Guinea. Univ Washington Press, Seattle, p. 175–221.

Brabant H (1963) Observations anthropologiques et odontologiques sur les dents des Hutu du Rwanda. Musée Royal de l'Afrique Centrale – Tervuren, Belgique Annales – Serie In–8 – Sciences Humaines 47: 1–30.

Brace CL (1976) Tooth reduction in the Orient. Asian Perspectives 19: 203–219.

Brace CL (1978) Cit. in Frayer DW (1978)

Brace CL (1979) Krapina, "classic" Neanderthals, and the evolution of the European face. J Hum Evol 8: 527–550.

Brace CL, Nagai M (1982) Japanese tooth size: Past and present. Am J Phys Anthrop 59: 399–411.

Brace CL, Xiang-Ging S, Zhen-Biao Z (1984) Prehistoric and modern tooth size in China. In: Smith FH, Spencer F (ed) The Origins of Modern Humans. Liss, New York, p. 485–516.

Bräuer G (1988) Osteometrie. In: Knussmann R (ed) Anthropologie, Band 1: Wesen und Methoden der Anthropologie. Fischer, Stuttgart, p. 186–189.

Butler PM (1939) Studies of the mammalian dentition – Differentiation of the postcanine dentition. Proc Zool Soc London, Ser B, 109: 1–36.

Campbell TD (1925) Dentition and Palate of the Australian Aborigines. Hassall Press, Adelaide.
Chamla M (1980) Études des variations métriques des courrones dentaires des Nord-Africains, de L'Épipaléolithique à l'époque actuelle. L'Anthropologie 84: 254–271.
Cormack RM (1971) A review of classification. J R Statist Soc A 134: 321–367.
Corruccini RS, Handler JS, Mutaw RJ, Lange FE (1982) Osteology of a slave burial population from Barbados, West Indies. Am J Phys Anthrop 59: 443–459.
Creel N (1968) Die Anwendung statistischer Methoden in der Anthropologie. Beitrag zur Erklärung der Entwicklungsprozesse europäischer Populationen. Nat Diss, Tübingen.
Dahlberg AA (1960) The dentition of the first Agriculturists (Jarmo, Iraq). Am J Phys Anthrop 18: 243–256.
Deichsel G, Trampisch HJ (1985) Clusteranalyse und Diskriminanzanalyse. Fischer, Stuttgart.
DeSmet RM (1966) Observations anthropologiques sur la denture des Indiens Xingu (Brésil). Bull Group Int Rech Sc Stomat 9: 401–414.
Dockrell RB (1956) Tooth size in Irish (Aran Island) families. Europ Orthodont Soc Reports 32: 200–216.
Doran GA, Freedman L (1974) Metrical features of the dentition and arches of populations from Goroka and Lufa, Papua New Guinea. Hum Biol 46: 583–594.
Doyle WJ, Johnston O (1977) On the meaning of increased fluctuating dental asymmetry: A cross population study. Am J Phys Anthrop 46: 127–134.
Drennan MK (1929) The dentition of a bushman tribe. Ann S Afr Mus 24: 61–87.
Ebeling CF, Ingervall B, Hedegard B, LEWIN T (1973) Secular changes in tooth size in Swedish men. Acta Odontol Scand 31: 141–147.
Eid AA, El-Nemarawy MA (1984) Mesiodistal widths for permanent and deciduous teeth in a population of Cairo school children. Egypt Dent J 30: 89–96.
Eismann D (1983) Zahngrößendiskrepanzen als additive Faktoren bei der Ätiologie von Gebißanomalien. Fortschr Kieferorthop 44: 248–255.
Forsberg CM (1988) Tooth size, spacing, and crowding in relation to eruption or impaction of third molars. Am J Orthod 94: 57–62.
Frayer DW (1978) The Evolution of the Dentition in Upper Paleolithic and Mesolithic Europe. Univ of Kansas Publ in Anthrop 10, Lawrence, KS.
Garn SM, Lewis AB, Kerewsky RS (1964) Sex differences in tooth size. J Dent Res 43: 306.
Garn SM, Lewis AB, Kerewsky RS (1967) The relationship between sexual dimorphism in tooth size and body size as studied within families. Archs Oral Biol 12: 299–301.
Garn SM, Lewis AB, Kerewsky RS (1968) Relationship between buccolingual and mesiodistal tooth diameters. J Dent Res 47: 495.
Ghose LJ, Baghdady VS (1979) Analysis of the Iraqi dentition: Mesiodistal crown diameters of permanent teeth. J Dent Res 58: 1047–1054.
Goaz PW, Miller III MC (1966) A preliminary description of the dental morphology of the Peruvian Indian. J Dent Res 45: 106–119.
Gonda K (1959) On the sexual difference in the dimensions of the human teeth. J Anthrop Soc Nippon 67: 151–163.
Goose DH (1976) The dental condition of Chinese living in Liverpool. In: Dahlberg AA, Graber TM (eds) Orofacial Growth and Development. Aldine, Chicago, p. 183–194.
Gordon AD (1987) A review of hierarchical classification. J R Statist Soc A 150: 119–137.
Haeussler AM, Irish JD, Morris DH, Turner CG II (1989) Morphological and metrical comparison of San and Central Sotho dentitions from South Africa. Am J Phys Anthrop 78: 115–122.
Hanihara K (1976) Statistical and Comparative Studies of the Australian Aboriginal Dentition. The University Museum, The University of Tokyo Bulletin No 11. Univ of Tokyo Press, Tokyo.

Harris EF, Nweeia MT (1980) Tooth size of Ticuna Indians, Colombia, with phenetic comparisons to other Amerindians. Am J Phys Anthrop 53: 81–91.
Harris EF, Rathbun TA (1991) Ethnic differences in the apportionment of tooth size. Adv Dent Anthrop 3: 121–142.
Hjelman, G (1928) Morphologische Beobachtungen an den Zähnen der Finnen. Acta Societas Medicorum Fennicae "Duodecim" 11, 1–132. Cit. in Pedersen PO (1949) The East Greenland Eskimo dentition. Meddedelser om Grønland 142: 1–256
Hosaka T (1936) Statistische Untersuchungen über Zähne bei Chinesen. J Oriental Med 25: 348–368, 24: 1230–1251, 1065–1090.
Janzer O (1927) Die Zähne der Neu-Pommern. Vierteljahrschrift für Zahnheilkunde 43: 289–319, 401–434.
Jonas I, Schienle R (1984) Veränderungen der Zahnmorphologie unter dem Einfluß von fluoridiertem Trinkwasser. Schweizer Mschr Zahnmed 94: 399–408.
Kaul V, Prakash S (1984) Crown dimensions of deciduous and permanent teeth of Jats from Haryana (India). Ann Hum Biol 11: 351–354.
Keene HJ (1971) Epidemiologic study of tooth size variability in cariesfree naval recruits. J Dent Res 50: 1331–1345.
Keene HJ (1979) Mesiodistal crown diameters of permanent teeth in male American Negroes. Am J Orthod 76: 95–99.
Kieser JA, Groeneveld HT, Preston CB (1985a) A metric analysis of the South African caucasoid dentition. J Dent Ass South Africa 40: 121–125.
Kieser JA, Groeneveld HT, Preston CB (1985b) An odontometric analysis of the Lengua Indian dentition. Hum Biol 57: 611–620.
Kieser JA, Cameron N, Groeneveld HT (1987) Evidence for a secular trend in the Negro dentition. Ann Hum Biol 14: 517–532.
Kirveskari P (1978) Racial traits in the dentition of living Skolt Lapps. In: Butler PM, Joysey KA (eds) Development, Function and Evolution of Teeth. Academic Press, London, p. 59–68.
Kolakowski D, Bailit HL (1981) A differential environmental effect on human anterior tooth size. Am J Phys Anthrop 54: 377–381.
Koyoumdjisky-Kaye E, Steigman S, Gudelevitch B (1978) A comparative study of tooth and dental arch dimensions and sexual dimorphism in Israeli children of Cochin and North-African descent. Z Morph Anthrop 69: 32–42.
Koyoumdjisky-Kaye E, Zilberman Y, Hazan O (1977) A comparative study of tooth and dental arch dimensions and sexual dimorphism in Circassian and Druse children. Z Morph Anthrop 68: 298–306.
Koyoumdjisky-Kaye E, Zilberman Y, Zeevi Z (1976) A comparative study of tooth and dental arch dimensions in Jewish children of different ethnic descent. Am J Phys Anthrop 44: 437–444.
Lavelle CLB (1972) Maxillary and mandibular tooth size in different racial groups and in different occlusal categories. Am J Orthod 61: 29–37.
Lavelle CLB (1973) Odontometric comparisons between maxillary premolars and molars of different ethnic groups. Hum Biol 45: 123–135.
Lavelle CLB (1984) A metrical comparison of maxillary first premolar form. Am J Phys Anthrop 63: 397–403.
Lin PM (1976) A factor analysis of the dentition of three Tlaxcaltecan populations. In: Crawford MH (ed) The Tlaxcaltecans: Prehistory, Demography, Morphology and Genetics. Univ of Kansas Publications in Anthropology 7. Lawrence, KS, p. 93–119.
Liu KL (1977) Dental condition of two tribes of Taiwan aborigines Ami and Atayal. J Dent Res 56: 117–127.
Lombardi AV (1975) A factor analysis of morphogenetic fields in the human dentition. Am J Phys Anthrop 42: 99–104.

Lundström A (1944) Fortidiga mjölktandsförluster och tandställningen. Svensk tandläkaretidskrift 6. Cit. in Selmer-Olsen R (1949).
Lysell L, Myrberg N (1982) Mesiodistal tooth size in the deciduous and permanent dentition. Europ J Orthodont 4: 113–122.
Macko DJ, Ferguson FS, Sonnenberg EM (1979) Mesiodistal crown dimensions of permanent teeth of Black Americans. J Dent Child 46: 314–318.
Mahalanobis PC (1936) On the generalized distance in statistics. Proc Nat Sc India 2: 49–55.
Mijsberg WA (1931) On sexual differences in the teeth of the Javanese. Proc K Ned Akad Wet Sect Sci 34: 1111–1115.
Miyabara T (1916) An anthropological study of the masticatory system in the Japanese. I. The teeth. Dental Cosmos 58: 739–749.
Moorrees CFA (1957) The Aleut Dentition. A Correlative Study of Dental Characteristics in an Eskimo People. Harvard Univ Press, Cambridge, MA.
Nelson CT (1938) The teeth of the Indian of Pecos Pueblo. Am J Phys Anthrop 23: 261–293.
O'Rourke DH, Crawford MH (1980) Odontometric microdifferentiation of transplanted Mexican Indian population: Cuanalan and Saltillo. Am J Phys Anthrop 52: 421–434.
Ozaki T, Satake T, Kanazawa E (1987) Morphological significance of root length variability in comparison with other crown dimensions. J Nihon Univ Sch Dent 29: 233–240.
Paul G (1990) Geschlechtsbestimmung von Skeletten mit Diskriminanzfunktionen für bleibende Zähne. Med Diss, Ulm.
Pedersen PO (1949) The East Greenland Eskimo dentition. Meddedelser om Grønland 142: 1–256.
Penrose LS (1954) Distance, size and shape. Ann Eugen 18: 337–343.
Potter RHY, Alcazaren AB, Herbosa FM, Tomaneng J (1981) Dimensional characteristic of the Filipino dentition. Am J Phys Anthrop 55: 33–42.
Prakash S, Kaul V, Kanta S (1979) Observations on Bhutanese dentition. Hum Biol 51, 23–30.
Radnzic D (1987) Comparative study of mesiodistal crown diameters and arch dimensions between indigenous British and Pakistani immigrant populations. Am J Phys Anthrop 72: 479–483.
Rahman NA (1962) On the sampling distribution of studentized Penrose measure of distance. Ann Hum Genet 26: 97–106.
Richardson ER, Malhota SK (1975) Mesiodistal crown dimension of the permanent dentition of American Negroes. Am J Orthod 69: 157–164.
Ritchie SG (1923) The Dentition of the Western and Central Eskimos. In: Report of the Canadian Arctic Expedition 1913–18, Vol 12: The Copper Eskimos. Part C, p. 59–66. Acland, Ottawa. Cit. in Pedersen PO (1949).
Rosenzweig KA, Smith P (1971) Dental variability in isolates. J Dent Res 50: 155–160.
Rosenzweig KA, Zilberman Y (1967) Dental morphology of Jews from Yemen and Cochin. Am J Phys Anthrop 26: 15–22.
Rosenzweig KA, Zilberman Y (1969) Dentition of Bedouin in Israel, II. Morphology. Am J Phys Anthrop 31: 199–204.
Rosenzweig KA, Mass E, Smith P (1969) La denture des Samaritains. Bull Group Int Rech Sc Stomat 12: 95–106.
Rösing FW (1990) Qubbet el Hawa und Elephantine. Zur Bevölkerungsgeschichte von Ägypten. Fischer, Stuttgart.
Rottstock F (1985) Odontologische Untersuchungen – Ein Beitrag zur Verschiedenheit und Einheitlichkeit rezenter Populationen. Med Diss, Berlin.
Rottstock F, Noack C, Thiele B (1983) Anwendung statistischer Auswertungsmethoden an einem Beispiel aus der Odontologie, 3. Mitteilung: Faktorenanalyse. Stomatol DDR 33: 472–475.

Schnutenhaus S (1994) Weltverteilung der Zahngrößen. Überprüfung der Eignung von Zahnmaßen für eine Taxonomie des Menschen. Med Diss, Ulm.
Schwidetzky I, Rösing FW (1992) Vergleichend-statistische Untersuchungen zur Anthropologie von Neolithikum und Bronzezeit. Homo 43: 16–56.
Seipel CM (1946) Variations of tooth position: A metric study of variation and adaptation in the deciduous and permanent dentition. Svensk Tandläkare-Tidskrift 39: Suppl. cit. in Moorrees CFA (1957).
Selmer-Olsen R (1949) An Odontometrical Study on the Norwegian Lapps. Skifter utgitt av Det Norske Videnskaps-Akademi Oslo I. Mat-Naturv Klasse No 3.
Sharma JC (1983) Dental morphology and odontometry of the Tibetan immigrants. Am J Phys Anthrop 61: 495–505.
Sharma JC, Kaul V (1977) Dental morphology and odontometry in Punjabis. J Indian Anthrop Soc 12: 213–226.
Shaw JCM (1931) The Teeth, the Bony Palate and Mandible in Bantu races of South Africa. Bale Sons and Davidsson, London.
Smith P (1982) Dental reduction. Selection or drift? In: Kurten B (ed) Teeth: Form, Function and Evolution. Columbia Univ Press, New York, p. 366–379.
Smith P, Brown T, Wood WB (1981) Tooth size and morphology in a recent Australian aboriginal population from Broadbeach, South East Queensland. Am J Phys Anthrop 55: 423–432.
Sofaer JA, Bailit HL, MacLean CJ (1971) A developmental basis for differential tooth reduction during hominid evolution. Evol 25: 509–517.
Sokal RR, Sneath PHA (1963) Principles of Numerical Taxonomy. Freeman, London.
Steinhausen D, Langer K (1977) Clusteranalyse. Einführung in Methoden und Verfahren der automatischen Klassifikation. De Gruyter, Berlin.
Thomsen S (1955) Dental morphology and occlusion in the people of Tristan da Cunha. Results of the Norwegian Scientific Expedition to Tristan da Cunha 1937–38 No. 25. Det Norske Videnskap-Akademi, Oslo.
Townsend GC (1983) Tooth size in children and young adults with Trisomy 21 (Down) Syndrome. Archs Oral Biol 28: 159–166.
Townsend GC, Brown T (1979) Family studies of tooth size factors in the permanent dentition. Am J Phys Anthrop 50: 183–190.
Turner CG II, Swindler DR (1978) The dentition of New Britain West Nakanai Melanesians. Am J Phys Anthrop 49: 361–372.
Wajeman G, Levy G (1979) Crown variations in the permanent teeth of modern man. J Hum Evol 8: 817–825.
Wei B (1996) Die Reduktion der Zahngröße in China und die Evolution des Menschen. Med Diss, Ulm.
Yamada E (1932) The anthropological study of the Japanese teeth. J Nippon Dental Assoc 25: 15–46, 101–132, 177–208, 255–286, 329–344, 450–481, 528–560, 609–656, 710–744, 774–812. Cit. in Moorrees CFA (1957).

7.2 Kinship Studies in Skeletal Remains – Concepts and Examples

Kurt W. Alt and Werner Vach

Introduction

Cultural, technological and economic changes in prehistoric times are determined primarily by ecological and sociological parameters. In order to understand such processes, knowledge of both population and social structures is essential. The importance and function of social structures, the origins of social stratification and the development and change of social organization are concerns of fundamental importance in this context (Friedman and Rowlands 1978; Steuer 1982; Renfrew 1984).

Archeological methods and theory permit the evaluation of data on the cognitive, social, economic and political organization of populations, on gender-specific variation in grave goods (weapons, ornaments), and the development of models of social structures (Jørgensen 1992). Yet the prehistoric sciences are unable to reconstruct biological structures and dynamics. In the absence of such data, reconstructions of social structure of former populations must necessarily be fragmentary in nature (Härke 1989). For the detection and evaluation of genetic relationships in former communities, archaeology must rely on anthropology and its methods.

The Importance of Knowledge About Genetic Relationships in Past Populations

The identification of biologically determined family structures in burial grounds is elementary to research on the social structures in early communities (Härke 1995). Kinship was the main constructional element of human societies, as kinship systems represented the basic element of household and economic units far into the Middle Ages. The determining factors in society were bound to kinship structures (clanship or lineage solidarity). Ancient social structures can therefore only be inferred using a kinship-oriented approach which aims to reconstruct genetically determined relationships of the members of local groups represented by skeletons in burial grounds. Relationships of this kind can then be used to suggest social structures of a higher order.

Related groups of individuals may be represented as multiple burials or by the arrangement of individual graves in burial grounds. Archeologically, the simultaneous burial of several individuals in a single grave is often interpreted to imply a familial relationship (e.g., parents/offspring). Such assumptions are undoubtedly justified in many cases. The type of grave goods, the sex of the individuals and further archeological findings additionally expand the field of speculation. Double burials are often suspected to contain a husband and wife, or perhaps a very important individual and a follower. In the case of multiple burials, the diversity of possible combinations complicates a solution by archeological means.

In collective graves or grave fields containing hundreds of burials an attempt is made to discriminate one or several groups of related individuals. Such an evaluation is complicated by the presence of individuals whose affinities with the group are determined by extra-biological factors (e.g., marriage partners, social affinities).

In cultural anthropology such mixed communities are interpreted as solidarity and subsistence groups, or residental units. The kinship terminology used to describe recent traditional communities is too extensive to be used in practical applications – Malinowski (1930) speaks of "kinship algebra"' in this regard. In this work, a residental unit is defined as comprising a "joint family" which includes individuals related by both consanguinity and affinity as well as persons belonging to the group by social assignment (followers, clients, servants, slaves). Kinship analysis must be regarded against the background of this complexity of social relations in former human communities.

Kinship analysis is most frequently applied in archeological contexts: to multiple burials, mass or collective graves, cemeteries or to the identification of eminent historical persons. Kinship analysis can be successful with regard to specific socio-historical questions or to problems of population dynamics. A good preservation of the investigated skeletal remains is of major importance to the outcome of this analysis. In legal medicine, knowledge of biological relationships may be used to identify unknown bodies, victims of crimes and natural or civil mass disasters, and of armed conflict or political terror (Alt and Vach 1995).

The Phenotypic Approach to Kinship Analysis

The basic task of kinship analysis is the detection of genetically related individuals. Access to the actual genotype of an individual is not yet available from skeletal remains. Recent progress in the analysis of ancient DNA (aDNA) hints of future change, but much work remains (Herrmann and Hummel 1993; Hummel 1994). Hence kinship analysis is based on the detection of similarities in phenotypical traits. Our methods are restricted to the use of discrete traits of the skull, especially of teeth and jaws. Metric traits seem to be unsuited for family reconstruction (Rösing 1990). In order to conclude genetic relationship from the joint occurrence of traits, those traits must fulfil basic requirements. First, the traits must be determined mainly by genetic factors – exogeneous influences should not play a major role in their occurrence. The occurrence of carious lesions in two individuals, for example, shows only that both individuals suffered from caries, but does not mean that they were related. Traits controlled by simple genetic (Mendellian) laws are of great value in kinship analysis. However, most discrete traits are polygeneously controlled, and the exact mechanisms of their heredity are known only to a certain degree. In traits exhibiting genetic polymorphisms and continuous variability the phenotypic variability is mostly based on genotypic variability (Sperlich 1988; Vogel and Motulsky 1996). An investigation of the heritability of such traits allows the estimation of genetic variance (Falconer 1984). The second basic requirement condition traits for kinship analysis have to fulfil concerns the general frequence of the traits. If the trait occurs with a frequency of perhaps 50%, there is a 25% chance that two non-related individuals will each show the trait. Hence only rare traits can contribute significantly to kinship analysis. The third requirement is that the traits be genetically independent from each other, i.e. that the occurrence of one trait is not conditioned on the occurrence of another.

Procedures used in kinship analysis are similar to those of morphological paternity determination (Knussmann 1988), with a basic difference. Morphological paternity expertises aim to detect the agreement of two individuals (usually father/child) in morphological traits with regard to the morphology of a third person (usually the mother). In kinship analysis of skeletal remains we are usually searching for a group of individuals showing joint occurrence of several traits. Hence, while each pair of individuals may show some similarity, it is additionally essential that all pairs show some similarity in identical traits. Furthermore, kinship analysis can only identify possible members of a family, but cannot conclude a specific relationship between individual family members. Specific relationships can only be determined in cases where extraordinary circumstances supply additional information (see example 1; Alt et al. 1992 and 1997).

The Use of Odontological Traits

In the past, nonmetric traits of the human cranium were used for kinship analyses in the majority of cases (Ullrich 1969; Sjøvold 1973; Stuchlikova et al. 1985; Zhong Pei 1985; Rösing 1986, 1990), but these studies have been subject to criticism because information about important features of many traits (e.g., heredity) is lacking (Reinhard and Rösing 1985; Hauser and De Stefano 1989; Saunders 1989). Some problems concerning epigentic variants could be solved by verification using skeletal samples where age, sex, and kinship relationships of the individuals are known (Sjøvold 1973; Corruccini 1974).

Odontologic traits fulfill the requirements for kinship analysis to a much higher degree than do non-metric traits of the cranium (Alt 1991, 1997a, b). Odontologic traits are easily validated in living populations, and information about the heredity of many traits is available (Sofaer 1969; Scott 1973; Harris 1977; Sharma 1986).

A further advantage is that in skeletal material teeth and jaws are generally in a better state of preservation than are other anatomical regions.

Alt (1991, 1997a) has catalogued more than 100 traits suitable for use in kinship analysis (see also Alt and Vach 1994). The list consists of four types of traits: variants of tooth crowns and roots, ontogenetic disturbances of the shape, number, size, structure, position and number of teeth, selected nonmetric traits of cranium and jaws, and congenital malformations and syndromes involving jaws and teeth. For the majority of these traits a macroscopical investigation suffices; few need to be diagnosed roentgenologically. Some of the traits in the catalogue can only be scored in adult individuals as they are not developed in children (e.g., the roots of the permanent teeth).

The first step in the practice of kinship analysis is the scoring of traits. In archeological skeletal material, the state of preservation usually prevents the registration of all traits in every individual from a burial site, but if even one trait is observable the individual is included in the analysis. All the traits are expressed bilaterally, and some traits on the jaws occur asymmetrically. In most cases, each trait can either be present, absent, or indiscernible (i.e., if the specific skeletal part is missing or insufficiently preserved, e.g., if the occlusal surface of a tooth shows attrition). Some traits are not scored by exclusive alternatives only, such as the fissure patterns of occlusal surfaces or the number of tooth cusps or roots. In these cases, the corresponding variant or the actual number of cusps or roots is recorded. In all, about 1000 entries are scored for each individual.

In a second step it is necessary to group traits representing one piece of genetic information into new complexes of multivariate traits. Most traits are scored bilaterally, but if they occur on the same tooth in both the right and left halves of the jaw they are evaluated as contributing just one piece of genetic information. Additionally, if a trait is present on only one side, we regard the genetic information as being present. If a trait is absent on one side and indiscernible on the other side, a problem results, since it could have been present on both sides. In such cases, we regard a complex trait as being present if at least one component is present, we regard it as indiscernible if it is indiscernible in all components, otherwise we regard it as absent. Some complex traits consist of quite a large number of single traits, if each single trait represents a single piece of genetic information. An example is aplasia within the group of hypodont teeth, i.e. the third molars, the second premolars, the second incisors in the maxilla and the first incisors in the lower jaw (Schulze 1987), all of which represent just one piece of genetic information. Hence the data matrix to be analyzed is constructed mainly of complex, multivariate traits. In the matrix the single components are binary, but some values may also be missing.

To interpret the joint occurrence of traits in several individuals to be a hint of familial relationship, it is useful that there be no genetic intercorrelation between traits. However, the complexity of odontological traits unfortunately results in the occurrence of intercorrelated traits. The problems are caused by scoring several traits on the same tooth which are genetically interdependent. An interdependency may be of a functional nature if the presence of one trait depends on the presence of another. Such pairs or sets of intercorrelated traits require special treatment in the analysis.

Depending upon on the state of preservation of skeletal material in a burial complex, many traits may be indiscernible. This does not mean that kinship analysis is impossible but that there are limitations to the results. It may still be possible to detect hints of genetic relationships among some individuals, but it often becomes impossible to make a further investigation of two individuals of special interest, because they have too few discernible traits in common.

Approaches for the Detection of Families

The general task of finding a family represented by a group of traits typical for this family can be solved by different approaches. The usefulness of the various approaches depends mainly on the size of the group or population studied.

Consider first the analysis of a small "population", e.g., individuals from a multiple burial (Rösing 1990). If this group includes a family, the frequency of traits typical for this family should be highly increased within the group in comparison to their frequency in a large population. Hence the key to a significant hint of the existence of a family is the identification of typical family traits by their increased frequency. Once suspicious traits are identified, individuals showing several may be found and hypothetical family constructed.

The second approach is based on similarity measures for pairs of individuals. In the simplest case, such a measure may simply count the number of common traits present. A family may then be identified by a high similarity between all pairs of its members. However, this approach is only useful if the total number of traits considered is small. If the number of traits is large and if the family is characterized by few traits, similarity measures need not be distinctly increased because the few relevant traits are superimposed by traits reflecting random noise.

In the third approach the data matrix is searched for a group of individuals and a corresponding set of traits, i.e., a group where each trait is exhibited by several individuals, and where the relative frequency of each trait within the group is increased in comparison to the population not belonging to the group. This means that – in contrast to the first approach – the population itself is used as a reference for the frequency of traits. This approach only works for large populations.

In order to operate with one of the above concepts, additional considerations are necessary to provide appropriate tools for the analysis. With the second approach, for example, we need a similarity measure, with the third we need to develop a search strategy. Such tools are described in the next section of our paper. A further difficulty of the above concepts is that we will always get results: We will find a trait with a maximal increase in its relative frequency, we will find a pair of individuals showing maximal similarity, and a search procedure will always identify a suspicious group of individuals with a corresponding block of traits. In order to decide whether a result really indicates a hint of a family, we need to find measures for the relevance or "significance" of the result indicating that the identified hint cannot be explained by random variation in the data. Such "significance measures" are also discussed in the next section.

So far our concepts have been motivated by the fact that if a family is present in the population considered, there is some likelihood that we will detect this family. For our concepts to be reasonable, we must be able to show that any hint detected (and marked "significant") presents an indication for the existence of a family. This is a serious problem for the first approach, where we need relative frequencies from a comparable population, i.e., a contemporary, neighbouring population. Unfortunately, frequencies from adequate comparable populations are often unavailable. Other references may be used, but any difference found may only indicate that the considered small sample is from a population which shows different frequencies for certain traits than does the reference population, and may be no indication that the individuals are genetically related at all.

Statistical Procedures

Comparison of Frequencies

Let us consider first a single trait, which is discernible in N individuals. If the relative frequency in a comparable population is p, under random conditions we must expect about $p \times N$ individuals to have the trait. "Random conditions" mean that any of the individuals shows the trait with probability p, and that these events are independent. Contrarily, if there is a family for which the trait is typical, we expect more than $p \times N$ individuals to have the trait. If we observe K such individuals, the question remains whether K is large enough to reject the assumption of "random conditions", i.e., to conclude that a family must exist to explain such a large value. To answer this question, we compute the p-value, which is the probability of observing at least K individuals with a present trait under random conditions. This computation is based on the binomial law (see for example Ihm 1978, 55). If the p-value is small, then the observed result is highly unlikely under random conditions, and we can conclude that we have a significant hint of the existence of a family.

Using this procedure for many traits, we can order them by their p-values and identify those most significant. However, the more traits we search, the larger is the probability of finding at least one small p-value. To decide whether the smallest p-value gives enough evidence to state the existence of a family, we compute the probability of observing at least such a small p-value within L traits, if for each trait the random conditions are valid. This probability can be approximated by multiplying the smallest p-value and L. This is called the Bonferoni procedure, and the resulting value is called the Bonferoni-adjusted p-value.

For moderate sizes of L, this approach is adequate. If we have several hundred traits, then it is more appropriate to look at the number of traits with a p-value smaller than a prespecified value p_0. Examples are given in Alt et al. (1996, 1997).

So far we have used the term "relative frequency" in a naive manner. As most traits consist of several components, some of which may be indiscernible, we are never able to find the exact number of individuals which have the trait. A further discussion of this topic is found in Alt and Vach (1991).

Analysis Based on Similarity Measures

Similarity measures based on binary traits can be constructed in different manners (cf., Bock (1974). For an appropriate measure in kinship analysis two properties are important. First, the measure should be based on the joint presence of traits, not on the joint absence, and second the measure must take into account the different state of preservation of the individuals. In Alt et al. (1995) the following suggestions were made: For each pair of individuals the number of traits present in both individuals is divived by the number of traits present in at least one individual and discernible in both individuals. This measure is interpreted to be an estimate of the probability that one of the individuals shows a trait if it is present in the other. However, if the numerator of the ratio is too small, this estimate is rather imprecise, hence we recommend including only those pairs in the analysis where the nominator is larger than 4.

Once similarity measures have been computed, further investigation can be based on well-known methods of cluster analysis (Everitt 1980). Hierarchical clustering methods are a little bit out of place because the families of a population do not build a hierarchy; methods for the construction of overlapping clusters are preferred. It is also possible to compute measures for the significance of clusters (see Dubes and Jain 1979). In our example we will not use such sophisticated methods, but try to compare the relationships indicated by a high similarity with a pregiven grouping defined by grave-mounds.

Search Strategy for Blocks

The data matrix may look as shown in Table 1. We are interested in finding subblocks which look like that in Table 2. A computer searching for a subblock can compute for any set *T* of traits a measure of how suspicious these traits are typical of a family. In Vach and Alt (1993) we suggest computing the measure *P*(*T*). Consider a statistic *S* summing up the number of traits in *T* present in each individual with at least two traits in *T*. For example, if *T* is of size 4, and if two individuals have two traits and

Table 1. Upper left corner of a hypothetical data matrix

individuals/ traits	*t*1	*t*2	*t*3	*t*4	*t*5	*t*6	*t*7	*t*8	*t*9	*t*10	*t*11	...
*i*1	– –	?–	–?	??	?	++++	– –	– –	– –	?	????	
*i*2	+?	– –	??	– –	–	– – ??	??	??	??	–	– – – –	
*i*3	–?	??	??	??	–	+ – ??	??	– –	??	?	??– –	
*i*4	?–	– –	– –	+ –	–	– – ??	– –	+?	?–	–	– – – –	
*i*7	??	??	– –	– –	?	????	– –	– –	+?	?	??– –	
*i*8	??	?–	??	??	?	????	?–	??	??	?	????	
*i*9	– –	++	?+	– –	–	– – ++	– –	–?	– –	?	?–?–	
*i*10	– –	?–	?–	??	?	– – – –	–?	??	++	?	????	
*i*11	–?	– –	– –	– –	–	– – ??	++	+?	?–	–	– – – –	
...												

Table 2. An example of a suspicious subblock of the hypothetical data matrix

individuals/traits	*t*16	*t*34	*t*72	*t*111	*t*115	*t*123
*i*12	++	?–	+	????	?	– –
*i*23	– –	??	?	++++	?	++
*i*34	– –	++	?	??++	–	??
*i*56	++	– –	?	??– –	+	??
*i*77	??	– –	?	++++	?	+?
*i*114	–??	++	–	????	+	++
*i*117	++	+?	+	– – – –	?	– –
*i*158	– –	??	+	++??	+	– –
*i*161	– –	+?	+	– – – –	?	++
*i*172	++	– –	?	++++	+	– –
*i*203	++	?+	–	++++	+	– –
*i*211	??	– –	+	++++	+	++

two individuals have three, then $S = 2 + 2 + 3 + 3 = 10$. $P(T)$ is defined as the probability of observing at least the observed value of S under the assumption of a random allocation of the traits to the individuals. This probability, which is similar to the p-values above, takes into account the missing patterns of the individuals and the frequency of each trait. The same value of S gives a smaller value for $P(T)$ if all the traits of T are rare rather than if all the traits of T are more frequent.

The use of $P(T)$ allows comparison of different sets of traits and the search for sets with a minimal value of $P(T)$. However, there are too many sets of traits to compute this value for each. Instead we use an agglomerative procedure: First, $P(T)$ is computed for all pairs of traits and a certain number of pairs with the smallest values is selected. To each pair is added each other trait. From these triples a certain number with the smallest values is again selected. Add to each of these triples another trait and so on, up to sets with eight traits. Through this search procedure can be solved the problem that some pairs of traits are known to be dependent, either by definition or by a genetic link. We restrict our search procedure to sets of traits which do not include two interdependent traits.

The search procedure always produces a result: the most suspicious set T of traits. However, the corresponding value of $P(T)$ gives no information as to how suspicious this finding is. To evaluate the significance a further measure $G(T)$ is computed. $G(T)$ is defined as the probability of finding at least one set of traits of the same size as T which shows a smaller value than $P(T)$, assuming independence between traits. A high value of this global measure indicates that the result of the search procedure can be explained by random variation, whereas low values indicate that a systematic cause must exist, probably a true family. The global measure $G(T)$ can be computed approximatively by Monte Carlo simulation.

Further Steps in Kinship Analysis

When the above data analysis procedures suggest a genetically related group, this must be regarded as a hypothesis. The computation of a global measure of significance is a first step in validating this hypothesis. A second step is a thorough examination of the chosen traits and their distribution within the suggested family and the whole population. Here it is necessary to focus on the independence of the traits and possible knowledge of their heredity. If, for example, all traits are on the same teeth or if all traits belong to the same subgroup of traits the result may reflect an unknown genetic relationship between the traits. On the other hand, if some traits are known to be highly heritable or to follow a Mendelian law, these traits become more valuable than others.

A further step is to consider the distribution of the anthropological parameters of sex and age within the suggested family. At this point we realize that rules determining the social phenomena of a family may imply certain selections with respect to the genetically-related family members buried. The key issue is that about half of the adult members of a family may change their social family by marriage. For example it is not necessary to expect a balanced sex distribution of the members of a genetic family observable in a burial ground because the members of one sex may have left the family more often than those of the other sex. However, a suggested family with members of only one sex is not very convincing (see Rösing 1990). Similarly a suggested family where all members died very young is not plausible. Archeological information on the chronological order of the graves can also be used to check the plausibility of a suggested family. In particular, large chronological gaps within a family are not probable also.

So far we have considered validation procedures allowing only negative results, i.e., a suggested family may show some unexpected features. Archeological finds such as weapons and ornaments may also allow a positive validation. The members of a suggested family may correspond with respect to a special archaeological finding, e.g., the female members may show a special decoration in their ornaments. Archeological findings may also hint to membership in a social class such as nobility, or to a special trade (Jørgensen 1992). Coincidences among the members of a suggested family in such findings can also contribute to validation, as social positions may be inherited.

Finally, in a burial complex with spatial distribution of the graves, close proximity of some members of a suggested family may suggest a family-oriented burial practice and hence may contribute to validation. As it is difficult for the human eye to distinguish systematic spatial patterns from those due to random variation, statistical procedures can be used to achieve an objective decision (Alt and Vach 1992).

In the comparison of the suspected family with archeological findings or spatial distribution we always use a second hypothesis about possible representation of a family in the findings available. A hypothesis such as family-oriented burial practices need not be true, hence if the spatial distribution is not suspicious this is not an argument against the hypothesis that the suggested family is a true one. Furthermore the different types of information used for validation are often not independent, e.g., the spatial distribution of graves often reflects their chronological order. If we then find members of a suggested family to be restricted to a chronological subperiod, spatial closeness may not be an additional argument validating the hypothesis.

To this point we have considered only the search for a family within the complete population of a burial site. It is often wise to restrict the search to a subpopulation. If for example a burial site was occupied during several centuries, it is unlikely that a family and its typical traits appear for the whole span. Moreover, often hypotheses suspecting members of a family within a specified group exists based on archeological information. Such a group may be the founding generation of a burial site or perhaps individuals identified as members of the noble class of a society (Rösing 1990). In this case the statistic S counts only individuals in the prespecified group I. The measure $PI(T)$ and $GI(T)$ are then defined in an analogous manner, cf., Vach and Alt (1993).

Examples of Kinship Analysis Application

Kinship Analysis of Victims from a Late Roman Mass Grave

The skeletal material investigated derives from Regensburg-Harting, a village in southern Germany. The excavation unearthed a Roman estate (villa rustica). On the territory of the estate were two wells in which the skeletal remains of thirteen individuals were discovered. They were identified as local Roman inhabitants of the villa. Each individual's skull exhibited cut and blow marks, part of them showed signs of scalping. The find was considered to be archeological proof of human sacrifice among German tribes in the Late-Roman provinces (Schröter 1984).

A standard anthropological analysis of these individuals detected the incidence of metopism in five of the thirteen individuals. The occurrence of this rare trait initiated the consideration of 70 additional traits in order to evaluate genetic relationships among the individuals (Alt et al. 1992). As this investigation was done prior to the establishment of Alt's catalogue (1991), the set of traits included 63 bone variants and only 7 complex odontological traits. The state of preservation allowed the investigation of twelve of the thirteen individuals.

In addition to metopism two further traits show a strikingly increased frequency in comparison to incidences known from recent reference populations. Table 3 shows the contrast of the frequencies and Bonferoni-adjusted p-values, indicating that the shown discrepancies cannot be explained by random variation (cf. subsection *comparisons of frequency*).

Table 3. Traits with striking frequency in skeletal remains from Regensburg/Harting, Germany. n^+: number of bearers of the trait; n^-: number of non-bearers of the trait; $n^?$: number of individuals with indiscernable traits; $n := n^+ + n^- + n^?$; rf $:= n^+/n$ relative frequency to observe the trait in the study population; q: comparative frequency (frequency in recent reference populations); p-value: p-value of test whether the probability for presence of the trait is larger than q; adjusted p-value: p-value after adjustment for multiple testing by the BONFERONI procedure

Trait	n^+	n^-	$n^?$	rf (%)	q (%)	p-value	Adjusted p-value
Metopism	5	4	3	42	5	0.0002	0.014
Aplasia of wisdom teeth	5	1	6	42	20	0.07	1.00
Amelogenesis imperfecta	4	6	2	33	<1	0.0001	< 0.007

The next suggestion of familial relationships among these individuals is given by the correspondence of four individuals in all traits (Tab. 4).

Table 4. Regensburg/Harting, Germany. Distribution of the three traits with increased frequency

	i113	i110	i115	i116	i079	i112	i231	i227c	i238	i227a	i227b	i114
Metopism	–	+	+	–	+	+	?	?	?	–	–	+
Aplasia of the M3	?	+	+	–	+	+	?	+	?	?	?	?
Amelogenesis imp.	?	+	+	–	+	+	–	–	–	–	–	?

Hence at least for these four individuals we can assume a strong genetic relationship. A fifth individual (i114) shows metopism, but is indiscernible with respect to aplasia of the wisdom teeth and amelogenesis imperfecta. In this special situation where all individuals probably died at the same time we can also suggest parent/offspring relationships between individuals simply by comparing ages at death. The results suggest that the five individuals represent three generations, and the genealogical tree of Fig. 1 represents a hypothetical reconstruction of this

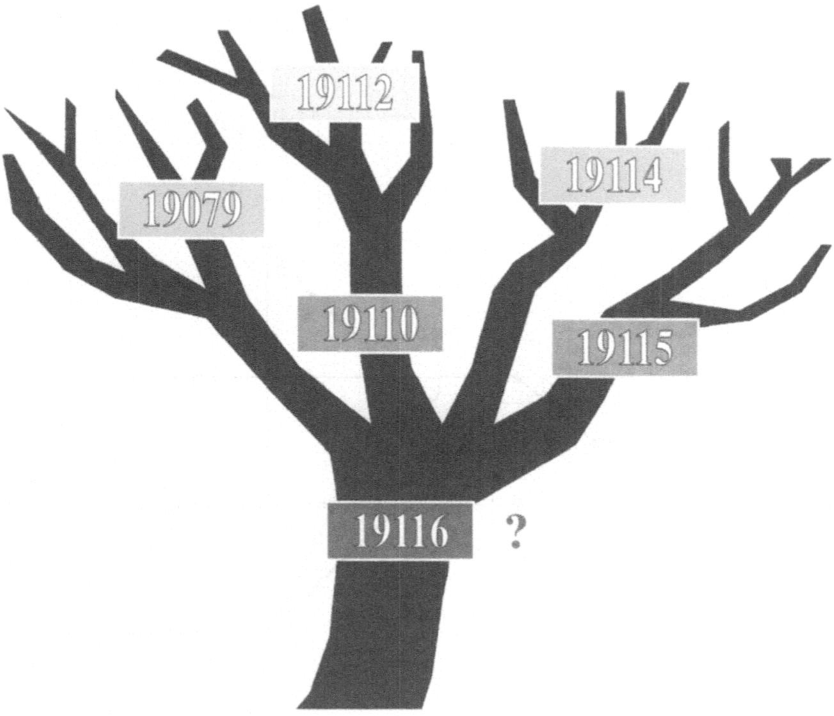

Fig. 1. Regensburg/Harting, Germany: a putative genealogical tree for five individuals of the mass grave. Individual 19116 possibly belong in the same kinship group (19116 shows microsymptoms of aplasia [impacted teeth 23 and 28, 31 reduced size] and a torus palatinus [together with 19115]).

"family" (Alt et al. 1992). The reasons why the rest of the former inhabitants of the villa rustica do not exhibits the same traits could be that either conspicuous traits were partly indiscernible or that there are natural causes for the absence of the traits: married couples are from different gene-pools (exogamy), and farmhands and slaves also lived in the villa; moreover incomplete heritabilities exist for all of these traits.

Kinship Analysis of Individuals from four Grave-Mounds of the Hallstatt Period (Iron Age)

In Dattingen (District Hochschwarzwald, Southern Germany) skeletal remains of 28 individuals placed in the segments of five circles were excavated. Hence the existence of five burial mounds is assumed, which date to the Hallstatt period (Dehn 1986). Twenty-two of the 28 individuals from four of the mounds allowed investigation of odontological traits according to Alt (1991). Of these traits, 53 are present in at least two individuals. For such a number of traits an analysis based on the similarity measure described in the subsection *analysis based on similarity measures* seems to be appropriate. Excluding all pairs of individuals with less than 5 traits present for at least one individual and discernible for both, there are 24 pairs with a similarity larger than 25%. These pairs are confronted with their membership

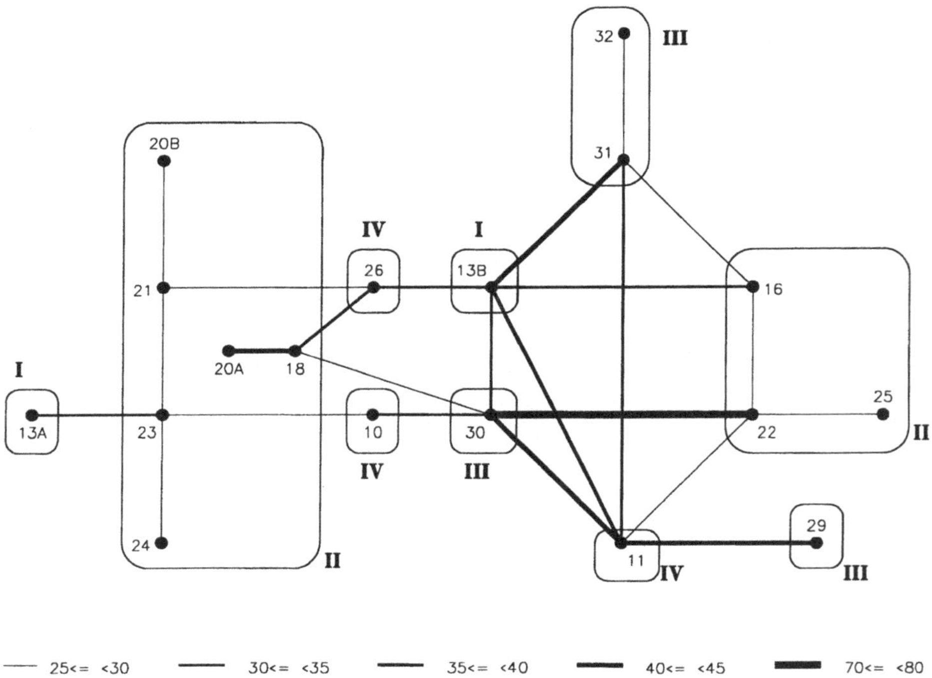

Fig. 2. Dattingen, Kr. Breisgau-Hochschwarzwald, Germany: pairs of individuals with a similarity larger than 25% compared with their membership to four grave mounds.

to the four grave-mounds in Fig. 2. Most pairs with a high similarity involve individuals from different grave-mounds, e.g., from 13 pairs with a similarity larger than 30% only one pair (i18 – i20A) is located in the same grave-mound. This indicates the existence of genetic relationships among the grave-mounds. Hints of genetic relationships within the grave-mounds mainly result from pairs with a similarity between 25% and 30%.

Further steps of our analysis investigated the single grave-mounds, see Alt et al. (1995). Here we mention one further result to demonstrate that it may also be possible to exclude a genetic relationship between two individuals. In the first grave-mound there was a double burial with one male and one female juvenile; the hypothesis of brother and sister is straightforward. However, although the state of preservation of both individuals was quite good, among 20 traits present in at least one individual and discernible in both only one trait was present in both individuals. This result indicates that the above hypothesis is rather unlikely.

Kinship Analysis in the Burial Ground of Kirchheim

The Alemannic burial ground of Kirchheim/Ries was used between the 6th and 8th centuries A.D. 460 individuals with skeletal remains suitable for kinship analysis were found (see Alt and Vach 1995). We focus on a set of 14 individuals from 19 graves containing repoussé sheet brooches. The skeletal remains from the other five graves with repousseé sheet brooches could not be evaluated because they were in a too bad preservation. The search procedure described in the subsection *search strategies for blocks* leads to the discovery of a set of seven traits such that three of the 14 individuals show three of the traits, one shows four and one shows five. Additionally two individuals out of the group with repoussé sheet brooches (graves 213 and 337A) also show three or four traits. The global measure *GI*(*T*) could not be shown to be different from 0.0 in 250 simulation experiments.

Table 5 shows the subblock of the matrix representing the suggested family with seven members. The upper part compares the frequency of traits within and outside of the family. Four traits are rather rare in the complete population except within the suggested family. Three traits show a frequency larger than 10% outside of the family, but these traits are always present within the family if they are discernible.

Additionally, the spatial distribution of the individuals supports the hypothesis of a family: five of the seven individuals are buried close together (Fig. 3). Contrary to this, the distribution of the 19 graves with disk brooches covers the complete burial ground (Fig. 4).

This example also demonstrates the problems of a kinship analysis based on similarities. For the two individuals in double burial 208/209, Neuffer-Müller (1983) suggests that "probably a mother is buried together with her daughter". In our analysis these two individuals are placed in the same family, but they have only one of seven traits in common. A comparison of pairs would probably be unable to detect the relationship, because five of the seven traits are indiscernible

Table 5. Hypothesis of a family *F* from the burial ground of Kirchheim/Ries, Germany based of the analysis of individuals with repoussé sheet brooches. Set of traits *T* typical for the hypothetical family *F*

	Frequency within *F** (%)	Frequency out of *F*** (%)
*t*1: median lingual ridge (upper canines)	100	17,7
*t*2: fissure pattern (first upper premolars)	50	6,4
*t*3: accessory ridge (first upper premolars)	50	2,1
*t*4: mesial paracone tubercle (first upper molars)	100	16,5
*t*5: abnormal size of roots (all premolars)	67	5,4
*t*6: concretio dentium (upper third molars)	40	1,7
*t*7: paramolar tubercle (lower third molars)	100	10,4

t : trait; * : frequency of occurence of the individual traits t in the hypothetical family *F*; **: frequency of occurence of the indivdual traits *t* in the complete cemetery

Suggested family *F*: Individuals with at least three traits in *T*

Grave	*t*1	*t*2	*t*3	*t*4	*t*5	*t*6	*t*7	Age	Sex anthrop./archaeol.	Chronological period
208	??	++	++	??	–	?	++	20–30	f/f	V
209	++	?–	?+	++	?	?	??	7–14	?/f	V
213	?+	– –	– –	?+	–	–	–+	15–20	m/?	IV
358	??	–+	– –	??	+	–	++	20–30	f/f	IV–V
232	??	?–	?+	??	+	+	?+	>20	f/f	V
337A	++	–+	– –	??	+	–	+–	>20	?/?	?
240	+?	??	??	++	+	+	++	15–20	?/?	IV

m: male; f: female ; +: trait present; –: trait absent; ?: trait indiscernible; P(*T*) = 1,07 10–12; G(*T*) ≫ 0,00

in at least one of the two individuals. Further results for kinship analysis in the burial ground of Kirchheim/Ries are demonstrated in Alt et al. (1998).

Discussion

Kinship analysis is an important tool to improve our understanding of prehistoric societies. Detection of genetic relationships among individuals of burial sites can help us to understand burial practices, mating patterns the constitution of social families and the heredity of social positions.

Kinship analysis based on phenotypic traits is possible, but restricted by limitations. It is based on the assumption of the existence of traits typical for a family. Some families may have sufficient typical traits, some may have none or perhaps only one typical trait and hence remain undetected. Some members of a family may not be detected because the typical traits are indiscernible. The boundaries of a family are always fuzzy, hence we can only expect to find the "core family". In most instances we can only produce hints of the existence of genetic relationships. Proof of the absence of genetic relationship among individuals is only possible in rare cases.

Fig. 3. Cemetery of Kirchheim/Ries, Germany: spatial distribution of a suggested “family” preselected by grave goods: repoussé sheet brooches. ○: grave. ●: grave containing repoussé sheet brooches (member of the hypothetical family; numbers refer the grave numbers assigned in the excavation of the cemetery.

Fig. 4. Cemetery of Kirchheim/Ries, Germany: spatial distribution of grave goods: repoussé sheet brooches. ○: grave. ●: grave containing repoussé sheet brooches.

The approaches to kinship analysis considered in this paper can generate hypotheses about genetic relationships which must be further validated. A first step is the computation of a measure of significance to demonstrate the unlikelihood of observing a specific finding under purely random conditions. However, computations of this type are restricted to well-defined procedures investigating one specific aspect. Of greater importance is the validation by coincidences of hints from independent sources, for example, the coincidence of phenotypic similarities and archeological findings. Hence kinship analysis in prehistoric burial sites can only be successful if it is based on a cooperation of anthropology and archeology. Recent developments in the preparation and analysis of ancient DNA may indicate that much more powerful tools for kinship analysis may become available in the future (Zierdt et al. 1996).

Conclusions

The role and importance of anthropology in the interpretation of social structures in burial complexes is obvious in view of the nature of the material sources. However, archeology by itself is an inadequate tool. The social structures of ancient populations can only be reconstructed through the correlation of available data by continuous reception and discussion, and by interdisciplinary research. The realization of the first applications of our newly developed method of kinship analysis have demonstrated the importance of the integration of anthropology, archeology, dentistry, and statistics.

References

Alt KW (1991) Verwandtschaftsanalyse an Skelettmaterial. Methodenentwicklung auf der Basis odontologischer Merkmale. Habil. Schrift, Freiburg.

Alt KW (1997a) Odontologische Verwandtschaftsanalyse. Fischer, Stuttgart.

Alt KW (1997b) Odontologische Binnenanalyse von Bevölkerungen. Individuum – Familie – Population. In: Alt KW, Türp JC (eds) Die Evolution der Zähne. Quintessenz, Berlin, p. 689–704.

Alt KW, Vach W (1991) The reconstruction of "genetic kinship" in prehistoric burial complexes – problems and statistics. In: Bock H, Ihm P (eds) Classification, Data Analysis, and Knowledge Organization. Springer, Berlin Heidelberg New York, p. 299–310.

Alt KW, Vach W (1992) Non-spatial analysis of "genetic kinship" in skeletal remains. In: M Schader (ed) Analyzing and Modeling Data and Knowledge. Springer, Berlin Heidelberg New York, p. 247–256.

Alt KW, Vach W (1994) Rekonstruktion biologischer und sozialer Strukturen in ur- und frühgeschichtlichen Bevölkerungen. Innovative Ansätze zur Verwandtschaftsanalyse in der Archäologie. Prähist Z 69: 56–91.

Alt KW, Vach W (1995) Odontologic kinship analysis in skeletal remains – Concepts, methods, and results. Forensic Sci Int 74: 99–113.

Alt KW, Munz M, Vach W (1995) Hallstattzeitliche Grabhügel im Spiegel ihrer biologischen und sozialen Strukturen am Beispiel des Hügelgräberfeldes von Dattingen, Kr. Breisgau-Hochschwarzwald. Germania 73: 281–307.

Alt KW, Vach W, Pichler S (1992) "Familienanalyse" an kaiserzeitlichen Skelettresten aus einer Villa rustica bei Regensburg-Harting. Bayer Vorgeschichtsbl 57: 261–276.

Alt KW, Pichler S, Vach W, Huckenbeck W, Stloukal M (1996) Early Bronze age family burial from Velké Pavlovice. Homo 46: 256–266.
Alt KW, Pichler S, Vach W, Klíma B, Vlček E, Sedlmeier J (1997) 25,000 year-old triple burial from Dolní Vestonice – an ice age family? Am J Phys Anthrop 102: 123–131.
Alt KW, Jørgensen L, Vach W (1998) Verwandtschaftsanalyse im alamannischen Reihengräberfeld von Kirchheim/Ries. PNM Publications from the National Museum, Studies in Archaeology and History (in prep.).
Bock HH (1974) Automatische Klassifikation. Vandenhoeck & Ruprecht, Göttingen.
Corruccini RS (1974) An examination of the meaning of cranial discrete traits for human skeletal biological studies. Am J Phys Anthrop 40: 425–446.
Dehn R (1986) Ein Gräberfeld der Hallstattzeit von Dattingen, Stadt Müllheim, Kreis Hochschwarzwald. Archäol Ausgr Bad.-Württemberg, p. 96–98.
Dubes R, Jain K (1979) Validity studies in cluster methodologies. Pattern Recognition 11: 235–254.
Everitt B (1980) Cluster Analysis. Halsted Press, New York.
Falconer DS (1984) Einführung in die Quantitative Genetik. Ulme, Ulm.
Friedman J, Rowlands MJ (eds) (1978) The Evolution of Social Systems. London.
Härke H (1989) Die Anglo-amerikanische Diskussion zur Gräberfeldanalyse. Archäol Korrespondenzbl 19: 185–194.
Härke H (1995) Zur Bedeutung der Verwandtschaftsanalyse aus archäologischer Sicht. Germania 73: 307–316.
Harris E (1977) Anthropologic and genetic aspects of the dental morphology of Solomon Islanders, Melanesia. PhD Thesis, Arizona State University, Tempe.
Hauser G, De Stefano GF (1989) Epigenetic Variants of the Human Skull. Schweizerbart, Stuttgart.
Herrmann B, Hummel S (1993) Ancient DNA. Springer, Berlin Heidelberg New York.
Hummel S (1994) DNA aus alten Geweben. In: Herrmann B (ed) Archäometrie. Springer, Berlin Heidelberg New York, p. 87–100.
Ihm P (1978) Statistik in der Archäologie. Rheinland-Verlag, Köln.
Jørgensen L (1992) The early medieval family in Europe: Studies of family burial practices in North-Germanic, Merovingian and Lombard societies (6th–8th centuries). In: Dickinson T, James E (eds) Death and Burial. Dept Archaeol, Univ York, p. 23–28.
Knussman R (Hrsg) Anthropologie, Bd. I. 1. Wesen und Methoden der Anthropologie. Fischer, Stuttgart.
Malinowski B (1930) Kinship. Man 30: 19–29.
Neuffer-Müller C (1983) Der alamannische Adelsbestattungsplatz und die Reihengräberfriedhöfe von Kirchheim am Ries. Forschungsber Vor- und Frühgesch Bad.-Württemberg, Bd. 15. Stuttgart.
Reinhard R, Rösing FW (1985) Ein Literaturüberblick über Definitionen diskreter Merkmale/anatomischer Varianten am Schädel des Menschen. Selbstverlag, Ulm.
Renfrew C (1984) Approaches to Social Archaeology. Univ Press, Edinburgh.
Rösing FW (1986) Kith or kin? On the feasibility of kinship reconstruction in skeletons. In: David AR (ed) Science in Egyptology. Manchester Univ Press, p. 223–237.
Rösing FW (1990) Qubbet et Hawa und Elephantine: Zur Bevölkerungsgeschichte von Ägypten. Fischer, Stuttgart.
Saunders SR (1989) Nonmetric skeletal variation. In: Işcan MY, Kennedy KAR (eds) Reconstruction of Life from the Skeleton. Liss, New York, p. 95–108.
Schröter P (1984) Skelettreste aus zwei römischen Brunnen von Regensburg-Harting als archäologische Belege für Menschenopfer bei den Germanen der Kaiserzeit. Das archäologische Jahr in Bayern, p. 118–120.
Schulze C (1987) Anomalien und Mißbildungen der menschlichen Zähne. Quintessenz, Berlin.

Scott GR (1973) Dental Morphology: A Genetic Study of American White Families and Variation in Living Southwest Indians. PhD Thesis, Arizona State University, Tempe.
Sharma K (1986) Heritability of morphological traits in a Punjabi population of India. Z Morph Anthrop 77: 87–93.
Sjøvold T (1973) The occurrence of minor non-metrical variants in the skeleton and their quantitative treatment for population comparisons. Homo 24: 204–233.
Sjøvold T (1984) A report on the heritability of some cranial measurements and nonmetric traits. In: van Vark GN, Howells WW (eds) Multivariate Statistical Methods in Physical Anthropology. Reidel Publ Comp, Dordrecht, p. 223–246.
Sofaer JA (1969) The genetics and expression of a dental morphological variant in the mouse. Arch Oral Biol 14: 1213–1223.
Sperlich D (1988) Populationsgenetik. Fischer, Stuttgart.
Steuer H (1982) Frühgeschichtliche Sozialstrukturen in Mitteleuropa. Phil Habil Schrift. Abhandl Akad Wiss Göttingen, Phil-Hist Kl 3, Folge 128.
Stuchlikova J, Stuchlik S, Stloukal M (1985) Ein Veterov-Massenbegräbnis aus Velké Pavlovice. Anthropologie (Brno) 23: 51–68.
Ullrich H (1969) Interpretation morphologisch-metrischer Ähnlichkeiten an ur- und frühgeschichtlichen Skeletten in verwandtschaftlicher Hinsicht. Z Archäol 3: 48–88.
Vach W, Alt KW (1993) Detection of kinship structures in prehistoric burial sites based on odontological traits. In: Andresen J, Madsen T, Scollar I (eds) Computing the Past – Computer Applications and Quantitative Methods in Archaeology. Aarhus Univ Press, p. 287–292.
Vogel F, Motulsky AG (1996) Human Genetics. Problems and Approaches. Springer, Berlin.
Zhong-Pei Z (1985) The social structure reflected in the Yuanjunmiao cemetery. J Anthrop Archaeol 4: 19–33.
Zierdt H, Hummel S, Herrmann B (1996) Amplification of human short tandem repeats from medieval teeth and bone samples. Hum Biol 68: 185–199.

Index

The manufacturer's authorised representative in the EU is Springer Nature Customer Service Centre GmbH, Europaplatz 3, 69115 Heidelberg, Germany. If you have any concerns regarding our products, please contact ProductSafety@springernature.com

Printed and bound by CPI Group (UK) Ltd, Croydon, CR0 4YY
22/09/2026
02220988-0012